Chromatography today

Colin F. Poole and Salwa K. Poole

Department of Chemistry, Wayne State University, Detroit, MI 48202, U.S.A.

ELSEVIER
Amsterdam — Oxford — New York — Tokyo   1991

ELSEVIER SCIENCE PUBLISHERS B.V.
Sara Burgerhartstraat 25
P.O. Box 211, 1000 AE Amsterdam, The Netherlands

Distributors for the United States and Canada:

ELSEVIER SCIENCE PUBLISHING COMPANY INC.
655, Avenue of the Americas
New York, NY 10010, U.S.A.

CHEMISTRY

ISBN 0-444-88492-0 (hard bound)
ISBN 0-444-89161-7 (paperback)

CONTENTS

CHAPTER 9. HYPHENATED METHODS FOR IDENTIFICATION AFTER CHROMATOGRAPHIC SEPARATION

PREFACE

The knowledge base of chromatography has continued to grow rapidly through the 1980s due to the many applications of chromatographic methods to problems of contemporary interest in chemistry, biology, medicine, pharmacy, clinical chemistry, industry and environmental sciences. Summarizing this information in a usable form for a diverse group of professional chromatographers has become increasingly difficult in a single, comprehensive text. The present book stemmed from the desire to revise an earlier work, "Contemporary Practice of Chromatography", published in 1984. It was soon realized that a simple revision would not provide the desired result of a contemporary picture of the practice of chromatography in the 1990s. The only workable solution was to start a fresh, maintaining the same general philosophy and concept of "Contemporary Practice of Chromatography", while creating essentially a new book. We hoped to capture the pulse beat of chromatography for the modern chromatographer without loosing sight of earlier developments, which sometime in the future, are destined to be the basis of the next generation of new ideas.

In writing this book we had in mind that it should present a comprehensive review of modern chromatographic techniques at a level commensurate with the needs of a text book for teaching post-baccalaureate courses in the separation sciences and as a self-study guide for professional chromatographers wishing to refresh their background in this rapidly expanding field. When using the book to teach graduate level courses, it is necessary to select a group of topics that can be conveniently taught in a one semester course. Individual teachers should be able to select that material which fits their desired emphasis for concentrated study while leaving other sections for the student to browse through for general perspective. The book is written in a modular fashion to accommodate this approach without loss of continuity. The practicing chromatographer should be easily able to identify subject areas of interest, and when further details are required, to use the extensive bibliography to access key references for further perusal.

Colin F. Poole
Salwa K. Poole

Detroit, March 1991

CHAPTER 1

FUNDAMENTAL RELATIONSHIPS OF CHROMATOGRAPHY

1.1   INTRODUCTION

The Russian botanist M. S. Tswett is generally credited with
the discovery of chromatography around the turn of the century
[1]. He used a column of powdered calcium carbonate to separate

green leaf pigments into a series of colored bands by allowing a solvent to percolate through the column bed. Since these early experiments by Tswett many scientists have made substantial contributions to the theory and practice of chromatography [2,3]. Not least among these is A. J. P. Martin who received the nobel prize in 1952 for the invention of partition chromatography (with R. L. M. Synge) and in the same year with A. T. James he introduced the technique of gas-liquid chromatography. On account of the pioneering efforts of these scientists chromatography is now an important tool used in all branches of the chemical and life sciences.

Chromatography is essentially a physical method of separation in which the components to be separated are distributed between two phases one of which is stationary (stationary phase) while the other (the mobile phase) percolates through it in a definite direction. The chromatographic process occurs as a result of repeated sorption/desorption acts during the movement of the sample components along the stationary bed, and the separation is due to differences in the distribution constants of the individual sample components.

A distinction between the principal chromatographic methods can be made in terms of the properties of the mobile phase. In gas chromatography the mobile phase is an inert gas, in supercritical fluid chromatography the mobile phase is a dense gas (fluid) which is generally above its critical temperature and pressure, and in liquid chromatography the mobile phase is a liquid of low viscosity. The stationary phase is generally a porous, granular powder in the form of a dense homogeneous bed packed into a tube (column) able to withstand the operating pressures normally employed. The sorbents are usually porous solids of high surface area, a similar solid modified by bonding a ligand to its surface or used as a support for a thin film of liquid, or an inert sorbent of controlled pore size. Alternatively, the stationary phase can be distributed as a thin film or layer on the wall of an open tube of capillary dimensions leaving an open passageway through the center of the column. For thin-layer chromatography the sorbent is spread as a thin, homogeneous layer on a flat glass or similar inert backing plate. In this case the mobile phase moves through the layer by the action of capillary forces, or if special equipment is used, by forced flow operation.

Today elution development has become synonymous with the word chromatography itself. This method is characterized by the introduction of a small volume of the sample to be analyzed into the flowing mobile phase (eluent) and the observation of the various components of the sample as they leave the column bed in the form of concentration bands separated in time. Other methods of development, largely of historic interest, are frontal and displacement chromatography. In frontal chromatography the sample is continuously fed onto the column during development. Each solute is retained to a different extent as it reaches equilibrium with the stationary phase until, eventually, the least retained solute exits the column followed by other bands in turn, each of which contains several components identical to the solutes in the zone eluting before it. Ideally the detector output will be comprised of a series of rectangular steps of increasing height. Displacement chromatography is a version of elution chromatography applicable to strongly retained solutes. It is assumed that the sample components distribute themselves on the column in zones according to their ability to interact with the stationary phase. To develop the chromatogram another substance, the displacer, is introduced into the mobile phase. The displacer must have a higher affinity for the stationary phase than the sample. The displacer then drives the adsorbed components progressively along the column, each component displacing the one in front, until they are eluted in the same order in which they were adsorbed on the column; the least strongly retained being eluted first. Displacement chromatography is sometimes used in preparative chromatography (section 4.7.3) and frontal chromatography in some physicochemical applications.

The information obtained from a chromatographic experiment is contained in the chromatogram, a record of the concentration or mass profile of the sample components as a function of the movement of the mobile phase. Information readily extracted from the chromatogram includes an indication of sample complexity based on the number of observed peaks, qualitative identification of sample components based on the accurate determination of peak position, quantitative assessment of the relative concentration or amount of each peak, and an indication of column performance. The fundamental information of the chromatographic process that can be extracted from the chromatogram and its associated vocabulary form the subject of this chapter [4-6]. A more detailed discussion of the theoretical basis and thermodynamic principles of the chromatographic process can be found in standard texts [7-29].

1.2 RETENTION

During their passage through the column, sample molecules spend part of the time in the mobile phase and part in the stationary phase. All molecules spend the same amount of time in the mobile phase. This time is called the column dead time or holdup time (t_m) and is equivalent to the time required for an unretained solute to reach the detector from the point of injection. The solute retention time (t_R) is the time between the instant of sample introduction and when the detector senses the maximum of the retained peak. This value is greater than the column holdup time by the amount of time the solute spends in the stationary phase and is called the adjusted retention time (t_R'). These values lead to the fundamental relationship, equation (1.1), describing retention in gas and liquid chromatography.

$$t_R = t_R' + t_m \tag{1.1}$$

Retention is usually measured in units of time for convenience. Volume units are more exact, Table 1.1, after suitable corrections have been applied [26]. Under average chromatographic conditions liquids can be considered incompressible, but not so for gases, and in gas chromatography elution volumes are corrected to a mean column pressure by multiplying them by the gas compressibility factor, j, equation (1.2)

$$j = 3/2 \ [(P^2 - 1)/(P^3 - 1)] \tag{1.2}$$

where P is the relative pressure (P_i/P_o), P_i the column inlet pressure and P_o the column outlet pressure. The column inlet pressure is usually measured with a pressure gauge at the head of the column. The gauge actually reads the pressure drop across the column; thus, the inlet pressure used for calculating P in equation (1.2) is the value read from the gauge plus the value for P_o. It is also common practice to measure flow rates in gas chromatography with a soap-film meter. For accurate measurements it is necessary to correct the measured value of the flow rate for the vapor pressure of the soap film (assumed to be the same as

TABLE 1.1

RETENTION EXPRESSED IN TERMS OF VOLUME
W_L weight of liquid phase in the column

Term	Symbol	Definition	Method of Calculation
Column Void Volume	V_m	Retention volume corresponding to the column holdup time	$V_m = t_m F_c$
Retention Volume	V_R	Retention volume corresponding to the retention time	$V_R = t_R F_c$
Adjusted Retention Volume	V_R'	Retention volume corresponding to the adjusted retention time	$V_R' = t_R' F_c$ $= V_R - V_m$
Corrected void Volume	V_m°	V_m corrected for mobile phase compressibility	$V_m^\circ = j V_m$
Corrected Retention Volume	V_R°	Retention volume corrected for mobile phase compressibility	$V_R^\circ = j V_R$
Net Retention Volume	V_N	Adjusted retention volume corrected for mobile phase compressibility	$V_N = j V_R$ $V_N = V_R^\circ - V_m^\circ$
Specific Retention Volume	V_g	Net retention volume at 0°C for unit weight of stationary phase	$V_g = \dfrac{V_N \, 273}{W_L T_c}$

that of pure water) and also for the difference in temperature between the column and flow meter, as indicated in equation (1.3)

$$F_c = F_a \, [T_c/T_a][1 - (P_w/P_a)] \tag{1.3}$$

where F_c is the corrected value of the carrier gas flow rate, F_a the flow rate at the column outlet, T_c the column temperature (K), T_a the ambient temperature (K), P_a the ambient pressure (Torr), and P_w the vapor pressure of water (Torr) at T_a. Appropriate values for P_w over a temperature range of 16-25.8°C are given in Table 1.2. For the most exact work it may be necessary to allow for non ideal behavior of the gas phase by applying a virial correction [9,10,31]. At moderate column pressure drops and for carrier gases that are insoluble in the stationary phase equation (1.4) is a reasonable approximation

$$\ln V_N = \ln V_N(0) + \beta P_o J_3^4 \tag{1.4}$$

$$\beta = (2 B_{12} - V_1)/RT_c$$

$$J_3^4 = 3/4 \ [\,(P^4 - 1)/(P^3 - 1)\,]$$

where $V_N(0)$ is the net retention volume at zero column pressure drop, B_{12} the second interaction virial coefficient of the solute with the carrier gas, V_1 the solute molar volume at infinite dilution in the stationary phase (commonly replaced by the bulk molar volume), and R the universal gas constant. Under normal operating conditions errors due to assuming ideality of the gas phase are not large, however, they increase with solute concentration, column pressure drop, and decreasing temperature. Virial corrections are usually made only when it is desired to calculate exact thermodynamic constants from retention volume measurements. Alternatively, high pressure gas chromatography can be used to calculate virial coefficients.

TABLE 1.2

VAPOR PRESSURE OF WATER IN TORR (mm Hg)

Temperature (°C)	0.0	0.2	0.4	0.6	0.8
16	13.634	13.809	13.987	14.166	14.347
17	14.530	14.715	14.903	15.092	15.284
18	15.477	15.673	15.871	16.071	16.272
19	16.477	16.685	16.894	17.105	17.319
20	17.535	17.735	17.974	18.197	18.422
21	18.650	18.880	19.113	19.349	19.587
22	19.827	20.070	20.316	20.565	20.815
23	21.068	21.324	21.583	21.845	22.110
24	22.377	22.648	22.922	23.198	23.476
25	23.756	24.039	24.326	24.617	24.912

The net retention volume and the specific retention volume, defined in Table 1.1, are important parameters for determining physicochemical constants from gas chromatographic data [9,10,32]. The free energy, enthalpy, and entropy of mixing or solution, and the infinite dilution solute activity coefficients can be determined from retention measurements. Measurements are usually made at infinite dilution (Henry's law region) in which the value of the activity coefficient (also the gas-liquid partition coefficient) can be assumed to have a constant value. At infinite dilution the solute molecules are not sufficiently close to exert any mutual attractions, and the environment of each may be considered to consist entirely of solvent molecules. The activity

coefficient and the specific retention volume are related by
equation (1.5)

$$V_g = (273 \ R) / (M_2 \gamma_1 P_1^0)$$ (1.5)

where M_2 is the molecular weight of the solvent, γ_1 the solute
activity coefficient at infinite dilution, and P_1^0 the saturation
vapor pressure of the pure solute at the given temperature.
Ideally, activity coefficients calculated from equation (1.5)
should be corrected for fugacity (solute-solute interactions),
imperfect gas behavior, and interfacial adsorption. The first two
corrections may introduce errors of ca. 1-5% in the value of the
activity coefficient depending on the circumstances of the
measurement; ignoring the importance of interfacial adsorption as
a retention mechanism may make values for the activity coefficient
completely meaningless. The implications of interfacial adsorption
as a retention mechanism in gas-liquid chromatography are
discussed in section 2.5.1. Typical infinite dilution activity
coefficients for nonionic solvents, used in gas chromatography,
have values in the range 0.3 to 50 [32]. Positive deviations from
Raoult's law ($\gamma > 1$) are common for the high molecular weight
solvents generally used in gas chromatography. Activity
coefficients much less than one indicate strong solute-solvent
interactions.

The gas-liquid partition coefficient is evaluated from the
specific retention volume using equation (1.6)

$$V_g = (273.2 \ K_L) / (T_c \rho_c)$$ (1.6)

where ρ_c is the liquid phase density at the column temperature, K_L
the gas-liquid partition coefficient (moles of solute per unit
volume of liquid/moles of solute per unit volume of gas phase).
More frequently, the gas-liquid partition coefficient is used to
correct the measured specific retention volume for contributions
to retention arising from interfacial adsorption. Also the partial
molar Gibbs free energy of solution for a solute at infinite
dilution in the stationary phase can be obtained directly from K_L.

$$\Delta G = -RT_c \ln K_L$$ (1.7)

where ΔG is the partial molar Gibbs free energy of solution. From
the slope of a plot of log (specific retention volume) against the

reciprocal of the column temperature over a small temperature range, 10-30 K, the enthalpy of solution is obtained. The entropy for the same process is obtained from a single value of the specific retention volume and the value of the enthalpy of solution calculated as just described [33-35]. Linearity of the above plots may not be preserved over a wide temperature range which is why the temperature interval used for measurements is small.

Gas chromatography is now a widely used technique for determining solution thermodynamic properties. Compared to classical static methods it has several advantages, namely, small sample size requirement, the ability to measure properties of impure samples, and provides easy variation of temperature. For the most exact measurements precise flow, pressure, and temperature control is needed that may require substantial modification to a standard analytical gas chromatograph [9,10]. Compared to gas chromatography liquid chromatography has been used far less for physicochemical measurements [32,36]. Inadequate knowledge of the true composition of the stationary phase and the absence of quantitative models for the accurate description of retention are the principal reasons for this.

For optimization of chromatographic separations the ratio of the time spent by the solute in the stationary phase to the time it spends in the mobile phase is more fundamentally important. This ratio is called the solute capacity factor and is given by equation (1.8)

$$k = t_R'/t_m = (t_R - t_m)/t_m \tag{1.8}$$

where k is the capacity factor. From its capacity factor, the retention time of any solute can be calculated from equation (1.9)

$$t_R = t_m (1 + k) = (L/u)(1 + k) \tag{1.9}$$

where L is the column length, and u the average mobile phase velocity.

The relative retention of two adjacent peaks in the chromatogram is described by the separation factor, α, given by equation (1.10).

$$\alpha = t_R'(B)/t_R'(A) = k_B/k_A \tag{1.10}$$

By convention, the adjusted retention time or the capacity factor of the later of the two eluting peaks is made the numerator in equation (1.10); the separation factor, consequently, always has values greater than or equal to 1.0. The separation factor is a measure of the selectivity of a chromatographic system. The separation factor is sometimes called the selectivity factor, selectivity or relative retention.

The gas-liquid partition coefficient is related to the capacity factor by equation (1.11).

$$K_L = \beta k \qquad (1.11)$$

where β is the phase ratio. For a wall-coated open tubular column the phase ratio is given by $(r_c - d_f)^2/2r_cd_f$ where r_c is the column radius, and d_f the film thickness for an open tubular column.

In gas chromatography the value of the partition coefficient depends only on the type of stationary phase and the column temperature. It is independent of column type and instrumental parameters. The proportionality factor in equation (1.11) is called the phase ratio and is equal to the ratio of the volume of the gas (V_G) and liquid (V_L) phases in the column. For gas-solid (adsorption) chromatography the phase ratio is given by the volume of the gas phase divided by the surface area of the stationary phase.

1.3 FLOW IN POROUS MEDIA

For an understanding of band broadening in chromatographic systems, the linear velocity of the mobile phase is more important than the column volumetric flow rate. The mobile phase velocity and flow rate in an open tubular column are simply related by

$$u_o = F_c / A_c \qquad (1.12)$$

where u_o is the mobile phase velocity at the column outlet, F_c the column volumetric flow rate, and A_c the column cross-sectional area available to the mobile phase. In a packed bed only a fraction of the column geometric cross-sectional area is available to the mobile phase, the rest is occupied by the solid (support) particles. The flow of mobile phase in a packed bed occurs predominantly through the interstitial spaces; the mobile phase trapped within the porous particles is largely stagnant [37-40].

The mobile phase velocity at the column outlet is thus described by the equation

$$u_o = F_c / \pi r_c^2 \epsilon_u \tag{1.13}$$

where r_c is the column radius and ϵ_u the interparticle porosity (typical value 0.4).

By definition, the experimentally determined average mobile phase velocity is equal to the ratio of the column length to the retention time of an unretained solute. The value obtained will depend on the ability of the unretained solute to probe the pore volume. In liquid chromatography, a value for the interstitial velocity can be obtained by using an unretained solute that is excluded from the pore volume for the measurement (section 4.4.4). The interstitial velocity is probably more fundamentally significant than the chromatographic velocity in liquid chromatography [39].

Under chromatographic conditions, the flow profile is usually laminar and therefore the mobile phase velocity can be described by Darcy's law

$$u(x) = (-K/\eta)(dP/dx) \tag{1.14}$$

where $u(x)$ is the velocity at some point x, K the column permeability, and η the mobile phase viscosity. As gases are compressible and liquids are not under average chromatographic conditions, equation (1.14) must be integrated differently for gases and liquids. For gas chromatography, the mobile phase velocity at the column outlet is given by

$$u_o = KP_o(P^2 - 1)/2\eta L \tag{1.15}$$

Equation (1.15) is valid for open tubular columns under all normal conditions and for packed columns at low mobile phase velocities. The average carrier gas velocity is calculated from the outlet velocity by correcting the latter for the pressure drop across the column, and is simply given by $u = ju_o$, where j is the gas compressibility correction factor, defined in equation (1.2).

In liquid chromatography, equation (1.14) can be integrated directly, neglecting the variation of viscosity with pressure and the compressibility of the mobile phase

$$u = \Delta P K_o d_p^2 / \eta L \qquad (1.16)$$

where K_o is the specific permeability coefficient, d_p the particle diameter, and ΔP the column pressure drop. These assumptions are valid for pressure drops up to about 600 atmospheres. The specific permeability coefficient has a value of ca. 1×10^{-3}, and can be estimated from the semi-empirical Kozeny-Carman equation [37]. The product $K_o d_p^2$ is the column permeability.

1.4 BAND BROADENING MECHANISMS

As a sample traverses a column its distribution about the zone center increases in proportion to its migration distance or time in the column. The extent of zone broadening determines the chromatographic efficiency, which can be expressed as either the number of theoretical plates (n) or the height equivalent to a theoretical plate (H or HETP). If the column is assumed to function as a Gaussian operator then the column efficiency is readily expressed in terms of the peak retention time and variance according to equation (1.17)

$$n = (t_R / \sigma_t)^2 \qquad (1.17)$$

where σ_t is the band variance in time units. In practice, various peak width measurements are frequently used based on the properties of a Gaussian peak profile, Figure 1.1 and equation (1.18)

$$n = a \ (t_R / w)^2 \qquad (1.18)$$

where w_i is the peak width at the inflection point when $a = 4$, w_h the peak width at half height when $a = 5.54$, and w_b the peak width at the base when $a = 16$. Alternatively the ratio of the peak height to the area of a Gaussian peak can be used to define n

$$n = 2\pi (t_R h / A)^2 \qquad (1.19)$$

where h is the peak height and A the peak area. The height equivalent to a theoretical plate is given by the ratio of the column length to the column plate count

$$H = L/n \qquad (1.20)$$

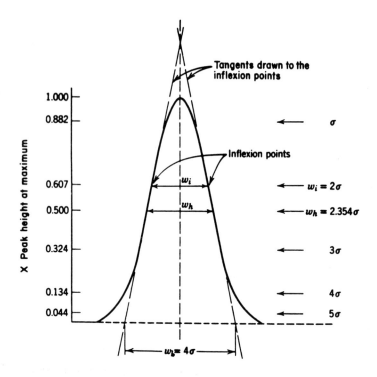

Figure 1.1 Characteristic properties of a Gaussian peak.

Column efficiency can also be measured as the number of effective theoretical plates (N) by substituting the adjusted retention time $(t_R - t_m)$ for the retention time in equation (1.18). The number of effective plates is considered more fundamentally significant than the number of theoretical plates since it measures only the band broadening that occurs in the stationary phase. The two measures of column efficiency are related by equation (1.21). For a weakly retained solute, for example one with k = 1, N will be only 25% of the value of n; however, for well retained solutes, k>10, N and n will be approximately equivalent as indicated in Figure 1.2 [41]. For useful column comparisons n and N should be determined for well retained solutes; at low k values n will be speciously high and misrepresent the actual performance that can be obtained from a particular column in normal use. Also for comparative purposes, it is general practice to normalize the value of n and N on a per meter of column length basis. For many of the relationships discussed in this chapter, n and N can be used interchangeably.

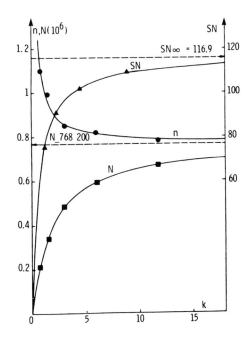

Figure 1.2 Plot of theoretical plate number (n), effective plate number (N) and separation number (SN) against the capacity factor for an open tubular column operated under isothermal conditions. (Reproduced with permission from ref. 41. Copyright Friedr. Vieweg & Sohn).

$$N = n \: [k/(1 + k)]^2 \tag{1.21}$$

An alternative measure of column performance is the separation number (SN), also known as the Trennzahl (TZ). It is defined as the number of component peaks that can be placed between the peaks of two consecutive homologous standards with z and z+1 carbon atoms and separated with a resolution of $R_s = 1.177$ (almost baseline resolved, see section 1.6).

$$SN - 1 = [t_{R(z+1)} - t_{R(z)}]/[w_{h(z)} + w_{h(z+1)}] \tag{1.22}$$

Normal alkanes or fatty acid methyl esters are generally used as the standard homologous compounds. The column separation number is dependent on the nature of the stationary phase, the column length, column temperature, and carrier gas flow rate [42-44]. Referring to Figure 1.2, at a sufficiently high capacity factor value either n, N, or SN provides a reasonable value for comparing the efficiencies of columns of similar length. The disadvantage of the separation number is that it varies with temperature, whereas n and N are reasonably temperature independent.

The separation number is the only column efficiency parameter that can be determined under temperature programmed conditions [45,46]. The critical parameters that must be standardized to obtain reproducible SN values for columns of different length are the carrier gas flow rate and the temperature program. The SN is widely used as part of a standardized test method to evaluate the quality of open tubular columns for gas chromatography (section 2.4.3).

The terms plate number and plate height have their origin in the plate model of the chromatographic process [7,13,47-49]. The plate model assumes that the column can be visualized as being divided into a number of volume elements or imaginary sections called plates. At each plate the partitioning of the solute between the mobile and stationary phase is rapid and equilibrium is reached before the solute moves onto the next plate. The distribution coefficient of the solute is the same in all plates and is independent of the solute concentration. The mobile phase flow is assumed to occur in a discontinuous manner between plates and diffusion of the solute in the axial direction is negligible (or confined to the volume element of the plate occupied by the solute). The plate model is useful for characterizing the efficiency of distillation columns and liquid extractors but its physical significance in chromatography is questionable. Axial diffusion contributes significantly to band broadening, the distribution constant is independent of concentration only over a narrow concentration range, and, quite obviously, the assumption that flow occurs in a discontinuous manner is false. Perhaps the largest shortcoming of the plate model is that it fails to relate the band broadening process to the experimental parameters (e.g., particle size, stationary phase film thickness, mobile phase velocity, etc.) that are open to manipulation by the investigator. Nevertheless, the measured quantities n and H are useful parameters for characterizing chromatographic efficiency and are not limited by any of the deficiencies in the plate model. The various rate models of the chromatographic process enable a similar expression for the theoretical plate to be derived [47-49].

The rate theory makes the following assumptions in its explanation of band broadening:

1. Resistance to mass transfer in both the stationary and mobile phase prevents the existence of an instantaneous equilibrium. Under most practical conditions this is the dominant cause of band broadening.

2. The flow velocity through a packed column varies widely with radial position in the column. Some molecules will travel more rapidly by following open pathways (channeling); others will diffuse into restricted areas and lag behind the zone center (eddy diffusion). These differing flow velocities will cause zone dispersion about the average velocity. For open tubular columns, the contribution of unequal flow velocities to the plate height is zero.

3. Longitudinal diffusion (molecular diffusion in the axial direction) leads to band broadening that is independent of the mobile phase velocity. Its contribution to band broadening increases with the amount of time the solute spends in the column.

The individual contributions to the band broadening mechanism are considered as independent variables except under some circumstances when the eddy diffusion term is coupled to the mobile phase mass transfer term. The above approach can be applied to gas or liquid mobile phases, although it is necessary to make allowances for the different physical properties of gases and liquids, Table 1.3 [50]. For liquid chromatography, the mobile phase is assumed to be incompressible. This is generally true in most practical situations, although at large pressure drops the solute diffusion coefficients, capacity factors, and the column plate height are all influenced by the column pressure drop [45]. Instrumental contributions to band broadening are considered in sections 1.7 and 5.2 and will not be discussed here. General reviews of the band broadening process which expand on the treatment presented below are available [7,13,19,25,26,46-63].

When a sample band migrates through a packed bed, the individual flow paths around the packing particles are of different lengths. These variations in the flow direction and rate lead to band broadening that should depend only on the density and homogeneity of the column packing. Its contribution to the total plate height is proportional to the particle size and can be described by

$$H_E = 2\lambda d_p \tag{1.23}$$

TABLE 1.3

APPROXIMATE VALUES OF CHARACTERISTIC PARAMETERS IMPORTANT IN
PREDICTING BAND BROADENING

Parameter	Gas	Supercritical Fluid	Liquid
Diffusion Coefficients (cm^2/s)	10^{-1}	$10^{-4} - 10^{-3}$	10^{-5}
Density (g/cm^3)	10^{-3}	$0.3 - 0.8$	1
Viscosity (poise)	10^{-4}	$10^{-4} - 10^{-3}$	10^{-2}
Reynold's Number	10		10^2

where H_E is the contribution to the total plate height from eddy diffusion, λ the packing factor, and d_p the average particle diameter. The packing factor is a dimensionless constant and usually has a value between 0.5 and 1.5. For an open tubular column the eddy diffusion term is zero. Band broadening due to stream flow maldistribution in packed beds can be minimized by employing packings of the smallest practical particle size with a narrow particle size distribution. The most practical particle size and column length will ultimately be determined by the column pressure drop. For commercial analytical instruments this corresponds to an average particle size of ca. 100 micrometers for gas chromatography and ca. 3 to 5 micrometers for liquid chromatography. Moreover, a column packed with particles ranging widely in size will give rise to band broadening characteristic of the properties of the largest particles while the pressure drop will be dictated by the particles of the smallest size. Both effects being detrimental to column performance.

The contribution to the plate height from molecular diffusion in the mobile phase arises from the natural tendency of the solute band to diffuse away from the zone center as it moves through the column [59,60,63,64]. Its value is proportional to the diffusion coefficient and the time the sample spends in the column. Its contribution to the total plate height is given by

$$H_L = 2\gamma D_m/u \qquad (1.24)$$

where H_L is the contribution to the plate height from longitudinal molecular diffusion in the mobile phase, γ the obstruction or tortuosity factor, D_m the solute diffusion coefficient in the mobile phase and u the average mobile phase velocity. The obstruction factor is a dimensionless quasi-constant that is not totally independent of the mobile phase velocity [65-67]. This dependence arises from the fact that the lowest flow resistance is offered by gaps or voids in the packing structure. Thus, at low velocities the value of the obstruction factor is averaged over tightly-packed and loosely-packed domains, while at high velocities it is weighted in favor of the loosely-packed domains where more flow occurs. Typical values for the obstruction factor are 0.6 to 0.8 in a packed bed and 1.0 for an open tubular column.

In liquid chromatography where diffusion coefficients are small, the contribution of H_L to the plate height is often negligible. Diffusion coefficients are much larger in gases and hence H_L is more important, particularly at low mobile phase velocities.

Mass transfer in either the stationary or mobile phase is not instantaneous and, consequently, complete equilibrium is not established under normal separation conditions. The result is that the solute concentration profile in the stationary phase is always displaced slightly behind the equilibrium position and the mobile phase profile is similarly slightly in advance of the equilibrium position. The combined peak observed at the column outlet is broadened about its band center, which is located where it would have been for instantaneous equilibrium, provided the degree of nonequilibrium is small. The stationary phase contribution to mass transfer is given by equation (1.25)

$$H_S = [2kd_f^2u]/[3D_s(1 + k)^2] \tag{1.25}$$

where H_S is the contribution to the plate height from the resistance to mass transfer in the stationary phase, d_f the stationary phase film thickness, and D_s the diffusion coefficient in the stationary phase. Equation (1.25) applies exactly to thin-film open tubular columns and is a reasonable approximation for packed column gas chromatography. For liquid chromatography the agreement is poor since there is no allowance made for the contribution of slow diffusion in the stagnant mobile phase.

When a liquid flows through a packed bed an appreciable fraction of the interstitial fluid is essentially stagnant with respect to the actual stream in the center region of the interparticle channels. The fluid space in the column is depicted as consisting of three domains: the free, streaming fluid space; the stagnant interstitial fluid space; and the intraparticle fluid space, which is also assumed to be stagnant. Diffusion is relatively slow in the stagnant mobile phase and its influence on band broadening in liquid chromatography is often significant. Thus equation (1.26) more adequately accounts for the contribution of mass transfer in the stationary phase to the total plate height in liquid chromatography than does equation (1.25)

$$H_S = [\theta(k_o + k + k_o k)^2 \, d_p^2 \, u_e]/[30 \, D_m k_o (1 + k_o)^2 \, (1 + k)^2] \qquad (1.26)$$

where θ is the tortuosity factor for the pore structure of the particles, k_o the ratio of the intraparticle void volume to the interstitial void space, and u_e the interstitial mobile phase velocity. With multicomponent eluents the value of k_o will vary with the mobile phase composition since the intraparticle space occupied by the stagnant mobile phase may change due to solvation of the stationary phase surface. In the derivation of equation (1.26), the influence of diffusion through the interparticle stagnant mobile phase has been neglected as it is generally very small compared to the value for the intraparticle stagnant mobile phase contribution. If a liquid, coated on a support, were used rather than a bonded-phase column packing, it would also be necessary to modify equation (1.26) to allow for mass transfer resistance at the liquid-liquid interface [68].

Mass transfer resistance in the mobile phase is more difficult to calculate because it requires an exact knowledge of the flow profile of the mobile phase. This is only known exactly for open tubular columns for which the contribution of mass transfer resistance in the mobile phase to the total plate height can be described by equation (1.27)

$$H_M = [(1 + 6k + 11k^2)/96(1 + k)^2][(d_c^2 u/D_m)] \qquad (1.27)$$

where H_M is the contribution to the plate height from the resistance to mass transfer in the mobile phase and d_c is the column diameter.

In a packed bed the mobile phase flows through a tortuous channel system and lateral mass transfer can take place by a combination of diffusion and convection. The diffusion contribution can be approximated by equation (1.28)

$$H_{M,D} = (wud_p^2)/D_m \qquad (1.28)$$

where $H_{M,D}$ is the contribution to the plate height from diffusion-controlled resistance to mass transfer in the mobile phase and w the packing factor function that corrects for radial diffusion (ca. 0.02 to 5). To account for the influence of convection, that is, band broadening resulting from the exchange of solute between flow streams moving at different velocities, the eddy diffusion term must be coupled to the mobile phase mass transfer term, as indicated below

$$H_{M,C} = 1/(1/H_E + 1/H_{M,D}) \qquad (1.29)$$

where $H_{M,C}$ is the contribution to the plate height resulting from the coupling of eddy diffusion and mobile phase mass transfer terms. In general, $H_{M,C}$ increases with increasing particle diameter and flow velocity and decreases with solute diffusivity. The packing structure, the velocity range, and the capacity factor value can significantly influence the exact form of the relationship.

In gas chromatography, the coupled plate height equation flattens out the ascending portion of the van Deemter curve at high mobile phase velocities in agreement with experimental observations. At flow velocities normally used the coupling concept appears to be unnecessary to account for the experimental results. In liquid chromatography the existence of a coupling term and its most appropriate form is a matter that remains to be finally settled [19,63].

Although the above listing of contributions to the column plate height is not comprehensive, it encompasses the major band-broadening factors and the overall plate height can be expressed as their sum, equation (1.30).

$$HETP = H_E + H_L + H_S + H_M \qquad (1.30)$$

A plot of HETP as a function of mobile phase velocity is a hyperbolic function (Figure 1.3) most generally described by the van Deemter equation (1.31).

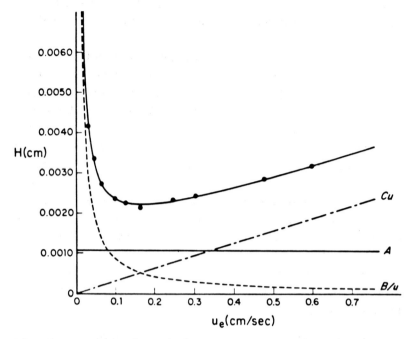

Figure 1.3   Relationship between band broadening and mobile phase velocity (van Deemter equation). (Reproduced with permission from ref. 48. Copyright Elsevier Scientific Publishing Co.)

$$HETP = A + B/u + (C_S + C_M)u \qquad (1.31)$$

The A term represents the contribution from eddy diffusion, the B term the contribution from longitudinal diffusion, and the C terms the contributions from mass transfer in the mobile and stationary phases to the total column plate height. By differentiating equation (1.31) with respect to the mobile phase velocity and setting the result equal to zero, the optimum values of mobile phase velocity (u_{opt}) and plate height ($HETP_{opt}$) can be obtained.

$$u_{opt} = [B/(C_M + C_S)]^{1/2} \qquad (1.32)$$

$$(HETP)_{min} = A + 2[B(C_M + C_S)]^{1/2} \qquad (1.33)$$

The highest column efficiency will be obtained at u_{opt}. In practice, higher values are frequently used to minimize the separation time. For gas chromatography a velocity of about 2 u_{opt}, called the practical operating gas velocity, is frequently recommended [67]. Provided that the ascending portion of the van Deemter curve is fairly flat at higher velocities than u_{opt}, then the loss in efficiency will be small and well worth the saving in separation time.

Optimization of kinetic chromatographic parameters comes from an understanding and interpretation of the coefficients of the van Deemter equation, [56,58,70-72]. For gas chromatography this is made more difficult by the compressibility of the gases used as the mobile phase. The local velocity and solute diffusivity depend on position along the column. Rearrangement in terms of the outlet pressure and velocity aids a general interpretation which is simple to do for the eddy diffusion term, longitudinal diffusion, and the mobile phase mass transfer term. The stationary phase mass transfer term, however, cannot be expressed explicitly independently of the column pressure drop.

Beginning with the most favorable case, band broadening in open tubular columns is satisfactorily described by the Golay equation, extended to situations of appreciable pressure drop by Giddings, equation (1.34)

$$\text{HETP} = f_1 [(2D_{m,o}/u_o) + (f_g(k)) (d_c^2 u_o/D_{m,o})] + f_2 [(f_s(k) d_f^2 u_o/D_s] \quad (1.34)$$

$$f_1 = 9/8 \ [(P^4 - 1) (P^2 - 1)] / [(P^3 - 1)^2]$$

$$f_2 = 3/2 \ [(P^2 - 1) / (P^3 - 1)]$$

$$f_g(k) = [(1 + 6k + 11k^2) / 96 (1 + k)^2]$$

$$f_s(k) = 2/3 [k / (1 + k)^2]$$

where $D_{m,o}$ is the mobile phase diffusion coefficient at the column outlet pressure, u_o the mobile phase velocity at the column outlet, d_f the stationary phase film thickness, and P the ratio of column inlet to column outlet pressure [72-75].

Open tubular columns in current use have internal diameters in the range 0.1 to 0.8 mm and film thicknesses from about 0.05 to 8.0 micrometers. Gases of high diffusivity, hydrogen or helium, are used to minimize mass transfer resistance in the mobile phase and, at the same time, minimize separation time [(HETP)$_{min}$ occurs at higher values of u_{opt} for gases of high diffusivity]. Narrow-bore columns are capable of higher intrinsic efficiency since they minimize the contribution from mobile phase mass transfer resistance to the column plate height, Figure 1.4. The C_M term increases successively with increasing column diameter and is also

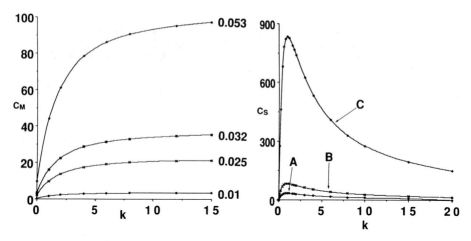

Figure 1.4 Variation of the resistance to mass transfer in the mobile phase, C_M, and stationary phase, C_S, as a function of the capacity factor for open tubular columns of different internal diameter (cm) and film thickness. A, d_f = 1 micrometer and D_s = 5 x 10^{-7} cm^2/s; B, d_f = 5 micrometers and D_s = 5 x 10^{-6} cm^2/s; and C, d_f = 5 micrometers and D_s = 5 x 10^{-7} cm^2/s.

influenced by the capacity factor value, particularly at low values of the capacity factor. When combined with the term describing the plate height contribution due to longitudinal diffusion, C_M is the dominant cause of band broadening for wide bore, thin-film columns. The stationary phase mass transfer term becomes increasingly important as film thickness increases, Table 1.4 (76). For thin-film columns ($d_f <$ 0.25 micrometers) the stationary phase mass transfer resistance term is generally only a few percent of the mobile phase term and, to a first approximation, can be neglected. In estimating the contribution of stationary phase mass transfer resistance to the plate height there is a strong dependence on the capacity factor value and the diffusion coefficient of the solute in the stationary phase, Figure 1.4. Diffusion coefficients in polar, gum and immobilized phases tend to be much smaller than those observed for phases which are not immobilized. Thick-film, polar open tubular columns tend to be substantially less efficient than similar apolar columns; the efficiency of both column types decreases with increasing film thickness.

The Golay equation is strictly applicable to open tubular columns with smooth walls but, with certain approximations, it can be extended to include support-coated [77] and whisker-walled [78]

TABLE 1.4

RELATIVE CONTRIBUTION (PERCENTAGE BASIS) OF MASS TRANSFER
RESISTANCE IN THE MOBILE AND STATIONARY PHASE TO COLUMN PLATE
HEIGHT FOR A SERIES OF 0.32 mm I.D. OPEN TUBULAR COLUMNS USING
UNDECANE AT 130°C AS THE TEST SOLUTE

Film Thickness (μm)	Capacity Factor	Phase Ratio	Mass Transfer Term (%)	
			C_M	C_S
0.25	0.56	320	95.2	4.8
0.5	1.12	160	87.2	12.8
1.00	2.24	80	73.4	26.6
5.00	11.2	16	31.5	68.5

open tubular columns. It can also be used to predict optimum
separation conditions in open tubular liquid chromatography
[79-87]. The main difference between gas and liquid chromatography
in open tubular column is that the diffusion coefficients in
liquids are roughly 10,000 times smaller than in gases and
therefore the last term in equation (1.34) can be neglected. For
high efficiency the column internal diameter must be reduced to a
very small size to overcome the diffusion disadvantage in open
tubular column liquid chromatography. This creates considerable
instrument and column technology constraints that limits the
practical utility of open tubular column liquid chromatography at
present.

Since the exact profile of the mobile phase flow through a
packed bed is unknown, only an approximate description of the band
broadening process can be attained. For packed column gas
chromatography at low mobile phase velocities, equation (1.35)
provides a reasonable description of the band broadening process
[70,82,83].

$$HETP = 2\lambda d_p + (2\gamma D_{m,o}/u_o) + [f_g(k)](d_p^2/D_{m,o})u_o + [f_s(k)](d_f^2/D_s)u \tag{1.35}$$

According to Scott the average linear velocity can be replaced by
$(4u_o/[P + 1])$ in equation (1.35) to permit evaluation entirely in
terms of the outlet velocity [84]. If $\lambda = 0$, $\gamma = 1$, and $d_p = r_c$ is
substituted into equation (1.35) then this equation can be used as
an alternative to equation (1.34) for evaluating the kinetic
column parameters of open tubular columns [58,84].

For packed columns small particles having a narrow size distribution and coated with a thin, homogeneous film of liquid phase are required for high efficiency. The particle size is controlled by the need to remain within limited pressure constraints; this results in the use of column packings with diameters of 120-180 micrometers in columns less than ca. 5 meters long. For heavily loaded columns, liquid phase loading of 25-35% w/w, slow diffusion in the stationary phase film is the principal cause of band broadening. With lightly loaded columns (less than 5% w/w), resistance to mass transfer in the mobile phase is no longer negligible. At high mobile phase velocities the coupled form of the plate height equation is used to describe band broadening.

When the mobile phase is a liquid a variety of equations can be used in addition to the van Deemter equation (1.31) to describe band broadening as a function of the mobile phase velocity, equations (1.36) to (1.39) [49,53,63,85-88].

$$HETP = A/[1 + (E/u)] + B/u + Cu \qquad (1.36)$$

$$HETP = A/[1 + (E/u^{1/2})] + B/u + Cu + Du^{1/2} \qquad (1.37)$$

$$HETP = Au^{1/3} + B/u + Cu \qquad (1.38)$$

$$HETP = A/[(1 + E/u^{1/3})] + B/u + Cu + Du^{2/3} \qquad (1.39)$$

A, B, C, D, and E are appropriate constants for a given solute in a given chromatographic system. Scott's comparison of these equations indicated a good fit with experimental data for all equations, but only equations (1.31), (1.36), and (1.38) consistently gave physically meaningful values for the coefficients A through E [48]. The van Deemter equation, expressed in form (1.40), was found to give the most reasonable fit for porous silica packings over the mobile phase velocity range of 0.02 to 1.0 cm/s

$$HETP = 2\lambda d_p + (2\gamma/u_e)D_m + [d_p^2 u_e][a + bk_e + ck_e^2]/[24(1 + k_e)^2 D_m] \qquad (1.40)$$

where k_e is the capacity factor determined for the interstitial column volume, and u_e the interstitial mobile phase velocity. The values of λ and γ may vary with the quality of the packing and, for a reasonably well-packed column, can be assumed to be 0.5 and

0.8, respectively; a, b, and c can be assigned values of 0.37, 4.69, and 4.04, respectively. At mobile phase velocities higher than those investigated by Scott, the coupled form of the plate height equation may be more appropriate. Equation (1.40) was derived with the assumption that diffusion in the mobile and stationary phases was similar and could therefore be represented by a single mass transfer term. Packings with other pore characteristics might necessitate the inclusion of a term to account for restricted diffusion of solutes in pores that are clogged with bonded phase and for diffusion of solutes along the bonded phase surface [88]. One general problem in verifying the correct plate height equation is that much of the early published studies contain an unacceptable contribution from extracolumn zone broadening that was not always separated from the column contribution [89,90]. More careful studies over a wider range of mobile phase velocities will be needed to finally settle the most exact form of the plate height equation for packed column liquid chromatography.

The highest efficiency in liquid chromatography is obtained using columns packed with particles of small diameter, operated at high pressures, with mobile phases of low viscosity. Both solute diffusivity and column permeability decrease as the mobile phase viscosity increases. For a fixed column pressure, the separation time will increase as the viscosity of the mobile phase is increased. Diffusion coefficients are much smaller in liquids than in gases and, although this means that longitudinal diffusion can often be neglected in liquid chromatography, the importance of mass transfer resistance in the mobile phase is now of much greater significance. The adverse effect of slow solute diffusion in liquid chromatography can be partially overcome by operating at much lower mobile phase velocities than is common for gas chromatography. This increase in efficiency, however, is paid for by an extended separation time.

1.5 PEAK SHAPE MODELS

The column is usually assumed to function as a Gaussian operator, broadening the sample plug into a Gaussian distribution as it passes through the column. In practice, chromatographic peaks are rarely Gaussian and significant errors can result from the calculation of chromatographic parameters based on this false assumption [90-92]. The Gaussian model is only appropriate when

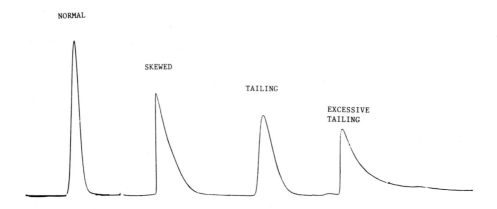

Figure 1.5 Representative peak profiles for different
interactions in column chromatography.

the degree of peak asymmetry is slight. Some examples of
asymmetric peaks frequently observed in column chromatography are
illustrated in Figure 1.5.

Peak asymmetry can arise from a variety of instrumental and
chromatographic sources. Chromatographic sources include
incomplete resolution of sample components, slow kinetic
processes, chemical reactions, and the formation of column voids
[93,94]. The influence of slow kinetic processes can be explained
by assuming that the stationary phase contains two types of sites.
Fast solute exchange between the mobile and stationary phases
occurs at normal sites. On the second type of site solute
molecules are only slowly sorbed and desorbed. If the time
constant for the desorption step of the slow process is greater
than half the standard deviation of the peak, then the peak will
not only be broadened but will also carry an exponential tail.
Examples of slow mass transfer processes include diffusion of the
solute in microporous solids, polymers, organic gel matrices, and
deep pores holding liquid droplets; interactions involving
surfaces with a heterogeneous energy distribution; and, in liquid
chromatography, interfacial mass transfer resistance caused by
poor solvation of bonded phases. Column voids formed by bed
shrinkage is usually a gradual process that occurs during the
lifetime of all columns and results in progressive peak broadening

and/or distortion. A void over the entire cross-section of the column near the inlet produces more peak broadening than asymmetry. However, voids occupying only part of the cross-section along the length of the bed can produce pronounced tailing or fronting, or even split all peaks into resolved or unresolved doublets. Partial voidage effects are due to channeling, i.e., different residence times in the two different flow paths, formed by the void and packed regions. Poor radial diffusion in liquids fails to relax the radial concentration profiles fast enough to avoid asymmetry or split peaks. In gas chromatography the phenomenon is far less significant because diffusion in gases is much faster.

Meaningful chromatographic data can be extracted from asymmetric peaks by digital integration or curve fitting routines applied to the chromatographic peak profile. Digital acquisition of chromatographic data by computer permits the rapid and easy calculation of the statistical moments of the peak profile by direct integration [90-92,95-97]. The statistical moments of a chromatographic peak in units of time are defined by the following equations

zeroth moment $\qquad M_0 = \int_0^\infty h(t) dt$ $\qquad\qquad$ (1.41)

first moment $\qquad M_1 = 1/M_0 \int_0^\infty t\, h(t) dt$ $\qquad\qquad$ (1.42)

higher moments $\qquad M_n = 1/M_0 \int_0^\infty (t-M_1)^n h(t) dt$ $\qquad$ (1.43)

where $h(t)$ is the peak height at time t after injection. The zeroth moment corresponds to the peak area, the first moment corresponds to the elution time of the center of gravity of the peak (retention time), and the second moment the peak variance. The column plate count is calculated from the first two moments using $n = M_1^2/M_2$. The third and forth statistical moments measure the peak asymmetry (skew) and the extent of vertical flattening (excess), respectively. For a Gaussian distribution, statistical moments higher than the second have a value of zero. A positive value for the skew indicates a tailing peak. A positive value for the excess indicates a sharpening of the peak profile relative to a Gaussian peak, while a negative value indicates a relative flattening of the upper portion of the peak profile.

Direct numerical integration of the peak profile may lead to many errors and uncertainties arising from the limits used in the integration, baseline drift, noise, and extracolumn contributions [90-92,96]. A slight error in determining the baseline will greatly influence the selected positions for the start and end of the peak resulting in a comparatively large error, particularly for the higher moments. To eliminate these inconsistencies curve fitting of peak profiles by computer or manual methods have been explored [98-102]. This has led to the general acceptance of the exponentially modified Gaussian function as an acceptable model for tailing peaks. The exponentially modified Gaussian function (EMG) is obtained by the convolution of a Gaussian function and an exponential decay function that provides for the asymmetry in the peak profile. The EMG function is defined by three parameters: the retention time and standard deviation of the parent Gaussian function and the time constant of the exponential decay function. By curve fitting portions of the peaks, these calculations can be conveniently performed on a small computer [99]. To make the EMG function more accessible to analytical chemists Foley and Dorsey have suggested a number of chromatographic figures of merit for ideal and skewed peaks [98]. These allow the calculation of such parameters as the observed column efficiency and the first through fourth statistical moments directly from a chromatogram. The column plate count, termed N_{sys} to indicate that the quantity measured refers to the combined column and instrumental contribution to zone broadening (as it always must do for experimental data) is given by equation (1.44).

$$N_{sys} = [41.7 \ (t_R/w_{0.1})^2]/[(A/B) + 1.25] \tag{1.44}$$

The width at 10% of the peak height ($w_{0.1} = A + B$) and the asymmetry function (A/B) are defined as indicated in Figure 1.6. For peaks with asymmetry factors ranging from 1.00 to 2.76 the percent relative error between equation (1.44) and the EMG function was -1.5 to 1.0%. Although useful data can be extracted from asymmetric peaks, it would seem to be preferable to eliminate the causes of the asymmetry in the first place. Another use of the EMG function is to indicate the magnitude of extracolumn effects assuming that the column behaves as a Gaussian operator [103].

1.6 PARAMETERS AFFECTING RESOLUTION

The separation factor, α, introduced in section 1.2 is a

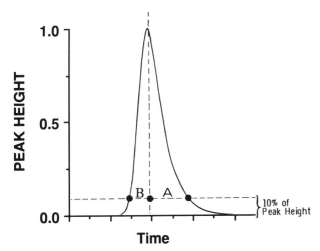

Figure 1.6 The 10% peak height definition of the asymmetry factor. The asymmetry factor is equal to the ratio A/B.

useful measure of relative peak separation. It is a constant for a given set of analytical conditions (stationary phase, temperature, etc.) and is independent of the column type and dimensions. The actual separation of two peaks in a chromatogram is not adequately described by the separation factor alone, however, since it does not contain any information about peak widths. The degree of separation between two peaks is defined by their resolution, R_s, the ratio between the separation of the two peak maxima (Δt) and the average base width of the two peaks, Figure 1.7, and equation (1.45)

$$R_s = 2\Delta t / (w_{b1} + w_{b2}) \tag{1.45}$$

$$\Delta t = t_{R2} - t_{R1} = t_{R2}' - t_{R1}'$$

A value of $R_s = 1.0$ corresponds to a peak separation on the order of 94% and is generally considered an adequate goal for an optimized separation. Baseline resolution corresponds to an R_s value of 1.5.

The resolution of two peaks is related to the adjustable chromatographic variables of selectivity, efficiency, and time by equation (1.46) [104-106]

$$R_s = [n^{1/2}/2][(\alpha - 1)/(\alpha + 1)][k_{Av}/(1 + k_{Av})] \tag{1.46}$$

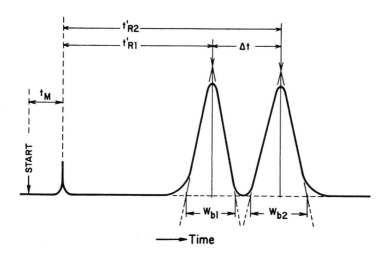

Figure 1.7 **Measurement** of resolution for two closely spaced peaks.

where n and α have their usual meaning and $k_{Av} = (k_1 + k_2)/2$ where k_1 is the capacity **factor of the first** eluting peak of a pair of peaks and k_2 the capacity **factor of the** second peak. Equation (1.46) can be derived directly from equation (1.45) without any assumptions. For two peaks eluting close together on a reasonably efficient column it can be assumed that the peaks have equal width and an approximate form of equation (1.46) derived.

$$R_s = [n^{1/2}/4][(\alpha - 1)/\alpha][k_2/(1 + k_2)] \tag{1.47}$$

The new equation (1.47) is widely used in optimization studies, although of course, it should be viewed as a special case of equation (1.46).

To a first approximation the three terms in equation (1.46) and (1.47) can be treated as independent variables. For a fixed value of n Figure 1.8 indicates the influence of the separation factor and capacity factor on the observed resolution. When the separation factor equals 1.0 there is no possibility of any separation. The separation factor is a function of the distribution coefficients of the solutes, that is the thermodynamic properties of the system, and without some difference in α, ($\alpha > 1$), separation is impossible. Increasing the

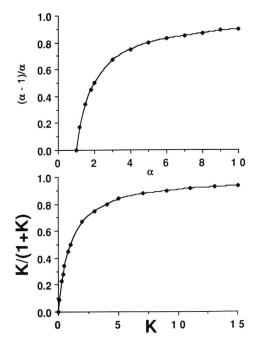

Figure 1.8 Influence of varying the separation factor and capacity factor on the observed resolution for two closely spaced peaks.

value of the separation factor initially causes a large change in resolution that levels off for large values of the separation factor. For values of α > 2 separations are generally easy. Since the separation factor has a large effect on the ease of achieving a certain resolution it is very important to optimize the chromatographic system, that is maximize α, as this will enable a separation to be achieved with the minimum value of n and/or the shortest possible separation time.

Again from Figure 1.8 it can be seen that resolution will initially increase rapidly with retention starting at k = 0. By the time k reaches a value of around 5, further increases in retention result in only small changes in resolution. The optimum resolution range for most separations occurs for k between 2 and 10. Higher values of k result in excessive separation time with little concomitant improvement in resolution.

From equation (1.47) it can be seen that resolution increases only as the square root of n. Thus, the column length must be increased four-fold to increase resolution by a factor of

TABLE 1.5

NUMBER OF THEORETICAL PLATES REQUIRED FOR A CERTAIN SEPARATION
WITH DIFFERENT SEPARATION FACTOR VALUES (k = 3 and R_s= 1.0)

Separation Factor	Required Number of Theoretical Plates
1.005	1,150,000
1.01	290,000
1.015	130,000
1.02	74,000
1.05	12,500
1.10	3,400
1.20	1,020
1.50	260
2.00	110

two. This will result in an approximately four-fold increase in
separation time and an increased column pressure drop if all other
parameters are held constant. For many analytical systems it may
be very difficult to obtain a four-fold increase in efficiency and
thus changing efficiency is the least rewarding factor for
increasing resolution.

Equation (1.47) can be rearranged to predict the number of
theoretical plates required to give a certain separation, equation
(1.48).

$$n_{req} = 16\ R_s^2[(\alpha/\alpha - 1)^2][(k_2 + 1)/k_2]^2 \qquad (1.48)$$

For a pair of solutes with a capacity factor of 3 for the later
eluting solute the number of theoretical plates required to give a
certain resolution, $R_s = 1$, for different values of the separation
factor can be calculated, Table 1.5 [50,107]. Practically all
chromatographic separations have to be made in the efficiency
range of 10^3-10^6 theoretical plates and from Table 1.5 it is
obvious just how important it is to optimize the separation factor
if a separation is to be achieved easily. Likewise, Table 1.6
indicates the number of theoretical plates required to achieve a
given resolution for two values of the separation factor at
different capacity factor values. At small capacity factor values
the number of theoretical plates required for the separation is
very high but falls rapidly as the capacity factor increases. If a
separation is to be carried out conveniently a minimum value for
the capacity factor is required. A common optimization strategy in

TABLE 1.6

NUMBER OF THEORETICAL PLATES REQUIRED TO GIVE A R_s = 1.0 AT
DIFFERENT CAPACITY FACTOR VALUES FOR SEPARATION FACTORS OF 1.05
AND 1.10

Capacity Factor (k)	α = 1.05	α = 1.10
0.1	853,780	234,260
0.2	254,020	69,700
0.5	63,500	17,420
1.0	28,220	7,740
2.0	15,880	4,360
5.0	10,160	2,790
10	8,540	2,340
20	7,780	2,130

difficult analyses is to fix the value of the capacity factor
between 1 and 3 for the most difficult pair to separate in a
mixture.

Figure 1.9 illustrates the relationship between resolution,
the separation factor, the average capacity factor and the column
efficiency for some real chromatographic peaks [108]. The central
portion of the figure illustrates how resolution increases with
the capacity factor for a fixed separation factor and column
efficiency. At first the resolution increases quickly as the
capacity factor is increased to 3 but in going from 3 to 8 the
increase is less dramatic. The peak pair at an average capacity
factor of 3 is also shown in Figure 1.9 with a separation factor
of 1.05 (top) and 1.25 (bottom) for a column with 1600 theoretical
plates. An acceptable separation is obtained for the larger
separation factor (R_s = 1.67) while the peaks are virtually
unresolved for the smaller separation factor. The peak pair at a
capacity factor value of 8 is shown in Figure 1.9 for three
different values of the column efficiency. It can be seen that the
improvement in resolution for increasing efficiency is not as
dramatic as for increasing selectivity.

The separating power of a column can be expressed as its
peak capacity defined as the number of peaks that can be resolved,
at any specified resolution level, in a given separation time. For
the general case it can be calculated using equation (1.49)

$$PC = 1 + \int_{t_m}^{t_R} (n^{1/2}/4t)\,dt \qquad (1.49)$$

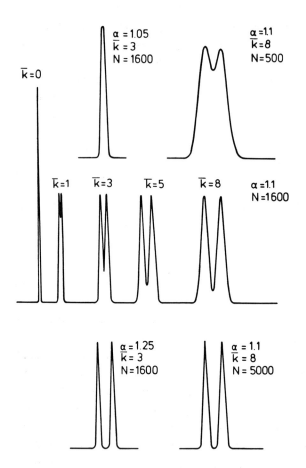

Figure 1.9 Observed change in resolution in a two peak chromatogram for different values of the separation factor or number of theoretical plates. The average capacity factor is indicated by k with a bar on top. (Reproduced with permission from ref. 108. Copyright Elsevier Scientific Publishing Co.)

where PC is the peak capacity, and t the separation time [109-111]. Assuming Gaussian peaks, a resolution of unity, and no dependence of the plate number on the analysis time (or on the capacity factor) equation (1.49) can be integrated to give equation (1.50).

$$PC = 1 + (n^{1/2}/4)\ln(t_R/t_m) \tag{1.50}$$

Equation (1.50) is a reasonable approximation for a packed column for which n shows only a weak dependence on separation time. The relationship between n and k is more complex for open tubular columns where n is a complex function of k, particularly at small values of k (see Figure 1.2). For long separation times (k > 10) equation (1.50) is a reasonable approximation for the peak capacity of open tubular columns but for small capacity factor values will give a value that is too low. Unfortunately, a more general solution to equation (1.50) that includes small capacity factor values is very complex [112]. The separation number, introduced in a previous section, is a special case of the peak capacity and is obtained when the time limits for the integral correspond to the retention time of two adjacent homologs. The peak capacity and separation number are related to each other as indicated below (at least for well retained solutes)

$$PC = 1.18(SN + 1) \qquad\qquad (1.51)$$

The peak capacity is an idealized approximation of the real resolving power of a chromatographic system. Real samples do not contain peaks that will emerge exactly at the correct retention time to fulfill the condition of unit resolution. The concept of the peak capacity can be combined with a statistical model that assumes that the component peaks of complex mixtures distribute themselves randomly along the elution axis and then solved to indicate the probability of peak overlap [113-118]. With random spacing between peaks the obtainable number of resolved peaks is at most 37% of the peak capacity [116]. However, more than half of these peaks may contain two or more sample components. The number of single, identifiable components, therefore, is only a small fraction of the calculated peak capacity for real samples.

The confident analysis of moderate to complex mixtures requires a very large peak capacity and, therefore, chromatographic systems that are able to maximize the information content of the whole, or parts, of the chromatogram of interest. A powerful approach to this problem is the use of multidimensional chromatographic techniques that combine distinctly different separation mechanisms for each component dimension [119-121]. Freeman has shown that the peak capacity for a multidimensional system is approximately multiplicative, while a similar number of columns coupled in series to enhance the total efficiency of the system, will only result in an increase in peak capacity approximately equal to the product of the square root of the number of coupled columns and their individual peak capacities

[122]. Thus, if two identical columns with a peak capacity of 25 are coupled in series, then the resultant peak capacity would be about 35, compared to a value of 625 if the same columns were used in the multidimensional mode. In many instances formidable technical problems must be solved to take full advantage of the potential of multidimensional systems (section 8.7).

1.7 OPTIMIZATION OF KINETIC COLUMN VARIABLES

The variables that control the extent of a chromatographic separation are conveniently divided into kinetic and thermodynamic factors. The thermodynamic variables control relative retention and are embodied in the selectivity factor in the resolution equation. For any optimization strategy the selectivity factor should be maximized (see section 1.6). Since this depends on an understanding of the appropriate retention mechanism further discussion will be deferred to the appropriate sections of Chapters 2 and 4.

The kinetic parameters are capable of a more general interpretation for gas and liquid chromatography. In most cases the kinetic variables are evaluated from the coefficients of the van Deemter equation and relate the column plate height to the properties of the mobile phase, particle size or column radius (open tubular columns) and film thickness. These in turn can be used to predict optimum conditions for maximizing resolution with a constant value for the selectivity factor, for minimizing the separation time, and for comparing the properties of columns of different types. In practice the separation time and column length are controlled by the available column pressure drop which must be considered as an important parameter in any optimization scheme.

1.7.1 Column Types in Gas Chromatography

Five types of columns are routinely used in gas chromatography: classical packed columns with internal diameters greater than 2 mm containing particles in the range 100 to 250 micrometers; micropacked columns having diameters less than 1 mm with a packing density similar to classical packed columns (d_p/d_c less than 0.3, where d_p is the particle diameter and d_c the column diameter); packed capillary columns have a column diameter less than 0.5 mm and a packing density less than classical packed columns (d_p/d_c = 0.2-0.3); SCOT columns (support-coated open

TABLE 1.7

REPRESENTATIVE PROPERTIES OF DIFFERENT COLUMN TYPES IN GAS
CHROMATOGRAPHY

Column Type	Phase Ratio	H_{min} (mm)	u_{opt} (cm/s)	Permeability $(10^7.cm^2)$
Classical Packed	4-200	0.5-2	5-15	1-50
Micropacked	50-200	0.02-1	5-10	1-100
Packed Capillary	10-300	0.05-2	5-25	5-50
SCOT	20-300	0.5-1	10-100	200-1000
WCOT	15-500	0.03-0.8	10-100	300-20000

tubular columns) where the liquid phase is coated on a surface
covered with a layer of solid support material, leaving an open
passageway through the center of the column; and WCOT (wall-coated
open tubular columns) in which the liquid phase is coated directly
on the smooth or chemically etched column wall. Some
characteristic properties of the various column types are given in
Table 1.7 [50,123-130]. The most significant difference among the
various column types is their permeability. The open tubular
columns offer much lower flow resistance and can therefore be used
in much longer lengths to obtain very high total plate counts. The
minimum plate height of the best packed column in gas
chromatography is about 2-3 particle diameters whereas that of an
open tubular column will be similar to the column diameter. So
that using a column packed with 10-micrometer particles and an
open tubular column of about 30 micrometers in diameter should
give a similar number of theoretical plates per unit length.
Jonker et al. [83] prepared a packed column with 10-micrometer
particles that yielded about 50,000 theoretical plates/m but had a
pressure drop of about 200 atmospheres/m limiting usable column
lengths to only a fraction of a meter (<25 cm). Whereas Schutjes
et al. [131] obtained approximately 1.1 million theoretical plates
from a 70-meter WCOT column of 50 micrometers internal diameter
operated at about 22 atmospheres. Thus, the intrinsic efficiency
of open tubular columns is not necessarily greater than that of a
packed column, but because of their greater permeability at a
fixed column pressure drop a greater total number of theoretical
plates may be obtained, since longer columns can be used.

Any optimization strategy that considered only efficiency is
inadequate to describe accurately resolution, which is a strong
function of the capacity factor at low capacity factor values. The

phase ratio (ratio of the volumes of gas to liquid phase in a column) for a number of typical columns is given in Table 1.8 [123]. At a constant temperature the partition coefficient will be the same for all columns prepared from the same stationary phase. Consequently, for a column with a large phase ratio the capacity factor will be small and vice versa (see equation 1.11). The number of plates required for a separation becomes very large at small capacity factor values (see Table 1.6). Columns with low phase ratios, that is thick-film columns, have a lower intrinsic efficiency than thin-film columns but provide better resolution of low boiling compounds, because they provide a more favorable capacity factor value. The opposite arguments apply to high boiling compounds that have long separation times because their capacity factor values are too large. Increasing the phase ratio lowers the capacity factor to a value within the optimum range so that there is little deterioration in resolution but a substantial saving in separation time. Phase ratios are usually lower (and analysis times longer at constant temperature) for packed columns than for WCOT columns, SCOT columns occupying an intermediate position. Since several combinations of film thickness and column radius can be used to generate the same phase ratio, there are other factors that need to be considered for selecting these variables for a particular separation.

In general, increasing the column radius and film thickness for a WCOT column, will lead to an increase in the column plate height and decrease in efficiency. However, the relationship between the variables is very complex (see section 1.4) and depends on the capacity factor since this term appears explicitly in the contribution of the resistance to mass transfer to the plate height. For thin-film columns (d_f < 0.25 micrometers) resistance to mass transfer in the stationary phase is small (frequently negligible) so that decreasing the radius of thin-film columns by minimizing the mobile phase mass transfer term leads to increased efficiency [131-134]. Narrow-bore, thin-film WCOT columns are the most intrinsically efficient provided that the inlet pressure is not limited (this does not mean that these columns will always provide the highest resolution since they may not provide optimum values of the capacity factor). If the column radius is held constant and the film thickness is increased the column efficiency will decline because the resistance to mass transfer in the stationary phase will eventually dominate the

TABLE 1.8

CHROMATOGRAPHIC PROPERTIES OF COMMERCIALLY AVAILABLE COLUMNS

Column Type	Length m	Internal Diameter mm	Film Thickness μm	Phase Ratio	Capacity* Factor	HETP mm	Total Plate Count	Plates Per Meter
Classical	2	2.16	10%(w/w)	12	10.4	0.55	3,640	1,820
Packed	2	2.16	5%(w/w)	26	4.8	0.50	4,000	2,000
SCOT	15	0.50		20	6.2	0.95	15,790	1,050
	15	0.50		65	1.9	0.55	27,270	1,820
WCOT	30	0.10	0.10	249	0.5	0.06	480,000	16,000
	30	0.10	0.25	99	1.3	0.08	368,550	12,285
	30	0.25	0.25	249	0.5	0.16	192,000	6,400
	30	0.32	0.32	249	0.5	0.20	150,000	5,000
	30	0.32	0.50	159	0.8	0.23	131,330	4,380
	30	0.32	1.00	79	1.6	0.29	102,080	3,400
	30	0.32	5.00	15	8.3	0.44	68,970	2,300
	30	0.53	1.00	132	0.9	0.43	70,420	2,340
	30	0.53	5.00	26	4.8	0.68	43,940	1,470

*undecane at 130 °C

TABLE 1.9

TYPICAL COLUMN EFFICIENCIES FOR WCOT COLUMNS (L = 25 m)

Column Diameter	H_{min} (mm)		UTE (%)	Film Thickness (μm)	u_{opt} (cm/s)
	Theoretical	Experimental			
0.10	0.069	0.09	77.1(*)	0.10	20.0
0.25	0.178	0.22	80.8	0.25	38.5
0.32	0.217	0.25	86.8	0.25	30.2
0.53	0.472	0.71	66.5	5.00	15.0
0.75	0.491	0.52	94.4	1.00	20.5

(*)This value is lower than expected perhaps reflecting extracolumn band broadening affects.

plate height equation. Some representative values for the efficiency of WCOT columns are given in Tables 1.8 and 1.9 [132,133]. In Table 1.9 the term utilization of the theoretical efficiency (UTE) is introduced; this term is often referred to as the coating efficiency in the older literature. This name is now considered inappropriate. The UTE of a WCOT column is calculated according to equation (1.52) [132-135].

$$UTE = [(HETP)_{Theor.} / (HETP)_{Exp.}] \times 100 \ (\%) \qquad (1.52)$$

$$(HETP)_{Theor.} = r_c [32 f_g(k)]^{1/2}$$

where $f_g(k)$ is defined in equation (1.34). The theoretical efficiency refers to a smooth wall WCOT column operated at its optimum velocity ignoring pressure gradients and resistance to mass transfer in the stationary phase as sources of additional band broadening. It is an idealized value that is a reasonable approximation for the anticipated performance of thin-film columns. In other cases it provides a window to view the relative contribution of mechanisms other than mobile phase mass transfer to the column plate height.

Representative column performance data for some micropacked and packed capillary columns are given in Table 1.10 [124,125]. As anticipated, the column efficiency increases as the particle diameter is decreased, but only with a simultaneous increase in the column pressure drop. Assuming a fixed available pressure, high total column efficiencies can be obtained by reducing the particle diameter (d_p/d_c < 0.03) or by using coarser particles and longer columns. In either case, it is not easy to exceed a limit of about 60,000 total number of theoretical plates without

TABLE 1.10

COMPARATIVE COLUMN PERFORMANCE DATA FOR SOME MICROPACKED AND PACKED CAPILLARY COLUMNS

Particle Diameter (mm)	Ratio of Particle Diameter to Column Diameter	H_{min} (mm)	Pressure Drop Per Meter(atms.)	L for n =10,000 (m)
0.175	0.70	1.00	1	10
0.150	0.17	0.26	3	2.6
0.140	0.18	0.50	1	5
0.125	0.39	0.60	0.5	6
0.113	0.23	0.16	0.8	1.6
0.030	0.03	0.10	7	1
0.030	0.03	0.10	8	1
0.025	0.067	0.09	25	0.9
0.010	0.008	0.02	240	0.2

resorting to unusually high operating pressures. Increasing the column length also increases the separation time, which can become long when a very large number of theoretical plates is needed for a separation. For high speed separations columns packed with small particles are preferred but because of pressure limitations the total number of plates available for the separation will be small, perhaps 10,000. Table 1.11 provides some representative column performance data for thick-film WCOT columns and micropacked columns of similar phase ratio [125]. The compared columns are similar in performance but the column pressure drop is at least an order of magnitude higher for the micropacked columns. Thus for difficult separations requiring a large number of theoretical plates with an optimum value for the capacity factor value thick-film WCOT columns would be clearly preferred. A lack of a commercial source of high quality columns, the attendant difficulties of operating gas chromatographs at high pressure, and the difficulty of uniformly packing long columns have conspired to make micropacked and packed capillary columns the least popular columns for analytical applications. For those applications that demand a large sample capacity or use of selective stationary phases that are difficult to coat on open tubular columns and/or a greater number of theoretical plates than is available for classical packed columns they might well be preferred.

1.7.2 Selection of the Mobile Phase in Gas Chromatography

The mobile phase in gas chromatography is generally

TABLE 1.11

COMPARISON OF WCOT AND MICROPACKED COLUMNS HAVING SIMILAR PHASE RATIOS

Column Type	Length m	Internal Diameter mm	Film Thickness μm	Phase Ratio	Capacity* Factor	H_{min} mm	U_{opt} cm/s	Column Plate Count	Plates Per Meter
Micro-	2.0	0.82	0.95(%)	154	17.2	0.46	10.1	4340	2170
packed	3.4	0.82	2.15(%)	67	36.4	0.46	10.5	7450	2170
	2.7	0.82	2.46(%)	58	42.2	0.63	8.0	4350	1590
WCOT	5.5	0.45	0.84	133	19.8	0.53	21.5	10380	1900
	3.6	0.45	1.97	57	42.5	0.67	16.9	5380	1490
	7.6	0.53	2.62	50	48.9	0.80	14.3	9500	1260

*ethyl undecanoate at 125 °C; d_p for micropacked columns (100-125 micrometers)

considered inert, in that it does not react chemically with the sample or stationary phase, and it does not influence the selectivity of the chromatographic system with the exception of a generally small contribution arising from nonideal gas behavior. It primarily provides the column transport mechanism by which the sample is moved through the column. The choice of carrier gas influences resolution through its effect on column efficiency, which arises from differences in the solute diffusion rates. It can also affect separation time, because the optimum carrier gas velocity decreases as the solute diffusivity decreases and also plays a role in pressure-limiting situations due to differences in gas viscosities. Other considerations for choosing a particular carrier gas might be cost, purity, safety, and detector compatibility. Taking all these factors into account, hydrogen, helium, nitrogen, and argon are the principal carrier gases used in gas chromatography. Occasionally organic vapors or other gases are added to these carriers to modify the column separation characteristics [136]. Typical examples are steam, ammonia, and formic acid. Since these gaseous mixtures do not behave ideally they may modify a separation by changing the relative retention of sample solutes, by masking the adsorption of polar solutes, and if appreciably soluble in the stationary phase, they may be used to create dynamic binary liquid phase mixtures whose selectivity can be adjusted by changing the partial pressure of the organic vapor in the carrier gas. In general, however, problems with detector incompatibility, the limited available temperature range for the desired effect to apply, and other inconveniences related to safety and equipment requirements have limited studies in this area.

The viscosity of the carrier gas determines the column pressure drop for a given velocity. For hydrogen, helium, and nitrogen it can be approximated by equations (1.53 to 1.55) in the temperature range of interest for gas chromatography

Hydrogen $\quad \eta_t = 0.1827\ t + 83.9899$ $\qquad\qquad\qquad$ (1.53)

Helium $\qquad \eta_t = 0.3993\ t + 186.6169$ $\qquad\qquad\qquad$ (1.54)

Nitrogen $\quad \eta_t = 0.3838\ t + 167.3534$ $\qquad\qquad\qquad$ (1.55)

where η_t is the carrier gas viscosity (micropoise) at temperature t (°C) [137,138]. For a given inlet pressure and temperature the

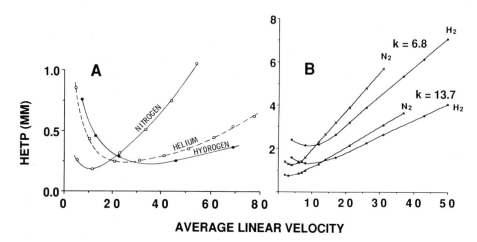

Figure 1.10   Van Deemter curves indicating the influence of the choice of carrier gas on the separation performance of a thin-film open tubular column (A), and a thick-film open tubular column for two solutes with different capacity factor vales (B).

average linear gas velocity for an identical column will be highest for hydrogen and the lowest for helium. Viscosity increases with temperature for gases, causing a decrease in the carrier gas velocity at constant pressure. The rate of change of viscosity with temperature is approximately the same for the three gases.

Solute diffusivity influences both the plate height and the position of the optimum linear velocity described by the van Deemter equation [138-141]. Comparing the van Deemter curves in Figure 1.10, nitrogen provides the lowest plate height but it occurs at an optimum gas velocity, which is rather low and leads to long analysis times. In the optimum plate height region the ascending portions of the curve are much shallower for hydrogen and helium. Thus for operation at carrier gas flow rates above the optimum in the van Deemter curve, the situation which applies in practice, hydrogen, and to a lesser extent helium, show only a modest sacrifice in column efficiency compared to nitrogen. In addition, because equivalent column efficiencies are obtained at higher average linear gas velocities, compounds exhibit lower elution temperatures during temperature programmed runs and, as the peaks are taller and narrower, sample detectability is improved.

The above observations are sound for thin-film (d_f < 0.5 micrometers) and low loaded packed columns. For these columns hydrogen is clearly the preferred carrier gas when efficiency and separation time are considered together. Different conclusions are reached for thick-film columns, since diffusion in the stationary phase contributes significantly to the band broadening mechanism. For thick-film columns the capacity factor, temperature, and solute diffusion coefficient in the stationary phase need to be taken into consideration when selecting a carrier gas for a particular analysis [142]. In going from a thin-film to a thick-film column (which corresponds to reducing the phase ratio) the efficiency of the columns, as measured by H_{min} decreases substantially and is dependent on the identity of the carrier gas. The optimum linear carrier gas velocity is reduced by perhaps a factor of two and also depends on the identity of the carrier gas, and the slopes of the ascending portion of the van Deemter curves at high velocity are much steeper for the thick-film columns and depend primarily on the liquid phase diffusion coefficient (Figure 1.10). In contrast to thin-film columns, for maximum efficiency thick-film columns, should be operated at their optimum linear velocity with nitrogen as the carrier gas. These conditions will maximize resolution while compromising analysis time.

For packed columns with normal loadings the liquid phase mass transfer term is always significant. In situations where the pressure drop is the limiting condition hydrogen is preferred, because its viscosity is only about half that of helium and nitrogen. In terms of efficiency nitrogen is slightly superior at low temperature and flow rates, while hydrogen seems to be superior at higher temperatures and at above optimum flow rates [140,141].

The only disadvantage to the use of hydrogen as a carrier gas is the real or perceived explosion hazard from leaks within the column oven. Experience has shown that the conditions required for a catastrophic explosion may never be achieved in practice. However, commercially available gas sensors will automatically switch off the column oven and carrier gas flow at air-hydrogen mixtures well below the explosion threshold limit [143].

1.7.3 Minimizing Separation Time in Gas Chromatography

The limitations on the time required for a particular separation can be conveniently derived by considering a simple model. This model assumes a three component mixture in which the optimum column length is dictated by the number of theoretical plates required to separate the two components most difficult to resolve and the total separation time by the time required for the last peak to elute from the column. It is further assumed that the temperature is constant. Under these conditions the retention time is described by equation (1.56).

$$t_R = (nH/u)(1 + k) \tag{1.56}$$

The number of plates required for a certain separation is given by equation (1.48) and can be substituted into equation (1.56) to give

$$t_R = 16\ R_s^2[\alpha/(\alpha - 1)]^2[(1 + k)^3/k^2][H/u] \tag{1.57}$$

When the object is to perform fast separations equation (1.57) indicates that we should not demand an overly large value for the resolution (R_s approximately 1), the separation factor (α) should be maximized for the pair most difficult to separate by careful selection of the stationary phase, the capacity factor should be minimized ($k = 1-5$) for the most difficult pair to separate, and the column should be operated at the minimum value of the plate height, H_{min}, corresponding to u_{opt} [107,131,144-146]. The ratio H/u is a complex function of the operating conditions (column pressure drop, film thickness, resistance to mass transfer terms and column dimensions). For thin-film WCOT columns having a limited number of theoretical plates (P approximately 1) the separation time in terms of the column variables is given by equation (1.58)

$$t_R = [f_g(k)][(1 + k)nd_c^2/2D_{m,o}] \tag{1.58}$$

(variables are defined in equation 1.34). For low plate number or wide-bore columns the separation time is directly proportional to the square of the column diameter. Increasing the column radius then has a large detrimental influence on the separation speed.

For columns with large plate numbers or operated with a vacuum outlet (as might be the case in GC/MS) such that P>>1 the optimum separation time is given by equation (1.59)

$$t_R = [4.5f_g(k)][(1 + k)(2\eta/P_oD_{m,o})^{1/2}n^{3/2}d_c] \tag{1.59}$$

where η is the carrier gas viscosity [146,147]. Since the separation time is directly proportional to the column diameter smaller diameter columns will provide faster separations, at least for thin-film columns. Also hydrogen should be used as the carrier gas because of its low $\eta/D_{m,o}$ ratio. Helium and nitrogen are respectively 50% and 250% slower than hydrogen. Operating a given column under minimum plate height conditions with a vacuum outlet always yields the shortest separation time but the maximum attainable plate number is decreased. However, the column can be lengthened to compensate for this loss while still reducing the separation time. The gain in speed of analysis by vacuum-outlet operation is reduced by decreasing column diameters and increasing maximum plate numbers. At the extreme of both terms attainable with narrow-bore columns the saving of time is very small indeed.

Scott has shown that for any given separation there is a unique (and predictable) column length, particle size or column radius, and film thickness which will achieve a given separation in the minimum separation time [58,84]. The derivation of the necessary equation for packed and open tubular columns combines equation (1.48) and (1.57) with the minimum values obtained by differentiating equation (1.35). The equations obtained for the optimum values are quite complex and will not be given here. Full details are contained in reference [84]. The conclusions can be summarized in outline, as follows. For columns of a fixed length and film thickness there is both an optimum particle diameter and an optimum radius that provides a minimum separation time. This minimum is quite sharp with radii less than optimum resulting in a very sharp increase in separation time whereas the increase in radii in excess of the optimum being less severe. An optimum film thickness exists at a fixed radius (particle diameter) which would produce a minimum separation time. A film thickness significantly less than the optimum causes a rapid increase in separation time whereas films thicker than optimum cause a much more gradual increase in separation time. The separation time increases approximately linearly with the required efficiency and it will be very difficult to obtain short separation times for difficult to separate mixtures compared to simpler mixtures.

High speed separations, particularly with narrow-bore WCOT columns of 100 micrometers or less, places special demands upon the instrument design [131,134,148-150]. The gas chromatograph must be capable of high pressure operation, sample introduction systems must be capable of delivering extremely small injection band widths, and the detector and data acquisition system must have extremely low dead volumes and a fast response. Ideally, injection band widths should be on the order of a few milliseconds, injection volumes a few nanoliters, detector dead volumes no more than a few nanoliters and a detector response time of a few milliseconds. These constraints are incompatible with the specifications of analytical instruments in general use necessitating the use of wider-than-optimum bore columns in practice. Packed columns share the above constraints, and because of their lower permeability, can only be used for separations requiring a relatively low number of theoretical plates [83]. They have the practical advantage of a higher sample capacity which is useful in trace analysis and can be designed to operate at higher flow rates relaxing some of the constraints placed on extracolumn void volumes.

1.7.4 Relationship Between Column Variables and Sample Capacity in Gas Chromatography

Column and detector properties determine the minimum amount of a component that can be reliably distinguished from the background noise. If we arbitrarily select a signal to noise ratio of 4 as the minimum value for the confident determination of a peak in a chromatogram then for a mass sensitive detector the minimum detectable amount is given by

$$Q_o = 4 (2\pi)^{1/2} (R_n/S) (t_R/n^{1/2}) \qquad (1.60)$$

where Q_o is the minimum detectable amount, R_n the detector noise level and S the detector sensitivity [135,146,151,152]. For a concentration sensitive detector the minimum detectable concentration is the product of Q_o and the volumetric gas flow rate through the detector. The minimum detectable amount or concentration is proportional to the retention time, and therefore, directly proportional to the column radius for large values of n. It follows, then, that very small quantities can be detected on narrow-bore columns.

The sample capacity Q_s, arbitrarily defined as the maximum amount of a component that can be injected on a column giving a limited (10%) increase in peak width, is given by

TABLE 1.12

SAMPLE CAPACITY (NG) FOR WCOT COLUMNS OF VARIOUS DIAMETERS AND FILM
THICKNESSES

| Internal | Film Thickness (micrometers) | | | |
Diameter (mm)	0.1	0.2	1.0	5.0
0.10	10	20		
0.25	25	50	250	1200
0.32	32	64	320	1600
0.53	53	130	530	2600

$$Q_s = \pi (MP_s^0/RT_c)(r_c - d_f)(1 + k)(3LH)^{1/2} \qquad (1.61)$$

where M is the molecular weight, P_s^0 the saturation vapor pressure
of the solute at T_c, T_c the column temperature, and R the universal
gas constant. The maximum sample size is proportional to both the
film thickness and to the column radius. Some typical values for Q_s
as a function of column radius and film thickness are given in
Table 1.12. The working range for the column is given by the ratio
of Q_s/Q_0. To a reasonable approximation for columns with large
plate numbers the working range is proportional to the radius
squared, a distinct advantage for wide-bore columns. The working
range should generally exceed the concentration range of the
components to be analyzed. If not the detector response for trace
components can only be distinguished from the noise level when the
column is overloaded for the main peaks. The selection of a column
radius, therefore, often implies a compromise between the speed of
analysis and the required working range.

Ettre has argued that the above definition of Q_s is too
severe since increasing the peak width by 10% will only reduce the
resolution by about 5%, a value that might be tolerable in many
cases [123]. As an alternative estimate of Q_s he proposed that the
volume corresponding to one theoretical plate be used. The peak
width is not significantly affected by the sample size as long as
the concentration of the solute vapor in the carrier gas plug
entering the column is smaller than the concentration of the
solute vapor in the gas volume corresponding to a single
theoretical plate. This is given by equation (1.62)

$$V_{max} = a_k V_G (1 + k)/n^{1/2} \qquad (1.62)$$

where V_{max} is the volume of one theoretical plate, a_k a factor that varies with the chromatographic conditions, and V_G the gas phase column volume. Since the factor a_k cannot be accurately defined equation (1.62) is best used in a relative rather than absolute sense. It predicts that the best approach to increasing the sample capacity is to increase the column radius and film thickness, but taking care that the ratio is decreasing. This can be accomplished by a higher relative increase in the film thickness than in the tube diameter. Although thick-film, wide-bore columns may approach the working range and sample capacity of packed columns used for analytical applications, it is always possible to prepare packed columns with a lower phase ratio that would be superior for preparative applications if a large number of theoretical plates are not needed for the separation.

1.7.5 Temperature and Flow Programming in Gas Chromatography

There are two general dependencies for retention data as a function of temperature in gas chromatography that influence resolution and separation time. These are most easily envisaged for compounds belonging to a homologous series. In this case there is usually an exponential correlation between the adjusted retention data (time, volume, specific retention volume, capacity factor, etc.) and the number of repeating methylene groups, solute saturation vapor pressure or boiling point under isothermal (constant temperature) conditions. A similar exponential correlation usually exists between adjusted retention data and the reciprocal of column temperature at a constant flow rate. Possible exceptions are the first few members of a homologous series. An example of the relationship between the adjusted retention time and column temperature for some homologous n-alkanes and n-alkylbenzenes is shown in Figure 1.11 [153]. The slopes of each member of the homologous series are approximately identical but different for the two homologous series. The lines for the two series intersect at various temperatures corresponding to the column temperatures at which there is no separation. At other than these intersection temperatures separation occurs. In isothermal gas chromatography there is an optimum temperature for a particular separation at which resolution will be a maximum. The range of temperatures over which a separation can be achieved depends on the properties of the individual solutes and the stationary phase. When the separation is easy to achieve and

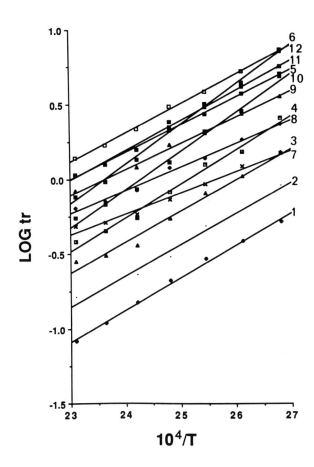

Figure 1.11 Plot of the logarithm of the adjusted retention time against the reciprocal of column temperature for two homologous series of n-alkane and n-alkylbenzenes (and ortho-xylene) on a 2 meter 20% (W/W) column of tetrabutylammonium 4-toluenesulfonate coated onto Chromosorb (W-AW). Solute identification, 1 = decane, 2 = undecane, 3 = dodecane, 4 = tridecane, 5 = tetradecane, 6 = pentadecane, 7 = benzene, 8 = toluene, 9 = ethylbenzene, 10 = propylbenzene, 11 = ortho-xylene, and 12 = butylbenzene.

properties of the sample and stationary phase allow, the highest effective column temperature should be used to minimize the separation time.

Since there is an approximate exponential relationship between retention time and solute boiling point under isothermal conditions, it is impossible to establish a suitable temperature for the analysis of mixtures with a boiling point range exceeding ca. 100°C. This is commonly referred to as "the general elution

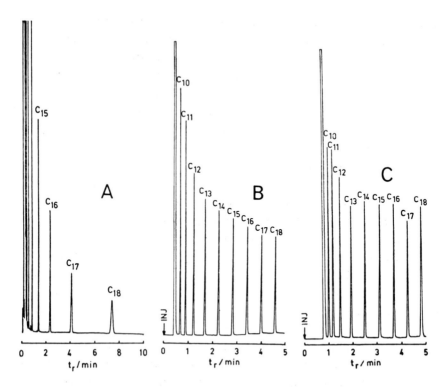

Figure 1.12 Comparison of isothermal (A), flow programmed (B), and temperature programmed (C) separation of C_{10} to C_{18} n-alkanes. (Reproduced with permission from ref. 154. Copyright Dr. Alfred Huethig Publishers).

problem" and is characterized by poor separation of early eluting peaks and poor detectability of late eluting peaks due to band broadening when a compromise temperature is chosen. The problem can be solved by either temperature or flow programming. Isothermal, temperature program, and flow program modes are compared in Figure 1.12 for the separation of a mixture of hydrocarbons covering a wide boiling point range [154]. Only in the programmed modes is a complete separation achieved in a reasonable time. The programmed modes tend to oppose the logarithmic dependence of retention on boiling point and, under certain optimized conditions, the relationship becomes approximately linear, with only a slow increase in peak width with increasing temperature or flow rate.

The temperature program mode is the most widely used separation technique in gas chromatography [155]. As well as reducing separation times for samples with a wide boiling point range and improving sample detectability temperature program techniques are the most useful approach for scouting the properties of an unknown sample and are compatible with the large volume injection modes employing low temperature solute refocusing used in trace analysis. A temperature program consists of a series of changes in the oven temperature and includes isothermal and controlled temperature rise segments that are selected by a mechanical or microprocessor controller. In essence, most programs are simple, consisting of an initial isothermal period, a linear temperature rise segment, a final isothermal period at the temperature reached at the end of the rise segment and a cool down period to return the oven to the starting temperature. The initial and final isothermal periods are optional, the temperature rise segment can be selected over a wide range (0.1 to ca. 30°C/min), non-linear changes in temperature may be used (extremely rare) and for complex mixtures, several linear programs may be used in sequence to optimize the separation. The initial oven temperature is selected with due consideration to the resolution of the earliest eluting peaks in the chromatogram. If the temperature chosen is too high the resolution of the initial peaks may be inadequate and, if it is too low, resolution may be acceptable but the separation time will be extended needlessly. The final temperature should be selected so that the termination of the temperature rise segment and elution of the last sample component coincide unless the last few eluting peaks are particularly difficult to separate and require an isothermal period. Peaks eluting after completion of the temperature rise segment will be wider than those eluted during the program. The selection of the heating rate represents a compromise between the necessity of maintaining a minimum acceptable resolution for the sample and the desire to reduce the separation time. This is governed largely by the complexity of the sample and its boiling point range. For samples containing components of different polarity temperature induced changes in selectivity make the prediction of the resolution of closely-spaced peaks a problem. Certain generalities can be made however. For the most difficult separations a slow heating rate will usually provide the optimum resolution. The separation time of weakly retained solutes is more readily adjusted by changing the flow rate of the carrier gas than the heating rate. For strongly retained solutes increasing the heating rate causes a proportional decrease in the separation time at a constant carrier gas flow rate. The retention time of well retained solutes are little affected by changing flow rates in temperature programmed gas chromatography. In general, heating rates are usually less than 5°C/min for open tubular columns and for narrow-bore columns, which are operated at low flow rates, it is particularly important to optimize the ratio of the heating rate to flow rate to minimize the separation time without compromising resolution [51,20,131].

The only practical limitations placed on the range of the temperature rise segment are the thermal stability of the stationary phase and flow changes. With the high temperature stationary phases recently introduced, the accessible upper temperature ranges have been dramatically increased. For constant-pressure-controlled instruments the flow rate will decrease with temperature, affecting the response of concentration-dependent detectors [156,157]. For this reason flow-controlled instruments are preferred for temperature programmed operation. Employing identical dual columns and simultaneous detection in the difference mode, the rapid baseline rise associated with flow and stationary phase bleed problems in packed column gas chromatography can be highly attenuated.

A rigorous theoretical treatment of temperature programmed gas chromatography is very difficult and the relationships proposed based on experimental data from a few, usually two, isothermal experiments do not have an exact solution. In linear temperature programmed gas chromatography the solute zones accelerate as they travel along the column in a manner that depends on the initial temperature, T_1, and the programming rate, r. If a zone center travels a distance dx in time dt then we can write that

$$dt = (dx/u_x)(1 + K_L/\beta) \tag{1.63}$$

where u_x is the linear velocity at point x, β the phase ratio, and K_L the gas-liquid partition coefficient (we will consider only a partition model here although other contributions to retention can easily be included) [158,159]. The partition coefficient can be expressed as

$$K_L = a.\exp(\Delta H_s/RT_c) \tag{1.64}$$

where a = exp($\Delta S_s/R$), ΔS_s is the partial molar entropy of solution, R the universal gas constant and ΔH_s the partial molar enthalpy of solution. In isothermal gas chromatography, the mean linear gas velocity is u = L/t_m (L being the column length and t_m the gas holdup time). If we substitute this relationship into equation (1.64), rearrange and integrate over the column length we arrive at

$$\ln k = \ln (a/\beta) + \Delta H_s/RT_c \qquad (1.65)$$

which is adequate for describing the influence of temperature on retention for isothermal operation. In temperature programmed gas chromatography the oven temperature increases during the analysis according to

$$T_c = T_1 + r.t \qquad (1.66)$$

where T_c is the column temperature, r the program rate (°C/min), and t the time from starting the program. Differentiation and substitution in equation (1.64) gives

$$dT/r = (dx/u_x)(1 + [a/\beta]\exp.[\Delta H_s/RT_c]) \qquad (1.67)$$

The integration of equation (1.67) is the point at which problems arise. After rearranging, the temperature component has to be integrated over the limits T_1 to T_E (the elution temperature) and the distance terms over the limits of zero to L, the column length. There is no exact analytical solution to the uncommon exponential integral and either numerical solutions using a computer [158-165] or graphical methods [166] have to be used, both of which frequently employ simplifying assumptions. Among the more common of these are that the column gas holdup time does not change with temperature (which it does) and that the mean gas velocity remains constant during temperature programming (which is untrue due to changes in gas viscosity and thermal expansion). The models thus take on a degree of inexactness although providing adequate insight into the interrelationships between the primary experimental variables. Solutions without approximations are more complicated but are intuitively more likely to be correct [159,164,165].

The lack of an exact mathematical model to describe temperature programmed separations makes computer simulation for

their rapid optimization difficult. Simplex optimization of
experimental variables [165] and a model based on the linear-
elution-strength approximation [167] are two approaches to the
problem that have been used successfully. Both methods are based
on analogies or approaches more commonly used for mobile phase
optimization in liquid chromatography (section 4.6). Simplex
methods are independent of any physical model but impose a fixed
experimental design on the data collection process. This approach
may require many experiments before converging to an optimum
(hopefully global optimum) set of experimental conditions. The
main advantage of the linear-elution-strength approach is that it
only requires experimental data from two temperature programmed
separations of a sample using different program rates. A series of
empirical equations are then employed to predict optimum
separation conditions using relative resolution maps. This latter
approach is very promising and is supported by inexpensive
commercial software; more studies, however, are required to
indicate its general usefulness.

Saxton has proposed that the emergence temperature (elution
temperature) in temperature programmed gas chromatography can be
used as a physical constant in the same way that boiling point is
used to identify individual solutes from tabulated reference data
[168,169]. This approach assumes a model akin to extractive
distillation in which solute retention times are the product of
two distinct sequences. In the first stage the solute is
immobilized at the head of the column during the time taken for
the column temperature to increase from the starting temperature
to the threshold temperature. The latter is defined as the
temperature at which a solute band, which was initially cold
trapped at the head of the column, begins to move through the
column. The second stage begins when the temperature is
sufficiently high that the stationary phase no longer affects the
retention time of that particular solute and hence the velocity of
the solute band becomes equivalent to that of the carrier gas. The
second stage ends with the emergence of the solute from the
column. A significant tenet of this model is that separations in
temperature programmed gas chromatography occur primarily at the
head of the column as a result of differences in threshold
temperatures which in turn depend on the strength of
intermolecular interactions between the solute and stationary
phase. Additional column length beyond that needed to trap the

compounds improves resolution, primarily by delaying emergence of the compounds from the end of the column while essentially maintaining the temporal spacing achieved at the head of the column. The delay leads to increased solute terminal velocities and, hence, narrower recorded peak width by allowing additional time for the column temperature to rise. Under standardized conditions the emergence temperature is largely independent of the column length, carrier gas flow rate, type of carrier gas, volume of liquid phase and heating rate. Provided that the initial temperature for the program is below the threshold temperature then the emergence temperature is independent of the start temperature or the length of any initial isothermal period. The proposed method relies upon arbitrarily assigning a specific emergence temperature to a suitable reference compound. Saxton suggested using n-eicosane assigned a value of 200°C for the primary standard and chlorpyrifos as 194°C as a secondary standard for use with element selective detectors. The standardized protocol then calls for arranging the experimental conditions such that the start temperature for the program is less than the threshold temperature of the sample components and the heating rate and flow rate are adjusted so that n-eicosane elutes with an emergence temperature of 200°C. How useful the emergence temperature becomes as an identification method will depend on whether the protocol becomes established as a standard method and on the availability of tabulated data similar to the compilations of retention index values available for isothermal gas chromatography.

Flow programming is not widely used in gas chromatography [154,170-173]. Many of its advantages tend to parallel those of temperature programmed gas chromatography, which is experimentally easier to perform. Exponential flow programs under isothermal conditions give peak distributions which are very similar to those obtained with temperature programming.

Short or wide-bore open tubular columns are most often used in flow programming because of their high permeability whereas packed columns are rarely used due to their limited dynamic flow range at normal operating pressures. The range of boiling points that can be covered in a flow program is relatively large but not as broad as with temperature programming.

The principal advantages of flow programming are that it shortens separation time for mixtures of wide volatility, while permitting the operation of the column at lower temperature. Since the analysis is performed at lower temperatures column bleed is less of a problem, and since the pressure of the gas in the column can be changed instantaneously, returning the instrument to the start conditions is very rapid. Flow programming might be preferred to temperature programming for the analysis of thermally labile samples. Disadvantages include a decrease in column efficiency for late eluting peaks, difficulty in calibrating flow sensitive detectors, and the inconvenience of instrument operation at high pressures.

1.7.6 Serially Connected Columns in Gas Chromatography

Increasing the effective length of a column by coupling identical column segments in series is not, in general, a very effective optimization strategy, except when the desired resolution can be achieved by a relatively modest increase in efficiency (see section 1.6). The analysis time and column pressure drop will increase proportionately with column length while resolution will increase only as the square root of the column length. Alternatively, serially coupling columns of different selectivity can result in dramatic changes in resolution. This optimization strategy is sometimes called selectivity tuning [121,174-176]. For convenience, and in keeping with experimental practice, we will consider two columns with different selectivities connected in series, although the arguments presented and general theory can be applied to as many columns as desired. Reversing the order of the columns in our two column tandem or changing the flow rate through the column system will result in changes in resolution and possibly elution order due to changes in selectivity of the complete system. Kaiser explained this finding elegantly by pointing out that the important parameter was not the retention time for the column system but the residence time of each solute in the different column segments [177]. The residence time, in turn, will depend on the interplay between the capacity factors characterizing each column unit and the compressibility effect of the carrier gas on local velocities and will, therefore, depend on the system operating variables, in particular, on the ratio of the column inlet to outlet pressure. In theory, the selectivity of the chromatographic system can be varied continuously over the selectivity range represented by the selectivity of the individual columns as limiting values. In practice, experimental constraints due to large flow rate differences make the extreme single column selectivity values difficult to reach without loss of column efficiency. However, most of the selectivity range is available for use, and of course, the columns can be used independently to generate the single column selectivity values. Selectivity tuning by series coupling of columns with different stationary phases is used almost exclusively with open tubular columns. This is in part due to the higher efficiency available for the separation of complex mixtures when open tubular columns are used and also because with packed columns an optimum selectivity is easily achieved by mechanically mixing packings in the desired proportions (see section 2.5.5). In contrast, mixed substrate open tubular columns are often difficult to prepare and series coupling of columns containing different phases becomes a more favorable option.

The residence time of the sample in the two columns can be adjusted by changing the relative column lengths, by adjusting the average linear velocity of the carrier gas in the two columns to preselected values, or by operating the columns at different temperatures in a dual oven gas chromatograph [174-178]. For practical convenience, the junction between the two columns is both pressure and flow controlled with also, for convenience of accurately setting the experimental conditions for the first column, the possibility of diverting a fraction of the sample from the first column to a detector. The residence time of the sample in the first column of the tandem will depend on the pressure drop between the column inlet and the column connection point. For the second column the residence time depends on the outlet pressure of the first column (pressure at the point of the column connection) and on the pressure at the column outlet of the second column, which will normally be atmospheric when the flame ionization detector is used. It is also possible to have a higher or lower flow in the second column than in the first. If a higher flow is required, carrier gas is added at the column connection point while, if a lower flow is desired, a part of the gas flow (and consequently, the sample) is vented.

The total retention of a given solute and the gas holdup time of an unretained solute eluting through two serially connected columns is the sum of the retentions in the individual column units, and thus we can write that

$$t_{ms} = t_{m1} + t_{m2} \tag{1.68}$$

and

$$k_s = (t_{m1}k_1 + t_{m2}k_2)/t_{ms} \tag{1.69}$$

where subscript s refers to the total system and 1 and 2 the first and second columns, respectively, k the capacity factor, and t_m the gas holdup time [174,179]. The ratio t_m/t_{ms} is called the relative retentivity and is given the symbol θ with subscripts to indicate the column in question. Substituting for θ in equation (1.69) leads to

$$k_s = \theta_1 k_1 + \theta_2 k_2 \tag{1.70}$$

and since by definition $\theta_1 + \theta_2 = 1$ then

$$k_s = \theta_1(k_1 - k_2) + k_2 \tag{1.71}$$

$$k_s = \theta_2(k_2 - k_1) + k_1 \tag{1.72}$$

The capacity factor for the system, therefore, is a linear function of the relative retentivity on either column and a plot of θ against k_s will have a slope equal to the difference in capacity factors on the two columns. Thus, the capacity factor of a solute for the two column system can be adjusted to any value between the capacity factor values measured on the individual columns by adjusting the relative retentivities.

In practice, it is more difficult to optimize resolution as a function of the relative retentivity than to optimize retention. Thus, unless the mixture is very complex or contains components that are particularly difficult to separate it may be possible to optimize a particular separation using the linear equation (1.72) as demonstrated by Ettre [177]. Figure 1.13 illustrates the relative change in peak position for a polarity test mixture with two identical, serially coupled open tubular columns, coated with a poly(dimethylsiloxane) and Carbowax 20 M stationary phases, as a function of their relative retentivity on the second column. The linear relationship predicted by equation (1.72) effectively predicts the relative peak positions and indicates that a nearly

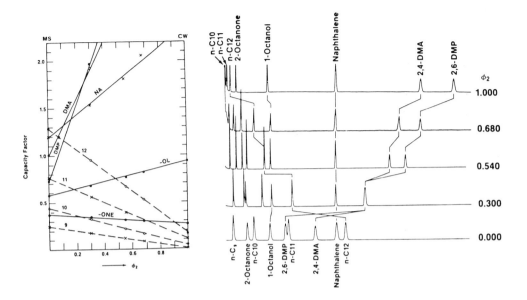

Figure 1.13   Separation of a polarity test mixture on two 25 m x
0.25 mm I.D. (d_f = 0.25 micrometers) serially coupled open tubular
columns, coated with a bonded poly(dimethylsiloxane), column 1,
and stabilized Carbowax 20 M, column 2, stationary phases. The
plot of k_s against θ is shown on the left and the experimental data
on the   right. Those probes identified by a number are n-alkanes,
2,6-DMP = 2,6- dimethylphenol and 2,4-DMA = 2,4-dimethylaniline.
(Reproduced with permission from ref. 174. Copyright Friedr.
Vieweg & Sohn).

optimal separation of the peak to peak positions would occur at a

θ value for the second column of about 0.40.

The trial and error approach to optimizing a separation
using serially connected columns can be very time consuming,

particularly for complex mixtures. Method development time can be reduced significantly by employing computer simulation procedures based on experimental data obtained in a relatively small number of preliminary experiments. Purnell and co-workers have described a method for the selection of the optimum column lengths to be coupled to provide a given separation based on the calculation of the relative column permeabilities, taking into account gas compressibility effects, and using the window diagram method to optimize selectivity [180,181]. An alternative approach, which is perhaps more versatile when a number of different mixtures are to be separated, is to fix the dimensions of the two columns and optimize the separation by varying the individual flow rates through the two columns by adjusting the coupling point pressure with a constant inlet and outlet pressure, or by independently adjusting the temperature of the two columns using a dual oven gas chromatograph. Computer assisted procedures for the prediction of the optimum selectivity by tuning the coupling point pressure [182-186] or column temperatures [187] for serially coupled column systems have been described. The optimization procedures are based on mathematical models in which the dependence of the retention indices of all solutes on the variable parameter (coupling point pressure, temperature) are described by polynomial equations or by using the variation in the system solute capacity factor as a function of the relative retentivity corrected for differences in the column permeability. A threshold optimization parameter based on the observed number of separated components and the total separation time has been proposed as an objective function for computer assisted optimization [184,185]. These methods are still evolutionary in character but promising for the future. There is no doubt that the development of predictive methods for the optimization of the separation of complex mixtures by serially coupled columns could have an enormous impact on the acceptance of this powerful tool beyond a small number of research laboratories that employ these techniques at present.

1.7.7 Column Types in Liquid Chromatography

Since many of the developments in modern liquid chromatography are of recent origin the nomenclature commonly used is less standardized than that of gas chromatography [188]. We have made an arbitrary selection of the terms we prefer to use in this book along with some other common alternatives in Table 1.13.

TABLE 1.13

NOMENCLATURE AND PHYSICAL CHARACTRISTICS OF ANALYTICAL LIQUID
CHROMATOGRAPHIC COLUMNS

Preferred Name	Other Names	Length(*)	Internal Diameter (mm)	Flow Rate (μl/min)
Conventional Packed		5-30 cm	4-5	1000-3000
Small Bore Packed	Microbore	10-100 cm	1-2	5-200
Packed Capillary	Slurry Packed Capillary	20-200 cm	0.1-0.5	0.1-20
Semi-Packed Capillary	Packed Capillary Semipermeable Capillary	1-100 m	0.02-0.1	0.1-2
Open Tubular	Microtubular Capillary	1-100 m	0.01-0.075	0.05-2

(*)In this case length refers to the longest single column normally
used.

The principal column type used in analytical liquid chromatography
and, thus referred to here as a conventional column, is a packed
column with an internal diameter of 4-5 mm and length of 5-30 cm
packed with porous particles of 3, 5, 7, 10, or 20 micrometers in
diameter. These columns are operated at reasonably high flow
rates and apart from the short, small particle diameter columns
can be used at close to their optimum performance with most
commercially available analytical liquid chromatographs. Wider
bore packed columns of similar length are used primarily for
preparative liquid chromatography where a large volume of column
packing is desirable to enhance sample capacity. Most recent
interest in analytical applications has focused on various types
of small bore packed and open tubular columns in the hope of
preparing columns with a higher intrinsic efficiency, in obtaining
greater economy in the use of column packing materials and a
reduction in mobile phase consumption, greater mass detection
through a reduction in peak volumes, and greater compatibility
with new detection principles, such as mass spectrometry and flame
based detectors that become possible when low volumetric mobile
phase flow rates are used [188-197]. Some of these advantages have

been realized for the small bore packed columns having internal
diameters of 1-2 mm and operated at volumetric flow rates of 5-50
microliters/min. To operate these columns at optimum efficiency a
low dispersion instrument is essential. Conventional instruments
may be modified for this purpose or small bore compatible
instruments purchased. Columns are also commercially available.
However, small bore columns have not been widely accepted for
reasons which will follow shortly. Packed columns of capillary
dimensions are also known and fall into two categories. Semi-
packed capillary columns have internal diameters less than 0.1 mm
and are loosely packed with particles of 10 to 30 micrometers in
diameter to give a d_c/d_p ratio of 2 to 6 for optimum stability
[198,199]. These columns are prepared by drawing glass tubes
filled with packing to capillary dimensions and in so doing some
of the packing becomes embedded in the column wall. This enhances
the mechanical stability of the packing and gives rise to the
characteristic open bed, zigzag structure of these columns. These
columns have not been widely used, are not commercially available
and demand extensive modification to existing instruments if they
are to be operated at close to their optimum conditions. Capillary
columns of slightly larger diameter, 0.1-0.5 mm, and higher
packing density have been prepared by slurry packing flexible
fused silica open tubular columns [190,200-204] or glass lined
stainless steel capillaries [205,206] with porous particles of 3
to 10 micrometers in diameter. These columns can be prepared in
comparatively long lengths and have been used primarily for the
analysis of complex mixtures were a large number of theoretical
plates was required and the separation time was of secondary
concern to performance. Extensive instrument modification is
needed to obtain optimum performance with these columns although
instrumental constrains are not as severe as they are for the
semi-packed capillary columns. These columns are commercially
available from a small number of manufacturers but have not been
used outside of a research setting so far. Open tubular columns of
internal diameters of 5-75 micrometers are truly experimental
systems whose optimum theoretical performance has rarely been
approached in the laboratory [191,196,207-214]. Severe technical
problems in column fabrication and instrument requirements limit
this column type to academic studies at the moment. Unlike the
situation in gas chromatography were open tubular columns are
clearly favored in terms of efficiency and separation time in

liquid chromatography, if realistic dimensions and operating conditions are considered, open tubular columns will only out perform packed columns when extremely large plate numbers are required for a separation [215,216].

1.7.8 Factors Affecting the Efficiency of Packed Columns in Liquid Chromatography

For a packed column if the mobile phase composition and velocity are held constant then the performance of various columns will depend on the column length, particle diameter and the available operating pressure. The intrinsic efficiency of a packed bed per unit length increases as the particle diameter is reduced, Figure 1.14 [3]. Also apparent from Figure 1.14 is the dramatic reduction in the contribution of mass transfer resistance to the column plate height as the particle size is reduced. For particles less than 5 micrometers in diameter the plate height versus mobile phase velocity curves are essentially flat in the region of the plate height minimum indicating that small-particle diameter columns can be operated at much higher linear velocities without appreciable loss in efficiency.

To place the above considerations into perspective it is necessary to consider the influence of the available operating pressure on column performance. From equation (1.16) it can be seen that the operating pressure required to maintain a constant mobile phase velocity is proportional to the ratio L/d_p^2. Since the available operating pressure is finite, the column length must be reduced as the particle diameter is decreased when the operating pressure is the column limiting factor. The theoretical efficiency (calculated assuming that the plate height equals twice the particle diameter) and relative column operating pressure referenced to a 25 cm column packed with 10-micrometer diameter particles for some typical packed column types are summarized in Table 1.14. Since these columns represent those most widely used in practice, we see that most separations are performed with about 5 to 20 thousand theoretical plates and that this range is largely independent of the particle diameter. However, since the retention time at a constant linear (optimum) velocity is proportional to the column length, equation (1.9), this arbitrary fixed number of plates is made available in a shorter time for columns packed with smaller diameter particles. Thus, the principal virtue of the use of small-diameter particle columns is that they permit a reduction

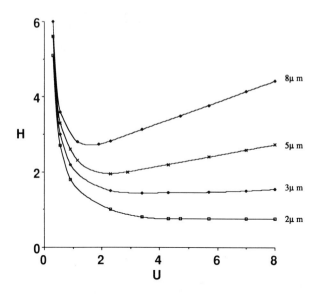

Figure 1.14   Plot of plate height (mm) as a function of the mobile phase velocity (ml/min) for a series of columns of 0.46 cm I.D. packed with a   spherical porous n-octadecylsilanized silica of different particle diameters. For the 8 and 5 micrometer diameter particles the column length was 25 cm, for the 3 micrometer diameter particles 15 cm and for 2 micrometer diameter particles 4 cm. (Reproduced with permission from ref. 3. Copyright Elsevier Scientific Publishing Co.)

in separation time for those separations that do not require an unusually large number of theoretical plates [217-220]. For separations requiring a very large number of theoretical plates long columns are needed and small-particle diameter packed columns are unsuitable for this purpose due to limits on length established by the available operating pressure [61,221]. The use of short columns packed with small diameter particles and operated at a high linear velocity for the separation of simple mixtures is sometimes referred to as high-speed liquid chromatography or fast liquid chromatography. A typical example is shown in Figure 1.15 for the separation of a seven component test mixture in less than 90 seconds [222]. Still faster separations are possible using specially designed instruments with low extracolumn dispersion and very fast detection systems [219]. Although the separation time is independent of the column diameter, narrow bore columns elute peaks in a much smaller volume of mobile phase than conventional bore columns placing still greater demands on the design characteristics of the instrument [223-227]. For this reason wider

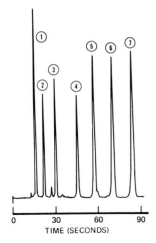

TIME (SECONDS)

Figure 1.15 Fast analysis of a test mixture on a 10 cm x 4.6 mm
I.D. column packed with 3 micrometer octadecylsilanized silica
with a mobile phase flow rate of 3.4 ml/min (acetonitrile-water
7:3) and operating pressure of ca. 340 atmospheres. Peaks: 1 =
uracil, 2 = phenol, 3 = nitrobenzene, 4 = toluene, 5 =
ethylbenzene, 6 = isopropylbenzene, and 7 = tert.-butylbenzene.
(Reproduced with permission from ref. 222. Copyright Friedr.
Vieweg & Sohn).

than normal bore columns, for example 8 cm x 0.62 cm I.D., have
been suggested for use with conventional liquid chromatographic
equipment [218]. Not all instruments are suitable for use in high-
speed liquid chromatography which has served to slow the growth of
this technique.

It is conventional practice to quote column performance as
the number of theoretical plates normalized to a length of one
meter. This approach was sensible for gas chromatography where
columns exceeding one meter in length are normally used. In liquid
chromatography, however, the same scale leads to an inflation of
expectations not born out in practice. For example, we see from
Table 1.14 that a column packed with 2 micrometer particles could
generate about 250,000 theoretical plates per meter.
Unfortunately, because of pressure limitations the longest column
that could actually be used in practice is about 7 cm (assuming an
operating pressure of about 400 atmospheres) which corresponds to
the more modest figure of about 17,500 theoretical plates [228].

In assembling the data in Table 1.14 it was assumed that the
packing density of the column is independent of particle size and
column length. In practice, this is not the case, and smaller bore

TABLE 1.14

THEORETICAL EFFICIENCY AND RELATIVE OPERATING PRESSURE FOR
CONVENTIONAL LIQUID CHROMATOGPHIC COLUMNS

Column Length (cm)	Particle Diameter (micrometers)	Column Plate Count	Plates Per Meter	Relative Operating Pressure
50	20	12,500	25,000	0.5
25	10	12,500	50,000	1.0
10	10	5,000	50,000	0.25
25	5	25,000	100,000	4.0
15	5	15,000	100,000	2.4
5	5	5,000	100,000	0.8
15	3	25,000	167,000	6.7
10	3	16,700	167,000	4.4
5	3	8,300	167,000	2.2
3	3	5,000	167,000	1.3
3	2	7,500	250,000	3.0

columns and smaller diameter particles are more difficult to pack
homogeneously than conventional bore columns with average size
particles [228,229]. These columns tend to be more unstable and
degrade faster than conventional columns. Faster dissolution rates
for the packing in the mobile phase (a saturator column is very
helpful), greater susceptibility to plugging by sample or wear
particles from the solvent delivery system, and limited pressure
shock stability (very fast valve switching or a bypass injector
should be used) all contribute to limiting the useful column life.
Sizing small particles to provide a narrow range of particle
diameters is also more difficult than for particles of larger
diameter.

There is also a practical limit for the longest column
length that can be homogeneously packed. This depends on the
particle size and the packing technique but is not much greater
than the largest dimension for each particle size given in Table
1.14 for conventional and small bore columns. The only practical
method of obtaining very high plate counts with these columns is
to couple several, short, well-packed columns together. Early
reports indicated that conventional bore columns could not be
coupled into longer lengths with additive efficiency [230]. This
is no longer accepted as true although from recent studies it is
obvious that not just any group of columns can be connected
together with additive efficiency [231-234]. The individual
columns to be concatenated must be efficient, have similar

efficiencies and stable packing structures and must not exhibit significant asymmetry for eluted peaks. Good columns and moderate columns when coupled together lead to results that are far from additive most probably due to the heterogeneous permeability of the coupled columns which results in an exaggerated multipath term increasing in importance as the column length is increased. For perspective, Snyder et al. [231] coupled nine 15 cm x 0.46 cm columns packed with 5 micrometer diameter octadecylsilanized silica in series and generated 76,000-85,000 theoretical plates (operating pressure not given). Longer columns packed with coarser particles will yield a greater number of theoretical plates. Halasz predicted that a 6.7 meter column packed with 8 micrometer particles operated at about 400 atmospheres would deliver about 400,000 theoretical plates with a holdup time of about 1 hour [232].

For long columns operated at high pressures and short, small particle columns operated at high flow rates a considerable amount of heat is generated due to frictional heating of the mobile phase during its passage through the column [217,218,232-236]. Heat is generated at a constant rate over the whole length of the column. However, as the thermal conductivity of the packing material is small most of the heat is carried away by the mobile phase. Thus there is a negative temperature gradient from the center of the column to the wall and a positive temperature gradient from the column inlet to the column outlet. Depending on the experimental conditions the radial temperature gradient may vary from 0.5 to several degrees celsius and the axial gradient by as much as 20 to 30°C. Such temperature gradients cause a non-uniformity in the migration velocity due to variation in the viscosity of the mobile phase and changes in the distribution constant of the solute between the mobile and stationary phases. The result of these effects is additional band broadening (loss of efficiency) and possibly peak asymmetry. Temperature gradients are less significant for small bore columns since lower volumetric flow rates are required to achieve the same mobile phase velocity and thus much less heat is generated and the reduced diameter of the columns permits heat to be conducted from it more readily. This allows long small bore columns operated at the maximum available operating pressure to be used to generate very large numbers of theoretical plates. For example, Scott and Kucera prepared an 18 m x 1.0 mm I.D. column packed with 8-micrometer diameter particles which when operated at its optimum linear velocity produced about one million theoretical plates [237]. Similarly, Menet et al. [234] prepared a 22 m x 1 mm I.D. small bore column packed with 8-micrometer diameter particles that generated about one million theoretical plates at a flow rate of 13-15 microliters/min and an operating pressure of 600-700 atmospheres. An example of the separation of a test mixture on this column and on a one meter column similar to the segments used to prepare the longer column are shown in Figure 1.16. The short column shown in this example is operated well above its optimum flow rate generating about 10,000 theoretical plates and completes the separation in about 6 minutes. At its optimum flow rate about 40,000-50,000 theoretical plates would be available but the separation time would be longer. The separation is much better on the long small bore column but now the separation time has been increased to about 25 hours. With very long columns the flow rate cannot be arbitrarily increased to reduce the separation time (pressure limited) so long separation times have to be accepted as the price to be paid for high efficiency. From a practical point of view about one million theoretical plates is the upper limit to performance that can be expected with a packed column using existing instrumentation and column packing technology. Such columns will only prove useful for the separation of very complex mixtures since for simple mixtures faster separations are possible using standard columns packed with smaller particles.

Although theoretically no more intrinsically efficient than the small bore columns discussed above, fused silica capillary columns seem to offer some advantages from a practical point of view [188,194,195,202-204]. They can be prepared in relatively long single lengths of several meters with small diameter particle packings. The columns are flexible, strong (provided that the protective polyimide outercoat is intact), easily coupled to the liquid chromatograph and since the column wall is made of fused silica, they facilitate on-column and in-column optical detection minimizing the detector contribution to extracolumn band broadening. There is some evidence that the small column diameter to particle diameter ratio of these columns influences the packing structure of the columns which require a lower mobile phase velocity for maximum efficiency and are more permeable than small bore columns. Both features favor the operation of longer columns with small particle diameter packings at a fixed pressure. An

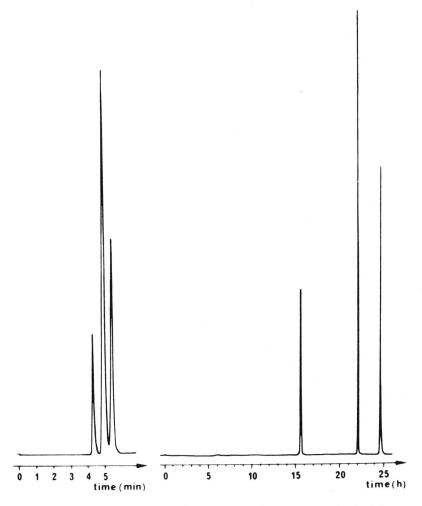

Figure 1.16 Separation of a test mixture by adsorption chromatography on a 1 m x 1 mm I.D. small bore column packed with 8 micrometer Zorbax B.P. Sil operated at a flow rate of 150 microliters/min (left) and a 22 m x 1 mm I.D. column of the same packing material prepared by series coupling of 1 m segments and operated at a flow rate of 15 microliters/min (right). (Reproduced with permission from ref. 234. Copyright American Chemical Society).

example of the separation of a complex mixture on a packed capillary column is shown in Figure 1.17 [238].

1.7.9 Relationship Between Column Diameter and Flow Characteristics in Packed Column Liquid Chromatography

The principal virtue of small bore and capillary packed

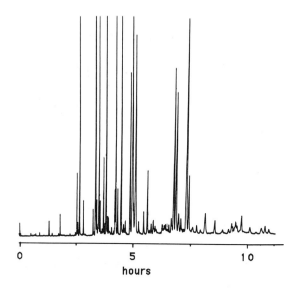

Figure 1.17 Separation of large ring polycyclic aromatic hydrocarbons extracted from carbon black on a 1.8 m x 0.2 mm I.D. fused silica capillary column packed with 3 micrometer spherical octadecylsilanized silica gel eluted with a stepwise solvent gradient at a flow rate of 1.1 microliters/min with an inlet pressure of about 360 atmospheres. Under isocratic conditions this column yielded ca. 225,000 theoretical plates. (Reproduced with permission from ref. 238. Copyright Friedr. Vieweg & Sohn).

columns is their potential for solvent savings. The volumetric flow rate, FR, in terms of column parameters is given by

$$FR = u\pi d_c^2 \epsilon_T / 4 \qquad (1.73)$$

where u is the linear velocity, d_c the column internal diameter and ϵ_T the total porosity (equal to ca. 0.75-0.85 for a well packed column) [57,239]. At a constant linear velocity the volumetric flow rate is directly proportional to the square of the column internal diameter, Table 1.15. Reducing the column diameter from 4.5 mm to 1.0 mm results in a 95 % decrease in solvent consumption which could be critically important if expensive or exotic mobile phases are employed for the separation. A further decrease in solvent consumption of ca. 16 fold is obtained by replacing a 1 mm small bore column with a 0.25 mm packed capillary column.

Another facet of small bore columns related to their low volumetric flow rate is their small peak volumes. For equal

TABLE 1.15

RELATIONSHIP BETWEEN COLUMN DIAMETER, SOLVENT CONSUMPTION, MASS
SENSITIVITY AND PEAK STANDRAD DEVIATION

L = 1 m, n = 40,000, k = 1, and detector sensitivity 1 μg/ml

Column Internal Diameter	Column Holdup Volume	Flow Rate	Relative Mass Sensitivity	Peak Standard Deviation
(mm)	(ml)	(ml/min)	(ng)	(μl)
4.5	11.6	1	232	116
2.0	2.2	0.19	44	22
1.0	0.55	0.047	11	5.5
0.5	0.14	0.012	2.8	1.4
0.25	0.03	0.003	0.7	0.3

injected mass the solute concentration at the column exit should
be greater for small bore columns due to the decreased volumetric
dilution. Assuming a Gaussian peak shape the peak volume, V_p, of a
solute as it leaves the column will be given by

$$V_p = \pi d_c^2 L \epsilon_T (1 + k) / n^{1/2} \qquad (1.74)$$

The peak volume is directly proportional to the square of the
column diameter and the column length and decreases with
increasing column efficiency (decreases for smaller particle
packings). The concentration at the peak maximum, C_{max}, is given by

$$C_{max} = m(8n/\pi^3)^{1/2} / \epsilon_T L d_c^2 (1 + k) \qquad (1.75)$$

where m is the mass of injected sample [193,194,239-242]. C_{max}
should increase in inverse proportion to d_c^2 corresponding to ca. a
20 fold increase in mass sensitivity for a reduction in the column
diameter from 4.5 mm to 1.0 mm. However, this claim which is
frequently made in the literature, ignores the reality of the
operating characteristics of the different diameter columns. The
sample size that can be loaded on a column without significantly
changing the performance of the column is directly proportional to
the column volume (d_c^2) and provided that the available sample is
not restricted then loading the maximum allowable sample size on
each column will result in identical values of C_{max}. This is a
realistic scenario since in normal practice to avoid column
overload injection volumes for small bore columns are typically
0.2-0.5 microliters. Thus the mass sensitivity will increase with

a decrease in column diameter only when the available sample size is limited.

An additional consideration dictated by the small peak volumes generated by small bore and capillary packed columns is the need to miniaturize detector volumes to minimize extracolumn band broadening. For example, to maintain acceptable column efficiency with spectroscopic detectors using 1 mm bore columns cell volumes ca. 1 microliter or less and a path length of 1 to 5 mm are generally used. Typical flow cells used with conventional bore columns have volumes of ca. 8 microliters and path lengths of ca. 10 mm. The reduced path length and increase in noise resulting from a reduction in the optical diameter of the low volume flow cells leads to a decrease of the detector sensitivity. Taking these factors into consideration the observed increase in mass sensitivity in changing from a 4.5 mm to a 1 mm diameter column is usually less than a factor of 5 and depends on the detector response characteristics.

The small peak volumes typical of samples eluted from small bore columns and short small diameter particle columns used in high-speed liquid chromatography place severe demands on the dispersion characteristics of all components of the liquid chromatograph. The standard deviation of a peak eluting from a column is given by

$$\sigma_{col} = V_m(1 + k)/n^{1/2} \tag{1.76}$$

where σ_{col} is the standard deviation of the eluted peak in volume units (peak volume is equivalent to 4 σ_{col}) and V_m is the column holdup volume. Using this equation some typical values of the peak standard deviation for columns of different dimensions and column packing are given in Table 1.16 [239]. If a 10% loss in column efficiency due to extracolumn dispersion is considered acceptable then the peak standard deviation due to extracolumn contributions should not exceed approximately 0.32 times the standard deviation of the peak volume [193,207,241-246]. The most severe case is represented by peaks eluting close to the holdup volume, requiring volume standard deviations less than one microliter for short small bore columns and not more than about 10 microliters for conventional bore columns. The sources of extracolumn band broadening include contributions from injection volumes, connecting tubes, detector volumes and electronic time constants.

TABLE 1.16

STANDRAD DEVIATION OF PEAK VOLUMES FOR TYPICAL CONVENTIONAL AND SMALL BORE COLUMNS

Column Length (cm)	Internal Diameter (mm)	Particle Size (μm)	Column Efficiency (n)	Column Holdup Volume (ml)	Peak Standard Deviation (μl)	
					k = 0	k = 5
25	4.6	20	6,250	3.0	38	228
25	4.6	10	12,500	3.0	27	161
25	5.0	10	6,000	3.7	50	286
25	4.6	5	25,000	3.0	19	114
10	5.0	5	6,000	1.5	20	116
6	3.0	3	8,000	0.32	4	22
100	1.0	20	25,000	0.7	4	27
100	1.0	10	50,000	0.7	3	19
50	1.0	10	12,500	0.3	3	16
100	1.0	5	100,000	0.7	2	13
50	1.0	5	40,000	0.3	2	9
25	1.0	5	18,000	0.15	1	13

A detailed discussion is given in section 5.2. However, instrument peak volume standard deviations for standard liquid chromatographs are typically in the range 20-50 microliters and for instruments designed specifically for small bore and high-speed liquid chromatography about 2-10 microliters. These values are already larger than those predicted as needed in Table 1.16 (about one third of the column peak volume standard deviation at k = 0) for commercially available columns operated at close to their optimum performance. For well retained peaks, represented by k = 5 in Table 1.16, the instrument peak volume standard deviation criterion is more easily met by commercially available equipment. Since the optimum value of the capacity factor for maximum resolution in a minimum separation time is generally in the range of 1 to 3, instrument limitations are a greater problem than available column technology at present.

Extracolumn dispersion is a major problem for the packed fused silica capillary columns with internal diameters less than 0.35 mm. Peak standard deviations will be in the submicroliter range and extensive equipment modification is required for operation under optimum conditions. A reasonable compromise is to employ injection volumes of a few hundred nanoliters or less with detector volumes of a similar or preferably smaller size. This demands considerable ingenuity on behalf of the analyst since, as was indicated in the previous discussion, this is well below the performance characteristics of commercial instruments. The modification of commercial instruments for use with packed fused silica capillary columns is discussed in Chapter 5.

1.7.10 Performance Characterization of Liquid Chromatography Columns Using Reduced Parameters

The major advantage of using the reduced parameters (h, ν, ϕ) rather than absolute parameters (HETP, u, K_o) is that they allow ready comparison of results obtained from columns containing packing materials of different sizes when using eluents of different viscosities and solutes of different diffusion coefficients. This approach also leads to a simple index of performance, the separation impedance, which can be used to judge both the relative performance of columns of a similar kind as well as the limit to performance of the different column types used in liquid chromatography. This general approach was largely developed by Knox and coworkers and is widely used for column testing (section 4.4) and for the kinetic optimization of column parameters [14,55,215,231,247-249]. The reduced plate height, h, is defined as the number of particles to a theoretical plate and is given by

$$h = HETP/d_p = (1/5.54)(L/d_p)(w_h/t_R)^2 \tag{1.77}$$

where w_h is the peak width at half height and t_R the retention time of a retained solute. The reduced velocity, ν, is the rate of flow of the mobile phase relative to the rate of diffusion of the solute over one particle diameter and is given by

$$\nu = ud_p/D_m = Ld_p/t_mD_m \tag{1.78}$$

where D_m is the solute diffusion coefficient in the mobile phase and t_m the column holdup time. When an accurate value for the diffusion coefficient is unavailable an approximate value can be found using the Wilke-Chang equation

$$D_m = A(\psi M_2)^{1/2}T/\eta V_1^{0.6} \tag{1.79}$$

where A is a constant depending on the units used (equal to 7.4 x 10^{-12} when D_m is in m^2/s), ψ is a solvent dependent constant (1.0

for unassociated solvents, 1.5 for ethanol, 1.9 for methanol, and 2.6 for water), T is the temperature in K, η the solvent viscosity, and V_1 the molar volume of the solute [11,248,250,251].

For mixed solvents the volume average value for the product (ψM_2) is used [248]. Typical values for low molecular weight solutes fall in the range 0.5-3.5 x 10^{-9} m^2/s. The higher value is typical of organic solvents of low viscosity such as hexane, and the lower value for polar aqueous solvents. The column flow resistance parameter, ϕ, provides a measure of the resistance to flow of the mobile phase and takes into account the influence of the column length, particle diameter, and eluent viscosity. It is given by

$$\phi = \Delta P d_p^2 t_m / \eta L^2 \tag{1.80}$$

where ΔP is the column pressure drop. Finally, the separation impedance, E, which represents the elution time per plate for an unretained solute times the pressure drop per plate, the whole corrected for viscosity is given by

$$E = (t_m/n)(\Delta P/n)(1/\eta) = h^2 \phi \tag{1.81}$$

The separation impedance represents the difficulty of achieving a certain performance and should be minimized for optimum performance. The highest performance is achieved by a column which combines low resistance to flow and minimum dispersion of chromatographic solute bands.

Some typical values for reduced parameters for the various column types used in liquid chromatography are summarized in Table 1.17. For conventional packed columns a reduced plate height of 2 is considered excellent with more typical values for good columns lying between 2 and 3. Larger values sometimes have to be accepted for small particle columns (packings less than 5 micrometers in diameter) due to difficulties in column packing [188]. Values less than 2 are rare and may be real or reflect errors in sizing the column packing. A limiting value of 2 for the reduced plate height is reasonable for an estimate of the best performance that can be obtained in practice and 3 as an acceptable value for a good column. The flow resistance parameter for conventional packed columns will normally lie within the range 500 to 1000. The separation impedance then has values between 2000 and 9000 for

TABLE 1.17

TYPICAL REDUCED PARAMETERS FOR DIFFERENT COLUMN TYPES

Column Type	Minimum Reduced Plate Height	Minimum Reduced Velocity	Flow Resistance Parameter	Separation Impedance
Conventional Packed	1.5-3.0	3-5	500-1000	2000-9000
Small-Bore Packed	1.5-3.0	3-5	500-1000	2000-9000
Packed Capillary	2.0-3.5	1-5	350-1000	3000-7000
Semi-Packed Capillary	2.0-3.0	3-5	70-250	500-4000
Open Tubular	0.5-30	4-30	32	8-80

good columns. The performance of packed columns should be independent of the column bore and this is reflected in the similar values for the separation impedance for conventional, small bore, and packed capillary columns. For packed capillary columns slightly lower flow resistance parameters than for wider bore columns have been recorded in a few cases but they were generally accompanied by an increase in the reduced plate height [188,200-202]. This probably reflects a reduction in the packing density leading to decreased efficiency accompanied by an increase in the column permeability. Since most conventional packed columns are rarely ideal in practice, these differences are probably not very significant.

The possibility of obtaining significant improvements in performance by using semi-packed and open tubular columns is clearly illustrated by the values for the separation impedance in Table 1.17. Variation of the reduced plate height with the reduced velocity for an open tubular column is given by equation (1.82), assuming that the resistance to mass transfer in the stationary phase can be neglected

$$h = HETP/d_c = (2/\nu) + [(1 + 6k + 11k^2)/96(1 + k)^2]\nu \qquad (1.82)$$

The reduced plate height is reasonably constant, independent of the capacity factor for a packed column, while equation (1.82) indicates that the reduced plate height, at least for small values of the capacity factor, will increase with the capacity factor for open tubular columns. Actual values of the minimum reduced plate height will vary from about 0.3 (corresponding to k = 0) to 1.0 (corresponding to k = infinity) for values of the reduced velocity in the range 4-14 [247]. The exceptional potential performance of open tubular columns cannot be explained entirely by the smaller values of the reduced plate height. The most significant difference compared to the packed columns is their greater permeability. The column flow resistance parameter can be calculated directly from the Poisseuille equation and is exactly 32 for open tubular columns [215]. The separation impedance is about two orders of magnitude lower for an open tabular column compared to a packed column. Table 1.17 indicates that if operation at the minimum separation impedance was possible then for a separation requiring a fixed number of plates the open tubular column would be 100 times faster at a constant inlet pressure. It can also be seen from Table 1.17 that semi-packed capillary columns cannot compete with open tubular columns on the basis of their separation potential [198,199,252]. Their plate height will always be higher and their permeability always lower than that of an open tubular column of the same internal diameter. The hydrodynamic properties of semi-packed capillary columns are somewhat unconventional due to their unique particle arrangements and increased flow permeability. Consequently, the classical plate height equations can no longer adequately describe their chromatographic behavior. Enhanced radial mass transfer in the semi-packed capillary columns resulting from turbulence may reduce the influence of slow diffusion in the mobile phase while at the same time some dispersion will undoubtedly result from multiple paths around the loosely packed particles. Yet another undesirable effect could be due to the presence of stagnant pools of mobile phase between particles. The separation impedance can be calculated independently of the reduced plate height and is thus unambiguous whereas further theoretical development will be needed to adequately account for the relationship between the column plate height and column parameters. The separation impedance for the semi-packed capillary columns is only three or four times lower than that of a packed column and probably offers little advantage in terms of separation speed. The semi-packed capillary columns are technically less developed than packed columns and their instrument requirements sufficiently demanding that it is not clear if they would be useful for all but special applications where their low volumetric flow rate would be their most important characteristic.

The optimum operating conditions for the various column types are summarized in Table 1.18 [14,55]. There are no serious problems when operating conventional packed columns under optimum conditions except perhaps for the first entry where particles of 1.6 micrometer diameter are considered. Small-bore packed columns place greater demands on instrumentation but again it should be feasible to meet the requirements for operation under optimum conditions for plate counts in excess of 10,000. There is virtually no possibility of operating semi-packed or open tubular columns under optimum conditions since the eluted peak volumes for the unretained solutes are so extremely small, even for high plate counts. Extracolumn contributions to dispersion should be about one-third of the volume standard deviation of an unretained solute to avoid a significant loss in performance for early eluting peaks. For an open tubular column exhibiting more than 1 million theoretical plates this corresponds to an extracolumn volume of less than 10 picoliters. Also, in practice, it may be difficult to achieve the optimum column internal diameter which for a column exhibiting 1 million theoretical plates is about 2.5 micrometers. The data in Table 1.18 were compiled at a fixed pressure of 200 bar. Increasing the inlet pressure reduces the separation time but leads to smaller optimum dimensions for the column diameter and column length which in turn leads to a smaller value for the peak standard deviation and a more hopeless case in terms of the practicality of operating open tubular liquid chromatographic columns at the minimum separation impedance.

Since packed columns can be operated at E_{min} = 2000 with plate numbers from 10,000 to 1 million, alternative column types will compete effectively only if they can be operated at separation impedance values below 2000. Enlarging the column diameter and using higher reduced velocities will increase the separation impedance for the column, but still below 2000, while simultaneously increasing peak volumes to relax the instrumental constraints for column operation. Optimization under conditions where extracolumn volumes are limiting can be handled by setting a minimum value for the peak standard deviation of an unretained solute. The smallest practical flow cell that can be devised is probably about 0.1 microliters [195]. Using on-column detection in

TABLE 1.18

OPTIMUM OPERATING CONDITIONS FOR DIFFERENT COLUMN TYPES
column pressure drop = 200 bar, viscosity = 1 x 10^{-3} N.s/m^2 and
diffusion coefficient = 1.0 x 10^{-9} m^2/s

Plate Count	Holdup Time (s)	Particle or Diameter (μm)	Length (m)	Flow Rate (μl/s)	Peak Standard Deviation at k = 0 (μl)
Conventional Packed Columns (5 mm internal diameter) ϕ = 500; h_{min} = 2; ν_{opt} = 5; E_{min} = 2000; total porosity = 0.75					
10,000	10	1.6	0.03	50	5
30,000	90	2.7	0.17	30	15
100,000	1,000	5	1.00	15	50
300,000	9,000	9	5.00	9	150
1,000,000	100,00	16	30.00	5	500
Small-Bore Packed Columns (1 mm internal diameter) ϕ = 500; h_{min} = 2; ν_{opt} = 5; E_{min} = 2000; total porosity = 0.75					
10,000	10	1.6	0.03	1.9	0.2
30,000	90	2.7	0.17	1.1	0.6
100,000	1,000	5	1.00	0.6	2
300,000	9,000	9	5.00	0.35	6
1,000,000	100,000	16	30.00	0.2	20
Semi-Packed Capillary Columns ϕ = 150; h_{min} = 2; ν_{opt} = 5; E_{min} = 600; d_c/d_p = 2.5; total porosity = 0.8					
10,000	2	0.87	0.017	1.7 x 10^{-5}	5 x 10^{-7}
30,000	27	1.5	0.090	3.0 x 10^{-5}	5 x 10^{-6}
100,000	3,00	2.7	0.550	5.4 x 10^{-5}	5 x 10^{-5}
300,000	2,700	4.7	2.9	9.0 x 10^{-5}	5 x 10^{-4}
1,000,000	30,000	8.7	17.0	1.7 x 10^{-4}	5 x 10^{-3}
Open Tubular Columns ϕ = 32; h_{min} = 0.8; ν_{opt} = 5; E_{min} = 20					
10,000	0.1	0.25	0.002	1.0 x 10^{-6}	1.0 x 10^{-9}
30,000	0.9	0.43	0.010	1.6 x 10^{-6}	0.8 x 10^{-8}
100,000	10	0.8	0.065	3.0 x 10^{-6}	1.0 x 10^{-7}
300,000	90	1.4	0.330	5.0 x 10^{-6}	0.8 x 10^{-6}
1,000,000	1,000	2.5	2	1.0 x 10^{-5}	1.0 x 10^{-5}
3,000,000	9,000	4.3	10	1.6 x 10^{-5}	0.8 x 10^{-4}
10,000,000	100,000	8	63	3.0 x 10^{-5}	1.0 x 10^{-3}

fused silica open tubular columns of internal diameter from 5 to
50 micrometers and a slit width of 0.3 cm would provide detection
volumes of 0.06 to 6.0 nanoliters [253]. Then for total instrument
dispersion a value of 0.1 microliters could probably be achieved

TABLE 1.19

DETECTOR LIMITED OPERATION OF OPEN TUBULAR COLUMNS
viscosity = 10^{-3} N.s/m^2; diffusion coefficient = 10^{-9} m^2/s; pressure
drop = 200 bar; ϕ = 32; mass transfer coefficient = 0.08

Column Plate Count	Column Holdup Time (s)	Column Length (m)	Reduced Plate Height	Reduced Velocity	Separation Impedance
d_c = 27 μm and peak standard deviation (k = 0) = 0.1 μl					
10,000	600	17	64	797	130,000
30,000	1,800	30	37	464	44,000
100,000	6,000	55	20	252	13,000
300,000	18,000	90	11	140	4,000
1,000,000	60,000	170	6.4	80	1,300
d_c = 9 μm and peak standard deviation (k = 0) = 1 nanoliter					
10,000	60	1.7	18.5	232	11,000
30,000	180	3	11.2	140	4,000
100,000	600	5.5	6.1	77	1,200
300,000	1,800	9	3.4	42	360
1,000,000	6,000	17			120

whereas 1 nanoliter might be possible but would be exceedingly
demanding. Thus values of 0.1 microliters and 1.0 nanoliter are
reasonable values for the limiting instrumental conditions. Taking
the above considerations into account the separation impedance can
be calculated for various sets of practical operating conditions
as shown in Table 1.19 [14]. For instrumental extracolumn volume
peak standard deviations of 0.1 microliters an open tubular column
will only show superiority over a packed column operated under
optimum conditions when the plate count is in excess of about
500,000. With a further reduction to 1 nanoliter the open tubular
column is superior to the packed column for a plate count in
excess of about 70,000. Thus the future of open tubular column
liquid chromatography will depend primarily on the development of
new technology not only to permit the fabrication of narrow bore
columns with sufficient stationary phase to provide reasonable
partition ratios but also instrumentation with injector and
detector volumes reduced to the nanoliter level. This is not a
trivial accomplishment and it would seem unlikely that open
tubular columns will be widely used in analytical laboratories in
the near future.

Similar conclusions to the above have been reached using different parameters for the optimization of open tubular columns in liquid chromatography. Guiochon considered the use of much higher pressure than those normally used in liquid chromatography [254]. He concluded that with an inlet pressure of 3000 bar a packed column and an open tubular column of the same length would have the same efficiency and separation time if the diameter of the open tubular column was about twice the particle diameter of the packed column when both are operated at their optimum linear velocity. Consequently, open tubular columns must be only a few micrometers in diameter just to be competitive with conventional packed columns. A view which could be considered a little more pessimistic than the calculations using the separation impedance. Jorgenson estimated the optimum diameter of an open tubular column constrained to operate within practical limits of pressure and time was only 1 to 2 micrometers [79]. In terms of resolving power Yang concluded that open tubular columns will have better resolving power than packed columns in a given separation time, if the column diameter for the open tubular column is the same as the particle diameter of the packed column [253]. In spite of some divergence in conclusions resulting from the different sets of limiting conditions, all studies to date are consistent in predicting that to be generally useful open tubular columns with internal diameters close to that of a particle diameter are needed. Detector and injection volumes around the nanoliter level are absolutely essential and smaller values would be preferable. Detectors will not only have to be very small but also extremely sensitive due to the limited sample capacity of the columns when operated under high resolution conditions with complex mixtures.

1.7.11 Temperature and Flow Programming in Liquid Chromatography

The vast majority of liquid chromatographic separations are carried out at ambient temperature for convenience and because ambient temperature provides reasonable column efficiency for low molecular weight solutes. Elevated temperatures improve the kinetic operating parameters of the column by increasing sample diffusivity and reducing mobile phase viscosity. Compared to ambient temperature a higher column efficiency, reduced separation time, and a lower pressure drop are obtained. Increasing the temperature will also change the column selectivity so that

predicting changes in resolution as a function of temperature can
be difficult. Subambient temperatures are reserved for the
separation of labile solutes where the inefficient column
operation, long separation times, and high column pressure drops
are justified by the need to protect the solute from
decomposition, denaturation, or conformational changes [255-257].

Many chromatographic systems show linear relationships
between the logarithm of the capacity factor and the reciprocal of
the column temperature (van't Hoff plots) [255,258-261]. In
thermodynamic terms the interaction of the solute with the
stationary phase can be described by

$$\log k = (-\Delta H^{\circ}/2.3RT) + (\Delta S^{\circ}/2.3R) + \log \beta \qquad (1.83)$$

where (ΔH°) is the standard enthalpy for the transfer of a solute
from the mobile phase to the stationary phase, (ΔS°) the associated
change in standard entropy, R the gas constant and β the column
phase ratio. For certain solutes the van't Hoff plots are
nonlinear and may have sharp discontinuities in their slopes.
Causes for this include retention by a mixed mechanism, a change
in solute conformation that influences binding, or the existence
of more than one form of the solute having different retention
characteristics. Plots of the standard entropy against enthalpy
can be used to establish the similarity of the retention mechanism
for a solute in different chromatographic systems [258-262]. The
slope of such plots is called the compensation temperature which
is approximately constant for solutes undergoing similar
interactions with the stationary phase.

Temperature programming is not widely used in liquid
chromatography [226,263-265]. Since the principal influence of
increasing temperature is to reduce retention, temperature
programming is used to improve front end resolution and to reduce
retention of later eluting peaks. The maximum permissible
operating temperature is established by the vapor pressure of the
mobile phase (a rule of thumb is that this temperature is about
20°C below the boiling point of the mobile phase when a back
pressure restrictor is used). In most cases this means that the
range of retention which can be controlled by changing temperature
is less than the range available by adjusting solvent strength.
With adsorbent stationary phases stripping of modifiers from the

column as the temperature increases and re-equilibrating the column to establish constant retention characteristics for the next analysis is a problem. Radial temperature gradients in normal bore columns may cause a loss of efficiency and the formation of asymmetric peaks. The smaller radius and reduced mass of microcolumns makes them more amenable to temperature programming techniques where this method of reducing separation times becomes more interesting on account of the greater difficulty in constructing gradient forming devices with very low mixing volumes [264,265]. The detection sensitivity of later eluting peaks in temperature program analysis is often enhanced as they elute in a smaller volume of mobile phase compared to isothermal analysis. This situation is opposite to flow programming where the sensitivity of later eluting peaks is diminished by volumetric dilution. Temperature programming can generally be used with spectrophotometric detectors after modification to minimize refractive index effects but not with temperature sensitive detectors such as the electrochemical detector.

The results of flow programming resemble those of temperature programming; it improves front end resolution and reduces the retention time of later eluting peaks [266,267]. However, flow is easier to change than temperature since it is controlled by the operating pressure of the pump, which can be changed rapidly to generate step flow gradients or varied continuously in a linear or exponential manner. Generally, a linear increase in inlet pressure and, hence, in flow velocity, results in a linear decrease in retention time, often with a decrease in resolution for later eluting peaks. Changing the mobile phase velocity within sensible ranges has only a small effect on column efficiency and selectivity. It is thus a less powerful way of changing separation characteristics than gradient elution. Disadvantages of flow programming are the limited range of pressure available for changing flow and volumetric dilution of later eluting peaks, which reduces sensitivity. Advantages lie in the fact that the separation conditions are easily returned to the initial conditions by depressurizing to the original pressure and that it may be used with both refractive index and spectrophotometric detectors. Differential refractometer detectors are likely to exhibit baseline drift as the flow velocity changes.

1.8 PRINCIPLES OF QUANTITATION IN COLUMN CHROMATOGRAPHY

This section reviews the basic performance characteristics of chromatographic detectors and the various methods of obtaining quantitative information from the chromatogram. The role played by standard substances used to improve the accuracy and precision of a chromatographic analysis will also be discussed.

1.8.1 Performance Characteristics of Chromatographic Detectors

The detector performance characteristics of interest to the chromatographer are sensitivity, minimum detectability, dynamic range, response linearity and noise characteristics [260-272]. Other properties of the detection system which indicate its suitability for a particular problem are flow sensitivity and response time. It is convenient to divide chromatographic detectors into two groups: concentration sensitive devices which respond to a change of mass per unit volume (g/ml) and mass sensitive devices which respond to a change in mass per unit time (g/s). Detector sensitivity can be defined as the signal output per unit mass or concentration of test substance in the mobile phase. For a concentration sensitive detector it is given by $S = AF/w$ and for a mass sensitive detector by $S = A/w$, where S is the sensitivity, A the peak area, F the flow rate through the detector, and w the sample amount. The sensitivity, however, is of little value unless the detector noise is also specified. The minimum detectability is defined as the amount of test substance that gives a detector signal equal to some selected multiple of the detector noise, usually taken to be 2 or 3 times the detector noise. When the test substance is also specified it can be used to compare the operating characteristics of different detectors under standard chromatographic conditions.

There are three characteristic types of noise (short term, long term, and drift) which may have different properties depending on whether they are measured under static or dynamic conditions (Figure 1.18). Static noise represents the stability of the detector when isolated from the chromatograph. Dynamic noise pertains to the normal operating conditions of the detector with a flowing mobile phase. Ideally, the static and dynamic noise should be very similar; the performance of the detector is otherwise being degraded by the poor performance of the mobile phase delivery system of the chromatograph. The noise signal is measured

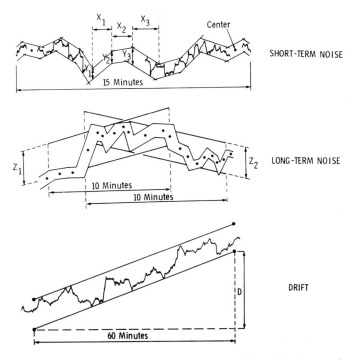

Figure 1.18 Methods for calculating short- and long-term noise and drift for chromatographic detectors.

over a period of time on a recorder with the detector amplifier set to the maximum usable sensitivity. Short-term noise is the maximum amplitude for all random variations of the detector signal of a frequency greater than one cycle per minute. It is calculated from the recorder trace by dividing the detector output into a series of time segments less than one minute in duration and summing the vertical displacement of each segment over a fixed time interval, usually 10 to 15 minutes. Long-term noise is the maximum detector response for all random variations of the detector signal of frequencies between 6 and 60 cycles per hour. The long-term noise is represented by the greater of Z_1 and Z_2 in Figure 1.18. The vertical distances Z_1 and Z_2 are obtained by dividing the noise signal into ten minute segments and constructing parallel lines transecting the center of gravity of the pen deflections. Long-term noise represents noise that can be mistaken for a late eluting peak. Drift is the average slope of the noise envelope measured as the vertical displacement of the pen over a period of 1 h. For spectrophotometric detectors, the signal response is proportional to the path length of the cell and noise values are normalized to a path length of 1 cm [271]. The dynamic range of the detector is determined from a plot of detector response or sensitivity against sample amount (mass or concentration). It represents the range of sample amount for which a change in sample size induces a discernible change in the detector signal (Figure 1.19). However, it is the linear range and not the dynamic range of the detector which is of most interest to the chromatographer. The linear range is the range of sample amount over which the response of the detector is constant to within 5%. It is usually expressed as the ratio of the highest sample amount determined from the linearity plot to the minimum detectable sample amount (Figure 1.19).

1.8.2 Quantitative Analysis

Quantitative analysis requires that a relationship between the magnitude of the detector signal and sample amount be established. The detector signal is measured by the peak height or area from the recorder trace or taken from the print-out of a data system. Manual methods for calculating peak areas include the product of peak height and width at half height, triangulation, trapezoidal approximation, planimetery and cut and weigh [270,273]. No single method is perfect and common problems include the difficulty of defining peak boundaries accurately, operator dependence on precision and the need for a finite time to make each measurement. A major disadvantage of manual measurements is the necessity that all peaks of interest must be completely contained on the chart paper (or adjusted to remain on the chart paper by varying the detector attenuation during the chromatographic run). This severely limits the dynamic range of solute composition that can be analyzed. For those methods that depend on the measurement of peak widths narrow peaks are usually difficult to measure with acceptable accuracy using a magnifying reticule or comparator unless high chart speeds are used to increase the peak dimensions. The product of the peak height and peak width at half height and the triangulation method can only be applied to symmetrical peaks and do not yield the total area for a Gaussian peak; the area measured corresponding to 93.9% and 96.9%, respectively [273]. This does not present a problem when the information is used for comparative purposes. For peak triangulation the tangents to the peak at the inflection points

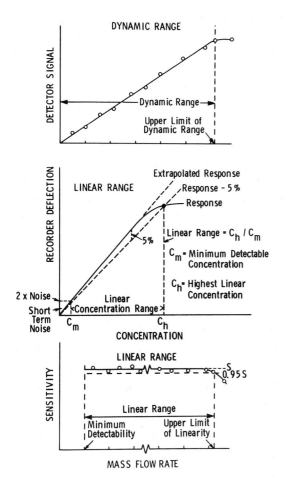

Figure 1.19 Methods for Calculating the dynamic and linear
response ranges for chromatographic detectors.

are drawn and the area of the triangle computed as the product of
half the constructed peak height and the peak width at base
(Figure 1.1). Clearly some prospects for bias exist resulting from
the subjective positioning of the tangent lines. The principal
virtue of the trapezoidal approximation is that it can provide
accurate area measurements of asymmetric peaks where the product
of the peak height and width at half height and triangulation
methods fail. The peak area by the trapezoidal approximation is
given by the product of the average of the peak widths at 15 and
85% of the peak height multiplied by the peak height.

Planimetry and cutting out and weighing of peaks makes no
assumptions about the shape of the peak profile and can be used to

determine the area of skewed peaks. The proper use of a planimeter (a mechanical device designed to measure the area of any closed plane by tracing out the periphery of the plane with a pointer connected by an armature to a counter) requires considerable skill and experience and, even so, obtaining accurate results requires repetitive tracing on each peak with the totals averaged. The cut and weigh procedure depends critically on the accuracy of the cutting operation. The homogeneity, moisture content and weight of the paper influence precision. Copying the chromatogram onto heavy bond paper, with expansion if possible, will preserve the original chromatographic record of the separation and enhance the precision of the weighings.

Electronic integrators and microcomputers are now commonly used for recording chromatograms [274,275]. They are capable of rapidly measuring and reporting peak heights and peak areas for even complex chromatograms. Since the manual methods discussed above are generally tedious and slow, few analysts presently use these methods but occasionally. For well behaved symmetrical peaks unperturbed by baseline noise most computing integrators are capable of very high accuracy and precision [274-281]. Examples of peaks often poorly treated by computing integrators are small peaks with large peak widths, peaks on the tail of larger peaks or the solvent front, and fused peaks. For fused peaks the accuracy of peak height or area measurements depends on the peak separation, peak ratio and peak asymmetry. The way in which different computing integrators treat fused peaks varies from system to system, but is generally based on the detection of inflection points in the signal being received. When such points have been detected peak dividing lines are computed and baselines set depending on the parameter setting used and the algorithms provided by the manufacturer. For computing peak areas the perpendicular drop method is commonly used for peaks of similar size and the tangent-skimming method for peaks of disproportionate size.

For all degrees of peak overlap it is well established that peak height is a more accurate measure of sample size than peak area for symmetrical peaks [277,279]. For either method the error increases for disproportionate peak sizes. For tailed peaks errors in either peak height or area can assume large proportions using the perpendicular drop method [279,281]. In general, the most serious errors in peak height and peak area measurements resulting from asymmetry effects occur when the first peak of an overlapping pair has a pronounced tail. When only tailing of the second peak is observed errors will be much smaller. In summary, there would seem to be few problems in using computing integrators for quantifying well behaved peaks in normal chromatograms but in problem cases no significant improvement in accuracy over manual methods can be anticipated. The fact that a computing integrator provides reproducible data should not be confused with the fact that the data may be highly inaccurate for the conditions identified in the above discussion.

A question frequently posed is whether peak height or peak area measurements provide the best representation of the quantitative information in a chromatogram. Leaving aside the special case of fused peaks, discussed above, there is no simple answer to this question since the precision and accuracy of peak height and area measurements depend on several chromatographic variables, including sample size, mobile phase composition, flow rate and column temperature [270,282-286]. From a purely theoretical viewpoint there is no doubt that when a mass sensitive detector is used quantitation should be based on peak area since precision of the measurement will be independent of flow rate, temperature stability and any other chromatographic factors that can alter either the elution volume or peak shape. In contrast when a concentration-sensitive detector is used the peak height should be independent of flow rate variations and be more precise than area measurements. From the perspective of liquid chromatography area measurements are preferred when the column flow can be controlled precisely even if the mobile phase composition shows some variability and vice versa as far as peak height measurements are concerned. Halasz has shown that as far as mobile phase flow rate variations are concerned, it is the average short-term variations that occur while the peak is in the detector which affects quantitative precision [283]. Consequently, even if an internal standard is included with the sample, it can not be expected to improve precision due to variations of the above kind. Quantitation in gradient elution chromatography requires careful control of total flow rate when peak areas are measured and gradient composition when peak heights are measured. To test which alternative is most applicable, variation in the retention time of early eluting peaks indicates poor flow precision and variation in the retention time of late eluting peaks suggests poor precision in the mobile phase composition. Consequently, whether peak height or peak area is selected for a particular analysis depends on system performance and not necessarily on sample composition. For modern instruments with precise control over flow, composition, and temperature the answer to our original question is that both peak area and peak height will provide acceptable precision. From the practical viewpoint, when making manual measurements peak heights are easier to determine than peak areas.

Four techniques are commonly used to convert peak height or area information into relative composition data for the sample. These are the normalization method, the external standard method, the internal standard method and the method of standard additions [12,15,270,287-289]. In the normalization method the area of all peaks in the chromatogram are summed and then each analyte is expressed as a percentage of the summed areas. This method will always lead to totals representing 100% whether or not all the sample is eluted from the column. If the detector response is not the same for all components then a response factor is required for normalization. Response factors can be calculated from a single measurement but the calibration method is preferred. Sample concentration (weight) is plotted against detector response for a minimum of three standards and the best straight line is drawn. The slope of the plot is the response factor. With moderately sophisticated computing integrators the response factors can be stored in memory and used to correct the detector response information for differences in response to individual sample components and the corrected normalized peak areas printed out.

In the external standard method standards are chromatographed separately alternating in order with samples for the highest precision. Ideally the concentration (weight) of standard injected will be similar to the concentration (weight) anticipated in the samples and the standards and unknowns will be identical so that response factors are not needed. The precision of the method depends very much on maintaining constant chromatographic conditions throughout the analysis and calibration and a high degree of precision in the sample volume injected. External standardization is frequently employed in quality control applications of raw materials, drugs and formulations, etc., where mostly the major components are analyzed and strict requirements on accuracy and precision of the method apply (0.5 to 2.0% RSD). The principal disadvantage of external calibration is the need to run a similar number of standards to samples which reduces the sample throughput. Also if the standard and sample solute are not identical or one standard is used to determine more than one sample component, then the appropriate response factors must be determined as described for the normalization method. The analysis time for the external standard method can be minimized by using the deferred standard method [21,290]. Injection of the standard is delayed until some time after the sample injection so that it will elute in some region of the chromatogram free of other components.

An internal standard is a substance that is added to the sample at the earliest possible point in an analytical scheme to compensate for sample losses occurring during sample preparation and final chromatographic analysis. The properties desired of an ideal internal standard are summarized in Table 1.20. Rarely will an internal standard meet all of these requirements. Substances most commonly used as internal standards include analogs, homologs, isomers, enantiomers, and isotopically labelled analogs of the analyte to be determined. Analogs and homologs are perhaps the most widely used substances simply because they are likely to be available. Isotopically labelled internal standards are frequently used in gas chromatography-mass spectrometry where the mass discriminating power of the mass spectrometer can be used to differentiate between the analyte and internal standard.

When using the internal standard method calibration curves are first prepared for all solutes to be determined and the internal standard to establish the relevant response factors. A constant amount of standard is added to each analyte, preferably at a concentration similar to the solutes of interest in the sample, and a calibration curve constructed from the ratio of the detector response to the analyte divided by the response to the internal standard plotted against the concentration (amount) of analyte. The ratio of the detector response to the sample unknowns and internal standard is then used for all quantitative measurements. The imprecision due to variation of the injection volume can largely be eliminated by use of an internal standard. This is frequently an important error in gas chromatography where injection volumes tend to be small but may not be the limiting factor for reproducibility in liquid chromatography using modern valve and autosampler injectors. The choice of an internal standard for an analytical procedure is often made in a too

TABLE 1.20

PROPERTIES REQUIRED OF AN IDEAL INTERNAL STANDARD

The internal standard should resemble the analyte as closely as possible in terms of chemical and physical properties.

The internal standard should not be a normal constituent of the sample.

The internal standard should be incorporated into the matrix in exactly the same way as the analyte. A situation which is rarely achieved.

The internal standard and analyte should be resolved chromatograph- ically to baseline (except for isotopically labelled samples when mass discrimination or radioactive counting are used for detection), elute close together, respond to the detection system in a similar way, and be present in nearly equal concentrations.

cavalier a fashion and may actually provide lower precision than external calibration. The successful use of an internal standard depends on the existence of a high correlation between the peak areas (heights) of the analytes and the internal standard for the complete analytical procedure and their being lower variability in the internal standard area (height) compared to those of the analytes [291,292]. External standardization will generally provide higher precision than internal standardization when the variation of the recovery of solutes and standards is sufficiently different and the standard deviation in the mean for repeated analyses of the internal standard is larger than that for the solute of interest.

The method of standard additions is the least widely used method of quantitation in chromatography because it requires several chromatographic separations to yield reliable results. A portion of the unknown sample is analyzed to estimate the concentration (amount) of analytes. Known amounts of the individual analytes are then added to individual portions of the sample to provide approximate incremental increases in detector response. Each sample is then reanalyzed individually and a pseudocalibration curve produced using the known amounts of analyte as the concentration (weight) axis yielding an accurate value for the unknown analyte concentration (amount) upon extrapolation. A lack of linearity in the pseudocalibration plot is a good indication of matrix effects.

Due to the growing importance of trace analysis in chromatography attempts have been made to provide a standard method to establish the limit of determination of an analytical procedure [293-297]. The limit of determination is generally defined as the quantity of analyte that can be detected with reasonable certainty for a given analytical procedure. This can be amplified in several ways, for example, it might be determined as the signal equal to three times the standard deviation of the gross blank signal. The limit of determination is a very useful concept in trace analysis made only less so by a lack of agreement as to the statistical approach to be used for the calculation. Until this becomes standardized a direct comparison of various analytical techniques using this term can lead to spurious conclusions. It should be noted that the limit of determination is different from the minimum sample detectability (Table 1.20) which refers only to the detection step whereas the limit of determination refers to the complete analytical procedure including the sample workup.

1.9 REFERENCES

1. V. G. Berezkin, Chem. Revs., 89 (1989) 277.
2. L. S. Ettre and A. Zlatkis (Eds.), "75 Years of Chromatography. A Historical Dialogue", Elsevier, Amsterdam, 1979.
3. F. Bruner (Ed.), " The Science of Chromatography", Elsevier, Amsterdam, 1985.
4. L. S. Ettre, J. Chromatogr., 165 (1979) 235.
5. L. S. Ettre, J. Chromatogr., 220 (1981) 29.
6. L. S. Ettre, J. Chromatogr., 220 (1981) 65.
7. B. L. Karger, L. R. Snyder, and C. Horvath, "An Introduction to Separation Science", Wiley, New York, NY, 1973.
8. C. J. O. R. Morris and P. Morris, "Separation Methods in Biochemistry", Wiley, New York, NY, 2nd Edn., 1976.
9. R. J. Laub and R. L. Pecsok, "Physiochemical Applications of Gas Chromatography", Wiley, New York, NY, 1978.
10. J. R. Conder and C. L. Young, "Physicochemical Measurement by Gas Chromatography", Wiley, New York, NY, 1979.
11. L. R. Snyder and J. J. Kirkland, "Introduction to Modern Liquid Chromatography", Wiley, New York, NY, 2nd Edn., 1979.
12. J. A. Perry, "Introduction to Analytical Gas Chromatography. History, Principles,and Practice", Dekker, New York, NY, 1981.
13. A. S. Said, "Theory and Mathematics of Chromatography", Huethig, Heidelberg, 1981.
14. C. F. Simpson, (Ed.), "Techniques in Liquid Chromatography", Wiley, New York, NY, 1982.
15. M. L. Lee, F. J. Yang, and K. D. Bartle, "Open Tubular Column Gas Chromatography: Theory and Practice", Wiley, New York, NY, 1984.
16. Z. Deyl (Ed.), "Seperation Methods", Elsevier, Amsterdam, 1984.

17. C. F. Poole and S. A. Schuette, "Contemporary Practice of Chromatography", Elsevier, Amsterdam, 1984.

18. R. L. Grob (Ed.), "Modern Practice of Gas Chromatography", Wiley, New York, NY, 2nd Edn. 1985.

19. J. A. Jonsson (Ed.), "Chromatographic Theory and Basic Principles", Dekker, New York, NY, 1987.

20. W. A. Jennings, "Analytical Gas Chromatography", Academic Press, New York, NY, 1987.

21. G. Guiochon and C. L. Guillemin, "Quantitative Gas Chromatography. For Laboratory Analyses and On-Line Process Control", Elsevier, Amsterdam, 1988.

22. R. M. Smith, "Gas and Liquid Chromatography in Analytical Chemistry", Wiley, New York, NY, 1988.

23. J. M. Miller, "Chromatography: Concepts and Contrasts", Wiley, New York, NY, 1988.

24. B. Ravindranth, "Principles and Practice of Chromatography", Ellis Horwood, Chichester, UK, 1989.

25. P. R. Brown and R. A. Hartwick (Eds.), "High Performance Liquid Chromatography", Wiley, New York, NY, 1989.

26. J. C. Giddings, "Unified Separation Science", Wiley, New York, NY, 1990.

27. R. E. Clement (Ed.), "Gas Chromatography. Biochemical, Biomedical, and Clinical Applications," Wiley, New York, NY, 1990.

28. E. Heftman (Ed.), "Chromatography, Part A and B", Elsevier, Amsterdam, 1991.

29. M. L. Lee and K. E. Markides (Eds.), "Analytical Super-critical Fluid Chromatography and Extraction", Chromatography Conferences, Inc., Provo, UT, 1990.

30. J. F. Parcher, J. Chem. Educ., 49 (1972) 472.

31. R. J. Laub, Anal. Chem., 56 (1984) 2110.

32. D. C. Locke, Adv. Chromatogr., 14 (1976) 87.

33. E. F. Meyer, J. Chem. Educ., 50 (1973) 191.

34. R. C. Castells, J. Chromatogr., 350 (1985) 339.

35. I. Ignatiadis, M. F. Gonnord, and C. Vida-Madjar, Chromatographia, 23 (1987) 215.

36. T. L. Hafkenscheid and E. Tomlinson, Adv. Chromatogr., 25 (1986) 1.

37. C. A. Cramers, J. A. Rijks, and C. P. M. Schutjes, Chromatographia, 14 (1981) 439.

38. G. Guiochon, J. Chromatogr. Revs., 8 (1966) 1.

39. G. Guiochon, J. Chromatogr., 185 (1979) 3.

40. A. Alhedai, D. E. Martire, and R. P. W. Scott, Analyst, 114 (1989) 869.

41. J. Krupcik, J. Garaj, G. Guiochon, and J. M. Schmitter, Chromatographia, 14 (1981) 501.

42. L. S. Ettre, Chromatographia, 8 (1975) 291

43. W. Jennings and K. Yabumoto, J. High Resolut. Chromatogr., 3 (1980) 177.

44. T. A. Rooney and M.J. Hartigan, J. High Resolut. Chromatogr., 3 (1980) 416.

45. K. Grob and G. Grob, J. Chromatogr., 207 (1981) 291.

46. L. A. Jones, C. D. Burton, T. A. Dean, T. M. Gerig and J. R. Cook, Anal. Chem., 59 (1987) 1179.

47. J. C. Giddings, "Dynamics of Chromatography", Dekker, New York, NY, 1965.

48. E. Katz, K. L. Ogan, and R. P. W. Scott, J. Chromatogr., 270 (1983) 51.

49. E. Grushka, L. R. Snyder, and J. H. Knox, J. Chromatogr. Sci., 13 (1975) 25.

50. C. F. Poole and S. K. Poole, Anal. Chim. Acta, 216 (1989) 109.
51. M. Martin and G. Guiochon, Anal. Chem., 55 (1983) 2302.
52. C. Horvath and H.-J. Lin, J. Chromatogr., 126 (1976) 401.
53. C. Horvath and H.-J. Lin, J. Chromatogr. 49 (1987) 43.
54. J. F. K. Huber, H. H. Lauer, and H. Poppe, J. Chromatogr., 112 (1975) 377.
55. J. H. Knox, J. Chromatogr. Sci., 18 (1980) 453.
56. J. H. Knox and H. P. Scott, J. Chromatogr., 282 (1983) 297.
57. S. J. Hawkes, J. Chem. Educ., 60 (1983) 393.
58. K. Ogan and R. P. W. Scott, J. High Reslout. Chromatogr., 7 (1984) 382.
59. V. R. Maynard and E. Grushka, Adv. Chromatogr., 12 (1975) 99.
60. N. A. Katsanos, "Flow Perturbation Gas Chromatography", Dekker, New York, NY, 1988.
61. R. P. W. Scott, J. Chromatogr., 468 (1989) 99.
62. D. P. Poe and D. E. Martire, J. Chromatogr., 517 (1990) 3.
63. P. Magnico and M. Martin, J. Chromatogr., 517 (1990) 31.
64. A. Berthod, F. Chartier, and J.-L. Rocca, J. Chromatogr., 469 (1989) 53.
65. R. Thumneum and S. J. Hawkes, J. Chromatogr. Sci., 14 (1981) 576.
66. G. Deininger, Chromatographia, 9 (1976) 251.
67. N. A. Katsanos and Ch. Vassilakos, J. Chromatogr., 471 (1989) 123.
68. J. P. Crombeen, H. Poppe, and J. C. Kraak, Chromatographia, 22 (1986) 319.
69. R. P. W. Scott, Adv. Chromatogr., 9 (1970) 193.
70. E. Bottari and G. Goretti, J. Chromatogr., 154 (1978) 228.
71. C. Gaget, D. Morel, and J. Serpinet, J. Chromatogr., 299 (1984) 119.
72. P. Sandra, J. High Resolut. Chromatogr., 12 (1989) 273.
73. A. K. Bemgard, L. G. Blomberg, and A. L. Colmsjo, Anal. Chem., 61 (1989) 2165.
74. P. A. Leclerc and C. A. Cramers, J. High Resolut. Chromatogr., 8 (1985) 764.
75. A. A. Clifford, J. Chromatogr., 471 (1989) 61.
76. L. S. Ettre, Chromatographia, 17 (1983) 553.
77. I. Brown, Chromatographia, 12 (1979) 467.
78. J. D. Schieke and V. Pretorius, J. Chromatogr., 132 (1977) 217.
79. J. W. Jorgenson and E. J. Guthrie, J. Chromatogr., 255 (1983) 335.
80. J. H. Knox and M. T. Gilbert, J. Chromatogr., 186 (1979) 405.
81. R. P. W. Scott, J. Chromatogr., 517 (1990) 297.
82. G. Guiochon, Adv. Chromatogr., 8 (1969) 179.
83. R. J. Jonker, H. Poppe, and J. F. K. Huber, Anal. Chem., 54 (1982) 2447.
84. E. D. Katz, K. Ogan, and R. P. W. Scott in "The Science of Chromatography", F. Bruner (Ed.), Elsevier, Amsterdam, 1985, p. 403.
85. J. F. K. Huber and J. A. R. J. Hulsman, Anal. Chim. Acta, 38 (1967) 305.
86. G. J. Kennedy and J. H. Knox, J. Chromatogr. Sci., 10 (1972) 549.
87. E. D. Katz and R. P. W. Scott, J. Chromatogr., 270 (1983) 29.
88. R. W. Stout, J. J. Destefano, and L. R. Snyder, J. Chromatogr., 282 (1983) 263.
89. J. F. K. Huber and R. A. Rizzi, J. Chromatogr., 384 (1987) 337.

98

90. A. L. Colmsjo and M. W. Ericsson, J. Chromatogr., 398 (1987) 63.
91. B. A. Bidlingmeyer and F. V. Warren, Anal. Chem., 56 (1984) 1583A.
92. J. V. H. Schudel and G. Guiochon, J. Chromatogr., 457 (1988) 1.
93. J. R. Conder, J. High Resolut. Chromatogr., 5 (1982) 341.
94. J. R. Conder, J. High Resolut. Chromatogr., 5 (1982) 397.
95. J. R. Conder, G. J. Rees, and S. McHale, J. Chromatogr., 258 (1983) 1.
96. D. J. Anderson and R. R. Walters, J. Chromatogr. Sci., 22 (1984) 353.
97. C. Vidal-Madjar, J. Chromatogr., 142 (1977) 61.
98. J. P. Foley and J. G. Dorsey, Anal. Chem., 55 (1983) 730.
99. J. P. Foley and J. G. Dorsey, J. Chromatogr. Sci., 22 (1984) 40.
100. W. Barber and P. W. Carr, Anal. Chem., 53 (1981) 1939.
101. D. Hanggi and P. W. Carr, Anal. Chem., 57 (1985) 2394.
102. N. S. Wu and M. Hu, Chromatographia, 28 (1989) 415.
103. W. M. A. Niessen, H. P. M. Van Vliet, and H. Poppe, Chromatographia, 20 (1985) 357.
104 . A. S. Said, Sepn. Sci. Tech., 13 (1978) 647.
105. K. Suematsu and T. Okamoto, J. Chromatogr. Sci., 27 (1989) 13.
106. P. Sandra, J. High Resolut. Chromatogr., 12 (1989) 82.
107. G. Guiochon, Anal. Chem., 50 (1978) 1812.
108. P. J. Schoenmakers, "Optimization of Chromatographic Selectivity: a Guide to Method Development", Elsevier, Amsterdam, 1986, p. 13.
109. E. Grushka, Anal. Chem., 42 (1970) 1142.
110. E. Grushka, J. Chromatogr., 316 (1984) 81.
111. G. Guiochon, L. A. Beaver, M. F. Gonnord, A. M. Souffi, and M. Zakaria, J. Chromatogr., 255 (1983) 415.
112. J. Krupcik, J. Garaj, P. Cellar, and G. Guiochon, J. Chromatogr., 312 (1984) 1.
113. J. M. Davis and J. C. Giddings, Anal. Chem., 57 (1985) 2168.
114. J .M. Davis and J. C. Giddings, Anal. Chem., 57 (1985) 2178.
115. D. P. Herman, M.-F. Gonnord, and G. Guiochon, Anal. Chem., 56 (1984) 995.
116. J. M. Davis and J. C. Giddings, Anal Chem., 55 (1983) 418.
117. J. M. Davis and J. C. Giddings, J. Chromatogr., 289 (1984) 277.
118. S. L. Delinger and J. M. Davis, Anal. Chem., 62 (1990) 436.
119. J. C. Giddings, Anal. Chem., 56 (1984) 1259A.
120. J. C. Giddings, J. High Resolut. Chromatogr., 10 (1987) 319.
121. H. Cortes (Ed.), "Multidimensional Chromatography: Techniques and Applications", Dekker, New York, NY, 1990.
122. D. H. Freeman, Anal Chem., 53 (1981) 2.
123. L. S. Ettre, Chromatographia, 18 (1984) 477.
124. G. Reglero, M. Herraiz, M. D. Cabezudo, E. Frenandez-Sanchez, and J. A. Garcia-Dominguez, J. Chromatogr., 348 (1985) 338.
125. T. Herraiz, G. Reglero, M. Herraix, R. Alonso, and M. D. Cabezudo, J. Chromatogr., 388 (1987) 325.
126. C. A. Cramers and J. A. Rijks, Adv. Chromatogr., 17 (1979) 101.
127. V. G. Berezkin and S. M. Volkov, Chem. Revs., 89 (1989) 287.
128. V. G. Berezkin, V. S. Gavrichev, and N. V. Voloshina, J. Chromatogr., 520 (1990) 91.
129. V. G. Berezkin, V. S. Gavrichev, and A. Malik, J. Liq. Chromatogr., 10 (1987) 1707.
130. L. S. Ettre and J. E. Purcell, Adv. Chromatogr., 10 (1974) 1.

131. C. P. M. Schutjes, E. A. Vermeer, J. A. Rijks, and C. A. Cramers, J. Chromatogr., 253 (1982) 1.
132. L. S. Ettre, J. High Resolut. Chromatogr., 8 (1985) 497.
133. W. Seferovic, J. V. Hinshaw, and L. S. Ettre, J. Chromatogr. Sci., 24 (1986) 374.
134. G. P. Cartoni, G. Goretti, B. Neri, and M. V. Russo, J. Chromatogr., 475 (1989) 145.
135. C. A. Cramers. F. A. Wijheijmer, and J. A. Rijks, Chromatographia, 12 (1979) 643.
136. J. F. Parcher, J. Chromatogr. Sci., 21 (1983) 346.
137. L. S. Ettre, Chromatographia, 18 (1984) 243.
138. W. Kimpenhaus, F. Richter, and L. Rohrschneider, Chromatographia, 15 (1982) 577.
139. L. Rohrschneider and E. Pelster, J. Chromatogr., 186 (1979) 249.
140. L. S. Ettre, Chromatographia, 12 (1979) 509.
141. M. Y. B. Othman, J. H. Purnell, P. Wainwright, and P. S. Williams, J. Chromatogr., 289 (1984) 1 .
142. F. David, M. Proot, and P. Sandra, J. High Resolut. Chromatogr., 8 (1985) 551.
143. B. Olufren, J. High Resolut. Chromatogr., 2 (1979) 579.
144. C. P. M. Schutjes, P. A. Leclercq, J. A. Rijks, C. A. Cramers, C. Vidal-Madjar, and G. Guiochon, J. Chromatogr., 289 (1984) 163.
145. C. A. Cramers, J. High Resolut. Chromatogr., 9 (1986) 676.
146. C. A. Cramers and P. A. Leclercq, CRC Revs. Anal. Chem., 20 (1988) 117.
147. P. A. Leclercq and C. A. Cramers, J. High Resolut. Chromatogr., 10 (1987) 269.
148. K. J. Hyver and R. J. Phillips, J. Chromatogr., 399 (1987) 33.
149. R. Tijssen, N. Vanden Hoed, and M. E. Van Kreveld, Anal. Chem., 59 (1987) 1007.
150. A. Van Es, J. Janssen, R. Bally, C. Cramers, and J. Rijks, J. High Resolut. Chromatogr., 10 (1987) 273.
151. T. Noy, J. Curves, and C. A. Cramers, J. High Resolut. Chromatogr., 9 (1986) 752.
152. Th. Noij, J. A. Rijks, A. J. Van Es, and C. A. Cramers, J. High Resolut. Chromatogr., 11 (1988) 862.
153. S. K. Poole, K. G. Furton, and C. F. Poole, J. Chromatogr. Sci., 26 (1988) 67.
154. S. Nygren, J. High Resolut. Chromatogr., 2 (1979) 319.
155. W. E. Harris and H. W. Habgood, "Programmed Temperature Gas Chromatography", Wiley, New York, NY, 1967.
156. G. H. Dodo, S. J. Hawkes, and L. C. Thomas, J. Chromatogr., 328 (1985) 49.
157. S. Wicar, J. Chromatogr., 298 (1984) 373.
158. D. W. Grant and M. G. Hollis, J. Chromatogr., 158 (1978) 3.
159. J. Curve, J. Rijks, C. Cramers, K. Knauss, and P. Larson, J. High Resolut. Chromatogr., 8 (1985) 607.
160. E. M. Sibley, C. Eon, and B. L. Karger, J. Chromatogr. Sci., 11 (1973) 309.
161. V. Bartu, S. Wicar, G.-J. Scherpenzeel, and P. A. Leclercq, J. Chromatogr., 370 (1986) 219.
162. V. Bartu, S. Wicar, G.-J. Scherpenzeel, and P. A. Leclercq, J. Chromatogr., 370 (1986) 235.
163. P. Y. Shrotri, A. Mokashi, and D. Mukesh, J. Chromatogr., 387 (1987) 399.
164. E. E. Akporhonor, S. Levent, and D. R. Taylor, J. Chromatogr., 405 (1987) 67.
165. E. V. Dose, Anal. Chem., 59 (1987) 2414 and 2420.

166. A. S. Said, Sepn. Sci., 12 (1977) 29.
167. D. E. Bautz, J. W. Dolan, W. D. Raddatz, and L. R. Snyder, Anal. Chem., 62 (1990) 1560.
168. W. L. Saxton, J. High Resolut. Chromatogr., 7 (1984) 245.
169. W. L. Saxton, J. Chromatogr., 312 (1984) 59.
170. L. S. Ettre, L. Mazor, and J. Takacs, Adv. Chromatogr., 8 (1969) 271.
171. S. Nygren and S. Anderson, Anal. Chem., 57 (1985) 2748.
172. S. Nygren and A. Olin, J. Chromatogr., 366 (1986) 1.
173. J. R. Larson, S. W. Barr, and R. A. Bredeweg, J. Chromatogr., 405 (1987) 163.
174. J. V. Hinshaw and L. S. Ettre, Chromatographia, 21 (1986) 561 and 669.
175. P. Sandra, F. David, M .Proot, G. Diricks, M. Verstappe, and M. Verzele, J. High Resolut. Chromatogr., 8 (1985) 782.
176. R. Villalobos and R. Annino, J. High Resolut. Chromatogr., 13 (1990) 764.
177. R. E. Kaiser, R. I. Rieder, L. Leming, L. Blomberg, and P. Kusz, J. High Resolut. Chromatogr., 8 (1985) 580.
178. H. T. Mayfield and S. N. Chesler, J. High Resolut. Chromatogr., 8 (1985) 595.
179. J. H. Purnell, M. Rodriguez, and P. S. Williams, J. Chromatogr., 358 (1986) 39.
180. J. H. Purnell, J. R. Jones, and M.-H. Wattan, J. Chromatogr., 399 (1987) 99.
181. J. R. Jones and J. H. Purnell, Anal. Chem., 62 (1990) 2300.
182. T. Maurer, W. Engewald, and A. Sternborn, J. Chromatogr., 517 (1990) 77.
183. W. Engewald and T. Maurer, J. Chromatogr., 520 (1990) 3.
184. D. Repka, J. Krupcik, E. Benicka, T. Maurer, and W. Engewald, J. High Resolut. Chromatogr., 13 (1990) 333.
185. E. Benicka, J. Krupcik, D. Repka, P. Kuljovsky, and R. E. Kaiser, Anal. Chem., 62 (1990) 985.
186. T. Hevesi, J. Krupcik, and P. Sandra, J. Chromatogr., 517 (1990) 161.
187. D. Repka, J. Krupcik, E. Benicka, P. A. Leclercq, and J. A. Rijks, J. Chromatogr., 463 (1989) 243.
188. M. Verzele and C. Dewaele, J. High Resolut. Chromatogr., 10 (1987) 280.
189. M. Novotny, Anal. Chem., 53 (1981) 1294A.
190. F. J. Yang, J. High Resolut. Chromatogr., 6 (1983) 348.
191. D. Ishii and T. Takeuchi, Adv. Chromatogr., 21 (1983) 131.
192. R. P. W. Scott, Adv. Chromatogr., 22 (1983) 247.
193. R. P. W. Scott, (Ed.), "Small Bore Liquid Chromatography Columns: Their Properties and Uses", Wiley, New York, NY, 1984.
194. P. Kucera, (Ed.), "Microcolumn High-Performance Liquid Chromatography", Elsevier, Amsterdam, 1984.
195. M. V. Novotny and D. Ishii (Eds.), "Microcolumn Separations: Columns, Instrumentation and Ancillary Techniques", Elsevier, Amsterdam, 1985.
196. D. Ishii (Ed.), "Introduction to Microscale HPLC", VCH Publishers Inc., New York, NY, 1988.
197. K. Jinno, Chromatographia, 25 (1988) 1004.
198. T. T. Suda and M. Novotny, Anal. Chem., 50 (1978) 271.
199. T. T. Suda, I. Tanaka, and G. Nakagawa, Anal. Chem., 56 (1984) 1249.
200. J. C. Gluckman, A. Hirose, V. L. McGuffin, and M. Novotny, Chromatographia, 17 (1983) 303.
201. C. Borra, S. M. Han, and M. Novotny, J. Chromatogr., 385 (1987) 75.

202. F. Andreotini, C. Borra, and M. Novotny, Anal. Chem., 59 (1987) 2428.
203. R. T. Kennedy and J. W. Jorgenson, Anal. Chem., 61 (1989) 1128.
204. K. E. Karlsson and M. Novotny, Anal. Chem., 60 (1988) 1662.
205. D. Ishii, K. Watanabe, H. Asai, Y. Hashimoto, and T. Takeuchi, J. Chromatogr., 332 (1985) 3.
206. M. Konishi, Y. Mori, and T. Amano, Anal. Chem., 57 (1985) 2235.
207. R. Tijssen, J. P. A. Bleumer, A. L. C. Smit and M. E. Van Kreveld, J. Chromatogr., 218 (1981) 137.
208. M. Krejci, K. Tesarik, M. Rusek, and J. Pajurek, J. Chromatogr., 218 (1981) 167.
209. P. Kucera and G. Guiochon, J. Chromatogr., 283 (1984) 1.
210. W. M. A. Niessen, H. P. M. Van Vliet, and H. Poppe, Chromatographia, 20 (1985) 357.
211. S. Folestad, B. Galle, and B. Josefsson, J. Chromatogr. Sci., 23 (1985) 273.
212. S. Folestad, B. Josefsson, and M. Larsson, J. Chromatogr., 391 (1987) 347.
213. P. P. H. Tock, C. Boshoven, H. Poppe, J. C. kraak, and K. K. Unger, J. Chromatogr., 477 (1989) 95.
214. P. P. H. Tock, P. P. E. Duijslers, J. C. Kraak, and H. Poppe, J. Chromatogr., 506 (1990) 185.
215. J. H. Knox and M. T. Gilbert, J. Chromatogr., 186 (1979) 405.
216. G. Guiochon, Anal. Chem., 53 (1981) 1318.
217. N. H. C. Cooke, B. G. Archer, K. Olsen, and A. Berick, Anal. Chem., 54 (1982) 227.
218. R. W. Stout, J. J. DeStefano, and L. R. Snyder, J. Chromatogr., 261 (1983) 189.
219. E. D. Katz and R. P. W. Scott, J. Chromatogr., 253 (1982) 159.
220. F. Erni, J. Chromatogr., 282 (1983) 371.
221. E. D. Katz, K. L. Ogan, and R. P. W. Scott, J. Chromatogr., 289 (1984) 65.
222. J. L. DiCesare, M. W. Dong, and L. S. Ettre, Chromatographia, 14 (1981) 257.
223. N. Sagliano, H. Shih-Hsien, T. R. Floyd, T. V. Raglione, and R. A. Hartwick, J. Chromatogr. Sci., 23 (1985) 238.
224. P. J. Nash, D. P. Goulder, and C. V. Perkins, Chromatographia, 20 (1985) 335.
225. R. Gill and B. Law, J. Chromatogr., 354 (1986) 185.
226. H. M. McNair and J. Bowermaster, J. High Resolut. Chromatogr., 10 (1987) 27.
227. R. P. W. Scott, P. Kucera, and M. Munroe, J. Chromatogr., 186 (1979) 475.
228. C. Dewaele and M. Verzele, J. Chromatogr., 282 (1983) 341.
229. N. D. Danielson and J. J. Kirkland, Anal. Chem., 59 (1987) 2501.
230. P. Kucera and G. Manius, J. Chromatogr., 216 (1981) 9.
231. L. R. Snyder, J. W. Dolan, and Sj. Van der Wal, J. Chromatogr., 203 (1981) 3.
232. I. Halasz and G. Maldener, Anal. Chem., 55 (1983) 1842.
233. P. Roumeliotis, M. Chatziathanassiou, and K. K. Unger, Chromatographia, 19 (1985) 145.
234. H. G. Menet, P. C. Gareil, and R. H. Rosset, Anal. Chem., 56 (1984) 1770.
235. H. Poppe and J. C. Kraak, J. Chromatogr.,282 (1983) 399.
236. E. D. Katz, K. L. Ogan, and R. P. W. Scott, J. Chromatogr., 260 (1983) 277.
237. R. P. W. Scott and P. Kucera, J. Chromatogr., 169 (1979) 51.

238. A. Hirose, D. Wiesler, and M. Novotny, Chromatographia, 18 (1984) 239.
239. V. R. Meyer, J. Chromatogr., 334 (1985) 197.
240. P. Kucera, J. Chromatogr., 198 (1980) 93.
241. R. P. W. Scott, J. Chromatogr. Sci., 23 (1985) 233.
242. F. M. Rabel, J. Chromatogr. Sci., 23 (1985) 247.
243. H. A. Claessens, C. A. Cramers, and M. A. J. Kuyken, Chromatographia, 23 (1987) 189.
244. K. A. Cohen and J. D. Stuart, J. Chromatogr. Sci., 25 (1987) 381.
245. F. W. Freebairn and J. H. Knox, Chromatographia, 19 (1985) 37.
246. K. Hupe, R. J. Jonker, and G. Rozing, J. Chromatogr., 285 (1984) 253.
247. J. H. Knox, J. Chromatogr. Sci., 15 (1977) 352.
248. P. A. Bristow and J. H. Knox, Chromatographia, 10 (1977) 279.
249. A. Berthod, J. Liq. Chromatogr., 12 (1989) 1169 and 1187.
250. C. R. Wilke and P. Chang, Amer. Inst. Chem. Engr. J., 1 (1955) 264.
251. V. Huss, J. L. Chevalier, and A. M. Siouffi, J. Chromatogr., 500 (1990) 241.
252. V. L. McGuffin and M. Novotny, J. Chromatogr., 255 (1983) 381.
253. F. J. Yang, J. Chromatogr. Sci., 20 (1982) 241.
254. G. Guiochon, Anal. Chem., 53 (1981) 1318.
255. D. E. Henderson and D. J. O'Connor, Adv. Chromatogr., 23 (1984) 65.
256. D. E. Henderson, D. J. O'Connor, J. F. Kirby, and C. P. Sears, J. Chromatogr. Sci., 23 (1985) 477.
257. D. E. Henderson and C. Horvath, J. Chromatogr., 368 (1986) 203.
258. F. D. Antia and C. Horvath, J. Chromatogr., 435 (1988) 1.
259. GY. Vigh and Z. Varga-Puchony, J. Chromatogr., 196 (1980) 1.
260. C. Viseras, R. Cela, C. G. Barroso, and J. A. Perz-Bustamante, Anal. Chim Acta, 196 (1987) 115.
261. L. C. Sandra and S. A. Wise, Anal. Chem., 61 (1989) 1749.
262. C. A. Chang and C. S. Huang, Anal. Chem., 57 (1985) 997.
263. W. R. Biggs and J. C. Fetzer, J. Chromatogr., 351 (1986) 313.
264. K. Jinno, J. B. Phillips, and D. P. Carney, Anal. Chem., 57 (1985) 574.
265. J. Bowermaster and H. M. McNair, J. Chromatogr. Sci., 22 (1984) 165.
266. L. R. Snyder, J. Chromatogr. Sci., 8 (1970) 692.
267. H. Wiedemann, H. Engelhardt, and I. Halasz, J. Chromatogr., 91 (1974) 141.
268. R. P. W. Scott, "Liquid Chromatographic Detectors", Elsevier, Amsterdam, 1986.
269. M. Dressler, "Selective Gas Chromatographic Detectors", Elsevier, Amsterdam, 1986.
270. E. Katz (Ed.), "Quantitative Analysis using Chromatographic Techniques", Wiley, New York, NY, 1987.
271. T. Wolf, G. T. Fritz, and L. R. Palmer, J. Chromatogr. Sci., 19 (1981) 387.
272. C. A. Dorschel, J. L. Ekmanis, J. E. Oberholtzer, F. V. Warren, and B. A. Bidlingmeyer, Anal. Chem., 61 (1989) 951A.
273. M. F. Delaney, Analyst, 107 (1982) 606.
274. A. N. Papas, CRC Crit. Revs. Anal. Chem., 20 (1989) 359.
275. N. Dyson, "Chromatographic Integration Methods", Royal Society of Chemistry, London, UK, 1990.
276. R. E. Pauls, R. W. McCoy, E. R. Ziegel, T. Wolf, G. T. Fritz, and D. M. Marmion, J. Chromatogr. Sci., 24 (1986) 273.

277. R. J. Hunt, J. High Resolut. Chromatogr., 8 (1985) 347.
278. L. Binsheng and L. Peichang, J. High Resolut. Chromatogr., 10 (1987) 449.
279. J. P. Foley, J. Chromatogr., 384 (1987) 301.
280. J. P. Foley, Anal. Chem., 59 (1987) 1984.
281. A. N. Papas and T. P. Tougas, Anal. Chem., 62 (1990) 234.
282. S. R. Bakalyar and R. A. Henry, J. Chromatogr., 126 (1976) 327.
283. I. Halasz and P. Vogtel, J. Chromatogr., 142 (1977) 241.
284. J. H. Park, A. Hussan, P. Couasnon, and P. W. Carr, Microchem. J., 35 (1987) 232.
285. W. Kipiniak, J. Chromatogr. Sci., 19 (1982) 332.
286. P. J. Naish, R. J. Dolphin, and D.P. Goulder, J. Chromatogr., 395 (1987) 55.
287. F. I. Onuska and F. W. Karask, "Open Tubular Column Gas Chromatography in Environmental Sciences", Plenum, New York, NY, 1984.
288. J. Novak and P. A. Leclerq, "Quantitative Analysis by Gas Chromatogray" Dekker, New York, NY, 1988.
289. H. Engelhardt, "Practice of High Performance Liquid Chromatography. Applications, Equipment, and Quantitative Analysis" Springer-Verlag, New York, NY, 1986.
290. C. L. Guillemin, J. C. Gressin, and M. C. Caude, J. High Resolut. Chromatogr., 5 (1982) 128.
291. P. Haefelfinger, J. Chromatogr.,218 (1981) 73.
292. R. E. Pauls and R. W. McCoy, J. High Resolut. Chromatogr., 9 (1986) 600.
293. J. E. Knoll, J. Chromatogr. Sci., 23 (1985) 422.
294. J. P. Foley and J. G. Dorsey, Chromatographia, 18 (1984) 503.
295. S. Ebel, H. Kuhnert, and W. Muck Chromatographia, 23 (1987) 934.
296. E. L. Inman and E. C. Richard, J. Chromatogr., 447 (1988) 1.
297. P. L. Bonate, J. Chromatogr. Sci., 28 (1990) 559.

CHAPTER 2

THE COLUMN IN GAS CHROMATOGRAPHY

2.1 INTRODUCTION

Numerous papers have been written about gas-liquid chromatography since the first description by James and Martin [1]. Its impact on modern analytical chemistry has clearly been immense; this technique has been used to solve a large number of significant problems in medicine, biology and environmental sciences, as well as finding an additionally impressive number of industrial applications. In spite of developments in spectroscopy and liquid chromatography, gas chromatography remains the most widely used separation tool in analytical chemistry and is likely to remain so in the foreseeable future. No other analytical technique can provide equivalent resolving power and low-concentration, sample detection for volatile organic compounds [2]. To be suitable for gas chromatographic analysis the sample, or some convenient derivative of it, must be thermally stable at the temperature required for volatilization. The thermal stability of the sample and column materials then represent the fundamental limit for the technique. In contemporary practice an upper temperature limit of about 400°C and a molecular weight less than 1000 is indicated, although higher temperatures have been used and higher molecular weight samples have been separated in a few instances.

In gas chromatography samples are separated by distribution between a stationary phase and a mobile phase by adsorption, partition, or a combination of the two. When a solid adsorbent serves as the stationary phase the technique is called gas-solid chromatography (GSC). When a liquid phase is spread on an inert support or coated as a thin film onto the wall of a capillary column the technique is designated gas-liquid chromatography (GLC). The separation medium may be a coarse powder, coated with a liquid phase through which the carrier gas flows. This is an example of packed column gas chromatography. If the adsorbent, liquid phase, or both, are coated onto the wall of a narrow bore column of capillary dimensions, the technique is called wall-coated open tubular (WCOT), support-coated open tubular (SCOT), or porous-layer open tubular (PLOT) column gas chromatography. A distinction between these column types will be made later.

In this chapter we will adhere closely to the historical development of gas chromatography, describing packed columns before open tubular columns. In recent years spectacular developments in column fabrication and instrument design have made the open tubular column the standard for most analytical applications. This trend will surely continue. However, it would be inappropriate to ignore packed columns entirely. Packed columns have been used for many of the theoretical and practical developments in gas chromatography, they are less expensive, require little training in their use, are better suited to isolating preparative-scale quantities, and can better tolerate samples containing involatile or thermally labile components. Only a limited number of stationary phases have been immobilized successfully in open tubular columns; thus the range of selectivity values is limited compared to the large number of phases available for use in packed columns. Packed columns may not be used as often as open tubular columns in contemporary practice, but they have not fallen into disuse either, and are always likely to find a significant number of applications in spite of the trend towards open tubular columns as the first choice for analytical separations.

2.2 PACKED COLUMN GAS-LIQUID CHROMATOGRAPHY

Separations occur in gas-liquid chromatography because of differences in vapor pressure and because of selective interactions between the solute and the stationary liquid phase. At the molecular level the principal intermolecular forces that occur between a solute and a solvent are dispersion, induction, orientation, and donor-acceptor interactions [3-7]. Dispersion (or London) forces arise from the electric field generated by rapidly varying dipoles formed between nuclei and electrons at zero-point motion of the molecules, acting upon the polarizability of other molecules to produce induced dipoles in phase with the instantaneous dipoles forming them. Dispersion forces are universal and independent of temperature. Induction (or Debye) forces arise from the interaction of a permanent dipole with a polarizable molecule. Orientation (or Keesom) forces arise from the net attraction between molecules or portions of molecules possessing a permanent dipole moment. Induction and orientation forces decrease with increasing temperature and at a sufficiently high temperature disappear entirely as all orientations of the dipoles become equally probable. Complementing the above physical interactions are donor-acceptor interactions of a chemical nature. Donor-acceptor complexes involve special chemical bonding interactions that arise from the partial transfer of electrons from a filled orbital on the donor to a vacant orbital on the acceptor molecule. Important examples in GLC are hydrogen bonding interactions and coordination forces between pi-electron rich systems and metal ions. Although the above forces provide a suitable framework for the qualitative understanding of the separation process, the concert of interacting forces between polyatomic molecules is far too complex for a quantitative description.

For nonpolar solutes (e.g., hydrocarbons) dispersive forces are the sole forces of interaction in the pure liquid state. Dispersive forces are non-selective and increase approximately with solute molecular weight. The elution of nonpolar species from any chromatographic column occurs, to a first approximation, therefore, in order of increasing boiling point. Dispersive forces are not considered important for the separation of polar solutes of similar boiling point. Here the important interactions are orientation and induction forces, which depend on the polarizabilities, ionization potentials, and dipole moments of the solute and liquid phase, and specific electron donor-acceptor interactions of a chemical nature. The sum of the various intermolecular forces between a particular solute and liquid phase is a measure of the polarity of that phase with respect to that solute. The magnitude of the individual interaction energies is a measure of phase selectivity. Differences in selectivity are important chromatographically, for these interactions enable two solutes of equal polarity to be separated by a selective liquid phase. Although not specifically discussed above, when fine tuning a separation, it may be necessary to take molecular size and shape into account as well.

2.2.1 Frequently Used Liquid Phases for Packed Columns

The properties desired of an ideal liquid phase are contradictory and a compromise must be reached between theoretical and practical considerations [8,9]. It is generally desirable for the liquid phase to have a wide temperature operating range. Ideally, this range would include all temperatures from the lowest to the highest used in GLC, approximately -60°C to 400°C. No phase meets this requirement but several popular phases are stable liquids for temperatures of 100 to 300°C above their melting point or glass transition temperature. The liquid phase should have a low vapor pressure and be chemically stable. For high temperature operation this generally dictates the use of high molecular weight polymeric liquids. As polymers lack a well-defined chemical structure their composition is subject to batch-to-batch variations and they cannot be considered ideal liquid phases. Practical considerations dictate that the liquid phase must have reasonable solubility in some common volatile organic solvent and adequately wet the supports used in column fabrication. The phases themselves must show reasonable solvent properties (i.e., dissolving power) for the solutes to be separated and phases with varying selectivities are needed to effect difficult separations of polar samples.

Prior to the development of suitable techniques for liquid phase characterization numerous phases were in general use, leading to what has been called, "stationary phase pollution" [10]. Gradually, many of those phases with poor chromatographic properties, those that simply duplicated the properties of others, and those available as general industrial materials whose composition varied from the same or different sources, have passed into disuse. Today most separations can be performed on a handful of preferred phases, often specially synthesized for GLC purposes [8-13]. These and some additional popular or novel phases will be described in the following paragraphs [8,9,14,15]. Shape-selective liquid crystal and chiral phases for the separation of structural isomers and enantiomers are discussed in section 8.15.

From the inception of GLC high molecular weight hydrocarbons such as squalane, 2,6,10,15,19,23-hexamethyltetracosane, and Apiezon greases have been used as nonpolar stationary phases [16,17]. Squalane is obtained by the complete hydrogenation of squalene isolated from shark liver oil. As a natural product its chemical purity is sometimes doubtful with squalene and batyl alcohol being the principal impurities. The Apiezon greases are prepared by the high temperature treatment and molecular distillation of lubricating oils. The Apiezon greases are of ill-defined composition and the commercial product contains unsaturated hydrocarbon groups as well as residual carbonyl and carboxylic acid groups. Both squalane and Apiezon should be purified by chromatography over charcoal and alumina before use

[17,18]. Commercially available Apiezon greases are usually orange colored prior to cleanup and colorless afterwards. Exhaustive hydrogenation can be used to saturate aromatic and olefinic groups to produce a more stable product, Apiezon MH, having properties more typical of a saturated hydrocarbon phase [18].

The dominant retention mechanism is dispersion for these phases with a variable contribution from induction possible when separating polar solutes. Accordingly, nonpolar solutes are eluted in order of boiling point while polar solutes elute largely according to their hydrophobicity (size of nonpolar portion). These phases are mainly used for the separation of mixtures of hydrocarbons. The principal limitation of squalane is its low maximum allowable operating temperature limit of about 120°C. The Apiezon greases provide higher maximum allowable operating temperature limits, up to about 300°C. Kovats has prepared a synthetic, high temperature hydrocarbon phase ($C_{87}H_{176}$, Apolane-87), 24,24-diethyl-19,29-dioctadecylheptatetracontane, $(C_{18}H_{37})_2CH(CH_2)_4C(C_2H_5)_2(CH_2)_4CH(C_{18}H_{37})_2$, with a molecular weight of 1222 and a maximum operating temperature of 280-300°C [17,19-21]. Its composition is fixed by its chemical formula and this is the most reproducible of the hydrocarbon phases commercially available.

Hydrocarbon phases are widely used as nonpolar reference phases in schemes designed to measure the selectivity of polar phases. Both Rohrschneider and McReynolds selected squalane for this purpose. Apolane-87 and Apiezon MH have been suggested as substitutes due to their greater thermal stability, and in the case of Apolane-87, greater chemical purity [6,17,22]. Retention index values on Apiezon MH and Apolane-87 are virtually identical and, in both instances, slightly larger than on squalane; this difference being attributed to the molecular weight difference between the less volatile phases and squalane. The reproducibility of retention index values on different batches of squalane is adversely affected by chemical impurities difficult to remove from the phase [17]. Schemes for measuring stationary phase selectivity are discussed in section 2.5.4.

All hydrocarbon phases are susceptible to oxidation by reaction with the small quantities of oxygen normally present in the carrier gas, particularly at elevated temperatures, or from the adsorption of air during removal of the column from the chromatograph and its subsequent storage [17,23,24]. Oxidation alters the chromatographic properties of these phases by introducing polar, oxygenated functional groups and in extreme cases may cause scission of the carbon backbone resulting in a high level of column bleed. Oxidation of squalane, which is fully hydrogenated, arises from the presence of tertiary hydrogen atoms that react with oxygen to form thermally unstable hydroperoxides that in turn yield hydroxylic and carbonyl derivatives. Apolane-87 has a much lower concentration of tertiary hydrogens and is therefore more resistant to oxidation. The use of chemical traps to remove oxygen from the carrier gas is a good general practice with these phases.

Sporadic accounts of the use of fluorocarbon liquid phases have appeared almost from the inception of GLC [25,26]. Perfluorocarbon phases have been used to separate substances of high chemical reactivity such as metal fluorides, halogens, interhalogen compounds, and the halide compounds of hydrogen, sulfur, and phosphorus. These compounds tend to destroy more conventional phases. Highly-fluorinated phases also show greater selectivity for the separation of isomeric perfluorocarbons and Freons. Attempts to use these phases for general applications were less successful due to a combination of poor support wetting characteristics, low column efficiencies and low upper operating temperature limits. Intermolecular forces in perfluorocarbon solvents are significantly weaker than in equivalent hydrocarbon solvents, thus making the separation of high molecular weight or thermally labile samples possible at lower temperatures than on conventional nonfluorinated phases [27-29]. These same weak intermolecular forces are responsible for the poor support wetting characteristics that can only be overcome by modifying the structure of the phase to contain anchor groups capable of strong support interactions. Typical of these modified phases are the poly(perfluoroalkyl ether) Fomblin oils, and the fluorinated alkyl esters [30,31]. Fomblin YR, with a molecular weight of 6,000-7,000, provides columns of high efficiency that can be used at temperatures up to 250°C. Above this temperature the stationary phase film is no longer homogeneous and the column efficiency declines. Fomblin YR provides some of the weakest nonpolar interactions exhibited by any moderately thermally stable phase. The fluorinated alkyl esters, Fluorad FC-430 and FC-431, are generally useful, medium polarity liquid phases that strongly deactivate diatomaceous earth supports. Low concentrations of

TABLE 2.1

PROPERTIES OF SOME COMMERCIALLY AVAILABLE POLYSILOXANE PHASES

Name	Type	Structure		Density (g/ml)	Viscosity (cP)	Average Molecular Weight
OV-1	Dimethylsiloxane gum	CH_3		0.975		$>10^6$
OV-101	Dimethylsiloxane fluid	CH_3			1500	3×10^4
OV-7	Phenylmethyl dimethylsiloxane	80% CH_3	20% C_6H_5	1.021	500	1×10^4
OV-17	Phenylmethyl-siloxane	50% CH_3	50% C_6H_5	1.092	1300	4×10^3
OV-25	Phenylmethyl diphenylsiloxane	25% CH_3	75% C_6H_5	1.150	>100,000	1×10^4
OV-210	Trifluoropropyl-methylsiloxane	50% CH_3	50% $CH_2CH_2CF_3$	1.284	10,000	2×10^5
OV-225	Cyanopropyl methylphenyl-methylsiloxane	50% CH_3 25% C_6H_5	25% C_3H_6CN	1.096	9000	8×10^3
OV-275	Dicyanoalkylsiloxane*				20,000	5×10^3

* mixture of 2-cyanoethyl and 3-cyanopropyl groups.

underivatized amines, phenols, and carboxylic acids can be separated on these phases. Since they are only partially fluorinated, they are less effective at reducing retention compared to Fomblin YR.

By far the most important group of liquid phases are the polysiloxanes [8,9,32-35]. Their high thermal stability, wide liquid operating range and availability in different polarities, spanning the range from nonpolar to some of the most polar phases presently available, contribute to their popularity. The basic siloxane backbone can be represented by $-(R_2SiO)-$, in which R can be either methyl, vinyl, phenyl, 3,3,3-trifluoropropyl, 2-cyanoethyl or 3-cyanopropyl groups. Many polymers contain mixtures of the above functional groups, as indicated in Table 2.1. The polysiloxane polymers are prepared by the acid hydrolysis of appropriately substituted dichlorosilanes or dialkoxysilanes, or by the catalytic polymerization of cyclosiloxanes in the presence of a small amount of hexamethyldisiloxane as a chain stopper. Materials of high purity, particularly with regard to the presence of residual catalyst, and of narrow molecular weight range are required for chromatographic purposes.

If there was such a commodity as a standard phase for gas chromatography then based solely on frequency of use it would have to be the poly(dimethylsiloxane) phases. These are only slightly more polar than the hydrocarbon phases so that most separations occur according to boiling point differences. They are chemically and thermally stable and can be used from subambient temperature up to about 350°C. Batch to batch reproducibility is not normally a significant problem although retention differences associated with molecular weight difference among the poly(dimethylsiloxanes) are known [36]. Thermal stability differences have been associated with the presence of low molecular weight oligomers or depolymerization in the presence of acid/base and Lewis acid catalysts [15,34-37]. Although statements are often made to the contrary, there is no fundamental reason why the higher molecular weight gum poly(dimethylsiloxane) phases should produce differences in column efficiency to the lower molecular weight and less viscous oils [38]. The greater solubility of the poly(dimethylsiloxane) oils in volatile organic solvents, however, facilitates column preparation compared with the poorly soluble gums.

After the poly(dimethylsiloxane) phases the second most widely used family of stationary phases are the poly(methylphenylsiloxane) phases. These phases retain the favorable chemical and thermal stability of the poly(dimethylsiloxanes) while introducing some additional selectivity due to the presence of the polarizable phenyl group. These phases also possess a low level of hydrogen-bond acceptor capacity [39]. Their properties are complementary to those of the poly(dimethylsiloxanes) and, since they are available in a wide range of phenyl composition, they allow optimization of the chromatographic selectivity at least over the polarity range associated with the phenyl group. Commercially available phases may be contaminated by lower molecular weight oligomers which can affect the thermal stability of the column packings [40].

Polysiloxanes containing α-cyano and α- or β-fluorine substituents are thermally labile at comparatively low temperatures and are not used in gas chromatography. Polysiloxane phases containing 3,3,3-trifluoropropyl groups are thermally stable up to about 250-275°C and possess rather unique selectivity, particularly for electron-donor solutes, such as ketones and nitro groups. The electronegative trifluoropropyl group should also induce some significant dipolar character to the phases. Mixed-phase packings containing poly(3,3,3-trifluoroproplymethylsiloxanes) have been widely used to separate pesticides and steroid isomers. The poly(cyanoalkylsiloxane) phases with a high percentage of cyanoalkyl groups are some of the most polar phases in common use. The presence of the polar cyano group with its large dipole moment provides strong orientation and induction interactions with dipolar solutes and unsaturated hydrocarbons. These phases are widely used for the resolution of saturated and unsaturated fatty acid methyl esters. Increased thermal stability is obtained when a phenyl group is bonded to the same silicon atom as the cyanopropyl group and commercially available phases can usually be used to 250 or 275°C. Batch-to-batch reproducibility of the poly(cyanoalkylsiloxanes) is not as good as other polysiloxane phases in part due to the introduction of an artifact substituent, a carboxamide group, during the synthesis [41]. Analysis of several commercially available poly(cyanoalkylsiloxane) phases indicated carboxamide groups occupying between 0 to 5% of the total substituent groups. The poly(cyanoalkylsiloxanes) are also susceptible to oxidation in the presence of air or oxidizing agents, particularly at high temperatures, which probably contributes to their changing properties in use. Hydrolysis of the cyano group is another possible cause of variability. The poly(cyanoalkylsiloxane) phases should be protected from oxygen (air) and moisture during use and storage.

The meta-linked poly(phenyl ethers) are useful, chemically well-defined, moderately polar liquid phases. Their low volatility is exceptional for their molecular weight. The five and six ring poly(phenyl ethers) are stable to 200°C and 250°C, respectively. The flexibility imparted to the chain by the ether linkage is responsible for their low viscosity. A high molecular weight poly(phenyl ether), in this instance a true polymer containing an average of 20 rings, was used to separate solutes of high boiling point with column temperatures in the range 125-400°C [42]. Poly(phenyl ethers) containing polar functional groups were less useful due to poor column efficiencies and low thermal stability [43]. The five and six ring poly(phenyl ethers) can be copolymerized with diphenyl ether-4,4'-disulfonyl chloride to produce a polymer of undefined molecular weight and chemical structure, containing repeating units of poly(phenyl ether) joined together by diphenyl ether sulfonyl bridging groups [44]. Commercially available as polyphenyl ether sulfone or, Poly-S-179, it has been used to separate a wide range of high-boiling polar solutes at temperatures in the range 200-400°C.

Phthalate esters $C_6H_4(COOR)_2$ are well-characterized, moderately polar liquid phases [8]. As might be expected, the polarity of the phases declines as the alkyl (R) group increases in size, while their volatility decreases. High volatility compared to other available liquid phases has reduced their importance in recent years. Tetrachlorophthalate esters and 2,4,7-trinitro-9-fluorenone have been used to separate unsaturated hydrocarbons and aromatic hydrocarbons (electron-donor solutes) by π-charge transfer complex formation [45-47]. Since such interactions are perturbed by steric factors, these phases are often able to separate isomeric mixtures that are difficult to resolve on other phases.

The term polyester is used to describe a wide range of resinous composites derived from the reaction of a polybasic acid with a polyhydric alcohol [48,49]. The most frequently used polyester phases for GLC are the succinates and adipates of ethylene glycol, diethylene glycol and butanediol. In particular, poly(ethylene glycol succinate) EGS, $HO(CH_2)_2[OOCCH_2CH_2COO(CH_2)_2]_nOH$, and poly(diethylene glycol succinate), DEGS, $HO(CH_2)_2O(CH_2)_2[OOCCH_2CH_2COO(CH_2)_2O(CH_2)_2]_nOH$, have been frequently used. However, changes in polymer composition upon column conditioning, low tolerance to oxygen and water (particularly above 150°C), and solute exchange reactions involving the polyester functional groups and alcohols, acids, amines and esters have contributed to their diminished use in recent years [15,50]. The presence of residual acid and hydroxyl groups in the polyester structure precludes the use of these phases for the analysis of isocyanates, anhydrides, acid chlorides, etc., which may react with these groups. For many applications they have been replaced by the poly(cyanoalkylsiloxanes).

Polyethers such as the Poly(ethylene glycols) of general formula, $HO(CH_2CH_2O)_nCH_2CH_2OH$ have been used in GLC for the separation of oxygenated compounds (particularly alcohols) and other polar solutes [51]. The most popular of these phases is Carbowax 20M which has an average molecular weight of 14,000 [52]. Carbowax 20M has a melting point of about 60°C and is stable up to about 225°C. The specially prepared and purified poly(ethylene glycols) Superox-4 (MW 4,000,000) [53] and Superox-20M [53,54] can be used at temperatures of up to 250-275°C (intermittently to 300°C). Pluronic phases are similar to the Carbowax phases but are of much lower polydispersity [55]. They are prepared by condensing propylene oxide with propylene glycol. The resulting chain is then extended on both sides by the addition of controlled amounts of ethylene oxide until the desired molecular weight is obtained. The pluronics most useful as liquid phases have average molecular weights in the range 2000 to about 8000, with maximum allowable operating temperatures in the range 220-260°C. Condensing Carbowax 20M with 2-nitroterephthalic acid produces a new phase, FFAP, that has been recommended for the separation of underivatized organic acids [15]. This phase, which has an upper column temperature limit of about 250°C, is not suitable for the analysis of basic compounds or aldehydes with which it reacts. The poly(ethylene glycols) are rapidly degraded by oxygen and moisture at high temperatures. Strong acids and Lewis acid catalysts may also degrade these polymers. The poly(ethylene glycols) have terminal hydroxyl groups and should not be used for the analysis of compounds that might react with this functional group.

The liquid organic salts are a new class of polar phases for gas chromatography [56,57]. They differ from the phases considered so far in that they contain ions that provide strong interactions associated with the Coulombic fields. Some representative examples of typical liquid organic salts and their chromatographic properties are given in Table 2.2 [28,58-64]. The minimum column operating temperature is generally established by the melting point of the salt. The maximum column operating temperature by the vapor pressure or decomposition temperature of the salt. For the low molecular weight tetraalkylammonium salts the lowest available melting points are often found for the tetrabutylammonium salts. Weak nucleophilic anions such as the sulfonate and tetrafluoroborate anions are generally the most stable with their decomposition temperature being above the vapor pressure limit for the salt. The tetraalkylphosphonium salts are generally more thermally stable to nucleophilic attack by the anion than the analogous tetraalkylammonium salts providing wider liquid ranges for some anions that cannot be studied as the tetraalkylammonium salts [62]. The column efficiencies observed for packings prepared with liquid organic salts are not obviously different to those prepared with conventional non-ionic phases. Chemical

TABLE 2.2

COLUMN OPERATING CHARACTERISTICS FOR SOME LIQUID ORGANIC SALTS
r.t. = room temperature

Salt Cation	Anion	Liquid Range (°C)
Tetrabutylammonium	Perfluorooctanesulfonate	r.t.-220
	3-(N-Morpholino)propane-sulfonate	r.t.-180
	4-Toluenesulfonate	55-200
	Tetrafluoroborate	162-290
	Trifluoromethanesulfonate	112-240
	Picrate	90-200
	tris(Hydroxymethyl)methyl-aminopropanesulfonate	110-210
Tetrapentylammonium	4-Toluenesulfonate	55-190
Tetraethylammonium		85-190
Diethylammonium		105-210
Triethylammonium		78-180
Ethylammonium		121-220
Tributylammonium		82-180
Tetrabutylphosphonium	Chloride	83-230
	Bromide	103-230
	Nitrate	65-220
	4-Toluenesulfonate	44-230
Tributylbenzyl-phosphonium	Chloride	165-240
Ethylpyridinium	Bromide	110-160
Stearylmethyldipoly-oxethyl(15)ammonium	Chloride	r.t.-300

transformations of polar solutes in the liquid organic salts are not commonly found [56,65]. Nucleophilic displacement is important for alkyl halides, degradation of alkanethiols occurs on some salts, and proton transfer and other acid/base reactions can affect the recovery of amines. The unique selectivity of the liquid organic salts is shown by the strong orientation and induction interactions, which generally exceed those of the non-ionic polar phases, and a variable capacity for proton acceptor interactions, that vary from moderate to extremely strong depending on the identity of the anion. For example, Figure 2.1 shows a separation of a polar test mixture on identical columns of tetrabutylammonium 4-toluenesulfonate and the poly(cyanoalkylsiloxane) phase OV-275. The increased retention time of the polar test solutes on the liquid organic salt column is indicative of the strong polar interactions between sample and

118

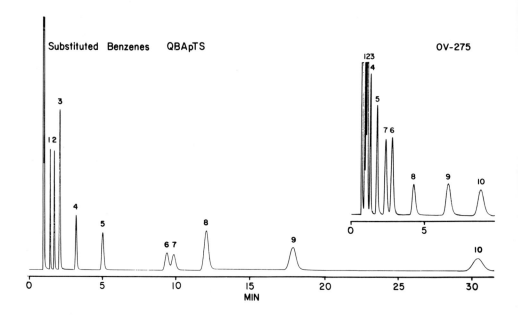

Figure 2.1 Separation of a polarity test mixture on matched
columns of tetra-n-butylammonium 4-tolenesulfonate (QBAPTS) and
OV-275. Each column was 3.5 m x 2 mm I.D. containing 10% (w/w) of
phase on Chromosorb W-AW (100-120 mesh) with a carrier gas flow
rate of 15 ml/min and column temperature of 140°C. Peak
assignments : 1 = benzene; 2 = toluene; 3 = ethylbenzene; 4 =
chlorobenzene; 5 = bromobenzene; 6 = iodobenzene; 7 = o-
dichlorobenzene; 8 = benzaldehyde; 9 = acetophenone; and 10 =
nitrobenzene.

the salt compared to those with OV-275. Many of the retention
properties of the salts can be correlated with the structure of
the anion or cation thus permitting the possibility of designing
phases for specific separations based on a consideration of sample
characteristics [56,58,60,61]. Also, since the liquid organic
salts have a well defined chemical structure, reproducible via
synthesis, they should not be plagued by the problems of batch-to-
batch irreproducibility typical of polymeric, non-ionic phases and
meet the chemical requirement for use as standard polar reference
phases.

2.2.2 Supports

The ideal support would behave like a sponge holding the
liquid phase as a thin film without playing any further role in

the separation process. To eliminate solute interactions with the support its surface activity should be low, yet contrarily, the surface must have sufficient energy to both hold the liquid phase stationary and to cause it to wet the surface in the form of a thin film. The ideal support would also have a large surface area to weight ratio to allow for the use of high liquid phase loadings and a regular shape, with a narrow range of cross-sectional diameters, so that columns of high efficiency can be prepared. It should also be a good conductor of heat and mechanically and thermally stable to avoid changes in properties on handling and while in use. No such ideal support exists and a compromise between those properties which are desirable and those which can be obtained practically must be reached. By far the most important support materials are the diatomaceous earths with glass beads, fluorocarbon powders, graphitized carbon and dendrite salt being used occasionally for some special applications [66-69].

Diatomite (diatomaceous earth, Kieselguhr) is composed of the skeletons of diatoms, single-celled algae, which have accumulated in large beds in various parts of the world. The skeletal material is essentially microamorphous silica with small amounts of alumina and metallic oxide impurities. The porous nature of the diatomite with its associated secondary structure gives the material a high surface area to weight ratio, approximately 20 m^2/g. As mined, the diatomite skeletons are too small and fragile to be used as chromatographic supports; further processing agglomerates and strengthens the natural material. This is achieved by calcining at temperatures in excess of 900°C. The crude material is ground to a powder and shaped into the form of a fire brick (perhaps with the addition of a small amount of clay binder) before being calcined. The calcination serves to fuse the ground particles together with some of the silica being converted to cristobalite (crystalline silica) and the mineral impurities forming complex oxides or silicates. The presence of complex iron oxides gives this material its characteristic pink color. Close inspection of the processed diatomite material reveals a compact mass of broken diatomite fragments with a portion of the secondary pore structure still intact so that the material retains a relatively large surface area. If the diatomite is first broken into chunks and then calcined in the presence of a small amount of sodium carbonate flux, a useful modification of its properties can be achieved. This material is white, as the metal impurities are

TABLE 2.3

PHYSICAL PROPERTIES OF CHROMOSORB SUPPORTS

Property	P	W	G	A	750	T
Color	pink	white	oyster white	pink	white	white
pH	6.5	8.5	8.5	7.1	8.0	- - -
Free fall density (g/ml)	0.38	0.18	0.47	0.40	0.37	0.42
Packed density (g/ml)	0.47	0.24	0.58	0.48	0.36	0.49
Surface area (m^2/g)	4.0	1.0	0.5	2.7	0.5-1.0	7-8
Surface area (m^2/ml)	1.9	0.3	0.3	1.3	- - -	- - -
Maximum useful liquid phase loading (%)	30	15	5	25	12	20

converted to colorless sodium metal silicates, and close inspection reveals a mass of large diatomite fragments held together by sodium silicate glass with most of the secondary pore structure destroyed. Thus the white material has a smaller surface area when compared to the pink and the use of sodium carbonate flux imparts a slightly basic character to the material compared to the natural acidic character of the pink material. The chemical composition of the two types of diatomaceous earths are very similar (SiO_2 88.9-90.6%, Al_2O_3 4.0-4.4%, Fe_2O_3 1.6%, CaO 0.6%, MgO 0.6%, and Na_2O + K_2O 1.0-3.6%) [66]. The balance is largely moisture and other minor component impurities. The higher Na_2O and K_2O content of the white material reflects the use of sodium carbonate flux in its preparation.

The physical properties of the popular series of Chromosorb supports are summarized in Table 2.3. The pink supports are relatively hard, have a high packing density, large surface area and high liquid loading capacity. They are used in both analytical and preparative gas chromatography. The white supports are more friable, less dense, have a smaller surface area and liquid loading capacity, and are used in analytical gas chromatography. Chromosorb G is a specially prepared white support which is harder, more robust, and inert than Chromosorb W. It has a low liquid loading capacity but is more dense than the ordinary white supports so that a given column contains approximately 2.5 times the amount of Chromosorb G, and therefore liquid phase, as

Chromosorb W of the same nominal percentage weight per weight coating. Chromosorb A, a pink support developed specifically for preparative gas chromatography, possesses the mechanical strength and high liquid phase capacity of pink supports with a reduced surface activity approaching that of white supports. Chromosorb 750 is a white support which is chemically treated after processing and is the most inert of the Chromosorb series.

Commercial gas chromatographs are designed to operate at column inlet pressures of a few atmospheres. The column pressure drop is inversely proportional to the square of the particle diameter, whereas the column efficiency is directly proportional to the particle diameter. For analytical purposes supports with a mesh range of 80-100 or 100-120 are a reasonable compromise. The highest efficiencies are obtained with particles of a narrow mesh range. Traditionally, particle size range is quoted in terms of mesh and the designation 100-120 mesh means that the particles passed through a sieve of 100 mesh but not a sieve of 120 mesh. The mesh range in units of micrometers is given in Table 2.4.

In the early development of gas chromatography, the crude diatomaceous supports performed successfully as long as experimental work was limited to hydrocarbons and other such nonpolar samples. The extension of gas chromatography into the biomedical and environmental fields demonstrated that the crude diatomaceous supports were not sufficiently inactive for these applications. Severe tailing of polar molecules, sample decomposition, structural rearrangements and even complete sample adsorption were often observed. These undesirable interactions were associated with the presence of metallic impurities and silanol groups at the surface of the diatomaceous supports. Acid and/or base washing to remove metallic impurities and silanization of surface silanol groups are presently the most widely used methods of support deactivation.

Surface metallic impurities, principally iron, are removed by soaking for several hours in 3 N HCl [70], by Soxhlet extraction with 6 N HCl until no further coloration is extracted [71-73], or by heating for one to three days at 850-900°C while passing a mixture of nitrogen and hydrogen chloride gases through the support [73,74]. The above treatments remove 95-98% of the iron present at the surface of the support and reduce the concentrations of sodium and aluminum as well. Reduction of metal oxides, particularly iron oxide, by high temperature treatment

TABLE 2.4

PARTICLE SIZE RELATIONSHIPS FOR DIATOMACEOUS SUPPORTS

Mesh Range	Top Screen Opening (μm)	Bottom Scree Opening (μm)	Spread (μm)	Range Ratio
10-20	2000	841	1159	2.38
10-30	2000	595	1405	3.36
20-30	841	595	246	1.41
30-40	595	420	175	1.41
35-80	500	177	323	2.82
45-60	354	250	104	1.41
60-70	250	210	40	1.19
60-80	250	177	73	1.41
60-100	250	149	101	1.68
70-80	210	177	33	1.19
80-100	177	149	28	1.19
100-120	149	125	24	1.19
100-140	149	105	44	1.42
120-140	125	105	20	1.19
140-170	105	88	17	1.19
170-200	88	74	14	1.19
200-230	74	63	11	1.19
230-270	63	53	10	1.19
270-325	53	44	9	1.20
325-400	44	37	7	1.19

with hydrogen has also been claimed to improve the properties of diatomaceous supports [75].

Surface silanol groups are converted to silyl ethers by reaction with dimethyldichlorosilane (DMCS), hexamethyldisilazane (HMDS), trimethylchlorosilane (TMCS), or a combination of these reagents, Figure 2.2 [76-81]. Even after exhaustive silanization some unreacted silanol groups will remain. The number of silanol groups is 12-24 times greater on the surface of the pink supports compared to the white, so that even after silanization pink supports show approximately the same adsorptive activity as untreated white supports. Silanization changes the hydrophilic character of the support to hydrophobic and, consequently, it is no longer completely wet by polar liquid phases. Silanized supports should not be used to prepare packings with polar liquids. The treatment that the support has undergone is normally stated on the manufacturer's label, for example, Chromosorb W-AW-DMCS would indicate that the white diatomite support was deactivated by acid washing and silanization with dimethyldichlorosilane.

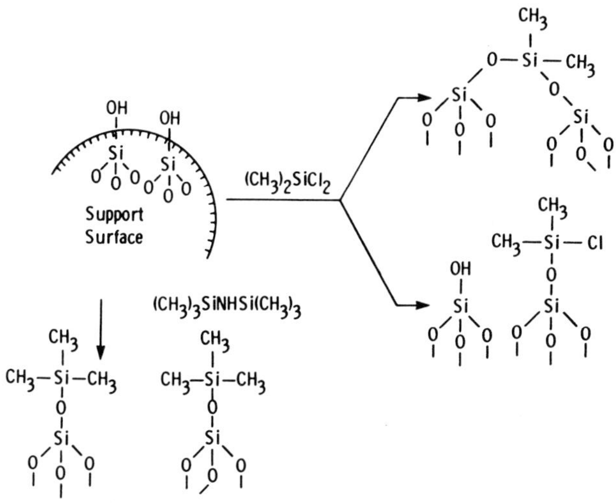

Figure 2.2 Support deactivation by silanization.

To improve the support wetting characteristics of diatomaceous support by nonpolar phases surface treatment with cyclic siloxanes [82,83], or silanization by bulky reagents such as octadecyldimethylchlorosilane has been recommended [20,84-86]. None of these treatments yields ideal coating characteristics for nonpolar phases; however, they all provide higher column efficiencies and improved mass transfer characteristics compared with conventional silanization, indicating that some improvement in the support wetting characteristics has been realized. Precoating diatomaceous support with a nonextractable film of Carbowax 20M or a polyester has been used to both deactivate and improve the support coating characteristics of polar phases [82,84,87,88]. This treatment provides good deactivation characteristics and is more appropriate than silanization procedures for polar phases.

The most extensive chemical treatment and silanization of diatomaceous supports cannot completely remove those active centers which cause tailing of strongly basic or acidic components. When compounds of this type are analyzed, the addition of small quantities of "tailing reducers" is necessary. Tailing reducers are coated onto the support in a manner similar to that for the liquid phase and, to be effective, they must be stronger acids or bases than the compounds to be chromatographed. For amines, the tailing reducer could be a few percent of potassium hydroxide or poly(ethyleneimine) [67,85,89]. For acidic compounds, suitable tailing reducers are phosphoric acid, the stationary phase FFAP, or commercial corrosion inhibitors such as trimer acid [67]. It must be remembered that these active substances will also act as subtractive agents and thus an acidic tailing reducer will remove basic substances from the chromatogram and vice versa. The phase itself must also be compatible with the reagents. For example, potassium hydroxide and phosphoric acid catalyze the depolymerization of polyesters and polysiloxane phases.

More than 90% of the separations using packed column GLC have probably been performed on columns prepared with diatomaceous supports. The only other support materials of any importance are various fluorocarbon powders, glass beads, graphitized carbon black and dendrite salt [66-69]. The most important of the fluorocarbon supports are: Teflon-6, Chromosorb T, Fluoropak 80 and Kel-F. Teflon-6 is a tetrafluoroethylene polymer prepared by the agglomeration of Teflon emulsions which are sieved to a 40-60 mesh fraction. The individual particles are composed of fragile aggregates possessing a porous structure, providing a large surface area with little surface energy. This results in a very inert surface but presents problems with polar liquid phases, as these cannot wet the surface of the particles. The Teflon supports develop electrostatic charges easily and are difficult to coat and pack. For high column efficiencies the Teflon particles should be handled at temperatures below 19°C (solid transition point), at which temperature they are rigid and free flowing. Kel-F is a hard, granular chlorofluorocarbon molding powder that can be handled like diatomaceous supports but generally gives columns of low efficiency. Fluoropak-80 is a granular type of fluorocarbon resin with a sponge-like structure of low surface area and, therefore, low capacity. The properties of the fluorocarbon supports are summarized in Table 2.5 [90].

Glass beads are available as spheroids in narrow mesh ranges. They can be used to prepare very efficient columns. As the surface energy of the beads is low, liquid phase loadings are limited to less than 0.5% w/w. At higher loadings the column efficiency is reduced due to the formation of pools of liquid phase at the contact points of the beads. The glass surface is not completely inert and tailing is observed with polar solutes [91,92]. In addition to the silanol groups on the surface of the

TABLE 2.5

CHROMATOGRAPHIC PROPERTIES OF FLUOROCARBON SUPPORTS

Property	Kel-F	Fluoropak-80	Teflon-6
Surface area (m^2/g)	2.2	1.3	10.5
Optimum liquid phase loading (%)	15-20	2-5	15-20
Maximum temperature (°C)	160	275	250

bead, Na^+ and Ca^{2+} ions are also present. These ions may function as Lewis acids and provide adsorption sites for electron-donor molecules such as ketones and amines. Silanization of the beads reduces the tailing caused by silanol groups but also diminishes the support capacity. To increase the capacity of the beads, chemical etching of the surface [91] or coating with a thin layer of an intermediate material of high surface area (such as bohemite or graphitized carbon black) have been used. Glass beads are used mainly in theoretical studies because of their controlled shape and size and for very rapid analyses where their low liquid loading capacity is not a disadvantage.

2.2.3 Bonded Liquid Phases

Four approaches have been used to create column packings with immobilized liquid phases [93-95]. Halasz described the preparation of estersils, support bonded alcohols or urethanes, prepared by the condensation of an alcohol or isocynate reagent with porous silica beads [96,97]. Due to the specific nature of the reaction only a monomeric film is possible. As long as the alkyl groups are relatively short they are probably orientated normal to the surface and closely packed. Consequently, the speed of mass transfer is considerably higher than for bulk liquid phases and estersil packings can be used at high flow rates for fast separations while retaining reasonable efficiency [98]. The estersils are easily hydrolyzed by water, alcohols and other proton-donor substances. At high temperatures oxidative degradation by traces of oxygen in the carrier gas can be a problem. The estersils are available commercially under the trade name Durapak, Table 2.6 [99].

The second approach involves the reaction of a monofunctional or multifunctional alkylsilane or cyclosiloxane

TABLE 2.6

PERMANENTLY BONDED LIQUID PHASES (ESTERSILS)

Support(*)	Stationary Phase(%w/w)	Polarity	Isothermal Temperature Limit (°C)	Recommended for the Separation of
Porasil C	3-Hydroxy-propio-nitrile (3.60)	Medium	135	Low molecular weight hydrocarbons, aromatics, and oxygenated compounds.
Porasil C	Carbowax 400 (7.86)	Nonpolar	175	Low molecular weight alcohols.
Porasil C	n-Octanol	Polar	175	Low molecular weight alcohols and hydrocarbons.
Porasil C	Phenyl isocyanate	Polar	60	C_1-C_3 hydro-carbons and their isomers.
Porasil S	Carbowax 400 (16.75)	Nonpolar	230	
Porasil S	Carbowax 4000	Polar	230	Aromatics and chloroaromatics.
Porasil F	Carbowax 400 (1.41)	Nonpolar	230	Higher molecular weight substances: waxes, steroids, polynuclear aromatics.

(*)The Porasils are spherical porous silicas that differ in surface area and pore diameter. Porasil C (100 m^2/g, 30 nm), Porasil F (10 m^2/g, 300 nm), and Porasil S (300 m^2/g).

reagent with a porous silica or diatomaceous support [100,101]. Chemical attachment via formation of siloxane bonds and polymerization are probably both involved in bonding the liquid phase to the support. Aue proposed a third approach to bonding a liquid phase to a support by the high temperature treatment of a coated support in the presence of a slow flow of carrier gas [102,103]. After treatment the packing was exhaustively extracted with appropriate solvents to leave a fraction of the original phase permanently bonded to the support. The nonextractable liquid phase loading corresponds to about 0.2% w/w. Modifications and improvements to the original preparation procedure have been described but in all cases thermal treatment is used for the bonding process [88,104-107]. Phases which have been bonded in this way include Carbowax 20M, polyester and polysiloxanes. They are commercially available under the trademark Ultra-bond [108]. Finally, immobilized polysiloxane phases have been prepared by peroxide-induced or ozone-induced free radical crosslinking of coated packings at elevated temperatures [109,111]. By reloading and performing sequential crosslinking reactions up to 20% w/w of liquid phase can be immobilized over the support surface. This type of immobilization reaction is analogous to the widely used procedure for preparing open tubular columns described in section 2.3.6. In general, non of the bonded-phase packings described above are widely used compared to conventionally coated liquid phases.

The extent of solute retention on estersils depends on the size and functionality of the organic chain, the degree of surface bonding, and the size and polarity of the solute [99,112-114]. The most important mechanism of solute retention is adsorption on either the solid support or the bonded organic group. These groups dominate the retention mechanism for those bonded phases containing short-chain alkyl groups. For bonded phases containing polar groups, adsorption at the surface of the bonded organic group tends to dominate the retention mechanism. This accounts for the contrary observation that the octane-bonded phase (prepared by reaction with octanol) is more polar than the Carbowax-bonded phase, Table 2.6, the exact opposite of results obtained with conventionally coated liquid phases. Long chain bonded alkyl groups show less spatial ordering than their short chain analogs. Their structure shows more liquid-like behavior and the possibility that partitioning contributes to the retention mechanism is more likely.

Less is known about the retention mechanism of solutes on the nonextractable phases [115]. It is probably similar to the estersil phases except that the liquid film may be present, in part, as "dense patches" and, therefore a greater contribution from partitioning might be anticipated. The principal chromatographic properties of the nonextractable phases are low retention, very low column bleed and higher column efficiencies and maximum operating temperatures than conventionally coated phases. The packings are also more inert and can be used for the separation of polar solutes. Oxidative degradation can occur at high temperatures.

2.2.4 Column Preparation and Evaluation

Supports are usually coated with the liquid phase using one of several evaporation methods. In outline, the liquid phase in a suitable solvent is mixed with the support, the solvent is removed, and the dried packing is then added to the empty column with the aid of pressure or suction [57,116]. The coating solvent should be a good solvent for the phase and of sufficient volatility for easy evaporation. High purity solvents without stabilizers should be used so as not to contaminate the packings. Recommended solvents for coating are usually indicated by the manufacturer of the phase. Although these are not the only solvents that can be used to prepare the column packing, it should be noted that a poor choice of solvent can lead to uneven coating and packings of below average efficiency. Conventionally, column packings are prepared on a weight per weight basis and quoted as percentage of liquid phase. Thus, to prepare 10 g of a 10% w/w column packing, 9 g of support and 1 g of liquid phase would be used. Typical phase loadings range from 0.5 to 30% w/w depending upon the type of separation. The maximum useful phase loadings for various supports were given in Tables 2.3 and 2.5.

The most common coating procedure is the rotary evaporator technique. The support is coated by adding it slowly to a fluted flask (Morton flask or powder drying flask) containing the liquid phase in sufficient solvent to completely wet the support and form a layer above it. The support is conveniently added to the hand swirled solution through a fluted paper with a small hole at its center. The solvent is then removed in the usual way on the rotary evaporator. Because of the fragile nature of the diatomaceous supports, it is necessary to avoid rapid rotation of the flask or violent bumping of the damp solid. Alternative coating techniques include filtering a slurry of the liquid phase and solid support through a Buchner funnel; packing the support into a column containing a glass frit and passing a solution of the liquid phase through the bed of support aided by overpressure or suction; or by the pan-dry method, in which the slurry is gently stirred in a large flat glass dish under an infrared heating lamp or stream of nitrogen to aid evaporation [116-118]. In all cases it is important to treat the packings gently to avoid the formation of fines.

After coating the damp support may be air dried, oven dried or dried in a fluidized bed dryer. Fluidized bed drying is faster and may lead to more efficient and permeable column packings, since fines developed during the coating procedure are carried away by the gas passing through the packing [57,116,119,120]. The dried packing may be mechanically sieved to remove fines. Dry sieving is easy to perform but may also cause extensive attrition of the fragile diatomaceous supports. Mechanical sieving prior to coating and fluidized bed drying after coating are the preferred methods of minimizing the deleterious effects of fines on column performance. Unopened bottles of support may contain from 30 to 50% fines by volume, thus the need for a preliminary screening process.

For general purposes it is assumed that the ratio of liquid phase to support does not change during the coating process. However, the amount of liquid phase coated on the support is frequently less than the nominal amount weighed out due to losses of phase to the walls of the coating vessel and from uneven coating which tends to result in heavily loaded particles adhering to the walls of the coating flask that remain behind after the coated support is removed. For physicochemical measurements the weight of liquid phase in the column must be accurately known. This is usually determined by either combustion [121], evaporation [122,123] or Soxhlet extraction [121,124,125] of a sample of the column packing. Soxhlet extraction is generally more convenient. The packing, ca. 0.5-1.0 g, accurately weighed into a fine porosity glass sintered crucible with a small piece of filter paper on top, is placed in a standard Soxhlet apparatus. Glass beads are used to raise the level of the crucible above that of the siphon arm. The packing is then extracted at rate of about 10-15 cycles per hour with, generally, the same solvent used for coating. Quantitative extraction of the liquid phase normally requires about 4 hours but overnight extraction is generally used to ensure completion. After extraction the packing is oven dried to constant weight, cooled, and accurately weighed. As a check on the procedure the weight of extracted liquid phase should also be determined. If the evaporation method is used then blanks should be run with each packing to allow for the weight loss from the support when raised to a high temperature.

Copper, aluminum, stainless steel, nickel, or glass tubes bent into various shapes to fit the dimensions of the column oven provide the container for column packings [126]. Neither copper nor aluminum tubing is recommended as both metals are readily oxidized; active, oxide-coated films formed on the inner walls promote decomposition or tailing of labile and polar solutes. Stainless steel is adequate for nonpolar samples but its catalytic activity precludes the analysis of labile solutes. Nickel, after acid passivation, and glass are the most inert column materials, and are preferred for the analysis of labile samples. Teflon or plastic tubing is also used occasionally, often in the separation of chemically-reactive substances for which Teflon supports are also required. Low upper temperature limits and permeability to some solutes restricts their general use.

Columns of 0.5-3.0 m in length and internal diameters of 2-4 mm can be conveniently packed by the tap-and-fill method, aided by suction. One end of the column is terminated with a short glass wool plug and attached via a hose to a filter flask and water pump aspirator [57,116]. A small filter funnel is attached to the other end and the packing material is added in small aliquots. The column bed is consolidated as it forms by gentle tapping of the column sides with a rod or with the aid of an electric vibrator. Over vibration is detrimental to the preparation of an efficient column and should be avoided. When sufficient packing has been added the funnel is removed and a glass wool plug is inserted. The length of the plug is usually dictated by the design of the injector. Columns longer than 3.0 m are difficult to pack and require a combination of vacuum suction at the free end and pressure applied to the packing reservoir at the other end [127].

The packed column requires a temperature conditioning period before use. The column is installed in the oven with the detector end disconnected. The column temperature is then raised to a value about 20°C higher than its intended operating temperature while a low flow of carrier gas is passed through it. In all cases the temperature used for conditioning should not exceed the maximum operating temperature of the phase. The conditioning period is complete when a stable detector baseline is obtained.

After conditioning the column a few preliminary tests are performed to ensure that the performance of the column is adequate. A measure of column efficiency is made with a test sample, for example, a mixture of hydrocarbons, Figure 2.3. An average value of about 2000 theoretical plates per meter is acceptable for a 10% w/w loaded column on a 80-100 mesh support. Higher values can be expected for columns with lower loadings and with finer mesh supports. Columns with less than 1500 theoretical plates per meter are of doubtful quality. An unusually high column inlet pressure also signals problems associated with attrition of the column packing.

After testing the efficiency of the column, some test of column activity and resolving power may be made. A polarity test

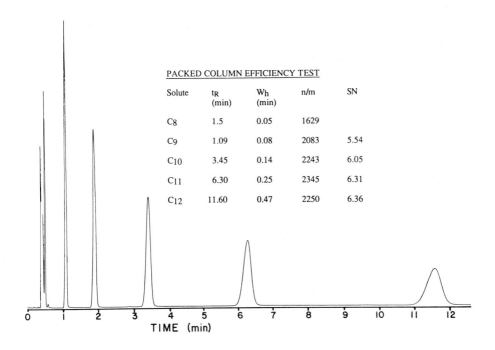

PACKED COLUMN EFFICIENCY TEST

Solute	t_R (min)	W_h (min)	n/m	SN
C_8	1.5	0.05	1629	
C_9	1.09	0.08	2083	5.54
C_{10}	3.45	0.14	2243	6.05
C_{11}	6.30	0.25	2345	6.31
C_{12}	11.60	0.47	2250	6.36

TIME (min)

Figure 2.3   Column efficiency test for a 1.5 m x 2 mm I.D. packed column coated with 10% (w/w) OV-101 on chromosorb P-AW (100-120 mesh) at 100°C with a nitrogen carrier gas flow rate of 30 ml/min. The test sample is a mixture of n-alkanes.

mixture reflecting the polar functional groups of the samples to be separated can be used. Excessive tailing or adsorption of the test probes indicates residual column activity. Polar interactions with support silanol groups can often be reduced by on-column silanization using one of the many commercially available column conditioning reagents. Residual sorptive activity is not unusual for nonpolar phases on diatomaceous supports. To test the resolving power a mixture that reflects the intended use of the column should be selected.

Micropacked column, which have internal diameters less than 1 mm and a packing density similar to the columns described above (see section 1.7.1), are packed by special methods, since they are generally used in longer lengths and operated at higher inlet pressures than conventional packed columns [128,129]. The coiled column is placed in an ultrasonic bath with its axis vertical, one end of the column is attached to a vacuum pump and the other to a pressurized reservoir. During the packing procedure the inlet pressure is slowly increased to maintain a constant fill rate for the column. Virtually all packings used in classical packed columns can be used to prepare micropacked columns. If fine particle supports are used then higher pressures are required for column packing. For small particles of about 10 micrometers in diameter liquid slurry packing is used, analogous to column preparation procedures in high pressure liquid chromatography [130].

2.3 OPEN TUBULAR COLUMNS

In his seminal paper published in 1957, Marcel Golay showed theoretically that an open tube of small internal diameter coated with a thin film of liquid would be capable of realizing efficiencies a hundred-fold or so greater than a packed column [131]. Thus the concept of open tubular column gas chromatography was born and the next quarter century was an era of struggles and turmoil devoted to developing the column technology necessary to make the technique widely accepted [132-134]. In the past column preparation was often described as an art rather than a science, with many practitioners but only a few grand masters. Today it is possible to purchase excellent open tubular columns whose efficiency, thermal stability and inertness surpass the capabilities of packed columns.

The theoretical efficiency of packed columns is limited by the intrinsic dispersion of solute molecules associated with the multiplicity of flow paths through the packing, and by differences in retention resulting from the uneven distribution of liquid phase within the particles and at the contact points between individual particles. By contrast, the efficiency of an open tubular column is governed solely by the rate of mass transfer and diffusion; flow heterogeneities due to the presence of packing are, of course, eliminated. The general improvement in efficiency of an open tubular column over a packed column can best be demonstrated by example. Consider the separation of the solvent extract of a river water sample shown in Figure 2.4 [135]. Clearly the complexity of the sample is better represented in the open tubular column chromatogram than in the packed column case. Both theory and practice indicate that open tubular column gas

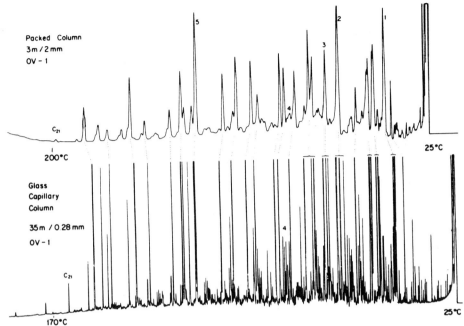

Figure 2.4 Comparison of a typical separation on a packed and a open tubular column. The sample (solvent extract of river water) is the same in both cases. (Reproduced with permission from ref. 135. Copyright Ann Arbor Science Publishers).

chromatography is capable of better separation efficiencies and greatly improved sample detectability for a given separation time than any packed column. The principal disadvantages of open tubular columns are that they are more demanding of instrument performance, less forgiving of poor operator technique and possess a lower sample capacity than packed columns.

Before discussing column preparation procedures a few comments on nomenclature are in order. Open tubular columns are also widely known as capillary columns. The characteristic feature of these columns is their openness, which provides an unrestricted gas path through the column. Thus open tubular column rather than capillary column is a more apt description. However, both descriptions appear frequently in the literature and can be considered interchangeable. The type of columns discussed so far are also known as wall-coated open tubular columns (WCOT). Here the liquid phase is deposited directly onto the column wall without the inclusion of any additive that might be considered as a solid support. Alternatives to the WCOT columns are the porous-layer open tubular columns (PLOT) and support-coated open tubular columns (SCOT). PLOT columns are prepared by extending the inner wall of the column by such substances as silica or elongated crystal deposits. In SCOT columns the liquid phase is coated on a surface that has been covered with some type of solid support material.

Continuous improvements in column technology have made many of the early studies on column preparation obsolete. We will present only a brief account of these older methods in the following sections. For a more extensive view of the evolution of column technology standard texts [136-142] and review articles [35,143-146] should be consulted.

2.3.1 Drawing Columns With Capillary Dimensions

Numerous materials have been used to fabricate open tubular columns. Most early studies were conducted using stainless steel tubing and later nickel tubing of capillary dimensions [147-149]. These materials had rough inner surfaces (leading to non-uniform stationary phase films), metal and oxide impurities at their surface which were a cause of adsorption, tailing, and/or decomposition of polar solutes and because their walls were thick, enhanced thermal inertia that prevented the use of fast temperature programming. None of these materials are widely used today.

At present, virtually all open tubular columns are prepared from either soda-lime, borosilicate, or fused silica glasses [137,140,150]. Fused silica, prepared by oxidizing pure silicon tetrachloride in an oxygen electric plasma torch, is essentially pure silica, containing less than 1.00 ppm of metal impurities [134]. Since it has a relatively high melting point, both special equipment and precautions are required in fabricating columns from this material. The surface of fused silica glass is relatively inert, containing primarily silanol and siloxane groups. The presence of silanol groups is responsible for the residual acidic character of the glass. High temperature treatment in an inert atmosphere can reduce the concentration of silanol groups, and hence activity, to very low levels [151,152].

Glasses with lower working temperatures than silica are prepared by adding metal oxides to the silica during manufacture. Thousands of glasses with a variety of compositions are known, but for chromatographic purposes only soda-lime (soft) and borosilicate (hard, Pyrex, Duran) glasses are important. Typical values of the bulk compositions of soda-lime glasses are SiO_2 67.7%, Na_2O 15.6%, CaO 5.7%, MgO 3.9%, Al_2O_3 2.8%, BaO 0.8%, and K_2O 0.6% and for borosilicate glass SiO_2 81%, B_2O_3 13%, Na_2O 3.0%, Al_2O_3 2.0% and K_2O 1.0%. However, for chromatographic purposes, the surface composition of the glass is more important than its bulk composition, and may vary substantially from that of the ratio of the constituents given for the bulk [153]. Soda-lime glasses are slightly alkaline, due to the high content of Na_2O, while the borosilicate glasses are somewhat acidic as a result of the presence of B_2O_3. The adsorptive and catalytic activity of the column materials can be attributed to the silica surface (silanol and siloxane groups) and to the presence of metal impurities at the surface which can function as Lewis acid sites. The metal impurities act as adsorption sites for electron donor molecules, such as ketones and amines [143].

Open tubular glass capillary columns with internal diameters of 0.05 to 1.0 mm and outer diameters of 0.3 to 1.5 mm can be prepared from glass stock, typically 4.0 to 10.0 mm outer diameter and 2.0 to 6.0 mm internal diameter, with a glass drawing machine [154,155]. The glass stock tube is fed into a softening furnace that is electrically heated to a temperature of 650 to 850°C and drawn out at the lower end by passage through a pair of motor-driven rollers. Capillary tubes of different dimensions are drawn by maintaining a fixed differential between the feed rate of the stock tube to the furnace and the pull rate of the softened tube, at the exit of the furnace. The capillary tubing emerging from the softening oven is coiled by passage through a heated bent pipe maintained at 550 to 625°C. The actual temperatures used for the softening oven and bent pipe depend on the wall thickness and composition of the glass. Higher temperatures are required for borosilicate than soda-lime glasses.

The flexible fused silica columns of the type introduced by Dandeneau and Zerenner in 1979 cannot be drawn on standard glass drawing machines [134,156-158]. These thin-walled, polymer-coated columns are drawn by techniques similar to those used for manufacturing optical fibers on equipment that is not normally found in analytical laboratories. A schematic diagram of an idealized drawing machine is shown in Figure 2.5. The drawn tubing is inherently straight but sufficiently flexible to be coiled on a

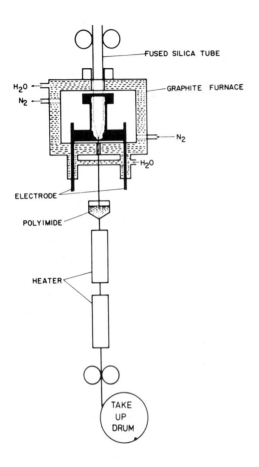

FUSED SILICA TUBE

GRAPHITE FURNACE

H_2O

N_2

N_2

H_2O

ELECTRODE

POLYIMIDE

HEATER

TAKE UP DRUM

Figure 2.5 Apparatus for drawing polymer-clad, flexible, fused-silica open tubular columns.

spool for collection purposes. The thin-walled tube must immediately be protected from moist air or dust particles that promote the growth of fissures or cracks, making the capillary weak and friable, by coating it with a protective film of polymer or aluminum. For this reason the capillary emerging from the softening oven is immediately passed through a reservoir containing a solution of a high temperature polymer, usually a polyimide, and then through an electrically-heated drying oven prior to collection on a spool. Typical drawing rates for fused silica capillary columns are around 1 m/s, substantially higher than the rate used to draw glass capillaries.

During the drawing process the fused silica is subject to various temperature changes that cause hydration and dehydration, yielding a material with an uncontrolled silanol concentration. Residual silicon tetrachloride, trapped in the fused silica preform, and oxides of nitrogen, formed at the high temperatures required to melt the fused silica, can result in the production of acidic impurities (HCl and HNO_3) on hydrolysis. Conditioning and hydrothermal treatment are generally required to remove acid impurities and provide a uniform and defined silanol concentration for column preparation [133,140,141,146,159]. These procedures are usually performed in the laboratory and are essential in minimizing batch-to-batch variation in the properties of the fused-silica columns, which can lead to failure of optimized column deactivation and coating procedures found in the literature.

Strength, flexibility, relative inertness, and the inherent straightness of column ends recommend fused silica as the column material of general choice for column preparation. However, as we shall see later, not all polar stationary phases form stable films on fused-silica surfaces, and columns with internal diameters greater than about 0.6 mm are not available. For these phases and for the preparation of wide bore columns, borosilicate glass is normally the choice. Greater availability of the preform in the desired size, greater uniformity of composition and increased mechanical strength tend to dictate the use of borosilicate glass over soda-lime glass in contemporary practice, except perhaps for the separation of basic solutes.

2.3.2   Film Formation on Glass Surfaces

To achieve a high separation efficiency in any type of open tubular column it is essential that the stationary phase be deposited as a smooth, thin and homogeneous film. This film must also maintain its integrity without forming droplets when the column temperature is varied. The ability of a liquid phase to wet a glass surface depends on whether the critical surface tension of the glass, a measure of the free energy of the surface, is greater than the surface tension of the liquid [160-163]. The stability of the film depends on the viscosity of the liquid and its resistance to disruption by forces tending to minimize the gas-liquid interface, Rayleigh instability. More viscous films provide greater stability whenever the time needed for the film to rearrange becomes long compared to the duration of the disturbance causing it to rearrange [163,164]. This is because the film is allowed insufficient time to break-up or form ripples under unstable conditions, and because the viscosity of the film acts as a dampening mechanism in opposition to the disturbance. To maintain temperature stability, phases showing little changes in viscosity with temperature are preferred. Many of the common oil-like phases developed for packed column gas chromatography have too low a viscosity, particularly at high temperatures, to enjoy equal popularity in contemporary open tubular column technology. Specially synthesized analogs of high viscosity, many of them gums, are widely used instead (see section 2.3.6). All liquid films on curved surface are unstable in that the liquid film shows a natural tendency to decrease the interfacial energy at the gas-liquid interface leading to droplet formation [165]. The logarithmic growth rate of Rayleigh instability is proportional to the surface tension of the liquid and to the third power of the film thickness, and inversely proportional to the viscosity of the liquid and to the fourth power of the column diameter. Again, liquid phases of high viscosity minimize Rayleigh instability. For similar reasons, wide bore and thin film columns show inherently greater stability and longevity than narrow bore and thick film columns.

The ability of a liquid to wet a solid surface is determined by the contact angle, that is, the angle formed between the tangent to the liquid drop and the solid surface as drawn through the liquid. For a liquid that completely wets the surface the contact angle is zero. When the cohesive forces of the liquid are greater than the adhesive forces between the liquid and the solid the contact angle increases. The contact angle, therefore, is an inverse measure of the wettability of the solid by the liquid. The surface tensions of typical stationary phases are in the range 30-50 dyne/cm. Clean fused silica and glass are high energy surfaces and should be wet by most stationary phases. However, adsorption of moisture and other impurities generally lowers the critical surface tension to about 25-50 dyne/cm. Also, columns coated without deactivation of the surface often show undesirable activity towards sensitive analytes. Deactivation, then is a common column preparation procedure, performed to maintain or enhance the wettability of the surface.

Solutions to the above problem are required if efficient open tubular columns are to be prepared. The energy of the smooth glass surface can be increased by roughening or chemical modification, or the surface tension of the stationary phase can be lowered by the addition of a surfactant. Roughening and/or chemical modification are the most widely used techniques for column preparation; the addition of a surfactant, although effective, modifies the separation properties of the stationary phase and may also limit the thermal stability of columns prepared with high temperature stable phases.

Roughening, achieved by chemical treatment of the glass surface, enhances the wettability of the glass surface by increasing the surface area over which interfacial forces can act and dissipate the cohesive energy of the drop. Acid leaching or hydrothermal treatment restores the high energy of the glass surface, 40-50 dyne/cm, and provides a stable and uniform surface to achieve maximum deactivation. Deactivation is usually achieved by creating a bonded organic layer by silylation or a non-extractable polymeric film by the thermal degradation of a polymeric substrate. Exhaustive silylation with a trialkylsilyl reagent forms highly deactivated surfaces but of low energy, about 21-24 dyne/cm, which can only be coated with nonpolar liquid phases. For more polar phases silylation reagents containing functional groups similar or identical to those in the stationary phase are used to maintain the desired degree of wettability. Formation of a non-extractable film of polar polymer, such as Carbowax 20M, by thermal treatment results in a high energy surface, about 45 dyne/cm, which can be coated with many polar phases. In subsequent sections the chemistry of the above processes will be discussed in more detail.

2.3.3 Surface Modification Reactions

Surface modification reactions are used to improve the wettability of glass surfaces by polar stationary phases and to improve the extent of deactivation by silylation [138-146,166]. Numerous procedures have been investigated but only a few are in common use. Of these, the most important reactions are etching by gaseous hydrogen chloride, leaching with aqueous hydrochloric acid, formation of whiskers and solution deposition of a layer of solid particles. Because of the high purity and thinness of the column wall, leaching is the only treatment generally used with fused silica columns.

The static or dynamic etching of soda-lime glass capillaries results in the formation of a regularly spaced array of sodium chloride crystals on the inner surface of the column. Borosilicate glasses contain a lower concentration of alkali ions and are little changed in appearance by hydrogen chloride treatment, unlike soda-lime glasses which become opaque. Hydrogen chloride etching is considered essentially a technique for roughening the surface of soda-lime capillaries, although similar treatment of borosilicate glass often improves its wetting and deactivation characteristics as well.

To etch soda-lime capillary columns the dry column is filled with hydrogen chloride gas and then sealed at both ends with a microtorch. The column is then heated at 300 to 400°C for between 3 to 12 h depending on the method followed [160,167-169]. Dynamic etching with a continuous flow of hydrogen chloride yields a surface covered with a large number of relatively small particles of dissimilar size and shape. In the presence of excess hydrogen chloride (the static method) recrystallization is relatively slow compared to particle growth, accounting for the differences in the surfaces generated by the two methods. The static method provides the most uniform surface and is preferred for column preparation. A detailed study of the mechanism of sodium chloride crystal formation has been given by Franken et al. [168]. The most important limitations of the hydrogen chloride etching procedure are: the reproducibility of the microroughened surface depends on the surface composition of the glass; the solubility of sodium chloride in various solvents can present a problem during the rinsing and coating of columns; thin film columns show adsorptive behavior resulting from the weak Lewis acid character of the sodium ions; and the high alkali concentration on capillary surfaces increases the catalytic decomposition of some stationary phases [143].

Acid leaching with solutions of hydrochloric acid produces a different effect than acid etching in the gas phase. Controlled acid leaching removes metallic cations from the glass surface to form a silica-rich surface. The new surface layer greatly minimizes the effects of glass variety on subsequent treatments and leads to a higher degree of reproducibility in column preparation. Lewis acid sites are removed, leaving highly adsorptive silanol groups; the latter are easily deactivated, for example, by silylation. Thus acid-leached borosilicate and soda-lime glasses are very inert after deactivation and can be coated with thin films of nonpolar phases. As fused silica contains only traces of metal impurities deionization is not required but leaching is useful to maximize the concentration of surface silanol groups to enhance deactivation and possibly bonding in subsequent column preparation procedures [140,159]. The experimental conditions for the hydrothermal treatment (equivalent to leaching) of fused silica is often milder than that required for other glasses.

Borosilicate and soda-lime glass capillary columns can be leached with aqueous hydrochloric acid using the static [140,170] or dynamic [171-173] method. In the static method the column is filled with 20% v/v hydrochloric acid, a small portion of the column emptied and then evacuated, the ends sealed and the column heated overnight at 140 to 170°C. The column is partially emptied and evacuated to provide an expansion zone for the heated liquid. For borosilicate columns it is recommended that the evacuated portion of the column should correspond to 7% of the column length and the column should be heated at 170°C for 12 to 16 h. For soda-lime columns the evacuated portion of the column should correspond to 4% of the column length and it should be heated at 140°C for 12 to 16 h. After leaching the columns are cooled and then rinsed with 1% v/v hydrochloric acid to remove the dissolved metal ions. One to two column volumes at a flow rate not greater than 2 cm/s are usually sufficient for the rinse step. To avoid readsorption of dissolved metal ions by the surface hydrogel, the entire rinsing process should occur under strong acid conditions (distilled water should not be used). Also, short (20 meter) columns are much easier to rinse out without readsorption of metal ions than longer columns. After rinsing, the swollen surface hydrogel is dehydrated by applying simultaneously a vacuum to both ends of the column, or by applying alternately the vacuum to either end of the column for 10 to 20 minutes while the column is heated. Recommended conditions for 15 to 20 m column lengths are 2 h at 300°C, and 3 h at 300°C for 30 to 50 m column lengths.

For dynamic leaching the column is mounted in an oven with both ends extending to the outside. One end is connected to a reservoir and a solution of 20% v/v hydrochloric acid is forced through the column at a temperature of 100°C. Generally, long lengths (e.g., 90 m) of capillary tubing are leached with 50 ml of acid for approximately 48 h. The leaching step may need to be repeated after an intermediate drying stage to produce the best results. After leaching the column is rinsed with distilled water and then dried by purging with dry nitrogen at 150°C for about 12 h. Venema et al. have recommended slightly different conditions for the dynamic leaching of soda-lime capillary columns [173].

A variety of reagents and methods have been used for extending the surface area of borosilicate or soda-lime glasses by the preparation of whisker surfaces. A high density light surface etch can be obtained by flushing the capillary column with either an aqueous solution of potassium bifluoride [174] or a methanolic solution of ammonium bifluoride [175]. Treatment of borosilicate glass with ammonium bifluoride results in a thin, uniform etch with a depth of 0.05 to 0.4 micrometers and a distance of 2 to 6 micrometers between whiskers. This is a relatively light surface modification compared to whiskers grown at high temperatures. However, the above methods are simple to carry out and, after acid leaching and deactivation, are suitable for preparing columns coated with nonpolar and moderately polar stationary phases.

Whisker columns of high density are prepared by the action of hydrogen fluoride gas on the glass surface at high temperatures. The gas itself is rarely used because of safety considerations (toxicity) and because whisker growth is often non-uniform [176]. The preferred method is to generate hydrogen fluoride in situ by the thermal decomposition of 2-chloro-1,1,2-trifluoroethyl methyl ether [177-183] or ammonium bifluoride [59,184,185]. The mechanism of whisker formation is not known with certainty but it is assumed that the hydrogen fluoride generated reacts with the glass to form silicon tetrafluoride, which is then converted into silicon dioxide and deposited in the form of whiskers. The whisker surface has a surface area a thousand times or so higher than the smooth glass surface and consists of a densely packed layer of filamentary projections whose size and length depend on the experimental conditions. Under optimum conditions whiskers of 4 to 5 micrometers in length are obtained. With the 2-chloro-1,1,2-trifluoroethyl methyl ether, the important experimental variables are reagent concentration, temperature and reaction time. At temperatures below 250°C whisker formation does not occur. The optimum temperature for whisker formation is in the region of 400°C. Whisker length and surface density increase with the growth period and a time of 24 h is considered optimum. After whisker formation carbon deposits formed by decomposition of the ether must be removed by flushing with oxygen. Typical conditions involve passing a continuous flow of oxygen through the column at 450°C for 8 to 12 h. The ammonium bifluoride method has the advantage that only volatile products (HF and NH_3) are produced by thermal decomposition. Column preparation time is also considerably reduced. For optimum whisker growth the reagent is dynamically coated and the sealed column is then heated at 450°C for 3 h. for borosilicate glass and 350°C for 3 h. for soda-lime glass.

The large surface area of whisker columns enables any stationary phase to be coated efficiently without droplet formation and their sample capacity is much higher than WCOT columns. On the debit side, whisker surfaces are extremely active and very difficult to deactivate, and, because of the high degree of roughening, they are less efficient than WCOT columns.

Glass surfaces can also be roughened by deposition of a thin homogeneous layer of particle material from solution. A dense layer of sodium chloride crystals can be deposited by dynamically coating columns with a stable sodium chloride suspension prepared by the addition of a saturated solution of sodium chloride in methanol to 1,1,1-trichloroethane or methylene chloride [186,187]. During passage of the suspension through the column, particles of sodium chloride deposit spontaneously on the column wall, building up a layer of sodium chloride which gradually reaches a maximum density. Coating with polar stationary phases produces efficient and thermally stable columns. The barium carbonate procedure developed by Grob is probably the most widely evaluated of the particle deposition methods [140,188-191]. The general procedure consists of dynamically coating the glass surface with barium hydroxide solution using carbon dioxide gas to push the plug of hydroxide solution through the column and to generate the barium carbonate layer. From a saturated solution of barium hydroxide, a rather dense crystal layer is obtained that is suitable for coating with polar phases. From gradually diluted coating solutions the crystals become more distant while the glass surface between the crystals remains covered by a very thin and smooth layer of barium carbonate. When the solution is diluted 50- to 100-fold the crystals disappear, leaving the smooth surface cover only. This thin film is suitable for coating with nonpolar phases where the barium carbonate layer functions more as a deactivating agent for surface active sites than as a roughening agent. The preparation of the barium carbonate layer is influenced by a large number of experimental variables including the glass surface structure, the crystallization temperature, the addition of stationary phase modifiers (surfactants), and the method used for deactivation. Careful attention to the experimental details enables columns to be prepared with a fairly high success rate, although column efficiency is usually lower than with other methods. General problems are the difficulty of completely deactivating the barium carbonate layer and the poor thermal stability of some phases when coated onto this layer.

2.3.4 Surface Deactivation Methods

Surface modification reactions used to improve the wettability of glasses by the stationary phase and to deionize some glasses result in an increase in the activity of the glass surface. Without subsequent deactivation, columns coated with nonpolar and moderately polar stationary phases exhibit undesirable chromatographic properties, such as tailing, incomplete elution of polar solutes and in extreme cases, sample decomposition. Although polar stationary phases may act as their own deactivating agents, column pretreatment is still advisable to ensure complete deactivation and to improve the thermal stability of the stationary phase film.

The effectiveness of a particular deactivating reagent depends on the properties of the glass, the identity of the stationary phase and the sample to be analyzed, as well as on the method of surface roughening. No universal method of deactivation exists, but some techniques have emerged as more useful than others. The most widely used methods include coating with the thermal degradation products of stationary phases [192-194], high temperature silylation [195] and reaction with silicon hydride polysiloxanes [196,197].

Carbowax 20M, polysiloxanes, and N-cyclo-3-azetidinol are the most widely used substances for the thermal degradation method [143,180,192-194]. In the case of the Carbowax treatment deactivation can be carried out in either of two ways. The column can be dynamically coated with a solution of Carbowax 20M in a volatile solvent, excess solvent evaporated with a stream of nitrogen, the column ends sealed and the column heated at about

260-280°C for 12-16 h. After the column is cooled, excess Carbowax and other byproducts are removed by rinsing with suitable solvents. Also, the column may be deactivated by vapor phase treatment [194]. For this purpose a short precolumn, containing Carbowax 20M coated onto a support, is inserted into a separate heater and connected directly to the column to be deactivated, which is housed in a separate oven. The precolumn in maintained at a temperature of about 260°C, 5-10°C higher than the capillary column. Deactivation is completed by allowing the degradation products to bleed through the capillary column for about 16 h with a slow flow of carrier gas. The mechanism of Carbowax deactivation remains unknown, although it is believed that the thermal degradation products bond chemically to the surface silanol groups at the high temperatures employed. Carbowax pretreatment is reasonably successful at masking the influence of silanol groups but less so at masking Lewis acid sites associated with metal ions; the latter are better removed by acid leaching. Carbowax deactivation methods have been used to prepare both nonpolar and polar columns and reduced activity toward many sensitive compounds has been demonstrated. The deactivated layer is thermally stable under continuous use at temperatures up to about 220°C and for intermittent use up to about 250°C. Columns coated with nonpolar stationary phases may show mixed liquid phase retention behavior. Carbowax pretreatment is inadequate for deactivating whisker columns. Thermal treatment with an azetidinol reagent is the preferred method in this case [180,181].

The thermal degradation products of polysiloxane stationary phases have been used by Schomburg to deactivate glass and fused silica columns prior to coating the columns with those same phases [192,198]. An acid-leached column is dynamically coated with a solution of the polysiloxane phase, the solvent evaporated with an inert gas, the column ends sealed and the column heated to between 300 and 450°C (depending on the glass used) for several hours. Afterwards, excess material is removed by solvent extraction or by high temperature conditioning with a continuous flow of inert gas. It is postulated that under the experimental conditions described above the polysiloxane phase decomposes, releasing degradation products which bond chemically to the glass surface. The mechanism may be very similar to that of silylation with cyclic siloxanes. Columns prepared in the above manner exhibit good deactivation of silanol groups with thermal stabilities that endure above 300°C.

High temperature silylation with disilazanes, disiloxanes, or cyclosiloxanes, thermal degradation of polysiloxanes, and the dehydrocondensation of silicon hydride polysiloxanes are now established as the most effective procedures for masking silanol groups on deionized glass and fused silica surfaces for subsequent coating with nonpolar and moderately polar polysiloxane stationary phases [140,146,195-198]. Some typical reagents are shown in Figure 2.6 in which the R group is varied to change the wettability of the glass surface and is generally the same as the substituent attached to the polysiloxane backbone of the stationary phase [199-203].

Under typical reaction conditions the capillary column is dynamically coated with the reagent or a solution of the reagent followed by removal of solvent, the column ends sealed, and the column heated to 400-450°C for a time that depends on the reagent used. The column ends are then broken and the column is purged with nitrogen at a high temperature to remove any residual reagent. At the high temperatures required for effective deactivation the polyimide external coating of the fused silica capillary columns must be protected with an inert atmosphere, since atmospheric oxidation decreases its flexibility and transparency. Narrow bore capillary columns can be difficult to deactivate by high temperature silylation using cyclic siloxanes, or polysiloxanes due to the formation of solid deposits that partially block the columns and make subsequent coating difficult [164,204]. These problems can be solved using the silicon hydride polysiloxane reagents which require lower deactivation temperatures, 250-300°C, and produce hydrogen as the only byproduct of their reaction with silanol groups.

Not all aspects of the high temperature silylation mechanism for glass surfaces are fully understood. There probably a large commonalty in the way the various reagents react, although they were introduced and optimized separately before a more general picture emerged. Careful hydroxylation and dehydration of the glass surface is an essential prerequisite for maximizing the number of silanol groups and for establishing a fixed concentration of water which will be consumed during the silylation reaction. At low and moderate temperatures the steric bulk of the trimethylsilyl group prevents complete reaction of all silanol groups resulting in a surface which retains some residual activity that is not useful for preparing inert, nonpolar columns.

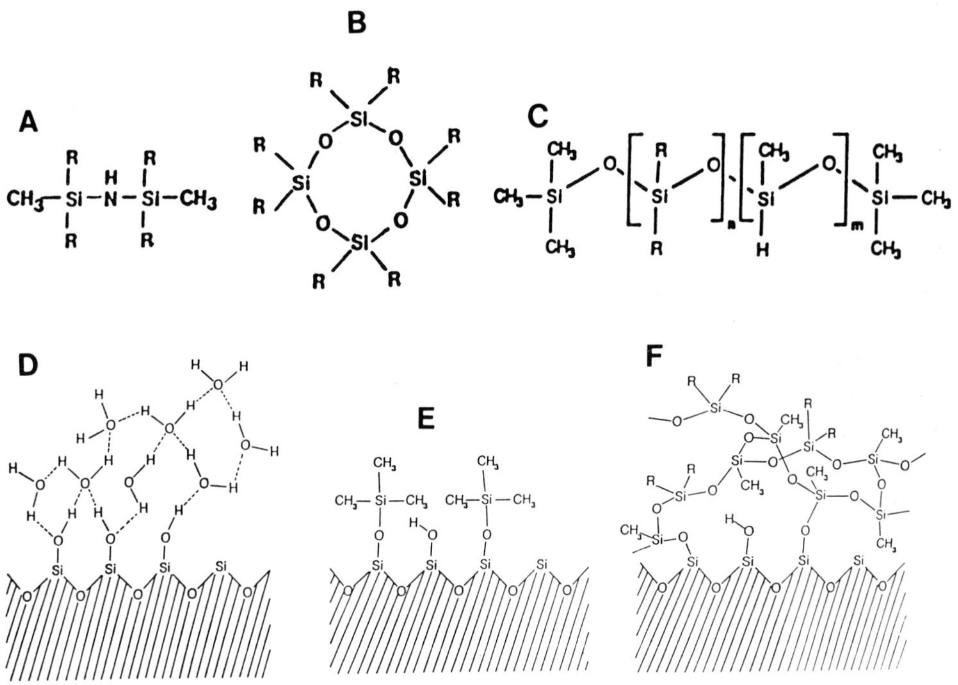

Figure 2.6   Reagents used for the deactivation of silanol groups
on glass surfaces. A = disilazanes, B = cyclic siloxanes, and C =
silicon hydride polysiloxanes in which R is usually methyl,
phenyl, 3,3,3-trifluoropropyl, 3-cyanopropyl, or some combination
of these groups. The lower portion of the figure provides a view
of the surface of fused silica with adsorbed water (D), fused
silica surface after deactivation with a trimethylsilylating
reagent (E), and fused silica surface after treatment with a
silicon hydride polysiloxane (F).

Above about 400°C several alternative reactions to simple
silylation can also occur. Condensation of neighboring surface
silanol groups to form siloxane groups, nucleophilic displacement
of the organic group from the bonded reagent by a neighboring
silanol group to form short cyclic siloxanes bonded to the surface
by two or three bonds and reactions with water, forming short
chain polymers bonded to the surface [195,205]. These reactions
result in a decrease in the number of unreactive silanol groups on
the deactivated surface and the formation of a protective

umbrella-type film, diminishing access of the analyte to the remaining unreacted silanols. In the case of the disilazane reagents the byproduct of the reaction with silanol groups, ammonia, can react with siloxane groups on the glass surface to generate additional silanol groups, which are subsequently silylated, yielding a high density of bonded organic groups. If the experimental conditions are less than optimal, an active surface may result because of the production of silanol groups that are subsequently inadequately reacted or shielded at the completion of the deactivation reaction [206]. The thermal decomposition of polysiloxane phases occurs with the liberation of cyclic siloxanes and this explains the equivalence of results observed for deactivation by thermal degradation of polysiloxanes and the direct application of cyclic siloxane reagents. Deactivation with silicon hydride polysiloxanes occurs primarily by dehydrocondensation with silanol groups on the fused silica surface, establishing surface-to-polymer bonding [196,197,207-209]. In addition, physically adsorbed water hydrolyzes silane bonds to silanols, which further dehyrocondense with silane bonds or condense with other silanol groups to form a highly crosslinked resin film as illustrated in Figure 2.6.

2.3.5 Procedures for Coating Open Tubular Columns

The object of all coating procedures is the distribution of the stationary phase as a thin, even film that completely covers the glass surface [139,140,143,146]. Dynamic coating and static coating methods, with several individual variations, are used for this purpose.

For dynamic coating, a suitable coating reservoir is charged with a solution of the stationary phase, 5-15% w/v, which is forced by gas pressure into the capillary column to be coated. When about 25% of the column volume is filled the column is withdrawn from the coating solution and gas pressure is used to force the liquid plug through the column at a linear velocity of 1-2 cm/s. To maintain a constant coating velocity a buffer capillary column, which is about 25% of the length of the column being coated, is attached to the outlet end of the capillary column. When all the coating solution has left the capillary column it is disconnected from the buffer column and the gas pressure is increased to evaporate the solvent. For coating long lengths of narrow bore capillary columns a high pressure glass-lined stainless steel reservoir is required [164,210]; otherwise, coating reservoirs are usually made from any convenient screw-capped PTFE sealed vial, bottle or test tube [140,211]. The dynamic coating procedure is comparatively fast but suffers from the general weakness that it is difficult to accurately predict the stationary phase film thickness prior to preparing the column. Monitoring the change in volume of the plug of coating solution during coating or using approximate equations defining the film thickness in terms of the experimental conditions are two possibilities for obtaining columns of known film thickness [212-214]. It is more likely, however, that accurate values of the film thickness can only be obtained by comparing the chromatographic properties of a newly prepared column with columns having an accurately known film thickness.

The thickness and uniformity of the stationary phase film produced by dynamic coating is dependent on the concentration and viscosity of the coating solution, the surface tension of the coating solution, the velocity and constancy of the velocity of the coating solution, temperature and the rate of solvent evaporation. Orienting the coil of the column horizontally when coating and drying can minimize the influence of drainage of the coating solution to the lowest parts of the coil. Fast column flow rates, normally between 4-10 ml/min, are preferred for drying as this minimizes the column preparation time but will usually result in the formation of thicker films [215,216]. Slow drying speeds do not disturb droplets which are pushed out of the column, while high drying speeds destroy droplets and prevents stationary phase transportation. Discarding the first 5% of the column length after drying may produce higher overall efficiency in the final column. According to Parker and Marshall the principal cause of low column efficiency provided that the stationary phase wets the column wall, is the formation of "lenses" (small plugs of liquid left behind the main plug during coating) [160]. Small temperature fluctuations may cause condensation of solvent vapors, forming a lens which is able to bridge the bore of the capillary column. Since the lens consists of pure solvent, it will dissolve the film already laid down. Lens formation can be eliminated by raising the temperature of the column during coating, although this is rarely done in practice.

To improve the success of the dynamic coating method, Schomburg suggested a modification of the above method, known as the mercury plug dynamic coating procedure [180,216-220]. In this method a mercury plug is interposed between the plug of coating solution and the driving gas. Mercury, which has a high surface tension, wipes most of the coating solution off the surface of the column wall as it moves through the column, leaving a more even and thinner film behind. More concentrated coating solutions (10-50% v/v) are used in this procedure, producing films which are more resistant to drainage during the drying process.

The static coating method yields efficient columns of predictable film thickness with gum or solid phases [140,180,221-226]. Phases of low viscosity, however, cannot be successfully coated by the static method due to their propensity to flow and accumulate in the lowest portions of the capillary column during the relatively long time required for solvent evaporation. For static coating the column is filled entirely with a dilute solution, 0.02-4.0% v/v, of the stationary phase in a volatile solvent which is then evaporated by sealing one end of the column and attaching the other to a vacuum source, elevating the temperature of the column and performing the evaporation step at about atmospheric pressure or by screwing the open end of the filled capillary into a high-temperature oven to force solvent evaporation. The vacuum technique remains the most widely used variation. In all cases the solvent is evaporated under quiescent conditions, leaving behind a thin film of liquid phase of thickness given by $rc/200$, where r is the capillary column radius, and c the concentration of the coating solution on a v/v basis.

The principal problem with the static coating method is the breakthrough of gas bubbles when the column is placed under vacuum. These bubbles generally originate at the solvent/seal interface or are trapped in microcavities in the glass surface. Breakthrough can eliminate all or part of the coating solution from the capillary column. The creation of a solvent/seal air-free interface is very important. Numerous methods have been described for this purpose and can be grouped into two categories: mechanical methods include clamping, crimping, freezing, or placing a stopper in a short length of Teflon or silicone tubing attached to the column end [224,227,228]; and chemical sealants such as waxes [229], Apiezon grease [230], epoxy resins [185,231,232], or water glass [221], etc., which are sucked into the end of the column and allowed to harden before evacuation is begun. It may be necessary to degas some sealants prior to use by placing them under vacuum. Air bubbles trapped in microcavities or scratches, etc., in the glass surface can be dissolved by allowing the column to stand several hours before evaporation is commenced or by pressurizing the filled column to several atmospheres for a short time. Rapid, even small, changes in temperature result in uneven film deposition and columns should be thermostated during the coating process by immersion in a large volume water or oil bath.

Certainly, one of the less desirable features of the static coating method is the long time required for column preparation, particularly for long and/or narrow bore columns. Only volatile solvents, such as pentane, dichloromethane, and Freons provide reasonable column preparation times at ambient temperatures. For example, a 20 m x 0.3 mm I.D. capillary column filled with dichloromethane requires about 15 h for complete evacuation; pentane requires only about half this time and is the preferred solvent for those phases having sufficient solubility [230,233]. Mixtures of pentane-dichloromethane can also be used to decrease the column preparation time of medium polar stationary phases. Static coating of narrow bore capillary columns at temperatures above the solvent boiling point has been used successfully to minimize column preparation time [164,230]. The choice of solvent and temperature cannot be viewed as independent variables in the coating process on account of the natural tendency of the film to rearrange and form droplets, due to Rayleigh instability (see section 2.3.2) [165,225]. The properties of the stationary phase, column surface, coating solvent, temperature and vacuum conditions are all interrelated. Further, the Rayleigh instability is magnified for narrow bore columns reducing the success rate for coating these columns. The properties of the coating solvent and temperature have a marked effect on the viscosity and surface tension of the stationary phase solutions at all concentrations during evaporation. High solution viscosities in the later stages of solvent removal and fast coating speeds are desirable to avoid the formation of inhomogeneities in the stationary phase film. Temperature and solvent composition will frequently have to be established as a compromise and will be different for different stationary phases.

Evaporation at atmospheric pressure, called the free release static coating method, has not been as widely used or evaluated as the vacuum method [146,226,234]. The experimental arrangement is

similar to that for the vacuum method except that no vacuum source
is required and a dampening capillary column is normally attached
to the open end of the column to increase the restriction to vapor
transport and minimize the occurrence of bumping. Temperatures
above the solvent boiling point can be used and coating is
comparatively rapid, particularly for long and/or narrow bore
columns, requiring very long coating times using the vacuum
technique.

2.3.6 Gum and Immobilized Stationary Phases

The gradual evolution of column technology revealed that the
low to medium viscosity oils used as stationary phases for packed
column gas chromatography were inadequate for preparing thermally
stable and efficient open tubular columns. New phases of much
higher viscosity were required to resist film disruption at
elevated temperatures. For poly(dimethylsiloxanes) this was
relatively easy to achieve by increasing the molecular weight. For
poly(dimethylsiloxanes) containing bulky or polar functional
groups, the regular helical conformation of the polymers is
distorted resulting in a greater change in viscosity with
temperature. For poly(dimethylsiloxanes) increasing temperature
causes the mean intermolecular distance to increase but at the
same time expansion of the helices occurs, tending to diminish
this distance. As a result, the viscosity appears to be only
slightly affected by temperature. A second breakthrough was
achieved when it was realized that crosslinking of the gum phases
to form a rubber provided a means of further stabilizing the
polysiloxane films without destroying their favorable diffusion
characteristics. The unmatched flexibility of the silicon oxygen
bond imparts great mobility to the polymeric chains, providing
openings that permit diffusion, even for crosslinked phases.
Modern stationary phases for open tubular column preparation are
characterized by high viscosity, good diffusivity, low glass
transition temperatures and are often suitable for crosslinking.
For the present the above properties are only apparent for the
polysiloxane and poly(ethylene glycol) phases, and this limits the
achievable range of selectivities.

The polysiloxane phases with various substituent groups are
generally prepared by either the catalytic hydrolysis of
dichlorosilane or dimethoxysilane monomers or by the platinum
catalyzed addition of an alkene to a poly(alkylhydrosiloxane)

polymer (the addition of an alkene to a Si-H bond) [35,235,236]. The dimethoxysilanes are preferred for the preparation of polysiloxanes containing cyanopropyl substituents to avoid hydrolysis of the cyano functional groups and trapping of hydrogen chloride, generated by hydrolysis of chlorosilanes [41]. Copolymers containing vinyl, tolyl or n-octyl groups to assist in the crosslinking of the more polar polysiloxanes are prepared by hydrolyzing and polymerizing the appropriate amounts of the relevant monomers. Eventually, a high molecular weight polysiloxane gum with terminal silanol groups results, which is suitable for use after fractionation for some column preparation procedures. Alternatively, the polymer can be endcaped by treatment with 1,3-divinyltetramethyldisilazane and chlorosilane catalyst. Endcaping introduces an additional functionality into the polymer to aid crosslinking, and by terminating the polymerization process it ensures a narrow molecular weight distribution for the polymer. Thermal curing of the silanol terminated polymer is a common secondary route to increase the average molecular weight of the polymer by additional condensation [237-239]. Prior to use the synthesized polymers are purified by solvent precipitation to remove low molecular weight oligomers and finally by size-exclusion chromatography to isolate a well defined molecular weight fraction. To obtain inert columns it is essential that the final product is free of all catalysts and byproducts and to obtain high thermal stability and reproducibility in the column preparation procedure that the structure and molecular weight of the phase is well defined.

Two different procedures have been used to immobilize polysiloxane phases during column preparation. Thermal condensation of nonpolar and medium polar silanol terminated polysiloxanes at high temperature with silanol groups on the glass surface results in bonding of the phase to the surface, yielding inert and thermally stable columns. Columns prepared with poly(dimethylsiloxanes) and poly(methylphenylsiloxanes) are easily made by static coating of the silanol terminated phase, which is then heated by temperature programming up to 300-370°C and maintained at that temperature for about 5 to 15 h with a slow flow of carrier gas. Bonded, nonpolar polysiloxane phases are suitable for high temperature gas chromatography with isothermal operation at 400-425°C [240-242]. Medium polar phases are more difficult to bond and require careful surface preparation to avoid film disruption during the bonding process and generally yield columns of lower thermal stability [243]. As well as bonding, additional crosslinking using a trifunctional crosslinking reagent or by conventional free radical crosslinking with azo-tert.-butane can be used for additional immobilization of cyanopropylsiloxanes [238,244].

An alternative approach to stationary phase immobilization, and the most popular method in contemporary practice, is the free radical crosslinking of the polymer chains, using peroxides [201,202,245-248], azo-compounds [202,249-251], ozone [252], or gamma radiation [253-255] as free radical generators. In this case, crosslinking occurs through the formation of (Si-C-C-Si) bonds as shown below:

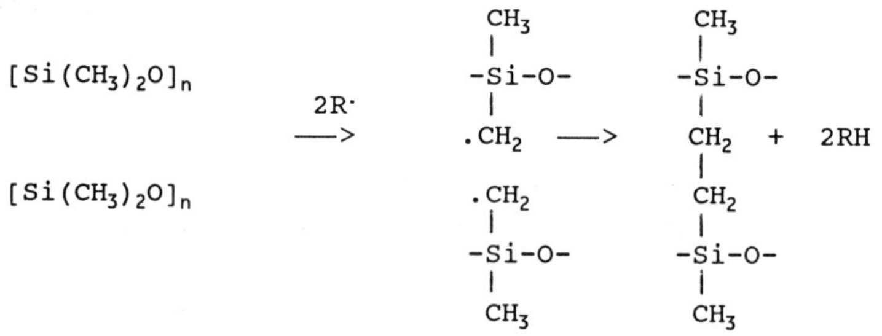

$$
\begin{array}{cccc}
& & CH_3 & CH_3 \\
& & | & | \\
[Si(CH_3)_2O]_n & & -Si-O- & -Si-O- \\
& 2R\cdot & | & | \\
& \longrightarrow & \cdot CH_2 \longrightarrow & CH_2 \quad + \quad 2RH \\
& & | & | \\
[Si(CH_3)_2O]_n & & \cdot CH_2 & CH_2 \\
& & | & | \\
& & -Si-O- & -Si-O- \\
& & | & | \\
& & CH_3 & CH_3 \\
\end{array}
$$

Very little crosslinking (0.1-1.0%) is required to render insoluble polysiloxanes with long polymer chains. Of the reagents indicated above dicumyl peroxide and azo-tert.-butane have emerged as the most widely used. Radiation-induced crosslinking has the advantage that no extraneous groups are introduced into the phase during crosslinking. On the debit side gamma ray sources, usually ^{60}Co, are not generally available in analytical laboratories and radiation induced crosslinking is not successful for polar phases.

An advantage of the crosslinking reaction is the relatively simple column preparation procedure. The deactivated glass surface is statically coated with a freshly prepared solution of the stationary phase containing 0.2 to 5% w/w of the peroxide free radical generator. It is very important that the stationary phase is deposited as an even film upon coating since the film will be fixed in position upon crosslinking and no improvement in film homogeneity can be obtained after fixation. The coated column is sealed and slowly raised to the curing temperature for static curing. For dynamic curing a slow flow of carrier gas is used. The curing temperature is a function of the thermal stability of the

free radical generator and is selected to give a reasonable half-
life for the crosslinking reaction. Temperatures in the range 80
to 250°C are common with the reagents presently used. Once the
column reaches the curing temperature, remaining at that
temperature for a short while is usually all that is required to
complete the reaction. After curing, the phase which was not
immobilized is rinsed from the column by flushing with appropriate
organic solvents.

A general problem with the use of peroxide free radical
generators is the formation of byproducts which may be difficult
to rinse from the column giving rise to phase degradation and
column activity. Certain susceptible functional groups, such as
tolyl or cynopropyl groups may be oxidized to more reactive
functional groups [250]. Azo-tert.-butane decomposes to neutral,
volatile byproducts that do not oxidize susceptible functional
groups. The column is saturated with azo-tert.-butane vapors
entrained in nitrogen gas at room temperature for about 1 h. Both
ends of the column are then sealed and the column temperature
slowly raised to above 200-230°C and held there for about 1 h
[250,251,256].

Poly(dimethylsiloxane) phases are relatively easy to
crosslink, but with increasing substitution of methyl by bulky or
polar functional groups the difficulty of obtaining complete
immobilization increases. For this reason moderately polar
polysiloxane phase are prepared with various amounts of vinyl,
tolyl or octyl groups that increase the success of the
crosslinking reaction. Even so, it has proven impossible to
immobilize poly(cyanopropylsiloxane) phases with a high percentage
of cyanopropyl groups completely [41,238,244,257-259].
Commercially available poly(cyanopropylsiloxane) columns are
either completely immobilized and contain, generally, less than
33% cyanopropyl groups as substituents or they are "stabilized",
indicating incomplete immobilization with a higher incorporation
of cyanopropyl groups. Stabilized columns usually have greater
thermal stability than physically coated columns but are not
resistant to solvent rinsing or as durable as immobilized columns.
Immobilized polysiloxane phase with p-cyanophenyl- and p-
cyanophenylethyl- substituents are easier to prepare and are
stable up to 300°C [260]. Poly(ethylene glycol) phases are also
available as bonded or stabilized phases, since, as in the case of
the poly(cyanopropylsiloxane) phases, it has proven difficult to

immobilize them successfully by using free radical crosslinking [256,261,262]. Polysiloxane phases containing oligoethylene oxide or crown ether substituents are easier to crosslink and have been suggested as alternatives to the poly(ethylene glycol) phases [263,264].

Another general advantage of immobilization is that it permits the preparation of open tubular columns having very thick films. Normal film thicknesses in open tubular column gas chromatography are 0.1-0.3 micrometers. It is very difficult to prepare conventionally-coated columns with stable films thicker than about 0.5 micrometers. However, thick-film columns of 1.0-8.0 micrometers can easily be prepared by immobilization. Thick-film columns permit the analysis of volatile substances with reasonable retention times without the need for subambient temperatures. They also permit the injection of much larger sample volumes without loss of resolution. This factor is important for identifying trace components using hyphenated techniques, such as GC/FTIR and GC/MS, where the small sample sizes tolerated by conventional columns may preclude positive identification. Immobilization also improves the resistance of the stationary phase to phase stripping by large volume splitless or on-column injection and allows solvent rinsing to be used to free the column from non-volatile sample byproducts or from active breakdown products of the liquid phase. Bonding and crosslinking procedures also provide a useful increase in the upper column operating temperature limit. Thus bonding and immobilization techniques have been useful not only in advancing column technology but they have also contributed to advances in the practice of gas chromatography by facilitating injection, detection and other instrumental developments that would have been difficult without them.

2.3.7 Porous-Layer Open Tubular (PLOT) and Support-Coated Open Tubular (SCOT) Columns

Porous-layer open tubular (PLOT) and support-coated open tubular (SCOT) columns are prepared by extending the inner surface area of the capillary tube. A layer of particles can be deposited on the surface or the column wall can be chemically treated to create a porous adsorbent layer. Obviously some of the wall-modified open tubular columns discussed in section 2.3.3 could be

~ ~~idered examples of SCOT and PLOT columns. In fact, distinction
  ·n certain kinds of WCOT, PLOT, and SCOT columns is quite
  ·d and we will attempt no such decisive division here.

PLOT and SCOT columns are usually prepared by an adaptation
of one of three general methods [265,266]. The porous layer can be
formed by chemical treatment of the inside tube wall, the inside
tube wall can be coated with a layer of porous particles, or the
porous particles can be partially embedded in the tube's inside
wall as the capillary tube is drawn. In the latter case a large
diameter tube is loosely packed with sorbent or coated with a thin
film of sorbent and then drawn out to a capillary tube in the
conventional manner (section 2.3.1) [265-268]. During the drawing
and coiling process the solid particles are partially embedded in
the inside wall of the capillary tube. The selected adsorbents
must be available in a narrow size range and have a melting point
higher than the softening point of the glass to be useful. SCOT
columns can then be prepared by coating the wall-modified
capillary columns in the usual way. These columns generally
exhibit low efficiency and this method of column preparation is
not widely used.

Open tubular columns for gas-solid chromatography are
generally prepared by chemical etching of the inside column wall
(glass for silica and aluminum for alumina) or by dynamic or
static coating of the capillary column with a suspension of
micrometer or sub-micrometer-sized adsorbent particles [269-273].
Porous polymer layers are prepared by coating with a porous
prepolymer of the styrene-divinylbenzene type that is subsequently
stabilized by in situ polymerization [274]. Glass that is etched
by heating with aqueous ammonia, hydrochloric acid or sodium
hydroxide for various times produces a porous silica layer of
average thickness 5-100 micrometers, depending on the reaction
conditions. Dehydration of this layer by conditioning at a high
temperature produces a surface suitable for gas-solid
chromatography or as a support for metal salts or stationary phase
modifiers. Good quality alumina, silica, and molecular sieve PLOT
columns can be prepared by dynamic or static coating of the
capillary column with a suspension of colloidal adsorbent or with
a stabilized suspension of micrometer-sized particles. Stable
suspensions are prepared using a balanced density slurry, by the
addition of a surfactant to the coating solution, or by adding
colloidal particles of the same material to the suspension.

Thixotropic suspensions are generally required for static coating due to the long column preparation time.

The popularity of SCOT columns has declined in recent years replaced largely by columns prepared with immobilized phases that are more inert and have a similar sample capacity. SCOT columns are prepared in a one or two step coating procedure employing a fumed silica, diatomaceous earth or graphite dispersion to stabilize the stationary phase layer [59,275-284]. The mechanism of film stabilization is most likely explained by the rheological behavior of the structured dispersion, since the solid material is dispersed in the liquid phase and cannot adhere to the wall through surface tension forces, nor can the stability of the film-surface relationship be explained in terms of surface roughness [278]. For nonpolar phases Silanox 101 (trimethylsilylated silica with a primary particle size of 7 nanometers) or similar dispersant is widely used. Unsilanized fumed silica, graphite, or kaolin are more commonly used with polar phases. In the two step coating procedure a suspension of the dispersant, stationary phase, and in some cases a surfactant as well, is used to dynamically coat the capillary column. After which the solvent is evaporated by increasing the nitrogen flow and subsequently the dried column dynamically coated with a more concentrated solution of stationary phase in a solvent that does not disturb the initial layer adhering to the glass surface. This enables the stationary phase film to be built up to the desired thickness. Also the initially deposited layer can be statically coated with a solution of the stationary phase containing a peroxide crosslinking agent to immobilize the stationary phase [279].

2.4 EVALUATION OF THE QUALITY OF OPEN TUBULAR COLUMNS

Whether using homemade or commercial columns, the assessment of column quality is of importance to all chromatographers. Periodic retesting using a standardized quality test is the best method of monitoring changes in column properties. It may then be possible to modify sample preparation procedures to extend the useful column lifetime. The quality of a column is a rather loosely defined concept that embodies such criteria as efficiency, inertness and the thermal stability of the stationary phase film. By inertness we signify the absence of adsorptive and catalytic interactions that can lead to incomplete sample elution, peak tailing and degradation for labile and polar analytes. All the above properties depend upon the method and materials chosen for column preparation as well as the care taken to exactly reproduce the critical steps in that procedure.

2.4.1 Activity Tests for Uncoated Columns

Activity tests for uncoated columns are important for studying the effectiveness of surface preparation and column deactivation procedures [180,198,208,285-287]. Even when using a well-tried method something may go wrong or the deactivation may be incomplete, and an intermediate test prior to coating is very useful for recognizing problems of this kind. The test method proposed by Schomburg is based on the principle that if a symmetrical solute band passes through an inert and uncoated capillary column it should emerge from that column unchanged except for symmetrical band broadening, and its retention time in that column should be equivalent to the column holdup time. To perform the test a short coated capillary precolumn of high quality is connected to the injector at one end and a short length of the capillary column to be tested at the other. The free end of the test capillary is connected to the detector. The test substances, after separation by the precolumn, enter the test capillary as perfectly shaped elution bands. The procedure generates nearly Gaussian peaks, from which symmetry distortions may be mathematically evaluated using symmetry factors, peak area/peak height ratios, etc. A practical example is shown in Figure 2.7 for the separation of a basic test mixture on a piece of deactivated fused silica capillary column compared to a piece of column prior to deactivation [208]. Almost any of the popular polarity test mixtures can be used to estimate the nature and extent of the activity of the test capillary Column. The Grob test mixture (section 2.4.3), the basic test mixture identified in the legend to Figure 2.7, and an acid test mixture containing n-decane, 1-octanol, 2,6-dimethylphenol, 1,6-hexanediol, octanoic acid, n-dodecane, 2,4,5-trichlorophenol, 4-nitrophenol and heptadecane or selected components from the above mixtures have been widely used in practice. Anthony et al. have shown that simple test mixtures containing aldehydes or ketones can be used to probe the concentration of water and bonded silanol groups on the surface of uncoated fused silica capillaries using the same experimental procedure as for the activity test [287,288].

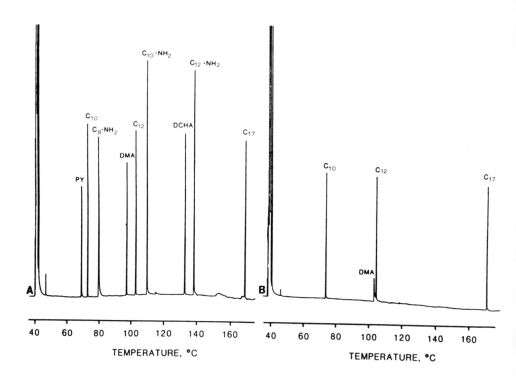

Figure 2.7 Activity test of an uncoated fused silica capillary after deactivation with poly(phenylmethylhydrosiloxane), (A), and before deactivation, (B). Precolumn: 15 m x 0.20 mm I.D. coated with SE-54. Test columns 10 m x 0.20 mm I.D. The column tandem was programmed from 40 to 180°C at 4°C/min after a 1 min isothermal hold with a hydrogen carrier gas velocity of 50 cm/s. The test mixture contained C_{10} = n-decane, C_8-NH_2 = 1-aminooctane, PY = 3,5-dimethylpyrimidine, C_{12} = n-dodecane, C_{10}-NH_2 = 1-aminodecane, DMA = 2,6-dimethylaniline, DCHA = N,N-dicyclohexylamine, C_{12}-NH_2 = 1-aminododecane, and C_{17} = n-heptadecane. (Reproduced with permission from ref. 208. Copyright Dr. Alfred Huethig Publisher).

2.4.2 Efficiency as a Column Quality Test

The efficiency of an open tubular column can be measured in several ways; the most widely used methods are the number of theoretical plates (n), the number of effective theoretical plates (N), the height equivalent to a theoretical plate (HETP) or effective plate, the coating efficiency and the separation number (SN). No single method is ideal, standardization is lacking and the terms are often used loosely in the literature without an adequate definition of the experimental conditions. All the above methods measure the total efficiency of both the column and instrument; if the values obtained are to be meaningful, it must be assumed that instrumental contributions to band broadening are negligible.

The number of theoretical or effective plates, usually normalized per meter of column length, is easily measured from a test chromatogram. However, the values measured will vary with retention, column dimensions, carrier gas velocity and stationary phase film thickness [141,289,290]. If the column is very long, as open tubular capillary columns frequently are, then the holdup time may become a sizeable fraction of the total retention time and the resultant number of theoretical plates speciously high. For this reason some authors consider the number of effective theoretical plates to be a better measure of column efficiency than the total number of theoretical plates, as the former is less influenced by the column holdup time. If a well-retained solute is used to measure column efficiency then either n or N would be an adequate measure. This can be seen from Figure 1.2 (section 1.4) where both terms converge to an approximately constant value at a sufficiently large value of the capacity factor [289]. In fact, as n levels off faster than N with increasing values of the capacity factor, this might be considered as the more reliable measure of column efficiency.

The plate height, and thus the total number of theoretical or effective plates, depends on the average linear carrier gas velocity (van Deemter relationship) and, for a particular carrier gas, the efficiency will maximize at a particular flow rate. Only at the optimum carrier gas flow rate are n, N, and HETP independent of the column length. The efficiency will also depend on the column diameter (see section 1.7.1) where typical values for n, N, and HETP for different column types can also be found. Values for n, N, and HETP are reasonably independent of temperature but may vary with the substance used for their determination, particularly if the test substance and stationary phase are not compatible.

The separation number, section 1.4, is dependent on the nature of the stationary phase, the column length, column temperature, and carrier gas flow rate. The separation number tends to increase as the column temperature decreases. At a constant temperature the separation number will approach a maximum and remain nearly constant as the capacity factor increases. The separation number is the only column efficiency parameter that can be determined under temperature programmed conditions [291,292]. The critical parameters that must be standardized to obtain reproducible separation numbers for columns of different lengths are the carrier gas flow rate and the temperature program rate (details are given in the next section). A maximum separation number is reached when the first compound of a homologous pair elutes with a capacity factor value of at least ten under optimized carrier gas flow and temperature program rates.

2.4.3 Standardized Quality Test for Coated Open Tubular Columns

The comprehensive column test procedure devised by Grob is now universally used by both column producers and column users [140,293,294]. The principal advantage of the Grob test is that it provides quantitative information about four important aspects of column quality: separation efficiency, adsorptive activity, acidity/basicity and the stationary phase film thickness. The standard test conditions are optimized for the measurement of efficiency and, since it is not feasible to standardize all parameters simultaneously in a multi-purpose test, these conditions are not necessarily optimum for measuring column activity. The experimental conditions are optimized for columns with a medium range of film thicknesses (0.08-0.4 micrometers) and column internal diameters of 0.25-0.35 mm. Although the separation number values obtained for wide bore or thick film columns may be slightly below their maximum values, for convenience the same experimental conditions are generally used for all column types.

The composition of the Grob test mixture is given in Table 2.7. To prepare a working solution, 1.0 ml of each standard solution is added to a 10.0 ml vial and 1.0 ml of this secondary standard is diluted to 20.0 ml with hexane. The resultant mixture can be stored in a refrigerator and is normally stable for many months. Reaction between nonanal and 2,6-dimethylaniline to form a Schiff base derivative occurs on prolonged storage resulting in reduced peak areas for these two compounds [295]. For nonpolar columns nonanal can be omitted as this compound is rarely adsorbed on high-temperature silylated surfaces. For polar phases the use of two test mixtures with nonanal in one and 2,6-dimethylaniline

TABLE 2.7

COMPONENTS OF THE CONCENTRATED GROB TEST MIXTURE

Test Substance	Abbreviation	Amount Dissolved in 20.0 ml solvent(a) (mg)
Methyl decanoate	E_{10}	242
Methyl undecanoate	E_{11}	236
Methyl dodecanoate	E_{12}	230
n-Decane	10	172
n-Undecane	11	174
n-Dodecane(b)	12	176
1-Octanol	ol	222
Nonanal	al	250
2,3-Butanediol	D	380
2,6-Dimethylaniline	A	205
2,6-Dimethylphenol	P	194
Dicyclohexylamine	am	204
2-Ethylhexanoic Acid	S	242

(a)Hexane, except for 2,3-butanediol, which is dissolved in chloroform. (b)Used in place of n-undecane to reduce the possibility of peak overlap on nonpolar stationary phases.

in the other is recommended. The concentrated standards should be stored in a freezer and are stable indefinitely.

A stepwise guide to the test method is given in Table 2.8. The test is performed under optimized conditions of carrier gas flow rates and temperature program rates, which are adjusted for column length and carrier gas viscosity, Table 2.9. To obtain a correct value for the gas holdup time for thick-film columns (d_f ca. > 0.7 micrometers) it should be measured at 100°C (methane is considerably retained at 25°C). As the viscosity of the carrier gas is greater at this temperature, the gas holdup time should be corrected by adding 10% of the measured time for hydrogen and 15% for helium. Some instruments do not allow the temperature program rate to be changed continuously to meet all conditions given in Table 2.8. In this case it is necessary to correct the speed of the run to an available temperature program rate by selecting a gas holdup time corresponding to this program rate.

The column efficiency is determined by the separation number obtained for the methyl esters of decanoic, undecanoic, and dodecanoic acids (E_{10}, E_{11}, E_{12}). As the relative difference in the molecular sizes of the homologous pairs decreases with increasing molecular size, the first pair of methyl esters (E_{10}/E_{11}) provides a separation number value that is about 8% higher than that of the

TABLE 2.8

STEPWISE PROCEDURE FOR PERFORMING THE GROB TEST

- Cool the column oven to below 40°C.
- Adjust the flow rate by measuring the gas holdup time for methane. Adjust the time to the standard time ($\pm$ 5%) Table 2.9.
- Adjust the temperature program rate to the appropriate value given in Table 2.9.
- Inject the test mixture under conditions that allow ca. 2 ng of a single test substance to enter the column (e.g., 1 microliter with a split ratio of 1:20 to 1:50, depending on injector design).
- Immediately after injection, heat the oven to 40°C (for very thin films to 30°C) and start the temperature program.
- Within the temperature range in which the third ester is eluted (on most columns, 110-140°C), make two marks on the recorder chart noting the actual oven temperature.
- At the end of the run, inter- or extrapolate the elution temperature of the third ester.
- Draw the "100% line" over the two alkanes and the three esters.
- Express the height of the remaining peaks as a percentage of the distance between the baseline and the 100% line.
- Determine SN as an average of SN E_{10}/E_{11} and SN E_{11}/E_{12}.
- Determine the film thickness using the nonogram (see text).

TABLE 2.9

STANDARD EXPERIMENTAL CONDITIONS FOR PERFORMING THE GROB TEST

Column	Hydrogen		Helium	
Length (m)	CH_4 elution (s)	Temperature Program (°C/min)	CH_4 elution (s)	Temperature Program (°C/min)
10	20	5.0	35	2.5
15	30	3.3	53	1.65
20	40	2.5	70	1.25
30	60	1.67	105	0.84
40	80	1.25	140	0.63
50	100	1.0	175	0.5

second pair (E_{11}/E_{12}). The average of the two values is normally used as a measure of the column separation efficiency.

An easy way to quantify the adsorptive and acid/base characteristics of a column is to measure the peak height as a percentage of that expected for complete and undisturbed elution. The two alkanes and three fatty acid methyl esters (non-adsorbing peaks) are connected at their apexes to provide the 100% line as shown in Figure 2.8 [190]. The column activity is then quantified

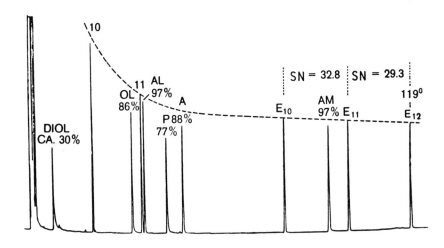

Figure 2.8   Test chromatogram of an open tubular column according
to the method of Grob. A line is drawn over the peaks of the non-
adsorbed solutes. The peak height of the remaining peaks is
determined as a percentage of the ideal peak height. In the
absence of adsorption all peaks should reach the dotted line.

by expressing the height of the remaining peaks as a percentage of
the distance between the baseline and the 100% line. It accounts
for all types of peak distortion that occur in practice: peak
broadening, peak tailing, irreversible adsorption and degradation.
The alcohols, octanol and 2,3-butanediol, are used to measure
adsorption by a hydrogen bonding mechanism. Acid/base interactions
are assessed from the adsorptive behavior of 2,6-dimethylaniline,
2,6-dimethylphenol, dicyclohexylamine, and 2-ethylhexanoic acid.
Probes with sterically-hindered functional groups are used for
this purpose in order to avoid adsorption by hydrogen bonding,
which can complicate the interpretation of the interaction. 2-
Ethylhexanoic acid and dicyclohexylamine provide much more
stringent tests of acid/base behavior than the phenol and aniline.
On nonpolar phases, even less than 1.0 ng of 2-ethylhexanoic acid
may cause column overloading, resulting in a distorted (leading)
peak. In this case, peak area as opposed to peak height must be
used to quantify column interactions.

The film thickness is calculated from the elution
temperature of methyl dodecanoate (E_{12}). However, quantitation of
film thickness requires a calibration of elution temperature

against stationary phase thickness for all stationary phases of interest. A nonogram can then be constructed and the stationary phase film thickness can be obtained from the elution temperature of E_{12} [293]. The elution temperature of E_{12} is reproducible to within $\pm$ 1°C if the standard conditions are kept within reasonable limits; in terms of film thickness this corresponds to a variation of 5% or less.

Saito used the Grob test to compare the activity of several columns coated with the poly(dimethylsiloxane) phase OV-101 after different surface pretreatments, Figure 2.9 [152]. The untreated fused silica surface shows a lower level of activity than the pyrex glass column even after deactivation with Carbowax 20M. Columns prepared by high temperature silylation are more inert than those shown in Figure 2.9 showing greater recovery of the diol, 2-ethylhexanoic acid and dicyclohexylamine. Incomplete silylation of fused silica usually results in a surface showing mild acidity with partial adsorption of amines and, possibly, normal recovery of 2,6-dimethylphenol and 2-ethylhexanoic acid.

The limitations of the Grob test are worthy of mention. It cannot be used to test columns coated with liquid phases of high melting point. The elution order of the test mixture is not the same on all stationary phases and the occurrence of peak co-elution cannot be eliminated entirely. The elution order of the test mixture on many common phases has been given by Grob [293,294]. For phases not previously tested, or in cases where the elution order is in doubt, individual standards must be injected for peak identification. In the case of peak co-elution, the test mixture should be divided into groups of separated components and each group injected independently. Finally, the test is biased towards the measurement of adsorptive as opposed to catalytic activity [296,297]. Catalytic activity causes time-dependent, concentration-independent losses of sample components. With increasing column temperature, adsorption decreases whilst catalytic/thermal decomposition increases. Although at high column temperatures adsorption may be of little importance compared to catalytic activity, the latter may not be observed in the Grob test. As catalytic activity is a fairly specific influence on particular solutes it may be necessary to customize an activity test for it based on the intended use of the column.

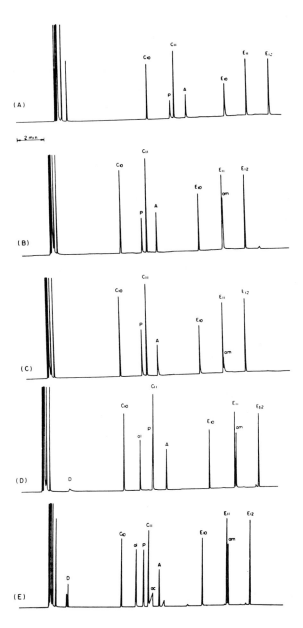

Figure 2.9 Use of the Grob test mixture to compare the activity of various glass surfaces coated with OV-101. Surface types: A = Untreated pyrex glass, B = pyrex glass deactivated by thermal degradation of Carbowax 20M, C = SCOT column, prepared with Silanox 101, D = pyrex glass column coated with a layer of barium carbonate and deactivated as in (B) , and E = untreated fused silica. Components are identified in Table 2.7 with ac = 2-ethylhexanoic acid. (Reproduced with permission from ref. 152. Copyright Elsevier Scientific Publishing Co.)

2.4.4 Column Thermal Stability

In quite general terms the amount of volatile decomposition products formed in the column per unit time depends on the stationary phase, temperature, surface properties of the support, surface area covered by the stationary phase and film thickness [143,298,299]. The mass flow of volatile products at the column outlet depends on the column length, the carrier gas flow rate and the rate of formation of the volatile products. The bleed products from polysiloxane phases consists primarily of low molecular weight cyclic siloxane impurities or thermally induced cyclic siloxanes [300]. For polar polysiloxanes the cyclic siloxane bleed products are often enriched in functional groups causing a decrease in polarity with column aging.

Standardization of the measurement of column bleed from different columns presents several problems. The influence of detector sensitivity on the results obtained can be eliminated by injecting a standard compound as part of the test, Figure 2.10. The amount of stationary phase bleed is represented by a rectangle whose area is defined by the change in recorder response between two set temperatures, one of which is selected to represent conditions of low column bleed, and the section of the abscissa that corresponds to unit time (minutes). By comparison of this area with that of a test compound, usually an n-alkane, the loss of liquid phase can be expressed as micrograms of n-alkane per minute. It is assumed that the response factors of the column bleed products and the test compound are identical. Although this is required as a working hypothesis, it is not necessarily the case.

The above method can be used to study the influence of the column preparation procedure on the stability of the liquid phase film. Surface effects are particularly noticeable in capillary columns since the ratio of the surface area to the amount of stationary phase is high. In general, glass capillary columns made from alkali glass exhibit much greater column bleed than borosilicate glasses unless the concentration of alkali ions is reduced by acid leaching. Fused silica columns show very low levels of thermally-induced catalytic phase decomposition. The method is also suitable for observing the influence of different methods of deactivation on stationary phase decomposition.

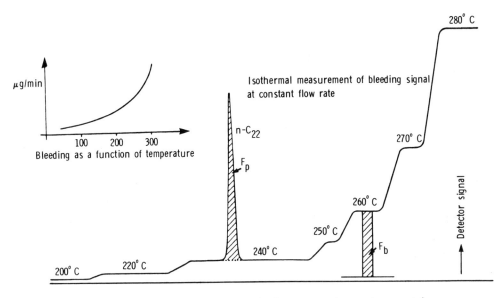

Figure 2.10 Standardized bleed test for gas chromatographic columns (see text for details). (Reproduced with permission from ref. 298. Copyright Elsevier Scientific Publishing Co.)

2.5 RETENTION IN GAS-LIQUID CHROMATOGRAPHY

The volume of gas required to elute a solute from a column in gas chromatography is determined by the experimental conditions and solute specific stationary phase interactions. In theory the specific retention volume (equation. 1.6 and Table 1.1) depends only on the measurement temperature and could be used for solute identification on a specified phase or as a measure of specific phase interactions with a specified solute [301]. In practice neither of these options is usually exploited because of the difficulty of measuring absolute retention volumes with adequate accuracy in the absence of specially designed instruments with performance characteristics exceeding those of analytical gas chromatographs in common use [121,301-304]. The accurate measurement of retention volumes is governed by the combination of individual errors in the absolute measurement of the retention time, carrier gas flow rate, column temperature, weight of column packing and/or liquid phase, column pressure drop and the absence of thermal gradients in the oven. The largest error is usually associated with the absolute determination of the weight of liquid phase [121,303]. A difference in temperature of 0.3 K is sufficient to change retention by about 1% [305]. Non-ideality of the gas phase should also be considered as a source of error in the most exact work (section 1.2). From this viewpoint the determination of specific retention volumes is much more of a problem than the determination of relative retention data. The usefulness of the latter, however, will depend on the development of a standardized protocol.

The general use of retention volume data for transferring information between laboratories for identification purposes was subject to too many systematic errors to be useful. This problem was solved by using the retention index system (section 2.5.2) in which the retention data was obtained in a standardized protocol with respect to a series of internal standards. Relative retention data are more reproducible and to some extent compensates for absolute errors in the experimental procedure. The retention index values rather than retention volumes are used to correlate physical and chemical properties of solutes to their chromatographic properties (section 2.5.3) and, in some procedures, to characterize the solvent properties of individual stationary phases (section 2.5.4). The interpretation of these properties depends on having an adequate model of the retention process in gas-liquid chromatography which is a convenient starting point for our discussion of retention.

2.5.1 A General Model for Retention in Gas-Liquid Chromatography

Early attempts to understand retention in gas-liquid chromatography were based on a partition model. Since the supports commonly used for column preparation were not entirely inert, it was realized that adsorption at the support interface could also contribute to retention. In most cases this was looked upon as an inconvenience and a cause of peak tailing and low sample recovery for polar solutes. Careful support deactivation can minimize the undesirable effects of tailing and irreversible sample adsorption but the significance of adsorption at the support interface as a general contribution to retention cannot be entirely ignored. When Martin suggested that adsorption at the gas-liquid interface might make a substantial contribution to retention in gas-liquid chromatography this was met, at first, by a great deal of skepticism [306]. However, when a liquid is spread upon a support of high surface area, the surface-to-volume ratio becomes very large and amplifies any properties of the solutes subject to variation due to differences between the bulk liquid and surface concentration of the solute. Adsorption at the gas-liquid interface is likely to become increasingly significant as the dissimilarity in properties between the liquid phase and solute increases, and is likely to be very important in the retention of hydrocarbons on polar liquid phases [307,309].

Any model devised to explain retention in general term must also take into account the distribution of the liquid phase on the support surface [310-313]. At low phase loadings with liquids that do not readily wet the support, the liquid phase will be present as droplets or pools of liquid located primarily at the outer grain surface with little penetration into the pores. At higher phase loadings coalescence from microdroplets to a continuous film will occur. For example, for squalane coated onto silanized Chromosorb P coalescence was observed to occur at a stationary phase loading of about 7.2% (w/w) [313]. Prior to coalescence the solute is exposed directly to a large support area and a very small liquid area, a situation that is reversed after coalescence. For liquids with good support wetting characteristics it is generally assumed that at low liquid phase loadings the liquid phase is first adsorbed as a monomolecular and multimolecular layer over the entire support surface. As the phase loading is increased it collects, initially, in the fine pores first, and then progressively appears in large cavities at the same time as the adsorbed layer thickens up. For diatomaceous earth supports with typical surface areas from 1-3 m^2/g a 0.01 to 2% (w/w) liquid phase loading should be adequate to cover the support with a monolayer, depending on the orientation of the adsorbed molecules. Liquid phase molecules near the support surface exist in a regular arrangement with an entropy of solution that is lower than that of the bulk liquid. It is unlikely that surface influences become completely non-existent beyond the first monolayer or so, for although the forces themselves are of very short range, they can be transmitted by successive polarization of adjacent molecules to a considerable depth in the liquid. The propagation of these forces is opposed by the thermal motion of the molecules and so falls of with distance after the strongly attracted first monolayer. The structured layer adjacent to the support surface, therefore, may be of considerable thickness and will likely dominate the retention characteristics of the liquid phase at low phase loadings. The characteristic properties of the liquid surface will be different for the two types of liquid films, structured and bulk, and the extent of the liquid surface area will change in a nonlinear manner with increasing phase loading until, at high phase loadings, it asymptotically approaches a limiting value, approximately equivalent to the support surface area less the area of its narrow pores and channels.

Taking the above considerations into account Nikolov proposed a general model to describe retention in gas-liquid chromatography given by equation (2.1)

$$V_N{}^* = V_L K_L + \delta V_L (K_S - K_L) + (1 - \delta) A_{LS} K_{DSL} + A_{GL} K_{GL} + A_{LS} K_{GLS} \qquad (2.1)$$

where V_N* is the net retention volume per gram of column packing, V_L the volume of liquid phase per gram of packing, K_L the gas-liquid partition coefficient, δ a constant constrained to have values of 1 when the film thickness is less than or equal to the thickness of the structured layer (d_s) and zero when the film thickness exceeds the thickness of the structured layer, K_S the gas-liquid partition coefficient for the structured liquid layer, A_{LS} the liquid-solid interfacial area per gram of packing, $K_{DSL} = d_s (K_S - K_L)$, A_{GL} the gas-liquid interfacial area per gram of packing, K_{GL} the adsorption coefficient at the gas-liquid interface, A_{LS} the liquid-solid interfacial area per gram of packing and K_{GLS} the coefficient for adsorption at the liquid-solid interface [314]. Although equation (2.1) provides a general description of the retention process it is rather unwieldy. Equation (2.1) contains five unknowns (K_L, K_S, K_{DSL}, K_{GL}, and K_{GLS}) which requires data for the phase characteristics (V_L, A_{GL}, and A_{LS}) of a minimum of five column packings prepared from the same support at different liquid loadings if it is to be solved numerically. There is generally no exact method of defining the thickness of the structured layer, and therefore δ, which must be set by intuition and trial and error. Nikolov applied equation (2.1) to the system Apiezon M and Carbowax 4000 on Celite at phase loadings of 1-18% (w/w). An abbreviated summary of Nikolov's data is presented in Table 2.10. Gas-solid adsorption of polar solutes contributes substantially to retention on the Apiezon columns. Even at high liquid phase loadings, adsorption at the support surface may contribute between 4 and 25% of solute retention. For

TABLE 2.10

RELATIVE CONTRIBUTION (%) TO THE NET RETENTION VOLUME BY DIFFERENT
SOLUTE RETENTION MECHANISMS
$V_S K_S$ = contribution due to partition with the structured
liquid phase layer

Solute	Phase	Inter-action	Percent Liquid Phase Loading					
			1	2.9	4.7	9.1	13	17
Octane	Apiezon M	$V_L K_L$	100	100	100	100	100	100
	Carbowax	$V_L K_L$	58	81	88	93	95	97
	4000	$A_{GL} K_{GL}$	42	19	12	7	5	3
Benzene	Apiezon M	$V_L K_L$	100	100	100	100	100	100
	Carbowax	$V_S K_S$	89	96	98	56	37.5	28.2
	4000	$V_L K_L$				43	61.7	71.3
		$A_{GL} K_{GL}$	11	4	2	1	0.8	0.5
1-Butanol	Apiezon M	$V_L K_L$	13	30	42	59	67	275
		$A_{LS} K_{GLS}$	87	70	58	41	33	25
	Carbowax	$V_S K_S$	86	95	97	60	40	30
	4000	$V_L K_L$				38	59	69
		$A_{GL} K_{GL}$	14	5	3	2	1	1
2-Butanone	Apiezon M	$V_L K_L$	52	77	84	92	94	96
		$A_{LS} K_{GLS}$	48	23	16	8	6	4
	Carbowax	$V_S K_S$	82	93	96	60	40	30
	4000	$V_L K_L$				38	58	69
		$A_{GL} K_{GL}$	18	7	4	2	2	1
Chloro-	Apiezon M	$V_L K_L$	100	100	100	100	100	100
benzene	Carbowax	$V_S K_S$	96	99	99	62.8	41.8	31.4
	4000	$V_L K_L$				36.8	57.9	68.4
		$A_{GL} K_{GL}$	4	1	1	0.4	0.3	0.2

Carbowax 4000, gas-liquid phase adsorption contributes only
moderately to the retention of all solutes, decreasing rapidly
with increasing liquid phase loading. At low phase loadings
adsorption at the gas-liquid interface may contribute between 3
and 42% to the retention of the solutes investigated

If retention measurements are made at higher phase loadings
so that the contributions from the structured liquid phase layer
can be neglected, then equation (2.1) can be simplified and
rearranged to give

$$V_N^*/V_L = K_L + (A_{GL} K_{GL} + A_{LS} K_{GLS})(1/V_L) \qquad (2.2)$$

from which the gas-liquid partition coefficient can be determined
from a plot of V_N^*/V_L vs $1/V_L$ for a series of columns and
extrapolating the data to give an intercept corresponding to $1/V_L =$
0 [6,307,315-317]. The shape of the plot is dependent on the
relative importance of adsorption at the gas-liquid and gas-solid
interfaces. Generally, one of these terms will tend to dominate

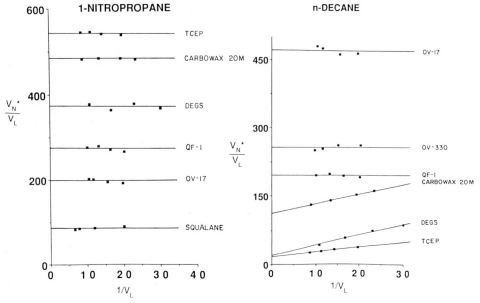

Figure 2.11   A plot of V_N^*/V_L vs $1/V_L$ for n-decane (left) and 1-nitropropane (right) on a series of stationary phases of different polarity at 80.8°C. OV-17 = poly(phenylmethylsiloxane), QF-1 = poly(trifluoropropylmethylsiloxane), Carbowax 20M = poly(ethylene glycol), DEGS = poly(diethylene glycol succinate) and TCEP = 1,2,3-tris(2-cyanoethoxypropane). (Reproduced with permission from ref. 318. Copyright Elsevier Scientific Publishing Co.)

and in most cases a linear extrapolation can be used. Some typical examples are shown in Figure 2.11 for the solutes n-decane and 1-nitropropane on several phases of different polarity [318]. For solutes retained solely by gas-liquid partitioning, as was found for 1-nitropropane, the plots have no slope. For n-decane a mixed retention mechanism is observed for some phases such as Carbowax 20M, DEGS, and TCEP. Interfacial adsorption is an important retention mechanism on the two most polar phases as indicated by the slope of the line and the small gas-liquid partition coefficient obtained by extrapolation to infinite phase volume. Equation (2.2) would also permit the measurement of the gas-liquid adsorption coefficient if the liquid surface areas were known for each column packing. Experimentally this is very difficult to determine and no reliable method has been forthcoming [6].

There are certain assumptions which must be fulfilled if equations (2.1) and (2.2) are used to calculate partition coefficients. It is assumed that the individual retention mechanisms are independent and additive. This will only be true under conditions where the infinite dilution and zero surface coverage approximations apply or, alternatively, at a constant concentration with respect to the ratio of sample size to amount of liquid phase. The infinite dilution/zero surface coverage approximation will apply to very small samples where the linearity of the various adsorption and partition isotherms are unperturbed and solute-solute interactions are negligible. These conditions can generally be met in gas-liquid chromatography using sensitive detectors such as the flame-ionization detector. The constancy of the solute retention volume with variation of the sample size for low sample amounts and the propagation of symmetrical peaks is a reasonable indication that the above conditions have been met. For asymmetric peaks, however, the constant concentration method of Conder and Purnell must be employed if reliable gas-liquid partition coefficients are to be obtained [319,320]. It is generally difficult to state absolutely the conditions for which contributions to retention from the structured liquid phase layer can be neglected. This will occur for some minimum phase loading that depends on the support surface area, the liquid phase support wetting characteristics and possibly the column temperature. For any set of experimental conditions the properties of the stationary phase will resemble those of the bulk liquid at high phase loadings. Occasionally this will be true for phase loadings below 5% (w/w) on typical diatomaceous earth supports but is more likely to be generally true at phase loadings exceeding 10% (w/w). The most convincing final proof that bulk liquid properties can be determined from gas-liquid chromatographic experiments is the excellent agreement between the solution thermodynamic properties measured chromatographically and those determined by conventional static and calorimetric methods [321].

There are surprisingly few studies of the retention mechanism for open tubular columns but the theory presented for packed columns should be equally applicable. For normal film thicknesses open tubular columns have a large surface area/volume ratio and the contribution of interfacial adsorption to retention should be significant for those solutes that exhibit adsorption tendencies. Interfacial adsorption has been shown to affect the reproducibility of retention for columns prepared with nonpolar phases of different film thicknesses [322-324]. The poor reproducibility of retention index values for columns prepared from polar phases was demonstrated to be due to interfacial adsorption of the n-alkane retention index standards [309]. Interfacial adsorption was shown to contribute to the retention of polar solutes on poly(dimethylsiloxane) phases which increased slightly when the bulk liquid phase was crosslinked [325]. Further studies are required to adequately elucidate the contribution of surface preparation procedures and crosslinking reactions on the retention mechanisms for immobilized phases.

2.5.2 Retention Index Systems

Since the early days of gas chromatography an enormous effort has been devoted to standardizing methods used to characterize retention data to make possible the wider use of collected published results. From the many proposals a general consensus has emerged in favor of the use of relative retention data expressed with respect to the retention of a standard substance. However, since gas chromatographic separations are performed over a wide range of temperatures with phases of different polarities, no single substance would fulfill the role of a universal standard. Also, in those cases where the retention of the sample and standard are grossly different accuracy would be impaired. This problem was finally solved by Kovats who proposed the retention index scale, in which the retention behavior of any substance was expressed with respect to the retention properties of a series of closely related standard substances [326-331]. The standard substances were the n-alkanes, each one of which was assigned an index value of 100 times its carbon umber (100 for methane, 200 for ethane, etc.) at any temperature and on any phase. For a homologous series the logarithm of the adjusted retention time values on a given column at a given temperature are linearly related to the carbon number if the first few members of the series are ignored. This provides the form of the relationship relating adjusted retention times to the fixed points of the retention index scale. The retention index of any substance can then be defined as equal to 100-times the carbon number of a hypothetical n-alkane which would have the same adjusted retention time as the substance of interest. The retention index is represented by the symbol I and is calculated under isothermal conditions using equation (2.3)

$$I = 100z + 100[\log t_{R'(x)} - \log t_{R'(z)}]/[\log t_{R'(z+1)} - \log t_{R'(z)}] \quad (2.3)$$

where t_R' is the adjusted retention time, z the carbon number of the n-alkane eluting immediately before the substance of interest denoted by x, and $z + 1$ the carbon number of the n-alkane eluting immediately after substance x. Thus the retention index of a substance is expressed on a uniform scale with increased precision in the determination due to the use of two closely eluting, bracketing, standards for the experimental measurement. For completeness, the retention index is usually given with the stationary phase identified as a superscript, the sample as a subscript, followed by the temperature of the measurement in parenthesis, for example $I_{benzene}^{squalane}(75°C) = 644.1$ index units. Indicating that the retention index value for benzene on squalane at 75°C is identical to that of an n-alkane (hypothetical) with 6.44 carbon atoms and that with the stationary phase squalane and the stated temperature benzene will elute between n-hexane ($z = 6$) and n-heptane ($z + 1 = 7$). In theory the retention index value of a substance depends only on the temperature and the identity of the stationary phase and is independent of other instrumental and experimental parameters. In practice, this is an over simplification, as we shall see later.

Equation (2.3). is the most useful of the expressions for the retention index, in practice, although the general theoretical basis for its correctness can be established by casting the equation in terms of the chemical potentials or Gibbs free energy of solution [318,328,332]. The partial molar Gibbs free energy of solution for a methylene group is reasonably constant for all members of a homologous series except for the first few members of the series [333,334]. Inaccuracies in the index values are likely to be greatest when the carbon number for the n-alkane standards is small, less than about five. Also, on polar phases the partial molar Gibbs free energy of solution for a methylene group is smaller than for a nonpolar phase compressing the scale in the direction of increasing polarity. This results in lower relative retention of the n-alkane standards and a decrease in the relative measurement precision. The reliability of the retention index value is strongly dependent on the retention mechanism, section 2.5.1, and equation (2.3) is independent of the column characteristics only when retention occurs solely by gas-liquid partitioning. For polar phases the n-alkanes are retained largely by interfacial adsorption and their retention will vary with the phase loading and support surface area in an unpredictable manner [6,307,309,318,335-337]. Under these circumstances the retention index values are unsuitable for qualitative identification and may well show variations between different laboratories running to hundreds of index units. Interfacial adsorption is frequently overlooked as a source of error in measuring retention index values but is easily determined through the use of equation (2.2) and plots of the kind shown in Figure 2.11. Substituting the appropriate values of the gas-liquid partition coefficients for the adjusted retention times in equation (2.3) enables a comparison between the experimental and the theoretical values of the retention index for a partition mechanism to be made.

The retention index value is temperature dependent and when an index value is required at another temperature it can be obtained by interpolation using an Antoine-type hyperbolic function

$$I(T) = A + B/(T + C) \tag{2.4}$$

where I(T) is the retention index at any temperature T(K) and A, B, and C are experimentally derived constants [338]. The curve can nevertheless have a significant linear portion, particularly for substances of low polarity on nonpolar stationary phases. Alternatively, the relationship

$$I(T) = I(T_1) + [(T - T_1)/(A + BT)] \tag{2.5}$$

can be used where $I(T_1)$ refers to the retention index value at some reference temperature T_1 and A and B are experimentally determined constants [339].

For mixtures of wide boiling point range the determination of retention indices under isothermal conditions would be time consuming and unnecessarily restrictive. Under temperature program conditions an approximately linear relationship exists between the elution temperature of n-alkanes and their carbon number, provided that the initial column temperature is low compared to the elution temperature of the first standard and that only a relatively limited range of carbon numbers are considered [340,341]. An expression equivalent to equation (2.3) can be given for linear temperature program conditions by replacing the adjusted retention time by the elution temperature, equation (2.6)

$$I = 100z + 100[T_{R(x)} - T_{R(z)}]/[T_{R(z+1)} - T_{R(z)}] \qquad (2.6)$$

where T_R is the elution temperature (K) and the subscripts x, z, z + 1 are identified in equation (2.3). Temperature programmed retention indices are more sensitive to the chromatographic conditions than isothermal indices and are generally of lower precision. The temperature program indices are influenced by the time and temperature of any isothermal period prior to programing, the temperature program rate (faster programing rates lead to higher elution temperatures), and the method of carrier gas flow control, since the viscosity of the mobile phase increases with temperature. A considerable effort has been devoted to predicting temperature program indices from the more readily available isothermal retention indices and their temperature dependence [342-348]. However, the unique dependence of the solute partition coefficient on temperature and the varying column holdup times have prevented any simple means of converting isothermal retention indices to programmed indices (and vice versa) from being developed.

Sources of error in determining retention indices are due primarily to poor control of fluctuations in temperature and carrier gas flow rates, non-ideal behavior of the mobile phase, uncertainties in the measurement of retention times and the column holdup time, sample size effects and variations caused by interfacial adsorption and impurities in the stationary phase [328-330,349-352]. Proper instrument design should minimize fluctuations in temperature and carrier gas flow rates. A significant systematic error can be caused by differences in the set point temperature of different instruments which might, unknowingly, differ by several degrees. For normal carrier gases and typical operating pressures non-ideal behavior should not be a significant source of error (section 1.2). Stationary phase purity speaks for itself and only those phases of sufficiently high purity can be expected to produce reproducible retention data. If interfacial adsorption of solutes or standards is significant then wide variations in the retention index values can be expected. The retention index values, in this case vary with the sample size, particularly the concentration ratio of the test substance to the n-alkanes, and with the phase loading, since the available liquid surface area declines in a nonlinear manner with increasing phase loading. Rather than treat these observations as exceptional, the case frequently made in the literature, they can be seen to be a general consequence of the general retention model discussed in section 2.5.1. Under favorable circumstances the reproducibility of retention indices between different laboratories is within one index unit for nonpolar phases and within a few index units for polar phases; the precision for temperature programmed indices is usually lower. The reproducibility of retention indices within the same laboratory is generally on the order of 0.2 index units for manual measurements and 0.05-0.1 index units with computer-controlled data acquisition.

The adjusted retention times used to calculate the retention index require that an accurate determination of the gas holdup time be made. This simple measurement is deceptively difficult to perform accurately and a bewildering array of experimental, graphical, statistical and iterative methods have been proposed for this measurement [340,350-355]. Experimentally, the gas holdup time is determined by the time required for an infinitesimal amount of non-adsorbed gas to pass through the chromatographic system under conditions identical to those used for the sample. It is generally agreed that neon, and to a lesser extent air, are unretained on most phases and are suitable standards for determining the gas holdup time. Unfortunately, these gases are not detected by the flame-ionization detector. On the other hand, methane is easily detected with a flame-ionization detector but is known to be retained, albeit slightly, on some phases. Methane retention is particularly noticeable on nonpolar stationary phases and at relatively low temperatures (<100°C). Whether methane is an acceptable standard for an unretained solute will depend on the accuracy required in the measurement and on the experimental conditions used for the measurement. Whichever approach is used the experimental accuracy may be limited by the method used to determine the elution time of the rapidly eluting peak.

Mathematical methods for determining the gas holdup time are based on the linearity of the plot of adjusted retention time against carbon number for a homologous series of compounds. Large errors in this case can arise from the anomalous behavior of early members of the homologous series (deviation from linearity in the above relationship). The accuracy with which the gas holdup time is determined by using only well retained members of a homologous series can be compromised by instability in the column temperature and carrier gas flow rate [353,357]. The most accurate estimates of the gas holdup time are obtained by regression procedures after testing the linearity of the homologous series plot and selecting an appropriate number of standards to give the required statistical accuracy in the calculation procedure [353,354].

The concept of the retention index system is not limited to the use of n-alkanes as index standards and any homologous series that shows a linear relationship between the logarithm of the adjusted retention time and carbon number can be used. To achieve a reasonable standardization of retention data the n-alkanes should be the first choice but there are two compelling reasons to use different standards under some circumstances. Interfacial adsorption of n-alkanes on polar phases yields unreliable retention index values. Standards of intermediate polarity such as 2-alkanones, alkyl ethers, alkyl halides and alkanoic acid methyl esters are less likely to be retained by adsorption on polar phases and will give more reliable index values [328,331,335,336,358-362]. In other cases standards have been selected because they are normally present in the samples to be analyzed or are of a similar structure to the compounds to be determined and should have similar properties in the chromatographic system. For example, fatty acid methyl esters are used in lipid analysis; androstane and cholestane for the calculation of steroid numbers; and naphthalene, phenantherene, chrysene and picene for polycyclic aromatic compounds [342,363-366]. A second reason for choosing index standards other than n-alkanes is to obtain compatibility with a selective detector such as the electron-capture, thermionic ionization, and photometric detectors that do not respond to the n-alkanes. Suitable standards in this case include n-alkyltrichloroacetates, n-bromoalkanes, dichlorobenzyl alkyl ethers and nitroalkanes for use with the electron-capture detector [367-372]; tri-n-alkylamines and dipropylalkylamines with the thermionic ionization detector [373]; dialkylsulfides with the flame photometric detector [374]; and multifunctional standards for use with several detectors, such as alkylbis(trifluoromethyl)phosphine sulfides [375].

2.5.3 Relationship Between Retention Indices and Solute Properties

There has been a great deal of interest in the relationship

between solute properties and retention for several reasons [328-330,376]. If predictive relationships could be developed then a method would exist to test the consistency of retention data, to predict the retention index of an analyte that was unavailable for study, or to predict the value of certain physical properties of an analyte that could be correlated through its retention index. This would make the retention index system and retention index libraries more useful as a qualitative identification method.

One such predictive relationship is the basis of the retention index system itself, that is that the members of a homologous series differ by 100 index units per methylene group (the first few members of the series excepted) [377,378]. Interpolation or extrapolation of the chromatographic data between knowns and unknowns establishes the probability of a substance in the chromatogram belonging to a particular homologous series. For compounds of low polarity on nonpolar phases the retention increment due to the presence of a functional group relative to the parent hydrocarbon can be used to predict the retention index of structurally related compounds with reasonable accuracy [377,379]. If on a nonpolar stationary phase the retention index of a substance of the type alkyl chain-functional group is determined then subtraction of the value for the corresponding hydrocarbon having the same carbon number as the substance of interest gives the contribution for the functional group. The retention index value for a compound on the same stationary phase can then be predicted by addition of the functional group increments to the base value predicted from its carbon number. For multifunctional compounds second order interactions may lower the predictive accuracy and the practical utility of the method is generally limited to nonpolar compounds on nonpolar phases.

More general approaches to retention index prediction are based on the generation of molecular descriptors which are subsequently fitted to the retention index using multiple linear regression methods [376]. The molecular descriptors are generally solute or stationary phase interaction parameters, topological indices, or some combination of both. One of the earlier successful computational approaches was that developed by Rohrschneider which is also the basis of the retention index method for stationary phase characterization discussed in the next section [380,381]. Ideal solutions do not exist in gas chromatography and equations used to predict retention should contain terms to account for specific and nonspecific interactions. Bermejo et al. [382-386] have applied a multiparameter approach of the general form

$$I = m + nT_B + pP + qQ + rR + etc. \tag{2.7}$$

where T_B is the solute boiling point; P, Q, R, etc., are independent but complementary solute parameters to account for the different possible retention mechanisms; and m, n, p, q, r, etc., are regression coefficients which describe the sensitivity of the retention process to the solute boiling point and the different solute-stationary phase interaction mechanisms. For nonpolar solutes on nonpolar phases biparameter equations including solute boiling point combined with the van der Waals volume, molar volume, or the first-order molecular connectivity index have been quite successful. For polar solutes and/or polar phases additional terms must be included. To account for contributions from polarizability and induction forces functions of the refractive index are generally used and for orientation forces functions of the dielectric constant or permanent dipole moment. Energy and charge parameters calculated by quantum chemical methods were found to result in better correlations for some compounds [376,381]. The above methods often provide reasonable results for compounds belonging to a homologous series but variable accuracy for heterogeneous mixtures. The prediction of retention indices is generally not sufficiently accurate for substance identification at present.

Several methods are available to represent the structural formula of solutes by a topological model [376]. The structural formula of a compound may be viewed as a molecular graph, where the vertices represent atoms and the edges represent covalent bonds. More than twenty different topological indices have been described in the literature with the molecular connectivity indices being the most popular [387-390]. Since topological indices basically reflect molecular shape and size (thus accounting for dispersive interactions) reasonable predictive agreement for retention indices of nonpolar solutes on nonpolar phases has been obtained. For polar phases, it seems that a combination of topological indices with interaction parameters, similar to those discussed above, are more rewarding. The accuracy obtained, however, is very variable.

2.5.4 Methods for Characterizing Stationary Phase Interactions

The characteristics of a liquid phase of general interest to the chromatographer are its operating temperature range, film forming properties, mass transfer characteristics, solvent strength and solvent selectivity. Film forming and mass transfer properties primarily influence peak shapes and column efficiency. Only those liquid phases that at least meet average standards are useful in gas chromatography. The minimum operating temperature for a stationary phase is usually established by its melting point or the temperature at which the column efficiency reaches a plateau region, if higher. Some phases have been used below their melting point as selective adsorbents for the separation of positional isomers which were not resolved at temperatures above the melt [391,392]; however, this is the exception rather than the rule. As shown in Figure 2.12, the performance of a column below the melting point of the phase is generally low and declines rapidly as lower temperatures are reached [58]. The retention characteristics of the phase may also change dramatically on melting, particularly for solutes retained predominantly by a partition mechanism [391,393]. Some polymers and liquid crystals exhibit multiple phase transitions, each one associated with a change in retention corresponding to a change in surface area, volume, or viscosity above the transition temperature [391].

The upper temperature operating limit for a liquid phase is established as the highest temperature the phase can be maintained at without decomposition or significant bleed from the column. Some phases are polydisperse substances containing low molecular weight oligomers. These oligomers may be selectively evaporated during column conditioning, leaving a much lower amount of phase on the column than expected. This is less of a problem for chemically defined phases and for phases of low polydispersity. In this case, the maximum allowable operating temperature of the phase is defined as the highest temperature that the phase can be maintained at for 24 h without changing the retention and performance characteristics of the components in a test chromatogram, obtained at a lower temperature before and after the conditioning period [31,43,394]. Alternatively, the maximum allowable operating temperature can be assumed to be that

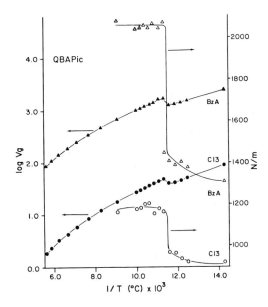

Figure 2.12 Plot of the logarithm of the specific retention volume and column efficiency (plates per meter) as a function of column temperature for benzaldehyde (BZA) and n-tridecane (C_{13}) on the stationary phase tetra-n-butylammonium picrate. (Reproduced with permission from ref. 58. Copyright Elsevier Scientific Publishing Co.)

temperature at which the liquid phase exhibits a vapor pressure of 0.5 Torr.

The solvent properties of a stationary phase are usually characterized by its strength and selectivity [6,15,303,328,395-397]. These terms can be defined in several ways without universal agreement as to their exact meaning. The solvent strength is synonymous with the notion of polarity and is a measure of the capacity of the stationary phase to enter into all intermolecular interactions: dispersion, induction, orientation and donor-acceptor complexation (including hydrogen bonding). The solvent selectivity determines the relative capacity of the stationary phase for each individual interaction. The solvent strength (polarity) criterion of a liquid phase is the least satisfactory measure of the properties of a phase, as it is difficult to define rigorously. Since polarity is not a unique property of a molecule but a composite expression for several different interactions, there is no single substance that can be defined as polar. Any scale or ranking of polarity that depends on the measured properties of individual substances must be largely allegorical but can still be useful for the qualitative classification of phases. The two most common measures of polarity are based on the reluctance of a stationary phase to dissolve a hydrocarbon and the sum of the retention index differences for the solutes benzene, butanol, 2-pentanone, nitropropane and pyridine on the polar phase and squalane (P_M value of McReynolds) [6,398]. The P_M value is widely quoted in supply catalogs but has little theoretical support, as will be discussed subsequently when the McReynolds system of stationary phase characterization is considered in detail. The reluctance of a stationary phase to dissolve an n-alkane is at least qualitatively in agreement with the general notion of polarity, in that it is widely accepted that hydrocarbons are poorly soluble in polar solvents. The reluctance of a stationary phase to dissolve an n-alkane can be expressed in several ways, probably the most useful are the partial molar Gibbs free energy or excess free energy of solution for a methylene group or by the solvent strength parameter (SSP), defined as the partial molar Gibbs free energy of solution for a methylene group per unit solvent volume [5,6,303,378,390,399,400]. These terms have the advantage that they are easily determined. The partial molar Gibbs free energy of solution for a methylene group is obtained directly from the slope of a plot of the logarithm of the partition coefficient against carbon number for the members of a homologous series; generally n-alkanes on nonpolar phases and 2-alkanones or fatty acid methyl esters on polar phases (n-alkanes may be retained largely by interfacial adsorption on polar phases, section 2.5.1). The free energy for a methylene group is simply the product of the slope value and $-2.3RT_c$ where R is the gas constant and T_c the column temperature. The solvent strength parameter is obtained by dividing the partial molar Gibbs free energy of solution for a methylene group, determined as indicated above, by the density for the stationary phase at the column temperature. Some typical values for the two solvent strength parameters of some common stationary phases are summarized in Table 2.11. Both scales produce a similar, but not identical, order in the ranking of stationary phases. The solvent strength parameter is less disturbed by molecular weight and size differences and, at least intuitively, provides a more logical ranking of phases [303,400]. The partial molar Gibbs free energy of solution per methylene group reflects a combination of solvent-

TABLE 2.11

SOLVENT STRENGTH CHARACTERISTICS OF SOME COMMON STATIONARY PHASES
AT 121°C
structures of phases identified in section 2.2.1.

Stationary Phase	$\Delta G°(CH_2)$ (cal/mol)	SSP (cal.cm^3/g.mol)
Squalane	-521	-728
Apolane-87	-518	-669
SE-30	-463	-578
Di-n-decylphthalate	-511	
OV-7	-467	-504
OV-17	-470	-463
Poly(phenyl ether)	-487	-436
OV-225	-418	-410
OV-25	-431	-396
Carbowax 20M	-400	-387
Tetra-n-butylammonium-4-toluenesulfonate	-377	-377
QF-1	-393	-337
DEGS	-324	-275
1,2,3-Tris(cyano-ethoxypropane)	-280	-273
OV-275	-265	-243

solvent bond breaking to form a cavity in the solvent suitable for
a methylene group and methylene group and solvent dispersion
interactions. These interactions are not necessarily an indirect
measure of the specific interactions that would occur between two
polar substances and there can be no tacit assumption that the
potential of a given phase for polar interactions is accurately
represented by its relative position in Table 2.11. These features
only serve to emphasize the difficulties and inadequacies of
defining a single parameter scale for a general term like solvent
strength and it seems inevitable that anomalies and exceptions
will arise.

Solvent selectivity is a measure of the relative capacity of
a solvent to enter into specific solute-solvent interactions,
characterized as dispersion, induction, orientation and
complexation interactions. Unfortunately, fundamental approaches
have not advanced to the point where an exact model can be put
forward to describe the principal intermolecular forces between
complex molecules. Chromatographers, therefore, have come to rely
on empirical models to estimate the solvent selectivity of
stationary phases. The Rohrschneider/McReynolds system of phase
constants   [6,15,318,327,328,380,397,401-403],   solubility

parameters [6], the solvent selectivity triangle of Snyder [336], Hawkes polarity indices [404], dispersion and selectivity indices [331,405], universal indices [406] and various thermodynamic approaches [303,397,404,407] have been the most widely used. Non of these methods, however, can be considered ideal and all can be faulted in some respect.

The McReynolds system of phase constants has become the most widely used systematic approach to solvent selectivity characterization and virtually all popular phases have been characterized by this method. In spite of its popularity the approach is fundamentally flawed and the phase constants are an unreliable indication of phase properties. The basic approach, however, has influenced the development of other methods of selectivity characterization, and although these methods have inherited many of the deficiencies of their parent, a brief description of the McReynolds approach is worthwhile to indicate the general limitations of methods based on retention index differences.

The McReynolds approach, which was based on earlier theoretical considerations proposed by Rohrschneider, is formulated on the assumption that intermolecular forces are additive and their individual contributions to retention can be evaluated from differences between the retention index values for a series of test solutes measured on the liquid phase to be characterized and squalane at a fixed temperature of 120°C. The test solutes, Table 2.12, were selected to express dominant intermolecular interactions. McReynolds suggested that ten solutes were needed for this purpose. This included the original five test solutes proposed by Rohrschneider or higher molecular weight homologs of those test solutes to improve the accuracy of the retention index measurements. The number of test solutes required to adequately characterize the solvent properties of a stationary phase has remained controversial but in conventional practice the first five solutes in Table 2.12, identified by symbols X' through S' have been the most widely used [6]. It was further assumed that for each type of intermolecular interaction, the interaction energy is proportional to a value a, b, c, d, or e, etc., characteristic of each test solute and proportional to its susceptibility for a particular interaction, and to a value X', Y', Z', U', S', etc., characteristic of the capacity of the liquid phase

TABLE 2.12

INTERACTIONS CHARACTERIZED BY MCREYNOLDS PROBES (ROHRSCHNEIDER
PROBES IN PARENTHESES)

Symbol	Test Solute	Boiling Point(°C)	Dipole Moment	Interaction Measured
X′	Benzene	80.1	0.03-0.1	Primarily dispersion with weak proton acceptor properties. Polarizable in induction interactions.
Y′	Butanol (Ethanol)	117.5	1.75	Orientation properties with both proton donor and proton acceptor capabilities.
Z′	2-Pentanone (2-Butanone)	102	2.82	Orientation with some weak proton acceptor but not proton donor capabilities.
U′	Nitropropane (Nitromethane)	103.5	3.59	Orientation with some weak proton acceptor capability.
S′	Pyridine	115.5	2.37	Strong proton acceptor with moderate orientation properties. Proton donor properties are absent.
H′	2-Methyl-2 pentanol	121.5	---	See butanol.
J′	Iodobutane	130	1.81	Moderate orientation interactions.
K′	2-Octyne	125	---	See benzene.
L′	Dioxane	101	0.45	Proton acceptor with weak orientation properties. Proton donor properties are absent.
M′	cis-Hydrindane	159(*)	---	Dispersive properties.

(*)boiling point is for trans isomer

to enter into specific intermolecular interactions. The retention
index difference, ΔI, is thus compiled of products as follows:

$$\Delta I = aX' + bY' + cZ' + dU' + eS' \tag{2.8}$$

$$I_{PH}^{P} = I_{SQ}^{P} + \Delta I \tag{2.9}$$

where I_{PH}^{P} is the retention index for test solute P on the stationary
phase to be characterized, PH, I_{SQ}^{P} the retention index for probe P
on the nonpolar reference phase squalane, SQ, and ΔI the retention

index difference equivalent to the contribution from polar interactions. To evaluate the phase constants numerically, McReynolds assigned values of 1 to each of the test solutes in turn. For example, for benzene, a = 1, b = 0, c = 0, d = 0, e = 0; therefore, X' is obtained from equation (2.9) being equal to ΔI when P = benzene. By repeating the above process for each test solute in turn, all the stationary phase constants X', Y', Z', U', and S' can be determined. Experimentally, this requires the determination of the retention indices of the five test solutes on the phase to be characterized and on a squalane column under identical and defined experimental conditions. The larger the numerical value of the McReynolds phase constant the stronger the interaction expressed by that constant. Although it is now known that these assumptions are false, reference to the use of McReynolds phase constants as the basis for selecting stationary phases for specific applications is still common in the contemporary literature.

Recent evidence has come to light that indicates that the approach used by McReynolds is unreliable for a combination of experimental and theoretical reasons [6,307,318,335,336,400,408]. These same problems are inherent in other methods of stationary phase characterization that employ retention index differences, for example, the solvent selectivity triangle, Hawkes polarity indices and dispersion and selectivity indices, etc. From an experimental point of view the poor retention of some test solutes at the recommended measuring temperature and the failure to account for interfacial adsorption as a significant retention mechanism are the major problems. In fact, on many common polar phases the n-alkane retention index standards are retained almost exclusively by interfacial absorption while the test solutes are retained largely by partitioning or by a mixed retention mechanism (section 2.5.1). Under these circumstances the phase constants derived from the retention index differences will be meaningless and will vary with the choice of column type and/or packing used to make the measurements. Other, less significant experimental objections are summarized in ref. 6. As a consequence of the above objections the large data collection of McReynolds, unfortunately, is unsuitable for re-evaluation by newer research methods proposed for stationary phase characterization.

Theoretical objections to the use of retention index differences for stationary phase characterization are based on the experimental and theoretical proof that the phase constants are composite values determined by the relative solubility of the n-alkane retention index standards in the compared phases, as well as by the magnitude of the selective interaction of the test solutes. In fact, in the majority of cases, the magnitude of the phase constants is determined almost entirely by the properties of the n-alkane retention index standards and a high level of correlation exists between the phase constants and various parameters describing the solubility of the n-alkanes in the stationary phase. This also explains why the McReynolds phase constants tend to increase monotonously with polarity rather than showing greater variation as predicted by chemical intuition and the results from analyzing test mixtures on different phases.

It has been suggested that approaches based on the differences of retention indices be abandoned for stationary phase characterization in favor of thermodynamic approaches which can be related to rigorous models describing the transfer of a solute from the gas phase to the stationary phase [409-413]. In these models the solvation process is considered to involve three steps: (1) the creation of a cavity in the solvent of a suitable size to accommodate the solute; (2) reorganization of the solvent molecules around the cavity (the free energy change for this process is probably small or zero); and (3) interaction of a solute molecule with the surrounding solvent molecules. Calculation of the free energy involved in cavity formation is difficult and most of the experimental parameters required are unavailable for common stationary phases. This requires that an empirical method be used to estimate the cavity formation term. One approach based on multiple linear regression and the principle of linear solvation energy relationships results in equation (2.10)

$$SP = SP_0 + l\log K_L^{16} + s\pi_2^* + d\delta_2 + a\alpha_2 + b\beta_2 \qquad (2.10)$$

where SP is some solute property linearly related to its energy of solvation (e.g., V_g, log k, etc.), SP_0 is a constant, π_2^*, α_2 and β_2 are solvatochromic parameters (see section 4.6), and l, s, a and b are the coefficients obtained by multiple linear regression which describe the susceptibility of the stationary phase for particular interactions. The parameter δ_2 is an empirical polarizability correction factor taken as zero except for polyhalogenated aliphatic compounds (0.5) and for aromatic compounds (1.0). The solvatochromic parameters are obtained spectroscopically and are a measure of the ability of a species to stabilize a neighboring dipole by virtue of its own dipolarity and polarizability (π^*), the solute hydrogen bond acidity (α), and the solute hydrogen bond basicity (β). The term log K_L^{16} is the partition coefficient of the solute from the gas phase to n-hexadecane at 293 K. It takes into account both solute phase dispersion interactions and the work needed to create a cavity in the condensed phase. The coefficients s, a, b and l, are a direct measure of the capacity of a stationary phase for orientation, hydrogen-bond basicity and acidity, and a combination of general dispersion interactions and cavity effects, respectively. These coefficients, then, perform the same function as the McReynolds phase constants, the greater their relative magnitude the stronger the interaction measured, but they are derived in an entirely different manner. Abraham et al. have summarized the coefficients for 77 stationary phases using experimental data obtained by McReynolds [410]. To obtain statistically valid values for the coefficients data for about 30 different solutes are required. This limits the available data sets that can be used to test equation (2.10); unfortunately, the McReynolds data set is known to be experimentally invalid which diminishes the current usefulness of the determined coefficients. Carr et al. note that using a "double-reference" column approach improves the statistical accuracy of the coefficients but greatly complicates their chemical interpretation [411]. There are some problems that still need to be solved with this approach but it appears very promising.

If it is assumed that the total free energy for the transfer of solute X from the gas phase to the stationary phase (with molecular interactions characteristic of infinite dilution) is the linear sum of the individual free energy contributions to the transfer process then a general expression for the solution process, equation (2.11), can be written as follows

$$\Delta G_S^{SOLN}(X) = \Delta G_S^{CAV}(X) + \Delta G_S^{NP}(X) + \Delta G_S^{P}(X) \tag{2.11}$$

where $\Delta G_S^{SOLN}(X)$ is the partial Gibbs free energy of solution for the transfer of solute X from the gas phase to the stationary phase S, $\Delta G_S^{CAV}(X)$ the partial Gibbs free energy of cavity formation for solute X, $\Delta G_s^{NP}(X)$ the partial Gibbs free energy of interaction of the nonpolar contribution of solute X with the surrounding solvent and $\Delta G_s^P(X)$ the partial Gibbs free energy of interaction of the polar contribution of solute X with the surrounding solvents. However, non of the free energy terms on the right hand side of the equation can be calculated directly. A practical solution can be found if a few simplifying assumptions are made. The polar contribution to the free energy of solution of solute X is assumed to be equal to the difference between the free energy of solution of solute X in a stationary phase and the free energy of solution for a hypothetical n-alkane with the same van der Waals volume as solute X in the same stationary phase. The induction component to $\Delta G_s^P(X)$ can be removed by assuming that it is equivalent to the polar contribution to the free energy of solution of solute X in a nonpolar hydrocarbon solvent, such as squalane. A second reason for using a reference hydrocarbon solvent is to correct, at least partially, for the fact that the hardcore van der Waals volume is a poor estimate of the size of the cavity and its accessible surface for solvent interactions for aromatic and cyclic solutes. The solvent accessible surface area would logically be the preferred parameter for the cavity term but is very difficult to calculate while the van der Waals volume is readily accessible. With the above approximations the solvent interaction term for polar interactions, $\Delta G_s^{INT}(X)$, can formally be defined as [413].

$$\Delta G_s^{INT}(X) = \Delta G_s^P(X) - \Delta G_{SQ}^P(X) \qquad (2.12)$$

where the subscripts S and SQ refer to the polar phase to be characterized and squalane, respectively. The solvent interaction term for polar interactions can formally be defined as equivalent to the solute-solvent interactions of solute X that exceed those interactions typified by an n-alkane of identical van der Waals volume in solvent S reduced by the identical interactions of solute X in a hydrocarbon solvent, squalane. The free energy contributions to solution from cavity formation, dispersion, induction and reorganization entropy changes should be largely eliminated making $\Delta G_s^{INT}(X)$ the most logical term to probe the importance of solute orientation and hydrogen-bonding interactions. $\Delta G_s^{INT}(X)$ is evaluated entirely from accessible experimental data but its chemical meaning depends on the selection of the solute X and the properties expressed by X. Nitrobenezene and n-octanol were identified as suitable solutes for stationary phase orientation and proton acceptor interactions. No suitable probe was identified for stationary phase proton donor capacity [412,413]. Those solutes that are most useful for characterizing the complementary interactions in a solvent should have a single, dominant type of interaction expressed against a weak background for other interactions. In fact solutes of this kind are extremely rare, since all hydrogen bonding solutes, for example, contain a dipole moment and simultaneously enter into orientation interactions. This may require a more sophisticated approach to dissect the individual contributions of intermolecular forces for the test solutes or the expression of phase parameters in terms of the interactions experienced by substances belonging to different chemical groups. A useful form of visualizing the data from the above method is in the form of a dendrogram, Figure 2.13, based on the similarity of $\Delta G_s^{INT}(X)$ interactions for 20 polar test solutes. The phases that are most similar are next to each other and are connected. Connections at the extreme left side of the dendrogram have a similarity of 1, representing duplicates, and those at the extreme right a similarity of zero, and have no significant features in common. The dendrogram clusters phases with similar properties that are less likely to change the selectivity in a separation and allows the identification of unique phases, or groups of phases, suitable for selectivity optimization. For example, in Figure 2.13, OV-11 and OV-17 are identified as being very similar phases with nearly identical properties while QF-1 [a poly(trifluoropropylmethylsiloxane) phase] is identified as being uniquely different to all other phases in Figure 2.13.

The last two approaches represent promising beginnings for new methods to characterize stationary phase selectivity. The methods are evolutionary and not fully developed at present. Their future prospects are quite good and should eventually evolve into a standardized protocol for phase characterization. This is urgently required to make both the selection of stationary phases from those currently available and the rationale synthesis of new phases a logical process.

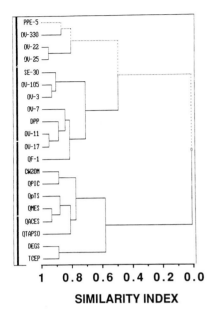

SIMILARITY INDEX

Figure 2.13   The use of cluster analysis to display the similarity
of stationary phases with the single linkage furthest neighbor
dendrogram.

2.5.5 Predicting Retention for Binary Stationary Phase
Mixtures

Consider the situation where a sample is incompletely
resolved on two different liquid phases and the components
unseparated on each phase are not the same. The natural conclusion
would be that a complete separation could probably be obtained if
the two phases were combined in appropriate proportions. Given the
separations on the two pure liquids how can one calculate the
exact composition of a binary liquid phase mixture that will
provide complete resolution of the sample? A method for this
purpose, based on the theory of diachoric solutions, was developed
by Purnell and Laub [414-417]. It is commonly known as the window
diagram method and has been applied with success to a number of
practical problems.

The basis of the window diagram approach is that the
relative retention of a solute on a mixed phase depends only on
the volume fractions of the individual phases and the partition

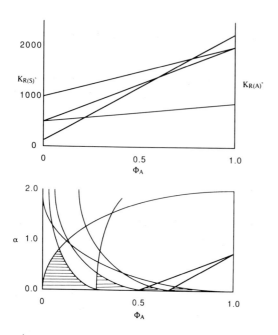

Figure 2.14 Plot of K_R against the volume fraction of A for four hypothetical solutes based on equation (2.14) and of the separation factor against the volume fraction of A based on equation 2.15. The best separation is predicted to occur at a volume fraction of 0.12.

coefficients of the solute in the pure liquids. For a binary mixture of two phases, A and S, this can be expressed as

$$K_R = \phi_A K_{R(A)}{}^\circ + \phi_S K_{R(S)}{}^\circ \qquad (2.13)$$

where K_R is the infinite dilution gas-liquid partition coefficient for a solute in the mixed phases (A + S), $K_{R(A)}{}^\circ$ the infinite dilution gas-liquid partition coefficient in phase A, $K_{R(S)}{}^\circ$ the infinite dilution gas-liquid partition coefficient in phase S, ϕ_A the volume fraction of A and ϕ_S the volume fraction of S. If equation (2.13) is rearranged to equation (2.14), it can be seen that a plot of K_R against the volume fraction of A will be linear, Figure 2.14.

$$K_R = K_{R(S)}{}^\circ + \phi_A (K_{R(A)}{}^\circ - K_{R(S)}{}^\circ) \qquad (2.14)$$

Furthermore, for any pair of solutes (1 and 2) whose retention behavior is described by equation (2.13), the separation factor, α, is simply the ratio of the two partition coefficients, which for computational purposes can be expressed in the form of equation (2.15).

$$\phi_A = [K_{R(S)1}{}^\circ - \alpha K_{R(S)2}{}^\circ] / [\alpha \Delta K_{R2}{}^\circ - \Delta K_{R1}{}^\circ] \qquad (2.15)$$

where $\Delta K_R{}^\circ = K_{R(A)}{}^\circ - K_{R(S)}{}^\circ$. Computer programs are available for performing the above calculations [418,419]. A plot of the separation factor against the volume fraction of A for all solute pairs provides a window diagram, Figure 2.14. For n solutes there are n!/2(n-2) solute pairs. To simplify the calculation all solute pairs with a separation factor greater than 1.2 are ignored, since their separation represents a relatively trivial problem. Also, for the purpose of visualizing the data, when plots of $K_R{}^\circ$ against the volume fraction of A cross for any solute, the calculated value of the separation factor is inverted so that the separation factor is greater than or equal to 1 at all times. The resultant window diagram is a series of approximate triangles rising from the volume fraction axis that constitute the windows within which complete separation for all solutes can be achieved. The optimum phase composition for the separation corresponds to the volume fraction of A with the largest value of the separation factor provided that the most difficult to separate pair are reasonably well retained. When this is not the case the largest value for the separation factor may not be the optimum choice and a different criteria is required to optimize the stationary phase composition [420]. Further, the number of theoretical plates required for complete separation can be calculated from equation (1.48). Thus, knowing the values of n attainable with the pure liquid phase columns, the required length of the mixed-phase column can be calculated with reasonable accuracy, given the predicted value of the separation factor. Finally, reference to the straight line diagram of K_R against the volume fraction of solvent A corresponding to the optimum phase mixture yields the elution order of the solutes.

The window diagram method can also be used to optimize the separation of mixtures when the number and identity of the components are unknown [421-423]. Two liquid phases, A and S, of different selectivity are chosen. Trial chromatograms are run on the two columns at some common temperature arrived at by visual optimization. A standard solute for which $K_{R(A)}°$ and $K_{R(S)}°$ are accurately known is included in the trial chromatograms to allow retention time data to be converted into $K_R°$ values. The sample, including standard, is then run at the selected temperature on columns containing pure A, pure S, and on 2:1, 1:1, and 1:2 v/v mixtures of the two liquid phases. For any given solute eluted from A, S and A + S mixed phases the corresponding ratio of the partition coefficient/volume fraction of A must lie on a straight line. This consideration enables the corresponding data points on each column to be distinguished and a plot similar to Figure 2.13 to be constructed. Once this has been completed the window diagram can be constructed as described previously. With this approach the presence of overlapping peaks on either column or changes in the elution order between the two columns does not present a problem. However, the predictive ability for polar phases is not always that good, most likely due to retention occurring by a mixed mechanism (see section 2.5.1) [424].

It is not necessary to blend liquid phases prior to coating, since it has been shown that mechanically mixed column packings of the pure phases provide the same results [424-426]. The window diagram approach can be used to optimize the temperature for a separation, but the method itself is intended primarily for use in isothermal gas chromatography. An extension to temperature programmed separations is not straightforward. The window diagram approach has not been widely used in open tubular column gas chromatography, since the greater intrinsic efficiency of such columns allows the separation of mixtures with low separation factor values and, in addition, the number of calculations required for an average capillary column run is prohibitive [426,427]. The method has been applied to the design of specific stationary phases for particular applications such as the prediction of the optimum composition of poly(cyanopropylphenylmethylsiloxane) gum phases for the resolution of purgeable halocarbon compounds found in contaminated water [428,429]. Mechanically mixing the desired volume fraction of liquid phases for preparing open tubular columns is not always successful due to the possibility that the phases may be immiscible and that finding a deactivation procedure compatible with the requirements of both phases simultaneously may be difficult. In unfavorable circumstances, both effects lead to poor film stability and a tendency for droplet formation as the temperature is varied. A more generally successful approach to selectivity optimization in open tubular column gas chromatography is the use of serially coupled columns (see section 1.7.6).

2.6 GAS-SOLID CHROMATOGRAPHY

Gas-solid chromatography (GSC) was in existence prior to the development of gas-liquid chromatography but has never attained the same status as a separation technique. Like all techniques its growth has been cyclic and at the moment it is enjoying an upward trend catalyzed by a maturing of the commercialization of column technology in gas-liquid chromatography, the commercialization of new sorbents and porous-layer open tubular columns for gas-solid chromatography and the realization that modified adsorbents provide greater scope for adjusting selectivity than can be obtained using bulk liquids [8,272-274,430-434]. Some of the reasons for the slow growth of gas-solid chromatography can be summarized as follows. Adsorption isotherms in GSC are often non-linear, even for small sample sizes. This produces several detrimental effects: solute retention volumes vary with sample size, peaks are asymmetric and the recovery of some samples from the column is incomplete. Many adsorbents in common use have large surface areas and/or energies leading to excessively long retention times, particularly for high boiling and polar molecules. With a few exceptions GSC is usually limited to the separation of solutes containing fewer than twelve carbon atoms and having a boiling point less than about 200°C [430]. The sample molecular weight range for which GSC is suitable is thus more restricted than for gas-liquid chromatography. Adsorbents are generally more difficult to standardize and prepare reproducibly than liquid phases. The slow kinetics of mass transfer in GSC can lead to columns of low efficiency compared to gas-liquid chromatography. However, this is not always the case, and in favorable circumstances the rate of mass transfer in solid sorbents can be faster than the rate of diffusion in a liquid film yielding columns of higher efficiency.

GSC does enjoy some advantages over gas-liquid chromatography and these have been sufficient to maintain some interest in the technique. Adsorbents are generally stable over a wide temperature range and are often insensitive to attack by oxygen. Column bleed is virtually nonexistent, so high sensitivity detectors such as the helium ionization detector can be used. The selectivity of GSC is usually much greater than gas-liquid chromatography for the separation of geometric and isotopic isomers. GSC is also suitable for the separation of inorganic gases and low molecular weight hydrocarbons, for which gas-liquid chromatography generally shows little selectivity.

Gas-solid chromatography is a suitable technique for studying changes in surface morphology and heterogeneity of solids, the shape of adsorption isotherms and the thermodynamics of adsorption at either zero or finite surface coverage [435-440]. Interpretation of the data is not as straightforward as for gas-liquid partitioning since isotherms are frequently non-linear, even for the case of zero surface coverage, and the choice of a standard state for the adsorbate, that of a two dimensional gas, must follow an arbitrary convention. Several references contain a good discussion of the problem [316,441-444].

The principal adsorbents used in GSC are silica, alumina, graphitized carbon blacks, porous polymer beads, zeolites and cyclodextrins [8,430,431,445]. The bonded phase sorbents discussed in section 2.2.3 could also be considered as modified adsorbents in many respects.

2.6.1 Inorganic Adsorbents

Silica gel adsorbents are available for GSC in the form of rigid, incompressible, spherical beads that can be dry packed into empty columns. These macroporous adsorbents, Table 2.13, are prepared from ordinary silica gel by either simple high temperature treatment (700 to 950°C) or by hydrothermal treatment with steam in an autoclave. Hydrothermal treatment causes pronounced pore widening, leading to a reduction in the specific surface area. Thus silica adsorbents with a wide range of specific surface areas, pore diameters and pore volumes can be readily prepared [446-448]. Silica and alumina adsorbents are recommended for the separation of low molecular weight saturated and unsaturated hydrocarbons, halogenated hydrocarbons and derivatives of benzene. Polar solutes interact too strongly with the highly adsorptive surface to be separated in a reasonable time or with reasonable peak shape.

Retention on silica or alumina adsorbents is a function of the specific surface area, the degree of surface contamination (particularly that of water), the prior thermal conditioning of

TABLE 2.13

INORGANIC ADSORBENTS FOR GAS-SOLID CHROMATOGRAPHY

Type	Name		Specific Surface Area (m^2/g)	Pore Diameter (nm)
Silica	Spherosil	XOA 400	300-500	8
		XOA 200	140-230	15
		XOB 075	75-125	30
		XOB 030	37-62	60
		XOB 015	18-31	125
		XOC 005	5-15	300
	Porasil	B	125-250	10-20
		C	50-100	20-40
Graphitized Carbon Black	Carbopack	C	12	
		B	100	
	Carbosieve		1000	1.3
	Spherocarb		1200	1.5

the adsorbent and the ability of the solute to undergo specific
interactions, such as hydrogen bonding, with surface silanol
groups. Alumina adsorbents with specific surface areas comparable
to silica adsorbents show similar retention properties but
different selectivity. This is due to the presence of Lewis acid
sites associated with surface aluminum ions. In both cases a
dramatic reduction in the retention of nonpolar solutes can be
obtained by the uptake of water vapor by the adsorbents. Mixing
the adsorbents with an inert solid diluent such as glass beads or
a diatomaceous support, or coating a support with a layer of fine-
particle adsorbent ("dusted columns"), reduces retention and
simultaneously improves column efficiency [449]. An alternative
approach to reduce retention volumes, modify column selectivity
and/or to improve the efficiency of silica and alumina adsorbents
is to coat the adsorbent with a small quantity of an involatile
liquid [450-453] or an inorganic salt [430,454-456]. It is assumed
that the liquid film is selectively adsorbed by the most energetic
adsorption sites, which are then no longer available to the
solute. The heterogeneous distribution of active sites is
consequently reduced, providing a more homogeneous surface for
interaction with the solute. As a result, both retention and peak
asymmetry are dramatically reduced. As the amount of liquid phase
is increased gas-liquid adsorption and, at still higher levels,
gas-liquid partitioning can be expected to contribute to the
retention mechanism. For the low levels of liquid phases commonly

used in GSC it is assumed that gas-solid interactions dominate the retention mechanism.

Commonly used inorganic salt modifiers generally fall into two categories. Alkali metal salts are used to reduce the surface heterogeneity of the adsorbent in a manner somewhat similar to polar liquid phases. Complex-forming salts such as silver nitrate and cuprous chloride deactivate the surface and can also adjust column selectivity by their ability to selectively retain some solutes. With low salt loadings, active sites on the original adsorbent surface are covered by the salt in the form of ions and/or ion pairs. With increasing salt loadings a new surface layer of salt gradually covers the original adsorbent surface. After a complete monolayer is formed additional salt is deposited as a thin, crystalline film whose thickness depends on the amount of salt added. At this point the specific surface area will decline due to a reduction in the pore volume now occupied by salt crystals. Therefore, at high salt loadings solute retention is determined chiefly by adsorption onto the surface of the salt film. At low salt loadings retention is determined primarily by interactions with the adsorbent surface. At intermediate salt loadings both retention mechanisms may be important, depending on the nature of the solute and its ability to displace the salt modifier from active sites on the adsorbent surface. Thus optimum salt loadings for a particular sample are generally found by trial and error and may vary from about 0.5% w/w up to about 30% w/w. An example of a separation of a mixture of hydrocarbons on a potassium chloride modified alumina porous-layer open tubular column is shown in Figure 2.15 [430].

Graphitized carbon blacks are prepared by heating ordinary carbon blacks to about 3000°C in an inert gas atmosphere [445,452,457,458]. This drives off volatile, tarry residues and induces the growth of graphite crystallites (particles in which graphite crystals are arranged in the form of polyhedrons). At the same time various functional groups originally present on the carbon black surface are destroyed. The surface of the graphitized carbon blacks is almost completely free of unsaturated bonds, lone electron pairs, free radicals and ions. Because of the high surface homogeneity of the adsorbent, solute-solute interactions on the adsorbent surface can be important. At moderate sample sizes this can result in a decreased linearity of the adsorption isotherms. The flat surfaces of graphitized carbon blacks are

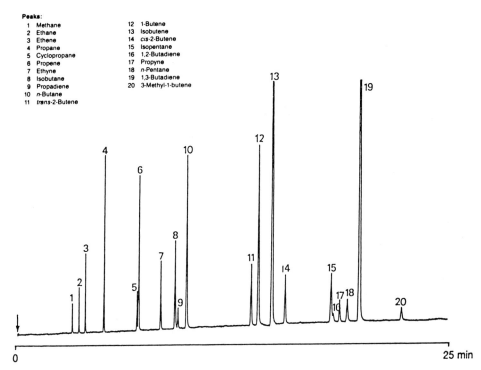

Peaks:

1	Methane	12	1-Butene
2	Ethane	13	Isobutene
3	Ethene	14	*cis*-2-Butene
4	Propane	15	Isopentane
5	Cyclopropane	16	1,2-Butadiene
6	Propene	17	Propyne
7	Ethyne	18	*n*-Pentane
8	Isobutane	19	1,3-Butadiene
9	Propadiene	20	3-Methyl-1-butene
10	*n*-Butane		
11	*trans*-2-Butene		

Figure 2.15 Separation of a mixture of saturated and unsaturated hydrocarbons on a 50 m x 0.32 mm fused silica porous-layer open tubular column coated with alumina modified with potassium chloride. The separation was performed by temperature programming from 70 to 200°C at 3°C/min. (Reproduced with permission from ref 430. Copyright Preston Publications, Inc.)

particularly well suited for the separation of molecules which differ only in geometric structure, i.e., structural isomers and geometric stereoisomers.

The retention behavior of solutes on graphitized carbon blacks can be explained on the basis of the availability of two types of adsorption sites [452,459]. The vast majority of the surface sites are nonpolar and correspond to the graphite-like array of carbon atoms. These sites show no tendency to interact preferentially with molecules carrying functional groups and dispersive interactions dominate retention behavior. Polar adsorption sites are few in number but they can establish specific, strong interactions with polar solutes. Preliminary treatment of the adsorbent may be used to reduce their number. For example, heating to 1000°C in a stream of hydrogen has been used to minimize those active sites associated with the presence of surface oxygen complexes [435,436,459] while washing with perchloric or phosphoric acid eliminates basic carbonium-oxygen complexes and sulfur present as sulfide [460]. However, even after such treatment further modification of the surface by the addition of small quantities of liquids that are capable of establishing strong interactions with residual active sites is generally required [459-461]. At low surface concentrations the modifying agents act primarily as tailing inhibitors and retention volume reducers. Liquids which are capable of entering into specific interactions with the sample may be used as selectivity modifiers. An example of this is shown in Figure 2.16, in which a low concentration of picric acid is added to the adsorbent to enhance the separation of a mixture of saturated and unsaturated hydrocarbons [462]. At the surface concentration of liquid modifier required to build up a densely-packed monolayer, retention will occur primarily by adsorption on this layer and the selectivity of the column will be different to that of a partially complete monolayer. This film, however, will not show properties characteristic of the bulk liquid as its properties are influenced by those of the adsorbent with which it is in contact. At high liquid phase loadings the adsorbent behaves as a support for the bulk liquid [463,464]. Thus quite different column selectivities can be expected over the range of liquid phase modifier added to the adsorbent. In most cases quite low levels are used to exploit the full advantages of gas-solid chromatographic interactions. Very efficient micropacked [465,466] and porous-layer open tubular columns [432-434,467] have been prepared with liquid-modified graphitized carbon blacks for separations requiring large numbers of theoretical plates.

Other forms of carbon-based adsorbents include Carbosieves [468] and porous glassy carbon beads [469]. Carbosieves are comprised of very small crystallites crosslinked to yield a disordered cavity-aperture structure. They are microporous and of high surface area, Table 2.13 , with a very pronounced retention of organic compounds. They are used primarily for the separation of inorganic gases and, in particular, for the separation of small polar molecules such as water, formaldehyde, and hydrogen sulfide [470,471]. Porous glassy carbon is produced by polymerization of a phenol-formaldehyde resin mixture in the pores of a silica or porous glass template which is subsequently converted to carbon by heating in an inert atmosphere to about 1000°C followed by dissolution of the template. Finally, the material is fired in an inert atmosphere at 2000-2800°C to anneal the surface, remove

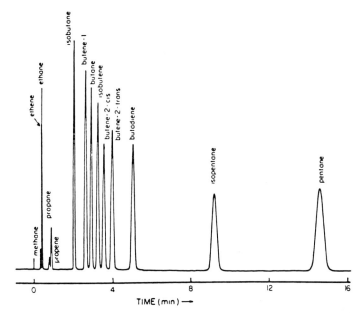

Figure 2.16   Isothermal separation of C_1 - C_5 hydrocarbons on a 2.2 m x 2 mm I.D. column of Carbopack C + 0.19% picric acid at 50°C. (Reproduced with permission from ref. 462. Copyright American Chemical Society).

micropores and, depending upon the temperature, produce some degree of graphitization. The particle size, shape, porosity and pore size are determined by the choice of the template material; the surface chemistry is determined by the final heat treatment and any subsequent chemical treatment. Porous glassy carbons have been made with surface areas varying from 20 to 400 m^2/g and have similar chromatographic properties to Carbopack B, but with somewhat less retention per unit surface area.

2.6.2   Porous Polymer Beads

Porous polymer beads are prepared by the process of suspension polymerization, in which a mixture of monomers and crosslinking reagents is polymerized in the presence of an inert solvent. The microstructure of the gel, formed within the droplet in the early stages, gradually grows into a sponge-like structure, with inert solvent filling the space between the microstructures [472]. Upon drying and evacuation, the porous structure remains, producing a uniform bead that is sufficiently rigid to be dry packed into a column. By adroit selection of monomer, crosslinking

TABLE 2.14

PHYSICAL PROPERTIES OF POROUS POLYMER BEADS

Porous Polymer	Type(a)	Physical Property			
		Free-fall Density (g/cm^3)	Surface Area(b) (m^2/g)	Average Pore Diameter (nm)	Temperature Limit (°C)
Chromosorb 101	STY-DVB	0.30	<50	300-400	275
Chromosorb 102	STY-DVB	0.29	300-400	8.5	250
Chromosorb 103	Polystyrene	0.32	15-25	300-400	275
Chromosorb 104	ACN-DVB	0.32	100-200	60-80	250
Chromosorb 105	Acrylic Ester	0.34	600-700	40-60	250
Chromosorb 106	Polystyrene	0.28	700-800	5	225
Chromosorb 107	Acrylic Ester	0.30	400-500	8	225
Chromosorb 108	Acrylic Ester		100-200	25	225
Porapak N	Vinylpyrolidone	0.39	225-350		200
Porapak P	STY-DVB	0.28	100-200		250
Porapak Q	EVB-DVB	0.35	500-700	75	250
Porapak R	Vinylpyrolidone	0.33	450-600	76	250
Porapak S	Vinylpyrolidone	0.35	300-450	76	250
Porapak T	EGDMA	0.44	250-300	9	200
Porapak PS	Silanized P				
Porapak QS	Silanized Q				250
Tenax-GC		0.37	18.6		375

(a) STY = styrene; DVB = divinylbenzene; ACN = acrylonitrile;
    EVB = ethylvinylbenzene; EGDMA = ethylene glycol dimethacrylate
(b) Values for surface area vary widely in the literature

reagent and inert solvent, porous polymer beads of different
properties can be prepared. Some examples are given in Table 2.14.
The properties of these beads vary with the chemical composition,
pore structure and surface area of the beads [473,474]. Porous
polymer beads with pore diameters less than 10 nm are used
primarily for the separation of gases while those with larger pore
diameters are used for the separation of higher-boiling organics.
The porous polymeric beads have found many applications in the
analysis of volatile inorganic and organic compounds, particularly
for samples with poor peak shapes on liquid-coated diatomaceous
supports. Notable examples are water, formaldehyde, carboxylic
acids and inorganic gases. Some general indications of the type of
samples which can be separated on porous polymer beads are given
in Table 2.15 [431,475-478]. Porous-layer open tubular columns
prepared with Porapak Q type polymers have been used for the
separation of alcohols, aldehydes, esters, ketones and halogenated
hydrocarbons, as well as permanent gases and light hydrocarbons

TABLE 2.15

SAMPLES SUITABLE FOR SEPARATION ON POROUS POLYMER BEADS

Porous Polymer	Recommended for the Separation of	Not Recommended For
Chromosorb 101 Porapak P and PS	Esters, ethers, ketones, alcohols, hydrocarbons, fatty acids, aldehydes and glycols.	Amines, anilines.
Chromosorb 102 Porapak Q	Light and permanent gases, low molecular weight acids, alcohols, glycols, ketones, hydrocarbons, esters, nitriles and nitroalkanes.	Amines, anilines.
Chromosorb 103	Amines, amides, alcohols, aldehydes, hydrazines and ketones.	Acidic substances, glycols, nitriles and nitroalkanes.
Chromosorb 104	Nitriles, nitro compounds, sulfur gases, oxides of nitrogen, ammonia and xylenols.	Amines and glycols.
Chromosorb 105 Porapak N	Aqueous mixtures of formaldehyde, acetylene from lower hydrocarbons and most gases.	Glycols, acids and amines.
Chromosorb 106 Porapak QS	Alcohols, C_2-C_5 carboxylic acids, alcohols and sulfur gases.	Glycols and amines.
Chromosorb 107 Porapak T	Formaldehyde from water and acetylene from lower hydrocarbons.	Glycols and amines.
Chromosorb 108	Gases, polar materials such as water, alcohols, aldehydes and glycols.	
Porapak S	Alcohols, ketones, and halocarbon compounds	Acids, amines, and nitro compounds.
Porapak R	Esters and ethers, nitriles and nitro compounds.	Glycols and amines.
Tenax-GC	High boiling polar compounds, diols, phenols, methyl esters of dicarboxylic acids, amines, diamines, ethanolamines, amides, aldehydes and ketones.	

[274,430]. Compared to inorganic adsorbents these columns are stable to aqueous injections.

The retention mechanism of organic solutes by porous polymer beads remains ambiguous [478]. At low temperatures adsorption dominates but at higher temperatures the polymer beads could behave as a highly extended liquid with solvation interactions. The evidence for a partition mechanism is not very strong and its importance, at present, remains speculative. Like other adsorbents it has proven possible to control retention and enhance efficiency by diluting porous polymers with an inert support material [479].

Various methods have also been attempted to standardize the selectivity of porous polymers using modified versions of the McReynolds' phase constants system, Synder's selectivity triangle, or the relative retention of ethylene, acetylene and carbon dioxide [477,480-482]. The applicability of these methods, however, has not been thoroughly tested. Among the Porapak polymers Porapak Q behaves as the least polar and Porapak N and R generally show the highest retention of polar compounds. Some of the porous polymers have been shown to react with certain samples. Porapak QS reacts with amines, Porapak Q and Chromosorb 102 are nitrated by nitrogen dioxide and Porapak S and Chromosorb 103 react with nitroalkanes [476,479,481]. Batch-to-batch variability of polymers is always present due to slight differences in the ratio of monomer and crosslinking reagents as well as changes induced by thermal aging while in use.

With the exception of Tenax-GC, the porous polymers are limited in application to those compounds with boiling points less than about 300°C; otherwise, retention times may be excessively long. Tenax-GC is unique in terms of its thermal stability. It may be used isothermally at temperatures up to 375°C. Tenax-GC is a linear polymer of p-2,6-diphenylphenylene oxide having a molecular weight of 5×10^5 to 10^6. A granulated powder of low surface area, it differs physically from the Porapak and Chromosorb polymers [483,484]. Its separation properties are similar to Porapak Q and similar porous polymers; its principal advantage is its greater thermal stability.

2.6.3 Molecular Sieves

Molecular sieves (zeolites) are artificially prepared aluminosilicates of alkali metals. The most common types for gas chromatography are molecular sieve 5A, a calcium aluminosilicate with an effective pore diameter of 0.5 nm, and molecular sieve 13X, a sodium aluminosilicate with an effective pore diameter of 1 nm. The molecular sieves have a tunnel-like pore structure with the pore size being dependent on the geometrical structure of the zeolite and the size of the cation. The pores are essentially microporous as the cross-sectional diameter of the channels is of similar dimensions to those of small molecules. This also contributes to the enormous surface area of these materials. Two features primarily govern retention on molecular sieves. The size of the analyte which determines whether it can enter the porous structure of the molecular sieve and the strength of adsorption interactions that take place on the surface both inside and outside the pores. Molecular sieves are used primarily for the separation of permanent gases such as hydrogen, oxygen, nitrogen, carbon monoxide, the inert gases and low molecular weight linear and branched chain hydrocarbons [8,273,431,485]. Water, carbon dioxide and other polar molecules are retained excessively and catalytic transformations of labile analytes are also common. The latter is a product of the high temperatures required for elution and the catalytic activity of the zeolites. To reduce retention thin layers of powdered molecular sieves coated on a support or directly on a capillary column wall can be used [273,430,485].

2.7 PREPARATIVE-SCALE GAS CHROMATOGRAPHY

Early in its development gas chromatography attracted considerable attention as a preparative as well as an analytical technique. Today liquid chromatography is more popular for preparative applications than gas chromatography except for the isolation of volatile compounds, which are not easily recovered in high yield from common solvents. Gas chromatography is also more suitable for the isolation of milligram amounts of pure substances from mixtures too complex to be separated by liquid chromatography using automated high resolution preparative gas chromatography [486-488]. With modern instrumentation infrared and mass spectra can usually be obtained on-line without prior isolation. However, for identification by nuclear magnetic resonance or to obtain sufficient material for biological testing or aroma classification, etc., isolation by repetitive accumulation of the desired components from several injections is normally required. The scale of preparative instruments varies tremendously. Pilot plant and industrial equipment capable of handling kilogram quantities per hour are in routine use [462]. For our discussion we will focus on more modest scale apparatus used to isolate substances in the microgram to gram range that are more likely to be found in an analytical laboratory.

The most common types of preparative-scale gas chromatographic instruments are based on packed column technology [489-491]. The primary objective in preparative-scale gas chromatography is to obtain a high sample throughput. An inevitable result of this goal is that either resolution or separation time, or both, must be compromised. The primary method of increasing the sample capacity of the column is to increase its size (i.e., the quantity of stationary phase available for the separation). Two solutions are suggested: increase the column diameter at constant length or increase the column length at constant diameter. For simple separations a short, wide packed column is usually used, for example a column of 1-3 m x 6-10 cm I.D. For more complex separations higher column efficiencies are required and long, narrow packed columns, for example 10-30 m x 0.5-1.5 cm I.D., recycle chromatography using a column of intermediate length [492] or multidimensional chromatographic techniques [493] are normally used. It is not possible to coil wide bore packed columns and, therefore, unless a purpose designed preparative gas chromatograph is available, the choice may be limited to long, narrow, coiled columns. Other critical instrument considerations include the injector, detector and method of sample collection. Compared to analytical operations the sample size is much greater and is generally increased to the maximum allowed. This maximum is limited by the requirement that a minimum resolution be maintained between critical sample components. Injection volumes in the range of 0.1 to 10.0 ml are common with packed columns and, therefore, normal injection techniques are not applicable. Injection by syringe is possible for small volumes (e.g., 100 microliters) of volatile samples provided that a slow injection rate is employed. For larger sample volumes an automated injector is used. Several types have been described employing pneumatic transfer from a reservoir through a capillary restrictor, a pneumatic piston pump, or by a syringe pump. The injection process is thus controlled by time, a necessity due to the limited thermal capacity of injection block heaters and their inability to flash vaporize large sample volumes. Larger samples usually require an evaporation device between the injector and column. This is often a tube heated separately from the column oven and packed with glass beads or metal spheres to increase the thermal capacity while reducing dead volume effects. The evaporation unit is usually maintained at a temperature 50°C above the boiling point of the least volatile sample component. The sample should be transferred to the column as a rectangular plug diluted in the minimum volume of carrier gas; if it is not completely vaporized then a homogeneous distribution of sample is not obtained over the cross-sectional area of the column and excessive band broadening may result. Vaporization problems may be circumvented by direct, slow, on-column injection without carrier gas flow. This is often the most practical solution for injecting large sample volumes using analytical instruments. The principal disadvantage is the possibility of stripping the stationary phase from the head of the column. Column stripping results in diminished efficiency due to the combination of sample interactions with bare support and the irregular distribution of liquid phase produced throughout the column. For the on-column injection of both vaporized and liquid samples a substantial pressure change may result from sample vaporization. Sample backflush into the carrier gas lines can be a problem and is normally solved by positioning a backflush or needle valve in the carrier gas line.

Most separations are usually carried out isothermally. Temperature programming is not possible with wide bore columns due to the uneven radial temperature gradients generated across the column. Temperature programming is only possible with long, narrow columns and then generally at slow rates. Detection of the separated sample is usually performed with a thermal conductivity or flame-ionization detector. As the FID is destructive, it is connected to the column via a splitter so that only a few percent of the total column flow is diverted to the detector. Although the thermal conductivity detector is a nondestructive detector, it is also usually operated with a splitter, since the filament lifetime may be shortened due to contamination by the high mass flow of material leaving the column.

The final operation, unique to preparative gas chromatography, is sample collection of the vapors in the carrier gas effluent. This process may be performed automatically or manually, and is sequenced by a signal from the detector. In a commercial preparative-scale gas chromatograph injection and sample collection are automated for unattended operation. When using an analytical instrument both processes will normally be performed manually. The exit from the detector splitter is usually led out through a side wall of the oven and maintained at a temperature sufficient to avoid premature condensation of the sample. This may involve the use of an auxiliary heating supply for the transfer line. There is no single solution to the problem of trapping the sample after it emerges from the transfer line; packed and unpacked cold traps, solution and entrainment traps, total effluent and adsorption traps, Volman traps and

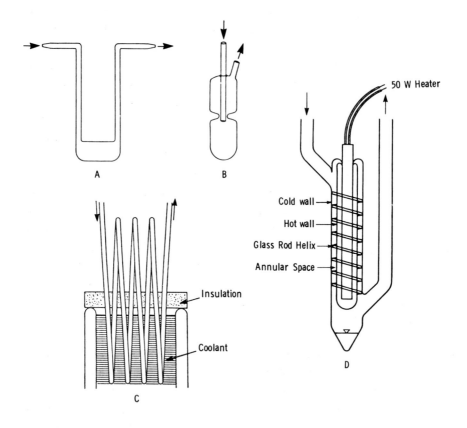

Figure 2.17 Traps for sample recovery from a gas chromatographic effluent. A, U-tube; B, simple trap; C, multiple temperature gradient trap; D, Volman trap.

electrostatic precipitators are among the more popular [494-499]. Some representative examples are shown in Figure 2.17. The minimum requirement for efficient sample collection is that the partial pressure of the solute entering the trap must be greater than its equilibrium vapor pressure with the condensed phase at the trap temperature. The equilibrium vapor pressure of the solute may be reduced by lowering the temperature, by dissolving the solute in a solvent, or by adsorbing it onto a material of high surface area. However, by far the most common approach is to reduce the temperature of the effluent, either in a trap filled with a material such as glass beads or column packing, or in a simple open tube trap. In practice the residence time of the solute in the trap may be too short to obtain equilibrium and the solute tends to pass directly through the trap, either as a supersaturated vapor or as a fog composed of small liquid droplets. The cold-trapping of samples with boiling points greater than about 150°C is almost inevitably accompanied by the formation of aerosols which are then swept through the trap. This can be minimized by cooling the effluent through a gradual temperature gradient rather than subjecting it to an abrupt temperature change. Also, turbulent flow within the trap facilitates the collision of liquid droplets with the wall, thereby inducing their retention. The Volman trap consists of a double-walled vessel within which turbulence is created by maintaining the walls at different temperatures. The Volman trap is particularly useful for recovering gram-scale quantities. Electrostatic precipitators are also used to trap large fractions. As aerosols contain large numbers of charged droplets, the trapping efficiency can be improved by passing the cooled effluent through a large electrostatic field. If a readily condensable gas such as argon or carbon dioxide is used as the carrier gas then the total effluent can be condensed, thereby entraining the small amount of solute in a much larger quantity of carrier gas. After condensation the sample and condensed carrier gas are separated by selective evaporation. The efficiency of the total effluent trap can be very high but the method is inconvenient and difficult in practice. For occasional sample trapping a simple packed or empty glass U-tube cooled in a dry ice-acetone or liquid nitrogen coolant usually suffices. For further details on all aspects of preparative-scale gas chromatography several reviews are available [490,491,500-502].

Instrumental constraints are established by the equipment available; the analyst, however, has control of the column selected. Analytical separations are usually performed initially to indicate the degree of difficulty of the separation. The stationary phase is then selected from a limited number of thermally stable phases. Preparative columns are expensive to prepare and generally only a few column types are available. The support material is coarse with a narrow particle size distribution, 35-40 mesh. High carrier gas flow rates, in the range 100-1000 ml/min depending on the column diameter, are used, so that a coarse column packing is needed to permit operation at a reasonable inlet pressure. As the stationary phase loading is greater and the column length may be longer than those used in analytical separations, operation at higher temperatures is also common. This can be a problem with thermally labile compounds. Further considerations for optimizing column conditions for a particular separation are summarized in references 503 and 504. Packing preparative columns reproducibly and with reasonable efficiency is, at best, never easy and increases in difficulty as the column diameter is increased [490,502,505-507]. The inferior chromatographic performance of preparative columns is due to the radial unevenness of the packing structure. During the packing of large diameter columns the solid support particles segregate preferentially according to their size, with the larger particles being closer to the wall. In addition, the packing in this region is less dense, due to the physical constraint of the wall. This leads to uneven flow profiles for the carrier gas and sample through the packed bed, inducing additional band broadening. To overcome this problem Verzele suggested using the shake, turn and pressure packing method whereby the column is shaken in the radial direction and is rotated along its long axis while being packed and periodically pressurized [505]. Reese has described a vacuum packing method with vertical tamping which is less time consuming and produces columns of good efficiency [507]. Since column permeability is an important consideration in preparative gas chromatography, the objective of any packing procedure can be defeated by the generation of fines; therefore robust packings and gentle handling are essential to obtain success.

Packed columns have a higher sample capacity than open tubular columns and have been more widely used for preparative-scale gas chromatography [508]. Open tubular columns are useful for isolating a few milligrams of components from complex mixtures under high resolution conditions. Automated sequencing of injection and fraction collection to accumulate sufficient sample is generally used [486-488,508]. Short lengths of wide bore capillary tubing coated with a thick film of stationary phase are efficient traps for room temperature sample collection [509,510]. The collected samples can be recovered by either solvent or thermal desorption. Because of narrow peak widths rapid switching between traps and/or waste is required to avoid cross contamination.

2.8 REFERENCES

1. A. T. James and A. J. P. Martin, Biochem. J., 50 (1952) 679.

2. C. F. Poole and S. K. Poole, Anal. Chim. Acta, 216 (1989) 109.
3. R. A. Keller, J. Chromatogr. Sci., 11 (1973) 49.
4. F. Vernon, in C. E. H. Knapman (Ed.), "Developments in Chromatography", Applied Science Publishers, London, UK, Vol. 1, 1978, p. 1.
5. R. V. Golovnya and T. A. Misharina, J. High Resolut. Chromatogr., 3 (1980) 4.
6. C. F. Poole and S. K. Poole, Chem. Revs., 89 (1989) 377.
7. H. Lamparczyk, Chromatographia, 20 (1985) 283.
8. G. E. Bailescu and V. A. Ilie, "Stationary Phases in Gas Chromatography", Pergamon Press, Oxford, UK, 1975.
9. K. K. Unger (Ed.), "Packings and Stationary Phases in Chromatographic Techniques", Dekker, New York, NY, 1990.
10. J. R. Mann and S. T. Preston, J. Chromatogr. Sci., 11 (1973) 216.
11. L. V. Semenchenko and M. S. Vigdergauz, J. Chromatogr., 245 (1982) 177.
12. J. A. Garcia-Domingues, J. Garcia-Munoz, V. Menendez, M. J. Molera, and J. M. Santiuste, J. Chromatogr., 393 (1987) 209.
13. J. A. Yancey, J. Chromatogr. Sci., 24 (1986) 117.
14. J. A. Yancey, J. Chromatogr. Sci., 23 (1985) 370.
15. J. A. Yancey, J. Chromatogr. Sci., 23 (1985) 161.
16. G. Castello, J. Chromatogr., 66 (1972) 213.
17. C. F. Poole, R. M. Pomaville, and T. A. Dean, Anal. Chim. Acta, 225 (1989) 193.
18. F. Vernon and C. O. E. Ogundipe, J. Chromatogr., 132 (1977) 181.
19. F. Reido, D. Fritz, G. Tarjan, and E. Sz. Kovats, J. Chromatogr., 126 (1976) 63.
20. L. Boksanyi and E. Sz. Kovats, J. Chromatogr., 126 (1976) 87.
21. J. K. Haken and D. K. M. Ho, J. Chromatogr., 142 (1977) 203.
22 J. K. Haken and F. Vernon, J. Chromatogr., 186 (1979) 89.
23. M. B. Evans and M. J. Osborn, Chromatographia, 13 (1980) 177.
24. M. B. Evans, M. I. Kawar, and R. Newton, Chromatographia, 14 (1981) 398.
25. R. M. Pomaville and C. F. Poole, Anal. Chim. Acta, 200 (1987) 151.
26. P. Varughese, M. E. Gangoda, and R. K. Gilpin, J. Chromatogr. Sci., 26 (1988) 401.
27. S. C. Dhanesar and C. F. Poole, Anal. Chem., 55 (1983) 1462.
28. R. M. Pomaville and C. F. Poole, J. Chromatogr., 468 (1989) 261.
29. R. M. Pomaville and C. F. Poole, J. Chromatogr., 499 (1989) 749.
30. W. W. Blaser and W. R. Kracht, J. Chromatogr. Sci., 16 (1978) 111.
31. S. C. Dhanesar and C. F. Poole, Anal. Chem., 55 (1983) 2148.
32. C. R. Trash, J. Chromatogr. Sci., 11 (1973) 196.
33. A. E. Coleman, J. Chromatogr. Sci., 11 (1973) 198.
34. J. K. Haken, J. Chromatogr., 300 (1984) 1.
35. L. G. Blomberg and K. E. Markides, J. High Resolut. Chromatogr., 8 (1985) 632.
36. F. G. Van Lenten, J. E. Conway, and L. B. Rogers, Sepn. Sci., 12 (1977) 1.
37. A. Venema, L. G. J. v. d. Ven and H. v. d. Steege, J. High Resolut. Chromatogr., 2 (1979) 405.
38. S. J. Hawkes, Anal. Chem., 58 (1986) 1886.
39. J. E. Brady, D. Bjorkman, C. D. Herter, and P. W. Carr, Anal. Chem., 56 (1984) 278.

40. E. F. Barry, P. Ferioli, and J. A. Hubball, J. High Resolut. Chromatogr., 6 (1983) 172.
41. B. A. Jones, J. C. Kuei, J. S. Bradshaw, and M. L. Lee, J. Chromatogr., 298 (1984) 389.
42. J. H. Beeson and R. E. Pecsar, Anal. Chem., 41 (1969) 1678.
43. S. C. Dhanesar and C. F. Poole, J. Chromatogr., 267 (1983) 293.
44. R. D. Schwartz, R. G. Matthews, S. Ramachandran, R. S. Henly, and J. E. Doyle, J. Chromatogr., 112 (1975) 111.
45. S. H. Langer, Anal. Chem., 44 (1972) 1915.
46. J. H. Purnell and O. P. Srivastava, Anal. Chem., 45 (1973) 1111.
47. C. L. de Ligny, Adv. Chromatogr., 14 (1976) 265.
48. D. G. Anderson and R. E. Ansel, J. Chromatogr. Sci., 11 (1973) 192.
49. F. Vernon, J. Chromatogr., 148 (1978) 397.
50. R. F. Kruppa and R. S. Henly, J. Chromatogr. Sci., 12 (1974) 127.
51. H. E. Persinger and J. T. Shank, J. Chromatogr. Sci., 11 (1973) 190.
52. Gy. Vigh, A. Bartha, and J. Hlavay, J. High Resolut. Chromatogr., 4 (1981) 3.
53. M. Verzele and P. Sandra, J. Chromatogr., 158 (1978) 111.
54. P. Sandra, M. Verzele, M. Verstappe, and J. Verzele, J. High Resolut. Chromatogr., 2 (1979) 288.
55. W. Blum, J. High Resolut. Chromatogr., 10 (1987) 32.
56. C. F. Poole, K. G. Furton, and B. R. Kersten, J. Chromatogr. Sci., 24 (1986) 400.
57. C. F. Poole, K. G. Furton, R. M. Pomaville, S. K. Poole, and B. R. Kersten, in R. J. Gale and D. G. Lovering (Eds.), Molten Salt Techniques, Plenum Press, London, UK, Vol. 4, 1990.
58. K. G. Furton and C. F. Poole, J. Chromatogr., 349 (1985) 235.
59. S. C. Dhanesar, M. E. Coddens, and C. F. Poole, J. Chromatogr., 349 (1985) 249.
60. K. G. Furton and C. F. Poole, Anal. Chem., 59 (1987) 1170.
61. R. M. Pomaville and C. F. Poole, Anal. Chem., 60 (1988) 1103.
62. R. M. Pomaville, S. K. Poole, L. J. Davis, and C. F. Poole, J. Chromatogr., 438 (1988) 1.
63. K. G. Furton, S. K. Poole, and C. F. Poole, Anal. Chim. Acta, 192 (1987) 49.
64. S. K. Poole, P. H. Shetty, and C. F. Poole, Anal. Chim. Acta, 218 (1989) 241.
65. S. K. Poole and C. F. Poole, J. Chromatogr., 435 (1988) 17.
66. D. M. Ottenstein, J. Chromatogr. Sci., 25 (1987) 536.
67. D. M. Ottenstein, J. Chromatogr. Sci., 11 (1973) 136.
68. J. F. Palframan and E. A. Walker, Analyst, 92 (1967) 71.
69. R. K. Gilpin, M. B. Martin, and S. S. Young, J. Chromatogr. Sci., 24 (1986) 410.
70. M. A. Kaiser and C. D. Batich, J. Chromatogr., 175 (1979) 174.
71. K. Lekva, L. Kardjieva, N. Hlebarova, and V. Natan, J. Chromatogr., 212 (1981) 85.
72. M. M. Daniewski and W. A. Aue, J. Chromatogr., 150 (1978) 506.
73. W. A. Aue, M. M. Daniewski, E. E. Pickett, and P. R. McCullough, J. Chromatogr., 111 (1975) 37.
74. J. A. Jonsson, L. Mathiasson, and Z. Suprynowicz, J. Chromatogr., 207 (1981) 69.
75. P. P. Wickramanayake and W. A. Aue, J. Chromatogr., 210 (1981) 133.

76. J. Bohemen, S. H. Langer, R. H. Perrett, and J. H. Purnell, J. Chem. Soc., (1960) 244.
77. D. M. Ottenstein, J. Gas Chromatogr., 6 (1968) 129.
78. Z. Suprynowicz, A. Waksmundzi, and W. Rudzinski, J. Chromatogr., 72 (1972) 5.
79. Y. Takayamo, J. Chromatogr., 178 (1979) 63.
80. J. Nawrocki, Chromatographia, 20 (1985) 308.
81. J. Nawrocki, J. Chromatogr., 362 (1986) 117.
82. G. V. Filonenka, V. A. Tertykh, V. V. Pavlov, G. Ya Guba, and A. N. Korol, J. Chromatogr., 209 (1981) 385.
83. P. P. Wickramanayake and W. A. Aue, J. Chromatogr., 195 (1980) 25.
84. C. Gaget, D. Morel, and J. Serpinet, J. Chromatogr., 244 (1982) 209.
85. Z. Kessaissia, E. Papier, and J.-B. Donnet, J. Chromatogr., 196 (1980) 481.
86. P. Claudy, J. M. Latoffe, C. Gaget, D. Morel and J. Serpinet, J. Chromatogr., 329 (1985) 331.
87. Z. Suprynowicz and J. A. Jonsson, Chromatographia, 14 (1981) 455.
88. K. Lekova, M. Bursukova, S. Ivanichkova, N. Bojkova, and R. Petkova, J. Chromatogr., 446 (1988) 31.
89. M. B. Evans, R. Newton, and J. D. Carni, J. Chromatogr., 166 (1978) 101.
90. J. J. Kirkland, Anal. Chem., 35 (1963) 2003.
91. C. Hista, J. P. Messely, and R. F. Reschke, Anal. Chem., 32 (1960) 1930.
92. L. S. Green and W. Bertsch, J. Chromatogr., 471 (1989) 113.
93. E. Grushka (Ed.), "Bonded Phases in Chromatography", Ann Arbor Science, Ann Arbor, MI, 1974.
94. V. Rehak and E. Smolkova, Chromatographia, 9 (1976) 219.
95. D. C. Locke, J. Chromatogr. Sci., 11 (1973) 120.
96. I. Halasz and I. Sebastian, J. Chromatogr. Sci., 12 (1974) 161.
97. G. E. Pollock, D. R. Kojiro, and F. H. Woeller, J. Chromatogr. Sci., 20 (1982) 176.
98. S. Boudah, K. Abdeddaim, and M. H. Guermouche, J. High Resolut. Chromatogr., 6 (1983) 281.
99. J. N. Little, W. A. Dark, P. W. Farlinger, and K. J. Bombaugh, J. Chromatogr. Sci., 8 (1970) 647.
100. J. Nawrocki and W. A. Aue, J. Chromatogr., 456 (1988) 337.
101. W. A. Aue and P. P. Wickramanayake, J. Chromatogr., 200 (1980) 3.
102. W. A. Aue, C. R. Hastings, and S. Kapila, J. Chromatogr., 77 (1973) 299.
103. W. A. Aue, C. R. Hastings, and S. Kapila, Anal. Chem., 45 (1973) 725.
104. M. M. Daniewski and W. A. Aue, J. Chromatogr., 147 (1978) 119.
105. R. F. Moseman, J. Chromatogr., 166 (1978) 397.
106. T. R. Edgerton and R. F. Moseman, J. Chromatogr. Sci., 18 (1980) 25.
107. N. D. Petsev, G. I. Pekov, M. D. Alexandrova, and C. Dimitrov, Chromatographia, 20 (1985) 228.
108. J. F. Suprock and J. H. Vinopal, J. Assoc. Off. Anal. Chem., 70 (1987) 1014.
109. L. Ghaoul, H. Shanfield, and A. Zlatkis, Chromatographia, 18 (1984) 11.
110. R. T. Ghijsen, F. Nooitgedacht and H. Poppe, Chromatographia, 22 (1986) 201.

111. C. H. Chuang, H. Shanfield, and A. Zlatkis, Chromatographia, 23 (1987) 169.
112. B. L. Karger and E. Sibley, Anal. Chem., 45 (1973) 740.
113. J. J. Pesek and J. E. Daniels, J. Chromatogr. Sci., 14 (1976) 288.
114. J. J. Pesek and J. A. Graham, Anal. Chem., 49 (1977) 133.
115. E. Papirer, H. Balard, Y. Rahmani, A. P. Legrand, L. Facchini, and H. Hommel, Chromatographia, 23 (1987) 639.
116. Standard Practice for Packed Column Gas Chromatography, Annual Book of ASTM Standards, E 260-85, The American Society for Testing and Materials, Philadelphia, PA., Vol. 14.01, 1990.
117. J. F. Parcher and P. Urone, J. Gas Chromatogr., 2 (1964) 184.
118. A. A. Spark, J. High Resolut. Chromatogr., 2 (1979) 577.
119. R. F. Kruppa, R. S. Henly, and D. L. Smead, Anal. Chem., 39 (1967) 851.
120. S. Melendez-R and W. C. Parker, J. High Resolut. Chromatogr., 2 (1979) 580.
121. R. J. Laub, J. H. Purnell, P. S. Williams, M. W. P. Harbison, and D. E. Martire, J. Chromatogr., 155 (1978) 233.
122. R. N. Nikolov, N. D. Petsev, and A. D. Stefanova, Chromatographia, 9 (1976) 81.
123. N. D. Petsev, V. H. Petkov, and C. Dimitrov, J. Chromatogr., 114 (1975) 204.
124. L. Mathiasson, J. Chromatogr., 174 (1979) 201.
125. E. F. Sanchez, J. A. G. Dominguez, J. G. Munoz, and M. J. Molera, J. Chromatogr., 299 (1984) 151.
126. D. C. Fenimore, J. H. Whitford, C. M. Davis, and A. Zlatkis, J. Chromatogr., 140 (1977) 9.
127. R. J. Laub and J. H. Purnell, J. High Resolut. Chromatogr., 3 (1980) 195.
128. C. A. Cramers and J. A. Rijks, Adv. Chromatogr., 17 (1979) 101.
129. G. Reglero, M. Herraiz, M. D. Cabezudo, E. Fernandez-Sanchez, and J. A. Garcia-Dominguez, J. Chromatogr., 348 (1985) 327.
130. R. J. Jonker, H. Poppe, and J. F. K. Huber, Anal. Chem., 54 (1982) 2447.
131. L. S. Ettre, J. High Resolut. Chromatogr., 10 (1987) 221.
132. L. S. Ettre, Anal. Chem., 57 (1985) 1419A.
133. S. R. Lipsky in F. Bruner (Ed.), "The Science of Chromatography", Elsevier, Amsterdam, 1985, p. 257.
134. D. H. Desty, J. Chromatogr. Sci., 25 (1987) 552.
135. K. Grob and G. Grob, in L. H. Keith (Ed.), "Identification and Analysis of Organic Pollutants in Water", Ann Arbor Science, Ann Arbor, MI, 1976, p. 75.
136. W. G. Jennings (Ed.), "Applications of Glass Capillary Gas Chromatography", Dekker, New York, NY, 1981.
137. W. G. Jennings, "Comparison of Fused Silica and Other Glass Columns in Gas Chromatography", Huethig, Heidelberg, 1981.
138. M. L. Lee, F. J. Yang, and K. D. Bartle, "Open Tubular Column Gas Chromatography. Theory and Practice", Wiley, New York, NY, 1984.
139. F. I. Onuska and F. W. Karasek, "Open Tubular Column Gas Chromatography in Environmental Sciences", Plenum Press, New York, NY, 1984.
140. K. Grob, "Making and Manipulating Capillary Columns for Gas Chromatography", Huething, Heidelberg, 1986.
141. W. G. Jennings, "Analytical Gas Chromatography", Academic Press, Orlando, FL, 1987.
142. C. F. Poole and S. A. Schuette, "Contemporary Practice of Chromatography", Elsevier, Amsterdam, 1984.

143. M. L. Lee and B. W. Wright, J. Chromatogr., 184 (1980) 235.
144. W. G. Jennings, Adv. Chromatogr., 20 (1982) 198.
145. L. G. Blomberg, J. High Resolut. Chromatogr., 7 (1984) 232.
146. B. Xu and N. P. E. Vermeulen, J. Chromatogr., 445 (1988) 1.
147. A. Zlatkis, C. F. Poole, R. S. Brazell, and S. Singhawangcha, J. High Resolut. Chromatogr., 2 (1979) 423.
148. V. Pretorius, E. R. Rohwer, P. J. Apps, K. H. Lawson, and J. Giesel, J. Chromatogr., 289 (1984) 17.
149. E. R. Rohwer, V. Pretorius, and G. A. Hulse, J. High Resolut. Chromatogr., 9 (1986) 30.
150. V. Pretorius and J. C. Davidtz, J. High Resolut. Chromatogr., 2 (1979) 703.
151. S. R. Lipsky and W. J. McMurray, J. Chromatogr., 217 (1981) 3.
152. H. Saito, J. Chromatogr., 243 (1982) 189.
153. M. L. Lee, D. L. Vassilaros, L. V. Phillips, D. M. Hercules, H. E. Azumaya, J. W. Jorgenson, M. P. Maskarinec, and M. Novotny, Anal. Letts., 12 (1979) 191.
154. J. L. Marshall and D. A. Parker, J. Chromatogr., 122 (1976) 425.
155. J. G. Schenning, L. G. van der Ven, and A. Venema, J. High Resolut. Chromatogr., 1 (1978) 101.
156. R. D. Dandeneau and E. H. Zerenner, J. High Resolut. Chromatogr., 2 (1979) 351.
157. S. R. Lipsky, W. J. McMurray, M. Hernandez, J. E. Purcell, and K. A. Billeb, J. Chromatogr. Sci., 18 (1980) 1.
158. V. Pretorius and D. H. Desty, Chromatographia, 15 (1982) 569.
159. M. W. Ogden and H. M. McNair, J. High Resolut. Chromatogr., 8 (1985) 326.
160. D. A. Parker and J. L. Marshall, Chromatographia, 11 (1978) 526.
161. K. D. Bartle, B. W. Wright, and M. L. Lee, Chromatographia, 14 (1981) 387.
162. M. W. Ogden and H. M. McNair, J. Chromatogr., 354 (1986) 7.
163. B. W. Wright, P. A. Peaden, and M. L. Lee, J. High Resolut. Chromatogr., 5 (1982) 413.
164. S. R. Sumpter, C. L. Wooley, E. C. Huang, K. E. Markides, and M. L. Lee, J. Chromatogr., 517 (1990) 503.
165. K. D. Bartle, C. L. Woolley, K. E. Markides, M. L. Lee, and R. S. Hansen, J. High Resolut. Chromatogr., 10 (1987) 128.
166. L. Blomberg, J. Buijten, K. Markides, and T. Wannman, J. Chromatogr., 279 (1983) 9.
167. G. Alexander and G. A. F. M. Rutten, J. Chromatogr., 99 (1974) 81.
168. J. J. Franken, G. A. F. M. Rutten, and J. A. Rijks, J. Chromatogr., 126 (1976) 117.
169. H. T. Badings, J. J. G. van der Pol, and D. G. Schmidt, Chromatographia, 10 (1977) 404.
170. K. Grob, G. Grob, W. Blum, and W. Walther, J. Chromatogr., 244 (1982) 197.
171. B. W. Wright, M. L. Lee, S. W. Graham, L. V. Phillips, and D. M. Hercules, J. Chromatogr., 199 (1980) 355.
172. B. W. Wright, P. A. Peaden, M. L. Lee, and G. M. Booth, Chromatographia, 15 (1982) 584.
173. A. Venema, J. T. Sukkel, and N. Kampstra, J. High Resolut. Chromatogr., 6 (1983) 236.
174. R. A. Heckman, C. R. Green, and F. W. Best, Anal. Chem., 50 (1978) 2157.
175. T. L. Peters, T. J. Nestrick, and L. L. Lamparski, Anal. Chem., 54 (1982) 2397.
176. F. I. Onuska and M. E. Comba, J. Chromatogr., 126 (1976) 133.

220

177. J. D. Schieke, N. R. Comins, and V. Pretorius, J. Chromatogr., 112 (1975) 97.
178. J. D. Schieke, N. R. Comins, and V. Pretorius, Chromatographia, 8 (1975) 354.
179. J. D. Schieke and V. Pretorious, J. Chromatogr., 132 (1977) 231.
180. P. Sandra and M. Verzele, Chromatographia, 10 (1977) 419.
181. P. Sandra, M. Verstappe, and M. Verzele, J. High Resolut. Chromatogr., 1 (1978) 28.
182. J. F. G. Clarke, J. High Resolut. Chromatogr., 2 (1979) 357.
183. K. Janak, K. Tesarik, P. Bauer, and V. Kalouskova, Chromatographia, 20 (1985) 379.
184. F. I. Onuska, M. E. Comba, T. Bistricki, and R. J. Wilkinson, J. Chromatogr., 142 (1977) 117.
185. S. C. Dhanesar and C. F. Poole, Anal. Chem., 56 (1984) 2509.
186. R. C. M. de Nijs, G. A. F. M. Rutten, J. J. Franken, R. P. M. Dooper, and J. A. Rijks, J. High Resolut. Chromatogr., 2 (1979) 447.
187. S. C. Dhanesar, M. E. Coddens, and C. F. Poole, J. Chromatogr. Sci., 23 (1985) 320.
188. K. Grob and G. Grob, J. Chromatogr., 125 (1976) 471.
189. K. Grob, G. Grob, and K. Grob, Chromatographia, 10 (1977) 181.
190. K. Grob, G. Grob, and K. Grob, J. High Resolut. Chromatogr., 1 (1978) 149.
191. R. F. Arrendale, L. B. Smith, and L. B. Rogers, J. High Resolut. Chromatogr., 3 (1980) 115.
192. G. Schomburg, H. Husmann, and H. Borwitzky, Chromatographia, 13 (1980) 321.
193. L. Blomberg and T. Wannman, J. Chromatogr., 148 (1978) 379.
194. R. C. M. De Nijs, J. J. Franken, R. P. M. Dooper, J. A. Rijks, H. J. J. M. De Ruwe, and F. L. Schulting, J. Chromatogr., 167 (1978) 231.
195. T. Welsch, J. High Resolut. Chromatogr., 11 (1988) 471.
196. C. L. Woolley, K. E. Markides, and M. L. Lee, J. Chromatogr., 367 (1986) 9.
197. M. Hetem, G. Rutten, B. Vermeer, J. Rijks, L. Van de Ven, J. De Haan, and C. Cramers, J. Chromatogr., 477 (1989) 3.
198. G. Schomburg, H. Husmann, S. Ruthe, and T. Herraiz, Chromatographia, 15 (1982) 599.
199. W. Blum, J. High Resolut. Chromatogr., 9 (1986) 120.
200. K. Grob, G. Grob, W. Blum, and W. Walther, J. Chromatogr., 244 (1982) 197.
201. K. Markides, L. Blomberg, J. Buijten, and T. Wannman, J. Chromatogr., 267 (1983) 29.
202. B. W. Wright, P. A. Peaden, M. L. Lee, and T. J. Stark, J. Chromatogr., 248 (1982) 17.
203. K. E. Markides, L. Blomberg, S. Hoffman, J. Buijten, and T. Wannman, J. Chromatogr., 302 (1984) 319.
204. R. C. Kong, C. L. Woolley, S. M. Fields, and M. L. Lee, Chromatographia, 18 (1984) 362.
205. M. Hetem, G. Rutten, L. Van de Ven, J. de Haan, and C. Cramers, J. High Resolut. Chromatogr., 11 (1988) 510.
206. K. Grob, G. Grob, and K. Grob, J. High Resolut. Chromatogr., 2 (1979) 31.
207. C. L. Woolley, K. E. Markides, and M. L. Lee, J. Chromatogr., 367 (1986) 23.
208. K. E. Markides, B. J. Tarbet, C. L. Woolley, C. M. Schregenberger, J. S. Bradshaw, M. L. Lee, and K. D. Bartle, J. High Resolut. Chromatogr., 8 (1985) 378.

209. K. E. Markides, B. J. Tarbet. C. M. Schregenberger, J. S. Bradshaw, M. L. Lee, and K. D. Bartle, J. High Resolut. Chromatogr., 8 (1985) 741

210. R. J. Laub, W. L. Roberts, and C. A. Smith, J. High Resolut. Chromatogr., 6 (1983) 44.

211. K. Grob, J. High Resolut. Chromatogr., 3 (1980) 525.

212. L. Blomberg, J. Chromatogr., 138 (1977) 7.

213. K. D. Bartle, Anal. Chem., 45 (1973) 1831.

214. J. Roeraade, Chromatographia, 8 (1975) 511.

215. L. Blomberg, Chromatographia, 8 (1975) 324.

216. G. Alexander and G. A. F. M. Rutten, Chromatographia, 6 (1975) 231.

217. G. Schomburg, H. Husmann, and F. Weeke, J. Chromatogr., 99 (1974) 63.

218. G. Schomburg and H. Husmann, Chromatographia, 8 (1975) 517.

219. T. Czajkowska, Chromatographia, 15 (1982) 305.

220. G. Redant, P. Sandra, and M. Verzele, Chromatographia, 15 (1982) 13.

221. J. Bouche and M. Verzele, J. Gas Chromatogr., 6 (1968) 501.

222. M. Giabbai, M. Shoults, and W. Bertsch, J. High Resolut. Chromatogr., 1 (1978) 277.

223. K. Janak, V. Kahle, K. Tesarik, and M. Horka, J. High Resolut. Chromatogr., 8 (1985) 843.

224. K. Grob and G. Grob, J. High Resolut. Chromatogr., 8 (1985) 856.

225. C. L. Woolley, B. J. Tarbet, K. E. Markides, J. S. Bradshaw, M. L. Lee, and K. D. Bartle, J. High Resolut. Chromatogr., 11 (1988) 113.

226. B. Xu and N. P. E. Vermeulen, Chromatographia, 18 (1984) 520.

227. H. F. Yin, A. J. Huang, and Y. L. Sun, Chromatographia, 28 (1989) 313.

228. P. Sandra and M. Verzele, Chromatographia, 11 (1978) 102.

229. K. R. Kim, L. Ghaoui, and A. Zlatkis, J. High Resolut. Chromatogr., 5 (1982) 571.

230. R. C. Kong and M. L. Lee, J. High Resolut. Chromatogr., 6 (1983) 319.

231. F. Janssen and T. Kaldin, J. High Resolut. Chromatogr., 5 (1982) 107.

232. C. Spagone and R. Fanelli, J. High Resolut. Chromatogr., 5 (1982) 572.

233. R. C. Kong and M. L. Lee, Chromatographia, 17 (1983) 451.

234. B. Xu, N. P. E. Vermeulen, and J. A. M. Smit, Chromatographia, 22 (1986) 213.

235. L. G. Blomberg, J. Microcolumn Sepns., 2 (1990) 62.

236. J. S. Bradshaw, N. W. Adams, R. S. Johnson, B. J. Tarbet, C. M. Schregenberger, M. A. Pulsipher, M. B. Andrus, K. E. Markides, and M. L. Lee, J. High Resolut. Chromatogr., 8 (1985) 678.

237. E. Geeraert and P. Sandra, J. High Resolut. Chromatogr., 7 (1984) 431.

238. F. David, P. Sandra, and G. Diricks, J. High Resolut. Chromatogr., 11 (1988) 256.

239. W. Blum, J. High Resolut. Chromatogr., 8 (1985) 718.

240. S. R. Lipsky and W. J. McMurray, J. Chromatogr., 289 (1984) 129.

241. S. R. Lipsky and M. L. Duffy, J. High Resolut. Chromatogr., 9 (1986) 725.

242. W. Blum and L. Damasceno, J. High Resolut. Chromatogr., 10 (1987) 475.

243. R. Aichholz, J. High Resolut. Chromatogr., 13 (1990) 71.

244. W. Blum, J. High Resolut. Chromatogr., 9 (1986) 350.

245. P. Sandra, M. Van Roelenbosch, I. Temmerman, and M. Verzele, Chromatographia, 16 (1983) 63.

246. S. R. Lipsky and W. J. McMurray, J. Chromatogr., 239 (1982) 61.

247. K. Grob and G. Grob, J. High Resolut. Chromatogr., 6 (1983) 153.

248. L. Blomberg, J. Buijten, K. Markides, and T. Wannman, J. High Resolut. Chromatogr., 4 (1981) 578.

249. K. J. Hyver and R. D. De Veaux, J. High Resolut. Chromatogr., 12 (1989) 208.

250. B. E. Richter, J. C. Kuel, J. I. Shelton, L. W. Castle, J. S. Bradshaw, and M. L. Lee, J. Chromatogr., 279 (1983) 21.

251. B. E. Richter, J. C. Kuel, N. J. Park, S. J. Crowley, J. S. Bradshaw, and M. L. Lee, J. High Resolut. Chromatogr., 6 (1983) 371.

252. J. Buijten, L. Blomberg, S. Hoffmann, K. Markides, and T. Wannman, J. Chromatogr., 289 (1984) 143.

253. G. Schomburg, H. Husmann, S. Ruthe, and M. Herraiz, Chromatographia, 15 (1982) 599.

254. E. F. Barry, G. E. Chabot, P. Ferioli, J. A. Huball, and E. M. Rand, J. High Resolut. Chromatogr., 6 (1983) 300.

255. V. Tatar, M. Popl, M. Matucha, and M. Pesek, J. Chromatogr., 328 (1985) 337.

256. J. C. Kuei, J. I. Shelton, L. W. Castle, R. C. Kong, B. E. Richter, J. S. Bradshaw, and M. L. Lee, J. High Resolut. Chromatogr., 7 (1984) 13.

257. A. Bemgard and L. G. Blomberg, J. Chromatogr., 395 (1987) 125.

258. K. Markides, L. G. Blomberg, S. Hoffmann, J. Buijten, and T. Wannman, J. Chromatogr., 302 (1984) 319.

259. L. M. Sidisky, P. L. Stormer, L. Nolan, M. J. Keeler, and R. J. Bartram, J. Chromatogr. Sci., 26 (1988) 320.

260. Z. Juvancz, M. A. Pulsipher, B. J. Tarbet, M. M. Schirmer, R. S. Johnson, K. E. Markides, J. S. Bradshaw, and M. L. Lee, J. Microcolumn Sepns., 1 (1989) 142.

261. C. A. Rouse, A. C. Finlinson, B. J. Tarbet, J. C. Pixton, N. M. Djordjevic, K. E. Markides, J. S. Bradshaw, and M. L. Lee, Anal. Chem., 60 (1988) 901.

262. L. Bystricky, J. High Resolut. Chromatogr., 9 (1986) 240.

263. B. J. Tarbet, J. S. Bradshaw, D. F. Johnson, A. C. Finlinson, C. A. Rouse, K. Jones, S. R. Sumpter, E. C. Huang, Z. Juvancz, K. E. Markides, and M. L. Lee, J. Chromatogr., 473 (1989) 103.

264. C.-Y. Wu, C.-M. Wang, Z.-R. Zeng, and X.-R. Lu, Anal. Chem., 62 (1990) 968.

265. L. S. Ettre and J. E. Purcell, Adv. Chromatogr., 10 (1974) 1.

266. J. G. Nikelly, Sepn. Purifn. Meths., 3 (1974) 423.

267. I. Halasz and E. Heine, Adv. Chromatogr., 4 (1967) 207.

268. P. Torline, G. du Plessis, N. Schnautz, and J. C. Thompson, J. High Resolut. Chromatogr., 2 (1979) 613.

269. T. Wishousky, R. L. Grob, and A. G. Zacchei, J. Chromatogr., 249 (1982) 1.

270. R. G. Mathews, J. Torres, and R. D. Schwartz, J. Chromatogr., 199 (1980) 97.

271. R. G. Mathews, J. Torres, and R. D. Schwartz, J. Chromatogr. Sci., 20 (1982) 160.

272. R. C. M. de Nijs and J. de Zeeuw, J. Chromatogr., 279 (1983) 41.

273. J. de Zeeuw, R. C. M. de Nijs, and L. T. Henrich, J. Chromatogr. Sci., 25 (1987) 71.

274. J. de Zeeuw, R. C. M. de Nijs, J. A. Peene, and M. Mohnke, J. High Resolut. Chromatogr., 11 (1988) 162.

275. P. Van Hout, J. Szafranek, C. D. Pfaffenberger, and E. C. Horning, J. Chromatogr., 99 (1974) 103.

276. J. Chauhan and A. Darbre, J. High Resolut. Chromatogr., 4 (1981) 11.

277. A. Liberti, G. Goretti, and M. V. Russo, J. Chromatogr., 279 (1983) 1.

278. C. A. Cramers, E. A. Vermer, and J. J. Franken, Chromatographia, 10 (1977) 412.

279. I. Santa Maria, J. D. Carmi, G. R. Haddad, and M. Cifuentes, J. Chromatogr., 329 (1985) 123.

280. O. Eddib, G. Nickless, and M. Cooke, J. Chromatogr., 368 (1986) 370.

281. S. L. Mackenzie and L. R. Hogge, J. Chromatogr., 147 (1978) 388.

282. R. G. McKeag and F. W. Hougen., J. Chromatogr., 136 (1977) 308.

283. J. J. Thieke, J. H. M. van den Berg, R. S. Deelder, and J. J. M. Ramaekers, J. Chromatogr., 160 (1978) 264.

284. G. Alexander, J. High Resolut. Chromatogr., 13 (1990) 65.

285. G. Schomburg, H. Husmann, and H. Behlau, J. Chromatogr., 203 (1981) 179.

286. G. Schomburg, J. High Resolut. Chromatogr., 2 (1979) 461.

287. A. D. Broske, J. High Resolut. Chromatogr., 13 (1990) 348.

288. L. J. Anthony, R. A. Holland, and S. A. Hellner, J. High Resolut. Chromatogr., 11 (1988) 167.

289. J. Krupcik, J. Garaj, G. Guiochon, and J. M. Schmitter, Chromatographia, 14 (1981) 501.

290. L. S. Ettre, Chromatographia, 8 (1975) 291.

291. K. Grob and G. Grob, J. Chromatogr., 207 (1981) 291.

292. L. A. Jones, S. L. Kirby, C. L. Garganta, T. M. Gerig, and J. D. Mulik, Anal. Chem., 55 (1983) 1354.

293. K. Grob, G. Grob, and K. Grob, J. Chromatogr., 156 (1978) 1.

294. K. Grob, G. Grob, and K. Grob, J. Chromatogr., 219 (1981) 13.

295. I. Temmerman and P. Sandra, J. High Resolut. Chromatogr., 7 (1984) 332.

296. M. Ahnoff and L. Johansson, J. Chromatogr., 279 (1983) 75.

297. H. F. Yin, A. J. Huang, and Y. L. Sun, Chromatographia, 25 (1988) 899.

298. G. Schomburg, R. Dielmann, H. Borwitzky, and H. Husman, J. Chromatogr., 167 (1978) 337.

299. G. Schomburg, H. Husmann, and H. Borwitzky, Chromatographia, 12 (1979) 651.

300. S. Schmidt. S. Hoffmann, and L. G. Blomberg, J. High Resolut. Chromatogr., 8 (1985) 734.

301. R. C. Castells, E. L. Aranciba, and A. M. Nardillo, J. Chromatogr., 504 (1990) 45.

302. O. Wicarova, J. Novak, and J. Janak, J. Chromatogr., 51 (1970) 3.

303. S. K. Poole and C. F. Poole, J. Chromatogr., 500 (1990) 329.

304. J. R. Conder and C. L. Young, "Physicochemical Measurement by Gas Chromatography", Wiley, New York, NY, 1979.

305. J. R. Conder, J. Chromatogr., 248 (1982) 1.

306. R. L. Martin, Anal. Chem., 33 (1961) 347.

307. B. R. Kersten and C. F. Poole, J. Chromatogr., 399 (1987) 1.

308. R. C. Castells, A. M. Nadrillo, E. L. Arancibia, and M. R. Delfino, J. Chromatogr., 259 (1983) 413.

309. A. Bemgard and L. Blomberg, J. Chromatogr., 502 (1990) 1.

310. V. G. Berezkin, Adv. Chromatogr., 27 (1987) 1.

311. A. N. Korol, J. Chromatogr., 67 (1972) 213.

224

312. J. Serpinet, Anal. Chem., 48 (1976) 2264.
313. J. R. Conder, N. K. Ibrahim, G. J. Rees, and G. A. Oweimreen, J. Phys. Chem., 89 (1985) 2571.
314. R. N. Nikolov, J. Chromatogr., 241 (1982) 237.
315. H.-L. Liao and D. E. Martire, Anal. Chem., 44 (1972) 498.
316. J. A. Jonsson (Ed.), "Chromatographic Theory and Basic Principles", Dekker, New York, NY, 1987.
317. K. Naito, T. Sagara, and S. Takei, J. Chromatogr., 503 (1990) 25.
318. S. K. Poole, B. R. Kersten, and C. F. Poole, J. Chromatogr., 471 (1989) 91.
319. J. R. Conder, D. C. Locke, and J. H. Purnell, J. Phys. Chem., 73 (1969) 700.
320. J. R. Conder, J. Chromatogr., 39 (1969) 273.
321. D. C. Locke, Adv. Chromatogr., 14 (1976) 87.
322. V. G. Berezkin and A. A. Korolev, Chromatographia, 21 (1986) 16.
323. V. G. Berezkin and A. A. Korolev, J. High Resolut. Chromatogr., 12 (1989) 617.
324. Z. Guoliang and C. Rixiao, Chromatographia, 29 (1990) 575.
325. M. Roth, J. Novak, P. David, M. Novotny, Anal. Chem., 59 (1987) 1490.
326. L. S. Ettre, Chromatographia, 6 (1973) 489.
327. J. K. Haken, Adv. Chromatogr., 14 (1976) 366.
328. M. V. Budahegyi, E. R. Lombosi, T. S. Lombosi, S. Y. Meszaros, Sz. Nyiredy, G. Tarjan, I. Timar, and J. M. Takacs, J. Chromatogr., 271 (1983) 213.
329. L. G. Blomberg, Adv. Chromatogr., 26 (1987) 229.
330. G. Tarjan, Sz Nyiredy, M. Gyor, E. R. Lombosi, T. S. Lombosi, M. V. Budahegyi, S. Y. Meszaros, and J. M. Tackacs, J. Chromatogr., 472 (1989) 1.
331. M. B. Evans and J. K. Haken, J. Chromatogr., 472 (1989) 93.
332. J. Novak and J. Ruzickova, J. Chromatogr., 91 (1974) 79.
333. R. V. Golovnya and D. N. Grigoryeva, Chromatographia, 17 (1983) 613.
334. S. J. Hawkes, Chromatographia, 28 (1989) 237.
335. B. R. Kersten, C. F. Poole, and K. G. Furton, J. Chromatogr., 411 (1987) 43.
336. B. R. Kersten and C. F. Poole, J. Chromatogr., 452 (1988) 191.
337. S. K. Poole, K. G. Furton, and C. F. Poole, J. Chromatogr. Sci., 26 (1988) 67.
338. S. J. Hawkes, Anal. Chem., 61 (1989) 88.
339. C. Bangjie and P. Shaoyi, Chromatographia, 25 (1988) 731.
340. E. Fernandez-Sanchez, J. A. Garcia-Dominguez, V. Menendez, and J. M. Santiuste, J. Chromatogr., 498 (1990) 1.
341. D. Messadi, F. Helaimla, S. Ali-Mokhnache, and M. Boumahraz, Chromatographia, 29 (1990) 429.
342. A. S. Said, A. M. Jarallah, and R. S. Al-Ameeri, J. High Resolut. Chromatogr., 9 (1986) 345.
343. J. Curvers, J. Rijks, C. Cramers, K. Knauss, and P. Larson, J. High Resolut. Chromatogr., 8 (1985) 607 and 611.
344. J. Lee and D. R. Taylor, Chromatographia, 16 (1983) 286.
345. A. Zhu, J. Chromatogr., 331 (1985) 229.
346. L. Podmaniczky, L. Szepesy, K. Lakszner, and G. Schomburg, Chromatographia, 21 (1986) 387.
347. C. Bangjie, G. Xijion, and P. Shaoyi, Chromatographia, 23 (1987) 888.
348. T. Wang and Y. Sun, J. Chromatogr., 390 (1987) 261, 269, and 275.
349. J. Klein and H. Widdecke, J. Chromatogr., 129 (1976) 375.

350. L. Mathiasson, J. A. Jonsson, A. M. Olsson, and L. Haraldson, J. Chromatogr., 152 (1978) 11.
351. F. Vernon and M. Rayanakorn, Chromatographia, 13 (1980) 611.
352. F. Vernon and J. B. Suratman, Chromatographia, 17 (1983) 597.
353. R. J. Smith, J. K. Haken, and M. S. Wainwright, J. Chromatogr., 334 (1985) 95.
354. R. J. Smith, J. K. Haken, M. S. Wainwright, and B. G. Madden, J. Chromatogr., 328 (1985) 11.
355. L.S. Ettre, Chromatographia, 13 (1980) 73.
356. H. Becker and R. Gnauck, J. Chromatogr., 366 (1986) 378.
357. M. S. Wainwright, C. S. Nieass, J. K. Haken, and R. P. Chaplin, J. Chromatogr., 321 (1985) 287.
358. U. Heldt and H. J. K. Koser, J. Chromatogr., 192 (1980) 107.
359. A. Yasuhara, M. Morita, and K. Fuwa, J. Chromatogr., 328 (1985) 35.
360. V. G. Berezkin and V. N. Returnsky, J. Chromatogr., 330 (1985) 71.
361. V. G. Berezkin and V. N. Returnsky, J. Chromatogr., 292 (1984) 9.
362. S. Bemgard and L. Blomberg, J. Chromatogr., 473 (1989) 37.
363. D. L. Vassilaros, R. C. Kong, D. W. Later, and M. L. Lee, J. Chromatogr., 252 (1982) 1.
364. J. Krupcik and P. Bohov, J. Chromatogr., 346 (1985) 33.
365. C. D. Bannon, J. D. Craske, and L. M. Norman, J. Chromatogr., 447 (1988) 43.
366. R. G. C. Wijesundera and R. G. Ackman, J. Chromatogr. Sci., 27 (1989) 399.
367. D. E. Wells, M. L. Gilles, and A. E. Porter, J. High Resolut. Chromatogr., 8 (1985) 443.
368. T. R. Schwartz, J. D. Petty, and E. M. Kaiser, Anal. Chem., 55 (1983) 1839.
369. H. J. Neu, N. Zell, and K. Ballschmiter, Fresenius Z. Anal. Chem., 293 (1978) 193.
370. R. Aderjan and M. Bogusz, J. Chromatogr., 454 (1988) 345.
371. F. Pacholec and C. F. Poole, Anal. Chem., 54 (1982) 1019.
372. F. Pacholec and C. F. Poole, J. Chromatogr., 302 (1984) 289.
373. G. L. Hall, W. E. Whitehead, C. R. Mower, and T. Shibamoto, J. High Resolut. Chromatogr., 9 (1986) 266.
374. L. N. Zotov, G. V. Golovkin, and R. V. Golovnya, J. High Resolut. Chromatogr., 4 (1981) 6.
375. A. Manninen, M.-L. Kuitunen, and L. Julin, J. Chromatogr., 394 (1987) 465.
376. R. Kaliszan, "Quantitative Structure Chromatographic Retention Relationships", Wiley, New York, NY, 1987.
377. L. S. Ettre, Chromatographia, 7 (1974) 41.
378. S. J. Hawkes, Chromatographia, 25 (1988) 313.
379. C. T. Peng, S. F. Ding, R. L. Hua, and Z. C. Yang, J. Chromatogr., 436 (1988) 137.
380. L. Rohrschneider, J. Chromatogr. Sci., 11 (1973) 160.
381. L. Rohrschneider, J. Chromatogr., 39 (1969) 383.
382. J. Bermejo and M. D. Guillen, Chromatographia, 17 (1983) 664.
383. J. Bermejo, C. G. Blanco, M. A. Diez, and M. D. Guillen, Chromatographia, 23 (1987) 33.
384. J. Bermejo and M. D. Guillen, Anal. Chem., 59 (1987) 94.
385. J. Bermejo and M. D. Guillen, Intern. J. Environ. Anal. Chem., 23 (1985) 77.
386. N. P. Dimov, J. Chromatogr., 360 (1986) 25.
387. L. Buydens, D. L. Massart, and D. Geerlings, J. Chromatogr. Sci., 23 (1985) 304.
388. J. Rayner, D. Wiesler, and M. Novotny, J. Chromatogr., 325 (1985) 13.

389. H. Tomkova, M. Kuchar, V. Rejholec, V. Pacakova, and E. Smolkova-Keulemansova, J. Chromatogr., 329 (1985) 113.

390. M. D. Hale, F. D. Hileman, T. Mazer, T. L. Shell, R. W. Noble, and J. J. Brooks, Anal. Chem., 57 (1985) 640.

391. P. R. McCrea, in J. H. Purnell (Ed.), "New Developments in Gas Chromatography", Wiley, New York, NY, 1973, p. 87.

392. L. Sojak and J. Krupcik, J. Chromatogr., 190 (1980) 283.

393. F. Pacholec and C. F. Poole, Chromatographia, 17 (1983) 370.

394. A.-M. Olsson, L. Mathiasson, J. A. Jonsson, and L. Haraldson, J. Chromatogr., 128 (1976) 35.

395. R. V. Golovnya and T. A. Misharina, J. High Resolut. Chromatogr., 3 (1980) 51.

396. R. V. Golovnya and B. M. Polanuer, J. Chromatogr., 517 (1990) 51.

397. L. S. Ettre, Chromatographia, 7 (1974) 261.

398. S. R. Lowry, H. B. Woodruff, T. L. Isenhour, J. Chromatogr. Sci., 14 (1976) 129.

399. J. Novak, J. Ruzickova, S. Wicar, and J. Janak, Anal. Chem., 45 (1973) 1365.

400. B. R. Kersten, S. K. Poole, and C. F. Poole, J. Chromatogr., 468 (1989) 235.

401 W. O. McReynolds, J. Chromatogr. Sci., 8 (1970) 685.

402. W. R. Supina and L. P. Rose, J. Chromatogr. Sci., 8 (1970) 214.

403. A. Hartkopf, J. Chromatogr. Sci., 12 (1974) 113.

404. E. Chang, B. de Bricero, G. Miller, and S. J. Hawkes, Chromatographia, 20 (1985) 293.

405. M. B. Evans and J. K. Haken, J. Chromatogr., 406 (1987) 105.

406. M. S. Vigdergauz and N. F. Belyaev, Chromatographia, 30 (1990) 163.

407. C. E. Figgins, T. H. Risby, and P. C. Jurs, J. Chromatogr. Sci., 14 (1976) 453.

408. C. F. Poole, S. K. Poole, R. M. Pomaville, and B. R. Kersten, J. High Resolut. Chromatogr., 10 (1987) 670.

409. M. H. Abraham, G. S. Whiting, R. M. Doherty, and W. J. Shuely, J. Chem. Soc. Perkin Trans. 2, (1990) 1451.

410. M. H. Abraham, G. S. Whiting, R. M. Doherty, and W. J. Shuely, J. Chromatogr., 518 (1990) 329.

411. J. Li, A. J. Dallas, and P. W. Carr, J. Chromatogr., 517 (1990) 103.

412. T. O. Kollie and C. F. Poole, J. Chromatogr., (1991) in press.

413. T. O. Kollie and C. F. Poole, J. Chromatogr., (1991) in press.

414. R. J. Laub, in T. Kuwana (Ed.), "Physical Methods of Modern Chemical Analysis", Academic Press, New York, NY, Vol. 3, 1983, p. 249.

415. J. H. Purnell, in F. Bruner (Ed.), "The Science of Chromatography", Elsevier, Amsterdam, 1985.

416. R. J. Laub and J. H. Purnell, J. Chromatogr., 112 (1975) 71.

417. G. J. Price, Adv. Chromatogr., 28 (1989) 113.

418. R. J. Laub, J. H. Purnell, and P. S. Williams, J. Chromatogr., 134 (1977) 249.

419. R. J. Laub, J. H. Purnell, and P. S. Williams, Anal. Chim. Acta, 95 (1977), 135.

420. M. Y. B. Othman, J. H. Purnell, P. Wainwright, and P. S. Williams, J. Chromatogr., 289 (1984) 1.

421. R. J. Laub and J. H. Purnell, Anal. Chem., 48 (1976) 1720.

422. W. Czelakowski, P. Kusz, and A. Andrysiak, Chromatographia, 22 (1986) 278.

423. R. J. Laub and J. H. Purnell, J. Chromatogr., 161 (1978) 59.

424. E. Fernandez-Sanchez, A. Frenandez-Torres, J. A. Garcia-Dominguez, J. Garcia-Munoz, V. Menendez, M. J. Molera, J. M. Santiuste, and E. Pertierra-Rimada, J. Chromatogr., 410 (1987) 13.

425. C.-F. Chien, R. J. Laub, and M. M. Kopecni, Anal. Chem., 52 (1980) 1402.

426. C.-F. Chien, M. M. Kopecni, and R. J. Laub, J. Chromatogr. Sci., 22 (1984) 1.

427. E. Matisova, E. Kovacicova, J. Garaj , and G. Kraus, Chromatographia, 27 (1989) 494.

428. P. Sandra, F. David, M. Proot, G. Diricks, M. Verstappe, and M. Verzele, J. High Resolut. Chromatogr., 8 (1985) 782.

429. R. R. Freeman and D. Kukla, J. Chromatogr. Sci., 24 (1986) 392.

430. L. Henrich, J. Chromatogr. Sci., 26 (1988) 198.

431. C. J. Cowper and A. J. DeRose, "The Analysis of Gases by Chromatography", Pergamon Press, Oxford, UK, 1983, p. 38.

432. V. G. Berezkin and S. M. Volkov, Chem. Revs., 89 (1989) 287.

433. F. Bruner, G. Crescentini, F. Mangani, and L. Lattanzi, J. Chromatogr., 473 (1989) 93.

434. F. Bruner, G. Crescentini, F. Mangani, and L. Lattanzi, J. Chromatogr., 517 (1990) 123.

435. F. J. Lopez-Garzon, I. Fernandez-Morales, and M. Domingo-Garcia, Chromatographia, 23 (1987) 97.

436. G. Crescentini, F. Managani, A. R. Mastrogiacomo, and P. Palma, J. Chromatogr., 392 (1987) 83.

437. S. P. Boudreau and W. T. Cooper, Anal. Chem., 59 (1987) 353.

438. F. J. Lopez-Garzon and M. Domingo-Garcia, Chromatographia, 21 (1986) 447.

439. M. Domingo-Garcia, F. J. Lopez-Garzon, R. Lopez-Garzon, and C. Moreno-Castilla, J. Chromatogr., 324 (1985) 19.

440. T. Paryjczak, "Gas Chromatography in Adsorption and Catalysis", Wiley, Chichester, 1986.

441. E. F. Meyer, J. Chem. Educn., 57 (1980) 120.

442. J. R. Conder and C. L. Young, "Physicochemical Measurements by Gas Chromatography", Wiley, New York, NY, 1979, p. 430.

443. J. A. Jonsson and P. Lovkvist, J. Chromatogr., 408 (1987) 1.

444. J. R. Conder, S. McHale, and M. A. Jones, Anal. Chem., 58 (1986) 2663.

445. F. Bruner, G. Crescentini, and F. Mangani, Chromatographia, 30 (1990) 565.

446. E. Smolkova, L. Feltl, and J. Zima, Chromatographia, 12 (1979) 463.

447. C. L. Guillemin, J. Chromatogr., 158 (1978) 21.

448. C. L. Guillemin, M. Deleuil, S. Cirendini, and J. Vermont, Anal. Chem., 43 (1971) 2015.

449. W. K. Al.-Thamir, J. H. Purnell, and R. J. Laub, J. Chromatogr., 188 (1980) 79.

450. G. Guiochon and C. L. Guillemin, "Quantitative Gas Chromatography for Laboratory Analyses and On-Line Process Control", Elsevier, Amsterdam, 1988, p. 213.

451. K. Naito, N. Ohwada, S. Moriguchi, and S. Takei, J. Chromatogr., 330 (1985) 193.

452. W. Engewald, J. Porschmann and T. Welsch, Chromatographia, 30 (1990) 537.

453. R. N. Nikolov, J. Chromatogr., 446 (1988) 221.

454. C. S. G. Phillips and C. D. Scott, in J. H. Purnell (Ed.), "Progress in Gas Chromatography", Wiley, New York, NY, 1968, p. 121.

455. K. Naito, M. Endo, S. Moriguchi, and S. Takei, J. Chromatogr., 253 (1982) 205.

228

456. Th. Noij, J. A. Rijks, and C. A. Cramers, Chromatographia, 26 (1988) 139.
457. W. R. Betz and W. R. Supina, J. Chromatogr., 471 (1989) 105.
458. N. V. Kovaleva and K. D. Shcherbakova, J. Chromatogr., 520 (1990) 55.
459. A. Di Corcia and A. Liberti, Adv. Chromatogr., 14 (1976) 305.
460. A. Di Corcia, R. Samperi, E Sabestiani, and C. Sererini, Chromatographia, 14 (1981) 86.
461. A. Di Corcia, A. Liberti, and R. Samperi, Anal. Chem., 45 (1973) 1228.
462. A. Di Corcia and R. Samperi, Anal. Chem., 47 (1975) 1853.
463. A. Di Corcia, A. Liberti, and R. Samperi, J. Chromatogr., 122 (1976) 459.
464. F. Bruner, G. Bertoni, and P. Ciccioli, J. Chromatogr., 120 (1976) 307.
465. A. Di Corcia and M. Giabbai, Anal. Chem., 50 (1978) 1000.
466. A. Di Corcia, A. Liberti, and R. Samperi, J. Chromtogr., 167 (1978) 243.
467. X. Min and F. Bruner, J. Chromatogr., 468 (1989) 365.
468. S. P. Nandi and P. L. Walker, Sepn. Sci., 11 (1976) 441.
469. M. T. Gilbert, J. H. Knox, and B. Knaur, Chromatographia, 16 (1982) 138.
470. R. E. Kaiser, Chromatographia, 3 (1970) 38.
471. A. Zlatkis, H. R. Kaufman, and D. E. Durbin, J. Chromatogr. Sci., 8 (1970) 416.
472. O. L. Hollis, J. Chromatogr. Sci., 11 (1973) 335.
473. M. Kraus and H. Kopecka, J. Chromatogr., 124 (1976) 360.
474. H. L. Gearhart and M. F. Burke, J. Chromatogr. Sci., 15 (1977) 1.
475. O. L. Hollis, Anal. Chem., 38 (1966) 309.
476. S. B. Dave, J. Chromatogr. Sci., 7 (1969) 389.
477. G. Castello and G. D'Amato, Chromatographia, 23 (1987) 839.
478. N. M. Djordjevic, R. J. Laub, M. M. Kopecni, and S. R. Milonjic, Anal. Chem., 58 (1986) 1395.
479. N. M. Djordjevic and R. J. Laub, Anal. Chem., 60 (1988) 124.
480. M. A. Hepp and M. S. Klee, J. Chromatogr., 404 (1987) 145.
481. G. Castello and G. D'Amato, J. Chromatogr., 243 (1982) 25.
482. G. Castello and G. D'Amato, J. Chromatogr., 254 (1983) 69.
483. R. van Wijk, J. Chromatogr. Sci., 8 (1970) 418.
484. J. M. H. Daeman, W. Kankelman, and M. E. Hendricks, J. Chromatogr. Sci., 13 (1975) 79.
485. T. G. Andronikashvili, V. G. Berezkin, N. A. Nadiradze, and L. Ya. Laperashvili, J. Chromatogr., 365 (1986) 269.
486. S. Nitz, F. Drawert, M. Albrecht, and U. Gellert, J. High Resolut. Chromatogr., 11 (1988) 322.
487. J. Roeraade, S. Blomberg, and H. D. J. Pietersma, J. Chromatogr., 356 (1986) 271.
488. B. V. Burger and Z. Munro, J. Chromatogr., 262 (1983) 95.
489. R. Bonmati, G. Chapelt-Letourneux, and G. Guiochon, Sepn. Sci. Technol., 19 (1984) 113.
490. A. Zlatkis and V. Pretorius (Eds.), "Preparative Gas Chromatography", Wiley, New York, NY, 1971.
491. E. Grushka (Ed.), "Preparative-Scale Chromatography", Dekker, New York, NY, 1989.
492. M. P. Zabokritsky, V. P. Chizhov, and B. A. Rudenko, J. High Resolut. Chromatogr., 8 (1985) 170.
493. G. Schomburg, H. Kotter, D. Staffels, and W. Reissig, Chromatographia, 19 (1985) 382.
494. D. A. Leathard and B. C. Shurlock, "Identification Techniques in Gas Chromatography", Wiley, New York, NY, 1970, p. 191.
495. G. Magnusson, J. Chromatogr., 109 (1975) 393.

496. A. A. Casselman and R. A. B. Bannard, J. Chromatogr., 90 (1974) 185.
497. H. T. Badings, J. J. G. van der Pol, and J. G. Wassink, Chromatographia, 8 (1975) 440.
498. A. B. Attygale and E. D. Morgan, Anal. Chem., 55 (1983) 1379.
499. S. T. Adam, J. High Resolut. Chromatogr., 10 (1987) 369.
500. G. W. A. Rijnders, Adv. Chromatogr., 3 (1966) 215.
501. K. P. Hupe, Chromatographia, 1 (1968) 462.
502. M. Verzele, in E. S. Perry (Ed.), "Progress in Separation and Purification", Wiley, New York, NY, 1 (1968) 83.
503. B. Roz, R. Bonmati, G. Hagenbach, P. Valentin, and G. Guiochon, J. Chromatogr. Sci., 14 (1976) 367.
504. J. R. Conder, J. Chromatogr., 256 (1983) 381.
505. J. Albrecht and M. Verzele, J. Chromatogr. Sci., 8 (1970) 586.
506. J. Albrecht and M. Verzele, J. Chromatogr. Sci., 9 (1971) 745.
507. C. E. Reese and E. Grushka, Chromatographia, 8 (1975) 85.
508. R. T. Ghijsen and H. Poppe, J. High Resolut. Chromatogr., 11 (1988) 271.
509. S. Blomberg and J. Roeraade, J. Chromatogr., 394 (1987) 443.
510. B. V. Burger, M. Le Roux, and W. J. G. Burger, J. High Resolut. Chromatogr., 13 (1990) 777.

CHAPTER 3

INSTRUMENTAL ASPECTS OF GAS CHROMATOGRAPHY

3.1   INTRODUCTION

The principal function of the gas chromatograph is to provide those conditions required by the column for achieving a separation without lowering the performance of the column in any way. As can be seen from Figure 3.1 this means providing a regulated flow of carrier gas to the column, an inlet system to vaporize and mix the sample with the carrier gas, a thermostated oven to optimize the temperature for the separation, an on-line

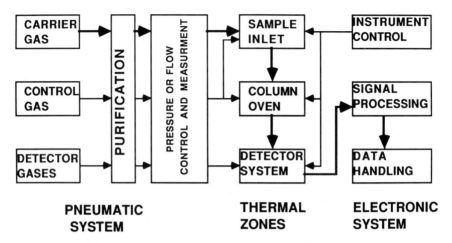

Figure 3.1 Schematic diagram of the principal components of a gas chromatograph. The bold line shows the path taken by sample and carrier gas resulting in the production of a chromatogram. The thin lines represent support and control functions.

detector to monitor the separation and associated electronic components to control and monitor instrument conditions, and to record, manipulate and format the chromatographic data [1-7]. Instruments differ in their degree of sophistication and automation, the tolerance within which the experimental condition can be controlled, and the provisions made for recording the chromatogram. Although the basic components remain the same, open tubular columns place greater demands on instrument performance than packed columns. This is due to the lower sample capacity, lower carrier gas flow rate, and faster detector response time required for optimized operation.

3.2 PNEUMATIC SYSTEMS

A supply of gases in the form of pressurized cylinders is required for the carrier gas and, depending on the instrument configuration, perhaps also for the detector, for operating pneumatic controls such as switching valves and for providing automatic cool-down by opening the oven door on command from a microprocessor. Each cylinder is fitted with a two-stage pressure regulator for coarse pressure and flow control. In most instruments provision is also made for secondary fine tuning of pressure and gas flow as well as for gas purification. The carrier gas flow is directed through a molecular sieve trap to remove moisture and low molecular weight hydrocarbons and possibly through an additional trap to remove oxygen. Oxygen causes degradation of some stationary phases, shortens filament lifetime for thermal conductivity detectors and yields unstable baselines with the electron-capture detector. Water shares some of the above properties as well as being a strong deactivating agent which can cause poor reproducibility of retention times in gas-solid chromatography. To minimize contamination high purity gases are used combined with additional chemical and/or catalytic gas purifying devices to maintain a contamination level at or below the one part per million level. Commercial gas purifying devices with plastic bodies and rubber O-ring seals may result in contamination of previously pure gases due to their permeability to atmospheric gases [8]. All metal or glass devices with metal fittings are preferred. The carrier gas then enters the pneumatic section of the gas chromatograph, thermostated to prevent drift, in which pressure regulators, flow controllers and, perhaps, additional gas purifying traps and particle filters are housed. The pressure regulators are usually of the diaphragm type and should have a metal diaphragm to minimize contamination of the carrier gas with organic impurities and to minimize the ingress of atmospheric air, particularly when open tubular columns are used [9].

Fuel gases required by flame-based detectors need only be coarsely controlled. Common arrangements include a pressure regulator combined with either a calibrated restrictor or needle valve. The carrier gas flow will normally be controlled to a higher precision using either pressure or flow controllers. Flow control is widely used for packed columns while pressure control is more common with open tubular columns. Poor retention reproducibility with a pressure-regulated system due to change in the column backpressure over time and from variations in atmospheric pressure can be a problem. Also, during temperature programmed separations the carrier gas velocity in a pressure-regulated system will decline with increasing temperature primarily due to increasing carrier gas viscosity. This may cause baseline drift and changes in response factors for concentration sensitive detectors. Column performance may be degraded and the separation time increased in server cases if the carrier gas velocity becomes too slow towards the end of a temperature programmed separation. The use of makeup gas at the column exit

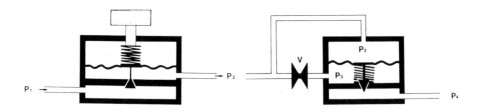

Figure 3.2   Schematic diagram of a pressure regulator and mechanical mass flow controller. P_1 and P_2 are the inlet and outlet pressure of the pressure regulator, respectively. P_2, P_3 and P_4 are the inlet, reference, and column head pressure for the flow controller and V is a needle valve used to set P_3 to the desired value.

with open tubular columns can improve the stability of concentration sensitive detectors, although sensitivity is reduced by sample dilution.

A simplified diagram of a pressure regulator and mechanical flow controller of the diaphragm type are shown in Figure 3.2 [5-7]. The pressure regulator maintains a constant pressure drop across an orifice by adjusting the flow of carrier gas through the orifice. The set point pressure is increased by increasing the tension on the spring by means of a threaded screw, which in turn depresses the diaphragm and increases flow through the orifice. An increase in the backpressure at the outlet will act on the diaphragm to reduce flow at the orifice and maintain a constant pressure differential across the orifice. The flow controller functions as a differential pressure regulator. At a constant inlet pressure and temperature it will deliver a constant mass flow that is independent of the system backpressure, P_4. For a given set point the mass flow rate through the orifice depends on the pressure differential (P_2 - P_3) pushing the membrane down and the opposing force due to tension in the spring. Any change in the downstream pressure, P_4, unbalances the diaphragm that controls the opening of the variable orifice to maintain a constant pressure differential across the needle value, V. This maintains a constant volumetric flow of carrier gas through the flow controller.

Electronic flow controllers are becoming more popular on modern instruments and are essential where chromatographic parameters are entered through a keyboard [4-6,10,11]. They are more complicated than mechanical flow controllers. One type uses a thermal flow sensor to compare temperature differences upstream and downstream of a heated filament in the gas. This signal is compared with a reference voltage equivalent to the desired flow and any imbalance is used to adjust a control valve. Other devices use semiconductor pressure sensors to monitor the pressure changes brought about by the laminar flow of a gas through a capillary. Electronic flow controllers with a range of 1-500 ml/min are capable of an accuracy of $\pm$1-2% full scale and a repeatability of the set point of $\pm$0.2% or better [6,10,11].

Soap-film flow meters have long been the popular method for measuring gas flow rates in the laboratory primarily because of their simplicity and low cost. The gas flow to be measured enters at the base of a calibrated glass tube where a soap film is introduced by squeezing a bulb containing soap solution. The time required for the film to move between two calibrated marks is noted and the flow rate calculated. Using a tube with three segments of increasing diameter permits the convenient measurement of flow rates in the 1-, 10-, and 100-ml/min ranges. The accepted accuracy of this type of flow meter is about $\pm$1%.

3.3 THERMAL ZONES

The column oven is generally a forced circulation air thermostat of sufficient size to allow comfortable installation of the longest columns normally used. In the design of a column oven it is important to ensure a uniform temperature throughout the column coil region. The temperature uniformity depends on the geometry of the oven, the location of the heater and sensor, and the pattern of mixing and circulation of air. A temperature stability of $\pm$0.1°C in time and $\pm$1.0°C in space are reasonable requirements, particularly for use with open tubular columns, although a number of commercial instruments do not meet this standard [3]. The error in the set point between various instruments may be much larger, for example, $\pm$3°C, although the precision of the set point for each instrument may be adequate. Poor column temperature stability has been identified as a distinct source of peak distortion or splitting with open tubular columns [12-14]. In severe cases, christmas-tree-like peaks are observed. When a column experiences oscillating temperatures, the distribution constants of solutes between the mobile and stationary phases will fluctuate, and this is believed to be responsible for the observed peak distortion. Splitting is most likely to occur when fused silica capillary columns are used, since their thin walls rapidly transmit temperature fluctuations from the oven to the phase boundary, allowing little time for cross-wall thermal averaging to dampen the oscillation. By keeping temperature fluctuations around 0.05% of the actual oven temperature, peak splitting is practically eliminated. Most modern gas chromatographs utilize microprocessors, sensors and proportional heating networks to maintain a stable isothermal temperature and to control the initial temperature lag, the linearity of the program rate, and the final temperature overshoot in temperature programmed gas chromatography.

The temperature operating range for column ovens is from about -100°C to about 450°C. For subambient temperature operation cryogenic cooling is required, using the boil over vapors from liquid nitrogen or carbon dioxide, which is mixed with air and then circulated at high velocity within the oven enclosure. Special instruments designed for high temperature gas chromatography are required for separations in the range 350-450°C. The range for normal temperature operation is 40 to 350°C. Typical linear temperature program rates are 0.1 to 1°C in 0.1°C/min increments and 1-30°C/min in 1°C/min increments. Rapid cool down of the column oven is achieved by opening the oven door to allow heat dissipation either manually or with an automated system. The actual heat-up and cool-down rates are limited by the power supply and the thermal mass of the oven.

Other thermal zones which should be thermostated separately from the column oven include the injector and detector ovens. These are generally insulted metal blocks heated by cartridge heaters controlled by sensors located in a feedback loop with the power supply. Detector blocks are usually maintained at a temperature selected to minimize detector contamination from condensation of column bleed or sample components and to optimize the response of the detector to the sample. The requirements for injectors may be different depending on the injector design and may include provision for temperature program operation.

In many cases it is possible to convert an older packed column gas chromatograph to accept open tubular capillary columns [15-18]. Normally this will involve changes to the pneumatic components to provide a stable pressure-regulated supply of carrier gas, the addition of splitter gas (if a splitter is installed) and detector makeup gas, a new injector for split, splitless or cold on-column injection and additional minor changes to plumbing, fittings, etc., to accommodate the added components. Before attempting such a conversion the response time of the detector electronics should be determined to ascertain the adequacy of time constants. The temperature stability of the oven in both time and space should also be determined to ensure the absence of temperature gradients and cycling or overshooting of preset or programmed temperatures, which would result in poor column performance even if the conversion was carried out successfully. Several companies offer kits for converting packed column instruments to capillary column use.

3.4 SAMPLE INLETS

The sample inlet provides the means by which the sample is volatilized and mixed with carrier gas prior to the start of the separation. Ideally the above processes should be achieved without degradation of the separation potential of the column; in the absence of thermal degradation, adsorption or rearrangement of sample components; without discrimination of sample components by boiling point, polarity or molecular weight; and with quantitative recovery for both trace and major sample components. It is also preferable that changes in the column operating conditions should not influence the sampling process. It is often difficult to meet all of these requirements for a wide range of samples and different column types used in gas chromatography. Frequently a compromise between desirable and less desirable properties has to be made in practice. For the same reason there is no single ideal sample inlet for gas chromatography and the choice of inlet system has to be based on the requirements for a particular analysis.

3.4.1 Syringe Handling Techniques

The most common method of introducing samples into a gas chromatographic inlet is by means of a microsyringe. Typically, this consists of a calibrated glass barrel with a close fitting metal plunger, which is used to dispense a chosen volume of sample by displacement through the syringe needle. Gas-tight syringes are available for injecting gases and vapors with Teflon-tipped plungers for improved sealing of the plunger with the syringe barrel against the backpressure created by the inlet pressure of the injector. However, microsyringes may cause problems that have frequently gone unnoticed. The accuracy of quantitative injection is known to depend on the rate of sample introduction, syringe dead volume, heating of the syringe needle by the injector and sample handling techniques. The most common syringe injection techniques are summarized in Table 3.1. Packed columns are very forgiving of poor injection techniques and most of the methods in Table 3.1 will work quite well, with the solvent flush method being the most popular. In this case injection volumes are typically on the order of 1 to 5 microliters; sufficiently large to minimize many of the problems mentioned above. Although large volume injection techniques have been devised for trace analysis using open tubular columns, the injection of comparatively small volumes, typically 0.1 to 2.0 microliters, is more common. In this range needle dead volumes, sample adhering to the outside wall of the needle, and backflushing of sample past the plunger can represent significant contributions to poor injection precision [19,29].

During the injection process the true injection volume may be difficult to determine with the required accuracy for calibration purposes. At the time of injection the sample volume delivered to the inlet is equivalent to the calibrated amount determined by the graduations on the syringe barrel plus some fraction of the sample volume retained in the needle. The correction for the needle volume is difficult to determine for vaporizing injectors, since it will depend on the column inlet temperature and pressure and in a more complex manner, on the sample concentration [21]. After injection, estimating the non-injected sample volume by drawing the remaining sample up into the syringe barrel is subject to error due to loss of vapors through the syringe needle while equilibration to ambient conditions occurs. For vapor samples adsorption on the syringe barrel can be a substantial source of low sample recovery. This problem can often be solved by the method of successive re-injection [22]. In this case the desired volume of sample is injected without removing the syringe from the injection port. Immediately afterwards an equal volume of carrier gas is withdrawn into the syringe and then re-injected some time later after the peaks in the first chromatogram have been integrated. In most cases two or

TABLE 3.1

SUMMARY OF METHODS USED FOR SAMPLE INTRODUCTION BY SYRINGE

Method	Principle
Filled needle	Sample is taken up into the syringe needle without entering the barrel. Injection is made by placing the syringe needle into the injection zone. No mechanical movement of the plunger is involved and the sample leaves the needle by evaporation.
Cold needle	Sample is drawn into the syringe barrel so that an empty syringe needle is inserted into the injection zone. Immediately the sample is injected by depressing the plunger. Sample remaining in the syringe needle leaves by evaporation.
Hot needle	Injection follows the general procedure described for the cold needle method except that prior to depressing the plunger the needle is allowed to heat up in the injection heater for 3-5 seconds.
Solvent flush	A solvent plug is drawn up by the syringe ahead of the sample. The solvent and sample may or may not be separated by an air barrier. The injection is usually made as indicated in the cold needle method. The solvent is used to push the sample out of the syringe.
Air flush	As for solvent flush, except that an air plug is used rather than a solvent plug.

three injections are sufficient to give adequate recovery of the sample by summing the integrated peak areas for each injection.

The volumetric accuracy of microsyringes when using the solvent flush injection technique is influenced by operating parameters, such as the speed of injection and viscosity of the sample solution, as well as design factors, such as the dead volume of the syringe needle and the tightness of the fit between the plunger and barrel [23-25]. During slow injection a film of sample is formed in the air segment between the sample and flush solvent. This film is taken up by the solvent plug and dispensed backwards and diluted by the flush solvent. During rapid injection the sample plug breaks up due to excessive shear forces and spreads as small droplets which are mixed with the solvent plug. For fast injections no improvement in accuracy can be expected by using the solvent flush method. The accuracy of the sample volume injected can only be improved by reducing the dead volume of the syringe.

A common problem encountered in sample introduction by
syringe into hot vaporizing injectors used with open tubular
columns is sample discrimination, as illustrated in Figure 3.3
[24]. The sample leaves the syringe and enters the vaporizer as a
stream of droplets, formed by the movement of the plunger and by
evaporation of the remaining sample from the syringe needle. It is
at this evaporation stage that discrimination is most likely to
occur; the solvent and more volatile sample components distill
from the syringe needle at a greater rate than the less volatile
components. Consequently, the sample reaching the column is not
identical in composition with the original sample solution; it
contains more of the most volatile and less of the least volatile
components of the sample. The portion of sample remaining inside
the syringe needle decreases as the sample volume injected
increases, and may be hardly noticed with packed columns, but it
may be critical with open tubular columns, where the sample volume
can be on the same order as that of the syringe needle. The "hot
needle" and "solvent flush" techniques are about equally effective
in minimizing discrimination and are generally preferred over the
other syringe handling methods denoted in Table 3.1

Adsorption or catalytic decomposition of labile substances
by the syringe needle can be a problem for some compounds using
hot vaporizing injectors [25]. For open tubular columns
deactivated fused silica syringe needles and cold on-column
injection techniques are used to minimize this problem.
Alternatively, syringes fitted with a needle shroud for cold-
needle injection can be used [26].

3.4.2 Packed Column Inlets

For packed columns injection of the sample in solution is made
with a microliter syringe though a silicone rubber septum into a
glass liner or the front portion of the column, which are heated
and continuously swept by carrier gas, Figure 3.4. When injection
is made in the on-column mode, the column is pushed right up to
the septum area and the column end within the injector is packed
with glass wool. Ideally, the tip of the syringe needle should
penetrate the glass wool filling and just reach the surface of the
column packing. For flash vaporization the sample is injected into
a low dead volume, glass-lined chamber, mixed with carrier gas,
and flushed directly onto the column. Whichever technique is used,
the injector must meet certain specifications. Firstly, it must

150

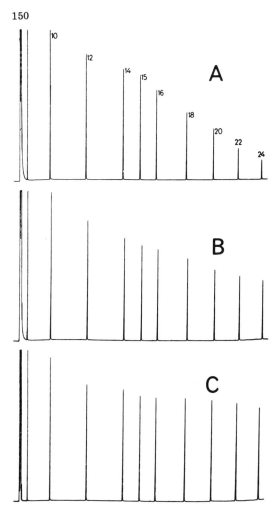

Figure 3.3 Discrimination of n-alkanes depending on injection technique. A, filled needle; B, hot needle; C, cold on-column. A and B were obtained using a 1:40 split. Sample: n-alkanes C_{10}-C_{24}, 1:10,000 in hexane. (Reproduced with permission from ref. 24. Copyright Dr. Alfred Huethig Publishers).

have sufficient thermal mass to rapidly vaporize the sample. The incoming carrier gas is usually preheated by directing its flow through a section of the injector block. This avoids possible condensation of the vaporized sample upon mixing with the cool carrier gas. The injector block should have variable temperature control and a temperature range of approximately 25-400°C. Unless dictated otherwise by the thermal instability of the sample, an injection temperature 50°C greater than the maximum column oven

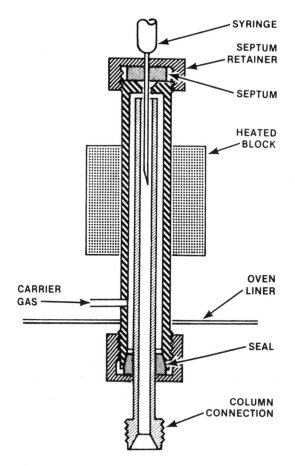

Figure 3.4 An example of a flash vaporization inlet for packed
column gas chromatography. (Reproduced with permission from
Hewlett-Packard Co.)

temperature used in the analysis is generally adopted. Because
high injection temperatures are frequently used, septum bleed may
give rise to an unstable baseline or the appearance of ghost peaks
in the chromatogram [27,28]. Various solutions to this problem are
available; low-bleed septa with good resealability, a finned
septum holder that allows cooling of the septum or a septum purge
device can be used. Many instruments incorporate a septum purge
particularly for injectors used with open tubular columns. Several
arrangements can be used for the septum purge, but in all cases a
portion of the carrier gas, or an auxiliary gas, is forced to flow
across the face of the septum and out through an adjacent orifice.

Volatile samples are injected by a gas-tight syringe in the normal way or more commonly by a gas sampling valve or by-pass injector [5,7,29-31]. Syringes are excellent for qualitative analysis and may give reliable quantitative results for large sample volumes. They should not used for quantitative analysis with small sample volumes unless a suitable internal standard of similar physical properties is available. Valves of the rotary type with external sample loops or piston or membrane valves are generally preferred for analyzing gases and volatile organic compounds. With the rotary valves, for example, the gas sample is introduced into a defined volume at a known temperature and pressure (usually ambient) and the carrier gas then sweeps the entire contents of this volume into the column by rotating the valve body, such that the carrier gas is rerouted through the sample loop. The defined volume (sample loop) can be changed, usually in the range 0.1 to 5 ml. With multifunctional valves continuous sampling is possible; the sample loops store individual samples which can be analyzed sequentially at a later time [30]. For the most accurate work it may be necessary to thermostat the injection valve in its own oven. Adsorption of the sample on the surface of the valve and sample loop can lead to systematic errors in quantitative analysis [29]. For corrosive samples a system based on a Deans switch which allows transfer of a known sample volume to the carrier gas stream without contact with the valve body or sample loop, may be beneficial [31].

Solid samples are usually dissolved in a suitable solvent and injected as described for liquids. Alternatively, the samples can be encapsulated in glass capillaries which are then pushed or dropped into the heated injection block and crushed by a mechanical device [32,33]. This form of injection is particularly useful for the analysis of trace volatiles which would be hidden in the solvent front with conventional injection techniques.

As well as the methods discussed above headspace samplers, pyrolyzers, thermal desorbers, and column switching devices could be considered as specialized sample inlet systems. These are treated separately in Chapter 8.

3.4.3 Open Tubular Column Inlets

The limited sample capacity and low carrier gas flow rates characteristic of open tubular column gas chromatography give rise to certain difficulties in sample introduction. Direct sample introduction of large volumes or concentrated samples can cause column overloading, leading to a decrease in column efficiency and/or the production of distorted peaks. The most difficult problems arise when mixtures spanning a wide concentration range or volatility range must be analyzed. No single injection method will solve all of the problems likely to be encountered. A number of different techniques are in routine use, the most important of which are split, splitless, on-column and programmed temperature vaporization injection [4,15,33-40].

For many applications split injection is the most convenient sampling method, since it allows injection of mixtures virtually independent of the choice of solvent for the sample, at any column temperature, with little risk of disturbing solvent effects or band broadening in time due to slow sample transfer from the injector to the column [34-36,41-44]. The classical hot split injector, Figure 3.5, is really an isothermal vaporizing injector, in which the evaporated sample is mixed with carrier gas and divided between two streams of different flow, one entering the column (carrier gas flow) and the other vented to the outside (split flow). The vaporizing chamber is usually constructed from a stainless steel tube lined with a removable glass or quartz liner to minimize sample contact with hot, catalytically active metal surfaces. The entire vaporizing chamber is heated to a temperature just sufficient to vaporize the least volatile components of the sample. The carrier gas at a constant pressure enters the inlet towards the bottom of the injector and passes into the inner liner at the top, flows by the column inlet (split point) and exits through the split vent. The liquid sample is introduced into the vaporizing chamber by syringe through a septum where it evaporates and mixes with the gas stream. A portion of the carrier gas and sample mixture then passes into the column inlet; the majority is routed out of the system through the split exit. A needle valve, restrictor, or a flow controller upstream of the splitter and a backpressure regulator at the split exit are used to control the split flow. An adsorption filter and/or buffer volume in the splitter line is used to prevent condensation of the sample vapors in the split line. An auxiliary flow of gas is used to purge septum bleed products and contaminants away from the vaporization chamber. Appropriate column loads are usually achieved by injecting sample volumes of 0.2 to 2.0 microliters with split

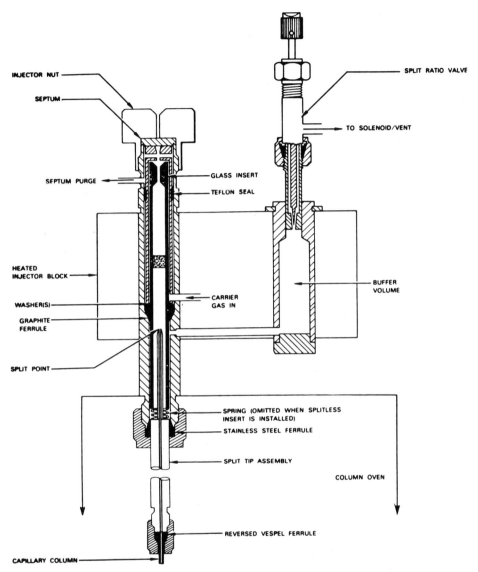

Figure 3.5 Cross-sectional view of a split-splitless injection assembly. (Reproduced with permission from Varian Associates).

ratios between 1:10 to 1:1000; 1:20 to 1:200 being the normal range of split ratios.

Unfortunately, the preset split flow ratio is only an approximate indication of the sample split ratio. The latter depends in a complex way on many parameters, including the range of sample volatilities, sample solvent, volume of sample injected,

TABLE 3.2

FACTORS AFFECTING THE REPRODUCIBILITY OF HOT SPLIT INJECTION

Parameter	Comments	Recommendations
Sample volume	The magnitude of the pressure wave depends on the sample size. It may also prolong the duration of the pressure wave. If the same amount of substance is injected in different solvent volumes, the peak areas will not necessarily be the same.	Reproduce the sample volume precisely for all injections.
Syringe handling	By slow movement of the plunger the pressure wave can be almost eliminated but discrimination will be very high.	Hot needle or solvent flush method. Rapid injection. Reproduce injection time as close as possible.
Distance between syringe needle tip and capillary inlet	Maximum amount of sample enters the column when the sample is released near the inlet of the column. This will depend on the design of the injector and the length of the syringe needle.	Reproduce penetration length of syringe needle into injector precisely.
Solvent	The molecular weight and density of the solvent influence the volume of the evaporated sample and hence the pressure wave. Solvent volatility may influence the distribution of sample between the vapor and droplet phases. The same sample amount dissolved in different solvents may produce different peak areas.	All samples should be dissolved in the same solvent for quantitative analysis.
Column temperature	Important because of the re-condensation effect. Particularly important when the column temperature is near the boiling point of the principal sample component or solvent.	Reproduce starting column temperature accurately.
Standards	Internal rather than external standards are recommended for quantitative analysis. If standard additions are used for calibration all the parameters listed in this table must be held constant.	

syringe handling technique, injector temperature and injector volume. For mixtures containing sample components of unequal volatility split injection discriminates against the less volatile sample components due to selective vaporization from the syringe needle (section 3.4.1) and from incomplete sample vaporization and inhomogeneous mixing of sample vapors with the carrier gas in the vaporization chamber. When the sample is vaporized it generates an instantaneous pressure pulse and rapid change in the viscosity of the carrier gas sample mixture, altering the flow of gas between the column and split line in an irreproducible manner. The residence time of the sample in the vaporization chamber is insufficient for the transfer of sufficient heat to the sample to complete the vaporization process. In most cases the sample arrives at the split point only partially evaporated as a mixture of vapor and droplets of various sizes. The sample components are unlikely to be evenly distributed between the vapor and liquid phase, the latter being split to different extents, resulting in discrimination. Discrimination is the principal cause of difficulties in quantitative analysis. Reproducing all aspects of the injection process is very important for accurate quantitation and some guidelines for performing hot (classical) split injection are summarized in Table 3.2. In general, hot split injection has been used as the preferred sampling technique for qualitative analysis under high resolution conditions, and less frequently for quantitative analysis.

An explosion-like evaporation model has been proposed to further exemplify the processes occurring in a split injector [42-44]. According to this model, the amount of sample reaching the column is dependent upon the magnitude of the pressure wave, the time taken by the pressure wave to reach a maximum and the column temperature. The pressure wave, caused by sample evaporation, fills the capillary column with a portion of the sample vapor followed by a period when the pressure falls back to normal. During this second period little sample enters the column and most of the sample vapor is vented through the split exit. The real function of the preset split ratio is not the one expected; in actuality, it controls the sample split ratio by its influence on the magnitude of the pressure wave and the time to re-establish normal flow conditions. Column temperature influences the split ratio through a sample recondensation mechanism. Sample recondensation greatly reduces the volume of sample vapor in the cooled column inlet, creating a zone of reduced pressure that sucks in further amounts of sample vapor. This causes a decrease in the split ratio (i.e., an increase in observed peak areas) as the column temperature is reduced and is particularly important at column temperatures near the boiling point of the solvent. At column temperatures 50-80°C below the boiling point of the solvent, recondensation is virtually complete and further decreases in column temperature have little effect.

The hot splitless injection technique was devised to overcome some of the deficiencies of split sampling for the quantitation of trace components by the introduction of relatively large sample volumes into the column [4,15,34-38,45-49]. Classical splitless injection is a hot vaporization technique subject to some of the same problems as hot split injection, namely, discrimination due to selective vaporization from the syringe needle and thermal decomposition of labile substances due to prolonged heating at relatively high temperature and contact with catalytic surfaces, including the metal syringe needle. The velocity of gas flow through a splitless injector is relatively low, and the sample is introduced into the column over a comparatively long time, relying on cold trapping and/or solvent effects to refocus the sample at the column inlet. Consequently, solutes eluted before the solvent are not refocused and peaks are generally very distorted. The majority of hot split injectors, such as Figure 3.5, can be used in the splitless mode with minor modification; changing the injector insert and turning off the split flow (usually automated by incorporating a solenoid valve in the split line) are all that is generally required.

For splitless sampling, the sample is injected through a septum by a microliter syringe into a thermostated vaporization chamber, swept by the flow of carrier gas. The split flow is turned off at the start of the injection and is restarted only at the end of the sampling period. The transport of sample vapors to the column is relatively slow since the linear velocity of the carrier gas through the vaporization chamber is low and dictated by the separation conditions for the column. During sample vaporization, transfer of vapors into the column is negligible. Thus, the vaporization chamber must be large enough to hold the whole volume of vapor produced by the evaporated sample. Sample volumes are normally limited to 0.5-2.0 microliters, so as not to overfill the injector and samples are injected rapidly to minimize discrimination, resulting from selective vaporization of the sample from the syringe needle. A chamber volume of about 1.0 ml is required for a sample volume of 2.0 microliters. The exact chamber volume will depend on the properties of the solvent and injection conditions. If the volume of the vaporization chamber is too small, sample vapors will be lost by backflushing through the septum purge exit or by deposition in the carrier gas lines. Sample should be introduced at the bottom of the chamber, close to the column, so that it fills with vapor from the bottom up, displacing the carrier gas backwards. Sample transfer to the column should be virtually complete but requires a comparatively long transfer time, for example, from several seconds up to 2.0 minutes, and varies with the injector design and experimental conditions. As a rule of thumb, the sample transfer time is approximately equivalent to twice the time required to sweep out the volume of the vaporization chamber by the carrier gas. Complete sample transfer is difficult to achieve, since the sample vapors are continually diluted with carrier gas, and some sample vapors accumulate in areas poorly swept by the carrier gas. At the end of the sample transfer period, the split flow is re-established to purge the inlet of remaining solvent vapors. If the split flow is started too soon, sample will be lost; if too late, the solvent peak will trail into the chromatogram. When the conditions are correct there will be no significant sample loss and the solvent peak will be rectangular. The nature of the column inlet refocusing mechanism will be discussed below. However, these mechanisms are effective in maintaining the chromatographic efficiency with columns of internal diameters greater than 0.3 mm but are not very effective for columns of narrower bore. In practice, when the chromatographic conditions permit, higher than normal carrier gas flow rates are used to aid sample transfer even if this results in some loss of column efficiency.

During splitless injection the sample enters the column over a period of time considerably longer than typical chromatographic peak widths and is possibly distributed over a portion of the column that is long compared to typical terminal band lengths for normal peaks. Both processes will cause a degradation of column performance, as well as possibly peak distortion, unless an effective refocusing method is employed [36,50-53]. The refocusing methods applied in splitless injection are known as cold trapping and solvent effects.

The slow introduction of sample into the column causes band broadening in time in which all solute bands are broadened equally in terms of gas chromatographic retention time, Figure 3.6, in the absence of a refocusing mechanism. Band broadening in time can be defeated or minimized by temporarily increasing the retention power of the column inlet during sample introduction. For this to be effective, the migration velocity of the sample entering the column at the start of the injection period must be sufficiently retarded to allow portions of the sample entering at the end of the injection period to catch up with it. This may be achieved by lowering the column temperature, called cold trapping, or by a temporary increase of the film thickness of the stationary phase using the injection solvent as a liquid phase, called solvent effects. Cold trapping is usually performed in the absence of solvent effects by maintaining the column inlet at a temperature not less than 15°C below the solvent boiling point to minimize solvent recondensation in the column inlet. A minimum temperature difference of about 80°C between the inlet temperature and the elution temperature of any of the solutes of interest is normally required to ensure efficient cold trapping. Note that the efficiency of cold trapping is independent of the boiling points of the solvent and solutes. Instead, it is determined by the ratio of the migration speed of the important solutes at the temperature of injection and of elution. Cold trapping is frequently used when temperature programmed analysis is required.

A prerequisite for achieving refocusing of bands broadened in time by solvent effects is recondensation of solvent in the column inlet to form a temporary liquid phase film of sufficient retention power to delay migration of the sample. The column temperature during the splitless period must be low enough to ensure that the concentration of solvent vapors in the carrier gas entering the column exceeds the saturation vapor pressure of the solvent and that sufficient solvent is recondensed so that some liquid remains in the column inlet up to the end of the sample introduction period. In most cases, these conditions will be met if the column temperature is at least 20-30°C below the solvent boiling point. Also, the solvent should provide strong interactions with the sample to avoid partial solvent trapping, which can lead to distorted peaks similar to those illustrated in Figure 3.6, as well as wet the stationary phase, otherwise a stable film will not be formed. Distortion due to band broadening

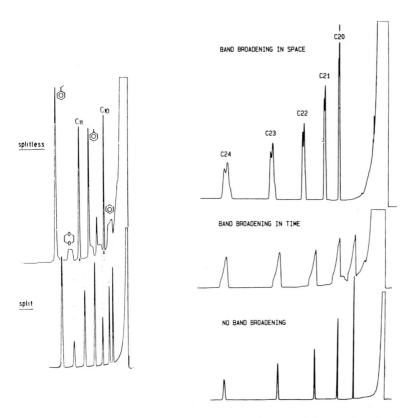

Figure 3.6 Different peak distortion problems due to band
broadening in time and band broadening in space observed during
hot splitless injection. Band broadening in space is characterized
by a broadening which grows proportionally with retention time and
may result in peak splitting that is poorly reproducible. Band
broadening in time is characterized by a constant broadening of
all peaks. Partial solvent trapping results in characteristic
chair and stool shaped peaks. (Adapted with permission from ref.
36 and 37. Copyright Dr. Alfred Huethig Publishers).

in space may be significant for solute bands eluted at
temperatures at least 50°C above the injection temperature. Band
broadening in space is discussed later for cold on-column
injection. It is frequently minimized by using retention gap
techniques.

The attractive features of splitless injection techniques
are that they allow the analysis of dilute samples without
preconcentration (trace analysis) and the analysis of dirty
samples, since the injector is easily dismantled for cleaning.
Success with individual samples, however, depends on the selection
of experimental variables of which the most important: sample

size, sample solvent, syringe position, sampling time, initial
column temperature, injection temperature and carrier gas flow
rate, often must be optimized by trial and error. These
conditions, once established, are not necessarily transferable to
a second splitless injector of a different design without
reoptimization. Also, the absolute accuracy of retention times in
splitless injection is generally less than that found for split
injection. For splitless injection the reproducibility of
retention times depends not only on chromatographic interactions
but also on the reproducibility of the sampling period and the
evaporation time of the solvent in the column inlet, if solvent
effects are employed. The choice of solvent, volume injected and
the constancy of thermal zones will all influence retention time
precision beyond those for split injection. For quantitative
analysis, with adequate injection control, the precision of
repeated sample injections is normally acceptable but the method
is subject to numerous systematic errors which may affect
accuracy. Internal standards are usually preferred to external
standards to improve accuracy.

Programmed temperature vaporization (PTV) injectors have
recently become available providing technical solutions to some of
the problems extant in the classical hot split and splitless
injectors discussed above [34,36-38,47,54-62]. They were
originally developed for large volume injections with solvent
elimination and for obviating discrimination arising from
selective sample volatilization from the syringe needle occurring
in hot vaporizing injectors. The latter problem can also be solved
by using special syringes with forced cooling of the syringe
needle [25,63]. In the PTV injector, for example Figure 3.7, the
liquid sample is introduced into the vaporizing chamber which is
maintained at a temperature below the boiling point of the
solvent. Compared to classical, hot vaporizing injectors the
vaporization chamber is much smaller, about one-tenth the size,
and is of lower mass to allow rapid heating and cooling by
circulated air. Typically, the injector can be raised from ambient
temperature to 300°C in about 20 to 30 s. The PTV can be operated
in three injection modes: split injection with a hot or cold
vaporization chamber, as a cold solvent split injector for solvent
elimination, and as a cold splitless injector for total sample
introduction. Heating of the injector is usually started a few
seconds after withdrawal of the syringe needle for split injection

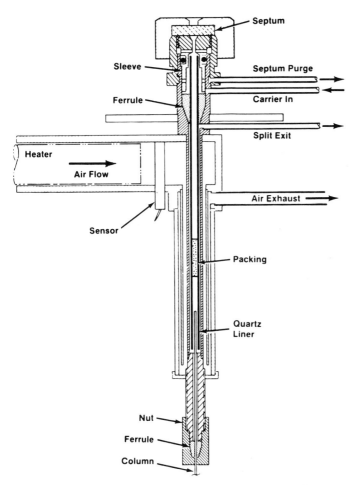

Figure 3.7 Schematic Diagram of the Perkin-Elmer PTV injector.
(Reproduced with permission from ref. 54. Copyright Dr. Alfred
Huethig Publishers).

and occurs ballistically to a temperature sufficiently high to
ensure rapid volatilization of the highest boiling sample
components. Relatively slow sample evaporation in the injector
aided by the presence of some packing material, usually glass
wool, avoids the formation of aerosols which are formed in hot
vaporizing injectors. Dilution of sample vapors with carrier gas
avoids strong recondensation effects in the column inlet. Also,
since the sample is introduced into a cold injector there is no
pressure wave effect. Compared to the classical hot vaporizing
injectors, there is far less discrimination, and the sample split
ratio and split flow ratio are usually similar. In the solvent

elimination mode, the sample is introduced into the cold injector at a temperature close to the boiling point of the solvent with the split vent open. The solvent is concurrently evaporated in the glass insert and swept through the split vent by the carrier gas. After completion of the solvent evaporation, the split vent is closed and the injector is ballistically heated to the temperature required to vaporize the sample and transfer it to the column. This injection technique permits the injection of large volumes of up to about 20 microliters rapidly and 300 microliter volumes slowly. The maximum sample sizes indicated are not definitive and depend on many factors that are not well understood. This injection technique is limited exclusively to high boiling solutes. Cold trapping and cosolvent trapping were found to be beneficial for enhancing the recovery of early eluting compounds when using the solvent elimination mode for large sample volumes [59]. For cold splitless injection the sample is introduced into the vaporization chamber at a temperature close to the solvent boiling point with the split vent closed. A few seconds after injection the injector is rapidly heated to the temperature required to complete the transfer of vapors into the column (30-90 s), the column temperature program is started, and the vaporizer is purged by opening the split vent to exhaust any solvent residues. As for hot splitless injection, cold trapping and solvent effects are employed as refocusing mechanisms. Since sample transfer takes place in two steps: transfer of solvent and volatiles followed by transfer of less volatile components at a later time, these refocusing mechanisms are generally most important for the separation of the volatile sample components.

Certainly, the precision and accuracy of the PTV techniques are generally superior to those of the classical hot split and splitless techniques and approach those obtained by cold on-column injection [34,36,37,54-57,62,64]. However, less is known concerning optimization of PTV injection and probably more parameters have to be considered than for cold on-column injection. The latter is, consequently, the preferred injection technique for most samples, except those contaminated by large amounts of involatile impurities and for headspace vapors.

Cold on-column injection differs from the vaporization techniques discussed above in that the sample is introduced directly as a liquid into the column inlet, where it is subsequently vaporized. In this way, discrimination is virtually eliminated and quantitation of mixtures of different volatilities is facilitated. Sample decomposition due to thermal and catalytic effects is minimized. An attractive feature of cold on-column injection is that the technique is easy to implement and can be optimized by controlling a few experimental variables. Its popularity has increased with the wider use of immobilized stationary phases that eliminate the problem of phase stripping. The technique can be automated by using wide bore precolumns connected to the analytical column. Also, it can be adapted to the introduction of large sample volumes by using the retention gap technique.

On-column injection requires special syringes with small external diameter stainless steel or fused silica needles, since at the point of sample release the needle must reside within the capillary column itself. The smallest needles available have an outer diameter of 0.17 mm and are used with 0.25 mm internal diameter columns. These needles are very fragile and not particularly easy to clean or fill. The larger 0.23 mm outer diameter (32 gauge) needles are more robust and easier to handle. They are suitable for use with 0.32 mm internal diameter columns. The injection device must provide a mechanism for guiding the fragile needle into the column, for positioning it at the correct height within the column and for sealing the needle passageway, or at least restricting the flow of carrier gas out through the syringe entry port. A number of different methods based on the use of valves, septa, or spring loaded O-ring seals are used for this purpose, Figure 3.8 [4,6,35,37,65,66]. Secondary cooling by circulating air is used in some designs to ensure that the temperature of the needle passageway can be controlled to avoid solvent evaporation. With efficient secondary cooling the oven temperature can be maintained well above the solvent boiling point by cooling the column inlet portion to below the oven temperature. This is important when using on-column injection in high temperature gas chromatography [34,37,67,68]. Alternatively, the column inlet can be housed in a separate injection oven which is temperature programmable and can be thermostated independently of the column oven [4].

One of the positive features of on-column injection is that it can be carried out successfully by adhering to a few simple guidelines. Sample volumes of about 0.2-2.0 microliters are injected rapidly into the column. For larger volumes a retention

256

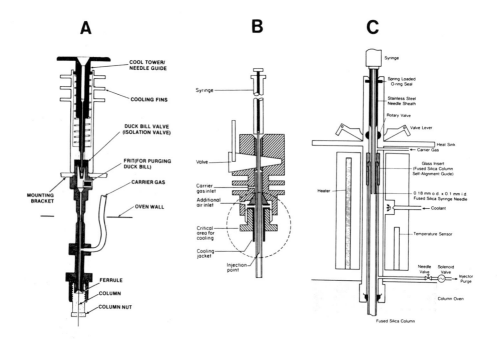

Figure 3.8 Alternative designs for cold on-column injectors. A, Injector with a duck bill valve (Hewlett-Packard), B, an injector with provision for secondary cooling of the column inlet (Carlo-Erba), and C, a temperature- programmable on-column injector with its own oven isolated from the column oven (Varian Associates).

gap, as discussed later, is required to avoid peak distortion by band broadening in space. The column temperature, or the thermostated temperature of the column inlet during injection, should be at or below the boiling point of the solvent at the carrier gas inlet pressure. After injection, the column oven should be abruptly raised to the operating temperature after a delay of 30 to 60 seconds to complete solvent evaporation, if the analysis temperature is very different from the solvent boiling point. Temperature programming is started from either of these two initial temperature settings depending on sample volatility. The primary disadvantage of on-column injection is that since the sample is introduced directly into the column, dirty samples will result in the build up of involatile residues that may affect peak shapes and the quantitative accuracy of the analysis.

The primary cause of peak distortion in on-column injection is band broadening in space. These effects are most common for large volume sample injections, for solvents that are incompatible with the properties of the sample or stationary phase, and for solutes which are eluted more than about 50°C above the injection temperature [37,69-71]. The plug of liquid leaving the syringe flows into the column inlet, closing the column bore. This plug is then pushed further into the column by the carrier gas until the liquid is spread over the surface of a considerable length of the column inlet. During the primary flow of liquid a layer of sample is left behind the plug, reducing the length of the plug until, eventually, the whole sample is spread as a stable film on the column wall. This layer of sample is called the flooded zone. Its length depends primarily on the volume of sample injected, whether the solvent wets the stationary phase or not, and the column temperature. High boiling or strongly retained solutes are distributed throughout the flooded zone with an initial band width corresponding to the zone length. At the injection temperature volatile solutes do not remain where they are deposited by the evaporating solvent but migrate with the unsaturated carrier gas. If the volatile solutes move as rapidly as, or more rapidly than the rear of the sample layer during solvent evaporation, they end up being reconcentrated at the place where the last portion of solvent evaporates from and form an initial sharp band. Such volatile solutes are not subjected to band broadening in space in the flooded zone. Solutes that migrate more slowly than the rate at which the rear of the sample layer withdraws, are only partially reconcentrated, since materials deposited at the rear of the flooded zone have migrated only part of the way to the front when the last portion of the solvent evaporates.

Provided the sample solvent wets the stationary phase film and the injected sample volume is less than 2 microliters, peak distortions attributed to band broadening in space are rarely a problem. Band broadening in space can be strongly reduced, or even eliminated, by making the injection at a column temperature slightly above the solvent boiling point at the carrier gas inlet pressure. This causes the sample vapor pressure to keep the sample plug from advancing into the column. However, quantitative results may not be as reliable as for the standard injection conditions and the peak widths of volatile sample components may be broadened due to the absence of solvent effects as there is no flooded zone. A more universal solution is the use of a retention gap [37,72-77]. A retention gap is a length of uncoated (but possibly deactivated) column, used as an inlet, which has a reduced retention power compared to the column used for the separation. Some advantages of retention gaps are that they function as guard columns to protect the analytical column from contamination by involatile residues; wide bore retention gaps permit the use of autoinjectors with regular syringe needles; and long retention gaps facilitate the injection of large sample volumes. As a general guide, typical retention gaps should be about 25-30 cm for each microliter of solvent injected and should be sufficiently inert to minimize adsorption and decomposition of the sample. It is also important that the solvent wet the retention gap surface in order to produce a film of sample liquid on the capillary wall. Most retention gaps are simply lengths of uncoated, fused silica tubing, deactivated by the same techniques as those used for column preparation, when required, to modify their solvent compatibility or activity. When a retention gap is used, refocusing of zones occurs because the retention power of the retention gap is 100 to 1000 times less than that of the column. Solutes migrate through the retention gap at temperatures well below those required to cause elution from the analytical column. Thus, they arrive at the entrance to the coated column at a temperature too low for significant migration and remain and accumulate there until a temperature is reached at which migration starts. Retention gaps are normally used when sample volumes greater than about 5 microliters and up to several hundred microliters are injected. Rotary loop injection valves with retention gaps can be used in place of syringes for injecting large sample volumes [78,79].

3.5 DETECTION DEVICES FOR GAS CHROMATOGRAPHY

Numerous methods have been described for detecting organic vapors in the effluent from a gas chromatograph [4,6,7,80-85]. Several have passed into common use and will form the backbone of this discussion. The principal methods of detection can be grouped under four headings: ionization, bulk physical property, optical and electrochemical detectors, according to the physical basis employed as the detection mechanism. Further division is possible based on the nature of the detector response. Detectors are broadly classified as universal, selective, or specific. These descriptions are applied loosely as no single detector exactly meets the dictionary obligations of these appellatives. The flame ionization and thermal conductivity detectors respond to the presence of nearly all organic compounds in the gas chromatographic effluent and are considered to be general or (near) universal detectors. Other detectors respond only to the presence of a particular heteroatom (e.g., the flame photometric, thermionic ionization, or atomic emission detectors) and are considered to be specific detectors, although element-selective detectors would be more descriptive. These detectors are able to discriminate between some property of the organic compound of interest, a heteroatom, and an organic compound lacking that property. The detection process is not specific, however, since at sufficiently high concentrations a response may be obtained from a compound lacking that property being monitored. The detector response is thus selective and may be described quantitatively by a selectivity ratio, the ratio of the detector sensitivities to two different compounds, compound classes, heteroatoms, etc. The response of detectors such as the electron-capture and photoionization detectors is also selective; these detectors are not element selective but rather structure selective. Their response range to organic compounds is broad, covering several orders of magnitude. The term "reaction detector" is occasionally applied to the electrolytic and chemiluminescence detectors, etc., to signify that prior to detection a chemical transformation of the organic compound into a species which can be detected is involved.

The signal from gas chromatographic detectors can be further characterized as mass or concentration dependent. For concentration-dependent detectors the most notable feature is that the detector response is dependent upon the flow rate through the detector (carrier gas and makeup gas, if any) and, therefore, the sensitivity of the detector is usually defined as the product of the peak area and flow rate divided by the weight of the sample. For mass-dependent detectors sensitivity is defined as the product of the peak area divided by the sample weight in grams or moles and is independent of flow rate through the detector.

Detectors are usually compared in terms of their operational characteristics defined by the minimum detectable quantity of standards, the selectivity response ratio between standards of different composition or structure, and the range of the linear portion of the detector-response calibration curve. These terms are widely used to measure the performance of different chromatographic detectors and were formally defined in section 1.8.1.

3.5.1 Ionization Detectors

At the temperatures and pressures generally used in gas chromatography the common carrier gases employed behave as perfect insulators. In the absence of conduction by the gas molecules themselves, the increased conductivity due to the presence of very few charged species is easily measured, providing the high sensitivity characteristic of ionization based gas chromatographic detectors [86]. Examples of ionization detectors in current use include the flame ionization detector (FID), thermionic ionization detector (TID), photoionization detector (PID), the electron-capture detector (ECD), and the helium ionization detector (HID). Each detector employs a different method of generating an ion current, but in all cases the signal corresponds to the fluctuations of this ion current in the presence of organic vapors which constitutes the quantitative basis of the detector operation.

3.5.1.1 Flame Ionization Detector

A nearly universal response to organic compounds, high sensitivity, long term stability, simplicity of operation and construction, low dead volume and fast signal response, and exceptional linear response range have contributed to the flame ionization detector (FID) being the most popular detector in current use. Only the fixed gases (e.g., He, Xe, H_2, N_2), certain nitrogen oxides (e.g., N_2O, NO, etc.), compounds containing a single carbon atom bonded to oxygen or sulfur (e.g., CO_2, CS_2, COS, etc.), inorganic gases (e.g., NH_3, SO_2, etc.), water and formic acid do not provide a significant detector response. Minimum sample detectability corresponds to about 10^{-13} g carbon/s with a linear response range of 10^6 to 10^7. McWilliam has provided an interesting historical account of the development of the FID on the occasion of the 25th anniversary of its invention (1983) [87].

The response of the FID results from the combustion of organic compounds in a small hydrogen-air diffusion flame which burns at a capillary jet, Figure 3.9. The carrier gas from the column is premixed with hydrogen and burned at a narrow orifice jet in a chamber through which excess air is flowing. A cylindrical collector electrode is located a few millimeters above the flame and the ion current is measured by establishing a potential between the jet tip and the collector electrode. To minimize ion recombination the potential is selected to be in the saturation region, that is, the region for which increasing the potential does not increase the ion current. Under normal conditions background currents of 10^{-14} to 10^{-13} Amperes, increasing to 10^{-12} to 10^{-5} Amperes in the presence of an organic vapor, are common. The small signal currents are amplified by a precision electrometer. The performance of the detector is influenced by experimental variables of which the most important are the ratio of air-to-hydrogen-to-carrier gas flow rates, the type (thermal conductance) of the carrier gas and individual detector geometry [88-91]. The optimum response plateau is usually fairly broad, permitting operation over a rather wide range of gas flow rates without incurring a large penalty in diminished response. The upper end of the linear response range is influenced primarily by the flame size, the bias voltage at the collector and detector geometry [92]. At increasing sample size, the sample becomes an additional fuel source to the flame, increasing the flame length. As the flame grows it eventually penetrates the interior of the collector electrode where the electric field is weaker and the ionization collection efficiency declines.

The mechanism of ion production in flames is complex and only poorly understood. The temperature and chemical composition

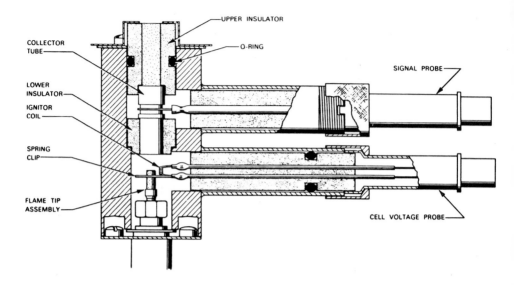

Figure 3.9   Cross-sectional view of a flame ionization detector.
(Reproduced with permission from Varian Associates).

of the flame is not uniform. In the hottest part of the flame a
temperature of about 2000°C is reached. The flame also contains a
large concentration of radicals (H, O, and OH, etc.) which in
combination with the high temperature, are primarily responsible
for the complete decomposition of the sample. The ionization
efficiency of the flame is not particularly high, perhaps a few
ions per million molecules are formed. This is consistent with the
fact that the thermal energy of the flame is too low to explain
the production of ions. It is generally believed that organic
compounds are ionized by a chemi-ionization mechanism, in which
energy released in strongly exothermic chemical reactions is
retained by one of the fragments as internal vibrational energy
and leads to ionization before thermal randomization of the energy
occurs. The ionization process would thus be a first order
reaction, and this explains the linear response of the FID. The
ionization mechanism would be placed in a low probability reaction
pathway, explaining the low ion yield. Two steps are thought to be
important in the above process: radical formation, requiring the
absence of oxygen, and chemical ionization of radicals formed by
excited atomic or molecular oxygen states. At the end of a chain
of reactions, in which methane or the methyl radical is thought to

TABLE 3.3

CONTRIBUTIONS TO THE EFFECTIVE CARBON NUMBER (ECN)

Atom	Type	ECN Contribution
C	Aliphatic	1
C	Aromatic	1
C	Olefinic	0.95
C	Acetylene	1.30
C	Carbonyl	0
C	Carboxyl	0
C	Nitrile	0.3
O	Ether	-1.0
O	Primary Alcohol	-0.5
O	Secondary Alcohol	-0.75
O	Tertiary Alcohol	-0.25
N	Amine	As for O in alcohol
Cl	On olefinic C	+0.05
Cl	Aliphatic C with 2 Cl atoms	-0.12 per Cl

be a key intermediate, the dominant ion-producing reaction is shown below:

$$CH\cdot + O* \longrightarrow CHO^+ + e^- \qquad (3.1)$$

As a consequence of the FID mechanism, each carbon atom capable of hydrogenation yields the same signal, and the overall FID response to the analyzed substance is proportional to the sum of these "effective" carbon atoms. The FID response is highest for hydrocarbons, being proportional to the number of carbon atoms, while substances containing nitrogen, sulfur, or halogens yield smaller responses, depending upon the heteroatom-carbon character and the electron affinity of the combustion products. The lower response can be explained by a complex series of recombination reactions and electron capture processes, resulting in a lower ionization current. The effective carbon number for a particular compound can be estimated by summation of the various carbon and heteroatom contributions, Table 3.3, which in turn can be used to predict relative response factors with reasonable, if not absolute, accuracy [93-95]. This simplifies the quantitation of complex mixtures where structures from mass spectrometry may be available but not standards.

The FID may be operated as an element-selective detector after minor modification. The hydrogen atmosphere flame ionization detector (HAFID) can be made selective towards organometallic compounds containing, for example, aluminum, iron, tin, chromium, and lead [96]. The detector employs a hydrogen-oxygen flame burning in a hydrogen atmosphere doped with a reagent such as silane to improve the consistency of its response. Conversely doping with an organometallic compound enables the HAFID to function as a selective silicon detector [97]. The detector is taller in height than a conventional FID and is operated with a negative potential at the collector electrode.

3.5.1.2 Thermionic Ionization Detector

The present generation of thermionic ionization detectors grew out of earlier studies on the properties of alkali-metal-doped flame ionization detectors [84,98]. Adding an alkali metal salt to a flame enhanced the response of the detector to compounds containing certain elements, such as nitrogen, phosphorus, sulfur, boron and the halogens, as well as some metals (e.g., Sb, As, Sn, Pb). However, in its earlier forms the detector was very noisy, the response depended critically upon experimental parameters and long term stability was poor. All these features contributed to making the detector difficult to use for quantitative trace analysis.

All modern thermionic ionization detectors (TID) employ a solid surface, composed of a ceramic or glass matrix, doped with an alkali metal salt, molded onto an electrical heater wire in one of two general detector configurations [84,99-101]. The detector originally proposed by Kolb, Figure 3.10, contains an electrically-heated rubidium silicate bead, situated a few millimeters above the detector jet tip and bellow the collector electrode [99]. The bead is maintained at a negative potential to minimize the loss of rubidium ions and to dampen the response of the detector in the flame ionization mode. The temperature of the source is controlled by an independent variable power supply to heat the bead to a dull red or orange glow (600-800°C). A plasma is sustained in the region of the bead by hydrogen and air support gases. The flow of hydrogen (1-5 ml/min) is too low to support a flame in the nitrogen-phosphorus mode. A normal hydrogen flow rate (approximately 30 ml/min) is used for the phosphorus-selective mode. To suppress the FID signal in the flame mode the detector jet is grounded and the negative potential of the bead deflects electrons to ground, and away from the collector electrode. The selective response to phosphorus-containing fragments occurs in

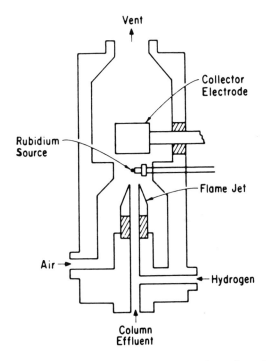

Figure 3.10 Thermionic ionization detector (Perkin-Elmer).

the region of the bead and electrons generated by this reaction are able to reach the collector electrode. In the flame mode the nitrogen response is diminished by at least an order of magnitude compared to the nonflame mode and the detector responds mainly to phosphorus-containing compounds. Both the selectivity and sensitivity of the detector are dependent upon experimental variables, principally the bead heating current, jet potential, choice of carrier gas, air and hydrogen flow rates and the bead position [84,101,102]. Operation at a fixed bead heating current will normally give poor long-term stability so that operation at a fixed background current is preferred.

The TID design proposed by Patterson consists of an alkali metal doped ceramic cylinder, containing an embedded heater surrounded by a cylindrical collector electrode [100]. The ceramic thermionic emitter is biased at a negative potential with respect to the collector electrode, and it is heated to a surface temperature of 400-800°C, depending on the mode of detection. The response of the detector to different elements depends on the electronic work function of the thermionic surface (i.e., the amount of energy required to emit a unit of electrical charge from the surface), the chemical composition of the gas environment immediately surrounding the thermionic surface, and the operating temperature of the thermionic surface. With only nitrogen as the plasma gas very specific ionization of compounds containing functional groups of high electron affinity, such as nitro, halogen, and thiol groups, etc., is obtained. This requires a thermionic source with a low work function (high cesium content) and the response is very structure dependent. With air or oxygen as the plasma gas and a relatively low source temperature (400 to 500°C) enhanced selectivity towards halogenated compounds is obtained. Using air mixed with a low flow of hydrogen as the plasma gases provides selectivity towards nitrogen and phosphorus containing compounds. The nitrogen/phosphorus response is observed when the thermionic emission surface is hot enough to ignite a dissociated hydrogen/air boundary layer in the immediate vicinity of the hot surface. Sample compounds are decomposed in this boundary layer, and nitrogen or phosphorus compounds which form electronegative decomposition products, are selectivity ionized by the thermionic surface.

Several models have been proposed to account for the selectivity of the TID response to nitrogen- and phosphorus-containing compounds [84,86,99,100,103-106]. They differ principally in whether the interaction between the alkali metal atoms and organic fragments occurs as a homogeneous reaction in the gas phase or is purely a surface phenomena. Kolb suggested that rubidium ions leave the bead as neutral atoms by accepting an electron from the plasma. While still in the vicinity of the bead, the rubidium atoms are excited and ionized by collision with plasma particles. The positive rubidium ions are immediately recaptured by the negatively charged bead. This mechanism occurs in a cyclic fashion, producing a steady state equilibrium. If this equilibrium is disturbed, for example by a process which results in ionization of rubidium atoms, more rubidium atoms will leave the source to re-establish the equilibrium, resulting in a concomitant increase in the ion current. The selectivity of the detector owes its origin to the increase in this ion current (according to the homogeneous mechanism theory) as only those radicals with electron affinities equal to or greater than the ionization potential of rubidium atoms will contribute to the ion current. Among the many fragments generated by the pyrolysis of organic compounds in the detector plasma, this criteria is only met by the CN, PO, and PO_2 radicals. It has proven difficult, however, to provide any conclusive evidence for this mechanism. Spectroscopic studies have indicated that alkali metal atoms are not lost from the bead by evaporation but rather by an exchange reaction with hydrogen in the plasma [104]. Sodium was found in the plasma in substantially larger amounts than rubidium and it was shown that rubidium was not essential for the selective detection of nitrogen and phosphorus compounds (its function may be to extend the lifetime of the bead). The signal response was claimed to result from the gas phase ionization of sodium atoms. The surface catalytic ionization model assumes that the nitrogen compound is adsorbed onto the bead surface, pyrolyzed in situ to among other fragments the cyano radical, accepts an electron from the bead, and then departs as an anion [103,105,106]. The charge-carrying species is the cyano anion in this case. The supply of electrons that supports the ionization process originates from the platinum wire used to heat the bead. The alkali glass is not a source since the rubidium atoms do not leave the glass; their role is that of surface catalysts for the electron transfer reaction. An alternative surface ionization model assumes that the principal role of the alkali metal in the source is to lower the work function of the surface [100]. To account for the influence of experimental variables it assumes an expanded role for the hydrogen plasma compared to the surface catalytic model. The high temperature of the source is sufficient to initiate decomposition of the hydrogen and oxygen gases supplied to the detector. Consequently, in the immediate vicinity of the source a hot, chemically-reactive gas boundary layer, containing radical species, is assumed to exist. A combination of the heat from the source and the reaction of sample molecules within this boundary layer gives rise to sample decomposition. Electronegative decomposition products from phosphorus- or nitrogen-containing compounds are then ionized by extracting an electron from the surface of the thermionic source. These negative ions produce the increase in ion current measured at the collector electrode.

Although, as indicated by the above studies, the detector response mechanism is poorly understood, the working limits of commercially available detectors are fairly well characterized and the detector is not particularly difficult to use. The minimum detectable quantity for nitrogen is about 10^{-13} g N/s and about 5 x

10^{-14} g P/s for phosphorus. The linear response range is about 10^4 to 10^5 and selectivity ratios of about 4×10^4 g C/g N, and 7×10^4 g C/g P may be obtained. The TID is widely used in environmental and biomedical research for determining pesticides, drugs, and profiling, where its high sensitivity and selectivity are useful in minimizing sample preparation requirements. Frequent checks on calibration should be made because of the adverse effect bead contamination has on response. Injection of excess silylation reagents should be minimized for the same reason.

3.5.1.3 Photoionization Detector

Photoionization is the process by which a molecule absorbs energy resulting in an electronic transition from a discrete low energy level of the molecule to the higher energy continuum of the ion. The energy required for such transitions is about 5-20 eV requiring photon excitation in the far ultraviolet. Early photoionization detectors operated at low pressures to maximize the photon intensity but this was not very convenient. The photoionization detector (PID) only became popular after the introduction of detectors that physically separated the source and ionization chamber from each other, allowing independent optimization of ion production and ion collection [84,107,108]. The photoionization efficiency is only about 0.001 to 0.1 %, consequently, the PID is often selected for those applications where a nondestructive detector is required and for the same reason can be used in tandem with other detectors as the sample exiting the detector is virtually identical with that entering. No auxiliary or support gases, beyond makeup gas for some applications, are required. Therefore, the PID can be used in environments where combustion gases may be considered hazardous or in portable instruments, where the additional weight of several gas bottles is undesirable. The detector selectivity can be varied by changing between sources of different energy. Ionization will occur only if the source energy is close to or exceeds that of the ionization potential of the sample. Unfortunately, the PID must be calibrated for every substance to be quantified, and as a concentration sensitive detector, its response is flow rate dependent.

A typical photoionization detector is shown in Figure 3.11. The UV source is a discharge lamp, containing an inert gas or gas mixture at low pressure, that emits monochromatic light of a

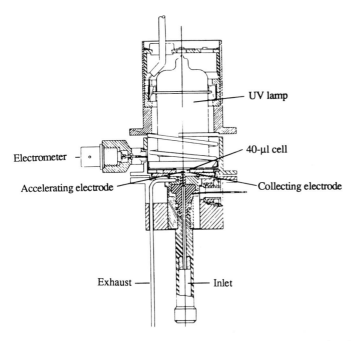

Figure 3.11 Cross-sectional view of a photoionization detector.
(Reproduced with permission from HNU Systems).

specific energy, depending on the choice of fill gases and the
window material. Sources of different energies (9.5, 10.0, 10.2,
10.9 and 11.7 eV) are available, while the 10.2 eV source is the
most widely used. Spectroscopic evidence suggests that the
absolute energy and monochromaticity of the sources may vary
widely from that quoted by the manufacturer, Table 3.4 [107].

The discharge compartment is mechanically separated from the
ionization chamber by an optically transparent window made of
metal fluoride. The effluent from the column passes through the
thermostated ionization chamber and between two electrodes,
positioned at opposite ends of the chamber. Detectors with
ionization chamber volumes of 40 and 175 microliters are available
for use with capillary columns and of 175 and 225 microliters for
packed columns. An electric field is applied between the
electrodes to collect the ions formed (or electrons, if preferred)
and the current amplified by a precision electrometer. It has been
shown that careful thermostating of the detector is required to
reduce baseline drift [107,109].

The processes occurring within the detector can be
represented by a series of equations (3.2-3.7) [110-123].

TABLE 3.4

CHARACTERISTIC LINES FOUND IN THE VACUUM UV OUTPUT OF HNU SOURCES

Lamp (eV)	Energy (eV)	Output (%)	Fill Gas
8.3	8.44	100	Xenon
9.5	10.88	0.03	Xenon
	10.58	0.01	
	10.40	0.18	
	9.92	0.05	
	9.57	2.1	
	8.44	97.6	
10.2	10.64	17.1	Krypton
	10.03	82.9	
11.7	11.82	26.2	Argon
	11.62	71.8	
	10.20	2.0	

$$AB + h\nu \longrightarrow AB^* \tag{3.2}$$

$$AB^* \longrightarrow AB^+ + e^- \tag{3.3}$$

$$AB^+ + e^- + C \longrightarrow AB + C \tag{3.4}$$

$$AB^* + C \longrightarrow AB + C \tag{3.5}$$

$$EC + e^- \longrightarrow EC^- \tag{3.6}$$

$$EC^- + AB^+ \longrightarrow AB + EC \tag{3.7}$$

Equation (3.2) leads to reaction (3.3) representing capture of a photon by a molecule AB followed by ionization resulting in the detector signal. Equations (3.4) to (3.7) represent competing reactions involving recombination or collisional de-excitation by a carrier gas molecule C or neutralization by reaction with an electron-capturing impurity EC. These reactions should be minimized. The use of electron-capturing solvents or impurities (e.g., O_2) in the carrier gas can lead to negative peaks in the chromatogram [111]. The choice of carrier gas also influences the response of the detector through the collision processes represented by equations (3.4) and (3.5), by its ability to influence the mobility of ions within the detector, and, in unfavorable cases, by absorption of some of the initial photon flux [110].

When competing reactions are minimized, the response of the detector can be described by equation (3.8) [112].

$$i = IF\eta\sigma NL[AB] \tag{3.8}$$

where i is the detector ion current, I the initial photon flux, F the Faraday constant, η the photoionization efficiency, σ the absorption cross-section, N Avogadro's number, L the path length and [AB] the concentration of an ionizable substance. Thus for a particular detector and source, the PID signal is proportional to the ionization yield, absorption cross-section and molar concentration of the analyte. The product ($\eta\sigma$) is the photoionization cross-section, which expresses both the probability that a molecule will absorb a photon and the probability that the excited state will ionize. The calculation of photoionization cross-sections is complex but it might be intuitively expected that a direct dependence of the photoionization cross-section on the photon energy and the ionization potential of the molecule exists. In practice, a fraction of the molecules with ionization potentials up to approximately 0.4 eV above the energy of the ionizing photons will be ionized, as some of these molecules will exist in vibrationally excited states.

The PID is nondestructive, relatively inexpensive, of rugged construction and easy to operate. With the 10.2 eV photon source most molecules are ionized; the exceptions are the permanent gases, C_1-C_4 hydrocarbons, methanol, acetonitrile and chloromethanes. The sensitivity for benzene is 0.3 Coulombs/g and the linear range approximately 10^7. The PID is 5 to 10 times more sensitive than the FID for alkanes and about 50 times so for aromatic compounds [107,108]. Freedman has shown that the ionization potential of the molecule is the most important single factor determining the PID response, the relative number of π-electrons having little significance [112]. This argument is well supported by the available experimental data, although other authors claim a much greater role for π-electrons in the ionization process [113]. The most comprehensive collection of response data for more than a hundred compounds, relative to benzene, has been compiled by Langhorst [114]. This study led to several empirical conclusions which are summarized in Table 3.5. A general increase in sensitivity as the carbon number increases was noted and at high carbon numbers the response was attributed mainly to the carbon chain with little influence from the presence of functional groups. These observations and those in Table 3.5

TABLE 3.5

RELATIONSHIP BETWEEN PHOTOIONIZATION DETECTOR RESPONSE (10.2 eV) AND MOLECULAR STRUCTURE

Sensitivity increases as the carbon number increases

Sensitivity for alkanes < alkenes < aromatics

Sensitivity for alkanes < alcohols < esters < aldehydes < ketones

Sensitivity for cyclic compounds > noncyclic compounds

Sensitivity for branched compounds > nonbranched compounds

Sensitivity for fluorine-substituted < chlorine-substituted < bromine-substituted < iodine-substituted compounds

For substituted benzenes, ring activators (electron-releasing groups) increase sensitivity and ring deactivators (electron-withdrawing groups) decrease sensitivity (exception: halogenated benzenes)

---

generally follow the trend in ionization potentials as discussed by Freedman [112].

3.5.1.4 Electron-Capture Detector

The structure-selective, electron-capture detector (ECD) is the second most widely used ionization detector [115-118]. It owes much of its popularity to its unsurpassed sensitivity to a wide range of toxic and biologically active compounds. Consequently, it is widely used in trace analysis for the determination of pesticides, herbicides and industrial chemicals in the environment, drugs and other biologically active compounds in biological fluids, and for determining the fate of volatile organic compounds in the upper atmosphere. As many of these applications have an impact on commerce, environmental quality and health, a great deal has been written on the properties of the ECD. However, it is quite correct to say that the ECD is one of the easiest to operate but least understood of the gas chromatographic detectors in common use.

In contrast to the other ionization detectors a decrease in the detector background current is measured rather than an increase in the number of ions or electrons generated. The detector standing current results from the bombardment of the carrier gas by beta electrons forming a plasma of positive ions, radicals and thermal electrons. By a series of elastic and inelastic collisions each beta electron may generate between one hundred and one thousand thermal electrons with mean energies of 0.02 to 0.05 eV. The application of a potential to the detector cell, either continuously or pulsed, allows the thermal electrons to be collected. This background current, in the presence of pure carrier gas, constitutes the detector standing current and the baseline value for all measurements. When an electron-capturing compound enters the cell it captures a thermal electron to produce either a negative molecular ion or fragment ion if dissociation accompanies electron-capture. The diminution in detector background current due to the loss of thermal electrons by formation of anions subsequently neutralized by positive ions or, due to a reduced drift velocity during the time the field is applied, constitutes the quantitative basis by which detector response is related to solute concentration.

The ideal source of ionizing radiation would produce a small number of ion pairs per disintegration to minimize the fluctuations in the ion current to be minimal. At the same time, the total ion pair formation should be large so that the resulting electron current during the passage of an electron-capturing substance can be measured conveniently without introducing other sources of noise. The best compromise among these demands appears to be low energy beta-emitting radioisotopes (minimum number of ion pairs per particle) at relatively high specific activities (maximum total ion pair formation) [119]. Commercial instruments use either ^{63}Ni or ^{3}H radioisotopic sources supported by a metallic foil. Of the two types, tritium would be preferred due to its lower energy beta emanation (0.018 MeV) compared to ^{63}Ni (0.067 MeV) and the fact that foils with higher specific activity are less expensive to manufacture. The principal advantage of ^{63}Ni sources is their high temperature operation stability (to 400°C) compared to 225°C for $Ti^{3}H_2$ and 325°C for $Sc^{3}H_3$. High temperature operation minimizes contamination from the chromatographic system and enhances the response of the detector to those compounds which capture electrons by a dissociative mechanism. Thus practical considerations and operational convenience dictate the use of ^{63}Ni for most purposes. Microwave discharges, thermionic emitters, and photoemission from thin metallic films have been suggested as altrnative thermal electron sources but have not been commercialized [120-123]. These sources are attractive since only electrons are produced, and those problems of a theoretical and practical nature due to the presence of positive ions in the ionization chamber, are eliminated.

Ionization chambers with parallel plate, coaxial cylinder and asymmetric (displaced coaxial cylinder) electrode geometry have been used in commercial detectors [115,117,124-128]. The parallel plate detector design was popular during the early stages of the development of the ECD, since it permitted adjustment of the relative position of the source and collector electrode to be easily made. It is rarely used nowadays. The majority of detectors in use today are of the coaxial cylinder or asymmetric design, Figure 3.12. The lower specific activity of the ^{63}Ni source, compared to ^{3}H, requires the use of a larger source area to obtain the same ionization efficiency. This is more easily attained in the coaxial design. Here the minimum spacing between the source, which surrounds the centrally located collector electrode, and the collector electrode is established by the penetration depth of the beta particles [129]. This distance should be great enough to ensure that all the beta particles are deactivated by collisions and converted to thermal energies without colliding with the anode.

By locating the anode entirely upstream from the ionized gas volume, collection of long range beta particles is minimized in the displaced coaxial cylinder design, and the direction of gas flow minimizes diffusion and convection of electrons to the collector electrode. However, the free electrons are sufficiently mobile that modest pulse voltages (e.g., 50 V) are adequate to cause the electrons to move against the gas flow and be collected during this time.

Miniaturization of the ionization chamber is important for use with open tubular columns where peaks may elute in 10-100 microliter gas volumes [118]. The effective lower limit of current designs is approximately 100-400 microliters, still too large to eliminate completely extracolumn band broadening [128]. The effective detector dead volume can be reduced by adding makeup gas at the end of the column to preserve column efficiency at the expense of some loss in detector sensitivity due to sample dilution. Since some electron-capture detectors designed for packed column use have cell volumes of 2.0 to 4.0 ml, they may not be suitable for use with open tubular columns.

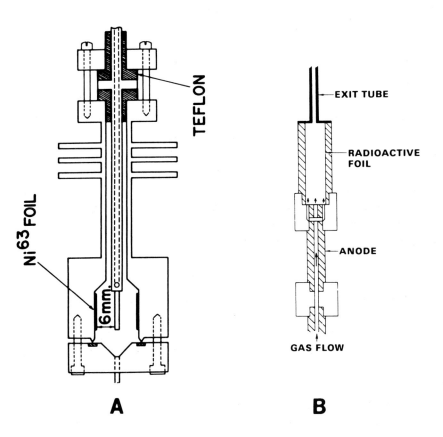

Figure 3.12 Schematic diagram of the coaxial cylinder, A, and the asymmetric (displaced coaxial cylinder), B, design.

The thermal electron concentration in the detector cell can be measured continuously (dc voltage) or discontinuously (pulsed voltage). The dc voltage mode has several disadvantages arising from space charge effects, contact potential effects, and the promotion of non-electron-capturing ionization processes, which can result in anomalous detector operation [117,130]. Pulse sampling techniques are used in all commercial instruments to minimize the above problems. The potential is applied to the cell as a square-wave pulse of sufficient height and width to collect all the thermal electrons and with sufficient time between pulses for the concentration of thermal electrons to be replenished by the ionizing beta radiation, and for their energy to reach thermal equilibrium. The signal from an ECD, operated with a long pulse period, can be described by equation 3.9 [117]

$$(I_b - I_e)/I_e = K [AB] \tag{3.9}$$

where I_b is the detector standing current, I_e the detector current measured at the peak height maximum, K the electron-capture coefficient and [AB] the sample concentration. By analog conversion the detector output can be linearized over about four orders of magnitude [131]. Without analog conversion and with pulse periods less than 1 millisecond, typical operating conditions under normal circumstances, the linear response range is approximately 100 to 1000. In the dc mode shorter linear operating ranges, approximately 10 to 100, are common.

The majority of the commercially available ECD's are designed for use with a modified version of the pulsed-sampling technique, termed the variable frequency constant-current mode [132]. Rather than measuring the cell current at a constant pulse frequency, the cell current is fixed with respect to a reference value, and the frequency of the pulse is changed so that the difference between the cell current and the reference current is zero throughout the chromatographic separation. Since pulse frequency is the variable quantity in this mode of operation, the detector signal is a voltage proportional to that frequency. The two principal advantages of this method are that the linear response range is approximately 10^4 to 10^5, much greater than for the constant frequency pulse sampling mode, and secondly, the detector operation is less disturbed by traces of interferents entering the detector cell. A disadvantage of this mode of operation is that for compounds with ultrafast electron attachment rate constants (e.g., between 2.8×10^{-7} and 4.6×10^{-7} ml/molecule.s) the detector response is non-linear [133-137]. Examples of compounds with ultrafast rate constants are: CCl_4, SF_6, $CFCl_3$, and CH_3I. The ECD is most responsive to these compounds and would normally be selected for their determination. For compounds with rate constants less than the above, the detector response is normally linear over the full operating range. This group includes most of the moderately strong and weak electrophores that make up the bulk of compounds determined with the ECD. It has been shown that operating the constant current ECD at moderately fast baseline frequencies will linearize the low concentration end of the response range for compounds with ultrafast rate constants at the expense of a reduction of the dynamic response range [136].

The choice of carrier gas for use with the ECD is limited to hydrogen, the noble gases and nitrogen. Hydrogen may strip tritium from the detector at high temperatures [125]. Pure argon and helium are unsuitable as they readily form metastable ions which can transfer their excitation energy by collision with solute vapors, resulting in undesirable ionization effects (Penning reaction). The addition of 5 to 10 percent of methane to argon removes the metastable ions as quickly as they are formed (by deactivating collisions) and also serves to reduce and maintain the secondary electron energy at a constant thermal level. Argon-containing methane and oxygen-free nitrogen are the most common carrier gases used with packed columns. For open tubular columns hydrogen or helium are usually used as carrier gases to maximize column efficiency and minimize separation time while argon-methane or nitrogen are used as makeup gases. Oxygen and water vapor traps should be used to purify all gases.

The response of the ECD to organic compounds covers approximately seven orders of magnitude, Table 3.7 [115,117,118]. The detector responds most strongly to compounds containing halogens or nitro groups, to organometallic compounds, and to conjugated electrophores. This latter group is structurally the least well defined, and is comprised of compounds containing two or more weakly electron-capturing groups, connected by some specific bridge that promotes a synergistic interaction between the two groups [138]. Examples of conjugated systems with a high detector response include conjugated carbonyl compounds (benzophenones, quinones, phthalate esters, coumarins), some polycyclic aromatic hydrocarbons, some sulfonamides and certain steroids. The response of the ECD to halocarbon compounds decreases in the order I > Br > Cl >> F and increases synergistically with multiple substitution on the same carbon atom. The identity and number of halogen substituents is more important than subtle variations in the geometric framework of the alkyl portion of the molecule, although even these small changes will have a measurable influence on the detector response in many cases [139,140]. The response of the ECD to haloaromatic and nitroaromatic compounds shows similar trends to the alkyl compounds. The position of electronegative functional groups (ortho, meta, para) has a measurable influence on the detector

TABLE 3.6

RELATIVE RESPONSE OF THE ELECTRON-CAPTURE DETECTOR TO ORGANIC COMPOUNDS

General Organic Compounds		Halocarbons	
Compound	Relative response	Compound	Relative response
Benzene	0.06	$CF_3CF_2CF_3$	1.0
Acetone	0.50	CF_3Cl	3.3
Di-n-butyl ether	0.60	$CF_2=CFCl$	100
Methylbutyrate	0.90	CF_3CF_2Cl	170
1-Butanol	1.00	$CF_2=CCl_2$	670
1-Chlorobutane	1.00	CF_2Cl_2	3×10^4
1,4-Dichlorobutane	15.00	$CHCl_3$	3.3×10^4
Chlorobenzene	75.00	$CHCl=CCl_2$	6.7×10^4
1,1-Dichlorobutane	111.00	CF_3Br	8.7×10^4
1-Bromobutane	280.00	$CF_2ClCFCl_2$	1.6×10^5
Bromobenzene	450.00	$CF_3CHClBr$	4.0×10^5
Chloroform	6×10^4	$CF_3CF_2CF_2I$	6.0×10^5
1-Iodobutane	9×10^4	CF_2BrCF_2Br	7.7×10^5
Carbon tetrachloride	4×10^5	$CFCl_3$	1.2×10^6

response but is less dramatic than the response variation due to the number and type of individual substituents. As the number of substituents is increased on polychlorinated and polybrominated compounds the detector response approaches its coulometric limit and the introduction of further substituents has little effect. This law of diminishing returns is observed for all organic compounds; the introduction of the first few electronegative substituents has a large impact on detector sensitivity, but further substitution has less influence.

Much of what is known about the structure response of the ECD is based on empirical observations. Clearly, the ability to correlate the response of the detector to fundamental molecular parameters would be useful. Chen and Wentworth have shown that the information required for this purpose is the electron affinity of the molecule, the rate constant for the electron attachment reaction and its activation energy, and the rate constant for the ionic recombination reaction [117,141,142]. In general, the direct calculation of detector response factors have rarely been carried out, since the electron affinities and rate constants for most compounds of interest are unknown.

The maximum response of the ECD to different organic compounds is markedly temperature dependent. For a single compound changes in response of 100- to 1000-fold may occur for a 100°C change in detector temperature [117,143,144]. This strong temperature dependence can be derived directly from the kinetic model of the ECD process. In simple terms, an electron-capturing solute (AB) may attach an electron, forming a stable molecular ion, or the molecular ion may be formed in a sufficiently excited state to dissociate, as shown in equations (3.10) and (3.11).

$$AB + e^- \longleftrightarrow AB^- \tag{3.10}$$

$$AB + e^- \longrightarrow A\cdot + B^- \tag{3.11}$$

The favored mechanism depends on the juxtaposition of the potential energy curves for the neutral molecule and the negative ion. In the case of the nondissociative mechanism the negative molecular ion is more stable than the neutral molecule, whereas in the dissociative case the potential energy curve for the negative ion crosses that of the neutral molecule at a level corresponding to a vibrationally excited state. An increase in detector temperature favors the populating of vibrationally excited states and thus the mechanism represented by equation (3.11). Conversely, the nondissociative mechanism is favored by low detector temperatures. The nondissociative mechanism is common among conjugated electrophores and the dissociative mechanism among halogen-containing (except fluorine) compounds. The two mechanisms can be distinguished from a plot of $\ln KT^{3/2}$ vs. $1/T$ where T is the detector temperature [117,134]. The calculation of K, the electron capture coefficient, is fairly involved and for routine use the peak area response for a fixed concentration of solute can be used instead. In most cases the plots will be linear with a positive slope for the nondissociative case and a negative slope for the dissociative case. In terms of maximizing the detector response, the optimum temperature will be either the maximum operating temperature of the detector for dissociative compounds, or the lowest advisable operating temperature for nondissociative compounds. Some situations may be more complex than represented above but an optimum detector temperature in the range indicated is found easily by trial and error.

Mathematical models of the electron-capture process are based on the stirred reactor model of Lovelock [145] and the kinetic model of Wentworth [117,142,146] as further modified by others [129,134-136,147-150]. The ionization chamber is considered to be a stirred reactor into which electrons are continuously introduced at a constant rate and electron-capturing solutes are added at a variable rate in a constant flow of carrier gas. The major consumption of cell electrons is via electron capture, recombination with positive ions, wall loss and ventilation. The model can be expanded to allow approximately for the presence of electron-capturing contaminants and can explain reasonably well the influence of different pulse sampling conditions on the detector response. However, an exact solution is not possible, as the importance of contaminants and the rate loss of positive ions by space charge diffusion and the corresponding magnitude of the positive and negative ion contribution to the ECD current are unclear.

When operated in the noncoulometric mode, the response of the ECD is concentration dependent. The signal response depends on the flow rate and purity of the carrier gas, specific detector design and operating characteristics (this includes detector temperature), and the contribution of background contamination from the column and chromatographic system. Response data is usually transposed into concentration terms by calibration and, since response factors cover a wide range for electron-capturing compounds, each substance to be quantified must have an individual calibration curve. For those compounds that are completely or reproducibly ionized it has been suggested that the ECD could be operated in an absolute, or coulometric mode, in which the time integral of the number of electrons captured during the passage of a chromatographic peak is equal, via Faraday's law, to the number molecules ionized [151-153]. This approach would be particularly useful for the quantitative analysis of certain substances at concentration levels where accurate calibration standards are difficult to prepare. Practically, it has proven difficult to construct devices that function reliably as coulometric detectors. Uncertainty of the role played by ions in the detector response mechanism no doubt contribute considerably to the problems of developing a coulometric ECD.

A technique known as selective electron capture sensitization has been used to increase the response of the ECD to weakly electron-capturing compounds [117]. In this mode a standard electron-capture detector is used with a supply of makeup gas doped with a specific sensitizing reagent such as oxygen or nitrous oxide. In this way the ECD functions as an ion-molecule reaction chamber. Since the dopants are present in significantly higher concentrations than the analytes, electron attachment takes place predominantly, if not exclusively, with the dopant species producing O_2^- ions in the case of oxygen and O^- and NO^- in the case of nitrous oxide. These ions then interact with the analyte to form stable negative ions. The instantaneous concentration of these ions must be coupled to the electron density. The sum of the chemical processes involved is the consumption of an electron, equivalent to direct electron capture except that the magnitude of the detector response is determined by the kinetic and thermodynamic characteristics of the ion-molecule reaction. Compounds with a high electron-capture response show only low or fractional enhancement values. For weakly electron-capturing compounds, such as chloromethane and polycyclic aromatic hydrocarbons and their derivatives, an enhancement in response covering three to four orders of magnitude have been obtained with oxygen added to the detector gases at about 0.3% (v/v) [117,154-156]. It was also found that the enhancement in response caused by the presence of oxygen was extremely sensitive to differences in molecular structure and the enhancement ratio could be used as a simple and practical method of compound verification and isomer identification. The addition of a few ppm of nitrous oxide to the ECD has been shown to enhance its response to a wide range of weakly electron-capturing compounds, such as vinyl chloride, carbon monoxide, methane and Freons [117,157]. An alternative approach is the use of a combination of a weakly electron capturing alkyl halide, such as chloroethane, in the presence of a co-additive of high proton affinity, such as triethylamine, which was used to enhance the detector response to polycyclic aromatic hydrocarbons [158]. The triethylamine was used to generate a stable population of positive ions with little reactivity towards the analyte. The chloroethane to form a stable population of chloride ions by electron transfer of the electrons captured initially by the analyte. The above processes resulted in a stable and reliable response enhancement of about two orders of magnitude for selected polycyclic aromatic hydrocarbons.

3.5.1.5 Helium Ionization Detector

The helium ionization detector (HID) is probably the least used of the ionization detectors. It is a universal and ultra sensitive detector with a reputation for unreliability and unparalleled difficulty in routine use. It is primarily used for the trace analysis of permanent gases and some volatile organic compounds that have a poor response to the FID and are present in too low a concentration for detection with a thermal conductivity detector [159].

Many aspects of the design of the HID are not unlike those of the ECD with which it shares a common parentage. The detector consists of a thermostated ionization chamber with a 100-200 microliter volume and a parallel plate or coaxial cylinder electrode configuration [159-161]. Tritium sources of a high specific activity (0.25-1.0 Ci) are used to maximize sensitivity. The detector current is measured by applying a fixed or pulsed voltage to the electrodes in either the saturation or multiplication region. Pulsed sampling seems to provide improved baseline stability but lower sensitivity than the fixed field mode. Likewise, voltages around 550 V (in the multiplication region), provide a greater response accompanied by increased background noise. The detector is usually more sensitive if less stable and more disturbed by contamination when operated in this region. The detector and column connections must be leak tight to avoid ingress of air which effects the stability and reliability of detector operation.

The principal mechanism for analyte response is ionization due to collision with metastable helium atoms. Metastable helium atoms are generated by multiple collisions with beta electrons from the radioisotopic source. Since the ionization potential of helium (19.8 ev) is higher than that of all other species except neon, then all species entering the ionization chamber will be ionized.

The response of the detector is notoriously dependent on the purity of the carrier gas (helium) and its flow rate, the applied field for electron collection, detector geometry, temperature, source activity, and the general level of contamination of the detector from column bleed and other instrument components. The polarity, linearity, and sensitivity of the detector response is influenced by detector contamination levels. The detector is generally operated with adsorption columns or partition columns at low temperatures to avoid detector contamination from column bleed. Overloading the detector produces M- or W-shaped peaks. Under favorable circumstances detection limits in the parts-per-billion range can be obtained and a linear range of about 10^4.

3.5.2 Bulk Physical Property Detectors

The bulk physical property detectors respond to some difference in the properties of the mobile phase due to the presence of the analyte. Usually, a large signal for some response characteristic of the mobile phase is desired to provide a reasonable working range, but for low concentrations of analyte the detector signal corresponds to a very small change in a large signal. The sensitivity of the bulk physical property detectors tend to be low compared to ionization detectors and are noise limited. The most important of the bulk physical property detectors are the thermal conductivity detector, gas density balance, and the ultrasonic detector. Of these, only the thermal conductivity detector is widely used.

3.5.2.1 Thermal Conductivity Detector

The thermal conductivity detector (TCD), also known as the hot-wire or katharometer detector, is a universal, non-destructive, concentration-sensitive detector that responds to the difference in thermal conductivity between pure carrier gas and carrier gas containing organic vapors. The TCD is generally used to detect permanent gases, light hydrocarbons and compounds which respond only poorly to the FID. For many general applications it has been replaced by the FID, which is more sensitive (100- to 1000-fold), has a greater linear response range, and provides a more reliable signal for quantitation. Detection limits for the TCD usually fall into the range of 10^{-6} to 10^{-8} g with a response that is linear over about four orders of magnitude.

In a typical TCD, the carrier gas flows through a heated thermostated cavity that contains the sensing element, either a heated metal wire or thermistor. With pure carrier gas flowing through the cavity the heat loss from the sensor is a function of the temperature difference between the sensor and cavity and the thermal conductivity of the carrier gas. When an organic solute enters the cavity, there is a change in the thermal conductivity of the carrier gas and a resultant change in the temperature of the sensor. The sensor may be operated in a constant current, constant voltage, or constant temperature compensation circuit as part of a Wheatstone bridge network. A temperature change in the sensor results in an out-of-balance signal, which is usually passed directly to a recorder without further amplification.

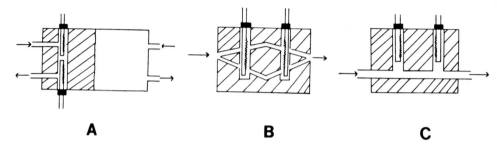

A **B** **C**

Figure 3.13 Cell designs for the thermal conductivity detector.
A, flow-through cell; B, semi-diffusion cell; C, diffusion cell.
(Reproduced with permission from ref. 163. Copyright Preston
Publications, Inc.)

The TCD has appeared in several different designs, some of
which have certain advantages for particular applications
[162,163]. They usually represent some variation of the three
basic geometries: the flow-through, semi-diffusion and diffusion
cells, Figure 3.13 [163]. The diffusion cell has a slow response
and is relatively insensitive; it is used mainly for preparative
chromatography. The semi-diffusion cell is used for packed column
analytical gas chromatography. It is less sensitive to flow
variations than the flow-through cell but, with a minimum detector
volume of 100 microliters (commercially available detectors have
volumes from 0.01-0.1 ml), it is not suitable for use with open
tubular columns unless operated at reduced pressure to overcome
dead volume effects [164]. Flow-through cells with volumes of 10
to 100 microliters are easily fabricated and can be used with open
tubular capillary columns [165]. A novel design by Hewlett-Packard
employs a single filament TCD and flow modulation to switch the
carrier gas between two channels, one of which contains a
filament. Every 100 ms a switching valve fills the filament
channel alternately with carrier gas and column effluent. No
reference column is required and, with an effective detector
volume of 3.5 microliters, it can be used with even narrow bore
capillary columns. Digital data processing and more reliable
temperature compensation provides greater sensitivity and
stability with either packed or open tubular columns than obtained
with conventional detector designs.

The sensing element in the TCD is usually a heated filament
or thermistor. Most filaments are made from tungsten, platinum,

nickel or alloys of these with other metals, such as iridium or rhenium. The desirable properties of a filament include a high temperature coefficient of resistance, relatively high electrical resistance, mechanical strength to permit forming into various shapes, a wide temperature operating range, and chemical inertness. Filaments can be coated with gold or PTFE to improve chemical inertness, particularly against oxidation by air. The filament is heated by a regulated power supply. The power setting, typically in the milliampere range (200-500 mA) at about 40 V, will affect both the sensitivity and the stability of the TCD response to different sample amounts [166]. Thermistors can be manufactured in the very small sizes needed to fabricate cells of low dead volume and have a much higher resistance than filaments. They can provide greater sensitivity than filaments when operated at low temperatures but, as they have a negative temperature coefficient, their resistivity and hence sensitivity declines with temperature; filament detectors are usually more sensitive above 60°C. Also, thermistors should not be used with hydrogen carrier gas to avoid reductive degradation of the metal oxides used in their manufacture.

Temperature gradients within the detector cavity can result in poor detector performance. At the detection limit temperature changes as small as 10^{-5}°C are encountered, presenting considerable problems in cell design. It is not possible to thermostat a cell to provide the necessary absolute thermal stability to measure these small temperature changes. The sensing element must be centrally located within a detector body having a large thermal mass of controlled temperature ($\pm$ 0.01°C). A difference signal must be used by incorporating a matched reference cell in the same environment as the sample cell. Some designs use two sampling and two reference cells to improve sensitivity and stability.

Carrier gases of low molecular weight and high thermal conductivity are required to maximize the response of the detector and to maintain a large linear response range. Consequently, helium and hydrogen are predominantly used. Heavier carrier gases such as nitrogen, as well as influencing sensitivity and linearity, may give rise to negative sample peaks or peaks with split tops. Theoretical models have been advanced to explain the response characteristics of the detector under different operating conditions [167]. These models take account of the effects of conduction, convection and radiation on the loss of heat from the sensor but do not lead to any simple mathematical expression to describe the operation of the detector. Several compilations of relative response data for the TCD are available [80,163,168]. These values are usually expressed on a weight or molar response basis relative to benzene. They depend on the nature of the carrier gas used for their determination but are generally sufficiently accurate for approximating sample concentrations. For precise quantitative analysis it is necessary to calibrate the detector for each substance determined. Alternatively, an internal standard can be used to determine accurate response ratios which may be subsequently used for quantitative analysis.

3.5.3 Optical Detectors

The use of flames as atom reservoirs for the spectroscopic determination of elements is a well established technique and is particularly valuable for metal analysis. Most non-metallic compounds, which account for the majority of samples analyzed by gas chromatography, have their principal emission or absorption lines in the ultraviolet region, where flame background contributions are troublesome. In addition, the diffusion flames used in gas chromatography lack sufficient stability and thermal energy to be useful atom reservoirs. For direct optical emission detection, microwave induced and inductively coupled plasmas provide more appropriate atom sources for organic compounds. However, the determination of phosphorus and sulfur by a flame photometric detector (FPD) is widely used in gas chromatography. For these elements, a hydrogen diffusion flame provides optimum excitation conditions. Since many industrial and pest control compounds contain sulfur or phosphorus, the FPD is widely used in environmental analysis. A number of chemiluminescent reaction detectors are also used in gas chromatography. They are based on pyrolytic or catalytic reactions, which release species that can be subsequently detected in a chemiluminescent reaction. The thermal energy analyzer is used as a selective detector for nitrosamines after low temperature pyrolysis to release nitric oxide, which is determined by its chemiluminescent reaction with ozone. In the redox chemiluminescence detector the effluent from a gas chromatographic column is dynamically mixed with a metered flow of dilute nitrogen dioxide and passed through a thermostated catalyst chamber. Compounds capable of reducing nitrogen dioxide to nitric oxide are subsequently detected by the chemiluminescent reaction between nitric oxide and ozone. Sulfur-containing compounds can be determined after combustion to sulfur monoxide by its chemiluminescent reaction with ozone.

3.5.3.1 Flame Photometric Detector

The flame photometric detector (FPD) uses a hydrogen diffusion flame to first decompose and then excite to a higher electronic state the fragments generated by the combustion of sulfur- and phosphorus-containing compounds in the effluent from a gas chromatograph. These excited molecules subsequently return to the ground state, emitting characteristic band spectra. This emission is monitored by a photomultiplier tube through either a 392 nm bandpass filter for sulfur or a 526 nm bandpass filter for phosphorus. It has been shown that by suitable changes in flame conditions the FPD will respond to elements other than phosphorus and sulfur; these include nitrogen, halogens, boron, selenium, germanium and several metals [84,169,170]. However, these applications do not represent the normal use of the detector, which is generally considered to be selective for phosphorus and sulfur [80,84,116,171].

In the most common detector design the carrier gas and air or oxygen are mixed, conveyed to the flame tip, and combusted in an atmosphere of hydrogen. With this burner and flow configuration interfering emissions from hydrocarbons occur mainly in the oxygen-rich flame regions close to the burner orifice, whereas sulfur and phosphorus emissions occur in the diffuse hydrogen-rich upper portions of the flame. To enhance the selectivity of the detector an opaque shield surrounds the base of the flame, preventing hydrocarbon emissions from reaching the photomultiplier viewing region. The extent and intensity of the various emitting regions of the flame are dependent upon the burner design and the gas flow rates. For any given burner design, the response of the FPD is critically dependent upon the ratio of hydrogen to air or oxygen flow rates, the type and flow rate of carrier gas and the temperature of the detector block [171-176]. Different optimum conditions are usually required for sulfur and phosphorus detection, for detectors with different burner designs, and perhaps also for different compound classes determined with the same detector. The sulfur response may decrease substantially at high carrier gas flow rates, more so with nitrogen than helium, but less response variation has been noted for phosphorus. As is

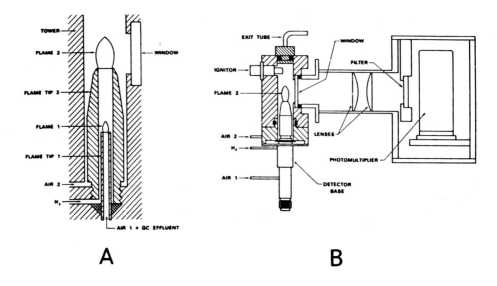

Figure 3.14 Dual-flame photometric detector (Varian). A, Schematic diagram of the dual-flame burner; B, Schematic diagram showing the relationship between the burner and the photometric viewing components. (Reproduced with permission from ref. 177. Copyright American Chemical Society).

obvious from the above comments, careful optimization of detector flow rates is required and these optimum values should be determined on a detector-to-detector basis and, in some cases, also on a compound-to-compound basis.

Problems with solvent flameout, hydrocarbon quenching and structure-response variations for different sulfur- and phosphorus-containing compounds can be partially solved by using a dual-flame burner, Figure 3.14 [177-179]. The lower flame is hydrogen rich and functions as a matrix-normalization reactor in which all compounds are decomposed to a highly reduced state (e.g., H_2S, S_2, H_2O, CH_4, etc). The combustion products from the first flame are swept into a second longitudinally-separated flame where the desired optical emission is generated under optimized flame conditions. The small detector volume, 170 microliters, and the high carrier and combustion gas flow rates in the passageways between the burners provide a very small effective detector dead volume, suitable for use with capillary columns. Although sensitivity for sulfur is generally higher for a single flame burner, the sample response is more likely to be dependent upon molecular structure and affected by the presence of coeluting hydrocarbons.

The response mechanism of the FPD is known superficially, even if the finer details remain obscure. The chemical processes occurring in flames are very complex involving a multitude of simultaneous and sequential reactions. In the case of the relatively low temperature (< 1000°C) and hydrogen-rich flames favored for use with the FPD, sulfur-containing compounds are combusted and interconverted by a large set of bimolecular reactions yielding species, such as H_2S, HS, S, S_2, SO and SO_2 in relative proportions that depend on the instantaneous and fluctuating flame chemistry. In the presence of carbon radicals in the flame, various carbon-sulfur containing species might also be anticipated. The presence of hydrocarbons coeluting with the sulfur compounds or variations in the hydrogen/oxygen ratio can change the relative concentration of either H or OH radicals in the flame and will also alter the distribution of sulfur species in the flame. Excited state S_2^* species could result from several two or three body collision reactions, such as those shown below

$$H + H + S_2 \longrightarrow S_2^* + H_2 \qquad (3.12)$$

$$S + S \longrightarrow S_2^* \qquad (3.13)$$

$$S + S + M \longrightarrow S_2^* + M \qquad (3.14)$$

where M is some third body. It is interesting to note that not all sulfur entering the flame is destined to be converted to S_2 species, and in fact, only a very small fraction may be so converted. In the case of phosphorus, incoming phosphorus-containing compounds are first decomposed to PO molecules, which are then converted into HPO^* by either of the reactions shown below

$$PO + H + M \longrightarrow HPO^* + M \qquad (3.15)$$

$$PO + OH + H_2 \longrightarrow HPO^* + H_2O \qquad (3.16)$$

A linear dependence between detector response and the amount of sample entering the detector is expected for phosphorus and is generally found. Deviations from the predicted detector response are more common with sulfur-containing than phosphorus-containing compounds [171,173]. The detector response in the sulfur mode can be described by equation (3.17)

$$I(S_2) = A [S]^n \tag{3.17}$$

where $I(S_2)$ is the signal intensity, A an experimental constant, [S] the mass flow rate of sulfur atoms, and n the exponential factor. The theoretical value for n is 2, but in practice, values between 1.6 and 2.2 are frequently observed. Non-optimized flame conditions, incomplete combustion, hydrocarbon quenching, competing flame reactions that lead to de-excitation, and sample structural variations all contribute to this deviation. Since deviations from the expected value are known to occur, the use of amplifiers with a linearization function should only be used when the value of n is known to be two [180]. It should also be noted that when measuring column performance parameters with the sulfur FPD, the equations derived for a linear detector response should be modified by incorporating the exponential factor into the denominator [181]. When n = 2, the response at one-quarter peak height yields a measure of peak width analogous to that obtained from a linear detector peak measured at half height. This discrepancy arises because the rate of elution into a detector with a square law response reaches half its maximum at one-quarter peak height.

Minimum detection limits and selectivity values for the FPD depend on the operating conditions of the detector, burner geometry and photomultiplier sensitivity. Typical detection limits are 5×10^{-13} g P/s and 5 to 50×10^{-12} g S/s. The linear range for phosphorus usually exceeds 1000 while the selectivity is more than 5×10^5 g C/g P. Sulfur selectivity depends upon the amount of sulfur present; it varies from 10^3 g C/g S at low sulfur amounts to greater than 10^6 g C/g S at high sulfur amounts. It is not always possible to take full advantage of the detector selectivity because of quenching, particularly of the sulfur response. The coelution of hydrocarbon compounds can quench the response of the detector due to a temporary change in flame conditions or due to de-exciting collisions or reactions in the flame-emitting zone [171,179,182,183]. Quenching is less of a problem with the dual-flame detector.

3.5.3.2 Chemiluminescence Detectors

Chemiluminescence is the term used to describe chemical reactions that result in the formation of vibrationally or electronically excited species that subsequently undergo emission of photons. Those reactions important for gas chromatography occur in the gas phase and are often combined with a reaction detector to thermally or catalytically liberate a species that is easy to detect [185]. By far the most important reaction is the release of nitric oxide which is subsequently combined with ozone to generate electronically excited nitrogen dioxide. The instrumentation is simple consisting of a reaction chamber and light sensor (photomultiplier tube) with associated amplifiers, a system to generate reagent gas, and possibly a vacuum pump to operate the reaction chamber at a reduced pressure. Reduced pressure operation has two advantages: it improves sensitivity by diminishing collisional deactivation of the excited reaction product and reduces the effective dead volume of the detector maintaining compatibility with the use of open tubular columns.

The thermal energy analyzer (TEA) is a selective detector for the determination of nitrosamines [185,186]. Low temperature catalytic pyrolysis (275-300°C) cleaves off the nitrosyl radical which is swept through a cold trap, used to condense interfering organic volatiles, and into an evacuated chamber (via a capillary restrictor) through which ozone is continuously bled. Detection limits in the region of 0.5 ng for the lower alkyl nitrosamines, a linear range up to 10^5, and selectivities greater than 10^3 against most amines and nitro compounds make this a very useful detector for the carcinogenic nitrosamines. By use of metal oxide catalyst for sample pyrolysis the detector can be made to respond to amines and nitro-containing compounds [187]. In the redox chemiluminescence detector (RCD) the effluent from the gas chromatography is first mixed with a gas stream containing 100 to 300 ppm of nitrogen dioxide and then passed through a thermostated catalyst chamber containing, usually, gold-coated glass beads [185,188,189]. In the catalyst chamber any compound that reduces nitrogen dioxide to nitric oxide, produces a surrogate pulse of nitric oxide, which is subsequently detected by reaction with ozone. The selectivity of the reaction is controlled by the choice of catalyst, temperature and the catalyst support material. Table 3.7 list some of the compounds that can be selectively detected using the RCD with a gold catalyst [185]. Under these conditions saturated hydrocarbons, fully chlorinated compounds, water, nitrogen, oxygen, carbon dioxide, and noble gases do not respond significantly. The RCD has been used to detect volatile compounds in environmental samples, especially those compounds with a poor

TABLE 3.7

COMPOUNDS DETECTED BY THE REDOX CHEMILUMINESCENCE DETECTOR

Alcohols	Aldehydes
Carboxylic acids	Phenols
Ketones	Ethers
Ammonia	Aromatic and aliphatic Amines
Nitrogen heterocyclics	
Olefins	Aromatic hydrocarbons
Sulfides	Disulfides
Mercaptans	Hydrogen sulfide
Carbonyl sulfide	Carbon disulfide
Sulfur dioxide	Hydrogen
Hydrogen peroxide	Carbon monoxide
Organic phosphates	Organic phosphonates
Organic phosphites	

response to the FID. Detection limits in favorable cases are in the sub-nanogram to a few nanograms range.

Sulfur-containing compounds can be determined at trace levels with high selectivity after combustion in a hydrogen diffusion flame to sulfur monoxide, which is detected by a chemiluminescent reaction with ozone [184,190-192]. Unlike the flame photometric detector the response to individual sulfur-containing compounds is nearly equal on a per gram of sulfur basis and is linearly related to the sulfur content. Hydrocarbon interferences are minimal with a discrimination ratio of 10^6 - 10^7 or greater on a weight-to-weight basis. The linear response range exceeds 10^3 with minimum limits of detection around 4×10^{-13} g S/s. Additional band broadening due to the detector dead volume is barely noticeable when packed columns and wide bore capillary columns are used. The sulfur chemiluminescent detector (SCD) does exhibit somewhat broader and more asymmetrical peaks than those obtained with the FID under the same conditions when narrow bore capillary columns are used [184,190].

The detector is based on the combustion of sulfur-containing compounds in a hydrogen rich air flame of a FID to form sulfur monoxide. The hydrogen/air flow rate ratio is the most critical parameter controlling the production of sulfur monoxide. Under optimum conditions sulfur monoxide may account for up to 20% of the sulfur species in the flame. Sulfur monoxide is a free radical and a very reactive species that is short lived; however, it can be stabilized in a vacuum, and a ceramic probe under reduced pressure can be used to sample it in the flame and transfer it to a reaction cell maintained at a pressure of a few Torr in which it can react with ozone to produce light. The reaction is monitored by a photomultiplier device in the range 260-480 nm selected by a blocking filter. The SCD operates according to the following series of reactions.

$$RS + O_2 \longrightarrow SO + Products \tag{3.18}$$

$$SO + O_3 \longrightarrow SO_2^* + O_2 \tag{3.19}$$

$$SO_2^* \longrightarrow SO_2 + light \tag{3.20}$$

3.5.3.3 Atomic Emission Detector

Plasma sources, and to a lesser extent flames and carbon furnaces, are readily coupled to a gas chromatograph [193-196]. The only interface required in most cases is little more than a direct connection with an auxiliary flow of makeup gas to minimize dead volume effects. Plasma sources are capable of exciting intense emission from the elements C,H,D,O,N,S,P and the halogens, and are thus uniquely suited for the analysis of organic compounds. The plasma consists of a mass of predominantly ionized gas at a temperature of 4,000-10,000 K. This state can be maintained directly by an electrical discharge through the gas (dc plasma) or indirectly via inductive heating of the gas. The latter is established by an electromagnetic field, using power generated at radio frequencies (inductively coupled plasma) or microwave frequencies (microwave induced plasma). When organic compounds enter the plasma molecular breakdown occurs due to absorption of energy from the plasma and emission spectra are produced which are characteristic of the atoms of the compounds. The intensity of these emissions is measured by a direct-reading optical emission spectrometer coupled to suitably placed photomultiplier tubes or a photodiode array. All elements are excited simultaneously and, at plasma temperatures, molecular breakdown is complete; the measured response for each element is thus proportional to the number of atoms in the plasma and independent of the structure of the parent compound. With a multichannel instrument as many elements as there are channels can be measured simultaneously. Ratioing the response of each channel during the passage of a chromatographic peak will, after response standardization with compounds of known elemental composition, enable the empirical formula of unknown compounds to be determined directly [197-202]. The precision of such measurements is limited by plasma instability, resulting mainly from fluctuations in the power input and plasma gas flow rates. In addition, accuracy is limited by the linearity of elemental responses and the possibility of incomplete compound destruction. Accurate values for oxygen and nitrogen can be particularly difficult to determine due to entrainment of atmospheric gases into the plasma. Good precision has been demonstrated under favorable circumstances, RSD of a few percent, which would indicate that empirical formulae can be determined with acceptable accuracy by emission spectroscopy on samples smaller than those needed for conventional combustion analysis. Elemental composition information derived from atomic emission data can be combined with molecular weight information obtained by mass spectrometry to derive molecular formulas [220]. If a high resolution value for the molecular mass is available then the molecular formula can be obtained directly from a limited set of elemental ratios. If only low resolution molecular weights are available, then usually all elements present in the molecule must be identified before a unique molecular formula can be determined.

Most early work was done with direct current or inductively coupled argon or low pressure helium plasmas. Inductively-coupled argon plasmas effectively decompose large amounts of organic materials but provide poor excitation efficiency for the nonmetallic elements that are of primary chromatographic interest. Helium is the preferred plasma gas, since it provides a simpler background spectrum, higher excitation energy, and an improved linear response range. Real interest in plasma based detectors outside the research laboratory started with the introduction of an automated multi-element atomic emission detector by Hewlett-Packard in about 1989 [202-205]. The atomic emission detector, AED, is able to determine up to 15 elements automatically and incorporates several novel design features, Figure 3.15 [202,204]. The AED is based on a microwave-induced plasma, and uses a movable photodiode array detector in a flat focal-plane spectrometer. The plasma is produced in a thin-walled, liquid-cooled silica discharge tube within a microwave "reentrant" cavity. Power is supplied by a magnetron, and coupled to the plasma through a waveguide. The power level is typically 50 watts. The plasma is generated in an atmospheric pressure flow of helium made up of the column flow and an additional makeup flow gas as

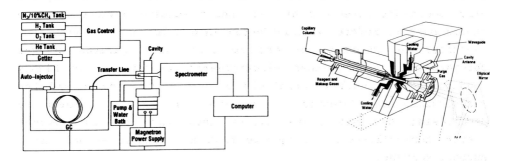

Figure 3.15 Schematic diagram of the Hewlett-Packard 5921A GC-AED system and cutaway view of the microwave cavity. (Adapted with permission from ref. 202 and 203. Copyright Dr. Alfred Huethig Publishers and American Chemical Society).

required. Depending on the elements being determined, low concentrations of various reagent gases are also added. There is also a purge flow in the exit chamber of the cavity, which is used to eliminate deposits on the window and to supply the flow for the solvent bypass mode, used to divert the column flow away from the plasma. The emission sensor consists of a polychromator and a movable photodiode array 12 mm long and containing 211 diodes. Prior to an analysis, the photodiode array is moved automatically to a position in the focal plane where the element(s) of interest have emission lines. The correct scavenger gases (H_2, O_2, or 10% CH_4 in N_2) are automatically turned on, and the required order shorter (optical filter) chosen by the software. In order to analyze elements simultaneously, they must have emission lines which fall within the wavelength range spanned by the photodiode array (10-30 nm depending upon its position in the focal plane) and require the same scavenger gases. For example, it is possible to obtain simultaneous chromatograms for C, S and N or C, H, Cl and Br, but under these conditions H, P, F and O would have to be run separately for additional measurements. Detection limits for individual elements range from 0.1 to 75 pg/s with a linear range of about 10^4.

3.5.4 Electrochemical Detectors

There are two general problems in the application of electrochemical detection to gas chromatography. First of all, few electrochemical detectors are gas phase sensing devices and the sample must therefore be transferred into solution for detection. Secondly, the majority of organic compounds separated by gas chromatography are neither electrochemically active nor highly conducting. The Hall electrolytic conductivity detector and the microcoulometric detector solve both of these problems by decomposing the gas phase sample into low molecular weight electrochemically-active fragments that are readily soluble in a support solvent.

Sample decomposition is carried out either by pyrolysis or by catalytic oxidation or reduction in a low-volume flow-through tube furnace. For oxidation or reduction, oxygen, air, or hydrogen is mixed with the carrier gas leaving the column and passed over a catalyst, usually a nickel or quartz tube with a nickel wire inside, maintained at a temperature of 500 to 1000°C. Organic compounds entering the furnace are decomposed into small molecular weight fragments, Table 3.8. To avoid a build-up of carbon deposits, the solvent is usually vented away from the furnace. A chemical scrubber is sometimes added to the flow system at the furnace exit to improve the selectivity of the detection process. Examples of chemical scrubbers include strontium hydroxide or potassium carbonate to selectively remove HX or SO_x, silver wire to remove HX or H_2S (X = halide ion), alumina to remove PH_3, and aluminum silicate to remove SO_x. With careful selection of pyrolysis conditions, mode of operation, and chemical scrubber, electrochemical detection permits highly specific and sensitive element detection.

3.5.4.1 Hall Electrolytic Conductivity Detector

The most widely used element-selective electrochemical detector is the Hall electrolytic conductivity detector (HECD) [98,116,206]. This is an improved version of an earlier design by Coulson [207,208]. In both detectors the reaction products are swept from the furnace into a gas-liquid contactor where they are mixed with an appropriate solvent. The liquid phase is separated from insoluble gases in a gas-liquid separator and then passed through a conductivity cell. The Coulson detector employed a

TABLE 3.8

PRODUCTS GENERATED BY THE CATALYTIC PYROLYSIS OF ORGANIC COMPOUNDS

Elemental Composition of Sample	Products Generated	
	Oxidative Mode	Reductive Mode
Carbon	CO_2	CH_4
Hydrogen	H_2O	H_2O
Halogen (X)	HX	HX
Nitrogen	NO_2 (low yield)	NH_3
Sulfur	SO_2/SO_3	H_2S
Phosphorus	P_4O_{10}	PH_3

siphon arrangement for the gas-liquid separator, which had a large dead volume and would occasionally loose prime during a separation [207-209]. The HECD employs a small-volume, concentric cylinder cell for mixing, separating, and monitoring the concentration of conducting species, Figure 3.16. The solvent and gaseous reaction products are combined in a small PTFE tee. The heterogeneous gas-liquid mixture thus formed separates into two smooth flowing phases upon contact with the inner surface of the stainless steel outer electrode. The driving forces of the cell are the downward force of the moving liquid phase, the attraction between the liquid phase and the detector surfaces, and the positive pressure of the liquid phase in the detector. Hall also improved the method of measuring sample conductance by employing an AC conductivity bridge in earlier designs and, more recently, a bipolar, pulsed, differential conductance circuit and cell design [116,210,211]. The bipolar, pulsed, differential detector employs a series conductivity cell (Figure 3.16) in which the conductivity of the solvent is monitored in the first portion of the cell while the conductivity of the solvent plus that of the reaction products is measured in the second part of the cell. The output of the two conductivity cells are differentially summed. Changes in the concentration of conductivity solvent and temperature fluctuations, which represent the principal source of daily response variations, are thus minimized [212,213]. Water was employed as the conductivity solvent by most early researchers but its use is now far less common. Organic solvents such as methanol and isopropanol, either alone or in admixture with water, are in common use today. These solvents provide higher sensitivity, greater selectivity for the species to be measured, and a greater

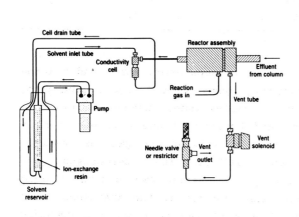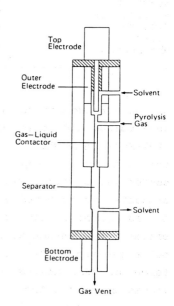

Figure 3.16   Schematic diagram of a Hall electrolytic conductivity detector with an expanded view of the bipolar pulsed differential conductivity cell.

linear response range than pure water. Most detectors recirculate the conductivity solvent through a closed system; a bed of ion-exchange resins purifies and conditions the replenished solvent. Conditioning may involve pH modification to improve the sensitivity and selectivity for a particular element. The selectivity and sensitivity of the detector response to individual elements can be improved by using an element-selective potentiometric electrode in place of the standard conductivity cell [214].

The practical problems most frequently encountered in operating the HECD are poor linearity and poor peak shape. Poor linearity is usually caused by neutralization of the conducting species when the pH of the conductivity solvent is incorrect and/or the scrubber is exhausted. Neutralization problems are readily recognized by a sharp dip in the baseline just prior to the peak, and a negative dip after the peak that gradually increases to the baseline. Peak tailing is often due to a contaminated scrubber, contaminated transfer line from the furnace to the cell, deactivated catalyst, or the presence of interfering, conducting reaction products that are not removed by the scrubber. The sensitivity of the detector is also influenced by the absolute flow rate and the flow rate stability of the conductivity solvent. Since it is a concentration-sensitive detector, a greater response is anticipated at lower solvent flow rates; this increase will, however, be accompanied by an increase in noise, resulting from fluctuations in the output of the solvent delivery system. Practical operating conditions are usually a compromise, in which the constancy and pulse-free nature of the solvent delivery system are extremely important.

The HECD is capable of high sensitivity and selectivity, although optimizing detector conditions and maintaining constant sensitivity at low sample levels can be troublesome. Detection limits on the order of 10^{-12} g N/s, 5×10^{-13} g Cl/s, and 10^{-12} g S/s can be obtained with a linear response range of 10^3 to 10^5. Detector selectivity values are variable, depending on the heteroelement and the detector operating conditions, but values in the range of 10^4 to 10^9 for nitrogen-, chlorine- or sulfur-containing compounds in the presence of hydrocarbons have been obtained. Compared to the FPD in the sulfur mode, the HECD has about the same or greater sensitivity and selectivity [210,211]. In comparison with the ECD, the HECD has a greater selectivity for chlorine-containing compounds and a more predictable response on a per gram of chlorine basis [215].

3.5.4.2 Microcoulometer Detector

The microcoulometer detector can be used to detect the same active species (e.g., SO_2, H_2S, HCl and NH_3) as the HECD [216-218]. It employs a four electrode electrochemical cell: two for generating the active species, one for sensing, and one as a reference electrode. A low concentration of a chemically active ion (e.g., Ag^+, H^+, or I_3^-) is generated in the cell and, when a reactive species enters the cell, this concentration falls. The concentration decrease is sensed by the working electrode and the resultant signal is fed via an amplifier to the generating electrodes, which then pass a current through the electrolyte until the original equilibrium ion concentration is restored. The energy consumed in regenerating the titrant ion is monitored with a precision resistor. This energy consumption is displayed as a voltage-time interval, integration of which reveals the total number of coulombs consumed in the reaction. The microcoulometer functions as an absolute detector, since the total number of coulombs consumed is related to the microequivalents of sample entering the cell by Faraday's law. The microcoulometer has a sensitivity of approximately 5×10^{-11} g/s of heteroelement, a linear range of about 100, and a selectivity ratio over hydrocarbons of about 10^6. However, the detector has a rather large dead volume and a very slow response time (ca. 20 s). Therefore, it is primarily used with packed columns for the analysis of relatively simple mixtures.

3.6 COLUMN CONNECTORS AND EFFLUENT SPLITTERS

Methods for interconnecting capillary columns are important for several common chromatographic procedures, such as joining two capillary tubes of identical or different diameters, fabrication of column effluent splitters and flow dividers, and for connection of the separation column to retention gaps [15,37,219]. The most common methods of making leak-tight, low-dead volume connections for capillary columns are press-fit connectors [220,222], heat shrinkable PTFE joints [223,224], butt connectors [15,225], and various procedures using glues and guiding tubes [219]. Some of these techniques demand considerable dexterity on the part of the user, others, however, are quite simple and make use of commercially available components. Low mass butt connectors, for example, are available from many vendors and achieve a stable joint using standard ferrules and screw-thread fittings. Problems may arise from ghost peaks caused by release of material from the ferrules, activity of the metal if the connection is not perfect, and from the lack of a simple visible test to check for misalignment of the capillary tubes or for tube breakage. Press-fit connectors are among the simplest and easiest to use connectors for fused silica capillary columns. Press-fit connectors are prepared by pushing polyimide-coated fused silica capillaries into a tapered seat in a glass connection tube. The thin layer of polyimide coating serves as a ferrule. The strong lateral forces exerted by pushing the capillary end into the conical seat provide both mechanical stability and tightness of fit. To connect a glass capillary to a fused silica capillary the glass capillary can be slightly opened using a heated, rotating, tungsten point to form the conical seat for insertion of the fused silica capillary [220]. Press-fit connectors for connecting fused silica columns of identical or different diameters are commercially available. Heat shrinkable PTFE tubing is one of the oldest methods of connecting glass or fused silica capillary columns. Connections are very easy to make by simply shrinking the PTFE tube over the two perfectly aligned columns or sleeve used to align the columns. A more stable connection is prepared by heat shrinking a second PTFE tube of large internal diameter over the first. The problems with PTFE connections include poor thermal stability above about 250°C, permeability to sample components and atmospheric oxygen that can cause degradation of some phases, and the adsorption of solvent when large volumes are injected [223]. The latter causes serious tailing of the solvent peak and baseline drift due to the slow release of the solvent.

It may be desirable to split the column effluent between two or more detectors operated in parallel to enhance the information content of the chromatogram, or to generate substance-characteristic detector response ratios, which aid compound identification. Ideally, an effluent splitter should provide a fixed split ratio that is independent of flow rate, temperature and sample volatility. It should also minimize extracolumn band broadening, be chemically inert, and for convenience, provide some mechanism for adjusting the split ratio over a reasonable range, e.g., 1:1 to 1:100, and be easy to clean [226]. For packed columns, T-splitters with fixed or valve-adjustable split ratios are generally used [227-229]. Dead volume effects are not usually a problem at packed column flow rates, although changes in the split ratio with temperature and the constancy of the split ratio with samples of different volatility may be. At the low flow rates used with capillary columns, dead volume effects can become significant unless the total volume of the splitter is reduced to the absolute minimum and all passageways are completely swept by carrier and/or makeup gas. Effluent splitters for columns have been fabricated from glass-lined stainless steel [226,230,231], platinum/iridium microtubing [232,233], glass capillary tubing [234-236] and zero-dead volume metal tees [237]. These may be in the form of simple T- or Y-splitters for dual detector operation or more complex roughs and stars for multiple detection and column switching, etc. [230,233,235]. Many of these devices can be constructed in the laboratory without special tools, although glass blowing and welding of capillary tubing requires practice and patience. Glass and silica are chemically inert; platinum/iridium tubing is relatively inert, although some catalytic activity remains even after tempering at red heat in a stream of oxygen [232]. As effluent splitters are constructed with fixed-diameter flow passages, the split ratio is varied by inserting wires or glass fibers of different diameters into one of the flow paths. To maintain a constant split ratio over a wide temperature range, as might be required during temperature programming, it is necessary to separately thermostat the splitter from the column oven, for example, by locating it in the detector oven block. The appreciable thermal mass of the commercially available zero-dead volume tees can cause problems due to cold trapping, and again, separate thermostating at a temperature above the column oven temperature is required [237]. One commercially available effluent splitter avoids this problem by employing a unique mixing and flow-stabilization chamber design [226].

The series coupling of detectors does not require an effluent splitter. This approach may be feasible when a nondestructive detector, such as the ECD or PID is used ahead of an FID, TID, HECD, etc. This technique has been demonstrated in a few instances for ECD/FID [238,239] and PID/ECD detection [240].

3.7 REFERENCES

1. B. E. Wenzel, J. Chromatogr. Sci., 28 (1990) 133.
2. J. N. Driscoll, CRC Crit. Revs. Anal. Chem., 17 (1987) 193.
3. J. Q. Walker, S. F. Spencer, and S. M. Sonchik, J. Chromatogr. Sci., 23 (1985) 555.
4. M. L. Lee, F. J. Yang, and K. D. Bartle, "Open Tubular Column Gas Chromatography: Theory and Practice", Wiley, New York, NY, 1984.
5. C. J. Cowper and A. J. DeRose, "The Analysis of Gases by Chromatography", Pergamon Press, Oxford, UK, 1985.
6. R. J. Grob, (Ed.), "Modern Practice of Gas Chromatography", Wiley, New York, NY, 1985.
7. G. Guiochon and C. L. Guillemin, "Quantitative Gas Chromatography for Laboratory Analysis and On-line Process Control", Elsevier, Amsterdam, 1988.

8. R. J. Bartram, W. J. Pinnick, and R. E. Shirey, J. Chromatogr., 388 (1987) 151.
9. K. Grob, J. High Resolut. Chromatogr., 1 (1978) 173.
10. B. W. Hermann, L. M. Freed, M. Q. Thompson, R. J. Phillips, K. J. Klein, and W. D. Snyder, J. High Resolut. Chromatogr., 13 (1990) 361.
11. S. Wicar, J. Chromatogr., 295 (1984) 395.
12. G. Schomburg, J. Chromatogr. Sci., 21 (1983) 97.
13. F. Munari and S. Trestianu, J. Chromatogr., 279 (1983) 457.
14. G. D. Reed and R. J. Hunt, J. High Resolut. Chromatogr., 9 (1986) 341.
15. W. Jennings, "Analytical Gas Chromatography", Academic Press, Orlando, Fl, 1987.
16. R. F. Severson, R. F. Arrendale, and O. T. Chortyk, J. High Resolut. Chromatogr., 3 (1980) 11.
17. H. J. Spencer, J. Chromatogr., 260 (1983) 164.
18. A. C. Mehta, J. Chromatogr., 494 (1989) 1.
19. K. Grob and H. P. Neukom, J. High Resolut. Chromatogr., 2 (1979) 15.
20. K. Grob and C. Gurtner, J. High Resolut. Chromatogr., 12 (1989) 335.
21. O. K. Guha, J. Chromatogr., 292 (1984) 57.
22. C. Dumas, J. Chromatogr., 329 (1985) 119.
23. J. Roeraade, J. Chromatogr., 441 (1988) 367.
24. K. Grob and G. Grob, J. High Resolut. Chromatogr., 2 (1979) 109.
25. J. Roeraade, G. Flodberg, and S. Blomberg, J. Chromatogr., 322 (1985) 55.
26. G. Schomburg and U. Hausing, J. High Resolut. Chromatogr., 8 (1985) 572.
27. D. M. Ottenstein and P. H. Silvis, J. Chromatogr. Sci., 17 (1979) 389.
28. J. E. Purcell, H. D. Downs, and L. S. Ettre, Chromatographia, 8 (1975) 605.
29. J. A. Jonsson, J. Vejrosta, and J. Novak, J. Chromatogr. 236 (1982) 307.
30. D. P. C. Fung and M. W. Channing, J. Chromatogr. Sci., 20 (1982) 188.
31. P. A. Hilling, P. A. Dawes, and W. David, J. High Resolut. Chromatogr., 11 (1988) 353
32. M. Ravey, J. Chromatogr. Sci., 19 (1981) 325.
33. A. B. Attygalle and E. D. Morgan, Angew. Chem. Int. Ed. Engl., 27 (1988) 460.
34. P. Sandra, (Ed.), "Sample Introduction in Capillary Gas Chromatography", Huethig, Heidelberg, Vol. 1, 1985.
35. G. Schomburg, U. Hausing, and H. Husmann, J. High Resolut. Chromatogr., 8, (1985) 566.
36. K. Grob, "Classical Split and Splitless Injection in Capillary Gas Chromatography", Huethig, Heidelberg, 1986.
37. K. Grob, "On-Column Injection in Capillary Gas Chromatography", Huethig, Heidelberg, 1987.
38. J. V. Hinshaw, J. Chromatogr. Sci., 25 (1987) 49.
39. J. V. Hinshaw, J. Chromatogr. Sci., 26 (1988) 142.
40. W. Jennings and M. F. Mehran, J. Chromatogr. Sci., 24 (1986) 34.
41. K. Grob, H. P. Neukom, and P. Hilling, J. High Resolut. Chromatogr., 4 (1981) 203.
42. K. Grob and H. P. Neukom, J. Chromatogr., 236 (1982) 297.
43. A. E. Kaufman and C. E. Polymeropoulos, J. Chromatogr., 454 (1988) 23.
44. J. Bowermaster, J. High Resolut. Chromatogr., 11 (1988) 802.

304

45. R. P. Snell, J. W. Danielson, and G. S. Oxborrow, J. Chromatogr. Sci., 25 (1987) 225.
46. K. Grob, M. Biedermann, and Z. Li, J. Chromatogr., 448 (1988) 387.
47. K. Grob, Th. Laubli, and B. Brechbuhler, J. High Resolut. Chromatogr., 11 (1988) 462.
48. K. Grob and M. Biedermann, J. High Resolut. Chromatogr., 12 (1989) 89.
49. J. V. Hinshaw and L. S. Ettre, J. High Resolut. Chromatogr., 12 (1989) 251.
50 K. Grob and B. Schilling, J. Chromatogr., 260 (1983) 265.
51. C. A. Saravalle, F. Munari, and S. Trestianu, J. Chromatogr., 279 (1983) 241.
52. K. Grob, J. Chromatogr., 279 (1983) 225.
53. K. Grob, J. Chromatogr., 324 (1985) 251.
54. J. V. Hinshaw and W. Seferovic, J. High Resolut. Chromatogr., 9 (1986) 69.
55. A. Tipler and G. Johnson, J. High Resolut. Chromatogr., 13 (1990) 365.
56. C. A. Saravalle, F. Munari, and S. Trestianu, J. High Resolut. Chromatogr., 10 (1987) 288.
57. H.-J. Stan and H. M. Muller, J. High Resolut. Chromatogr., 11 (1988) 140.
58. K. Grob and Z. Li, J. High Resolut. Chromatogr., 11 (1988) 626.
59. M. Termania, B. Lacomblez, and F. Munari, J. High Resolut. Chromatogr., 11 (1988) 890.
60. F. Poy, S. Visani, and F. Terrosi, J. Chromatogr., 217 (1981) 81.
61. W. Vogt, K. Jacob, A.-B. Ohnesorge, and H. W. Obwexer, J. Chromatogr., 186 (1979) 194.
62. M. Herraiz, G. Reglero, and T. Herraiz, J. High Resolut. Chromatogr., 12 (1989) 442.
63. G. Schomburg, U. Hausing, H. Humann, and H. Behlau, Chromatographia, 19 (1985) 29.
64. F. I. Onuska, R. J. Kominar, and K. Terry, J. Chromatogr. Sci., 21 (1983) 512.
65. F. Pacholec and C. F. Poole, Chromatographia 18 (1984) 234.
66. M. Galli, S. Trestianu, and K. Grob, J. High Resolut. Chromatogr., 2 (1979) 366.
67. M. Termonia, F. Munari, and P. Sandra, J. High Resolut. Chromatogr., 10 (1987) 263.
68. K. Grob and T. Laubi, J. Chromatogr., 357 (1986) 345 and 357.
69. P. Sandra, M. Van Roelenbosch, M. Verzele, and C. Bicchi, J. Chromatogr., 279 (1983) 279.
70. L. Ghaoui, F.-S. Wang, H. Shanfield, and A. Zlatkis. J. High Resolut. Chromatogr., 6 (1983) 497.
71. K. Grob, J. Chromatogr., 251 (1982) 235.
72. K. Grob and B. Schilling, J. Chromatogr., 391 (1987) 3.
73. K. Grob, G. Karrer, and M.-L. Riekkola, J. Chromatogr., 334 (1985) 129.
74. K. Grob, J. High Resolut. Chromatogr., 7 (1984) 461.
75. K. Grob, H. P. Neukom. and M.-L. Riekkola, J. High Resolut. Chromatogr., 7 (1984) 319.
76. K. Grob, H. P. Neukom, K. Chromatogr., 323, (1985) 237.
77. K. Grob, J. Chromatogr., 328 (1985) 55.
78. T. McCabe, J. F. Hiller, and P. L. Morabito, J. High Resolut. Chromatogr., 12 (1989) 517.
79. G. Hagman and J. Roeraade, J. High Resolut. Chromatogr., 13 (1990) 461.

80. D. J. David, "Gas Chromatographic Detectors", Wiley, New York, NY, 1974.
81. J. Sevcik, "Detectors in Gas Chromatography", Elsevier, Amsterdam, 1976.
82. E. R. Adlard, CRC Crit. Revs. Anal. Chem., 5 (1975) 1 and 13.
83. L. S. Ettre, J. Chromatogr. Sci., 16 (1978) 396.
84. M. Dressler, "Selective Gas Chromatographic Detectors", Elsevier, Amsterdam, 1986.
85. E. Katz, (Ed.), "Quantitative Analysis Using Chromatographic Techniques", Wiley, New York, NY, 1987.
86. P. L. Patterson, J. Chromatogr. Sci., 24 (1986) 466.
87. I. I. G. McWilliam, Chromatographia, 17 (1983) 241.
88. R. K. Simon, J. Chromatogr. Sci., 23 (1985) 313.
89. A. T. Blades, J. Chromatogr. Sci., 11 (1973) 251.
90. B. A. Schaefer, J. Chromatogr. Sci., 16 (1978) 211.
91. M. M. Thomason, W. Bertsch, P. Apps, and V. Pretorius, J. High Resolut. Chromatogr., 5 (1982) 690.
92. E. R. Colson, Anal. Chem., 58 (1986) 337.
93. J. T. Scanlon and D. E. Willis, J. Chromatogr. Sci., 23 (1985) 333.
94. A. D. Jorgensen, K. C. Picel, and V. C. Stamoudis, Anal. Chem., 62 (1990) 683.
95. H. Yieru, O. Qingyu, and Y. Weile, Anal. Chem., 62 (1990) 2063.
96. J. H. Wagner, C. H. Little, M. D. Dupuis, and H. H. Hill, Anal. Chem., 52 (1980) 1614.
97. A. R. Gholson, R. St. Louis, H. H. Hill, J. Chromatogr., 408 (1987) 329.
98. R. C. Hall, CRC Crit. Revs. Anal. Chem., 8 (1978) 323.
99. B. Kolb, M. Auer, and P. Pospisil, J. Chromatogr. Sci., 15 (1977) 53.
100. P. L. Patterson, J. Chromatogr. Sci., 24 (1986) 41.
101. G. R. Verga, J. Chromatogr., 279 (1983) 657.
102. I. B. Rubin and C. K. Bayne, Anal. Chem., 51 (1979) 541.
103. K. Olah, A. Szoke, and Zs. Vajta, J. Chromatogr. Sci., 17 (1979) 497.
104. P. Van de Weijer, B. H. Zwerver, and R. J. Lynch, Anal. Chem., 60 (1988) 1380.
105. C. S. Jones and E. P. Grimsrud, J. Chromatogr., 409 (1987) 139.
106. D. D. Bombick and J. Allison, J. Chromatogr. Sci., 27 (1989) 612.
107. J. N. Davenport and E. R. Adlard, J. Chromatogr., 290 (1984) 13.
108. J. N. Driscoll, J. Chromatogr. Sci., 23 (1985) 488.
109. P. A. Rodriguez, C. L. Eddy, and C. R. Culbertson, J. Chromatogr., 330 (1985) 365.
110. A. N. Freedman, J. Chromatogr., 190 (1980) 263.
111. G. I. Senum, J. Chromatogr., 205 (1981) 413.
112. A. N. Freedman, J. Chromatogr., 236 (1982) 11.
113. M. K. Casida and K. C. Casida, J. Chromatogr., 200 (1980) 35.
114. M. L. Langhorst, J. Chromatogr. Sci., 19 (1981) 98.
115. E. D. Pellizzari, J. Chromatogr., 98 (1974) 323.
116. S. O. Farwell, D. R. Gage, and R. A. Kagel, J. Chromatogr. Sci., 19 (1981) 358.
117. A. Zlatkis and C. F. Poole, (Eds.), "Electron Capture. Theory and Practice in Chromatography", Elsevier, Amsterdam, 1981.
118. C. F. Poole, J. High Resolut. Chromatogr., 5 (1982) 454.
119. G. R. Shoemake, D. C. Fenimore, and A. Zlatkis, J. Gas Chromatogr., 3 (1965) 285.

120. W. E. Wentworth, T. Limero, C. F. Batten, and E. C. M. Chen, J. Chromatogr., 468 (1989) 215.
121. W. E. Wentworth, E. D. D'Sa, C. F. Batten, and E. C. M. Chen, J. Chromatogr., 517 (1990) 87.
122. P. G. Simmonds, J. Chromatogr., 399 (1987) 149.
123. A. Neukermans, W. Kruger, and D. McManigill, J. Chromatogr., 235 (1982) 1.
124. D. C. Fenimore, P. R. Loy, and A. Zlatkis, Anal. Chem., 39 (1971) 1972.
125. C. H. Hartmann, Anal. Chem., 45 (1973) 733.
126. P. L. Patterson, J. Chromatogr., 134 (1977) 25.
127. G. Wells and R. Simon, J. High Resolut. Chromatogr., 6 (1983) 427.
128. G. Wells, J. High Resolut. Chromatogr., 6 (1983) 651.
129. J. Connor, J. Chromatogr., 210 (1981) 193.
130. J. E. Lovelock, Anal. Chem., 35 (1963) 474.
131. D. C. Fenimore and C. M. Davis, J. Chromatogr. Sci., 8 (1970) 519.
132. R. J. Maggs, P. L. Joynes, A. J. Davies., and J. E. Lovelock, Anal. Chem., 43 (1971) 1966.
133. E. P. Grimsrud and W. B. Knighton, Anal. Chem., 54 (1982) 565.
134. P. Rotocki and B. Drozdowicz, Chromatographia, 27 (1989) 71.
135. J. Connor, J. Chromatogr., 200 (1980) 15.
136. W. B. Knighton and E. P. Grimsrud, J. Chromatogr., 288 (1984) 237.
137. P. Rotocki and B. Drozdowicz, Chromatographia, 17 (1989) 71.
138. J. Vessman, J. Chromatogr., 184 (1980) 313.
139. C. A. Clemons and A. P. Altshuller, Anal. Chem., 38 (1966) 133.
140. D. A. Miller and E. P. Grimsrud, Anal. Chem., 51 (1979) 851.
141. E. C. M. Chen and W. E. Wentworth, J. Chromatogr., 217 (1981) 151.
142. E. C. M. Chen, W. E. Wentworh, E. Desai, and C. F. Batten, J. Chromatogr., 399 (1987) 121.
143. C. F. Poole, J. Chromatogr., 118 (1976) 280.
144. A. Zlatkis, C. K. Lee, W. E. Wentworth, and E. C. M. Chen, Anal. Chem., 55 (1983) 1596.
145. J. E. Lovelock and A. J. Watson, J. Chromatogr., 158 (1978) 123.
146. N. Hermandez-Gil, W. E. Wentworth, T. Limero, and E. C. M. Chen, J. Chromatogr., 312 (1984) 31.
147. P. L. Golby, E. P. Grimsrud, and S. W. Warden, Anal. Chem., 52 (1980) 473.
148. R. Simon and G. Wells, J. Chromatogr., 302 (1984) 221.
149. G. Wells, J. Chromatogr., 346 (1985) 1.
150. P. Rotocki and B. Drozdowicz, J. Chromatogr., 446 (1988) 329.
151. E. P. Grimsrud and S. W. Warden, Anal. Chem., 52 (1980) 1842.
152. S. W. Warden, R. J. Crawford, W. B. Knighton, and E. P. Grimsrud, Anal. Chem., 57 (1985) 659.
153. W. A. Aue, K. W. M. Siu, D. Beauchemin, and S. S. Berman, J. Chromatogr., 500 (1990) 95.
154. E. P. Grimsrud and D. A. Miller, Anal. Chem., 50 (1978) 1141.
155. J. A. Campbell, E. P. Grimsrud, and L. R. Hageman, Anal. Chem., 55 (1983) 1335.
156. J. A. Campbell and E. P. Grimsrud, J. Chromatogr., 291 (1984) 13.
157. M. A. Wizner, S. Singhawangcha, R. M. Barkley, and R. E. Sievers, J. Chromatogr., 239 (1982) 145.
158. C. A. Valkenburg, W. B. Knighton, and E. P. Grimsrud, J. High Resolut. Chromatogr., 9 (1986) 320.

159. F. Andrawes and R. Ramsey, J. Chromatogr. Sci., 24 (1986) 513.
160. R. S. Ramsey and R. A. Todd, J. Chromatogr., 399 (1987) 139.
161. F. Andrawes and P. Deng, J. Chromatogr., 349 (1985) 405.
162. J. Johns and A. L. Stapp, J. Chromatogr. Sci., 11 (1973) 234.
163. D. M. Rosie and E. F. Barry, J. Chromatogr. Sci., 11 (1973) 237.
164. M. G. Proske, M. Bender, H. Schirrmeister, and G. Bottcher, Chromatographia, 11 (1978) 715.
165. C. H. Lochmuller, B. M. Gordon, A. E. Lawson, and R. J. Mathieu, J. Chromatogr. Sci., 16 (1978) 523.
166. C. Roy, G. R. Bellemare, and E. Chornet, J. Chromatogr., 197 (1980) 121.
167. G. Wells and R. Simon, J. Chromatogr., 256 (1983) 1.
168. J. W. Carson, G. Lege, and R. Gilbertson, J. Chromatogr. Sci., 16 (1978) 507.
169. C. G. Flinn and W. A. Aue, J. Chromatogr. Sci., 18 (1980) 136.
170. C. G. Flinn and W. A. Aue, J. Chromatogr., 186 (1979) 136.
171. S. O. Farwell and C. J. Barinaga, J. Chromatogr. Sci., 24 (1986) 483.
172. T. J. Cardwell and P. J. Marriott, J. Chromatogr. Sci., 20 (1982) 83.
173. M. Maruyama and M. Kakemoto, J. Chromatogr. Sci., 16 (1978) 1.
174. M. Dressler, J. Chromatogr., 262 (1983) 77.
175. C. E. Quincoces and M. G. Gonzales, Chromatographia, 20 (1985) 371.
176. G. H. Liu and P. R. Fu, Chromatographia, 27 (1989) 159.
177. P. L. Patterson, R. L. Howe, and A. Abushumays, Anal. Chem., 50 (1978) 339.
178. P. L. Patterson, Chromatographia, 16 (1982) 107.
179. P. L. Patterson, Anal. Chem., 50 (1978) 345.
180. C. H. Burnett, D. F. Adams, and S. O. Farwell, J. Chromatogr. Sci., 15 (1977) 230.
181. P. J. Marriott and T. J. Cardwell, Chromatographia, 14 (1981) 279.
182. A. R. Baig, C. J. Cowper, and P. A. Gibbons, Chromatographia, 16 (1982) 297.
183. J. Efer, T. Maurer, and W. Engewald, Chromatographia, 29 (1990) 115.
184. K. K. Gaines, W. H. Chatham, and S. O. Farwell, J. High Resolut. Chromatogr., 13 (1990) 489.
185. R. S. Hutte, R. E. Sievers, and J. W. Birks, J. Chromatogr. Sci., 24 (1986) 499.
186. D. H. Fine, D. P. Rounbehler, E. Sawicki, and K. Krost, Environ. Sci. Technol., 11 (1977) 577.
187. D. P. Rounbehler, S. J. Bradley, B. C. Challis, D. H. Fine, and E. A. Walker, Chromatographia, 16 (1982) 354.
188. S. A. Nyarady, R. M. Barkley, and R. E. Sievers. Anal. Chem., 57 (1985) 2074.
189. N. Pourreza, S. A. Montzka, R. M. Barkley, R. E. Sievers, and R. S. Hutte, J. Chromatogr., 399 (1987) 165.
190. R. L. Shearer, D. L. O'Neal, R. Rios, and M. D. Baker, J. Chromatogr. Sci., 28 (1990) 24.
191. R. S. Hutte, N. G. Johansen, and M. F. Legier, J. High Resolut. Chromatogr., 13 (1990) 421.
192. A. L. Benner and D. H. Sledman, Anal. Chem., 61 (1989) 1268.
193. L. C. Ebdon, S. Hill, and R. W. Ward, Analyst, 111 (1986) 1113.

308

194. A. H. Mohamad and J. C. Caruso, Adv. Chromatogr., 26 (1987) 191.
195. P. C. Uden, Trends Anal. Chem., 6 (1987) 238.
196. P. C. Uden, Y. Yoo, T. Wang, and Z. Cheng, J. Chromatogr., 468 (1989) 319.
197. P. C. Uden, K. J. Slatkavitz, R. M. Barnes, and R. L. Deming, Anal. Chim. Acta, 180 (1986) 401.
198. K. J. Slatkavitz, P. C. Uden, and R. M. Barnes, J. Chromatogr., 355 (1986) 117.
199. A. L. P. Valente and P. C. Uden, Analyst, 115 (1990) 525.
200. D. B. Hooker and J. DeZwaan, Anal. Chem., 61 (1989) 2207.
201. P. L. Wylie, J. J. Sullivan, and B. D. Quimby, J. High Resolut. Chromatogr., 13 (1990) 499.
202. J. J. Sullivan and B. D. Quimby, J. High Resolut. Chromatogr., 12 (1989) 282.
203. B. D. Quimby and J. J. Sullivan, Anal. Chem., 62 (1990) 1027 and 1034.
204. P. L. Wylie and B. D. Quimby, J. High Resolut. Chromatogr., 12 (1989) 813.
205. P. L. Wylie and R. Oguchi, J. Chromatogr., 517 (1990) 131.
206. R. C. Hall, J. Chromatogr. Sci., 12 (1974) 152.
207. D. M. Coulson, Adv. Chromatogr., 3 (1966) 197.
208. M. L. Selucky, Chromatographia, 5 (1972) 359.
209. P. Jones and G. Nickless, J. Chromatogr., 73 (1972) 19.
210. B. J. Ehrlich, R. C. Hall, R. J. Anderson, and H. G. Cox, J. Chromatogr. Sci., 19 (1981) 245.
211. S. Gluck, J. Chromatogr. Sci., 20 (1982) 103.
212. R. K. S. Goo, H. Kanai, V. Inouye, and H. Wakatsuki, Anal. Chem., 52 (1980) 1003.
213. T. L. Ramus and L. C. Thomas, J. Chromatogr., 473 (1989) 27.
214. J. N. Driscoll, D. W. Conron, and P. Ferioli, J. Chromatogr., 302 (1984) 269.
215. T. L. Ramus and L. C. Thomas. J. Chromatogr., 328 (1985) 342.
216. E. M. Fredericks and G. A. Harlow, Anal. Chem., 35 (1964) 263.
217. R. L. Martin and J. A. Grant, Anal. Chem., 37 (1965) 644.
218. J. Sevcik, Chromatographia, 4 (1971) 102.
219. K. Grob, J. Chromatogr., 330 (1985) 217.
220. B. Schilling, K. Grob, P. Pichler, R. Dubs, and B. Brechbuhler, J. Chromatogr., 435 (1988) 204.
221. E. R. Rohwer, V. Pretorius, and P. J. Apps, J. High Resolut. Chromatogr., 9 (1986) 295.
222. A. D'. Amata, C. Bicchi, and M. Galli, J. High Resolut. Chromatogr., 12 (1989) 349.
223. V. Pretorius, P. J. Apps, E. R. Rohwer, and K. H. Lawson, J. High Resolut. cHromatogr., 8 (1985) 77.
224. G. Alexander and B. R. Gande, J. High Resolut. Chromatogr., 10 (1987) 156.
225. J. Roeraade, S. Blomberg and G. Flodberg, J. Chromatogr., 301 (1984) 454.
226. F. J. Yang, J. Chromatogr. Sci., 19 (1981) 523.
227. R. Digliucci, W. Averill, J. E. Purcell, and L. S. Ettre, Chromatographia, 8 (1975) 165.
228. H. A. McLeod, A. G. Butterfield, D. Lewis, W. E. J. Phillips, and D. E. Coffin, Anal. Chem., 47 (1975) 674.
229. P. L. Coduti, J. Chromatogr. Sci., 14 (1976) 423.
230. T. H. Parliament and M. D. Spencer, J. Chromatogr. Sci., 19 (1981) 432.
231. B. V. Burger and Z. Munro, J. Chromatogr., 262 (1983) 95.
232. F. Etzweiler and N. Neuner-Jehle, Chromatographia, 6 (1973) 503.

233. J. Roeraade and C. R. Enzell, J. High Resolut. Chromatogr.,
      2 (1979) 123.
234. P. Sandra, T. Saeed, G. Redant, M. Goderfroot, M. Verstappe,
      and M. Verzele, J. High Resolut. Chromatogr., 3 (1980) 107.
235. V. Pretorius, P. Apps, and W. Bertsch, J. High Resolut.
      Chromatogr., 6 (1983) 104.
236. F. Eltzweiler, J. High Resolut. Chromatogr., 7 (1984) 578.
237. M. F. Mehran, W. J. Cooper, M. Mehran, and R. Diaz, J. High
      Resolut. Chromatogr., 7 (1984) 639.
238. A. Sodergren, J. Chromatogr., 160 (1978) 271.
239. F. Poy, J. High Resolut. Chromatogr., 2 (1979) 243.
240. S. Kapila, D. J. Bornhop, S. E. Manahan, and G. L. Nickell,
      J. Chromatogr., 259 (1983) 205.

CHAPTER 4

THE COLUMN IN LIQUID CHROMATOGRAPHY

4.1 INTRODUCTION

The term "liquid chromatography" encompasses a lexicon of separation techniques with a single common feature, that of a liquid mobile phase. Compared to gases liquids provide a much greater variety of solvating capabilities with greater scope for selectivity optimization, while gases have more favorable kinetic properties yielding higher efficiencies and shorter separation times. Separations in liquid chromatography, consequently, are usually performed with a moderate number of theoretical plates at an optimized selectivity achieved by appropriate selection of the separation mode, stationary phase structure, and mobile phase composition. Understanding the complex relationship between the last three terms is the key to understanding how separations occur in liquid chromatography.

A number of distinct separation modes are employed in liquid chromatography. Separations based on the competitive adsorption of the sample and mobile phase by active sites on a solid adsorbent, such as silica gel, are referred to as liquid-solid chromatography (LSC). With the advent of chemically bonded phases in the late 1960s and early 1970s, the dominant position of inorganic oxide adsorbents as stationary phases declined. Today, about 15% of separations in liquid chromatography are performed by LSC using inorganic oxide or polar chemically bonded stationary phases and a relatively nonpolar mobile phase. Separations employing a relatively nonpolar hydrophobic sorbent as stationary phase and a polar mobile phase (in most cases an aqueous organic solvent mixture) are referred to as reversed-phase chromatography (RPC). This nomenclature reflects the historical precedent of the earlier development of LSC (also known as normal phase chromatography) followed by the explosive growth of RPC, in which the relative polarity of the stationary and mobile phases was reversed compared to LSC. Based on current usage it would be reasonable to call RPC normal phase chromatography, since about 65% of all separations by liquid chromatography are performed by RPC. As will be discussed later the current popularity of RPC is due to its unmatched versatility for the separation of neutral, polar and ionic samples encompassing a wide range of molecular weights, and the comparative ease of optimizing separations in this mode.

Bonded phase chromatography has largely replaced liquid-liquid chromatography (LLC), in which samples are separated by partitioning between two immiscible liquids, one of which is immobilized by coating onto a porous sorbent. A lack of genuinely immiscible solvent pairs, capable of forming stable separation systems, has been the primary reason for its demise. Separations employing a stationary phase containing immobilized charged groups to separate ionic or easily ionized samples are termed ion-exchange chromatography (IEC). The use of ion-exchange in liquid chromatography has grown rapidly in the last few years to about 15% of all liquid chromatographic applications due to increasing use in the life sciences and for the separation of inorganic ions by ion chromatography. Separations achieved by differences in molecular size using a non-sorptive stationary phase of controlled pore dimensions are referred to as size-exclusion chromatography (SEC). From an historical perspective, controlled porosity organic gels were used for molecular size separations, thus giving rise to the names gel permeation chromatography and gel filtration chromatography; in the latter case when aqueous mobile phases were employed. These terms are still commonly found in the modern literature but the general term size-exclusion chromatography is to be preferred for both the older soft gel packings and the modern rigid porous organic and inorganic gels.

4.2 COLUMN PACKING MATERIALS

Totally porous particles of relatively large particle sizes were widely used in low pressure liquid chromatography for many years. These column packings had good sample capacity but only limited efficiency accompanied by long separation times, due to the large size and unfavorable size distribution of the particles and the presence of relatively deep pores within the particles through which sample molecules diffused in and out of very slowly.

The same observations were true of porous polymeric beads used originally for ion-exchange packings that where instrumental to the development of the automated amino acid analyzer, probably the first example of a modern liquid chromatographic separation system. These packings were mechanically weak and could not be used at more than moderate pressure. The introduction of pellicular ion-exchange packings, developed by Horvath, to overcome the deficiencies of the soft porous polymers could be considered as the first introduction of a modern column packing [1]. These packings consisted of a glass bead core coated with a thin layer of organic polymer. The rigid core permitted operation at high pressures, but the packings produced less than desirable column efficiencies, due mainly to poor stationary phase mass transfer properties. These packings were soon replaced by superficially porous particles having improved mass transfer characteristics, Figure 4.1 [2,3], generally referred to as porous layer beads, or simply pellicular packings, that had particle diameters of 30 to 55 micrometers and consisted of a glass bead core to which was fused an intermediate porous silica or alumina layer 1-3 micrometers thick. The porous layer beads can be used without modification for liquid-solid chromatography, as a support in liquid-liquid chromatography, and in bonded phase and ion-exchange chromatography after chemical modification by reaction with organosilanes. These packings have a very small pore volume, providing rapid solute diffusion in and out of the stationary phase.

The importance of porous layer beads in modern liquid chromatography declined when totally porous silica microparticles with diameters less than 10 micrometers, and with narrow particle size ranges, became available (Figure 4.1) [3]. Indeed, when totally porous microparticles are employed, the most attractive feature of pellicular particles, namely their short diffusion pathways, becomes an intrinsic part of the support, and the use of an impervious core is unnecessary for particles with sufficient intrinsic strength. Compared to porous layer beads, totally porous silica microparticles can provide increases of up to an order of magnitude in column efficiency, sample capacity, and separation time, Table 4.1. Technical difficulties have prevented the production of porous layer beads in the same size ranges possible for totally porous microparticles. The current use of porous layer beads is largely limited to guard columns and extraction columns

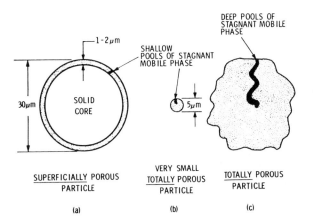

Figure 4.1   Types of particles for HPLC column preparation.

TABLE 4.1

PROPERTIES OF HPLC COLUMN PACKINGS

Property	Porous Layer Beads	Totally Porous Microparticles
Average Particle size (micrometers)	30-40	3-10
Best HETP values (mm)	0.2-0.4	0.01-0.03
Typical column lengths (cm)	50-100	10-30
Typical column diameters (mm)	2	2-5
Pressure drop (p.s.i./cm)	2	20
Sample capacity (mg/g)	0.05-0.1	1-5
Surface area (m^2/g)	10-15	400-600

where their ease of dry packing and lower cost are their principal virtues. In recent years pellicular packings have resurfaced in the form of nonporous microparticles of 1-3 micrometers diameter, for use in the separation of biopolymers [5,6]. Here, their principal advantages are high mechanical stability and improved mass transfer properties due to the elimination of slow diffusion of the high molecular weight biopolymers through the porous structure of conventional packings. Their surface areas and sample capacity are very low and only short columns can be used due to the high column backpressure, a consequence of their small particle size.

Totally porous microparticles can be prepared from a number of substances [7-10]. By far the most important are based on the hydrous oxides, silica gel and to a lesser extent alumina, and the porous organic polymer beads. Porous graphitic carbon packings have recently become available but have not been widely evaluated so far [8-11]. Packing materials based on inorganic phosphates (hydroxyapatite), cellulose, dextran, and agarose are used almost entirely for the separation of biopolymers [12-18].

4.2.1 Macroporous Silica and Alumina Packings

Macroporous silica gel is by far the most important adsorbent for liquid-solid chromatography and is also the material used to prepare most bonded phase packings, giving it a preeminent position in modern column technology [8-10,19-23]. Irregular silica gel microparticles are prepared by the sol-gel procedure [9,19,20,23]. A hydrosol is initially prepared by the addition of sodium silicate or tetraalkysilane to an acidic aqueous solution. Hydrolysis occurs spontaneously followed by condensation to poly(silicic acid) polymers which, depending on the processing conditions, form discrete nonporous spheres of 5-100 nm in diameter with a characteristic particle size distribution. Aging of the hydrosol results in gelling of the colloidal silica particles to larger aggregates accompanied by siloxane bond formation between particles and a strengthening of the gel framework. An insoluble cake is formed and eventually crushed into suitably sized portions, washed free of residual sodium salts, and finally dehydrated. A substantial shrinkage of the hydrogel lumps occurs upon dehydration, yielding a hard, porous xerogel that is subsequently milled and air sieved to the desired particle size range.

Spherical microparticles are more difficult to manufacture and can be prepared by several methods. One method prepares silica hydrogel beads by emulsification of a silica sol in an immiscible organic liquid [20,21,24,25]. To promote gelling a silica hydrosol, prepared as before, is dispersed into small droplets in a water immiscible liquid and the temperature, pH, and/or electrolyte concentration adjusted to promote solidification. Over time the liquid droplets become increasingly viscous and solidify as a coherent assembly of particles in bead form. The hydrogel beads are then dehydrated to porous, spherical, silica beads. An alternative approach is based on the agglutination of a silica sol by coacervation [25-27]. Urea and formaldehyde are polymerized at low pH in the presence of colloidal silica. Coacervated liquid droplets are formed having a spherical shape, which settle from the reaction medium and are subsequently harvested and dried. The polymer binder is then burned out of the particles forming porous silica microspheres. The mean particle diameter is controlled by the process variables and the pore diameter varied by starting from different sizes of silica sol in the range of 5-400 nm. As the pore diameter is increased the strength of the microparticles decreases. To compensate for this effect, and to maintain a minimum crush resistance, the larger pore size materials (50-400 nm) are processed at a higher temperature to increase the degree of sintering for strengthening purposes. Spherical silica microparticles can also be prepared by spray-drying techniques [20,25]. A silica sol with a defined concentration is dried in air at an elevated temperature by spraying from a nozzle into a centrifugal atomizer. During spray drying the liquid evaporates, leaving beads composed of compacted silica particles.

Different silica gels are characterized by their shape and mean particle size, specific surface area, mean pore diameter and specific pore volume [20,28,29]. Their chromatographic properties are also influenced by the type and number of surface functional groups, the presence of trace metal impurities, the surface pH, and solubility in mobile phases of different pH. For analytical applications microparticles of 5 to 10 micrometers in diameter are the most widely used, and to a lesser extent, 3 micrometer diameter microparticles. These sizes provide a reasonable compromise between column performance, stability, operating pressure and separation time. Smaller particles in the 1 to 2 micrometer diameter range (both porous and nonporous) have been introduced for the rapid separation of biopolymers, which because of their size, have poor diffusion characteristics [5,6,27,30]. Larger diameter particles, generally between 10 to 25 micrometers, are used in preparative-scale liquid chromatography. Still larger particles are used primarily for low pressure column separations. Although packings are sold in terms of a mean particle diameter, the material will contain particles of different sizes characterized by a size range. The particle-size distribution of packing materials can be determined in several ways, but the resulting data are presented for evaluation in one of two basic formats: as the mass (volume) distribution or the frequency (count) distribution. For the same material the cumulative statistical analysis of the data can lead to different mean particle diameters and size ranges. The frequency distribution is generally more sensitive to the presence of outsize particles (fines and boulders) which may cause poor chromatographic performance. Unfortunately, modern microparticle size analysis requires expensive, specialized equipment that is rarely found in analytical laboratories and, therefore, the analyst must usually rely on the quality control procedures of the various manufacturers for knowledge of the properties of the materials being used. With some skill particle size analysis can be performed by comparative microscopy [31] or by spectrometry using the time dependence of the turbidity of a solution of microparticles suspended in an appropriate liquid [32].

Spherical microparticles have been preferred in modern column technology since they form more homogeneous, stable and permeable column beds. Irregular microparticles are less expensive and still widely used, largely because in practice, it has not been shown that their properties are significantly inferior. Particle shape may become more important as the particle size is reduced, and spherical microparticles are considered superior for particle diameters less than 5 micrometers [33].

As well as shape and size, macroporous silica particles are characterized by their mean pore diameter, specific surface area, and pore volume. These parameters can be determined by well-established methods based on the physisorption of gases or vapors and the controlled penetration of fluids [20,30,33]. The specific surface area can be separated into two components: the surface area within the pores and the external surface area of the particle. The pore surface area is several orders of magnitude larger than the external surface area and, in general, the larger the surface area of the packing material the smaller will be the pore diameter. Some typical values for these parameters are summarized in Table 4.2 for Nucleosil silica gel.

Silica gels with mean pore diameters of 5-15 nm and surface areas of 150-600 m^2/g have been preferred for the separation of low molecular weight samples, while silica gels with pore diameters greater than 30 nm are preferred for the separation of biopolymers to avoid restricting the accessibility of the solutes to the stationary phase [15,16,29,34]. Ideally, the pore size distribution should be narrow and symmetrical about the mean value. Micropores are particularly undesirable as they may give rise to size-exclusion effects or irreversible adsorption due to

TABLE 4.2

PHYSICAL PROPRTIES OF NUCLEOSIL SILICA GEL (MACHEREY-NAGEL)

Designation	Mean Pore Diameter (nm)	Pore Volume (ml/g)	Specific Surface Area (m^2/g)
-50	5	0.8	500
-100	10	1.0	350
-120	12	0.7	200
-300	30	1.0	100
-500	50	0.9	35
-1000	100	0.75	25
-4000	400	0.70	10

the high surface energy existing within the narrow pores (micropores are considered to have a diameter < 3 nm, mesopores between 3-50 nm, and macropores > 50 nm; most types of chromatographic grade silicas are really mesoporous but are generally referred to as macroporous). It is important to note that of the above parameters only the specific pore volume, defined as the amount of liquid that fills the total volume of the pores per gram of adsorbent, has any real physical significance. The specific surface area and the pore diameter are calculated by reference to an appropriate model. The values obtained are only as accurate in absolute terms as the model can be said to be a perfect replica of the real situation. For porous silica particles a globular model is usually used to estimate the mean pore diameter. This model assumes that the porous system can be described by a regular arrangement of densely packed spheres. The pore volume is formed by the crevices between the spheres and is controlled by the size of the spheres and their packing density. The pore space can be envisioned as a matrix of alternating wide cavities and narrow constrictions. The effective pore size range can also be determined by size-exclusion chromatography and this value may be more meaningful for chromatographic studies than values based on mercury intrusion and gas condensation/evaporation methods interpreted using the globular model (see section 4.4) [35].

Commercial silica gels are available with a wide range of properties. The characteristics of some common chromatographic packings are summarized in Table 4.3 [36]. These data are taken mainly from the survey of Sander and Wise and are based on experimental determinations from a single batch of material. The

TABLE 4.3

PHYSICAL PROPERTIES OF SOME COMMERCIALLY AVAILABLE SILICA GELS

Brand Name	Nominal Pore Diameter (nm)	Specific Surface Area (m²/g)	Specific Pore Volume (ml/g)	Density (g/cm³)	pH of a 10% w/w Aqueous Suspension	Particle Type
LiChrosorb 60	6	398	0.66	0.44	7.2	Irregular
Polygosil	6	245	0.71	0.43	6.5	Irregular
Rsil	6	433	0.75	0.42	8.0	Irregular
Zorbax 60	6	432	0.50	0.57	5.6	Spherical
Econosphere	8	171	0.52	0.62	9.0	Spherical
Rosil	8	357	0.71	0.51	8.4	Spherical
Spherisorb	8	190	0.6		--	Spherical
Partisil	8	429	0.72	0.44	4.9	Irregular
Lichrosorb 100	10	297	1.11	0.36	6.7	Irregular
Zorbax 100	10	139	0.45	0.66	3.8	Spherical
Hypersil	12.5	149	0.70	---	8.2	Spherical
Zorbax 300	30	39	0.28	0.72	5.4	Spherical
Vydac TP	33	82	0.73	0.50	4.1	Spherical

values quoted are not necessarily identical to those found in the manufacturer's literature. For chromatographic applications the nominal mean pore diameter is used to determine the likelihood of size-exclusion effects. To a first approximation, the specific surface area indicates the strength of solute interactions (retention), but ideally should not influence selectivity, although this is not always true, since the composition of all silica surfaces are not necessarily identical. Therefore, to obtain similar retention on compared silicas it may be necessary to adjust the mobile phase composition. The apparent surface pH will be discussed later, but this is important in judging the suitability of different silicas for separating basic and acidic samples. The apparent density and specific pore volume of silica gel can be related to its mechanical strength and resistance to pressure deformation in HPLC [29]. Values of about 0.4 for density and 0.7 for pore volume are typical of stable silica gels. Wide pore materials have larger pore volumes and lower densities and are thus more fragile than smaller pore-size materials.

An understanding of the surface chemistry of silica is required to interpret its chromatographic properties. The silica surface consists of a network of silanol groups, some of which may be hydrogen bonded to water, and siloxane groups, as shown in Figure 4.2. A fully hydroxylated silica surface contains about 8

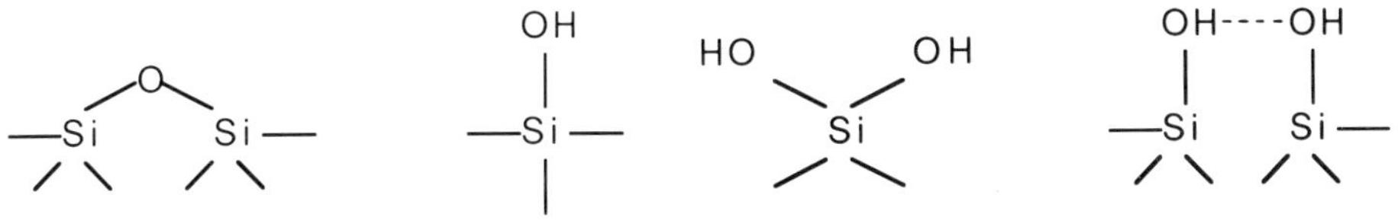

OH----OH
OH
HO   OH
| |
—Si   Si—   —Si—   \   /   —Si   Si—
/ \   / \ | Si   / \   / \
O   / \
SILOXANE   FREE SILANOL   GEMINAL   ASSOCIATED
                              SILANOLS   SILANOLS

Figure 4.2   Characteristic functional groups present on a silica surface.

micromol/m^2 of silanol groups (4.8 silanols per nm^2). Many types of commercial silica packings are not fully hydroxylated and have silanol group concentrations in the range 5.0-7.0 micromol/m^2 depending on the method of preparation and subsequent treatment [28,37,38]. The surface concentration of silanol groups can be adjusted to the maximum level by rehydroxylation, for example, by boiling in water (very slow) or, by treatment with dilute aqueous hydrofluoric acid or tetra-n-butylammonium hydroxide [25-27,38].

Spectroscopic evidence indicates that for a typical silica 68% of the silanol groups are of the single type and 32% are of the geminal type. By FTIR spectroscopy it has been shown that hydrogen-bond acceptor solutes are adsorbed preferentially on acidic, isolated silanols, whereas solutes with hydroxyl groups are adsorbed preferentially on associated silanols [37,39]. Contrary to popular belief, silica gels with a low adsorptivity for basic solutes should possess the highest possible concentration of associated silanol groups to ensure a minimum concentration of highly acidic, isolated silanols. Chromatographic grade silica gels usually contain 0.1-0.3% (w/w) metal oxide impurities (chiefly, Na, Ca, Al, Mg, Ti, and Fe) [28,37,40,41]. It has been suggested that the indirect influence of these metal impurities on adjacent silanol groups is responsible for their enhanced acidity. This would account for the small fraction of isolated silanols that react more strongly (perhaps by about 50-fold) with basic solutes than the average isolated silanol groups. Acid treatment of the silica gel can remove a sizable fraction of the metal impurities (but not all), resulting in a more homogeneous active surface.

The composition of silica depends on the method of preparation and thermal treatment [21,39,42]. Three temperature regions of importance in the thermal modification of silica include the region over which physically adsorbed water is removed (dehydration), a higher temperature region over which mainly associated silanols reversibly condense to siloxanes (dehydroxylation), and a third region, in which the temperature is high enough to induce lattice rearrangement, such that dehydroxylations become irreversible and sintering occurs. Thermal gravimetric analysis and FTIR spectroscopy indicate that physically adsorbed water is removed completely by heating to about 200°C, and is nearly complete by vacuum treatment at room temperature, or by heating at 115°C. Reversible dehydroxylation of silanols, resulting in an increased concentration of siloxane groups, commences at about 200°C but is most efficient over the temperature range 400-600°C. Thermal modification below 400°C generally occurs with little change in the surface area of the silica and is readily reversible by boiling in water. FTIR measurements indicate that the associated silanol groups are removed preferentially, particularly above 400°C, and that isolated silanol groups can remain intact until much higher temperatures are reached. At temperatures above 400°C, the siloxane groups, formed from the dehydration of associated silanols, rearrange to a more stable configuration that is apparently less prone to silanol reformation. At temperatures in the range 600-1100°C sintering of the silica increases progressively accompanied by a significant decrease in the pore volume and a decrease in the concentration of isolated silanols (the absolute temperature at which sintering occurs is strongly influenced by the presence of metal impurities). During sintering the silica becomes more fluid like and larger changes in physical and chemical properties are possible. A completely dehydrated silica surface is largely chemically and physically inactive and adsorbs moisture extremely slowly. It is not generally useful for chromatographic applications.

Theoretically, a fully hydroxylated silica surface should have a $pK_a = 7.1 \pm 0.5$ [28]. In practice, commercially available silica gels have an apparent pH which is very different from that predicted from theory, Table 4.2 [26,28,29,43,44]. The range of apparent pH values, determined from the pH of a suspension of silica in neutral, salt free water, range from 3.8 to 9.5. The most acidic and the most basic pH values are found for spherical silica microparticles, whereas relatively neutral values are common for irregular silica microparticles. These differences probably reflect differences in the manufacturing processes, the presence of surface impurities, and the different concentrations and types of surface silanol groups present. Although these pH values cannot be directly related to the pH of the silica used in a particular column separation employing a nonpolar eluent, it can, however, be demonstrated that the aqueous suspension pH value correlates well with the retention behavior of polar solutes. For polar solutes the surface pH affects such chromatographic parameters as relative retention, peak shape and column efficiency. It was observed that "acidic silicas" were more efficient at eluting proton donor solutes while proton acceptor solutes could not be eluted in an acceptable time with reasonable peak shapes. The opposite is true for "basic silicas". The tailing of polar compounds on silica gel can often be controlled by coating the silica packings with a buffer salt [45]. This can be done in a batch process prior to column packing or by an in situ coating procedure for packed columns. To be practically useful, the buffer should have a low solubility in organic solvents, for example, citrate, phosphate and borate buffers.

The equilibrium concentration of amorphous silica in water at room temperature is about 100 ppm [24]. This value does not change very much between pH 2-7. The solubility of silica increases exponentially above pH 8 due to the formation of the silicate anion. For mixed aqueous-organic eluents the solubility of silica depends on the aqueous content and type and concentration of pH buffer [46]. By using a high concentration of organic modifier, a saturator column before the injector, and ammonia as the source of hydroxyl ions, it was possible to use a silica column for a long time at a pH of 9.2 (given for the buffer solution). In another study, replacing ammonia by ethylenediamine permitted long term operation at pH 10.2 [47]. Normally, however, it is not recommended that silica-based packings be used at pH values greater than 8 and less than 1.0.

Alumina is not widely used in modern HPLC [48]. Porous gamma alumina is prepared by dehydration and thermal treatment of crystalline bayerite [8,49]. It is available in several types with pore diameters from 6-15 nm, surface areas 70-250 m^2/g and pore volumes 0.2-0.3 ml/g. After conditioning with acid or base its apparent surface pH can be adjusted between pH 3-9. The alumina surface is more heterogeneous than silica containing both hydroxyl groups and Lewis acid sites associated with partially coordinated aluminum ions. Its solubility at high pH is much less than that of silica and it can be used over the pH range 2-12.

4.2.2 Macroporous Chemically Bonded and Polymer Encapsulated Packings

Nearly all chemically bonded and polymer encapsulated column packings in use today are prepared from the same macroporous silica substrates discussed above. Bonded phases are prepared by reaction between the surface silanol groups of silica and reactive organosilanes to form siloxane bonds [7-9,15,20,24,28,50-52]. Polymer encapsulated phases are prepared by coating the silica substrate with a thin film of prepolymer which can be easily crosslinked to form an immobilized skin over the silica surface [53-56]. The first approach is better established and accounts for the majority of packing materials offered commercially. The latter approach is relatively new and is applicable to substrates other than silica, since it does not require chemical bonding to the surface for immobilization. The major disadvantage of silica-based bonded phases is their limited hydrolytic stability which restricts their use to the pH range 2-8. At higher pH their use is restricted by the solubility of silica and at lower pH by the hydrolysis of the siloxane bonded phase. Although alumina is more pH stable than silica, Al-O-Si bonded phases are less hydrolytically stable than bonded phases prepared from silica [49,57,58]. Similar deficiencies have been noted for zirconia bonded phases [59], although zirconia-clad silica substrates can be used to improve the stability of silica bonded phases at high pH, at least within the range pH 7-9 [60-62]. Polymer encapsulated phases based on alumina or zirconia macroporous supports are not limited by the hydrolytic stability of the bonds anchoring the organic ligands to the substrate and can be used over the full pH range dictated by the solubility of the substrate [61,63].

Bonded phases have been prepared by other general methods besides those indicated so far [64-66]. Reaction of silica with an alcohol or isocyanate resulted in the formation of silicate esters (estersils), but these phases were too hydrolytically unstable to be generally useful. Bonded phases with an Si-R or Si-NHR structure are more hydrolytically stable than the estersils but require a multi-step synthesis for their preparation, and the organometallic reagents used in their preparation often leave undesirable residues which are difficult to wash out of the final product. The greater simplicity and versatility of the direct reaction with reactive organosilanes has lead to their demise, except, perhaps for the preparation of alkyl bonded phases with substrates that do not yield stable siloxane bonded products [58].

Most commercially available bonded phase packings of the siloxane type are prepared by reacting an organosilane reagent with the surface silanol groups on a silica substrate [15,20-22,36,51,61,67]. The silica substrate is treated with aqueous acid to maximize the concentration of silanol groups. Silica gels containing about 8 micromol/m^2 of silanols produce a final product with a higher density of bonded phase, improved hydrolytic stability of the bonded ligands, increased mechanical strength, and a markedly lower adsorptivity for basic compounds [26,28,37,38,66,68]. The fully hydroxylated silica is finally washed to neutrality and dried under vacuum at less than 200°C to remove physically adsorbed water without dehydrating the surface silanol groups. The bonding reaction is performed by refluxing the silica in an inert solvent with various reactive organosilanes. Under anhydrous conditions monofunctional organosilanes, usually in the presence of a catalyst, result in the formation of a monomeric bonded phase, Figure 4.3. Using a difunctional or trifunctional organosilane under similar conditions will also produce a monomeric phase with each ligand bonded to the surface by one or two bonds to the reagent silicon atom. This material will always contain some unreacted reagent leaving groups, which must be endcapped during the workup. In the presence of a controlled amount of water multifunctional organosilanes react to form a polymeric bonded phase (Figure 4.3). In this case one or more of the leaving groups on the silane will hydrolyze and then react with other leaving groups to form a polymeric network extending out from the silica surface. This polymerization reaction may occur in solution before bonding to the silica surface, after the silane has already been bonded to the surface, or both [67,69]. Polymeric phases contain a higher concentration of bonded phase than monomeric phases, typically about twice as much, but they also contain a large number of silanol groups, many of which arise from the hydrolysis of the reagent and are not consumed in the polymerization process. After the bonding reaction

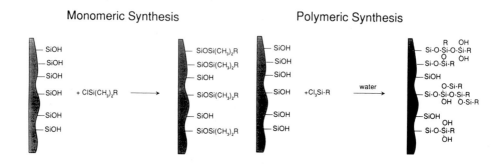

Figure 4.3 Synthesis of monomeric and polymeric siloxane bonded phases by reaction of organochlorosilanes with silica gel under different conditions.

both monomeric and polymeric phases are washed with a series of organic and aqueous solutions to remove adsorbed reagent and to hydrolyze the remaining reactive functional groups that are part of the skeleton of the bonded ligands. Since many undesirable chromatographic properties of bonded phases are associated with the presence of accessible silanol groups, their concentration is often minimized by reaction with various trimethylsilyl reagents in a subsequent reaction following the initial preparation of the bonded phase [52,66,70]. The replacement of accessible silanol groups in a bonded phase by trimethylsilyl groups is generally referred to as endcapping.

Comparing the synthetic procedures discussed above, monomeric phases are more straightforward to prepare and the reaction conditions should be more reproducible. The resulting monolayer coverage exhibits excellent mass transfer properties that produce high column efficiency for most solutes. Polymeric phases are more stable towards hydrolysis, but are not as easy to prepare reproducibly as are monomeric phases and may also exhibit lower column efficiency. Monomeric and polymeric bonded phases containing the same ligand substituent groups and prepared from the same silica substrate will, in many cases, exhibit different chromatographic selectivity. For example, consider the separation of the two test mixtures of polycyclic aromatic hydrocarbons on a monomeric and polymeric octadecylsiloxane bonded phase packing in Figure 4.4 [69]. The resolution and elution order of the different polycyclic aromatic hydrocarbons are both different and poorer for the monomeric packing compared to the polymeric material. It has also been suggested that the elution order of the hydrocarbons benzo[a]pyrene (BaP), tetrabenzonaphthalene (TBN) and phenanthro[3,4-c]phenanthrene (PhPh) is sufficiently characteristic on the two types of phase to be used to distinguish between different bonding chemistries [67,69]. These characteristic differences in selectivity result in one type of phase being preferred over the other for a particular separation rather than it being possible to claim general superiority for either type of phase. Similar comments can be made concerning endcapping. There are instances when undesirable silanol interactions can destroy a separation compared to the same phase after endcapping, Figure 4.5 [71]. This is particularly apparent in the separation of proton donor and acceptor solutes. By contrast, there are other cases where the presence of silanol groups contributed to the selectivity of the separation in a useful manner, and after endcapping the separation was no longer possible. These observations indicate that there is no ideal bonding chemistry and that variations in the properties of packings with the same nominal ligand substituent can be substantial.

A large number of factors influence the success of the bonding reactions and the quality of the final bonded phase product [72]. The rate and extent of the bonding reaction depends on the reactivity of the silane, choice of solvent, and catalyst, time, temperature and the ratio of reagents to substrate. Reactive organosilanes with Cl, OH, OR, $N(CH_3)_2$, $OCOCF_3$, and enolates as leaving groups have been widely used [52,66,72-74]. The dimethylamine, trifluoroacetate and enol ethers of pentane-2,4-dione are the most reactive leaving groups although economy, availability and familiarity result in the chlorosilanes and alkoxysilanes being the most widely used, particularly by commercial manufacturers. Initially, reactions may be almost stoichiometric but as the surface coverage approaches a maximum value the reaction becomes very slow. For this reason reaction times tend to be long, 12-72 hours, reaction temperatures

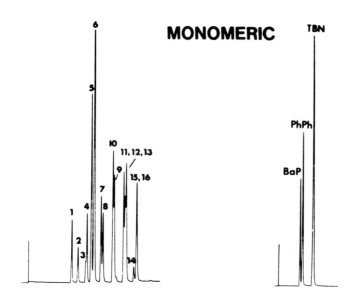

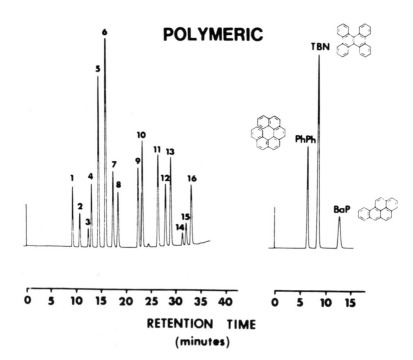

Figure 4.4 Separation of SRM 1647 and SRM 869 polycyclic aromatic
hydrocarbon test mixtures on a monomeric and polymeric reversed-
phase octadecylsiloxane bonded phases by gradient elution.
(Reproduced with permission from ref. 69. Copyright American
Chemical Society).

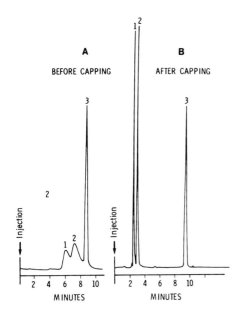

Figure 4.5 The influence of endcapping on peak shape and retention of some PTH-amino acids using a reversed-phase separation system. Peak identification: 1 = PTH-histidine, 2 = PTH-arginine and 3 = PTH-valine. (Reproduced with permission from ref. 71. Copyright Preston Publications, Inc.)

moderately high (in most cases around 100°C), and in the case of chlorosilanes, an acid acceptor catalyst (e.g., pyridine) is used. Some reagents, such as the alkylsilyl enolates and alkylsilyldimethylamines do not require additional catalyst, or even solvent, to carry out the reaction [73,75]. The most common solvents employed are toluene and xylene, although other solvents, such as carbon tetrachloride, trichloroethane and dimethylformamide have been recommended as being superior [72,74]. Since the bonding reactions are carried out by refluxing in an inert atmosphere, solvents are often selected based on their capacity to be a good solvent for the organosilanes and to attain the desired reaction temperature at reflux. Except for 3-cyanopropylsiloxane bonded phases, the high reactivity of chlorosilanes towards certain polar functional groups (e.g., NH_2, OH, CO_2H, etc.) precludes their use for the preparation of polar bonded phases. These are generally prepared from the less reactive alkoxysilanes [66,76,77]. Alkoxysilanes containing acidic or basic functional groups are autocatalytic and the bonded phases are usually prepared by refluxing the silane in an inert solvent at a temperature high enough to distill off the alcohol formed by the condensation reaction with the surface silanol groups. Bonding of neutral, polar ligands generally requires the addition of a catalyst, such as toluene-4-sulfonic acid or triethylamine in the presence of sufficient water to generate monolayer coverage of the silica. The presence of water speeds up the hydrolysis of the alkoxy groups of the adsorbed organosilane which tends to react with surface silanol groups rather than polymerize in solution. It seems to be a general problem in the preparation of polar bonded phases, that surface silanol groups are blocked by physically adsorbed organosilanes, giving rise to a lower bonded phase density after workup than the maximum theoretically predicted. The bonded phase density can be increased by repeating the reaction a second time or exposed silanol groups minimized by endcapping. Diol bonded phases are prepared by refluxing glycidoxypropyl-trimethoxysilane and silica in an aqueous suspension buffered to acidic or basic pH to promote bonding and simultaneous hydrolysis of the oxirane ring to a diol group [61,77]. The diol phase can be subsequently derivatized to introduce strong and weak ionogenic functional groups for ion-exchange chromatography [78]. Ion-exchange packings can also be prepared from appropriately substituted alkoxysilanes, or by subsequent reaction or coupling of ligands containing ionic groups to a prebonded phase containing reactive functional groups [79]. Some typical examples of bonded phases that can be prepared from reactive organosilanes are summarized in Table 4.4 [20,52,66]. Although most bonded phases are prepared from organosilanes containing a single functionalized ligand bonded to silicon with the remaining groups being leaving groups and/or methyl groups, more highly substituted organosilanes can also be used. Bifunctional organosilanes such as 1,3-dichlorotetraisopropyldisilazane are able to react with surface silanol groups at both ends of the chain forming a bonded phase that is more hydrolytically stable than bonded phases formed from conventional organosilanes [80]. The bidentate organosilanes have reactive sites that more closely match the spacing of the silanol groups on the silica surface and provides a higher bonded phase coverage than is achieved with dichlorosilanes with both leaving groups attached to the same silicon atom. For alkyldimethylsilanes increasing the length of the alkyl group increases the hydrolytic stability of the bonded phase relative to that of the

TABLE 4.4

STRUCTURES OF SILOXANE BONDED PHASES

Functional Group	Structure	Applications
Alkyl	$-CH_3$	Reversed phase
	$-C_4H_9$	
	$-C_8H_{17}$	
	$-C_{18}H_{37}$	
Phenyl	$-C_6H_5$	Reversed phase
Cyano	$-(CH_2)_3CN$	Normal and reversed phase
Amino	$-(CH_2)_3NH_2$	Normal, reversed phase and weak anion exchanger
Diol	$-(CH_2)_3OCH_2CH(OH)CH_2(OH)$	Normal phase and size exclusion
Amide	$-(CH_2)_3CONHCH_3$	Size exclusion
Sulfonic Acid	$-(CH_2)_3SO_3H$	Strong cation exchanger
	$-C_6H_4SO_3H$	
	$-(CH_2)_3SO_3H$	
	$-(CH_2)_3C_6H_4SO_3H$	
Carboxylic Acid	$-(CH_2)_3OCH_2COOH$	Weak cation exchanger
	$-(CH_2)_3COOH$	
	$-(CH_2)_3C_6H_4CH_2COOH$	
Dimethylamine	$-(CH_2)_3N(CH_3)_2$	Weak anion exchanger
Quaternary Amine	$-(CH_2)_3N^+(CH_3)_3$	Strong anion exchanger

trimethylsilyl bonded ligands [81]. Increasing the chain length of the methyl groups increases the hydrolytic stability of the bonded phase but reduces the phase coverage due to steric effects. The use of monofunctional organosilanes containing one or two bulky groups, for example, isopropyl or t-butyl on the silicon atom of the silane, may become more important in the preparation of bonded phases for use at low pH. The bulky alkyl groups provide better steric protection to the hydrolytically sensitive siloxane groups on the packing surface than does the methyl group. Mixed mode or zwitterionic siloxane bonded phases have been introduced for the separation of biopolymers [82,83]. These have been prepared by co-bonding hydrophobic ligands, such as octyl or octadecyl, with either cationic, or anionic ligands, in a single stationary phase. An alternative approach is to first bond a ligand with a reactive terminal group that can then be used to introduced the desired chemical functionalities in a second reaction. For example, silica gel bonded with aminopropyl groups can be partially acylated with a reagent containing an ester function. Selective hydrolysis of the ester group to form a carboxyl group yields a phase containing both surface bonded amine and carboxylic acid groups. Manipulation of the ionic strength and pH of the mobile phase enables neutral, anionic, cationic and zwitterionic molecules to be separated on the same stationary phase. Various mixed mode phases are commercially available, but they have not proven very popular.

There are three main factors that determine the accessibility of surface silanol groups for reaction with organosilane reagents [84]. A steric effect, due to the fact that the cross-sectional areas of most bonded ligands is greater than the average spacing between surface silanol groups, differences in the reactivity and possible clustering of silanol groups, and the silica morphology, particularly its porosity. From spectroscopic evidence, it has been shown that the geminal and isolated silanol groups are the most reactive towards chlorosilanes [38,39]. Typical bonded phase coverage for monomeric phases corresponds to about 3.0-3.9 micromol/m^2 for octadecyldimethylsilyl ligands rising to a maximum of about 4.4 micromol/m^2 for trimethylsilyl ligands. Thus even at maximum surface bonding densities between 30 and 50 percent of the surface silanol groups will remain unreacted. From isotope exchange studies it has been shown that access to all unreacted surface silanols is sterically hindered to different extents by the dense graft [85]. However, it is likely that all silanols are accessible in a time average fashion through occasional holes that appear in the graft, formed by the lateral vibrational motion of the bonded ligands. The concentration of bonded ligands also depends on the pore structure of the silica [86-89]. On passing from a flat to a concave surface the bonding density no longer depends solely on the size of the anchor group, but is determined by steric hindrance that occurs when the upper parts of the bonded ligands come into contact. This effect is of greater importance the longer the alkyl chain anchored to the pore surface. The extended conformation of the octadecylsilyl chain is

TABLE 4.5

INFLUENCE OF SILICA MORPHOLOGY ON THE PROPERTIES OF
OCTADECYLDIMETHYLSILOXANE BONDED PHASES

Parameter	Bare Silica			Bonded Phase		
	10	20	25	10	20	25
Nominal pore diameter (nm)	10	20	25	10	20	25
Specific surface area (m^2/g)	423	261	204	176	147	136
Pore volume (ml/g)	1.13	1.30	1.36	0.53	0.80	0.39
Mean pore diameter (nm)	10.5	18.8	25.0	8.5	16.2	21.6
Percent carbon bonded phase	---	---	---	19.2	15.47	13.2
Bonding density (micromol/m^2)	---	---	---	2.52	3.09	3.25

about 2.45 nm long (for methyl 0.3 nm, butyl 0.7 nm, and octyl 1.2 nm) and for pore diameters less than about 12 nm, assuming a cylindrical pore shape, the pore curvature will decrease the bonding density of the octadecyldimethylsiloxane ligands. Silica substrates with narrower pores would probably show a tendency to produce ink-bottle-shape pores after silanization as the initial silanization of the cylindrical pores at the edges would prevent other molecules of the octadecyldimethylsilane reagent from diffusing into the pores. A decrease in bonding density leads to alterations in the bonded layer thickness and to an increase in the concentration of unreacted surface silanol groups. The physical characteristics of the silica substrate are also changed by the bonding reaction, Table 4.5 [88,89]. A substantial reduction in the specific surface area, pore volume and mean pore diameter occurs upon siloxane bonding of organic ligands. There is a greater retention of the surface area and residual pore volume upon bonding as the pore diameter of the silica substrate increases. When there is no steric limitation on the observed bonding density the retention time decreases with increasing pore diameter as a consequence of the lower surface area, and therefore decreased quantity of bonded phase in the column. Selectivity differences, however, for phases based on a similar silica substrate are minimal [67]. More significant differences exist for polymeric phases, Figure 4.6 [36]. For the separation of polycyclic aromatic hydrocarbons differences in selectivity are apparent for different pore diameters that cannot be explained by

334

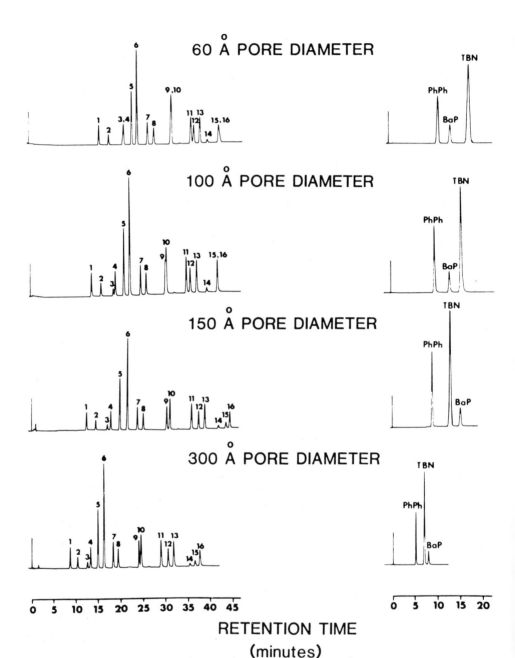

Figure 4.6 Separation of SRM 1647 and SRM 869 polycyclic aromatic hydrocarbon test mixtures on a series of polymeric octadecylsiloxane reversed-phase columns differing in pore diameter. (Reproduced with permission from ref. 36. Copyright Elsevier Scientific Publishing Co.)

size exclusion of the analytes. The differences in selectivity can be explained in terms of a size-exclusion mechanism that limits the extent of polymeric modification during bonded phase synthesis [36,67,68]. Under the conditions of the bonding reaction, a mixture of organosilane monomer and polymer exist together in solution. Both species are reactive and can bond to the silica substrate. However, because the organosilane polymer is larger than the monomer, size differentiation is possible within the pore network. For small-pore substrates, a phase with a higher monomeric character results with lower selectivity for the separation of the hydrocarbon test mixture. Greater polymeric character is possible for the larger-pore substrates because more of the pore volume of the substrate is accessible to the reactive polymer molecules.

Differences in the chromatographic properties of nominally similar column packings may result from differences in the phase coverage, carbon loading, extent of silanol deactivation, the type of silane used in the synthesis and even differences in the silica substrate. A number of chromatographic test methods have been developed to determine some of the above properties (section 4.4). Non-chromatographic methods are also in use and complement the information that can be obtained chromatographically [52]. The percent carbon loading for bonded phases is easily obtained by standard combustion analysis and is frequently quoted in the manufacturers literature [90]. The percent carbon loading increases with the size of the bonded ligand for the same surface coverage, but in the absence of specific knowledge of the surface area of the silica substrate, it is not a very useful parameter for comparison between phases. A more useful parameter is the surface coverage of the bonded phase calculated using equation (4.1)

$$N(\text{micromol/m}^2) = 10^6 P_c / [1200 n_c - P_c (M - 1)] S \qquad (4.1)$$

where P_c is the percent carbon loading of the bonded phase, n_c the number of carbon atoms in the bonded ligand, M the molecular weight of the bonded ligand, and S the surface area of the silica substrate [67,70,86]. For octadecylsiloxane bonded phases prepared from tricholorsilanes, $C_{18}H_{37}Si(OH)_2O-$ (M = 331) has been used as the representative unit for the phase structure. The oxygen atoms in the molecule are introduced as a result of hydrolysis during phase synthesis, and as such are considered part of the bonded

TABLE 4.6

REAGENTS USED IN THE PREPARATION OF OCTADECYLSILOXANE BONDED PHASES
IDENTIFIED BY ACID OR BASE CLEAVAGE

Phase	Manufacturer	Silane Reagent	Endcapped
Vydac 201-HS	The Separation Group	Monofunctional	Yes
Hypersil MOS	Shandon	Monofunctional	
Zorbax ODS	DuPont	Monofunctional	
μBondapak C_{18}	Waters	Difunctional	
LiChrosorb PR-18	Merck	Difunctional	
Partisil C_{18}	Whatman	Trifunctional	Yes
Nucleosil C_{18}	Macherey-Nagel	Trifunctional	Yes
Hypersil ODS	Shandon	Trifunctional	Yes
Spherisorb ODS	Phase Separations	Trifunctional	
Vydac 201 RP	The Separation Group	Trifunctional	

phase rather than part of the silica. The bonded phase coverage
can be used to indicate the completeness of the bonding reaction
for silicas with different pore structures and to estimate the
concentration of unreacted silanol groups for bonded phases
prepared from monofunctional organosilanes. The concentration of
unreacted silanol groups can also be determined by proton isotope
exchange or by reaction with an organometallic reagent such as
methyllithium [73,85,91,92]. The ligand type and method of bonding
to the surface can be determined by cleavage of the siloxane bonds
by alkali fusion in triglyme suspension [52,93] or by treatment
with methanolic aqueous hydrogen fluoride [94,95]. Alkali fusion
results in the formation of alkylsilanols and linear and cyclic
siloxane, which can be identified by gas chromatography. Treatment
with hydrofluoric acid produces stable, volatile
organofluorosilanes which are identified by gas chromatography.
The bonding chemistry used in the manufacture of some popular
octadecylsiloxane bonded phases has been identified by the
cleavage reactions given above, Table 4.6. This table illustrates
one reason why octadecylsiloxane bonded phases from different
sources often exhibit different chromatographic properties, since
they are prepared by different synthetic routes. Conventional FT-
NMR and CP-MAS NMR of ^{1}H, ^{13}C, and ^{29}Si is now routinely used to
identify the type of surface silanol groups, the level of
hydration, the type of surface silane bonding, and to gain insight
into the conformation of the bonded ligand in the dry state and in
solution [52,96-101]. FTIR can be used to study changes in the
type and number of silanol groups, conformation and order of the

bonded phase, and the presence of ligand functional groups [37-39,52]. Spectroscopic studies show that chain motion increases with distance from the surface, the chain mobility decreases with increasing surface coverage or after endcapping, and also decreases in the presence of water. Octadecylsiloxane bonded phases formed from trifunctional organosilanes show restricted motion in the region of the anchor group suggesting a brush-like or haystack-like structure. Octadecylsiloxane bonded phases prepared from monofunctional organosilanes can move freely suggesting a blanket structure with the top terminal part of the alkyl ligand bent to the surface of the silica substrate. Endcapping of these packings causes a change in order from a blanket structure to a brush or haystack configuration. Octadecylsiloxane bonded phases prepared from difunctional organosilanes exhibit structures intermediate between those for the mono- and trifunctional organosilanes, but generally close to those observed for monofunctional organosilanes. Also, it appears that the surface bonding is generally inhomogeneous with patches of high and low ligand density covering the surface.

In the last few years new bonded phase packings known as polymer coated or polymer encapsulated packings have been introduced, based on techniques used to prepare immobilized stationary phases for open tubular column gas chromatography [53-56,63,73,102]. In this case the phases are prepared by mechanically coating either bare silica or presilanized silica microparticles with a poly(siloxane) or poly(butadiene) prepolymer which is then immobilized by peroxide, azo-tert.-butane, or gamma radiation induced chemical crosslinking reactions. In a variation of this method the silica surface is first modified by reaction with vinyltrichlorosilane followed by polymerizing acrylic acid derivatives to and over the derivatized silica surface [53]. The availability of a large number of useful monomers and prepolymers has enabled a wide variety of reversed-phase, polar, and ion-exchange packings to be prepared using the same general reaction. Also, since the general approach does not depend on the chemistry of the underlying substrate, materials other than silica, for example, alumina and zirconia, can be modified and used under conditions for which silica is unsuitable, for example, with mobile phases outside the pH range 2-7.5. Returning to silica, presilanization decreases the number of active silanol groups which are then further shielded by the polymeric film anchored over the surface. In reversed-phase liquid chromatography these packings have shown improved chromatographic properties compared to monomeric chemically bonded phases for the separation of basic solutes. Polymer encapsulated packings have a film thickness of about 1 nm to maintain reasonable mass transfer characteristics. These films must be permeable, so presumably analytes can still interact with the substrate, and although one might anticipate a reduction in undesirable solute/substrate interactions using polymer encapsulated packings, it is difficult to envisage how these could be eliminated entirely by polymer encapsulation.

4.2.3 Macroporous Polymer Packings

Macroporous polymer packings have found increasing use for those applications that silica based packings are inadequate due to their hydrolytic instability or surface heterogeneity [9,12,103-106]. The polymeric packings are totally organic and have a more homogeneous surface devoid of strong bonding sites, and depending on their structure, can frequently be used at high and low pH where silica packings are degraded due to the formation of soluble silicates or hydrolysis of siloxane bonded ligands. They are often considered more suitable than silica bonded phases for the separation of certain ions, basic compounds and high molecular weight biopolymers that require relatively severe separation conditions using ion-exchange, reversed-phase or size-exclusion chromatography. The range of applications for some polymeric phases is limited, however, by a lack of rigidity, poor efficiency and a propensity to change dimensions (shrink or swell) with changes of solvent.

Macroporous polymers are generally prepared by a two-phase suspension polymerization of vinyl monomers, such as styrene, acrylic esters, vinyl acetate, etc., in a mixture of water, organic solvent, crosslinking agent, initiator and a suspension stabilizing agent. The suspension is rapidly stirred forming droplets in which the growing polymer chains form and eventually precipitate when they reach a critical size. The pore size is adjusted by using various ratios of a poragen, a solvent which is a good solvent for the monomer but a poor solvent for the polymer, in the presence of a crosslinking reagent, to form a rigid three dimensional structure. The resultant resin bead is approximately spherical and is comprised of many hard microspheres interspersed with pores and channels. Typical commercial products can be

TABLE 4.7

CHARACTERISTIC PROPERTIES OF MACROPOROUS POLYMERIC REVERSED-PHASE PACKINGS

Polymer	Particle Size (μm)	Pore Size (nm)	Surface Area (m^2/g)	Pore Volume (ml/g)
Hamilton				
PRP-1	5, 10 and 15-20	11	396	0.80
PRP-2		16	154	0.45
PRP-3		22	104	0.38
PRP-∞		--	1	
Polymer Laboratories				
PLRP-S	5, 8, 10 and 15-25	10	414	
		30	384	
		100	267	
		400	139	

prepared with particle sizes from 5-20 micrometers (if intended for use in HPLC), pore diameters from about 2-600 nm, and surface areas from about 2-500 m^2/g, Table 4.7 [103,107]. Resins containing less than about 8% by weight of crosslinking reagent yield soft gels unsuitable for high pressure operation. Semirigid packings used in HPLC may contain up to 50% or more of crosslinking reagent to permit high pressure operation and to minimize changes in particle size with changes in solvent [108-111]. Many of the undesirable chromatographic properties of polymeric packings based on poly(styrene)-divinylbenzene copolymers, the most popular type, has been attributed to the unavoidable incorporation of micropores into the polymeric structure [109-115]. For porous polymers the column permeability depends on the solvent and the efficiency depends strongly on the sample type, solvent and temperature and generally decreases with increasing retention time. The polymeric packings contain both macroporosity, due to pores among the microspheres and their agglomerates, and microporosity, due to pores within the microspheres. In good solvents, such as moderately polar organic solvents and aqueous mixtures containing acetonitrile or tetrahydrofuran, the micropores are swollen and column properties are most predictable. In poor solvents the microspheres are less swollen, the column efficiency lower, and the retention time greater. The poor efficiency is due to hindered diffusion within the micropores and increasing retention to the enhanced sorption effect due to the larger surface area contained

within the micropores. The presence of micropores in the polymeric
packings has been demonstrated by size-exclusion chromatography
[113-116].

The chromatographic properties of porous polymers can be
easily changed by choosing different vinyl monomers or by chemical
reaction of the preformed polymeric beads. Although styrene based
packings remain the most common, more hydrophilic packings based
on poly(vinyl alcohol), poly(alkyl methacrylate), poly(vinyl
pyridine) and poly(hydroxyalkyl methacrylate) are commercially
available [103,195,117]. Both poly(styrene) and poly(vinyl
alcohol) packings have been modified after polymerization to
incorporate an octadecyl or other alkyl group into the polymer
structure, to produce packings better able to mimic the properties
of silica-based materials, but having a wider pH operating range
[103,118]. A more common modification is the introduction of ionic
functional groups to produce packings for ion-exchange and ion
chromatography [9,15,16,104-106,119-121]. Virtually all
commercially available materials contain a sulfonate group (strong
cation exchanger), carboxylate group (weak cation exchanger),
tetraalkylammonium group (strong anion exchanger) or an amine
(weak anion exchanger) as the ionogenic functional group. Direct
reaction of the aromatic rings of poly(styrene) based packings
with oleum or chlorosulfonic acid, or chloromethylation with
chloromethyl ether and catalyst followed by amination, are widely
exploited reactions for introducing sulfonate and
tetraalkylammonium groups, respectively, into the column packing
materials, Figure 4.7. Poly(styrene)-divinylbenzene copolymer
packings containing one functional group per benzene ring would
have an ion-exchange capacity of about 4.5 mequiv/g [106].
Commercially available ion-exchange packings usually have
capacities of about 2.5-4.0 mequiv/g. For life science
applications ion exchangers have been prepared by covalently
bonding a poly(ethyleneimine) polymer to nonporous
poly(methacrylate) resin beads to obtain fast separations of
proteins [122]. Also, tentacle-like ion exchangers with ionic
groups exclusively located on linear polymer chains grafted on the
support surface, have been developed for biopolymer separations
[79]. This arrangement markedly reduces the contact between the
analyte and the matrix, thus suppressing nonspecific support
interactions. Further, this tentacle-like arrangement of the ionic
groups allows ionic interactions between the ion exchanger and the

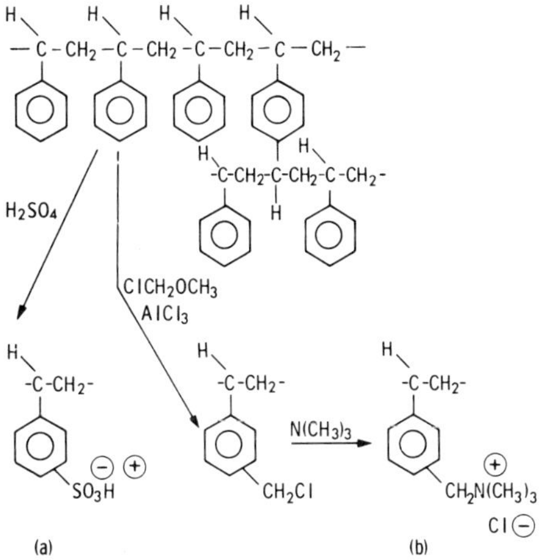

Figure 4.7 Methods for the preparation of ion-exchange resins.

analyte, which are impossible with standard-type ion exchangers for steric reasons. Low-capacity ion exchangers, 0.001-0.1 mequiv/g, are generally used in ion chromatography to facilitate sample elution by eluents of low ionic strength with conductivity detection [106,121,122,124]. These polymeric packings can be prepared by conventional techniques, modified to produce a low degree of functionalization. Alternatively, surface agglomerated pellicular packings have proven very popular for anion chromatography [106,121,124-128]. Modern versions of the original material consist of nonporous or porous poly(styrene) microparticles which are surface sulfonated and then coated with a layer of an oppositely charged latex resin. The core particle is about 5 micrometers in diameter and the latex particles much smaller, about 0.6-0.9 nm. The latex layer is very stable and held in place by electrostatic forces. This type of structure creates an anion-exchange pellicular layer within which the ion-exchange separation takes place and an oppositely charged underlayer which excludes anions through Donnan potential forces. These packings show higher efficiency than the macroporous packings due to the greater permeability and fast kinetics of the external layer, and exhibit low anion-exchange capacity because of the partial neutralization of the charge groups on the latex by the oppositely charged resin core.

4.2.4 Macroporous Graphitic Carbon Packings

Macroporous graphitic carbon packings are mechanically strong, pH stable and possess a homogeneous hydrophobic surface. They are prepared by the template replication method [8-11,129]. A wide pore, spherical silica microparticle packing is impregnated with a phenol-formaldehyde or phenol-hexamine mixture and polymerized, after which it is carbonized in an inert atmosphere at about 1000°C, and the silica template dissolved out by treating the carbon-silica composite with hot potassium hydroxide solution. Finally, the carbon skeleton is heated to above 2000°C to induce graphitization. This step eliminates the microporosity of the material but leaves the mesoporous structure intact. Typical aterials have surface areas of 100-300 m^2/g, pore volumes of 1.4-2.0 ml/g and a mean pore diameter that depends on the properties of the silica template. In reversed-phase chromatography the graphitic carbon packings behave as strong hydrophobic adsorbents with increased retention compared to typical octadecylsiloxane bonded silica packings: this might be a limitation for the separation of high molecular weight solutes. In the absence of unintentional oxidation of carbon during its thermal treatment, the graphite surface is chemically unreactive and cannot be derivatized directly. Possible methods of preparing chemically modified carbon based packings include physical adsorption of polymers onto the surface of the porous material or by crosslinking polymers over the carbon surface to generate polymer encapsulated packings in a manner similar to that discussed earlier for silica and alumina. Macroporous graphitic carbon packings were introduced relatively recently and are not fully evaluated. They are useful for the separation of basic solutes such as amines without tailing or irreversible adsorption, and exhibit novel stereoselectivity.

4.3 COLUMN PREPARATION

Few laboratories have developed their own facilities for packing conventional and small bore columns, preferring to rely on commercial manufacturers for their needs. This is understandable since packed columns containing any common phase can be obtained at an acceptable cost and with a guarantee of acceptable performance and stability. Column packing generally requires access to equipment not readily available in all laboratories and considerable experience of the packing process to guarantee success. Column packing has not been reduced to an exact science, and although good columns can be produced using a wide variety of different procedures, attempts to optimize the many variables and identify the best methods have met with limited success [130-132]. Similar packing materials often require an individual packing procedure to produce columns of optimum performance. A perusal of the contemporary literature of column packing techniques indicates many conflicting statements and generalizations that are not adequately tested, and can trap the unwary. If only a handful of columns are consumed per year it is unlikely that the investment in time for acquiring the necessary skills and experience in column packing can be recovered from the costs saved in not having to purchase columns. These arguments are irrelevant for those laboratories experimenting with new phases that are not commercially available, or working with packed capillary and open tubular columns that are not so common place that a viable commercial market has developed for them.

4.3.1 The Column Blank and Fittings

The column blank must be able to withstand the pressure at which the column is to be operated and be chemically resistant to a wide range of mobile phases. For the majority of applications seamless and polished 316 stainless steel meets the above requirements and is the preferred material. It is mechanically strong, available in a wide range of dimensions, and is inert to most solvents with the exception of aqueous solutions of halide salts, particularly at low pH, and certain organic solvents which can induce corrosion upon prolonged contact with the metal [4,133,134]. Alternatives to stainless steel include heavy-walled glass tubing, glass or polymer lined stainless steel tubing and plastic tubing. Heavy-walled glass tubing is used for cartridge columns and has many desirable features except for high pressure stability [135]. Columns with internal diameters of 3.0 mm are usually rated safe to about 200 atmospheres. Fused silica capillary columns, similar to those used in gas chromatography, with diameters up to about 0.5 mm are used to prepare packed capillary columns and have virtually replaced glass for this application due to their greater strength, flexibility and optical transparency for on-column detection [130,136-138]. Glass-lined stainless steel tubing combines the smoothness and inertness of glass with the strength of stainless steel and has been used to prepare conventional and narrow bore columns [139-143]. However, these columns do not always represent an ideal solution for all applications, since at high pH the glass becomes slightly soluble and it is difficult to attach compression fittings to narrow bore columns with thin metal walls to give an effective seal at high pressure without fracturing the inner glass liner [135,142]. Plastic walled columns are used in a unique way requiring a radial compression module [144-146]. The columns are prepared from heavy wall poly(ethylene) cartridges (packed with normal chromatographic packings) that are uniformly radially compressed by hydraulic pressure in a purpose-built hydraulic press. The column wall is forced to deform and molded to the shape of the internal packing structure. This stabilizes the packing structure and reduces the number of channels available to the mobile phase at the wall/packing interface. When compared to the performance characteristics of standard stainless steel columns some improvement in the packing homogeneity has been demonstrated but not a significant improvement in efficiency that might make these columns more popular than is presently the case.

Prior to packing, the interior wall of the empty stainless steel column must be cleaned and polished. It is particularly important that the column blank has a smooth inner surface and that residual grease and metal fines from the tube drawing process are removed. This is achieved by a thorough washing sequence involving aqueous and organic solvents. Four or five solvents of different polarity are usually sufficient. Dilute acids and detergent solutions may also be included in the washing sequence. The air-dried tube is then polished internally by passing a lint-free cloth attached to a nylon thread or a pipe cleaner through the column several times [140,147,148]. The inner wall of the polished column should be smooth and reflective when viewed against the light, and free from any indentations and/or burrs.

The packing material is retained and protected from the intrusion of particle matter by porous frits or screens. These may be either inserted into the column ends or made an integral part of the column end fittings. The frits are made from porous stainless steel, Hastaloy, titanium, poly(ethylene) or PTFE and have an average pore diameter less than the particle size of the column packing (typically 0.5, 2, or 10 micrometers). Column end fittings are not standardized, and although different fittings may look identical, they may have different thread sizes which makes them non interchangeable. The most common type are stainless steel internal column end fittings with female threads in the fitting body and a male nut. This arrangement tends to be more durable when reused frequently. Becoming more popular are finger-tightened fittings which rely on a polymeric ferrule to make the seal between the tubing and the fitting body. These can be used at pressures up to 2,000-6,000 p.s.i. depending on materials and design, and are commonly employed in cartridge column systems, since they can be reused many times. Other connections are usually made with zero-dead volume unions and short lengths of stainless steel capillary tubing. For metal-free systems, fittings, unions and capillary connecting tubing made from solvent and pressure resistant polymers such as Peek [poly(etheretherketone)] are employed.

4.3.2 Selection of the Column Packing Method

The method chosen for column packing depends mainly on the mechanical strength of the packing, its particle size, particle type, and column dimensions. Rigid particles, greater than 20 micrometers in diameter, can be dry packed, usually without much difficulty [149,150]. Particles of smaller diameter have a high surface energy to mass ratio and when dry packed result in an unstable packing structure with many voids, unsuitable for HPLC. For particles with diameters less than 20 micrometers, slurry packing techniques are used. Knox has summarized the critical points for the successful slurry packing of columns as follows [151]:

1. The particles must not sediment too fast during the procedure.
2. The particles must not agglomerate.
3. The particles must hit the accumulating bed with a high impact velocity.
4. Each particle should have time to settle in before it is buried by other particles.
5. The liquid used to support the slurry must be easily washed out of the packing and must not react with it.

These are broadly in line with the comments of others, except that some hold the view that to obtain the densest possible bed structure the impinging particle must hit the accumulating bed with sufficient speed to forcibly displace those particles already in position [130,131,152-154]. The slurry is displaced into the column under pressure and the accumulating bed grows by filtration. Independent of the method used for packing the ease of obtaining acceptable column performance decreases with decreasing particle size, decreasing column diameter, and increasing column length. The particles have to enter the column at high velocity in order to overcome frictional forces from viscous flow around the particles and at the wall. The packing speed is dependent on the applied pressure, viscosity of the slurry solvent, and the particle size. The wall effect should become more important the greater the column length to internal diameter ratio and minimized by using columns with smooth walls. There is little evidence to suggest that polishing the column wall has much impact on the packing structure of conventional bore columns of normal length [155] but it is widely believed that this is an important parameter for the successful preparation of small bore and packed capillary columns [33,140,147,148,156]. Since glass-lined stainless steel and fused silica columns have naturally smooth surfaces as well as reasonable bursting pressures, this has catalyzed interest in their use for preparing narrow bore columns (but not exclusively for small bore columns). The rationale being that the efficiency of the column is influenced by the packing density of the sorbent near the walls, which is less favorable in the case of unpolished tubes. The kinetic energy with which a particle hits the accumulating bed depends on both its mass (particle size) and velocity, as well as the particle shape and liquid viscosity. For large particles, about 10 micrometers, at a sufficiently high packing pressure optimum conditions are easily achieved over a reasonable range of experimental conditions. The choice of packing method and slurry type (Table 4.8) is not critical to success if experimental conditions are optimized. Reducing the particle size lowers the kinetic energy of the particles which must be compensated for by increasing the particle velocity. Higher pressures, lower viscosity slurry liquids, and upward packing will be more favorable for packing smaller diameter particles (2-5 micrometers) [33,131]. Similar reasoning indicates that frictional resistance to flow should be less for spherical particles which should be easier to pack than irregular particles having small particle diameters. With smaller particles, and especially with long columns, the backpressure that accompanies the accumulating bed increases to such values that the rate of packing declines below the critical value. In this case, upward

TABLE 4.8

OUTLINE OF SLURRY PACKING TECHNIQUES FOR RIGID MICROPARTICLES

Parameter	Column Type		
	Conventional	Small Bore	Packed Capillary
Column blank	Stainless steel Glass-lined stainless steel Glass	Stainless steel Glass-lined stainless steel	Fused silica
Treatment of column blank	Optional	Polished	Untreated or inner wall coated
Typical column internal diameter (mm)	4-5	0.5-3	0.1-0.5
Typical Column length (m)	0.05-0.3	0.1-1	0.2-2
Slurry type	Balanced density Viscous Low viscosity	Balanced density Viscous	Low viscosity
Slurry concentration (w/v)	5-30%	1-5%	20%
Packing technique	Up or down	Down	Down
Packing pressure (atms.)	300-1000	600-1800	300-500

packing is favored to avoid settling of the particles in the column, which would give a weak bed structure. Most columns have been packed at a constant pressure using widely available high pressure gas amplification pumps. To promote a more even packing structure both pressure programming [153,157] and constant velocity packing have been advocated [147,154,158]. One limitation of the constant velocity approach is the general lack of availability of pumps capable of maintaining the optimum flow at high pressure, at least for conventional diameter columns. Contemporary practice of column packing, at least in analytical laboratories, is dominated by constant pressure packing techniques. Even if not simple to achieve, it would seem desirable to arrange for columns to be packed with a constant pressure drop per unit column length to ensure a homogeneous packing density throughout the column. Also, there should be some minimum critical value for this parameter associated with the formation of a packing structure of maximum density. These conditions may be approached approximately at least for standard column lengths, conventional column diameters, and fairly large microparticles

(8-10 micrometers) by using a prepacked column as a restrictor down stream from the empty column blank to be packed [159].

The main requirements of the slurry packing liquid are that it must thoroughly wet the packing, provide adequate dispersion of the packing material, and be easily washed out of the column after packing. Since packings with very different surface properties and densities are used in HPLC, a wide range of common solvents and solvent mixtures have been used for this purpose. With particle diameters of about 10 micrometers, balanced density slurry techniques have been commonly used and also for packing small bore columns with smaller particles [140,147,160-164]. Balanced-density slurry solvents are usually mixtures of solvents, such as tetrabromomethane, carbon tetrachloride, methyl iodide and methanol mixed in various proportions to match the density of the column packing material. Particle segregation by sedimentation decreases as the density of the slurry liquid approaches that of the packing material. Consequently, the slurry liquid mixture must be tailored to the properties of the individual packing material and no universal recipe for all packings exists. The optimum slurry liquid combination is easily established by trial and error. After ultrasonic agitation, the balanced-density slurry should be stable at room temperature in a draft-free environment for at least 30 minutes without signs of sedimentation. Particle sedimentation towards the bottom of the flask indicates that the concentration of the denser slurry liquid should be increased. The opposite is true when sedimentation occurs against the influence of gravity. The brominated and iodated alkanes are somewhat expensive, toxic and may react with some packing materials, such as aminopropylsiloxane bonded phases. Settling out of the suspension can also be avoided by using slurry liquids of high viscosity, such as mixtures of isopropanol and glycerol [136,165,166]. However, solvents of high viscosity increase the resistance to flow through the column, so that the packing procedure is slow or requires a higher packing pressure than other methods. Packing materials of a narrow size range with diameters less than about 7 micrometers sediment very slowly in liquids with good dispersive properties which can be used for column packing if the packing procedure is carried out rapidly [33,167-170]. Typical solvents include carbon tetrachloride, chloroform, acetone, diethyl ether or methanol with low viscosity solvents being preferred, particularly for small particle sizes.

Packed capillary columns with internal diameters from 0.04 to 0.5 mm are packed in a similar manner to larger bore columns but are more permeable indicating a lower packing density. This increased permeability favors the packing of longer columns with smaller diameter particles than is possible for conventional diameter columns. In a way that is not completely understood the column inner diameter and packing density conspire to provide favorable kinetic efficiency under conditions that would be unfavorable for conventional diameter columns [138,171,172]. This effect is thought to be due to a decrease in the range of flow velocities for the packed capillary columns, more effective diffusional relaxation of retention variations due to flow heterogeneities, and a more uniform packing density brought about by the influence of the column wall. Reducing the column internal diameter increases proportionally the number of particles in contact with the wall which exerts a stabilizing influence on the packed bed to flow and pressure variations in spite of the lower packing density of the column. It has been suggested that the packing stability of packed capillary columns can be improved using fused silica columns with inner coated walls [173]. In this case the organic polymer film on the inner wall holds the packing in position because the particles are pressed into the elastic layer. Special frits are usually required to retain the packing in the column. Some examples include thin porous Teflon discs, porous ceramic plugs polymerized in the column end, glass wool or a thin sandwich of coarse packing held in position by a narrow tube glued to the capillary column [138,174,175]. In most cases columns have been packed using a concentrated slurry in a solvent of low viscosity, increasing the packing pressure in steps over time [138,153,176-179], although in one case columns were packed successfully using the constant flow method [172]. For reversed-phase materials better results were obtained by adding a small volume of a non-ionic surfactant to the slurry liquid. In all cases packing pressures tend to be lower than those used for small bore columns, Table 4.8, due to the lower bursting pressure of the fused silica capillaries compared to stainless steel.

4.3.3 Dry-Packing Column Procedures

Rigid particles with diameters greater than 20 micrometers can be dry-packed efficiently with relatively simple apparatus using the tap-fill method [149]. The empty column is held vertically with an end fitting at its lower end and a small amount of packing, equivalent to about 3-5 mm of column bed, is added from a funnel. The packing is consolidated by tapping firmly on a hard surface 2-3 times a second (80-100 times per packing increment) while lightly rapping on the side of the column at the approximate packing level and rotating the column slowly. Some authors state that vertical tapping during the packing process is detrimental to column performance and recommend only lateral tapping at the level of the column packing [150]. Further increments of packing are added and the process is repeated until the column is filled, further consolidated, and then the inlet fitting attached to the column without disturbing the column packing. The packed column is then attached to the liquid chromatograph and equilibrated with the mobile phase until no further bubbles appear at the column outlet and the column pressure and flow rate have stabilized.

It should be noted that incremental addition of the column packing, followed by consolidation of the packed bed, generally gives better results than bulk filling. Also, lateral tapping should be performed with a hard metal tool, vibration devices similar to those used to pack GC columns should not be used, since they tend to produce particle sizing across the diameter of the column. This results in a heterogeneous bed with large particles near the wall and the smaller particles at the center.

4.3.4 Down-Fill Slurry Column Packing

Slurry packing techniques are required for the preparation of efficient columns with rigid particles of less than 20 micrometers in diameter. The same general packing apparatus, Figure 4.8, can be used to pack columns by the balanced-density slurry, liquid slurry, or the viscous slurry techniques. Down-fill slurry packing is the method of choice for small bore columns and packed capillary columns.

Prior to the preparation of the slurry, some preliminary treatment of the packing material is required. Silica gel packings are heated in an oven to remove physically adsorbed water that tends to cause particle sedimentation during packing. Fines may be removed by repetitively suspending the material in a solvent, allowing the main fraction to settle, and pouring off the supernatant containing the fines. Aggregates can also be removed

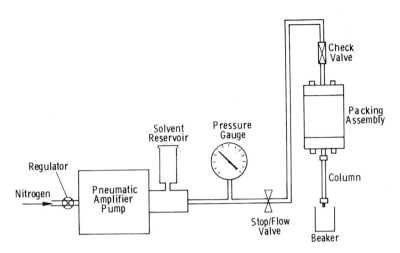

Figure 4.8 Down-fill slurry packing apparatus. (Reproduced with permission from ref. 168. Copyright Elsevier Scientific Publishing Co.)

in a similar manner. The larger particles settle fastest and can be removed at the bottom of an appropriate vessel.

Typical slurry concentrations are 1-30% (w/v) and have to be optimized for each solvent and packing combination. Before packing, the slurry should be thoroughly degassed and dispersed using an ultrasonic bath. The size of the packing reservoir should be scaled in proportion to the dimensions of the column blank and should not be overly large. Stainless steel bombs with finger-tightened connections are commercially available for packing conventional and semipreparative columns; for narrow bore columns sections of stainless steel tubing will serve as a slurry reservoir. Some reservoirs have a conical internal shape to promote a smooth passage of the slurry into the column [162].

Using Figure 4.8 as a guide, the down-fill slurry column packing procedure will be described [140,147,149,158-162,165-168]. A retaining porous frit or screen and standard Swagelok fitting are attached to the lower end of the precleaned column blank. The top end of the column is connected to a short length (3-5 cm) of connector tubing, which in turn is attached to the base of the slurry reservoir. The purpose of the connector tubing is to serve as a mini-reservoir; it guides the slurry into the column blank and ensures that the column is completely filled when the packing procedure is complete. It is important that the internal diameter of the connector tubing and the outlet fitting of the slurry reservoir are the same as that of the column blank. This will ensure a rapid and smooth delivery of the slurry into the column without subjecting the packing to undue sheer forces.

The column blank and connector tubing are filled with either the slurry liquid or a solvent of higher density. The side of the column is tapped with a metal rod to ensure that air bubbles are not left in the column during this step. The solvent is retained in the column by attaching either a stopcock or a terminator end fitting to the bottom of the column. The stop-flow valve is then opened to fill the reservoir and connector tubing with the packing solvent (the packing solvent may be the same as the slurry solvent but is often different). With the stop-flow valve closed the packing pressure is then established (see Table 4.8 for typical values). At this point the degassed slurry is rapidly added to the slurry reservoir which is then closed. Any air pockets are displaced with additional slurry solvent. The reservoir is then immediately attached to the pump line, the stopcock opened or the terminator removed, and the stop-flow valve opened. The packing process is rapid (except for the viscous slurry method); the actual filling of the column is usually completed within a few seconds. However, a constant pressure is maintained until about 20 column volumes of packing solvent has been pumped through the column. The stop-flow valve is then closed and the pressure allowed to bleed through the column until the solvent flow ceases. The column performance and stability is often improved if the freshly packed column is further consolidated without removing it from the packing assembly. The packing apparatus is pressurized to its original value, the stop-flow valve is opened, and a further 20 column volumes or so of solvent is pumped through the column. This process is repeated several times until the column permeability is unchanged, that is, at a constant pressure, the column flow rate does not change after repressurization. The column is then removed from the packing apparatus and excess packing material removed from the top of the column with a razor blade. Afterwards a screen or frit and a zero-dead volume fitting are attached to the column detector end and the column is ready for testing and use. Since the bottom frit can become plugged with fines during the packing process, it is sometimes recommended that it is replaced with a new frit before testing [180]. The column may require further conditioning to wash out the packing solvent or to control the activity of adsorbent packings.

4.3.5 Up-Fill Slurry Column Packing

The up-fill slurry packing method is the only packing technique that has been used to pack columns longer than 25 cm with particles of 5 micrometers in diameter or less [149]. It is generally considered unsuitable for packing small bore columns. It is easily adapted to the simultaneous packing of several columns [181]. Low viscosity slurry solvents are preferred for packing 2 and 3 micrometer diameter particles [33].

The principle of the packing method is based on the upward displacement of a stable slurry into the column blank from a reservoir whose contents are continuously diluted by incoming pressurized solvent [33,169,170,181,182]. Dilute slurries (1-10% w/v) and continuous stirring are used to minimize agglomeration of the packing slurry in the reservoir. This is important for particles larger than 5 micrometers in diameter to ensure that the particles remain suspended during the up-flow packing process. The viscosity of the slurry liquid determines the time required to pack the column and the impact velocity of the particles during the packing process. Therefore, low viscosity slurry solvents are preferred. The controlling factor for packing columns by the up-fill method is that the velocity of both the particles striking the bed and the solvent transporting them must be high enough to prevent the particles from falling back into the reservoir. Yet the required velocities are still relatively slow. As a result, the up-fill method allows the greatest opportunity for ordering of the bed to produce a densely packed structure.

To pack a column, the slurry is poured into the reservoir, Figure 4.9. It is stirred continuously to maintain a homogeneous suspension. With the lid of the reservoir securely in place and the outlet valve open, the slurry solvent is pumped gently into the attached column until the first drop of slurry appears at the end of the column. This purges trapped air from the system. At this point the outlet valve is closed and the packing pressure established (2,000-7,500 p. s. i.). The valve is then reopened, and the slurry is forced into the column. The packing process is complete when the flow rate of solvent through the column has stabilized. The pump and magnetic stirrer are stopped, the pressure is allowed to bleed away, and the column is disconnected

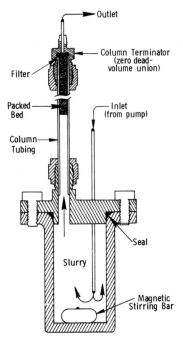

Figure 4.9   Up-fill slurry packing apparatus. (Reproduced with permission   from Micromeritics Instruments Corp.)

from the reservoir. A zero-dead volume fitting and frit are attached to the column inlet end and the column is ready for equilibration and testing.

4.3.6   Packing Columns with Semirigid Particles

Packing procedures for semirigid particles are similar to those used for rigid particles, except that the column packing pressure is limited by the lower mechanical strength of the packing. Ion-exchange resins with a polymeric backbone are usually packed by the salt-balanced-density slurry method. Organic resins must be swollen in the slurry solvent prior to packing. The resin is packed into the column at a pressure dependent on the mechanical strength of the swollen resin beads, usually less than 5000 p.s.i. for even the strongest beads. Semirigid organic gels are usually packed by the balanced-density slurry packing technique after first being swollen for several hours in the slurry medium. These gels are normally packed at pressures less than 3000 p.s.i. The organic gels have a low density compared to silica and require lower density solvents for slurry preparation.

4.3.7 Preparation of Open Tubular Columns

Slow diffusion of solutes in liquids requires that open tubular columns have extremely small physical dimensions to be competitive with packed columns in terms of performance and separation time. This, in turn, results in severe experimental difficulties in their use which has effectively limited their development outside of a handful of research laboratories (see section 1.7.10). Two types of microtubular columns have been developed, truly open tubular columns and semi-packed capillary columns [179,183]. Semi-packed capillary columns are prepared by filling a glass tube with sorbent, such as alumina or silica gel, and drawing it out to capillary dimensions with a conventional capillary glass drawing machine similar to those used to prepare open tubular columns for GC [184-186]. Under these conditions, the particles are partially welded to the capillary wall, thus giving the columns adequate mechanical stability at high operating pressures. Column internal diameters are typically 0.03-0.1 mm and particle diameters of 10 to 30 micrometers. Chemically bonded packings are unable to withstand the high temperature used in the drawing process and are prepared after the column is drawn by silanization of the sorbent with an appropriate reagent.

Open tubular columns are prepared from glass or fused silica capillaries having internal diameters between 5 to 75 micrometers. Fused silica open tubular columns can be coated with immobilized siloxane phases using peroxide crosslinking reactions developed for general use in GC. Short column lengths with internal diameters as small as 5 micrometers can be coated by the static coating method [187] or by precipitation coating [188]. Static coating is often limited by the unrealistically high pressures required to fill narrow capillaries with the coating solution and the slow evaporation rate of the solvent during the coating process. Precipitation coating relies on the difference in solubility of a polymer in a given solvent system at different temperatures. The capillary is filled with a hot solution of the polymer and then quickly cooled to room temperature. The remaining solvent is then pushed out by pressure leaving the polymer film behind for subsequent crosslinking. In both cases the volume of stationary phase that can be deposited is limited by the low surface area of the columns. Soda-lime and borosilicate glass capillaries can be chemically leached by treatment with aqueous base [179,183,189-192] or by electro-etching techniques [193,194], respectively, to selectively remove surface constituents producing a thin layer of porous silica with a greater wall surface area than the original capillary. This thin adsorbent layer can then be either dynamically coated with a liquid stationary phase or reacted with an appropriate silylating reagent to prepare bonded phases.

4.3.8 Scavenger Columns and Guard Columns

A scavenger column is a conventional column of various sizes placed between the pump and the injector. It removes unwanted contaminants from the mobile phase, acts as a filter to protect the analytical column from particle matter, pump-seal wear particles, etc., and as a saturator column when using aggressive mobile phases. Scavenger columns are most frequently used to saturate the mobile phase with silica gel when mobile phases of pH > 8 are employed [194]. The column is not located in the sample pathway and need not be identical to the analytical column in particle size or packing density. For reasons of economy, and to minimize the operating pressure, scavenger columns are usually packed with coarse particles but of a fairly narrow size range. For isocratic operation the size of the column is not too important, as long as it has sufficient capacity to perform its function. With gradient systems the scavenger column must be small to minimize the gradient delay volume. Scavenger columns are optional components used only when dictated by the characteristics of the mobile phase.

A guard column is a short column placed between the injector and the analytical column that is used to extend the life of the analytical column by protecting it from deterioration caused by sample contaminants and highly retained solutes, as well as from wear particles from the pump and injection valve [195-197]. It also functions as a saturator column to prevent dissolution of the stationary phase in the analytical column , although a scavenger column should be used for this purpose. Since it is located in the sample pathway, it must be packed with the same material and to the same packing density as that used for the analytical column. To maintain an adequate capacity for sample impurities without introducing excessive peak dispersion, the volume ratio of the guard column to that of the analytical column should be in the range 1:15 to 1:25. A well-designed guard column should not increase sample dispersion by more than 5-10%. When a greater loss of column efficiency can be tolerated, guard columns packed with pellicular packings can be used. These are less expensive, are easily repacked in the laboratory by the tap-and-fill method, but because of their low sample capacity offer less protection than slurry packed microparticle guard columns.

Guard columns are meant to be sacrificed and do not have a similar useful working life to the analytical column. They should be checked periodically and discarded as dictated by the nature of the samples analyzed.

4.4 COLUMN EVALUATION AND TEST METHODS

A new column, whether purchased or packed in the laboratory, should always be tested before use, and periodically retested while in use. In this way the analyst can be sure that the column meets reasonable specifications for general performance and has a reliable method to monitor changes in column properties as a function of time or the type of samples analyzed. Routine column tests are simple, they require the generation of a single isocratic chromatogram, yet reveal a great deal of information about the quality of the column packing and the success of the column packing process [9,27,50,52,71,180,198-201]. Other, more specific tests, are available to assess particular chromatographic properties, such as the column hydrophobicity, concentration of residual silanol groups and/or metal activity of chemically bonded phases, and the effective pore size distribution. A knowledge of these properties is often crucial for column selection, since notionally identical columns may have very different separation characteristics.

4.4.1 Routine Column Test Methods

No single set of conditions exists for testing the performance of all column types. The chemical nature of the column packing dictates the choice of test solutes and the mobile phase. Some general considerations for selecting test solutes are summarized in Table 4.9. Within these constraints, a wide range of test mixtures can be defined and should work adequately. Some examples of suitable test mixtures for evaluating the performance of normal and reversed-phase column packings are given in Table 4.10. Uracil/guanine/cytosine with the mobile phase 0.2 M $NH_4H_2PO_4$

TABLE 4.9

DESIRABLE PROPERTIES OF TEST SOLUTES AND MOBILE PHASES FOR COLUMN
TESTING

- Test solutes should be of low molecular weight to ensure rapid
  diffusion and easy access to the packing pore structure.

- Test mixture should contain components that correctly
  characterize the column in terms of both kinetic and
  thermodynamic performance.

- A value for the column dead volume is required in most
  calculations. It is convenient to have one component of the test
  mixture as an unretained solute.

- At least two components of the test mixture should have k values
  between 2 and 10.

- All measurements should be made with a mobile phase of simple
  composition, low viscosity, and under isocratic conditions.

- The sample volume and/or amount of test sample should not
  verload the column.

- It is convenient to use test solutes with a strong absorbance at
  254 nm. A solute giving 50-100% FSD with a detector setting of
  0.05 or 1.0 AUFS is convenient.

---

at pH 3.5 has been used to evaluate cation exchangers and
cytidine-5-monophosphate/uridine-5-monophosphate/guanosine-5-
monophosphate with 0.05 M KH_2PO_4 at pH 8 for evaluation of anion
exchangers. Details for testing size-exclusion columns are given
in section 4.5.9.

Having chosen the test mixture and mobile phase composition,
the chromatogram is run, usually at a fairly fast chart speed to
reduce errors associated with the measurement of peak widths,
etc., Figure 4.10. The parameters calculated from the chromatogram
are the retention volume and capacity factor of each component,
the plate count for the unretained peak and at least one of the
retained peaks, the peak asymmetry factor for each component, and
the separation factor for at least one pair of solutes. The
pressure drop for the column at the optimum test flow rate should
also be noted. This data is then used to determine two types of
performance criteria. These are kinetic parameters, which indicate
how well the column is physically packed, and thermodynamic
parameters, which indicate whether the column packing material
meets the manufacturer's specifications. Examples of such
thermodynamic parameters are whether the percentage of bonded

TABLE 4.10

TEST MIXTURES FOR EVALUATING THE PERFORMANCE OF NORMAL AND
REVERSED-PHASE PACKINGS

Test Mixture	Mobile Phase
a) Normal phase	
Toluene/nitrobenzene/ o-,m-,p-nitroaniline	Isooctane/ethanol/water (84.5/15/0.5)
Naphthalene/m-dinitrobenzene /o-nitroaniline	Hexane/methylene chloride/ 2-propanol (89.5/10/0.5)
Toluene/phenanthrene/nitrobenzene	Hexane/acetonitrile (99/1)
Toluene/nitrobenzene/acetophenone /2,6-dintrotoluene/1,3,5- trinitrobenzene	Hexane/methanol (99.5/0.5)
b) Reversed Phase	
Benzene/naphthalene/biphenyl	Methanol/water (70/30)
Dimethyl phthalate/nitrobenzene/ anisole/diphenylamine/fluorene	Acetonitrile/water (70/30)
Uracil/phenol/benzaldehyde/N,N- dimethyl-3-toluamide/toluene/ ethylbenzene	Acetonitrile/water (65/35)
Resorcinol/acetophenone/ naphthalene/anthracene	Acetonitrile/water (55/45)
Acetone/acetophenone/anisole/ benzene/toluene	Acetonitrile/water (60/40)
Benzamide/benzene/benzophenone/ biphenyl	Acetonitrile/40 mM sodium phosphate [pH = 6.3] (55/45)

phase is correctly stated, and whether this particular column will perform a separation similar to a column of the same type used earlier.

The peak asymmetry factor should be scrutinized first. In this discussion we refer to the peak asymmetry factor measured at 10% of the peak height (see section 1.5). Some column supply companies use the baseline measurement to specify peak asymmetry, leading to larger limiting values than those given here. Peak asymmetry, especially of unretained or weakly retained peaks ($k <$ 3), is typical of poorly packed columns (if instrumental contributions can be excluded). If only the unretained peak ($k <$ 1) is asymmetric and/or there is a significant difference (> 15%)

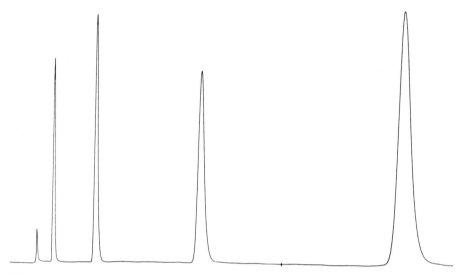

Figure 4.10 Typical routine column test chromatogram for a 30 cm
x 4.6 mm I. D. column packed with an octadecylsiloxane bonded
silica packing of 10 micrometers particle diameter. The test
mixture consisted of resorcinol (0.55 mg/ml), acetophenone (0.025
mg/ml), naphthalene (0.20 mg/ml) and anthracene (0.01 mg/ml) in
acetonitrile, 10 microliters injected. The separation was
performed isocratically at 23°C with acetonitrile-water (55:45) as
the mobile phase at a flow rate of 1.5 ml/min. Detection was by UV
at 254 nm (0.1 AUFS).

Test solute	k	Asymmetry factor	n/m	Separation Factor
Resorcinol	0.2	1.2	9,600	
Acetophenone	1.4	1.7	15,900	
Naphthalene	4.3	1.6	17,100	3.07
Anthracene	9.8	1.4	16,000	2.28

between the plate counts for the unretained and retained peaks, at
least for test solutes lacking strong, specific interactions with
the stationary phase, then extracolumn effects are implicated and
should be investigated prior to repeating the test. Extracolumn
effects can be dramatic for small bore columns due to improper
connection of the column to the instrument. As a general guide,
columns yielding peak asymmetry factors for solutes lacking strong
specific interactions greater than 1.2 are of poor quality and
those with values greater than 1.6 are unacceptable and should be
discarded.

If the test solutes show reasonable peak asymmetry with
lower than average plate count values for well retained solutes (k
> 2) the column is most probably poorly packed. The number of
theoretical plates, normalized to one meter column length by

convention, will correlate primarily with the average particle size of the packing. Some typical values are summarized in Table 1.14 [202]. These values are approximate upper limit values and plate counts exceeding 75% of the tabulated values are generally considered acceptable. The column plate count is not an ideal parameter for column comparison, since its value is influenced by a large number of experimental variables, including the column flow rate, composition of the mobile phase, temperature, choice of test solute, and extracolumn effects. Also, the plate count determined for asymmetric peaks will produce values having a large positive deviation from the true result unless the values are corrected by data analysis techniques that take the peak asymmetry into account (see section 1.5).

Under the test conditions, large changes in the capacity factor and separation factor values for the test solutes compared to those found previously for similar columns indicate a change in the chemical nature of the packing material. The separation characteristics of the column will be different from notionally similar columns used previously, and if a new column, it may not be capable of providing a similar separation to one obtained on a column of that type before. Significant changes in the column pressure drop between comparable columns measured under the same test conditions indicates that either the column or its fittings are partially blocked. If cleaning the fittings does not restore the pressure drop to the normal range the column should be discarded.

Knox introduced the concept of reduced parameters for column testing [50,180,201-203]. The reduced plate height, reduced velocity, and the flow resistance parameter are defined in section 1.7.10. Columns packed with materials of different sizes, eluted with solvents of different viscosities, and tested with solutes having different diffusion coefficients can be directly compared using the parameters introduced above. The experimental value for the flow resistance parameter provides information on how the chromatographic system as a whole is performing. It is the troubleshooting parameter that indicates whether the experimental conditions are appropriate for performing the test. Typical values for porous packings are given in Table 1.17. A very high value compared to the normal value, for example ten times the normal value, is indicative of a partial blockage in the chromatographic

system. This should be diagnosed and rectified before proceeding
further with the column evaluation.

According to chromatographic theory, the reduced plate
height is related to the reduced velocity by equation (4.2)

$$h = (B/\nu) + (A\nu^{1/3}) + (C\nu) \qquad (4.2)$$

The constant B reflects the geometry of the eluent in the column
and the extent to which diffusion of the solute is hindered by the
presence of the packing. For columns of acceptable performance, B
should be less than 4, and in practice, an approximate value of 2
is desired. It is responsible for the decrease in efficiency at
very low flow rates. The constant A is a measure of how well the
column is packed. A well-packed column will have a value of A
between 0.5 and 1.0 while a poorly packed column will have a
higher value, say between 2 and 5. The constant C reflects the
efficiency of mass transfer between the stationary phase and the
mobile phase. At high reduced mobile phase velocties the C term
dominates the reduced plate height value, and therefore column
efficiency. The value for C is close to zero for pellicular
packings, a reasonable value is 0.003, but has a greater value for
porous packings with a value of 0.05 being reasonable for the
latter. Packings with a polymeric stationary phase have a higher
value, perhaps approaching unity.

The constant A, B, and C can be calculated by curve fitting
from a plot of the reduced plate height against the reduced
velocity [201,203]. Accurate values for the constants can only be
obtained if a wide range of reduced velocity values are covered
and the data is of high quality. In particular, the method used to
calculate the plate height is very important, since the data are
easily distorted when inappropriate methods are used to treat
asymmetric peaks [203,204]. From the shape of a log-log plot of
reduced plate height against reduced velocity, Figure 4.11, the
important features of the column are easily deduced. If the
minimum is below 3 and occurs in the range 3 < reduced velocity <
10 then the column is well packed (low A). If the reduced plate
height is below 10 at a reduced velocity of 100 then the material
has good mass transfer characteristics (low C) and is probably
well packed as well (low A). If the curve has a high flat minimum
the column is poorly packed (high A).

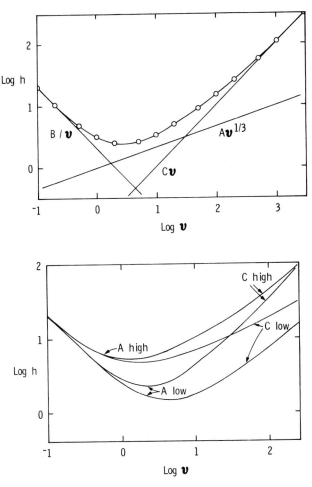

Figure 4.11 Plot of the reduced plate height against the reduced mobile phase velocity indicating the velocity region in which the three terms in equation (4.2) contribute to reduced plate height. The lower figure shows some typical results from column testing. The lowest curve (A and C low) illustrates ideal behavior. The plots for other columns can be considered to deviate from this curve due to high A (poorly packed column), high C (poor quality column packing), or from having both a high A and C value.

Because of the time required to develop sufficient data points to make a plot similar to that shown in Figure 4.11 it is useful to have a shorter method for assessing potential problems. For a good column the value of the reduced plate height should not exceed 3 or 4 at a reduced velocity of about 5 and 10 to 20 at a reduced velocity of about 100.

4.4.2 Column Activity Tests for Chemically Bonded Phases

Chemically bonded phases are generally prepared by reacting the silanol groups of a silica substrate with a reactive organosilane reagent. Differences in reagents, reaction conditions, and changes in the properties of the silica substrate lead to notionally similar packings having different concentrations of bonded ligand. Since for steric reasons not all silanol groups can react, these same variations result in packings having differences in their residual silanol concentrations. Also, it is now accepted that the population of silanol groups is inhomogeneous showing differences in their hydrogen bonding and ion-exchange capacity. Residual trace metal impurities in the silica substrate are also known to influence the chromatographic properties of bonded phases through either metal-solute interactions, or through their influence on the acidity of neighboring silanol groups. It can be anticipated, therefore, that column packings from the same or different manufacturers will vary in their hydrophobicity, silanophilicity, ion-exchange capacity, and activity induced by trace metal impurities. By designing methods to evaluate the contribution these interactions make to retention of specific compounds the variability in column packings from the same manufacturer over time and the similarity of packings prepared by different manufacturers can be judged.

To a first approximation the hydrophobicity of a reversed-phase column packing increases with carbon loading, for example, by increasing the chain length of the bonded ligand, increasing the bonding density, by endcapping the phase, and by increasing the surface area of the silica substrate [52]. The hydrophobicity of a column can be easily assessed, therefore, by determining the retention of a nonpolar solute under standard conditions. In practice, the difference in retention for two substances may show a smaller dependence on experimental variables and offer greater reliability. Some proposed test systems include: the ratio of k-anthracene/k-benzene (acetonitrile-water 63:36) [198]; k-pentylbenzene/k-butylbenzene (methanol-water 4:1) [200]; k-ethylbenzene/k-toluene (methanol-water 65:35) [205]; and k-anthracene [206,207]. Some typical values for the hydrophobicity index of different column packings using the ratio of k-anthracene/k-benzene are given in Table 4.11 [198]. Fairly large differences in the hydrophobicity index were observed among this

TABLE 4.11

HYDROPHOBICITY AND SILANOPHILIC INDICES FOR OCTADECYLSILOXANE
BONDED PHASES

Column Packing	Hydrophobicity Index(a)	Silanophilic Index(b)
Novapak C-18	5.10	0.50
Lichrospher C-18	5.88	0.63
Zorbax	4.77	1.36
Spherisorb ODS-2	4.74	0.82
Resolve C-18	4.43	1.07
μBondapak C-18	3.54	0.78
ASI-18	3.42	0.68
Partisil ODS-3	3.30	0.88

(a)based on the capacity factor for anthracene/benzene, (b)based
on the separation factor for N,N-diethyl-m-toluamide/anthracene.

group of octadecylsiloxane bonded phases reflecting the
differences in their preparation and also indicating the large
difference in retention that can be anticipated when
octadecylsiloxane phases from different manufacturers are selected
for the same application. Still larger variations are observed for
different ligands. The capacity factor for anthracene was found to
vary from about 0.5 to 6.5 on 31 reversed-phase packings that
included octadecyl-, octyl-, methyl-, cyanopropyl- and
phenylsiloxane bonded ligands with the same mobile phase [206]. It
is interesting that both the lowest and highest values observed
for k-anthracene were found for the octadecylsiloxane bonded
packings. Walczak et al. [208] developed a hydrophobicity scale
based on factor analysis using 23 reversed-phase packings with
different ligand substitutes for 63 test solutes. Although their
approach was different, their conclusions that large differences
in hydrophobicity exist among column packings of the same type, as
well as between different types, agrees with the other studies.
There are few published studies of the lot-to-lot variation in the
separation properties of column packing materials from a single
manufacturer [207]. The data in Table 4.12 for 21 lots of a
octadecylsiloxane bonded packing prepared over a 1.5 year period
is probably fairly representative of other manufacturers. For the
separation factor of anthracene/phenanthrene, the quality control
range of acceptable lots was established as 1.20-1.44. This means
that for two columns of equal efficiency prepared from packings
passing the quality control test, the resolution of
anthracene/phenanthrene could differ by as much as 125%. For

TABLE 4.12

QUALITY CONTROL RANGES FOR A REVERSED-PHASE PACKING MATERIAL

Parameter	Quality Control Ranges		
	Low	High	Mean
Anthracene (k)	3.59	4.39	3.99
Anthracene/Phenanthrene (α)	1.20	1.44	1.32
Androstenedione (k)	7.10	10.18	8.64
Androstenedione/Testosterone (α)	1.16	1.28	1.22

androstenedione/testosterone, resolution at constant column efficiency differed by as much as 75% between acceptable lots. For the same column materials, the resolution between naphthalene/biphenyl showed only small variation between lots, indicating that the choice of test compounds, and therefore the sample to be separated, influences column reproducibility.

Sander and Wise have proposed a test method to determine the bonding chemistry used to prepare octadecylsiloxane column packings based on the relative retention of three polycyclic aromatic hydrocarbons, benzo[a]pyrene (BaP), phenanthro-phenanthrene (PhPh), and 1,2:3,4:5,6:7,8-tetrabenzonaphthalene (TBN) eluted with the mobile phase acetonitrile-water (85:15) [52,67,199,210]. On monomeric phases the test solutes elute in the order BaP $\leq$ PhPh < TBN and on polymeric phases PhPh < TBN < BaP. For densely loaded monomeric or lightly loaded polymeric phases the elution order becomes PhPh < BaP < TBN. The separation factor for TBN and BaP was used to order a number of commercially available column packings into the three groups discussed above, Table 4.13. Similar conclusion were reached for the separation of triphenylene and o-terphenyl used to evaluate the shape selectivity of different octadecylsiloxane packings [200].

Residual silanol groups in chemically bonded phases have been associated with a number of undesirable interactions with polar solutes such as excessive peak tailing, irreproducible retention times, and excessively long retention times. These problems are particularly prevalent for amines and other strong bases. A large number of test systems have been proposed to characterize the concentration of residual silanol groups on bonded phase packings, and some representative examples are

TABLE 4.13

CHARACTERIZATION OF OCTADECYLSILOXANE BONDED PHASES BASED ON THE
SEPARATION FACTOR FOR TBN AND BaP

Polymeric (α 0.5-0.9)	Intermediate (α 1.0-1.7)	Monomeric (α 1.7-2.2)
Chromspher PAH Supelcosil LC-PAH Vydac 201TP Spherisorb PAH	LiChrospher 100 RP-18 Backerbond C-18 LiChrospher 60 RP-select B Partisil 5 ODS-2 Partisil 5 ODS Spherisorb ODS-1 Spherisorb ODS-2 Supelcosil LC-18-DB	Partisphere C-18 Zorbax ODS Partisil 5 ODS-3 Hypersil ODS (HP) Ultrasphere ODS Supelcosil LC-18 Hypersil ODS

TABLE 4.14

TEST MIXTURES FOR DETERMINING THE RESIDUAL SILANOL CONCENTRATION OF
REVERSED-PHASE COLUMN PACKINGS

Test Mixture	Mobile Phase	Parameter	Reference
Benzene/nitrobenzene	Heptane	α	211
Pyridine or 2,6-dimethylpyridine	Heptane-chloroform (3:1)	k	212
Aniline/benzene	Methanol-water (65.4:34.6)	α	213
Naphthalene/1-nitro- naphthalene	Methanol-water (60:40) containing 0.5% (w/v) sodium acetate	α	214
Aniline/phenol	Methanol-water (2:3)	α	52, 205
Cyclam/chrysene	Methanol	see text	215
Caffeine/phenol	Methanol-water (3:7)	α	200
N,N-diethyl-m- toluamide/anthracene	Acetonitrile	α	198

summarized in Table 4.14. The test systems fall into two kinds;
those employing normal phase conditions and those using either
reversed-phase or water-miscible solvents as the mobile phase.
Normal phase conditions require careful equilibration of the
column to be tested which can add considerably to the time
required for the experiment. These systems contain a polar
hydrogen bonding solute whose retention varies with the silanol

concentration, increasing with increasing silanol concentrations, and usually referenced to the retention of a nonpolar solute, which is not significantly influenced by silanophilic interactions. More popular are those test systems compatible with reversed-phase conditions. These may also provide a better assessment of silanophilic interactions under operating conditions close to those used in practice. The silanophilic index for some octadecylsiloxane bonded phases based on the observed separation factors for N,N-diethyl-m-toluamide/anthracene are summarized in Table 4.11. Confirming general opinion, the residual silanol concentration of different phases is shown to vary widely, reflecting differences in manufacturing practices. For the separation of basic solutes it is nearly always required to choose a packing of low silanol activity and/or to add a masking agent to the mobile phase. According to Verzele and Dewaele the separation factor for naphthalene/1-nitronaphthalene can be used to determine the success of deactivation of octadecylsiloxane bonded packings by endcapping [214]. A separation factor of at least 1.4 or more is characteristic of a well deactivated packing. Lower values around 1.1-1.2 are characteristic of phases without endcapping. Sadek and Carr used the macrocyclic tetraaza compound dimethyldiphenylcyclam, which forms strong complexes with silanol groups, to characterize the silanophilic interactions of a number of commercially available octadecylsiloxane bonded packings using methanol as the mobile phase and chrysene as a reference compound to account for different hydrophobic interactions for the packings [215]. Equation (4.3) was used to define a silanophilic index, R, for reversed-phase column packings.

$$R = [(k-cyclam) - (k-chrysene)]/[(k-chrysene)] \qquad (4.3)$$

The surface silanol groups are not homogeneous and a small proportion of these are strongly acidic and ionized even at low pH [207,216,217]. The small number of highly acidic sites can function as cation exchangers and are probably responsible for the anomalous behavior of solutes which are ionized under the conditions of elution. These ion-exchange sites are generally considered to contribute significantly to the low recovery and denaturing of proteins in reversed-phase chromatography. A test for this assessment has been developed using a series of synthetic peptides containing from 1 to 4 basic residues (lysine) [207]. A general test for ion-exchange capacity of reversed-phase packings has been described based on the separation factor for benzylamine/phenol in a mobile phase buffered to pH 2.7 [200]. At this pH normal silanol groups are undissociated and only highly acidic silanols contribute to the retention of the amine by ion exchange. The separation factor for benzylamine/phenol was found to vary from 0.01 to 1.43 for a number of column packings indicating a wide range of ion-exchange capacity. Acid treatment of the silica substrate prior to bonding seems to minimize the ion-exchange capacity of the packing, which may be due to the removal of metal impurities, thought to be activators for the low concentration of highly acidic silanol groups [37,38,41]. Metal impurities may also affect the chromatographic properties of the packing through their ability to chelate with certain functional groups of some solutes. The retention characteristics of 2,4-pentanedione in the mobile phase methanol-water (60:40) containing 0.5% (w/v) sodium acetate is a sensitive test of trace metal contamination of chemically bonded phases [40,214]. For metal-free packings 2,4-pentanedione will elute close to the void volume with acceptable peak shape. With increasing metal contamination 2,4-pentanedione shows increasing retention and deterioration in peak shape. It is not uncommon to observe complete retention of 2,4-pentanedione for some commercially available packings with a high metal content. Other sources of metal in the chromatographic system, such as frits and screens, may also contribute to the observed metal activity of the column.

It is necessary to use the test methods presented in this section with some caution. Not all of them have been adequately standardized or applied in a broad sense. They should prove useful for aiding column selection and improving the properties of column packings prepared in the laboratory. It should also be noted that columns may deteriorate with use depending on their treatment. An old column of a type that has been shown to perform well in a particular test in the literature may not live up to expectations, due to its altered state.

4.4.3 Effective Pore Size Distribution

The pore size distribution of a packing can influence the chromatographic behavior of the material through exclusion of high molecular weight solutes and excessive retention of solutes in micropores. The pore size distribution is also known to affect the extent of surface coverage during the bonding process (see section

4.2.2). It is common practice for manufacturers to quote the median pore size for the silica substrate for chemically bonded packings, which has two disadvantages. This provides no indication of the pore size distribution, and does not take into account the reduction of the pore diameter due to part of the pore volume being occupied by the bonded ligand groups, whose size can be of the same order as that of the pore diameter, and thus far from negligible. The effective pore size distribution under chromatographic conditions can be easily determined by size-exclusion chromatography using a series of hydrocarbon and poly(styrene) standards and a mobile phase, such as tetrahydrofuran, which is a strong solvent for the test solutes [35,218]. Each test solute has a characteristic random coil diameter in solution given by equation (4.4)

$$RCD = 0.062(MW)^{0.59} \qquad (4.4)$$

where RCD corresponds to the diameter of the smallest pore in nanometers allowing unhindered access to a poly(styrene) standard of a given molecular weight (MW). Normal hydrocarbon standards are required to ensure that sufficient data points are available for the small pores and their molecular weight must be converted to a poly(styrene) equivalent value by multiplying by 2.3 for insertion into equation (4.4). The elution volume of individual standards is converted to the parameter R, the percentage of the total pore volume for which a given poly(styrene) probe has access, using equation (4.5)

$$R = 100[(V_E - V_0)/(V_m - V_0)] \qquad (4.5)$$

where V_E is the elution volume of a standard able to explore some of the pore volume, V_0 the elution volume of a standard completely excluded from the pore volume, and V_m the elution volume of a standard able to explore the complete pore volume. The cumulative pore size distribution for the packing is then obtained by plotting R against log(RCD) and fitting the data to a polynomial function for interpolation. Values of RCD corresponding to R = 10, 25, 50, 75, and 90% are then obtained from the above plot. The value of RCD at R = 50%, RCD_{50}, gives the mean pore diameter, and the difference between two RCD values is used to express the range (for example, 80% of the pores have diameters between RCD_{90} and RCD_{10}). Some typical values for the effective pore size

TABLE 4.15

EFFECTIVE PORE SIZE DISTRIBUTION FOR COLUMN PACKINGS DETERMINED BY SIZE-EXCLUSION CHROMATOGRAPHY

Column Packing	Nominal Pore Size	Pore Size Distribution (nm)				
		RCD_{10}	RCD_{25}	RCD_{50}	RCD_{75}	RCD_{90}
Ultrasphere C-8	8.0	89.1	13.2	7.2	---	---
LiChrosorb RP-18	10.0	85.1	20.0	8.9	---	---
Zorbax ODS	7.5	---	---	5.7	3.7	2.9
Vydac 218TP	33.0	116.2	56.2	25.1	11.0	1.5
Spherisorb ODS	8.0	24.5	12.6	7.1	---	---
μBondapak C-18	12.5	41.7	21.4	10.0	5.0	3.0
μPorisil	12.5	55.0	25.1	11.0	5.0	2.7
Novapak C-18	9.0	263.0	75.9	6.8	---	---
LiChrosorb RP-8	10.0	77.6	19.5	9.5	---	---
Ultrastyragel 500A	---	15.8	9.3	4.6	2.4	1.6
Protesil 300 C-8	30.0	65.3	35.5	16.2	7.8	---

distribution of commercially available column packings determined by size-exclusion chromatography are given in Table 4.15 [35]. As would be expected large differences between the nominal pore size for the silica substrate and the effective pore size for the bonded phases are commonly found. Also the distribution of pore sizes can be very different for different packings, which will be reflected in their chromatographic behavior.

4.4.4 Determination of the Column Void Volume

In its broadest sense the void volume can be defined as the volume of the mobile phase contained within the chromatographic system between the sample injector and the detector. It can be determined by the elution volume of an unretained and unexcluded solute. This is the system void volume, the true column void volume being the system void volume less the extracolumn void volume. In a well-designed chromatographic system the extracolumn void volume should be negligible, if not, it must be determined experimentally (section 5.2). It would seem that the column void volume should be an easy parameter to determine in liquid chromatography, as it is for gas chromatography, but this is not the case. In liquid chromatography, employing a porous packing material, a problem arises from the difficulty of unambiguously defining the mobile and stationary phase volumes. The mobile phase is contained in the interstitial volume between the particles and within the pores, that is, in all those volumes not occupied by the fixed stationary phase. The interstitial volume can be divided into two parts, that portion of the volume containing the moving phase, and that volume close to the points of contact between the particles and away from the flowing stream that is essentially static. The mobile phase within the pores is stagnant. The composition of mixed mobile phases trapped in the pores is not of a constant composition. One component of the mobile phase can become selectively adsorbed or associated with the stationary phase resulting in a region of altered composition in the vicinity of the stationary phase. During preparation some of the smaller pores of the packing become blocked by the bonded ligand and can exclude solutes from the pore volume. For kinetic column characteristics, such as the linear velocity and band broadening phenomena, it is the volume of the moving mobile phase that is important. In this case the void volume is comprised of essentially the interstitial volume less the stagnant portion of the interstitial volume. The thermodynamic void volume is used to calculate the capacity factor, separation factor, etc., and unlike the kinetic void volume includes all those static portions of the mobile phase that have the same composition as the moving phase. Included in this case are the interstitial volume and that portion of the pore volume of identical composition to the interstitial mobile phase. The difficulty of defining the thermodynamic void volume arises from two sources. It is difficult to define the volume of solvated stationary phase along the pore wall. This volume should also depend on the composition of the mobile phase and, therefore, is not constant. Because of the exclusion effect solutes of different size will have available to them different phase volumes. In the thermodynamic sense a column may not have a single void volume, but a variable volume that must be matched to the solute size. It is not an easy task, however, to determine the various column volume elements, and generally kinetic and thermodynamic parameters have been determined for inadequately defined column systems and may not be correct in the absolute sense. For perspective, Figure 4.12, presents a diagrammatic summary of the various volume elements for a typical reversed-phase column and is instructive in indicating their relative magnitude [219].

A large number of experimental methods have been proposed to estimate the column void volume without any single method emerging

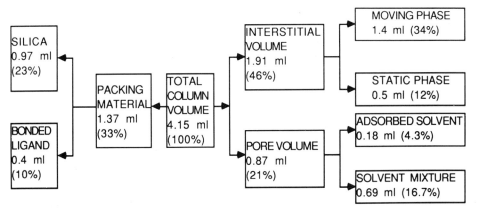

Figure 4.12   Summary of the various volume elements for a 25 cm x
4.6 mm I. D. column packed with Zorbax C$_8$ (5 micrometer particle
diameter and 330 m^2/g specific surface area) and methanol-water
(1:4) as mobile phase.

as   a   preferred   choice   [219-227].   Comments   and   opinions   on
comparative   studies   have   generally   resulted   in   contradictory
conclusions.   The   principal   experimental   methods   include   static
weighing procedures, the determination of the elution volume of a
solvent disturbance or system peak obtained by injecting a single
component of a mixed mobile phase, determination of the elution
volume of an unionized and unexcluded solute that gives the lowest
elution volume, determination of the elution volume of an easily
detected   ion,   determination   of   the   elution   volume   of   an
isotopically   labelled   component   of   the   mobile   phase,   and
mathematical procedures based on linear regression of the elution
volumes for members of a homologous series. The total porosity of
the column can be determined by successively filling the column
with two solvents of different density and then weighing it after
each filling [211,219,225]. This method is tedious and provides a
value for the maximum volume accessible to the mobile phase. It
ignores the possibility that the stationary phase is solvated by
the mobile phase reducing the column volume occupied by the mobile
phase. One of the simplest methods of determining the void volume
when mixed mobile phases are used is to inject one of the pure
mobile phase components and determine the elution volume of the
disturbance peak [221-223]. However, the disturbance peak may not
be easy to identify and its position may be dependent on the
mobile phase composition. The disturbance peak will correspond to
the column void volume only when the mobile phase components are

more or less equally sorbed by the packing. By far the most popular approach is to determine the elution volume of an ionic or organic compound that is notionally unretained, such as sodium nitrate, potassium bromide, potassium dichromate, uracil, acetone, dimethylformamide or fructose for reversed-phase columns [220-227]. These methods are basically limited by partial or complete exclusion from the interior of the porous packing by size or ion exclusion effects and by retention of some test solutes by the stationary phase. For neutral organic molecules retention by the stationary phase is the most common problem and manifests itself in the form of elution volumes that vary with the mobile phase composition. Ionic compounds are affected mainly by ion exclusion. Their elution behavior is influenced by the amount of sample injected and the presence of background electrolyte in the mobile phase. At low sample concentrations in unbuffered mobile phases, a salt is excluded from the pores of the packing, presumably due to electrical charges on the phase surface: with increasing electrolyte concentration in the mobile phase (or with injections of highly concentrated solutions of a salt) the ion exclusion effect is suppressed and the pores become accessible to the salt [219,223,224,227]. For sodium nitrate, one of the most commonly used ionic test solutes, it is recommended that the most dilute solution conveniently detectable be used to determine the interstitial void volume, and concentrated solutions, about 1 M or greater, be used to determine the interstitial and pore volume, assuming a 10 microliter injection (at least 3×10^{-6} moles of sodium nitrate injected). Alternatively, if the mobile phase is buffered with 0.1 M phosphate then consistent elution volumes for any sample size are obtained. For mobile phases other than aqueous methanol high salt concentrations may result in demixing preventing the use of ionic test solutes. The main difficulty in using neutral organic solutes as test probes, particularly under reversed-phase conditions, is ensuring that there is no retention by the stationary phase. This seems to occur, if only to a small extent, under a wide range of experimental conditions. In normal phase chromatography the elution volume of benzene with methanol as the mobile phase has been widely used to determine column void volumes, although Knox states that the correct criteria for choosing a test solute in liquid-solid chromatography is that the test solute should have the same elution strength as the mobile phase [228]. The use of isotopically labelled components of the mobile phase at first sight might appear ideal, but this is not the case for a single labelled component, due to the disturbance of the stationary phase solvation layer resulting in retention of the labelled component [221,223,226,229]. This can be accounted for by labelling all components used in the mobile phase and using their characteristic elution volumes to calculate the void volume [219,222]. This process is very slow and usually requires off-line detection by a scintillation counter. Mathematical methods of determining the void volume are based on the linear relationship between the logarithm of the capacity factor and carbon number for members of a homologous series [221,225,227]. This method requires very accurate elution data for at least 4 or 5 homologous standards to accurately define the appropriate relationship. However, it has been shown experimentally that the relationship obtained is not always linear and that different homologous series can give rise to different values of the void volume.

4.5 RETENTION MECHANISMS IN LIQUID CHROMATOGRAPHY

Retention in liquid chromatography is a complex process involving solute interactions in both the mobile and stationary phases that are difficult to describe exactly. Qualitative and approximate models have been advanced in lieu of more exact treatments, which at least provide some insight into the complex separation mechanisms involved, and in the most favorable cases, allow the rough prediction of retention and separation factors to aid the optimization of the separation of mixtures. The simpler models are most useful to the analyst, since they are described using terms derived from the chromatographic experiment, while more exacting models based on statistical thermodynamics may be more fundamentally significant, they are generally difficult to use in practice, due to a lack of knowledge of the molecular parameters incorporated into the models and the complex calculation procedures required. Thus, we will adopt the former approach whenever possible in this section.

4.5.1 Liquid-Solid Chromatography

Liquid-solid chromatography (LSC), sometimes referred to as normal phase or straight phase chromatography, is characterized by the use of an inorganic adsorbent or chemically bonded stationary phase with polar functional groups and a nonaqueous mobile phase consisting of one or more polar organic solvents diluted to the desired solvent strength with a weak solvent, such as hexane. A characteristic of these systems is the formation of an adsorbed layer of mobile phase molecules at the surface of the stationary phase with a composition that depends on the composition of the mobile phase, but is not necessarily identical to it, nor does its composition necessarily change linearly with changes in the composition of the mobile phase. The retention of a solute is essentially determined by the balance of interactions it experiences in the mobile phase and its competition with mobile phase molecules for adsorption sites at the surface of the stationary phase. Two retention models have been proposed to describe this process: the competition model (also called the displacement model) developed by Snyder and Soczewinski [230-233] and the solvent interaction model proposed by Scott and Kucera [8,21,231,234]. As we shall see, both models lead to similar conclusions for the description of retention as a function of solvent composition, but mechanistically are different. The Snyder-Soczewinski model is an adsorption model based on the displacement of solvent molecules by solute molecules from sites on the stationary phase, while the Scott-Kucera model treats the retention mechanism as a sorption process emphasizing solute mobile phase interactions.

In its simplest form the competition model assumes the entire adsorbent surface is covered by a monolayer of solute and mobile phase molecules. Under normal chromatographic conditions, the concentration of sample molecules will be small and the adsorbed monolayer will consist mainly of mobile phase molecules. Retention of a solute molecule occurs by displacing a roughly equivalent volume of mobile phase molecules from the monolayer to make the surface accessible to the adsorbed solute molecule. For elution of the solute to occur the above process must be reversible, and can be represented by the equilibrium depicted by equation (4.6)

$$X_m + nS_a \longleftrightarrow X_a + nS_m \qquad (4.6)$$

where X represents solute molecules, S solvent molecules, and the subscripts a and m designate molecules in the stationary (adsorbed) and mobile phases, respectively. The competition model thus assumes that a solute molecule replaces n solvent molecules in the adsorbed monolayer, where n is given by the ratio of the adsorption cross section of the solute to that of the solvent. While solute retention can be envisaged as a continuous competitive displacement process between solute and mobile phase molecules on the adsorbent surface, the net retention will depend on the relative interaction energy of the solute and mobile phase molecules with the adsorbent surface. Given the simplifying assumptions that the adsorbent surface is energetically homogeneous and that solute-solvent interactions in the mobile phase are effectively canceled by similar interactions in the adsorbed phase, the empirical equations which describe the variation of solute retention as a function of solvent strength (equations 4.7 and 4.8) and the variation of solvent strength with composition for a binary mobile phase mixture (equation 4.9) were derived.

$$\log k_1 = \log(V_a w/V_m) + \alpha'(S^0 - A_s \epsilon_1) \qquad (4.7)$$

$$\log(k_1/k_2) = \alpha' A_s (\epsilon_1 - \epsilon_2) \qquad (4.8)$$

$$\epsilon_{AB} = \epsilon_A + \log[N_B 10^Z + 1 - N_B]/\alpha' n_B \qquad (4.9)$$

$$Z = \alpha' n_B (\epsilon_B - \epsilon_A)$$

The terms k_1 and k_2 are the solute capacity factors in the mobile phase, V_a the volume of the monolayer of adsorbed solvent per unit mass of adsorbent, w the weight of adsorbent (gm), V_m the column void volume, α' the adsorbent activity parameter proportional to the adsorbent surface energy ($\alpha' = 1$ for a standard adsorbent), S^0 the dimensionless free energy of adsorption of the solute on a standard adsorbent of $\alpha' = 1$, A_s the adsorption cross section of solute X, ϵ_1 and ϵ_2 the solvent strength of mobile phase 1 and 2, ϵ_{AB} the solvent strength of the binary mobile phase AB, ϵ_A the solvent strength of the weaker solvent A, ϵ_B the solvent strength of the stronger solvent B, N_B the mole fraction of the stronger solvent B in the binary mobile phase, and n_B the adsorption cross section of solvent molecules B. The adsorbent activity parameter is a measure of the ability of the adsorbent to interact with adjacent molecules of solute or solvent and is constant for a given adsorbent (e.g., 0.57 for silica). The above equations are reasonably accurate for predicting separations involving solute and solvent molecules which are nonpolar, or of intermediate polarity and non-hydrogen bonding. In other cases, the model is less successful, since for polar solutes and solvents selective adsorption at high energy sites predominates (the adsorbent surface is not energetically homogeneous) and as solutes and solvents increase in polarity, their interactions with one another become stronger and more specific (solute-solvent and solvent-solvent interactions in the adsorbed and mobile phases do not cancel). These imperfections in the model can be accommodated by either adjusting the parameters in the above equations or by adding additional terms as discussed below.

For weak solvents there is little preference for adsorption on any given part of the surface and little tendency for weakly retained molecules to localize. For more polar solute and/or solvent molecules localized adsorption occurs in which molecules are centered over specific adsorption sites across the surface of the adsorbent and have a higher energy of adsorption than for the non-localized state [232,235-242]. Experiment shows that localization of the strong solvent B in a binary mobile phase AB will occur provided that the mole fraction of B in the adsorbed monolayer is less than about 0.75. Once most of the surface is covered by localized B molecules, the remaining spaces on the surface are unable to accommodate B molecules in the configuration and positioning required for the localizing effect. This factor is referred to as restricted-access delocalization. Restricted-access delocalization is predicted to occur for adsorbents with rigid surfaces and a high concentration of surface adsorption sites, such as silica gel, but should be much less important for chemically bonded phases due to the relatively low ligand density and the flexibility of the bonded chains. A practical consequence of restricted-access delocalization of polar molecules is that the solvent strength of the mobile phase will vary continuously with the mole fraction of polar solvent in the mobile phase and in the adsorbed monolayer. The separation of two compounds will vary according to equation (4.10)

$$\log(k_2/k_1) = X + Ym \qquad (4.10)$$

where X and Y are constants for a given combination of sample and solvents and m is the mobile phase localization parameter, defined in equation (4.11)

$$m = m^\circ f(\Theta_B) \qquad\qquad (4.11)$$

where m° is the value of m for the pure solvent B and $f(\Theta_B)$ is some function of the surface coverage of localizing solvent B and non-localizing solvent A. The function $f(\Theta_B)$ can have values from 0 to 1 but is not easy to calculate. Details of the calculation method for binary and ternary mobile phases, etc., are given in reference [243] and Palamareva has outlined a computer program for aiding the calculation method [244]. The net result of restricted-access delocalization is that the value for Θ_B in equation (4.9) progressively declines with the surface coverage of the localizing solvent B and the value for Θ_B must be adjusted for each mobile phase composition employed when using the competition model.

The retention of polar solute molecules is also affected by site-competition delocalization. A moderately polar non-localizing solvent molecule can interact laterally with sites upon which a solute molecule is localized. This added competition for the site by both the solute and solvent molecules weakens the net interaction of the solute with the surface such that for solvents of increasing polarity a greater decrease in the capacity factor with increasing polarity of the non-localizing solvent occurs than is predicted by the simple competition model. Site-competition delocalization is predicted to be important for adsorbents with active sites extending above the surface, such as silanol groups and chemically bonded ligands, but not for alumina (active sites buried below the surface). This effect can be quantitatively accounted for by assuming a larger value of A_s than is calculated from the molecular dimensions of the solute.

For polar solutes and solvents, particularly those capable of hydrogen bonding, secondary solvent effects due to the specific nature of solute-solvent interactions may also have to be included in the model, since the assumption that they are identical in the adsorbed and mobile phases, and therefore self-canceling, is no longer necessarily true. The addition of a secondary solvent term to equation (4.8), in addition to modifying the values for Θ and A_s due to localization effects, is required to improve the prediction of changes in relative retention of similar solutes [237,240,243]. With dioxane in the mobile phase the retention of proton donor solutes was found to increase compared to when 2-propanol was used as a mobile phase modifier due to coadsorption [245]. These differences were interpreted as a result of a dual adsorption mechanism for proton donor solutes: competitive adsorption on the silica surface and non-competitive adsorption on the monolayer of dioxane molecules. The bifunctional dioxane molecules acting as a bridge to the surface with the exposed ether oxygen as the active site for hydrogen bonding. At high dioxane concentration in the mobile phase, coadsorption tends to be counterbalanced by solvation interactions in the mobile phase.

For strong mobile phase modifiers ($\theta_B \gg \theta_A$) when $N_B \gg 0$ and θ_B does not vary with N_B, equation (4.9) can be simplified to the practically useful equation

$$\log k = \log k_B - (A_s/n_B) \log N_B \tag{4.12}$$

where k_B is the solute capacity factor for the pure strong solvent B [231,242,245-248]. The slope of a plot of log k vs. log N_B is equivalent to the number of solvent molecules displaced by the adsorption of the solute at an active site. When applicable, equation (4.12) is convenient for optimizing chromatographic separations.

The solvent interaction model differs from the competition model by proposing formation of solvent bilayers adsorbed onto the adsorbent surface. The composition and extent of bilayer formation depends on the concentration of polar solvent in the mobile phase. For weak solvents interacting with the adsorbent surface largely by dispersive forces, the solvent interaction model predicts results similar to the competition model with retention occurring by displacement. For binary solvent mixtures AB, the stronger solvent B is preferentially adsorbed by the hydrated silanol groups. As the concentration of strong solvent increases, polar interactions between the solvent molecules result in the formation of a bilayer. When the mobile phase contains a low concentration of the polar modifier B, the second layer of solvent molecules is incomplete, and solute retention can be explained by association of the solute within this layer and without the need for displacement. At high concentrations of the strong solvent B, the solvent bilayer is complete, and solute retention occurs by displacement of B solvent molecules from the second solvent layer. For solutes with high capacity factors (indicating that the polarity of the solute is equal to or greater than that of the B molecules), displacement of the primary solvent layer is possible by direct interaction of the solute with the adsorbent surface. The solvent interaction model thus provides a variety of mechanism for solute retention that depend mainly on the relative polarity of the solute and B solvent molecules and on the concentration of B molecules in the mobile phase. At high concentrations of solvent B the solute capacity factor can be described by equation (4.13)

$$1/k = D + EC_B \qquad\qquad (4.13)$$

where D and E are constants for a particular solute and polar solvent B and C_B is the concentration of strong solvent B (%w/v) in the binary mobile phase mixture [50,231,234].

The competition model and solvent interaction model were at one time heatedly debated but current thinking maintains that under defined conditions the two theories are equivalent, however, it is impossible to distinguish between them on the basis of experimental retention data alone [231,249]. Based on the measurement of solute and solvent activity coefficients it was concluded that both models operate alternately. At higher solvent B concentrations, the competition effect diminishes, since under these conditions the solute molecule can enter the interfacial layer without displacing solvent molecules. The competition model, in its expanded form, is more general, and can be used to derive the principal results of the solvent interaction model as a special case. In essence, it seems that the end result is the same, only the tenet that surface adsorption or solvent association are the dominant retention interactions remain at variance.

The above models provide a quantitative description of retention in LSC but the calculation of solvent strength parameters can be rather involved and a more empirical approach can often be justified for routine purposes or for the separation of simple mixtures. In these cases an estimate of solvent strength is required to provide adequate retention and then the selectivity is changed at constant solvent strength by using different solvent modifiers in binary or higher order mixtures. The solvent strength of a pure solvent in LSC according to the competition model is characterized by the solvent strength parameter, ϵ^o, determined empirically by the magnitude of the solute capacity factor for a solute of known cross-sectional area for a series of solvents referenced to pentane as having a zero solvent strength on an adsorbent of known activity [225,232,244,250]. The solvents can then be ordered by increasing magnitude of ϵ°, corresponding to increasing elution strength, known as an eluotropic series, Table 4.16 [8,232,237,240,241]. Although ϵ° is an empirical parameter, it can be correlated to more general solvent strength scales, such as the Kamlet-Taft constants [251] and Reichardt's $E_T(30)$ scale of solvent polarity [252].

The solvent strength parameter (ϵ°) is defined as the free energy of adsorption of the solvent per unit of surface area. Its magnitude, therefore, is dependent on the particular adsorbent used. The ϵ° values for silica are about 70% of those for alumina, but the general ranking of solvents remains the same. For the chemically bonded phases the ϵ° values are substantially smaller than for silica indicating much weaker retention of solutes by these phases, but their selectivity is anticipated to be different to that of the inorganic oxide phases. Solutes and solvents do not localize as strongly on the 3-aminopropylsiloxane bonded phase and restricted access delocalization is minimal, resulting in weaker retention compared to silica [235,247]. Since the energy of interaction of solute and solvent molecules with the adsorbent is substantially less for the chemically bonded phases than for silica and alumina, solute-solvent interactions in the mobile phase should be proportionately more important in explaining retention differences. In the simplified competition model the differences in elution strength of different solvents are assumed to arise exclusively from the different free energies of adsorption per unit area of these solvents. The effects of specific solute-solvent interactions in the mobile phase and on the adsorbent surface are unimportant. However, when the differences in the free energies of adsorption per unit area of different eluents become small, secondary solvent effects become important, and the order of eluting power of different solvents is then dependent on the nature of the solute. There can be no such thing then as a unique eluotropic series applicable to a majority of solutes, and the concept of an eluotropic series loses much of its usefulness. This seems to be the case for carbon and may well be the case, to a lesser extent, for the chemically bonded phases [8].

TABLE 4.16

ELUOTROPIC SERIES FOR DIFFERENT ADSORBENTS

Solvent	Solvent Strength Parameter					
	Alumina	Silica	Carbon	Aminopropyl	Cyanopropyl	Diol
Pentane	0.00	0.00				
Hexane	0.01	0.01	0.13-0.17			
Carbon tetrachloride	0.17	0.11		0.069		
1-Chlorobutane	0.26	0.20	0.09-0.14			
Benzene	0.32	0.25	0.20-0.22			
Methyl-tert. butyl ether	0.48			0.11-0.124	0.049-0.085	0.071
Chloroform	0.36	0.26	0.12-0.20	0.13-0.14	0.106	0.097
Dichloro- methane	0.40	0.30	0.14-0.17	0.13	0.120	0.096
Acetone	0.58	0.53		0.14		
Tetrahydrofuran	0.51	0.53	0.09-0.14	0.11		
Dioxane	0.61	0.51	0.14-0.17			
Ethyl acetate	0.60	0.48	0.04-0.09	0.113		
Acetonitrile	0.55	0.52	0.01-0.04			
Pyridine	0.70					
Methanol	0.95	0.70	0.00	0.24		

384

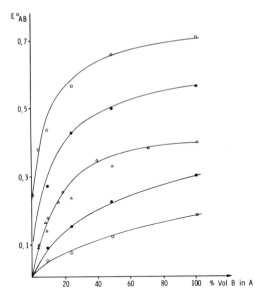

Figure 4.13 Solvent strength of eluent mixtures on alumina. (o) pentane-carbon tetrachloride, (■) pentane-n-propyl chloride, (Δ) pentane-dichloromethane, (•) pentane-acetone And (□) pentane-pyridine.

As can be gleamed from the various models describing changes in solvent strength as a function of mobile phase composition the change in solvent strength as a function of the volume fraction of the more polar solvent cannot be represented by a simple expression. At low concentrations of the polar solvent, small increases in concentration produce large increases in solvent strength; at the other extreme, relatively large changes in concentration of the polar solvent affect the solvent strength of the mobile phase to a lesser extent, Figure 4.13. Graphical methods, such as Figure 4.14, have been developed for obtaining the percent volume composition of binary mobile phases having similar solvent strength but different solvent constituents (selectivity). These are a useful aid in optimizing separations on silica and alumina stationary phases [20,244,253].

Retention and selectivity in LSC are dramatically influenced by the presence of even low concentrations of polar additives in the mobile phase, particularly water [20,22,253-255]. Their influence is most pronounced when the mobile phase is nonpolar. However, when used in controlled amounts (in which case they are

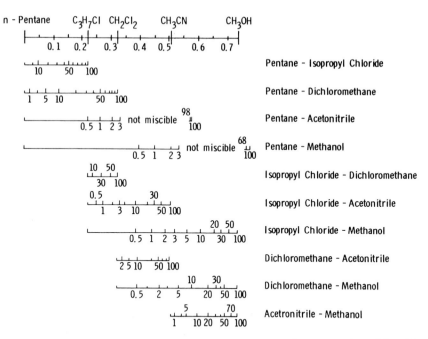

Figure 4.14 Solvent strength of eluent mixtures for liquid-solid chromatography.

called moderators or modulators), their use can be beneficial, resulting in less variation of sample retention from run-to-run and a substantial increase in sample capacity due to improved linearity of the adsorption isotherm. Higher column efficiencies, reduced band tailing, and a diminished tendency for sample decomposition may also be observed in some cases. These benefits result from the preferential adsorption of the moderator from the mobile phase by the most active sites on the surface of the adsorbent leading to a more energetically homogeneous adsorbent surface as seen by the sample.

When changing between mobile phases containing different concentrations of water, column equilibration is a slow process and several hours of adsorbent conditioning are usually required before constant retention volumes are obtained. The situation can be even more complex with gradient elution due to changing hydration levels during the course of the gradient, leading to the possible uptake of water by the stationary phase. This adversely affects the separation and leads to irreproducible separations and long column regeneration times. Some of these problems can be solved for isocratic separations by using isohydric solvents, that

is, solvents having the same hydration level as the adsorbent
[22,256,257]. The use of a 100% saturated solvent is undesirable
because such liquid-solid chromatographic systems are often
unstable. Under these conditions the pores of the adsorbent
apparently fill gradually with water, leading to changes in
retention with time and possibly also to a change in the retention
mechanism as liquid-liquid partitioning become more important.
When silica is the adsorbent, 50% saturation of the mobile phase
has been recommended for stable chromatographic conditions (25%
saturation in the case of alumina). Solvents with 50% water
saturation can be prepared by mixing dry solvent with a 100%
saturated solvent, or preferably, by using a moisture control
system [250]. The latter consists of a water-coated, thermostated
adsorbent column through which the mobile phase is recycled for
the time required to reach the desired degree of saturation.

A column which has been deactivated with water may no longer
show adequate separation properties. Restoring the activity of the
column by pumping a large volume of dry mobile phase through it is
slow and costly. Alternatively, reactivation can be accomplished
chemically using the acid-catalyzed reaction between water and
2,2-dimethoxypropane, the products of which, acetone and methanol,
are easily eluted from the column [259].

In addition to water, virtually any organic polar modifier
may be used to control solute retention in liquid-solid
chromatography. Alcohols, alkylamines, acetonitrile,
tetrahydrofuran and ethyl acetate in volumes of less than one
percent can be incorporated into nonpolar mobile phases to control
adsorbent activity. In general, column efficiency declines for
alcohol-moderated eluents compared to water-moderated eluent
systems. Many of the problems discussed above for water-moderated
eluents are true for organic-moderated eluents as well.

Liquid-solid chromatography on alumina or silica with non-
aqueous eluents is generally considered to be suitable for the
separation of non-ionic organic compounds soluble in organic
solvents. Retention results largely from the interaction of sample
polar functional groups with discrete adsorption sites on the
stationary phase. The strength of these polar interactions is
responsible for the selectivity of the separation with the general
elution order following the sequence: alkanes < aromatics with one
or two rings < halogenated alkanes and aromatics (F < Cl < Br < I)
< polynuclear aromatics < ethers < nitro compounds < nitriles <

most carbonyl-containing compounds < alcohols < phenols < amines < amides < carboxylic acids < sulfonic acids. Internal hydrogen bonding between functional groups and bulky alkyl groups adjacent to the polar functional group diminish retention. LSC is generally the method of choice for the separation of geometric isomers and for class separations [20,22,230,253,254]. The ability of LSC to separate geometric isomers has been attributed to a lock-key type steric fitting of solute molecules with the discrete adsorption sites on the silica surface. For maximum interaction the adsorbed molecule must orient itself parallel to the adsorption surface. The cis isomer of a compound, therefore, will be retained more strongly than the trans isomer. The unique ability of adsorbents to differentiate solutes based on functional group type, number of functional groups and polarity is the basis for the class separation of complex mixtures, where the isolation or measurement of the contribution of compound types to the overall sample is required, rather than a within-group separation (see section 8.5).

Some of the problems associated with the use of silica and alumina (irreproducible retention of polar compounds, the difficultly in maintaining a constant hydration level of the mobile phase, chemisorption, catalytic activity, etc.) can be avoided by using polar chemically bonded phases. Rapid equilibration with the mobile phase allows the use of gradient elution. In general, retention on polar chemically bonded phases is less than that for silica. The surface of the chemically bonded phases is heterogeneous with unreacted silanol groups contributing to retention. The retention of moderately polar solutes on 3-cyanopropylsiloxane bonded phases suggests that the adsorbent behaves like a deactivated (weak) silica gel with unreacted silanol groups as the dominant active sites [240,241,246,248,260]. The acidic character of the adsorbent can be eliminated by adding a small amount of a moderator, such as alkylamines, to the mobile phase. In fact, cyanopropylsiloxane bonded phases have been used for the separation of many classes of basic drugs using normal and reversed-phase mobile phases with an alkylamine or pH buffer added to reduce tailing [261]. With hydrogen bonding mobile phases the retention of compounds on silica and 3-cyanopropylsiloxane bonded phases become similar due to effective deactivation of silanol groups in both cases. The 3-aminopropylsiloxane bonded phase has properties complementary to those of silica and 3-cyanopropylsiloxane bonded silica [237,241,247,248]. The amino function imparts strong hydrogen bonding properties to the stationary phase as well as acid or base properties, depending on the nature of the solute. The 3-aminopropysiloxane bonded phase has been used for the class fractionation of polycyclic aromatic hydrocarbons and their alkylated analogs into groups containing the same number of rings [262,263]. Since amines are easily oxidized, degassing the mobile phase and avoiding solvents which may contain peroxides are recommended. Samples or impurities in the mobile phase containing reactive functional groups may react chemically with the amino group altering the separation properties of the column. For example, ketones and aldehydes can form Schiff's base complexes with the amino group which are only slowly dissociated by flushing the column with dilute acid [264]. The 1,2-dihydroxypropoxypropylsiloxane bonded phase (diol phase) has properties intermediate between those of the 3-aminopropyl- and 3-cyanopropylsiloxane bonded phases, exhibiting less retention of acidic and basic solutes than the former and lower retention of dipolar solutes than the latter [241,248,265,266].

4.5.2 Dynamically Modified Inorganic Oxide Adsorbents

A number of separation techniques have been developed to take advantage of the ion-exchange capacity of silica and alumina and its dynamic modification by interaction with ions in the mobile phase [44,49,267-271]. The isoelectric point pH for silica is about 2 while that for alumina varies from 3.5-9.2, depending on the method of manufacture and the composition of the buffered mobile phase [49,269-273]. Silica does not exhibit useful amphoteric behavior because of its reactivity in acidic solution and experiments are restricted to its cation-exchange properties at mobile phase pH > 2. Alumina, by contrast, is capable of exhibiting both anion (low pH) and cation (high pH) ion exchange behavior because of its chemical stability in both environments. The mobile phase pH determines the ion-exchange capacity of alumina and silica. The higher the pH of the mobile phase the greater the cation-exchange capacity of silica associated with increased dissociation of surface silanol groups. The higher specific surface area of silicas (for example, 500 m^2/g compared to 70 m^2/g for alumina), results in a higher retention capacity for silica, although the wide pH range stability of alumina allows both cations and anions to be separated, while silica is limited to cation separations.

Silica and alumina with reversed-phase type mobile phases have been used to separate basic compounds with a $pk_a \geq 8$ [47,269-273]. This includes many pharmaceutically active compounds and their metabolites, such as the example shown in Figure 4.15 [271]. Typical mobile phases are relatively simple consisting of an aqueous buffer mixed with an organic solvent. Neutral and acidic components are largely unretained while basic compounds are retained by a complex and multifunctional mechanism, of which ion-exchange interactions generally dominate. The retention behavior of basic compounds appears to be controlled predominantly by the mobile phase pH and the concentration and nature of the organic modifiers; the nature and concentration of the competing ions (buffer components) also exert some influence on retention. Typical of ion-exchange separations, retention normally increases with decreasing ionic strength of the mobile phase. Retention normally increases with solute pk_a and is diminished by steric crowding near the exchange site. At high organic modifier concentrations retention decreases due to a reduction in solute ionization, while at low modifier concentrations retention increases due to reduced solvation of the competing ions. Inorganic ions are less solvated than alkylammonium ions at all concentrations of organic modifier and, therefore, ion solvation is less important in this case. The optimum concentration of organic modifier is a compromise between the two competing effects. For effective retention on silica, mobile phase pH values in the range 7.5-10 (of buffer prior to addition of modifier) are commonly used. Columns are reported to be stable and retention reproducible within this range with the use of a saturator column between pump and injector [46,47,271].

Ion exchange is also the basis of the separation of organic ions using nonaqueous mobile phases [274,275]. Basic drugs were separated using mobile phases containing methanolic solutions of perchloric acid or ammonium perchlorate of an appropriate pH and ionic strength on silica. Retention depends primarily on the pk_a of the solute, the extent of protonation of the solute, and the degree of ionization of surface silanol groups. These parameters depend largely on the mobile phase pH (retention increases with pH) and ionic strength (retention is proportional to the reciprocal of the ionic strength). Organic solutions of tetraalkylammonium hydroxide have been used as mobile phases for the separation of acidic compounds on silica. Neutral and basic

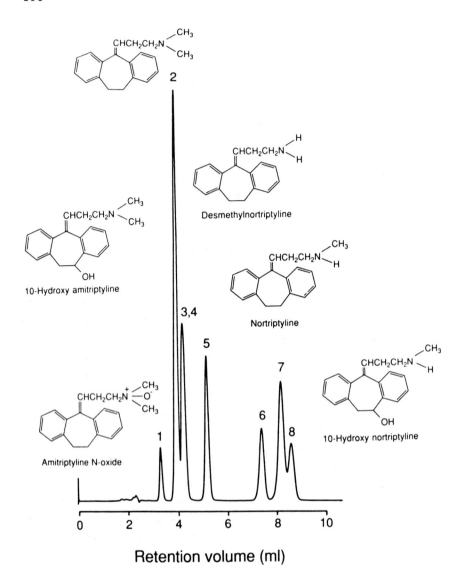

Retention volume (ml)

Figure 4.15 Separation of the tricyclic antidepressant amitriptyline and its major metabolites on a 10 cm x 4.6 mm I. D. column packed with Spherisorb S5W silica with methanol-aqueous ammonium acetate (9:1), pH 9.1, as mobile phase at a flow rate of about 1 ml/min. Peak identification: 1 = amitriptyline-N-oxide; 2 = amitriptyline; 3 = E-10-hydroxyamitriptyline; 4 = Z-10-hyroxyamitriptyline; 5 = desmethylnortriptyline; 6 = nortriptyline; 7 = E-10-hydroxynortriptyline; and 8 = Z-10-hydroxynortriptyline. (Reproduced with permission from ref. 271. Copyright Elsevier Scientific Publishing Co.)

compounds without proton donor groups are virtually unretained in this system. The tetraalkylammonium hydroxide in the mobile phase reacts with surface silanol groups on silica to form tetraalkylammonium salts. The surface concentration of tetraalkylammonium salts increases with increasing concentration of tetraalkylammonium hydroxide until a maximum surface coverage is reached. This point corresponds to maximum stationary phase retention of acid solutes through interaction with the silicate salts. At the same time as the concentration of tetraalkylammonium hydroxide in the mobile phase increases so will elution of acidic solutes due to interaction between the hydroxide and acid in the mobile phase, leading to a decrease in retention. The two interactions operate in the opposite sense, and for any separation a compromise mobile phase composition is required. The above interactions are strongly solute dependent and gradient elution is required for the elution of samples with a wide range of acidity. The general elution order of organic acids (alcohols < amides < phenols < carboxylic acids < multifunctional acids) is consistent with the ability of the respective acids to donate a proton to the basic tetraalkylammonium hydroxide-silica surface.

Reversed-phase separations on dynamically modified silica have been investigated extensively as an alternative to octadecylsiloxane bonded silica phases for the separation of a wide range of compounds, particularly organic bases and drugs in complex biological matrices [44,269,276-279]. The reproducibility of reversed-phase systems based on dynamically modified silica was found to be much better than that of chemically bonded phases and fewer problems were experienced in the separation of basic compounds [277]. The dynamic stationary phase system was generated by equilibrating a silica column with an aqueous organic mobile phase, buffered to pH > 5, and containing a certain amount of a long-chain quaternary ammonium compound, such as cetyltrimethylammonium bromide. The cetyltrimethylammonium ions are adsorbed on to the silica surface in such a way that the hexadecyl group points away from the surface, Figure 4.16 [44,279]. The cetyltrimethylammonium ions adsorbed on the silica surface form an apolar layer similar to that of a chemically bonded material except that the adsorbed layer is in dynamic equilibrium with the alkylammonium ions in the mobile phase. The amount of cetyltrimethylammonium ions adsorbed depends on the surface area of the silica (assuming an absence of size-exclusion by small pores) and the concentration of cetyltrimethylammonium ions in the mobile phase. The maximum adsorbed amount of

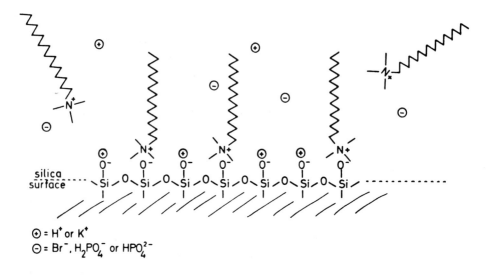

⊕ = H⁺ or K⁺
⊖ = Br⁻, H₂PO₄⁻ or HPO₄²⁻

Figure 4.16   A model for the surface of silica gel in equilibrium
with a mobile phase of (55:40:5) methanol-water-0.2 M potassium
phosphate buffer (pH 7.5) with the addition of 2.5 mM of
cetyltrimethylammonium bromide. (Reproduced with permission from
ref. 279. Copyright Pergamon Journals Ltd).

cetyltrimethylammonium ions coincides with the formation of the
critical micellar concentration of the cetyltrimethylammonium
ions. The concentration of organic modifier, nature of the buffer
ions, and the pH of the mobile phase also influence the
concentration of adsorbed ions through their influence on the
ionization of the silanol groups and competition with the
cetyltrimethylammonium ions for the ionized silanol groups.
Neutral and nonionized solutes are retained by reversed-phase
partitioning similar to chemically bonded phases. Anionic solutes
are retained as ion pairs by a reversed-phase mechanism. Cationic
solutes are retained largely by a reversed-phase mechanism, and
partly by cation exchange, depending on their ability to compete
with mobile phase ions, which are usually present in a much higher
concentration, for surface ionized silanol groups. The primary
disadvantages of dynamically modified silica systems are the
relatively long time required to reach equilibrium compared to
chemically bonded phase systems and the number of additional
parameters which must be optimized to achieve a given separation,
which can lead to lengthy methods development.

4.5.3 Liquid-Liquid Chromatography

Liquid-liquid chromatography (LLC) declined in popularity
with the advent of chemically bonded phases and almost passed into
disuse, until revived recently [50]. Renewed interest is due to
the development of solvent generated LLC systems which are both
stable and highly reproducible compared to conventional
preparation techniques that rely on loading the column from a
solvent in which the stationary phase is soluble followed by
displacing the solvent and excess stationary phase from the column
with mobile phase saturated with stationary phase [50,280-284].
Because of the slight mutual solubility of the two liquid phases
the latter systems tended to be unstable and required extensive
precautions for continuous operation, such as presaturation of the
mobile phase with stationary phase and thermostating of the mobile
phase and column to avoid fluctuations in the phase ratio and
displacement of the stationary phase from the column. Also,
because of the instability of most LLC systems the sample solvent
should have a composition similar to the mobile phase and gradient
elution is usually not feasible. LLC systems have some attractive
features, however, such as unambiguously defined and reproducible
distribution constants, higher column loadability, and a great
variety of liquids from which phase systems can be composed. These
features offer much greater variety and reproducibility than is
the case for chemically bonded phases, since it is much easier to
prepare and reproduce a bulk liquid phase with defined solvent
properties than it is a solid surface with defined adsorbent
properties. For these reasons interest in LLC remains, although
the number of applications published in recent years has not been
very great.

The problems of stationary phase erosion can be largely
overcome by solvent-generated LLC, where the stationary liquid
phase is generated dynamically by the mobile phase. In this
approach, one of the phases of an equilibrated liquid-liquid
system is applied as a mobile phase to a solid support which is
better wetted by the other phase of the liquid-liquid system. The
support is usually silica when the stationary phase is aqueous or
a polar solvent and a reversed-phase chemically bonded support
when the stationary phase is a nonpolar solvent. Under these
conditions a multimolecular layer is formed on the surface of the
solid support which has the properties of the liquid phase in

equilibrium with the mobile liquid phase. In most cases the pores will be completely filled with the stationary liquid phase. To adjust the phase ratio of the column, a support of different surface area and specific pore volumes is selected.

The selection of the support is critical for several reasons [285-287]. The surface of the solid support must be wetted by the stationary liquid phase better than by the mobile phase, otherwise a stable stationary liquid phase film will not be formed. Although sufficiently strong adsorptive properties are required to obtain wetting, some compromise is required, since the support should have negligible adsorptive properties for the components of the sample. In the absence of interfacial adsorption, retention in LLC is very simply defined by equation (4.14)

$$V_R = V_m + K_D V_s \qquad (4.14)$$

where V_R is the retention volume, V_m the volume of mobile phase, V_s the volume of stationary liquid phase, and K_D the liquid-liquid distribution constant. V_m and V_s are usually calculated from the intercept and slope, respectively, of the plot of V_R against K_D, where values of K_D for a number of test solutes are determined in a static liquid-liquid distribution experiment. Once values of V_m and V_s have been determined, values of K_D for different solutes are simply obtained from the elution volumes of the solutes in the same chromatographic system. Deviations from linearity in the plot of V_R against K_D are symptomatic of interfacial adsorption. Tailing of polar compounds is also symptomatic of interfacial adsorption.

4.5.4 Reversed-Phase Liquid Chromatography

The popularity of reversed-phase liquid chromatography (RPC) is easily explained by its unmatched simplicity, versatility and scope [15,22,50,52,71,149,288-290]. Neutral and ionic solutes can be separated simultaneously and the rapid equilibration of the stationary phase with changes in mobile phase composition allows gradient elution techniques to be used routinely. Secondary chemical equilibria, such as ion suppression, ion-pair formation, metal complexation, and micelle formation are easily exploited in RPC to optimize separation selectivity and to augment changes available from varying the mobile phase solvent composition. Retention in RPC, at least in the accepted ideal sense, occurs by non-specific hydrophobic interactions of the solute with the stationary phase. The near universal application of RPC stems from the fact that virtually all organic molecules have hydrophobic regions in their structure and are capable of interacting with the stationary phase. Since the mobile phase is polar, and generally contains water, the method is ideally suited to the separation of polar molecules which are either insoluble in organic solvents or bind too strongly to inorganic oxide adsorbents for normal elution. Many biological polymers fall into this category and the development of wide pore reversed-phase packings has had a profound effect on the separation of biopolymers in the life sciences and biotechnology [7,13-16,288]. RPC employing acidic, low ionic strength eluents has become a widely established technique for the purification and structural elucidation of proteins. However, the structure of biopolymers is very sensitive to mobile phase composition, pH and the presence of complexing species which can result in anomalous retention, peak splitting and even denaturing of proteins [291-294].

A general characteristic of reversed-phase systems is that a decrease in polarity of the mobile phase, that is increasing the volume fraction of organic solvent in an aqueous organic mobile phase, leads to a decrease in retention; a reversal of the general trends observed in liquid-solid chromatography or "normal phase" chromatography. It is also generally observed for RPC that for members of a homologous or oligomous series, the logarithm of the solute capacity factor is a linear function of the number of methylene groups or repeat units of the oligomeric structure, with the possible exception of the first few members of the series [22,295]. For monomeric bonded phases a small discontinuity in the plots may be observed at the point where the chain length of the sample exceeds that of the bonded ligand [296]. In this case it is assumed that once the length of the solute alkyl chain exceeds that of the bonded phase ligands the remaining solute methylene groups no longer penetrate into the bonded phase region and consequently undergo weaker dispersive interactions than those that can penetrate this region. It is also generally observed in RPC that branched chain compounds are retained to a lesser extent than their straight chain analogs and that unsaturated compounds are eluted before their fully saturated analogs. Polymeric bonded-phase packings have been shown to exhibit greater selectivity for the separation of polycyclic aromatic hydrocarbons (PAHs) than monomeric phases, presumably due to the more rigid structure of

the former [52,67]. A slot model was proposed to explain the
elution order of isomeric PAHs which was observed to depend on the
length-to-breadth ratio of the compound and its planarity.
Multisite interactions of biopolymers with the stationary phase
can result in anomalous retention behavior and peak splitting if
the stationary phase topography and that of the biopolymer are
incompatible. Multisite interactions, which require additional
time for the conformation of the molecule to attain the right
position on the stationary phase surface, also result in much
slower kinetics for the adsorption and desorption process, and
lower column efficiency than is generally observed for low
molecular weight solutes.

The most common method for varying the chromatographic
selectivity for neutral molecules in RPC is to change the type of
organic modifier in the mobile phase. In numerous studies using
binary mobile phases, equation (4.15) has been shown to describe
reasonably well the variation of solute retention with the volume
fraction of organic solvent in an aqueous-organic mobile phase

$$\log k = \log k_w + a\phi + b\phi^2 \tag{4.15}$$

where k is the solute capacity factor, k_w the solute capacity
factor with pure water as the mobile phase, ϕ the volume fraction
of organic solvent, and a and b are constants for a given solute
and eluent combination [297-302]. Deviations from equation (4.15)
occur, particularly at high and low water content (outside the
range ϕ = 0.1-0.9). For mobile phases of intermediate composition
equation (4.16) can often be used as a reasonable approximation
for the variation of retention with the volume fraction of organic
solvent in the mobile phase

$$\log k = \log k_w - S\phi \tag{4.16}$$

where S is a solute-dependent factor related to the solvent
strength of the organic solvent. Equation (4.16) is more
restricted in the range of solvent composition over which it
applies than equation (4.15), but provides some insight into the
selection of mobile phases of constant solvent strength for
methods development in RPC. The S-values determined from the slope
of log k vs. ϕ are similar for many substances (although

significant variations are also found), enabling these values to be used as descriptors, in a semiquantitative sense, of the solvent strength of the organic solvent. Some typical S-values for common solvents are methanol (2.6), acetonitrile (3.2), acetone (3.4), dioxane (3.5), ethanol (3.6), isopropanol (4.2) and tetrahydrofuran (4.5). The use of S-values to determine the composition of mobile phases for methods development is discussed in section 4.6.

Several reasons have been proposed for the poor description of experimental data by equations (4.15) and (4.16) for mobile phases representing the extremes of the solvent composition range. In nearly totally aqueous mobile phases changes in the conformation of the stationary phase are probably important. In this case it is assumed that the extended alkyl chains collapse onto the surface of the support and form a rigid surface in which solute penetration is limited and retention reduced compared to the extended conformation. At very low water content the properties of the mobile phase depend largely on the properties of the organic modifier and cease to be influenced by the hydrophobic properties of water. Also, when the water content of the mobile phase is low, silanophilic interactions with the column packing are likely to increase in importance and cause deviation from the results predicted from predominantly hydrophobic interactions at higher water content.

The intercept value, k_w, obtained by extrapolation based on equation (4.15) or (4.16) is useful for predicting parameters related to a solute's hydrophobicity, for example, the selection of a sorbent for solid-phase extraction and trace enrichment or for the prediction of log P values (octanol-water partition coefficients) used to model the absorption of substances by biological membranes [298-304]. The excessive retention of most solutes on RPC packings using only water as the eluting solvent prevents direct measurement of k_w. While extrapolation methods are effective for this purpose, they yield values of k_w that vary with the equation used to fit the data, the composition range studied, and the number of data points available. Log P values calculated from k_w must be used cautiously, but compared to the alternative methods of measuring log P by the shake flask method, RPC is faster, requires less sample, can handle impure samples, and can determine a wider range of values.

The complexity of the retention process in RPC has encouraged activity in non-chromatographic techniques to evaluate parameters appropriate for predicting retention as a function of mobile phase composition. Preliminary studies have indicated that solvatochromic methods can provide some useful insight into the retention process. The $E_{T(30)}$ scale of solvent strength, based on the position of the long-wavelength absorption of a standard betaine dye, was shown to give significantly better linearity for plots of log k against $E_{T(30)}$ polarity than similar plots against the volume fraction of organic modifier [305,306]. Solvatochromic measurements are made independently of the stationary phase and, therefore, the $E_{T(30)}$ scale allows the effects of changes in mobile phase composition to be examined independently of chromatographic retention experiments. In a different approach, based on the methodology associated with linear solvation energy relationships, a general equation has been proposed to predict retention from a knowledge of the solvatochromic parameters of the solutes of the form

$$\log k = A + mV/100 + s\pi^* + b\beta + a\alpha \tag{4.17}$$

where V is a parameter characteristic of the solute size (e.g., molar volume, intrinsic volume, etc.), π^* a measure of the solute's capacity for dipole/polarizability interactions, and α and β characterize the solute's hydrogen donor acidity and basicity, respectively [307-309]. The solute's solvatochromic properties are determined spectroscopically by the measurement of the absorption bands for a series of indicator compounds. The coefficients A, m, s, b, and a are related to the chemical nature of the mobile and stationary phases and are unique for every combination of mobile and stationary phase. The volume term in equation (4.17) is always positive in sign and, therefore, free energy concepts favor solute transfer from the cohesive mobile phase to the less cohesive stationary phase. Increasing the solute's size will lead to an increase in retention. The coefficients a, b, and s are usually negative in sign. Therefore, increasing the solute's capacity to enter into dipole and/or hydrogen bonding interactions will reduce retention. For aqueous mobile phases the solvatochromic comparison method indicates that the primary factors that influence solute retention are the solute size and capacity for hydrogen bonding interactions. Solute dipolarity/polarizability is considerably less important, but sometimes still significant.

The general retention mechanism in RPC is complex and difficult to describe in quantitative terms using readily accessible parameters. Several factors contribute to this difficulty. Typical stationary phases are intrinsically heterogeneous containing bonded ligands and various types of silanol groups. This promotes a series of general and specific interactions with solutes that is not easy to model. Selective absorption of one or more mobile phase constituents into the bonded phase and the concomitant variation of the phase ratio results in the composition of the stationary phase being poorly defined, and leads to difficulties in the quantitative interpretation of retention data. The solvated stationary phase is probably in some form of dynamic equilibrium with the mobile phase, and although the adsorbed solvent is unlikely to be identical to the composition of the mobile phase, it probably varies with changes in the composition of the mobile phase. Finally, there is the question of the retention mechanism itself. Whether adsorption on the exposed surface of the bonded phase ligands or partitioning into a stationary phase composed of the bonded ligands and associated solvent molecules taken up from the mobile phase is the dominant retention mechanism. It is possible that both mechanisms may operate alternately or concurrently depending on the solvent environment.

General retention models based on solvophobic theory [50,53,310-313] and lattice statistical thermodynamic theory [51,314-317] are quite successful at explaining many of the observations and generalities observed experimentally in RPC but, are mathematically complex, and often require the input of system variables that are unknown or difficult to calculate. The solvophobic theory assumes that aqueous mobile phases are highly structured due to the tendency of water molecules to self-associate by hydrogen bonding and that this structuring is perturbed by the presence of nonpolar solute molecules. As a consequence of the very high cohesive energy of the solvent, the less polar solutes are literally "squeezed out" of the mobile phase and are bound to the hydrocarbon portion of the stationary phase. This model emphasizes interactions in the mobile phase alone and suggests that interactions with the stationary phase are unimportant. If the solute contains polar functional groups then the dipolar or hydrogen bonding interactions of these groups with the mobile phase will oppose the solute transfer mechanism.

The solvophobic theory is formulated in terms of the free energy change required for dissolving a species from a hypothetical gas phase containing the species at atmospheric pressure. The free energy change for solute retention is the difference between the free energy for transferring the ligand-solute complex from the gas phase into solution and the free energy change in transferring the individual components into solution. For each species the free energy change resulting from placing the species into solution is equal to the difference between the free energy required for the creation of a cavity of a suitable size and shape to accommodate the species in the solvent and the free energy arising from the interaction between the species and the surrounding solvent molecules. The free energy for cavity formation is given by equation (4.18)

$$\Delta G_c = -N\gamma\Delta A - N\gamma A_s (X^e - 1) \qquad (4.18)$$

where ΔG_c is the free energy term associated with cavity formation, N is Avogadro's number, ΔA is the change in surface area due to formation of the ligand-solute complex, γ the surface tension of the bulk solvent, A_s the surface area of a solvent molecule, and X^e a factor which adjusts the macroscopic surface tension to the molecular dimensions. The free energy of interaction of the species with the solvent is composed of essentially two chemical terms due to dispersive interactions (van der Waals energy) and electrostatic interactions (dipole, hydrogen bonding, etc.) and an entropic term due to the reduction in free volume accompanying solution of each species from the gas phase. As before, the interaction energy for association of the solute and ligand is the difference in free energy for the ligand-solute complex and the solute and ligand independently, ΔG_{INT}, which is composed of similar terms for the difference in van der Waals energy, ΔG_{VDW}, electrostatic energy, ΔG_{es}, and the entropic term due to the net change in free volume. This can be expressed by equation (4.19)

$$\Delta G_{INT} = \Delta G_{VDW} + \Delta G_{es} + RT\ln(RT/PV) \qquad (4.19)$$

where V is the solvent molar volume and P the reference pressure of 1 atmosphere. The mathematical expressions representing these terms are quite complex and in many cases intractable unless it can be assumed that some terms are negligible [50,310,313,318]. In agreement with experiment, the solvophobic theory predicts that the capacity factor will increase with a decrease in surface tension of the mobile phase, and that an approximately linear relationship exists between the logarithm of the capacity factor and the volume fraction of organic modifier in the aqueous mobile phase. It also predicts a linear increase in the logarithm of the capacity factor with increasing chain length for a homologous series; an increase in the capacity factor for neutral solutes with addition of salts to the mobile phase; a reduction in the capacity factor for solutes that ionize; and a linear relationship between the logarithm of the capacity factor and the reciprocal of the absolute temperature (the slope is proportional to the enthalpy of binding).

The solvophobic theory considers only the transfer of solutes from the mobile phase to an assumed amorphous, disordered oil stationary phase. It does not predict changes in retention with changes in the surface density of the bonded ligands or as a function of increasing ligand chain length. Silanophilic interactions as a source of retention are also ignored. All of these effects can be accommodated in the lattice statistical thermodynamic model. In this model it is assumed that retention occurs by partitioning of the solute between the mobile phase and the solvated ligands (interphase) region of the stationary phase [51,314-317]. Solute retention involves the creation of a solute-size cavity in the stationary phase, the transfer of the solute from the mobile to the stationary phase, and the closing of the solute-size cavity in the mobile phase. The driving force for retention by the stationary phase is the chemical differences for the contacts of the solute with surrounding molecular neighbors in the stationary and mobile phases and the partial ordering of the bonded ligands which leads to an entropic expulsion of solute relative to the properties of an isotropic liquid. Using binary interaction parameters the lattice statistical thermodynamic model leads to the general expression for retention

$$(1/\phi_B)\ln(k/k_o) = (X_{SB} - X_{SA} - X_{AB}) + \phi_B(X_{AB}) \qquad (4.20)$$

where ϕ_B is the volume fraction of organic modifier, k_o the value of k when $\phi_B = 0$, and $(X_{SB} - X_{SA} - X_{AB})$ are the binary interaction parameters among the solute (S), water (A) and organic modifier (B) molecules. For regular solutions a plot of $(1/\phi_B)\ln(k/k_o)$ against ϕ_B should be linear for any value of X_{AB}. When the regular solution approximation cannot be applied, equation (4.20) predicts the experimentally observed quadratic dependence of log k against ϕ_B. If stationary phase interactions are negligible the lattice statistical thermodynamic model and the solvophobic model predict similar results. The strength of the lattice statistical thermodynamic model is that it can explain the shape selectivity observed for certain stationary phases and can accommodate silanophilic interactions.

So far the models proposed to explain retention in RPC have largely remained the province of the physical chemist. The mathematical difficulty of using these models and their lack of a simple conceptual picture of the retention process in familiar chromatographic terms has diminished interest in their use compared to simple empirical rules for trial and error optimization of separations.

4.5.5 Hydrophobic Interaction Chromatography

Hydrophobic interaction chromatography (HIC) is widely used for the separation and purification of proteins using either soft gels at low pressures [319] or modern chemically bonded microparticle packings based on silica or macroporous polymers at high pressures [15,16,320]. During separation in HIC proteins are induced to bind to a weakly hydrophobic stationary phase using a buffered mobile phase of high ionic strength and then selectively desorbed during a decreasing salt concentration gradient. Proteins are usually separated in HIC according to their degree of hydrophobicity, much as in RPC, but because of the gentler nature of the separation mechanism, there is a greater probability that they will elute with their conformational structure (biological activity) intact [294,321,322]. In RPC multiple peaks, peak distortions and changes in the elution volume of proteins result from conformational unfolding of proteins on the bonded phase surface as a consequence of the high interfacial tension existing between the mobile and the bonded stationary phases. These conditions are minimized in HIC by using stationary phases of lower hydrophobicity together with totally aqueous mobile phases, in general, since solvent strength is controlled by varying ionic strength rather than by increasing the volume fraction of an organic modifier [320,322-327]. Retention and selectivity in HIC depend substantially on the type of stationary phase. Retention increases for more hydrophobic ligands and with it the possibility of denaturing certain proteins. Some proteins are only satisfactorily handled on hydrophilic stationary phases. The ligand density and structure as well as the hydrophobicity of the stationary phase are the primary stationary phase variables that should be optimized for the separation of individual proteins.

Mobile phase parameters that have to be optimized are the salt concentration, salt type, slope of the salt gradient, pH, addition of surfactant or organic modifier and temperature [320-322,325,328-335]. In the absence of specific binding of the salt to the protein molecule and at relatively high salt concentration in the mobile phase, retention increases linearly with the salt molality and at constant salt concentration with the molal surface tension increment of the salt used in the aqueous mobile phase. Non-linearity in plots of log k against salt molality are due to salt selectivity arising from quantitative differences in salt-specific binding to proteins and/or differential hydration of both stationary phase and protein surfaces [327,328,333]. In general, higher resolution is attained with shallower salt gradients, but when the slopes are too gentle any effect may be insignificant, and very gentle gradients result in long retention times, increased peak dilution, and for some proteins, increased denaturation. The initial salt concentration primarily affects earlier eluting proteins with little effect on well retained proteins. The general trend is for increased retention accompanying an increase in the initial salt concentration, subject to individual variation for different proteins. Retention in HIC can often be predicted using the linear solvent strength gradient model (section 4.6.3) which predicts a linear change in log k per unit time for a descending linear gradient in salt concentration [321]. Selectivity in HIC can be varied conveniently by holding the surface tension of the eluent constant while exchanging one salt for another in single-salt gradients, or by varying the composition and type of salts in binary and ternary salt gradients [328]. Changes in pH do not affect retention as much as the stationary phase hydrophobicity and the nature of the salt, but changes in selectivity are noticeable due to changes in ionization (increased ionization generally reduces retention) and pH-induced changes in protein conformation. Also, separations may be improved by adding organic solvents, chaotropic agents (e.g., urea) or surfactants in small amounts to the mobile phase [320]. The strength of hydrophobic interactions generally increase with temperature [322,332]. Thus, proteins are usually retained longer at higher temperatures, or can be eluted with constant retention using a lower salt concentration. Not all proteins, however, are conformationally stable at elevated temperatures. Typical separation conditions in HIC make use of an ammonium sulfate gradient from 3 to 0 M over 30 min at close to neutral pH (0.1 M phosphate buffer), ambient temperature, and a 5-25 cm long column operated at a slightly lower flow rate than is usually employed for small molecules (0.5-1.0 ml/min for a 4.6 mm I. D. column).

Retention in HIC can be described in terms of the solvophobic theory, in which the change in free energy on protein binding to the stationary phase with the salt concentration in the mobile phase is determined mainly by the contact surface area between the protein and stationary phase and the nature of the salt as measured by its propensity to increase the surface tension of aqueous solutions [331,333-338]. In simple terms the solvophobic theory predicts that the logarithm of the capacity factor should be linearly dependent on the surface tension of the mobile phase, which in turn, is a linear function of the salt concentration. At sufficiently high salt concentration the electrostatic contribution to retention can be considered constant, and in the absence of specific salt-protein interactions, log k should depend linearly on salt concentration as described by equation (4.21)

$$\log k = \log k_o + mM_s \tag{4.21}$$

where k_o is the retention factor at zero salt concentration, m is a parameter reflecting the contact area between the protein and stationary phase, and M_s the molal salt concentration. Plots of log k vs M_s should be linear, as generally observed, but for a series of plots for different salts at the same surface tension, different values of m indicate that factors other than surface tension frequently influence protein retention. Salt-mediated changes in the surface tension depend on both the concentration and the nature (molal surface tension increment) of the salt. Consequently, changes in retention with salt concentration are generally not identical for different proteins, and the chromatographic selectivity will depend on both the nature and concentration of the salt. The magnitude of k_0 will depend on the net charge on the protein, which in turn depends on the pH of the mobile phase. Thus changes in retention should be observed with mobile phase pH, the magnitude and direction of which depends on the protein's isoelectric point. An increase in the hydrophobic character of the stationary phase will cause a proportionate increase in m resulting in an increase in retention of all species retained by solvophobic interactions. In the absence of specific salt-protein binding interactions the solvophobic model provides a reasonable description of the experimentally observed phenomena in protein HIC.

4.5.6 Secondary Chemical Equilibria in Reversed-Phase Liquid Chromatography

In liquid chromatography the primary equilibrium is the distribution of the solute between the mobile phase and stationary phase. All other equilibria in the mobile phase or stationary phase are termed secondary chemical equilibria (SCE) by convention. Some examples of SCE are ionization, ion pairing (section 4.5.7), metal complexation and solute-micelle association. Solutes participating in SCE can exist in more than one form, for example, a weak acid could exist in its fully protonated form or ionized form. The different forms of the solute usually have different retention characteristics in the chromatographic system, but since they are rapidly interconverted, at least on the chromatographic time scale, only one peak elutes from the column. If this was not the case, either broad and/or multiple peaks would be observed for each solute. The elution volume of a solute is a function of the weighted average of the various forms of the solute in coexistence and can be adjusted by changing the parameters which control the equilibrium composition of the solute. In this way SCE provides an additional mechanism for control of retention and selectivity for optimizing a separation. Although some types of SCE can be utilized in the normal-phase mode, the greater ease of application, better understanding of SCE in aqueous or partially aqueous solutions, and the rapid column equilibration times in RPC have resulted in the reversed-phase mode becoming dominant for practical applications.

The influence of SCE on the primary retention of a solute can be explained in general terms using the simple model depicted in Figure 4.17 [339-341]. The addition of an equilibrant (X) to the mobile phase introduces a secondary equilibrium that allows the solute to exist in two forms, free analyte (A) and the analyte-equilibrant complex (AX). The concentration of A and AX in the mobile phase also depends on the primary distribution of A and AX with the stationary phase. The equilibrant X is assumed to be unretained, which is a reasonable assumption for acid-base equilibria and some types of ion-pairing, complexation and micellar equilibria, but is not necessarily true for all cases. If the SCE is shifted completely to the left, then the solute will exist solely as species A with a capacity factor (k_A) given by the product of the distribution constant (K_A) and the phase ratio of the chromatographic system. If the SCE is shifted completely to the right, then the solute will exist solely as the complexed form AX with a capacity factor (k_{AX}) given in a similar manner by the product of the distribution constant (K_{AX}) and the phase ratio of the chromatographic system. The capacity factors k_A and k_{AX} are the limiting values for the retention of the solute which can elute at any value within the range k_A to k_{AX} depending on the value of K_{SCE}, the equilibrium constant for the SCE. Under intermediate conditions, the retention of solute A will be the weighted average of k_A and k_{AX},

$$k_{obs} = F_A k_A + F_{AX} k_{AX} \qquad (4.22)$$

where F_A and F_{AX} are the stoichiometric fraction of the solute in each of its two forms A and AX, and k_{obs} is the capacity factor at which the solute is observed to elute under the experimental conditions. F_A and F_{AX} can be expressed explicitly in terms of the equilibrium concentrations of A, X and AX and substituted into equation (4.22) to give

$$k_{obs} = K_{SCE}/([X] + K_{SCE})k_A + [X]/([X] + K_{SCE})k_{AX} \qquad (4.23)$$

From equation (4.23) it can be seen that k_{obs} depends on the concentration of the equilibrant [X], the equilibrium constant for the SCE process, K_{SCE}, and the limiting capacity factors for the two

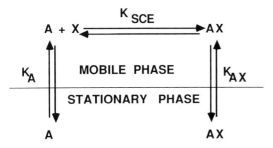

Figure 4.17 General phenomenalogical retention model for a solute that participates in a secondary chemical equilibrium in liquid chromatography. A = solute, X = equilibrant, AX = analyte-equilibrant complex, K_{SCE} = secondary chemical equilibrium constant, and k_A and k_{AX} are the primary distribution constants for A and AX, respectively, between the mobile and stationary phases.

possible forms of the solute A and AX. The effect of the concentration of equilibrant X on retention depends on the relative values of the limiting capacity factors; k_{obs} will increase with increasing [X] if $k_{AX} > k_A$, and vice versa. Since the limiting capacity factor values are subject to control by other chromatographic variables, such as the identity and concentration of organic solvent, stationary phase properties, ionic strength and temperature, the separation conditions in RPC using SCE can be quite complex in spite of the simplicity of equation (4.23).

The extent of ionization of weak acids or bases in an aqueous solution can be controlled by buffering the pH of the mobile phase. This operation is known as ion suppression and the mobile phase is generally selected to either optimize the separation of a series of ionizable substances with different dissociation constants or raised to such a value as to completely inhibit dissociation of all solutes [339,341,342]. Both neutral and ionizable solutes can be simultaneously separated using ion suppression in which, depending mainly on the identity of the buffer and its concentration, only the retention of the ionizable solutes is affected by the presence of the buffer. For strong acids and bases ionization control cannot be employed with silica-based bonded phases because of the instability of the packings outside of the pH range 2-8, and either ion-pair chromatography or porous polymeric column packings are generally used in this case. When the equilibrant is [H^+] and the complex AX is the acidic form of a weak acid, equation (4.23) can be written in the familiar form for the dissociation of an acid

$$k_{obs} = k_{HX}([H^+]/([H^+] + K_a)) + k_{X^-}(K_a/([H^+] + K_a)) \qquad (4.24)$$

The equivalent expression for polyprotic acids, zwitterions and weak bases are given in ref. 339. From equation (4.24) it can be seen that the relationship between k_{obs} and pH is sigmoidal with an inflection point at the pH corresponding to the pK_a of the acid. A small change in the pH near the pK_a value for the acid results in a large change in k_{obs} but at extreme values with respect to the pK_a of the acid (pH $\geq pK_a + 2$) changes in k_{obs} are largely dampened. In agreement with equation (4.24) it has been demonstrated experimentally that plots of $k_{obs}(1 + K_a/[H^+])$ vs. $K_a/[H]$ are linear and yield reasonable values for k_{HX} and k_{X^-}, the intercept and slope respectively [343]. For the separation of two weak acids, the optimum pH is given by

$$pH_{OPT} = pK_{AVG} + 1/2 \log (k_{HX}/k_{X^-}) \qquad (4.25)$$

where pK_{AVG} is half the sum of the pK_a values for the two acids. Equation (4.25) demonstrates that the optimum pH is dependent on chromatographic variables through k_{HX} and k_{X^-} as well as the ionization properties of the acids themselves. It should also be noted that equations (4.23) to (4.25) are exact as presented for aqueous mobile phases and a reasonable approximation for aqueous-organic solvent mixtures. For exact calculations the pH and pK_a values appropriate to the solvent composition employed should be used instead of those values obtained from tables for aqueous solution. Variation in retention due to changes in column pressure are also possible because of a shift in the pH of the mobile phase and changes in the extent of solute ionization [344]. These variations can be predicted from the volume changes for solute and buffer ionization in exact work. For equation (4.24) to be valid it is also necessary for the mobile phase ionic strength to be reasonably constant (typical buffer concentrations used are 20-60 mM).

Aqueous micellar solutions, i.e. solutions containing a surfactant at a concentration above its critical micelle concentration, have been studied extensively during the last decade, in part from curiosity, and because of the possibility of providing unique chromatographic selectivity compared to conventional RPC [345-349]. Above the critical micelle concentration individual surfactant molecules self-aggregate to form structures known as micelles which are microscopically inhomogeneous, and provide a microenvironment that is distinctly different from the bulk solvent. In particular, micelles are able to greatly enhance the solubility of insoluble, hydrophobic compounds in aqueous media through surface adsorption, pseudophase extraction (partitioning) and solute-surfactant coassembly (comicellization). Since micelles generally exhibit little affinity for the stationary phase, increasing the concentration of surfactant above the critical micelle concentration increases the concentration of micelles in the mobile phase, and reduces the retention of those solutes that are distributed between the mobile phase and stationary phase and the mobile phase and the micellar components of the mobile phase. Chromatographic selectivity and eluent strength can then be controlled by micelle concentration and surfactant type in the mobile phase. In the absence of significant interactions between the micelles and stationary phase equation (4.23) can be simplified (k_{HX} = 0) to give

$$k_{obs} = k_S[1/(1 + K_{SM}[M])] \hspace{3cm} (4.26)$$

where K_{SM} is the solute-micelle association constant and k_S the capacity factor of the unbound solute. In equation (4.26) [M] may represent either the concentration of surfactant or the concentration of micelles with the resulting association constant, K_{SM}, understood to be per surfactant molecule or per micelle, respectively. In agreement with equation (4.26) experimental plots of ($1/k_{obs}$) against [M] are frequently linear with a slope equal to K_{SM}/k_S and an intercept of $1/k_S$. The optimum micelle concentration required to separate two similar substances is given by analogy to equation (4.25) as

$$pM_{OPT} \approx \log K_{SM} - 1/2 \log (k_S/k_1) \hspace{2cm} (4.27)$$

where K_{SM} is substituted for $(K_{SM})_{average}$ assuming two similar substances (the interesting case for selectivity optimization) and k_1 is the solute capacity factor when the surfactant concentration is unity, since compared to equation (4.25) k_{HX} = k_{SM}, which is assumed to be zero in micellar LC and, therefore, does not yield a viable expression for optimization of the selectivity function.

The future development of micellar RPC will probably depend on the development of applications for which micellar mobile phases offer a significant advantage over the conventional use of aqueous-organic solvent mixtures and an increase in the efficiency of micellar liquid-chromatographic systems. The origins of the generally poor chromatographic efficiency observed in micellar liquid chromatography results from poor stationary phase mass transfer properties [346,350-352]. The characteristics of surfactant-modified bonded phases are changed compared to general RPC conditions due to adsorption of the surfactant which increases the effective film thickness, viscosity, and flexibility of the modified ligand-surfactant surface, all of which adversely impact on the column efficiency in micellar liquid chromatography. Reasonable efficiency may be obtained by operating at elevated temperatures (e.g., 40°C) and by adding a small volume (3% V/V) of an organic modifier, such as 1-propanol, to the mobile phase. The addition of alcohol modifiers to the mobile phase changes the characteristics of the stationary phase by decreasing the amount of surfactant sorbed onto the stationary phase compared to that possible in the absence of organic modifier (351,352). A further drawback of micellar liquid chromatography is that the mobile phase sometimes is not strong enough to elute compounds that are highly retained under reversed-phase conditions. In this case organic solvent can be added to the mobile phase to increase its elution strength but the amount added is limited to fairly low concentrations, generally, to prevent disruption of the aggregation process responsible for micelle formation. The relative effectiveness of organic modifier in micellar liquid chromatography appears to be directly related to its ability to partition and bind to the micellar pseudophase, and enhance the solubility of the solute in the organic micellar assembly. The organic modifier will also reduce the concentration of surfactant sorbed to the stationary phase changing both its kinetic and sorptive properties compared to similar use in the absence of surfactant. Also, it has been shown that aqueous micellar mobile phases provide improved gradient elution compatibility with electrochemical detection and an enhancement in response with fluorescence detection, which could prove practically useful for some analyses [346].

The addition of metal ions to the mobile phase frequently yields improved separations of solutes capable of forming complexes (conversely the addition of ligands to the mobile phase may allow the separation of metal ions based on differences in the distribution constants of the complexes between the mobile phase and stationary phase) [353-355]. A number of important applications of metal-modified chromatographic selectivity have been developed including the reversible formation of charge-transfer complexes between silver ions and solutes with π-donor groups, the resolution of enantiomers by ligand-exchange with chiral metal chelates, and the separation of nucleotides after chelation with different metal ions (section 8.15.3). Nucleotides are known to complex with alkali and alkaline earth metals, particularly calcium and magnesium, through the phosphate groups and with transition metals by these groups and nitrogen of the purine bases [353,356]. This complexation changes the electronic structure and the conformation of the compounds, which in turn alters solute-solvent interactions and retention behavior. Metal ions are essentially unretained under reversed-phase conditions and for a 1:1 complex the retention of the metal complex can be expressed in a form equivalent to equation (4.24)

$$k_{obs} = [M]/([M] + K_{ML})k_{ML} + K_{ML}/([M] + K_{ML})k_L \qquad (4.28)$$

where M is the metal, L the ligand, K_{ML} the formation constant for the metal-ligand complex in the mobile phase, and k_{ML} and k_L the limiting capacity factor values for the complex and ligand, respectively. In practice, the influence of pH on the formation constant and possible interactions between the bound ligand and metal ion result in a more complex relationship that contains additional equilibrium terms [353,354]. For ionic ligands, the ligand concentration available for complexation will depend on the dissociation constant of the ligand, as well as and the mobile phase pH. In general, the chromatographic selectivity involving metal-ligand complexation is controlled primarily by altering the pH (adjusts K_{ML} and possible [L]), the metal ion concentration [M] in equation 4.28), or by changing to a metal that forms either a weaker or stronger complex with the ligand (K_{ML} in equation 4.28). This is in addition to the primary chromatographic variables, such as the type and concentration of organic modifier, which affects k_L and k_{ML}, as well as the value of the formation constant, K_{ML}, and the apparent pH of the mobile phase.

4.5.7 Ion-Pair Chromatography

Ion-pair chromatography (IPC) is a further example of the use of secondary chemical equilibria to control retention and selectivity in liquid chromatography (section 4.5.6), given separate treatment here because of the many studies and popularity of this method compared to those discussed in the previous section. The origins of IPC can be traced to the use of ion-pairing reagents for the extraction of ionic compounds from aqueous solution by an immiscible organic solvent [357,358]. Schill and co-workers pioneered the adaptation of this principle to liquid chromatography in both the normal- and reversed-phase mode. Early studies in normal phase chromatography used silica or cellulose column packings coated with a buffered aqueous solution containing the ion-pairing reagent and a relatively nonpolar organic solvent as the mobile phase. The advantages of this system where that changes in retention and selectivity could be regulated by changing the composition of the organic mobile phase and the possibility of using detector-oriented counterions of high sensitivity to improve the detection of ions with little natural UV absorption or fluorescence properties [359]. In the latter case the extraction of an easily detected ion from the stationary phase by the detector transparent analyte ion of opposite charge resulted in the elution of an ion pair with favorable detection characteristics. The main limitation of normal phase systems is that the sample must be transferred to the organic phase prior to analysis; many ionic, hydrophilic substances have low solubility in moderately polar solvents. Later systems overcame this problem by using an aqueous buffer containing a low concentration of ion-pairing reagent as the mobile phase and a relatively nonpolar solvent, such as pentanol or octanol coated onto an inert support, as the stationary phase. However, both the normal-phase and reversed-phase liquid partition system have problems with instability due to the possible leaching of the stationary liquid phase from the support and have largely been superseded in contemporary practice by the use of chemically bonded phases. Since most samples have greater solubility in water than moderately polar organic solvents, reversed-phase chromatography dominates the current practice of IPC.

Reversed-phase IPC has quickly gained widespread acceptance as a versatile and efficient method for the separation of ionized and easily ionizable analytes [342,355,360-364]. It is complementary to ion-exchange (section 4.5.8) and ion chromatography (section 4.5.9) which are used to separate similar samples. Important advantages of IPC are its ability to

TABLE 4.17

VARIABLE PARAMETERS FOR ION-PAIR REVERSED-PHASE CHROMATOGRAPHY

Variable	Effect
Type of counterion	Retention increases with the ability of the counterion to form an ion pair.
Size of counterion	An increase in the size of the counterion will increase retention.
Concentration of counterion	Increasing concentration increases retention up to a limit.
pH	Effect is dependent upon the nature of the solute. Retention increases as pH maximizes concentration of ionic form of solute.
Type of organic modifier	Retention decreases with increasing solvent strength.
Concentration of organic modifier	Retention decreases with increasing concentration.
Temperature	Retention decreases as temperature increases.
Stationary phase	Retention increases with the hydrophobic character of the stationary phase.

simultaneously separate samples containing neutral and ionized molecules, it exploits conventional RPC columns which provide higher efficiency and greater versatility than fixed-site ion exchangers, and does not require any special or modified equipment. Method development in IPC is generally more flexible since the type and capacity of the stationery phase for ion interactions can be varied by changes in composition of the mobile phase.

Retention and selectivity in reversed-phase IPC are influenced by a large number of experimental variables, including the type and hydrophobicity of the counterion, the concentration of ion-pairing reagent, the type and concentration of the buffer, pH, ionic strength, concentration of organic modifier, temperature and the sorptive properties of the stationary phase [288,363,365-369]. Some general indications of the effects of these variables on retention are summarized in Table 4.17 [365]. To select a counterion for a particular separation, the most important consideration is charge compatibility. Some compounds commonly used as counterions are summarized in Table 4.18 [365].

TABLE 4.18

TYPICAL COUNTERIONS

Type	Major Applications
Quaternary amines, e.g., tetramethyl, tetrabutyl, palmityltrimethylammonium ions	For strong and weak acids, sulfonated dyes and carboxylic acids.
Tertiary amines, e.g., trioctylamine	Sulfonates.
Alkyl- and arylsulfonates, e.g., methane- or heptanesulfonate, camphorsulfonic acid	For strong and weak bases, benzalkonium salts and cathecholamines.
Perchloric acid	Forms very strong ion pairs with a wide range of basic solutes.
Alkylsulfates	Similar to sulfonic acids, but yields different selectivity.

For the separation of strong and weak acids, tetraalkylammonium salts in the halide or phosphate forms are commonly used. Protonated alkylamines are another possibility for the separation of strong acids, such as sulfonic acids. Inorganic ions, alkyl- or arylsulfonates and alkysulfates are generally used for the separation of cationic solutes. Hydrophobic amino compounds are best separated with small hydrophilic counterions, such as dihydrogenphosphate, bromide or perchlorate. For cations of intermediate polarity, hydrophobic counterions, such as naphthalenesufonate, picrate or bis-(2-ethylhexyl)phosphate are used. The reversed-phase separation of small or polar cations requires the use of hydrophobic counterions, such as alkyl- and arylsulfonates or sulfates. Generally, the retention of oppositely charged ionic solutes increases with increasing hydrophobicity (increasing chain length) of the pairing ions when used at identical mobile phase concentrations [370-372]. Solute retention depends primarily on the surface concentration of the sorbed ion-pairing reagent, which is coupled to the composition of the mobile phase, as well as the concentration of reagent. It has been shown that alkylsulfonate pairing ions of different chain length result in identical solute retention at similar surface concentrations [372]. With increasing concentration of ion-pairing reagent in the mobile phase, solute retention initially increases sharply but eventually reaches a plateau and may slowly decline at still higher concentration [373]. Since the mobile phase concentration of the pairing ion and the organic modifier both affect the surface concentration of the pairing ion, these variables should have the greatest effect on solute ion retention, and should be used for adjusting the mobile phase elution strength [374-376]. Once the plateau concentration region is reached increasing the pairing-ion concentration has little influence on the selectivity. To adjust selectivity in this region the pH should be varied first, followed by the concentration of organic modifier. To optimize a particular separation a reasonable starting concentration for the ion-pairing reagent is 5 mM for counterions with large alkyl groups and 10 mM for counterions with short alkyl groups or inorganic ions. Other considerations in selecting an ion-pairing reagent are availability in an adequate degree of purity, solubility in useful mobile phase compositions, and minimum interference with low-wavelength UV absorption detection [363,377].

Some guidelines for selecting the optimum pH for a given sample are summarized in Table 4.19. For chemically bonded reversed-phase columns, the working range is about pH 2-8. Outside of this range, physical damage to the column (e.g., corrosion, cleavage of the bonded phase, dissolution of silica, etc.) can occur. The pH of the mobile phase is chosen so that both the sample and counterion are completely ionized. Within the pH range 2-8 this is always the case for strong acids and bases but weak acids and bases are only partially ionized depending on the pH of the mobile phase. For these substances ion suppression also may be an effective method of reversed-phase separation by preventing dissociation (section 4.5.6). The pH is controlled by adding a buffer to the mobile phase. As well as its buffering capacity, the solubility of the buffer in a wide range of aqueous-organic solvent mixtures is of equal importance. Inorganic phosphate buffers are commonly used for pH control because of their wide pH range, but more recently organic buffers, such as citric acid-citrate and alkylamine-phosphate have become more popular due to their greater range of solubility in aqueous-organic solvent mixtures [368,374,378]. Organic buffers, such as triethylamine-phosphate are particularly successful in suppressing anomalous retention behavior and peak shape degradation caused by the interaction of basic solutes with unshielded silanol groups on the bonded stationary phase surface. Typical buffer concentrations are

TABLE 4.19

SELECTION OF pH

Type of Solute	Example	pH (for RP-IPC)	Comment
Strong acid ($pK_a < 2$)	Sulfonated dyes	2 - 7.4	These solutes are ionized throughout the pH range; actual pH selected is dependent upon other types of solutes present.
Weak acids ($pK_a > 2$)	Amino acids, Carboxylic acids	6 - 7.4	Solutes ionized; retention dependent upon the nature of the ion pair.
		2 - 5	Ionization of solutes is suppressed; retention dependent upon the nature of solute (not ion pair).
Strong bases ($pK_a > 8$)	Quaternary amines	2 - 8	Solutes are ionized throughout pH range; similar to strong acids.
Weak bases ($pK_a < 8$)	Catechol-amines	6 - 7.4	Ionization is suppressed; retention dependent upon the nature of the solute.
		2 - 5	Solutes are ionized; retention dependent upon the nature of the ion pair.

in the same general range as those of the ion-pairing reagent (0.5-20 mM) and contribute significantly to the ionic strength of the mobile phase. The solubility of the solutes and ion-pairing species in a given mobile phase is affected by changes in ionic strength due to salting-in and salting-out effects [363,371,374]. In addition, the buffer ions may compete with the counterions for the ionized solute or, alternatively, enhance the formation of ion-pairing complexes. The nature and concentration of organic modifier can influence retention in two ways. By changing the solvent strength of the mobile phase (as for RPC) and by modifying the concentration of ion-pairing reagent sorbed onto the stationary phase. Solvent selection, both type and range, is limited by the solubility of the ion-pairing reagent and buffer in the mobile phase. For this reason methanol has been the most widely used modifier with acetonitrile as the most common second choice. For gradient elution applications, it is important to ensure that the salts added to the mobile phase remain soluble throughout the full range of the gradient composition.

Although it would seem that there are many parameters to optimize, for simple separations this is not a great problem, since changes in just one or two of the parameters discussed above while maintaining the others within sensible ranges will yield the desired result. The primary variables of most importance are the concentration of the ion-pairing reagent, the concentration of the organic modifier and pH. Computer-guided strategies are available for varying these parameters simultaneously, and for using preselected gradient conditions to estimate reasonable starting conditions for samples of unknown origin [368,373-375,379]. For the separation of samples containing both neutral and ionic species the conditions are usually established by defining the mobile phase that provides adequate resolution in a reasonable time for the neutral species followed by adding the ion-pairing reagent and buffer to this mobile phase to adjust the separation of the ionic species. The buffer and ion-pairing reagent will, usually, have little influence on the retention of the neutral species allowing conditions to be selected to move the ionic species to open areas in the chromatogram. An example of a separation of antihistamines and decongestants by reversed-phase IPC is shown in Figure 4.18 [342]. Under the conditions of the separation maleic acid (peak one) and phenacetin (peak four) are non-ionic and their retention is unaffected by the presence of the ion-pairing reagent.

Several theoretical models, such as the ion-pair model [342,360,361,363,380], the dynamic ion-exchange model [342,362,363,375] and the electrostatic model [342,369,381-386] have been proposed to describe retention in reversed-phase IPC. The electrostatic model is the most versatile and enjoys the most support but is mathematically complex and not very intuitive. The ion-pair model and dynamic ion-exchange model are easier to manipulate and more instructive but are restricted to a narrow range of experimental conditions for which they might reasonably be applied. The ion-pair model assumes that an ion pair is formed in the mobile phase prior to the sorption of the ion-pair complex into the stationary phase. The solute capacity factor is governed by the equilibrium constants for ion-pair formation in the mobile phase, extraction of the ion-pair complex into the stationary phase, and the dissociation of the ion-pair complex in the

418

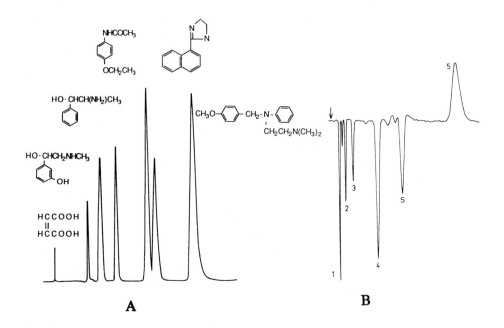

A

B

Figure 4.18   A, separation of antihistamine and decongestant drugs
by reversed-phase IPC. Mobile phase methanol-water (1:1)
containing 5 mM hexanesulfonate and 1 % acetic acid at a flow rate
of 3 ml/min. B, separation and indirect UV detection of carboxylic
acids by reversed-phase IPC. Components: 1 = acetic acid, 2 =
propionic acid, 3 = butyric acid, 4 = valeric acid, 5 = caproic
acid, and S = system peak. Mobile phase 0.3 mM 1-phenethyl-2-
picolinium in acetate buffer at pH 4.6.

stationary phase. Additional equilibria can be included in the
model by using a conditional equilibrium constant for the transfer
of the ion-pair complex to the stationary phase. The ion-pair
model is reasonably successful at explaining retention in liquid-
liquid chromatographic systems but is more circumspect when
applied to chemically bonded phases. The dynamic ion-exchange
model assumes that the ion-pairing reagent is first reversibly
sorbed onto the stationary phase through hydrophobic interactions
to form a dynamic ion-exchanger. Solute retention is governed by
ion-exchange of the oppositely charged solute ion with the reagent
ion sorbed onto the stationary phase combined with the equilibria
for dissociation of the analyte anion in the mobile phase and

sorption of the neutral form of the analyte onto the stationary phase. In this model retention of ionic solutes is governed mainly by ionic interactions, but by changing the pH of the mobile phase, it is possible to bring the solute into a non-ionized form and to increase the competition from non-ionic binding to the stationary phase.

The ion-pair model and the dynamic ion-exchange models are stoichiometric models, that is to say, a reaction scheme is constructed and the corresponding equilibrium constants express the interactions between the oppositely charged reagent ions and analyte ions in the system. The principal tenet of the electrostatic model is that this situation is physically unrealistic, since in the presence of long-range coulombic forces the equilibrium constants are no longer constant when the electrostatic field varies. The electrostatic theory of IPC assumes the formation of a surface potential between the bulk mobile phase and stationary phase resulting from the selective adsorption of ions with a higher affinity for the stationary phase compared to their counterions. For example, in the case of sodium alkylsulfonates, the alkylsulfonate anion will exhibit a greater affinity for association with the ligands of the chemically bonded stationary phase than sodium ions due to favorable hydrophobic interactions between the organic chains. To maintain electrical neutrality this will result in the formation of an electrical double layer and a difference in electrostatic potential between the bulk solvent and the stationary phase surface. The electrostatic theory then assumes that when the concentration of ion-pairing reagent is varied, changes in retention for an ionic solute are governed by the changes induced in the surface potential. The surface potential is calculated assuming a modified Langmuir isotherm for adsorption of the pairing ions and the Gouy-Chapman theory to derive the equation relating the amount of adsorbed ions to the electrostatic potential of the surface. These assumptions lead to the following relationship between the solute capacity factor of an analyte ion with a charge z at a fixed concentration of ion-pairing reagent in the mobile phase

$$k = k_0 \exp(-zF\psi_0/RT) \hspace{3cm} (4.29)$$

where ψ_0 is the difference in electrostatic potential between the surface of the stationary phase and the mobile phase, k_0 the capacity factor for the analyte at a reference composition of the mobile phase for which the surface potential is set to zero (i.e., in the absence of ion-pairing reagent in the mobile phase) and F the Faraday constant. The value for ψ_o depends primarily on the concentration of pairing ion on the stationary phase surface and the ionic strength and dielectric constant for the mobile phase. It is not, however, an easy term to evaluate mathematically. There is good experimental support for the robustness of the electrostatic theory of IPC to provide insight into the retention of analyte ions over a wide range of experimental conditions. It also explains the deficiencies in the stoichiometric models, but unfortunately, does not replace those models with a framework that is easy to manipulate for a practical understanding of the retention of different ions.

Indirect detection has become a widely used technique in IPC (also in ion-exchange and ion chromatography) for the detection of ions lacking chromophores or other suitable properties for their convenient detection by commonly used detectors [387-389]. The technique is so named because the detector response is due to the absence of the mobile phase ion, which determines the detector signal, rather than the presence of the analyte. A constant concentration of a detectable ion is added to the mobile phase and distributed to the stationary phase, such that the concentration of the detectable ion exiting the column is constant and yields a fixed detector response. Analytes injected into the chromatographic system give rise to a detector response as a result of their influence on the distribution of the detectable ion in the chromatographic system. Two types of peaks appear in the chromatogram, which furthermore can be positive or negative in direction with respect to the detector baseline, Figure 4.18. There should be one peak for each separated analyte in the chromatogram and a number of additional peaks that are characteristic of the chromatographic system (system peaks) that depends on the number of additives in the mobile phase. When the analytes and ion responsible for the detector signal are of opposite charge the analyte peaks eluted before the system peak are always negative with respect to the detector baseline and positive if they elute after the system peak. Neutral analytes and ions with the same charge as the ion detected show the opposite behavior, with positive peaks if eluted before the system peak and negative peaks if eluted after it. The direction of the detector response, therefore, depends only on the charge and retention of the solute relative to that of the ion monitored by the detector, and is independent of analyte concentration and the amount of other sample components. The magnitude of the detector response is proportional to the amount of solute injected and its retention relative to the system peak, but it is independent of the properties and the amount of other sample components. The response for an analyte eluting well before the system peak is very low. When the analyte and system peak elute close together there is an enhancement in response for both the system peak and the analyte, whereas when the capacity factor of the analyte exceeds that of the system peak, there is, at first, a decrease in response for the analyte, but at high values of the capacity factor (with respect to the system peak) the response levels out and becomes nearly constant. The system peaks appear with a retention that is dependent only on the properties of the chromatographic system and independent of the nature of the analyte, although their direction may change with the composition of the injected sample. Each mobile phase component, except for the main (weak) solvent of the mobile phase can give rise to a system peak.

The above properties of systems using indirect detection might seem confusing at first sight and are due to the mechanistic coupling of the chromatographic process with the detector operation. The properties are not anomalous and can be explained quantitatively by theoretical models [389-395]. When a sample is injected into a chromatographic system, the established equilibrium between the mobile and stationary phase will be perturbed due to the addition of new components (analytes) and deficiencies of other components related to differences in the composition of the mobile phase and sample solvent. The system will strive to reestablish the equilibrium at the point of injection in a rapid relaxation process that adjusts the concentration of all those species that are affected by the competition for access to the stationary phase. One of the species that is affected by the relaxation process is being monitored by the detector. The analyte zones come to a new equilibrium with the mobile phase in which the composition of the mobile phase that moves with each zone has a different composition from that entering the column. The resulting excess or deficit of the mobile phase components outside the analyte zones move through the column with a velocity that is characteristic for the distribution of each mobile phase component and give rise to the system peaks. The system peaks arise, therefore, from the relaxation process caused by the introduction of the sample in a solvent with a different composition from that of the mobile phase. The number of system peaks observed will depend on the complexity of the mobile phase and whether each individual component is coupled to the stationary phase equilibria that affects the instantaneous concentration of the monitored mobile phase ion. The concentration of analytes is generally low and their chromatographic properties are easily predicted assuming normal chromatographic behavior. The position of the system peaks depends on the concentration of the ion monitored by the detector, which is generally relatively high, and their behavior can only be predicted using nonlinear chromatographic relationships, that are more complex to solve mathematically [394,395]. The magnitude of the response changes resulting from the relaxation process are also predictable from the relative retention, charge type and fractional coverage of the stationary phase by the detected mobile phase ion, as described by Schill and co-workers [387,389]. An unusual feature of indirect detection is the operation of the detector at a high background signal to monitor small changes in this signal. The limit of determination is generally noise limited and will depend on the stability of the chromatographic system and that of the detector [387-390,396]. Widely different detection limits and linear response ranges can be found in the literature reflecting system properties rather than analyte properties. Response characteristics are certainly adequate for many problems.

4.5.8 Ion-Exchange Chromatography

Ion-exchange chromatography (IEC) is used mainly for the separation of ions and easily ionized substances (e.g., substances that form ions by pH manipulation or complexation) in which one of the principal contributions to retention is the electrostatic attraction between mobile phase ions, both sample and eluent, for immobilized ion centers of opposite charge in the stationary phase. The sample ions are separated based on differences in their relative affinity for the stationary phase ion centers compared to that of the mobile phase counterions in a dynamic exchange system, in which sample ions and eluent ions interact with multiple stationary phase ion centers as they pass through the column. Ion-

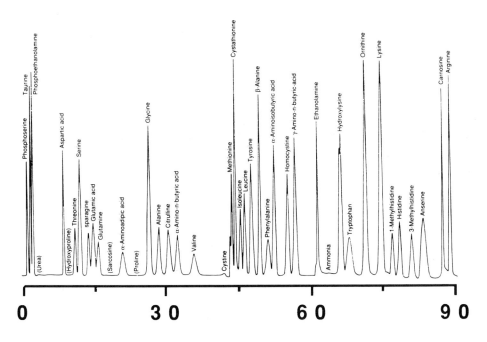

Figure 4.19   Ion-exchange separation of physiologically important amino acids using a strong cation-exchange column and a stepwise pH gradient with citrate and borate buffers. The temperature was increased from 38 to 60°C during the separation. (Reproduced with permission from ref. 104. Copyright Preston Publications, Inc.)

exchange finds applications in virtually all branches of chemistry [9,15,16,104,121,123,397-401]. In clinical laboratories ion-exchange has long been employed as the basis for the routine, automated separation of amino acids and other physiologically important amines used to identify metabolic disorders and to sequence the structure of biopolymers. Figure 4.19 provides an example of the ion-exchange separation of amino acids commonly found in physiological fluids [104]. Under typical conditions the amino acids are separated in the protonated form on a strong-acid, polymeric, cation exchanger with stepwise changes in pH (from ca. 3 to 9.6) using various combinations of citrate and borate buffers. The experimental conditions can be modified to provide rapid separation of simple amino acid mixtures or for the separation of complex mixtures requiring a longer separation time.

The amino acids are usually detected as their fluorescent products after a postcolumn reaction with ninhydrin or o-phthalaldehyde. Ion-exchange has also long been used for the isolation, purification and separation of peptides, proteins, nucleotides and other biological polymers on soft non-denaturing gels. During the 1980s there was a rapid development in the preparation of stable, wide pore, low capacity packings designed specifically for the high resolution separation of these samples [15,16,119,288,399]. This remains an active area of research with many types of stationary phases now available supported by a voluminous applications literature. The separation of carbohydrates by ion-exchange is another well-established method that has recently undergone a rapid technical development bringing many more modern packings and new separation techniques to the market place [400,402-405]. Probably the oldest and most common method of separating carbohydrate mixtures is by ligand exchange using polymeric cation-exchange columns loaded with metal ions, such as calcium (recommended for separating sugar alcohols) and silver (preferred for determining oligomeric distributions), using demineralized water or aqueous solutions of organic solvents as the mobile phase. Retention in this case is governed by a combination of steric exclusion and electrostatic attraction between the electronegative sugar oxygen atoms and electropositive metal cations. Carbohydrates have also been separated by anion-exchange chromatography of their stable, negatively charged complexes formed between sugars with vicinal diol groups and the borate anion [399-402,406]. Carbohydrates are weak acids ($pK_a \approx$ 11.9-12.5) and can be separated by anion-exchange chromatography using modern, polymeric, pellicular packings at high pH (> 12) and an aqueous solution of sodium hydroxide or sodium hydroxide/sodium acetate as eluents. The carbohydrates are expected to posses a single negative charge per saccharide unit (regardless of the degree of oligomerization) and elute in order of increasing size and degree of effective ionization. As well as new column packings, the development of the pulsed amperometric detector (PAD) has contributed to the increasing interest in carbohydrate analysis [400,403]. Carbohydrates, in general, lack useful chromophoric properties and are difficult to detect conveniently except with the relatively insensitive refractive index detector. The PAD provides an alternative and sensitive method of detection that is compatible with gradient elution techniques (section 5.6.3).

Ion chromatography is an ion-exchange based separation system introduced by Small and coworkers in the mid-1970s that has revolutionized the analysis of inorganic and organic ions with a $pK_a < 7$ [106,121-124,397,398,401,407-411]. Its rapid growth and acceptance was due in part to the difficulty of determining these ions by other methods; it has replaced many tedious wet chemical analyses with a simple, automated instrument that can determine several ions simultaneously in a single method. Ion chromatography, as originally practiced, was a synergistic development of high efficiency pellicular ion-exchange columns of low capacity, low concentration eluents with a high affinity for the ion-exchange packing, and universal, on-line detection with a flow-through conductivity detector. Modern ion chromatography still retains these features but now detection is frequently performed without use of a suppressor column (to minimize the background conductivity of the eluent) by indirect photometric detection and electrochemical detection. Typical of the ions that can be separated by ion chromatography are cations (particularly group I and II), NH_4^+, alkylammonium ions, halide anions, SO_4^{2-}, NO_2^-, NO_3^- and PO_4^{2-} as illustrated in Figure 4.20 [407]. Most of the metal cations in the periodic table have been separated at one time or another using ion exchange with acids or complexing agents as eluents, or by separation of their metal complexes with, generally, postcolumn reaction and spectrophotometric detection techniques employed for quantitation [106,121,398,401,412]. However, the major application of ion chromatography remains the separation of inorganic anions and organic cations in industrial, agricultural, food and environmental samples.

Ion-exclusion chromatography (also known as ion chromatographic exclusion) is a valuable technique for separating neutral and weakly acidic or basic substances from ionic compounds based on a Donnan exclusion mechanism rather than ion exchange [106,121,123,407,413-416]. An ion-exchange packing with the same immobilized formal charge as the sample ions is selected for this purpose and retention is dependent on the degree of access of the sample to the pore volume of the packing. Ions with the same charge as the packings are repelled and elute in the interstitial volume while nonionic and weakly ionic materials are retained and separated by partitioning between the solvent in the pore volume

426

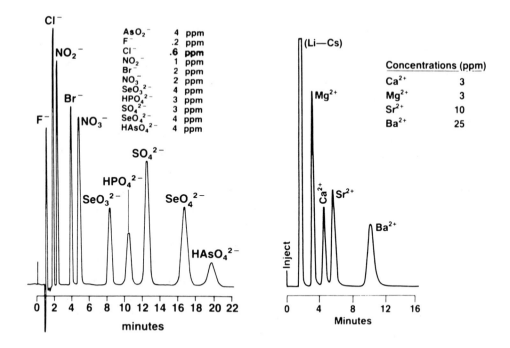

Figure 4.20 Separation of common anions and alkaline earth cations by ion chromatography using conductivity detection.

and the interstitial volume. Ion-exclusion chromatography is particularly useful for the separation of organic acids using a strong anion exchanger in the hydrogen form and an eluent containing a strong acid (e.g., 1-10 mM H_2SO_4, HCl, etc.) and, to a lesser extent, for the separation of sugars, alcohols, phenols and certain weakly ionized inorganic compounds. At a sufficiently low pH weak organic acids are undissociated, or weakly dissociated, and can diffuse into the resin pores largely uninhibited by the ionic groups on the stationary phase. The observed retention of organic acids in ion-exclusion chromatography depends primarily on the dissociation of the acid (they generally elute in order of their pK_a values) and the extent of sorptive (nonpolar and polar) interactions of the acid with the stationary phase. Also, it is generally found that dibasic acids are less strongly retained than monobasic acids, branched chain aliphatic carboxylic acids are more weakly retained than their straight chain homologs, and aromatic (also unsaturated aliphatic) acids are more strongly retained than saturated aliphatic acids.

From a global point of view, sample retention in ion-exchange chromatography can be envisioned as a simple exchange between the sample ions and those counterions originally attached to the charge-bearing functional groups of the stationary phase. However, this simple picture is a poor representation of the actual retention process. Retention in ion-exchange chromatography is known to depend on factors other than coulombic interactions. With organic ions and biopolymers, for example, sorptive interactions between the sample and the non-ionic regions of the support are important. In ion-exchange chromatography the mobile phase is often of high ionic strength; this favors hydrophobic interactions by "salting-out" the sample. For biopolymers size-exclusion effects may be important by preventing access of the sample to the total pore volume of the packing where the highest concentration of ion-exchange centers are located. Also, since Donnan membrane potentials are developed on charge-carrying supports, ions having the same charge as the support may be excluded from the pore volume by ion repulsion.

A retention model based on the coulombic interaction component of the retention mechanism is easily derived assuming that sample ions and eluent ions are in equilibrium with the available ionic sites on the stationary phase [9,106,123,124,126,417-422]. The equilibrium can be written in the form of equation (4.30) where A^X represents an analyte ion of charge x, E^Y a competing eluent ion of charge y, and M and S refer to ions in the mobile phase and the stationary phase, respectively.

$$yA_M^X + xE_S^Y \longleftrightarrow yA_S^X + xE_M^Y \qquad (4.30)$$

The equilibrium constant for this expression is given by

$$K_{A,E} = [A_S^X]^Y [E_M^Y]^X / [A_M^X]^Y [E_S^Y]^X \qquad (4.31)$$

To be absolutely correct the terms in the brackets should be activities and not concentrations, also, two of the terms involve stationary phase activities that cannot be easily evaluated. The weight distribution coefficient for the analyte A^X is designated as D_A and is given by:

$$D_A = [A_S^X] / [A_M^X] \qquad (4.32)$$

and is related to the capacity factor k_A for analyte A by

$$k_A = D_A w / V_m \tag{4.33}$$

where w is the weight of stationary phase and V_m is the volume of mobile phase. Combining equations (4.32) and (4.33) and substituting into equation (4.31), after rearrangement, we have

$$K_{A,E} = (k_A V_m / w)^Y ([E_M^Y] / [E_S^Y])^X \tag{4.34}$$

As a general condition in ion-exchange chromatography, the sample ion A^X is at a much lower concentration than E_S^Y, which can be replaced by Q/y, where Q is the ion-exchange capacity of the stationary phase. Substituting for E_S^Y in equation (4.34) and rearranging to express the other terms as a function of the capacity factor, gives

$$k_A = w / V_m (K_{A,E})^{1/Y} (Q/y)^{X/Y} [E_M^Y]^{-X/Y} \tag{4.35}$$

For a particular combination of column, eluent and analyte $K_{A,E}$, Q, w and V_m will be constant and equation (4.35), expressed in logarithmic form, simplifies to

$$\log k_A = C - (x/y) \log [E_M^Y] \tag{4.36}$$

where C is an experimental constant. From equation (4.35) the retention of A^X is determined by the selectivity coefficient, the ion-exchange capacity of the stationary phase, the ratio of stationary phase to mobile phase (w/V_m), and the concentration of competing ion in the eluent. Increasing $K_{A,E}$, Q or w/V_m leads to increased retention while increasing $[E_M^Y]$ leads to decreased capacity factors. Increased eluent charge, y, leads to decreased capacity factors, whilst increased analyte charge, x, leads to increased capacity factors. A plot of $\log k_A$ vs. $\log [E_M^Y]$ should be linear with a negative slope given by the ratio of the charge on the analyte and competing ions. A number of separation systems have been shown to obey equation (4.36), which can, therefore, be used to optimize the separation condition with respect to the eluent concentration. Exceptions are also known and occur when non-ionic interactions make significant contributions to the retention mechanism.

The above derivations assume the presence of a single competing ion in the mobile phase. For ion chromatography mixtures of polyprotic acids and their salts are commonly used, for example, sodium bicarbonate/sodium carbonate buffers for anion separations. For multiple eluent competing ions a modified approach is required [417,418,422]. The dominant equilibrium approach assumes that the divalent competing ion would be bound preferentially to the stationary phase during equilibrium and would exert a dominant effect on solute retention. Retention behavior can, therefore, be modified by considering the eluent to contain only a single (divalent) competing anion. The effective charge approach assigns equal importance to both the competing ions in the eluent and assumes that each will influence retention to the same degree. In equation (4.36) the effective charge y is replaced by $(a_1 + 2a_2)$ where a_1 and a_2 are the fractions of the total eluent species present as the monovalent and divalent forms, respectively. Both of the above approaches are gross approximations that can only be expected to fit the experimental data when the approximations made in their derivation reflect realistically the experimental conditions. The dual eluent species approach provides a more general model which considers both the concentration of the individual competing ions and their different affinities for the stationary phase [417,418]. Included in this approach are hydrogen ion concentration, acid dissociation constants and the ion-exchange selectivity coefficients, which must be added to the terms in equation (4.35) and (4.36). The final relationships are quite complex and will not be given here.

The equilibrium constant, $K_{A,E}$, described by equation (4.31), is called the selectivity coefficient, and is a measure of the preference of the stationary phase for one ion over another [123-127]. Selectivity series have been established for many common counterions, such as: $Li^+ < H^+ < NH_4^+ < K^+ < Cs^+ < Ag^+$; $Cu^{2+} < Cd^{2+} < Ni^{2+} < Ca^{2+} < Sr^{2+} < Pb^{2+} < Ba^{2+}$; and $F^- < OH^- <$ acetate , formate $< Cl^- < SCN^- < Br^- < I^- < NO_3^- < SO_4^{2-} <$ citrate. The absolute order depends on the individual ion exchanger but deviations from the above order are usually only slight for common ion exchangers. For weak-acid resins, H^+ is preferred over any common cation, while weak-base resins prefer OH^- over any of the common anions. Multiply-charged ions are, as a rule, sorbed more strongly than monovalent ions owing to their stronger electrostatic interactions with the fixed ionic sites on the stationary phase. For ions of the same charge, retention tends to decrease with increasing hydrated ion radius.

The retention of proteins by ion-exchange chromatography is more complex than is predicted by simple electrostatic models [15,16,119]. Protein retention depends on both the distribution of charge within a molecule and the number of charge sites interacting with the stationary phase, as well as on non-electrostatic interactions [14,423-427]. The net charge on a protein depends on its isoelectric point and solution pH. However, because of multiple sorptive interactions, different binding energies for each interaction, and the influence of charge asymmetry (the distribution of charge species may not be uniform throughout the protein structure), the net charge concept provides only a poor description of relative retention. Protein selectivity in ion-exchange chromatography can be manipulated through changes in the mobile phase pH and ionic strength which alter the electrostatic surface potential of the protein [424,428]. Also, the nature of the displacer ion and co-ion have been shown to significantly influence the retention of proteins through their effects on protein solubility and aggregation. Over a limited range of mobile phase composition, the capacity factor and concentration of the displacer ion, c, are sometimes linearly related, equation (4.37)

$$\log k = \log K + z_c \log [1/c] \qquad (4.37)$$

where K is the distribution coefficient which incorporates several terms including the binding constant for the equilibrium process, the column phase ratio, and the stationary phase ligand concentration and z_c the number of charges on a protein associated with the sorptive interaction. Horvath states that a major limitation of stoichiometric electrostatic models, such as the above, is their failure to consider explicitly stationary phase properties (and the implications of hydrophobic interactions which occur concomitantly with electrostatic interactions), as well as their adherence to a site-bound model [429]. Another way of looking at the problem is to assume that both the salt and protein are territorially bound (i.e., they are retained by the electrostatic field at the stationary phase surface, but remain free to move within a certain layer above it). With simplifying assumptions the influence of the salt concentration on the retention of proteins can be represented by a three-parameter equation

$$\log k = A - B\log(M_s) + C(M_s) \tag{4.38}$$

where M_s is the molality of the salt in the eluent, A is a constant encompassing all system characteristics, B the electrostatic interaction parameter (depends on the characteristic charge of the protein and the salt counterion and governs the change of retention with salt concentration according to the ion-exchange mechanism) and C the hydrophobic interaction parameter (depends on the hydrophobic contact area upon protein binding at the stationary phase surface and the properties of the salt as measured by its molal surface tension increment, see section 4.5.5. Plots of log k vs. $\log(M_s)$ are u-shaped, and at least for well behaved systems, show reasonable agreement with experimental data. At present none of the models are sufficiently sophisticated to explain the ion-exchange retention of proteins over a wide range of experimental conditions.

In analytical ion-exchange chromatography, once a selection of the column type has been made, sample resolution is optimized by adjusting the ionic strength, pH, temperature, flow rate and concentration of buffer or organic modifier in the mobile phase [50,398,430,431]. The influence of these parameters on solute retention is summarized in Table 4.20. Increasing the concentration of counterions in the mobile phase by either increasing the buffer concentration, or by the addition of a neutral salt, provides stronger competition between the sample and counterions for the exchangeable ionic centers and generally reduces retention. Since the ionic strength of the mobile phase has the greatest influence on solvent strength, this parameter is usually optimized first to obtain sample capacity factor values between 2-20. Selectivity is adjusted by varying the pH of the mobile phase. Changing the pH modifies both the character of the ion-exchange media and the acid/base equilibria and ionization of the sample. Also, a pH gradient can be used to control solvent strength. Slow column equilibration, however, can cause problems if the pH ranges over several pH units. Under these conditions the separation may not be reproducible. Ionic strength gradients are usually used for controlling retention whereas pH gradients of a narrow range are used to control selectivity. The operating pH range for a separation can be estimated from the pK_a values of the sample components. When sample pK_a values are unknown, approximate values can be estimated by considering the molecular structure and

TABLE 4.20

FACTORS INFLUENCING RETENTION IN ION-EXCHANGE CHROMATOGRAPHY

Mobile Phase Parameter	Influence on Mobile Phase Properties	Effect on Sample Retention
Ionic Strength	Solvent Strength	Solvent strength generally increases with an increase in ionic strength. Selectivity is little affected by ionic strength except for samples containing solutes with different valence charges. Nature of mobile phase counterion, $K_{A,E}$ value, controls the strength of the interaction with the stationary phase.
pH	Solvent Strength	Retention decreases in cation-exchange and increases in anion-exchange chromatography with an increase in pH.
	Solvent Selectivity	Samll changes in pH can have a large influence on separation selectivity.
Temperature	Efficiency	Elevated temperatures increase the rate of solute exchange between the stationary and mobile phases and also lower the viscosity of the mobile phase.
Flow Rate	Efficiency	Flow rates may be slightly lower than in other HPLC methods to maximize resolution and improve mass transfer kinetics.
Buffer Salt	Solvent Strength and Selectivity	Solvent strength and selectivity are influenced by the nature of the counterion i.e., its $K_{A,E}$ value. A change in buffer salt may also change the mobile phase pH.
Organic Modifier	Solvent Strength	Solvent strength generally increases with the volume percent of organic modifier. Its effect is most important when hydrophobic mechanisms contribute significantly to retention. In this case, changing the organic modifier can be used to adjust solvent selectivity as normally practiced in reversed-phase chromatography.
	Efficiency	Lowers mobile phase viscosity and improves solute mass transfer kinetics.

the number and type of functional groups present. Since only ionized solutes are retained by an ion-exchange mechanism, as a general rule of thumb, the optimum buffer pH should be 1 or 2 pH units below the pK_a of bases and 1 or 2 pH units above the pK_a of acids. When choosing a buffer salt two criteria must be met. The buffer must be able to establish the operating pH for the

TABLE 4.21

COMMON BUFFER SALTS USED IN ION-EXCHANGE CHROMATOGRAPHY

Buffer Salt	Useful pH Range
Citric acid	2.0 - 6.0
Ammonium phosphate	2.2 - 6.5
Potassium hydrogen phthalate	2.2 - 6.5
Disodium hydrogen citrate	2.6 - 6.5
Sodium formate	3.0 - 4.4
Sodium acetate	4.2 - 5.4
Triethanolamine	6.7 - 8.7
Sodium borate	8.0 - 9.8
Sodium perchlorate	8.0 - 9.8
Sodium nitrate	8.0 - 10.0
Ammonia	8.2 - 10.2
Ammonium acetate	8.6 - 9.8
Sodium dihydrogen phosphate	2.0 - 6.0
	8.0 - 12.0
Potassium dihydrogen phosphate	2.0 - 8.0
	9.0 - 13.0

separation, and the exchangeable buffer counterion must provide the desired eluent strength. Some common pH buffers and their usable ranges are summarized in Table 4.21. The solvent strength of the buffer can be estimated from the $K_{A,E}$ value for the counterion. Typical buffer concentrations are 1-500 mM.

Ion-exchange columns are generally less efficient than other column types used in HPLC. To improve solute diffusion and mass transfer, ion-exchange columns are often operated at elevated temperatures. As well as increasing column efficiency, an increase in column temperature usually results in a decrease in capacity factor values. Small changes in column temperature often result in large changes in selectivity, particularly for structurally dissimilar compounds. Also, ion-exchange columns are often operated at low mobile phase flow rates to maximize column efficiency. For most separations the mobile phase is usually completely aqueous. Water-miscible organic solvents may be added to the mobile phase to increase column efficiency and to control solvent strength. The presence of an organic modifier in the mobile phase has most influence on the separation when solute retention is at least controlled partly by a reversed-phase mechanism. When this is the case, solvent strength and selectivity can be adjusted as described for reversed-phase chromatography.

Eluent selection in ion chromatography is made more complex by the need to consider detector operating characteristics, particularly when conductivity detection is employed, as it is frequently, as well as other chromatographic properties [106,121,123,124,397,407,412,432,433]. Many common ions have poor responses using photometric detection and a method was needed that could detect the separated ions in the background of a highly conducting eluent. Since conductivity is a universal property of ions in solution and can be simply related to ion concentration, it was considered a desirable detection method provided that the contribution from the eluent background could be eliminated. The introduction of eluent suppressor columns for this purpose led to the general acceptance of ion chromatography [434]. Suppressor columns are now uncommon, being replaced at first by hollow fiber suppressors, and more recently, by micromembrane suppressors with superior chromatographic properties and eluent treatment capacity. It is now realized that the key factor for the detection of ions is not that the eluent should have a low background conductivity, but that the eluent ions should have a different equivalent conductance than the sample ions. This opens up many possibilities for direct detection with unsuppressed column systems resembling the general instrument configuration used in modern liquid chromatography. This has been referred to as single column ion chromatography (SCIC). Further possibilities for detection arise from the fact that the ion-exchange process is an example of a replacement process. Each sample peak eluting from the column is really composed of both sample and eluent ions. When solute ions elute from the ion-exchange column they replace in the eluent an equivalent number of eluent ions, so that electroneutrality is maintained. It follows, therefore, that if a property of the eluent is monitored by the detector, then changes in the detector signal will occur upon elution of a solute ion which has a different value of that same property. This is an example of indirect detection, which can be applied successfully to any detection principle, and not just conductivity, provided that there is a large enough difference in the values of the measured property between the eluent and solute ions. All of the above detection principles are now used widely in modern ion chromatography.

In its original form the ion chromatograph used two ion-exchange columns in series followed by a flow-through conductivity detector. The first column, the separator column, separates the ions in the injected sample while the second column, the suppressor column, served the dual purpose of reducing the conductivity of the eluent while simultaneously enhancing the conductivity of the separated ions by converting them to highly conducting species. By way of an example, consider the separation of a mixture of Li^+, Na^+ and K^+ on a cation-exchange column with dilute hydrochloric acid as the mobile phase. The three cations are separated on the first column according to their strength of interaction with the column packing and leave the column in the form of their chloride salts dissolved in the mobile phase. As the eluent enters the suppressor column two reactions occur as shown below:

$$HCl + Resin^+ OH^- \longrightarrow Resin^+ Cl^- + H_2O$$

$$MCl + Resin^+ OH^- \longrightarrow Resin^+ Cl^- + MOH$$

$$M = Li^+, Na^+, K^+$$

The eluent leaving the suppressor column contains the sample ions as their highly conducting hydroxide salts in a weakly conducting background of water. Anion suppression occurs in an analogous manner except that the eluent is usually a dilute solution of carbonic acid salts (e.g., $NaHCO_3/Na_2CO_3$) and the conducting species is now a strong acid (HX where $X = Cl^-$, NO_3^-, etc.) in a background of carbonic acid, H_2CO_3. Because of the high equivalent conductance of the H^+ and OH^- ions compared to other ion species this detection scheme achieves maximum detection sensitivity with limits of determination in the ppm to ppb range for standard injection volumes (100 microliters). The suppressor column is packed with an ion-exchange packing, opposite in type to the separator column, and of higher capacity, since it must neutralize a large volume of column eluent. The disadvantages of suppressor columns are that eventually the column becomes exhausted and must be regenerated (a process that needs to be repeated periodically) and the dead volume of the suppressor column reduces the chromatographic efficiency of the separator column (the Donnan exclusion phenomena prevents entry of the sample ions into the pore volume and only the interstitial particle volume need be considered normally). There should be no retention of ions in the suppressor column, which is true of strong acids and bases. Weak acids and bases, however, may be retained by several mechanisms and their retention volume will change with the degree of exhaustion of the suppressor column. In anion chromatography, ion-exclusion of carbonic acid in the unexhausted portion of the suppressor column gives rise to the so-called "carbonate dip", a negative excursion of the baseline using conductivity detection that can interfere in the determination of some ions, such as F^-, Cl^- and NO_2^-. The location of the "carbonate dip" in the chromatogram depends on the degree of exhaustion of the suppressor column and will change while in use. Some of the above problems were solved by the introduction of hollow fiber suppressors based on ion-exchange polymeric membranes [397,412]. The main advantage of the hollow fiber ion-exchange suppressor was that it allowed continuous operation of the ion chromatograph by exchanging ions across the membrane wall into a continuously regenerated solution of electrolyte bathing the fibers. Relatively large dead volumes, limited suppression capacity, and the fragile nature of the fibers made them less than ideal. Eluent suppression capacity was largely limited by slow mass transfer of ions to the membrane wall, and although second generation hollow fiber suppressors filled with beads or fibers and shaped into various helical forms to promote turbulence improved ion transport to the membrane wall, they lack equivalent performance to the micromembrane suppressors now in common use. The micromembrane suppressor, Figure 4.21, combines the high ion-exchange capacity of packed-bed suppressors with the constant regeneration feature of hollow fiber suppressors in a low-dead volume (< 50 microliter) configuration [397]. The effluent from the separator column flows between two thin ion-exchange membranes that are separated by an intermediate screen made from plastic with ion-exchange sites. A solution of electrolyte is pumped countercurrently and externally to the membrane in small volume channels that are partially filled with plastic screens. The purpose of the screens is to enhance the transport of ions to the ion-exchange membrane by generating turbulent flow and by providing a site-to-site path for transport of ions to the membranes (screens in the eluent chamber contain ion-exchange sites). The micromembrane suppressor has a higher suppression capacity than other suppressor designs. It is compatible with the use of higher buffer concentrations and higher eluent flow rates. One outcome of these properties has been the introduction of ion separations employing either composition or concentration gradients, that allow a wider range of ions to be separated in a single injection, as shown in Figure 4.21 [397,435-438].

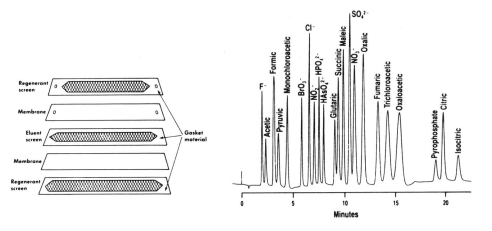

Figure 4.21 Exploded view of a micromembrane suppressor and gradient elution separation of a mixture of inorganic and organic anions by ion chromatography employing conductivity detection with a micromembrane suppressor.

Composition gradients are difficult to perform due to the slow reequilibration of the ion-exchange column with the weaker displacing eluent ion at the end of the gradient. Concentration gradients are the most popular, since they reduce retention without affecting selectivity for ions of the same charge, however, polyvalent ions are affected differently to monovalent ions, and samples containing ions of mixed valency exhibit significant selectivity differences with changing slope of the concentration gradient. Typical eluents for suppressed ion chromatography are of two general types: salts of weak acids or bases; or amphoteric ions that can be ionized at one pH and neutralized at another. Typical concentrations using a micromembrane suppressor are 1-100 mM. Some common eluents for anion analysis are sodium hydroxide, sodium phenate, sodium carbonate/bicarbonate mixtures and sodium borate. Hydroxide ion is an excellent choice of displacing ion because it is converted in the suppressor to water, regardless of its concentration. However, hydroxide has a low affinity for the stationary phase and must be used in high concentration to elute tightly bound ions. Such high concentrations only became practical with the introduction of micromembrane suppressor technology. Gradient elution with sodium hydroxide concentration gradients is now common practice. For polyvalent anions, such as sulfate and phosphate, the phenate ion is a good choice because of its high stationary phase affinity and conversion to the weakly conducting phenol in the suppressor system. With the older style suppressor columns, mixtures of sodium carbonate/bicarbonate became widely used due to the high affinity of the divalent carbonate for the stationary phase and the low conductivity of carbonic acid produced in the suppressor column. The most commonly used eluents in cation ion chromatography are dilute solutions of mineral acids (e.g., HCl, HNO_3, H_2SO_4, etc.), and mineral acid salts of organic amines, such as phenylenediamine, ethylenediamine and diaminopropionc acid. Mineral acids in the concentration range 1-10 mM readily elute alkali metals, ammonium and many organic amines. The hydronium ion is an excellent choice from a suppression standpoint, since its reaction product, water, is of low conductivity. The diamine salts are required for the elution of polyvalent cations of high stationary phase affinity.

In single column ion chromatography (SCIC) a stationary phase of low capacity (0.01-0.30 meq./g) is used with dilute solutions of electrolytes that have a strong affinity for the stationary phase and a significantly different conductivity to the sample ions for conductivity detection in the absence of a suppressor column [106,411,433]. In most cases 0.1-5 mM solutions of weak aromatic acids, such as phthalic, benzoic and salicylic acids, or their salts, have been used for anion analysis. Dilute solutions of strong acids and ethylenediammonium salts have been used for cation analysis. The advantage of SCIC is the simplicity of the apparatus and the higher chromatographic efficiency attainable. A wider choice of eluents is also possible as there is no restrictions related to the eluent suppression mechanism. In the SCIC mode, conductivity detection limits are usually poorer and the linear calibration range shorter than for suppressed ion chromatography. For many applications either mode of operation is acceptable.

4.5.9 Size-Exclusion Chromatography

In size-exclusion chromatography (SEC) a separation results from the distribution of the sample between the moving mobile phase and the stagnant portion of the mobile phase retained within the porous structure of the stationary phase [15,77,439-442]. Retention differences are controlled by the extent to which sample components can diffuse through the pore structure of the stationary phase, which depends on the ratio of molecular dimensions to the distribution of pore-size diameters. The distribution of pore sizes is the key factor for stationary phase selection for a particular separation, since no separation will result under conditions where the sample is completely excluded from the pore volume, or can completely permeate the pore volume; the essential requirement is that the fraction of the pore volume accessible to individual sample components must be different in a size-dependent manner. Compared to the other liquid chromatographic retention mechanisms discussed so far, a unique feature of SEC is that there are no enthalpic (attractive) stationary phase interactions involved. The driving force for retention is the difference in configurational entropy of the solute within the pore volume compared to that in the moving mobile phase.

The origins of modern size-exclusion chromatography can be traced to the introduction of crosslinked poly(dextran) and poly (saccharide) gels used for the size separation of water soluble biopolymers and of semirigid, porous crosslinked poly(styrene) gels for the separation of organic polymers [440,443,444]. These developments occurred in parallel with the separation of water soluble biopolymers being called 'Gel Filtration Chromatography' (GFC), largely by biochemists, and the separation of organic polymers with organic solvents 'Gel Permeation Chromatography' (GPC), largely by polymer chemists. Although both names remain widely used in the contemporary literature, they are redundant, and size-exclusion chromatography is the preferred term independent of the type of mobile phase employed. The properties of soft gels for conventional (low pressure) SEC will not be discussed here, although soft gels remain widely used for the isolation of polymer fractions intended for further

440

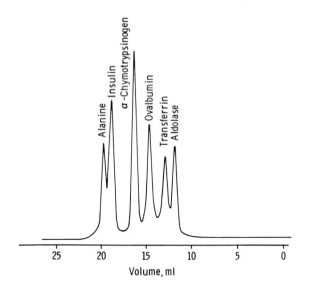

Figure 4.22 Separation of proteins on a small particle gel
column. Column 60 x 0.75 cm. TSK-Gel 2000 SW; mobile phase, 0.01 M
phosphate buffer (pH 6.5) with 0.2 M sodium sulfate; flow rate,
0.3 ml/min.

characterization, for desalting biological extracts, for studying
the formation of polymer association complexes, and for sample
preparation techniques in conjunction with gas and liquid
chromatography.

Modern SEC employs small particle, rigid, polymeric or
silica-based gels of controlled pore sizes to separate samples of
different molecular size, and to obtain average molecular weights
and molecular weight distribution information for polymers.
Polymers of an exact molecular weight, such as proteins, and
synthetic polymers of a narrow molecular weight distribution, are
separated as narrow bands as illustrated in Figure 4.22. For
reasons that will be discussed presently, the peak capacity of SEC
is low compared to other liquid chromatographic techniques, and
only a few separated bands can be accommodated within the
separation. Consequently, SEC is not the method of choice for
resolving complex samples into individual peaks. However, it can
be a powerful exploratory method for the separation of unknown
samples, since it provides quickly and conveniently an overall
view of sample composition within a predictable time and requires
little method development. Solvent optimization is not needed
beyond finding a solvent that is a good solvent for the sample and

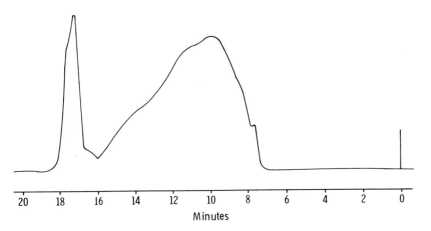

Figure 4.23 Molecular weight distribution of a carboxymethyl-cellulose sample. Column combination of LiChrospher Si-100 and Si-500; mobile phase 0.5 M aqueous sodium acetate, pH 6, and flow rate 0.5 ml/min.

is compatible with the properties of the stationary phase and the detector. The absence of interaction or reaction with the stationary phase ensures total sample elution, at least for the ideal case. Polymers of a broad molecular weight range contain many components of similar size that cannot be resolved into individual species, and are eluted as a continuous band with a molecular size profile that is characteristic of the polymer, for example, Figure 4.23. Gross differences in the profile obtained for similar polymers can often be related to changes in the synthetic procedure or physical and chemical properties of the finished product of value for indicating its performance in its intended application. Size-exclusion separations are not limited to high molecular weight polymers as new gels with mean pore diameters less than 10 nm are available for separating samples with a molecular weight below 500 [441,445,446]. These are useful for the separation of simple mixtures and for the analysis of small molecules in a high molecular weight matrix. In most cases, however, separations using small molecule SEC have been shown to depend (at least partially) on sorption interactions and cannot be rationalized completely within a general size-exclusion mechanism.

In SEC separation results from the distribution of the sample between the mobile phase inside and outside the pores of the column packing. In the ideal case there is no direct interaction with the stationary phase and a separation is complete when a volume of mobile phase equivalent to the column void volume has passed through it. In other liquid chromatographic techniques, retention is measured as the difference between the elution volume and the void volume, and is always greater than the void volume. Because of the fundamental difference in elution, the general methods of describing retention and resolution must be modified for SEC. In particular, the capacity factor and the separation factor have no real meaning in SEC. All peaks elute with a negative value of the capacity factor, using conventional terminology. Retention in SEC is better describing by the distribution function, K_{SEC}, which is related to the experimental parameters by equation (4.39)

$$V_E = V_o + K_{SEC}V_P \qquad\qquad (4.39)$$

where V_E is the elution volume of the solute, V_o the column interstitial volume (mainly the liquid volume between particles) and V_P the pore volume of the column packing, equivalent to (V_m - V_o), V_m being the column void volume as defined in section 4.4.4. K_{SEC} is constrained to values between 0 and 1, representing the extremes of complete exclusion and permeability of the pore volume by the solute. It is a function of the size and shape of the molecules being chromatographed with respect to the size and shape of the pores of the packing. K_{SEC} can be related directly to changes in the entropy of a macromolecule as it penetrates from the mobile phase into a stationary phase pore using appropriate thermodynamic terms and various models for the pore shape and distribution [447,448]. The various models provide a useful insight into macromolecular separations by SEC but are mathematically complex and not treated further in this text.

The conceptual idea of a theoretical plate can be used in SEC to measure column efficiency and to compare the performance of packed columns. For column comparisons it is usually measured with small molecules, such as toluene, acetone or benzyl alcohol, which can explore all of the pores of the packing (K_{SEC} = 1). Plate counts measured in this way produce HETP values lower than the actual values measured with monodisperse polymers and proteins. The plate count in this case can be expressed by equation (4.40)

$$n = 5.54 (V_E/w_h)^2 \qquad\qquad (4.40)$$

where w_h is the peak width at half height measured in the same units as V_E. With biopolymers having a defined structure it is legitimate to talk about an HETP value; however, most synthetic polymers are not single substances but mixtures with a mean molecular size distribution. In this case peak dispersion is comprised of chromatographic dispersion and the molecular size dispersion of the solute. Under these circumstances the HETP value is an inappropriate measure of column efficiency.

In SEC the HETP value is due almost entirely to the contribution from stagnant mobile phase dispersion and interparticle mobile phase mass transfer [77,449,450]. Contributions from longitudinal diffusion are insignificant, since the large polymeric molecules encountered in SEC have small diffusion coefficients. Resolution in SEC can be improved significantly by minimizing peak dispersion by using packings of small average particle size and working close to the optimum mobile phase velocity. Historically, packings of rather large particle size were used in SEC, but the modern trend is to use packings of 3 to 10 micrometers in particle diameter, similar in size to those generally used in HPLC. These are often packed in columns of greater internal diameter than conventional columns to avoid significant extracolumn dispersion effects, which because of low retention in SEC ($V_E < V_m$), are consequently more significant. Optimal mobile phase velocities for macromolecular solutes are up to an order of magnitude slower than those used for small molecule separations. Even comparatively long columns packed with small-particle packings can be used at modest column pressure drops.

Peak-to-peak resolution in SEC can be calculated by the ratio of peak separation at the peak maxima to the sum of the baseline peak widths. This general definition of resolution is less useful in SEC, where a measure of the ability of the column to separate solutes of different molecular weight is required. For this purposes, we define a new term, the specific resolution factor, R_{sp}, which relates peak resolution to sample molecular weight, assuming all measurements are made within the linear region of the molecular weight calibration curve, equation (4.41)

$$R_{sp} = R_s [1/\log (M_1/M_2)] \qquad\qquad (4.41)$$

where R_s is the chromatographic resolution defined by equation (1.45) and M_1/M_2 is the molecular weight ratio of two standards. Typically, two standards of narrow molecular weight distribution (polydispersity ≈ 1.1), differing by about 10-fold in average molecular weight, are selected to determine R_{sp} [451,452]. If a chromatographic resolution of unity is accepted as a desirable optimization factor (see section 1.6), then the ratio M_1/M_2 can be defined as the minimum molecular weight ratio, R_m. The minimum molecular weight ratio is a useful parameter for comparing column performance and can be used to relate resolution to the basic properties of the chromatographic system, equation (4.42)

$$\log R_m = 4mV_E/V_p n^{1/2} \tag{4.42}$$

where m is the slope of the linear region of the molecular weight calibration curve. Equation (4.42) indicates that in order to obtain increased resolving power (corresponding to a minimum R_m value) for a particular column, the column efficiency and the internal pore volume of the packing must be maximized and the slope of the calibration curve minimized. All other things being equal, using small particles with a narrow particle size distribution, particles with a narrow pore size distribution, particles with a larger pore volume, optimum (low) mobile phase velocities and increased column temperatures will improve resolution. Mobile phases of high viscosity and increased sample molecular weight (due to a decrease in diffusion coefficients) will usually lead to diminished resolution.

Another approach to defining the separation capacity of a column is by its peak capacity (the number of peaks than can be resolved at any specific resolution, usually $R_s = 1$, in a given separation time). For SEC the peak capacity, PC_{SEC}, is given approximately by

$$PC_{SEC} = 1 + 0.2n^{1/2} \tag{4.43}$$

Typical values for PC_{SEC} are 20-40, considerably smaller than can be obtained with other HPLC systems, and results primarily from the limited retention range characteristic of SEC.

Non-ideal behavior in SEC, due to the contribution of nonsize effects to retention, particularly in the case of the separation of polyelectrolytes in aqueous SEC, can render erroneous the calculation of molecular weight and molecular weight distributions from the chromatographic results [440,452-459]. Non-ideal behavior results from sorptive interactions with the stationary phase, solvent-dependent conformational changes of the solute, and from poor control of experimental variables. Since solute-solvent interactions play a critical role in controlling the hydrodynamic volume of a macromolecule, mixed mobile phases may lead to deviations from ideal behavior due to preferential solvation of either the stationary phase or the solute by one of the solvents. In the case of preferential solvation of the stationary phase, the composition of the mobile phase trapped in the porous support and that in the interstitial volume, are no longer identical. This may cause the hydrodynamic volume of the polymer to change when it enters the packing leading to unexpected elution behavior. Polyelectrolytes exhibit size-dependent changes that depend on the mobile phase ionic strength. In low ionic strength solvents, electrostatic repulsive forces among neighboring ionic sites expand the polymer chain, increasing its hydrodynamic volume. With the addition of electrolyte to the solvent, these electrostatic forces are diminished and the polymer contracts. The extent of contraction can be expected to depend on the degree of dissociation and ionic substitution on the polyelectrolyte, as well as by the uniformity and location of ionic groups on the polymer chain. Also, at low mobile phase ionic strength, polyelectrolytes may exhibit expansion as the polymer concentration is decreased. This phenomenon is caused by a decrease in concentration of closely associated counterions surrounding the polyelectrolyte in proportion to the decrease in polyelectrolyte concentration. This results in decreased electrostatic screening among ionic sites on the polymer, leading to an increase in the hydrodynamic volume of the polymer. It is usually necessary to use mobile phases of high ionic strength, typically 100-200 Mm, to obtain correct molecular weight data for polyelectrolytes, since then the thickness of the double layer becomes negligible. Stationary phase interactions that affect the retention of polyelectrolytes include ion exchange, ion exclusion and ion inclusion, as well as various sorptive interactions that depend on nonionic interactions and apply to all macromolecules. Ionic interactions for polyelectrolytes are caused by the presence of dissociated silanol groups on silica-based packings, carboxylic acid groups on polymer packings, or other ionized groups intentionally or otherwise introduced into the packing to enhance biocompatability. Ion-exchange interactions are minimized by optimizing the pH, ionic strength and type of buffer salt used for the separation. Also, sites on the support may cause ion exclusion, the process whereby a charged solute is prevented from exploring the pore volume due to electrostatic repulsion by stationary phase ions of like charge. This can be overcome by adding electrolyte to the mobile phase to screen the electrostatic repulsions. Ion inclusion can result in the appearance of an additional peak in the chromatogram when polyelectrolytes are separated in a mobile phase containing a salt. In this case, the stationary phase acts as a semipermeable membrane (permeable to small ions but impermeable to macroions), permitting the establishment of a so-called Donnan equilibrium. As the zone of eluent containing polyelectrolyte solute travels down the column, the activity of small ions outside of the pores becomes higher because the overall ionic concentration is higher. Some of the small ions are thus driven into the pores by the difference in activity. They become eluted later than the polyelectrolyte of low molecular weight. Sorptive interactions with the stationary phase are usually controlled by tailoring the properties of the mobile phase so that it can compete effectively with the solute for active sites on the stationary phase. The most general type of interactions, hydrophobic interactions, can be eliminated by adding an organic modifier to the mobile phase and/or by decreasing the ionic strength of the mobile phase. It should be noted that many of the problems discussed above are prevalent for the separation of polyelectrolytes but for nonionic polymers, soluble in common organic solvents, the difficulties are much less.

Experimental variables such as temperature, flow rate, sample concentration and mobile phase composition can cause changes in the elution volume of a polymer [439,457,460-464]. Chromatographic measurements made with modern equipment are limited more by the errors in the absolute methods used to characterize the molecular weight of the calibration standards than any errors inherent in the measurements themselves, since the determination of molecular weights by SEC is not an absolute method and is dependent on calibration [462]. The influence of temperature on retention in SEC is not very great, since no strong sorptive interactions are involved in the retention mechanism. Temperature differences between the column and solvent delivery system are important, however, because temperature gradients cause variations in the elution volume of the sample due to thermal expansion or contraction of the solvent volume [461]. It is generally observed that elution volumes are independent of flow rate within the normal flow rate range used in SEC, except perhaps at high column backpressures where solvent compressibility or other factors may cause changes in the effective hydrodynamic volume of the solute [464]. Organic polymers insoluble in common solvents, such as toluene, chloroform or tetrahydrofuran at room temperature, are often analyzed using dimethylformamide, dichlorobenzene, or m-cresol at higher temperatures. Polymer degradation by thermal or chemical reactions is an increasing possibility. Mixed solvents and more exotic solvents, such as hexafluoroisopropanol, may be better choices for sensitive polymers [403]. High molecular weight polymers separated on efficient microparticle columns may undergo shear degradation caused by the hydrodynamic forces exerted on the polymer in the sample flow path (column and instrument components) [457,459]. This has the effect of increasing the concentration of low molecular weight components and broadening the molecular weight distribution observed for the polymer. Adverse concentration effects for high molecular weight polymers are manifested by increased elution volumes and, if the viscosity of the injected solution is significantly higher than that of the mobile phase, by peak distortion. The former effect is known as macromolecular crowding and results in a reduction of the hydrodynamic volume of the polymer caused by volume constraints imposed by neighboring polymers. At high polymer concentrations, a critical concentration is approached whereby segmented chain motion becomes somewhat restricted by chain overlap. In addition, macromolecular crowding decreases the conformational entropy of the polymer in the interstitial volume, leading to an increase in K_{SEC}. Viscosity differences between the sample and mobile phase cause viscous fingering, resulting in peak distortion that arises from the perturbation of the velocity streamlines as the mobile phase moves through the zone occupied by the sample plug.

The most critical decision in size-exclusion chromatography is column selection. This is also a fairly simple decision, since the molecular weight separating range of the column is defined by the pore size and pore size distribution of the packing. The molecular size separating range of a packing is described by the

448

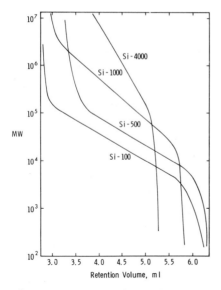

Figure 4.24 Calibration curves for LiChrospher column.
Calibration standards poly(styrene) and mobile phase
tetrahydrofuran.

shape of its calibration plot. Some representative calibration
plots for LiChrospher macroporous silica packings are shown in
Figure 4.24. The flat central portion of the curve, the
fractionation range, represents the usable molecular weight
separating range of the packing. The sharp breaks in the
calibration plot that occur at either end of the fractionation
range correspond to the molecular weight region in which the
sample is either totally excluded or free to explore the total
pore volume of the packing. In these regions the separation
properties of the packing are poor. To separate two components of
different molecular size, a column packing for which the two
components elute in the middle of the fractionation range should
be selected. All other things begin equal, the column with the
shallowest gradient for the fractionation range will provide the
highest resolution. Once the optimum column has been selected,
resolution can be increased by coupling two or more columns of the
same pore size to increase the chromatographic efficiency of the
system.

For separating samples with a relatively narrow molecular
weight range (< 500), packings with a single pore size are used.
To separate sample with a broad molecular weight distribution,

columns containing packings with a wide pore size range are used. This is achieved by coupling columns having nonoverlapping but closely adjacent fractionation ranges, or by intimately mixing the packings prior to filling the column [465,466]. A bimodal pore size configuration provides better resolution and a wider linear calibration range than can be obtained by the series coupling of several column types with overlapping molecular weight ranges. Columns containing only two discrete pore sizes and a narrow size distribution, differing by approximately a factor of ten in pore size and of approximately equal pore volume, are coupled for this purpose. A linear molecular weight separating range of four orders of magnitude or more can be obtained in this way.

The information often sought from an analysis by SEC is the sample molecular weight or molecular weight distribution. The information yielded by the experiment is a separation based on molecular dimensions, size and shape - a parameter that is not necessarily synonymous with molecular weight. Even monodisperse samples of the same molecular weight do not necessarily have the same size. The size of a molecule in solution is a function of sample-solvent interactions and depends on the nature and strength of intermolecular forces. Biopolymers adopt different conformations and, therefore, different molecular sizes, when dissolved in various solvents. Some solutes are hydrated or otherwise associated with small molecules in solution; such interactions change their hydrodynamic size and separation characteristics. For monodisperse samples careful column calibration with standards of similar structural properties is required to obtain molecular weight information. For polydisperse samples there is no absolute, well-defined value for the molecular weight. Many synthetic polymers are polydisperse systems characterized by some distribution of species of different absolute molecular weight around an average or central value. The molecular weight of a polydisperse sample must be described by a statistical function, such as the number-average molecular weight ($M_n = \Sigma N_i M_i / \Sigma N_i$), the weight-average molecular weight ($M_w = \Sigma N_i M_i^2 / \Sigma N_i M_i$) or the z-average molecular weight ($M_z = \Sigma N_i M^3) / \Sigma N_i M_i^2$) where N_i is the number of molecules having a molecular weight M_i and i is an incrementing index over all molecular weights present. The various molecular weight averages are calculated by digitizing the peak and using the relationships given above [77,439,457].

This may be done manually but is quite tedious and error prone. Many integrator-based and computer-based data handling systems have optional software for automatically performing these calculations. The polydispersity of a broad molecular weight sample is characterized by the ratio (M_w/M_n). For comparison with absolute molecular weight methods, osmometry yields values for M_n, light scattering M_w, and ultracentrifugation M_z. However, there is no preferred choice of molecular weight value for correlating polymer properties to an intended application, and most often the value that shows the most sensible change with the property being correlated is arbitrarily selected.

The first step in extracting molecular weight information from a SEC chromatogram is to establish a calibration curve relating the retention volume (or K_{SEC}) to the molecular weight of the polymer sample [77,439,448,457,459]. Molecular weight, rather than the size of the polymer, is used as the calibration parameter, since the former is independent of the experimental conditions. For monodisperse samples and polymers of low polydispersity (≈ 1.1) the primary method of calibration is the peak position calibration method. A series of different molecular weight standards are chromatographed under constant experimental conditions and their elution volume measured. A plot of molecular weight against elution volume constitutes the calibration plot as shown in Figure 4.25. The linear portion of the calibration curve is usually described by an equation of the type

$$\log M = A - mV_E \tag{4.44}$$

where M is the molecular weight corresponding to an elution volume V_E and A and m are system constants. The peak position calibration method is limited in scope by the availability of suitable standards. Apart from poly(styrene), poly(ethylene oxide) and dextrans, very few other narrow molecular weight polymer standards are available for characterizing synthetic polymers. The accuracy of the peak position calibration method depends on whether the solution conformations of the standards and samples are similar, the condition required for the molecular weight and molecular size to be uniquely correlated. Large errors in estimating the sample molecular weight can result when calibration curves prepared from narrow molecular weight standards of one polymer are used to characterize polymers of a different type. For this reason,

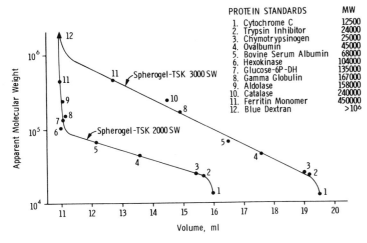

PROTEIN STANDARDS	MW
1. Cytochrome C	12500
2. Trypsin Inhibitor	24000
3. Chymotrypsinogen	25000
4. Ovalbumin	45000
5. Bovine Serum Albumin	68000
6. Hexokinase	104000
7. Glucose-6P-DH	135000
8. Gamma Globulin	167000
9. Aldolase	158000
10. Catalase	240000
11. Ferritin Monomer	450000
12. Blue Dextran	>10⁶

Figure 4.25 Protein calibration curves for Spherogel TSK-SW 2000 and TSK-SW 3000 column. Mobile phase: phosphate buffer 0.2 M, pH 6.8, flow rate 1.0 ml/min.

various attempts have been made to determine a universal calibration parameter to characterize the effective dimensions of macromolecules and various polymers. The hydrodynamic volume is one such parameter that has been widely applied for this purpose [77,439,459,467]. The hydrodynamic volume is equal to the product of the intrinsic viscosity and the molecular weight of the polymer. The intrinsic viscosity of the polymer is an experimental quantity derived from the measured viscosity of the polymer in solution or calculated using Mark-Houwink constants. A plot of the logarithm of the hydrodynamic volume against the elution volume provides a calibration curve that is approximately valid for all polymers.

The peak position and universal calibration methods rely on peak position calibration with known polymers of narrow molecular weight distribution. Several other calibration procedures requiring only a single broad molecular weight standard have been proposed [77,439]. These procedures are quite complex and have a major drawback in that, unlike the peak position methods, instrumental peak broadening must be accounted for correctly if accurate results are to be obtained.

The use of a molecular weight detector and a concentration detector in tandem provides data that can be used to make absolute molecular weight determination [439,457,459,468-470]. Light scattering and viscometric detectors offer fast responses to changes in molecular weight, have small mixing volumes, and are sufficiently sensitive to determine molecular weights at concentrations typical of those found in chromatographic effluents. The intensity of light scattering is proportional to both molecular weight and concentration using the low angle laser light scattering detector. In the case of the viscometric detector the pressure drop across a capillary is used to continuously monitor the intrinsic viscosity of the column eluent. Using either detector, the molecular weight can be evaluated across the whole chromatogram if a concentration-sensitive detector, usually a refractive index detector, is connected in-line, eliminating the need for a molecular weight calibration for the instrument. Although this is a very elegant approach towards calibration, cost and availability limit its use to a few specialized polymer laboratories.

4.5.10 Precipitation Liquid Chromatography

Separation of polymers by size-exclusion chromatography (SEC) provides information about the hydrodynamic volume (molecular weight) of the sample but gives no information about the chemical composition distribution of copolymers which is often essential for understanding their properties. Precipitation liquid chromatography, generally referred to as high performance precipitation liquid chromatography (HPPLC), is used to provide information about the chemical composition of polymers and is a complementary technique to SEC [471-475]. It is often used in tandem with SEC to determine the chemical composition of fractions isolated by SEC. Typically in HPPLC, polymer species are first precipitated at the head of the column by injection into a mobile phase that is a poor solvent for the polymer. Individual polymeric species are eluted in a solvent gradient of a good solvent added continuously in increasing amounts to the poor solvent. Each polymeric species is dissolved during gradient elution at an eluent composition characteristic of its molecular weight and chemical composition [473,476,477]. In the case where the elution strength of the mobile phase is sufficient to exclude adsorption on the stationary phase, the polymers are eluted in a solvent mixture corresponding to the cloud point in almost quantitative agreement with observations on polymer solubility made by turbidimetric titrations. The concentration of good solvent required to dissolve the sample increases with the sample size resulting in an increase in elution volume with increasing sample size. When the dissolved polymers are additionally adsorbed on the stationary phase they are eluted in the gradient at a higher concentration of the strong solvent than predicted from polymer solubility data. Since with porous stationary phases the velocity of the excluded polymer fractions is faster than that of the eluent front (solvent is not excluded from the pore volume), the polymers are eluted as colloidal solutions which can cause problems with photometric detection resulting from simultaneous absorption and light scattering. This results in an increased photometric response with increasing polymer molecular weight and should be avoided for quantitative analysis.

4.6 METHOD DEVELOPMENT STRATEGIES FOR LIQUID CHROMATOGRAPHY

The process of method development in liquid chromatography is often difficult and time consuming [478-484]. This is due primarily to the diversity of retention mechanisms available, the large number of interdependent parameters associated with the practice of each method, and the fact that several chromatographic separations are generally required to arrive at the optimum experimental conditions for the separation. The expert chromatographer uses the sum of his/her experiences to resolve the many questions that arise in the decision making steps to develop a suitable method, but most likely in a manner that is not simply systematized for the benefit of the novice chromatographer. The best that can be hoped for in this section is to provide some logical guidelines as to how a method is generally developed, but we cannot provide a panacea for all problems. At least by eliminating those decisions or experiments from the method development process that are least likely to yield useful information the process of method development should be both faster and more logical.

The general steps in developing an acceptable analytical method in liquid chromatography are summarized in Figure 4.26. Method development starts with a clear definition of the needs of the analysis. How many detectable components are present in the sample? Are all peaks equally relevant? In the first case all peaks must be resolved and the difficulty of providing the desired result will increase with the number of components in the sample.

Figure 4.26 An outline of the steps involved in developing an analytical method by liquid chromatography.

In the second case the problem is truncated to the requirement that the relevant peaks are adequately separated from each other and from the sample matrix (irrelevant peaks) as a group. Any method development strategy will be simpler if the number of relevant sample components is known in advance, so that the completeness of the separation can be easily established. The availability of pure standards is helpful for aiding the separation optimization. The concentration range of relevant peaks and the detection characteristics of the sample components dictates the choice of detector and minimizes the choice of mobile phases to those that are compatible with the detector. General sample properties of interest are its solubility in different solvents, component molecular weights and chemical structure or general class of sample components. These considerations are very important for mode selection, that is matching the sample properties to the retention mechanism (stationary phase) most suitable for the separation. Also, the separation time may be an implied constraint for the final method. All of the above factors are important in making decisions that affect all subsequent identifiable steps in Figure 4.26.

To select a method, the chromatographer usually relies heavily on personal insights and experiences gained with similar samples in the past, information taken from the literature describing the separation of the same or similar samples, and/or the immediate availability of certain columns and equipment. The last approach is based on expediency rather than scientific rigor and is not considered here. Some preliminary information, easily obtained by a few simple experiments, such as the molecular weight range of the sample components, relative solubility in organic solvents and water, and whether or not the sample is ionic can be used to select a suitable retention mechanism using the flow chart

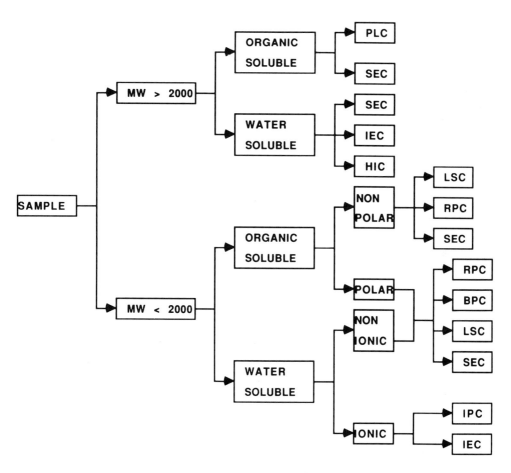

Figure 4.27 Flow chart for column selection based on sample type
(MW = molecular weight). PLC = precipitation-liquid
chromatography; SEC = size-exclusion chromatography; IEC = ion-
exchange chromatography; HIC = hydrophobic interaction
chromatography; LSC = liquid-solid chromatography; RPC = reversed-
phase liquid chromatography; BPC = (polar) bonded-phase
chromatography; and IPC = ion-pair chromatography.

given in Figure 4.27. The molecular weight cut off at 2000 is
quite arbitrary and reflects the fact that size-exclusion packings
are readily available for the separation of higher molecular
weight solutes. Packings for small molecule size-exclusion
separations are available but have a limited peak capacity
restricting their use to the separation of simple mixtures.
Likewise, wide pore packings permit the separation of polymers
with a molecular weight exceeding 2000 by reversed-phase and ion-
exchange chromatography, for example. Some separation techniques

have greater scope than others for the separation of certain types of samples. For example, if the information sought is the molecular weight distribution of a sample, size-exclusion chromatography (section 4.5.9) would be an obvious choice. Oligomer distributions or the separation of members of a homologous series are generally best accomplished using reversed-phase chromatography (section 4.4.4). Liquid-solid chromatography is usually the preferred method for the separation of mixtures of low molecular weight geometric isomers (section 4.5.1). The technique of ion suppression is preferred for the separation of weak acids and bases by reversed-phase chromatography (section 4.5.6). Strong acids and bases are usually separated by ion-pair (section 4.5.7) or ion-exchange chromatography (section 4.5.8). In many cases several different approaches may lead to an acceptable separation and method development is not particularly difficult, in other cases the selection of the optimum combination of stationary and mobile phases is the key step to obtaining an acceptable separation.

Having selected the separation method to be used the separation itself must be optimized to meet the goals of the analysis [478,480,481,485-490]. Based on the number of components and their polarity range either an isocratic (section 4.6.2) or gradient elution (section 4.6.3) separation may be employed. A general strategy in isocratic liquid chromatography is to change the composition of the mobile phase to obtain satisfactory retention and then optimize the chromatographic selectivity while maintaining approximately similar separation times. The relatively modest number of theoretical plates available in conventional liquid chromatography makes efficiency optimization far less powerful than selectivity optimization (section 1.6). In the isocratic mode the peak capacity of a typical conventional column with a separation time of about 20 minutes is around 50 [486,487]. Statistically there is a reasonable probability of being able to resolve 7 peaks without extensive optimization. This probability rises to 13-15 after selectivity optimization. For more complex samples the probability of success is low and isocratic separations will be very difficult to achieve, even with extensive optimization.

A number of manual and computer-aided strategies are available as a guide to selectivity optimization. The relative merits of these approaches are discussed in section 4.6.2. The desired result of the selectivity-optimization process is a chromatogram in which all the relevant peaks are separated and evenly distributed throughout the chromatogram. The problem is that it is not a simple task, in most cases, to accurately predict the changes in retention that accompany changes in the experimental variables and, therefore, fairly sophisticated experiment-based optimization procedures are required.

Once the selectivity is optimized, a system optimization can be performed to improve resolution or to minimize the separation time. Unlike selectivity optimization, system optimization is usually highly predictable, since only kinetic parameters are generally considered (see section 1.7). Typical experimental variables include column length, particle size, flow rate, instrument configuration, sample injection size, etc. Many of these parameters can be interrelated mathematically and, therefore, computer simulation and expert systems have been successful in providing a structured approach to this problem [480,482,491-493].

The final step in the method development chain is to demonstrate the applicability of the method by validation [494-496]. Full validation of the method involves testing different aspects of its performance such as accuracy, precision, sensitivity, selectivity and limitations using statistical methods. The level of method validation required depends on the scope of the method. For example, if it is intended to analyze a few samples by a single method without changing the operating conditions, then provided that the method is accurate, sensitive enough, and repeatable, that would probably be sufficient. On the other hand, if a method is to be used in other laboratories with different equipment, columns and operators, etc., the ruggedness of the method will be of equal importance. This would include the effects of changing parameters away from the determined optimum values and describing the influence on the performance of the method. A method that is rugged is able to tolerate small, expected variations in experimental parameters while meeting the required specifications of the analysis. If a method is not rugged re-optimization may be required to develop a method that is. The validation testing may indicate those parameters which are responsible for the inadequate performance of the method and thus provide useful insight into how to re-optimize the procedure. Again, expert system can be very useful in aiding the selection of the best statistical and experimental design for the method validation process [495-497].

The remainder of this section deals primarily with selectivity optimization in isocratic liquid chromatography and with gradient elution optimization. Before entering these subjects proper, however, a discussion of the relevant chromatographic properties of solvents is in order as a framework for the intuitive selection of the preferred solvent or solvent mixtures for selectivity optimization.

4.6.1 Selection of Solvents for Liquid Chromatography

Some fairly broad generalizations can be made about the selection of certain preferred solvents for liquid chromatography from the relatively large number of liquids that might be employed as solvents. A suitable solvent will preferably have a low viscosity, be compatible with the detection system, be readily available in a pure form, and if possible, have a low flammability and toxicity. Since detection in HPLC occurs on-line, compatibility with the detection system employed is very important although, of course, not related to the function of the solvent as a mobile phase. Solvents of low viscosity enhance column performance and minimize the column pressure drop for a given column length and/or particle size [498]. In practice, any useful solvent must be able to completely dissolve the sample without reacting with it chemically.

An abbreviated list of useful solvents for HPLC is given in Table 4.22. More extensive compilations are available [149,499,500]. Special grades of solvents are available for HPLC; the majority have been carefully purified to remove contaminants, UV-absorbing impurities and particulate matter [501]. Some solvents contain a preservative or antioxidant which can influence their chromatographic properties or detector compatibility. These may have to be removed prior to use [4,501-503]. In general, solvent impurities cause a drift in the detector baseline and diminished sensitivity under isocratic conditions, and large baseline fluctuations and spurious interfering peaks when gradient elution is used. If the impurities are not eluted from the column they will accumulate in the stationary phase, resulting in a change in solute retention. By a combination of forward and reverse gradients, it is possible to establish the level of solvent detectable impurities in either the water or organic

TABLE 4.22

PHYSICAL PROPERTIES OF SOME COMMON SOLVENTS USED IN LIQUID CHROMATOGRAPHY (20°C)

Solvent	Boiling Point (°C)	Viscosity (cP)	Refractive Index	UV Cut-off (nm)	Dielectric Constant	Dipole Moment (D)	Surface Tension (dyne/cm)
Hexane	69	0.30	1.375	195	1.89	0.08	18.4
Carbon Tetrachloride	77	0.97	1.460	263	2.24	0.0	26.8
Benzene	80	0.65	1.501	278	2.28	0.0	28.9
Diethyl Ether	35	0.24	1.352	218	4.30	1.15	17.1
Methyl tert.-Butyl Ether	55	0.27	1.369	210	----	1.32(a)	19.4(a)
Dioxane	101	1.37	1.422	215	2.25	0.45	34.5(a)
Tetrahydrofuran	66	0.55	1.407	212	7.58(a)	1.75(a)	26.4(a)
Ethyl Acetate	77	0.45	1.372	256	6.02(a)	1.88(a)	23.8
Chloroform	61	0.57	1.446	245	4.81	1.15(a)	27.2
Dichloromethane	40	0.44	1.424	233	9.08	1.14(a)	28.1
Acetone	56	0.36	1.359	330	20.70(a)	2.69	23.3
Acetonitrile	82	0.36	1.344	190	37.50	3.44	19.1
2-Propanol	82	2.40	1.377	205	18.30(a)	1.66(b)	21.8(c)
Methanol	65	0.55	1.328	205	33.60	2.87	22.6
Acetic Acid	118	1.30	1.372	---	6.15	5.15	27.8
Water	100	1.00	1.333	<190	80.37	3.11	73.0

component of a binary solvent mixture in reversed-phase liquid
chromatography [504].

The properties of a solvent which determine its capacity to
dissolve a sample are its strength, or polarity, and its
selectivity. The solvent strength is a measure of the capacity of
a solvent for all intermolecular interactions, for nonelectrolyte
solvents this includes dispersion, orientation, induction and
hydrogen bonding interactions. The selectivity of a solvent is a
measure of its relative capacity to enter into specific
intermolecular interactions, such as orientation or hydrogen
bonding. Solvents can have similar solvent strength but different
selectivity resulting in substantial differences in solvating
power for individual substances.

A number of models have been proposed to describe the
solution formation process [505-509], some of which can be
extended to include chromatographic processes and other solvent-
dependent phenomena. In terms of chromatographic applications the
most useful are the solubility parameter concept, solvatochromic
parameters and Snyder's solvent strength and selectivity
parameters. The Hildebrand solubility parameter, δ_T (total
solubility parameter), is a rough measure of solvent strength, and
is easily calculated from the physical properties of the pure
solvent. By definition it is equivalent to the square root of the
solvent vaporization energy divided by its molar volume [507,510].
The original solubility parameter concept was developed from
assumptions of regular solution behavior in which the principal
intermolecular interactions were dominated by dispersive forces.
Several workers have extended the concept to polar solvents by
assuming that the various interaction energies in solution are
additive [510-513]. One such representation is given in equation
(4.45)

$$\delta^2_T = \delta^2_d + 2\delta_{in}\delta_d + \delta^2_0 + 2\delta_a\delta_b \qquad (4.45)$$

where δ_d is a measure of the ability of a solvent to participate in
dispersive interactions, δ_0 to participate in orientation
interactions, δ_{in} to induce a dipole moment in surrounding
molecules, δ_a to function as a proton donor, and δ_b to function as
a proton acceptor. The partial solubility parameters (δ_d, δ_0, δ_a,

and δ_b) are measures of the ability of the solvent to enter into selective interactions; the larger the value, the stronger the interaction being characterized. Maximum solvency occurs when the partial solubility parameters for the solute and solvent are matched, that is, δ_d, δ_o and δ_{in} have similar values and δ_a and δ_b have complementary values. Some typical values of the solubility parameters for common solvents are summarized in Table 4.23. The polar partial solubility parameters must be considered approximate, since there is no general agreement as to the correct method for their calculation, reflected in the discordance between literature values for the same solubility parameter. This, and the fact that the model is only approximate for polar substances ($\delta_T \gg \delta_d$) has contributed to the diminished interest in this approach. Most frequent current use is in studies of the solubility of polymers, particularly those of moderate polarity, where the solubility parameters are useful for predicting solvents for size-exclusion and precipitation chromatography.

Solvatochromic parameters, so called because they were initially derived from solvent effects on UV/visible spectra, have been applied subsequently with success to a wide variety of solvent-dependent phenomena and have demonstrated good predictive ability. The $E_{T(30)}$ scale of solvent polarity is based on the position of the intermolecular charge transfer absorption band of Reichardt's betaine dye [506]. $E_{T(30)}$ values are available for over 200 common solvents and have been used by Dorsey and co-workers to study solvent interactions in reversed-phase liquid chromatography (section 4.5.4) [305,306]. For hydrogen-bonding solvents the magnitude of the $E_{T(30)}$ value is largely governed by the solvent's Lewis acidity. The most comprehensive solvatochromic treatment of solvent selectivity are the π^*, α and β parameters of Kamlet and Taft, Table 4.23 [507,508,514,515]. The π^* value is an index of solvent dipolarity/polarizability, normalized to dimethyl sulfoxide = 1, which measures the ability of a solvent to stabilize a charge or dipole by virtue of its dielectric effect. The α scale of hydrogen bond donor acidities measures the ability of a solvent to donate a proton in a solvent-solute hydrogen bond, normalized to a value of 1.0 for methanol. The β scale of hydrogen bond acceptor basicities measure the ability of the solvent to

TABLE 4.23

SOLVENT STRENGTH AND SOLVENT SELECTIVITY PARAMETERS FOR SOME COMMON SOLVENTS USED IN LIQUID CHROMATOGRAPHY.

Solvent	Solubility Parameters						Solvatochromic Parameters				Snyder's Parameters			
	A	B	C	D	E	F	G	H	I	J	K	L	M	N
Hexane	7.7	7.7					31.0	-0.04	0.0	0.0	-0.14			
Carbon Tetrachloride	9.2	8.2					32.5	0.29	0.0	0.0	1.56	0.26	0.40	0.34
Benzene	9.7	8.5					34.3	0.59	0.0	0.1	3.19	0.27	0.28	0.45
Diethyl Ether	7.9	7.4					34.5	0.27	0.0	0.47	3.15	0.53	0.13	0.34
Dioxane	10.7	8.1	1.2	3.4	6.2	2.0	36.0	0.55	0.0	0.37	5.27	0.37	0.23	0.40
Tetrahydrofuran	9.9	8.0	3.3	0.3	3.5	1.5	37.4	0.58	0.0	0.55	4.28	0.41	0.19	0.40
Ethyl Acetate	9.6	7.6	3.6	1.4	3.5	1.0	38.1	0.55	0.0	0.45	4.24	0.36	0.22	0.42
Chloroform	9.9	8.1					39.1	0.58	0.44	0.0	4.31	0.31	0.35	0.34
Dichloromethane	10.7	8.0	4.4	1.1	4.0	1.0	41.1	0.82	0.30	0.0	4.29	0.27	0.33	0.40
Acetone	10.5	8.0					42.2	0.72	0.08	0.48	5.10	0.35	0.23	0.42
Acetonitrile	13.2	7.3	5.8	0.2	10	4.0	45.6	0.75	0.19	0.31	5.64	0.33	0.25	0.42
2-Propanol	12.4	7.6					48.4	0.48	0.76	0.95	3.92	0.57	0.17	0.26
Methanol	15.9	7.2	3.9	0.1	17.1	5.4	55.4	0.60	0.93	0.62	5.10	0.48	0.22	0.31
Acetic Acid		7.2					51.9	0.64	1.12		6.13	0.41	0.30	0.29
Water	25.5	7.2			21.7	14.2	63.1	1.09	1.17	0.4	10.2	0.37	0.37	0.25

$A = \delta_T$; $B = \delta_d$; $C = \delta_o$; $D = \delta_{in}$; $E = \delta_a$; $F = \delta_b$; $G = E_{T(30)}$; $H = \pi^*$; $I = \alpha$; $J = \beta$; $K = P'$; $L = X_e$; $M = X_d$; $N = X_n$

to accept a proton (donate an electron pair) in a solvent-solute hydrogen bond, normalized to a value of 1.0 for hexamethylphosphoramide. The relative proton-sharing acidity and basicity (α and β) differ significantly from the corresponding proton transfer acidity and basicity (pK_a and pK_b). In many cases proton transfer properties are not relevant to the solubility of nonelectrolyte solutes, while the ability to participate in hydrogen bonding interactions frequently is. Correlation equations using π^*, α and β parameters as variables have been used to predict retention in reversed-phase liquid chromatography (section 4.5.4) [307-309] and to explain the relationship between the experimental solvent strength parameter, ϵ^0, and retention in liquid-solid chromatography [516]. The values of α and β quoted in Table 4.23 refer to bulk solvent properties. For solvents that self-associate in the bulk liquid state (e.g., alcohols) the values of α and β require modification when the solvent is used as a test solute to probe the properties of other solvents, and a different series of monomer values should be used [514-516].

The solvent triangle classification method of Snyder is the most common approach to solvent characterization used by chromatographers [510,517]. The solvent polarity index, P', and solvent selectivity factors, X_i, which characterize the relative importance of orientation and proton donor/acceptor interactions to the total polarity, were based on Rohrschneider's compilation of experimental gas-liquid distribution constants for a number of test solutes in 75 common, volatile solvents. Snyder chose the solutes nitromethane, ethanol and dioxane as probes for a solvent's capacity for orientation, proton acceptor and proton donor capacity, respectively. The influence of solute molecular size, solute/solvent dispersion interactions, and solute/solvent induction interactions as a result of solvent polarizability were subtracted from the experimental distribution constants by first multiplying the experimental distribution constant by the solvent molar volume and then referencing this quantity to the value calculated for a hypothetical n-alkane with a molar volume identical to the test solute. Each value was then corrected empirically to give a value of zero for the polar distribution constant of the test solutes for saturated hydrocarbon solvents. These residual values were supposed to arise from inductive and entropy effects. Poppe and Slaats proposed two modifications to the above procedure to eliminate the entropy contribution to the distribution constant arising from differences in molecular size between the solute and solvent (Flory-Huggins correction factor) and introduced an improved method to estimate the contribution of a hypothetical n-alkane with the same molar volume as the solute to the experimental distribution constant [518]. This resulted, however, in only minor changes to the P' and X_i values calculated by Snyder. Rutan et al. [519] have provided a revised tabulation of Snyder's constants based on an improved experimental procedure for calculating the gas-liquid distribution constants; some of these values are summarized in Table 4.23, and in certain cases differ significantly from the original values given by Snyder [517]. In general, the Snyder constants do not correlate very well with the solvatochromic parameters for the same solvents. It has been suggested that the original choice of test solutes, nitromethane, ethanol and dioxane are inappropriate as their solubility is due to multiple and not single interactions. Thus, for example, ethanol exhibits significant orientation and weak proton acceptor capacity, as well as strong proton donor capability. It will show significant solubility in dipolar solvents due to favorable orientation interactions in the absence of contributions from hydrogen bonding interactions. However, if it is assumed that the solubility of ethanol is predominantly governed by hydrogen bonding interactions then an inflated value for the proton acceptor capacity of the solvent will be obtained. Since it is impossible to fined a test solute which is a strong hydrogen bond donor or acceptor and is not dipolar, it may not be possible to characterize intermolecular interactions based on the solubility properties of individual test solutes. Multiple linear regression methods may prove more useful for identifying and quantifying the relative contribution of specific intermolecular interactions involved in solution formation as represented by the linear solvation energy relationship approach [507,508,514,515].

However, not withstanding the above objections, further discussion of the Snyder solvent triangle classification method is justified by its common use in many solvent optimization schemes in liquid chromatography. The polarity index, P', is given by the sum of the logarithms of the polar distribution constants for ethanol, dioxane and nitromethane and the selectivity parameters, X_i, as the ratio of the polar distribution constant for solute i to the total solvent polarity (P'). The sum of the three values for X_i will always add up to 1.0, but for non-identical solvents X_e (i = ethanol), X_d (i = dioxane) and X_n (i = nitromethane) the solvents will have different individual values for X_i. When the individual X_i values for the solvents are arranged in diagrammatic form, Figure 4.28, Snyder was able to show that the many solvents available could be grouped into eight classes with distinctly different selectivities. Solvents within the same selectivity group exhibit similar separation properties compared to solvents of similar strength from different groups. As a general strategy for solvent selection, it should be more profitable to evaluate solvents from different groups, than to try several solvents from within the same group. The effect on the measured property whether solubility or peak separation in liquid chromatography, etc., should be greatest in the first instance and relatively minor in the latter. Some typical examples of selective solvents and their group assignments are given in Table 4.24. The underlined solvents are those generally preferred for solvent optimization in liquid chromatography. The trialkylamines are arbitrarily placed in Group I although they are significantly more basic than the other solvents in this group. Chloroform, a moderately strong hydrogen bond donor solvent, is only loosely assigned to group VIII where the other members of the group consist of stronger proton donors, such as fluoroalkanols and phenols.

Very few separations in liquid chromatography are performed using single solvents because of the rather limited possibilities for simultaneously optimizing solvent strength and solvent selectivity within the range of common solvents employed for chromatography. A more logical and flexible approach is to use binary or higher-order solvent mixtures. In the case of binary solvents, mixing a strength-adjusting solvent with various volume fractions of a strong solvent enables the complete solvent strength range between the extremes represented by the pure solvents to be covered. The strength-adjusting solvent is usually a non-selective solvent, such as water for reversed-phase chromatography and hexane for liquid-solid chromatography using an inorganic oxide stationary phase. The solvent strength of a mixed mobile phase is the arithmetic average of the solvent strength weighting factors adjusted according to the volume fraction of each solvent, equation (4.46)

466

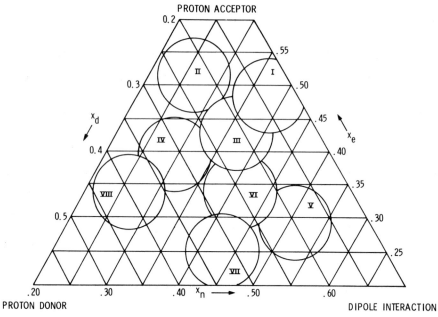

Figure 4.28 Selectivity triangle for solvents. (Reproduced with permission from ref. 517. Copyright Preston Publications, Inc.)

TABLE 4.24

CLASSIFICATION OF SOLVENT SELECTIVITY ACCORDING TO SNYDER

Group Designation	Solvents
I	Aliphatic ethers, <u>methyl t-butyl ether</u>, tetramethylguanidine, hexamethyl phosphoric acid amide (trialkylamines)
II	Aliphatic alcohols, <u>methanol</u>
III	Pyridine derivatives, <u>tetrahydrofuran</u>, amides (except formamide), glycol ethers, sulfoxides
IV	Glycols, benzyl alcohol, <u>acetic acid, formamide</u>
V	Ethylene chloride
VI	a) Tricresyl phosphate, aliphatic ketones and esters, <u>dioxane</u>, polyesters
	b) Sulfones, nitriles, <u>acetonitrile</u>, propylene carbonate
VII	Aromatic hydrocarbons, <u>toluene</u>, halosubstituted aromatic hydrocarbons, <u>nitro</u> compounds, <u>methylene chloride</u>, aromatic ethers
VIII	Fluoroalcohols, m-cresol, water, (<u>chloroform</u>)

$$S_T = \Sigma_i S_i \phi_i \qquad (4.46)$$

where S_T is the total solvent strength of a solvent mixture, S_i the solvent strength weighting factor of solvent i and ϕ_i the volume fraction of solvent i. For normal phase chromatography the solvent strength weighting factor, S_i, is the same as the polarity index, P', given in Table 4.23. For reversed-phase chromatography a series of empirically determined weighting factors, calculated from the average slopes of a plot of log k against ϕ for several solutes (see section 4.5.4), is used. Some typical examples are methanol S_i = 2.6, acetonitrile S_i = 3.2, tetrahydrofuran S_i = 4.5 and water S_i = 0. For a binary solvent mixture containing 60% methanol and 40% water, the solvent strength of the mixed solvent is calculated as follows

$$S_T = S_m \phi_m + S_w \phi_w$$

$$S_T = (2.6)(0.6) + (0)(0.4) = 1.56$$

where the subscripts m and w identify the parameters for methanol and water, respectively.

Several mobile phase optimization strategies in liquid chromatography are based on the use of isoeluotropic solvents, that is, solvent mixtures of identical strength but different selectivity. Suitable binary (and higher order) solvent mixtures can be selected from the solvent selectivity triangle discussed previously. To maximize the differences in selectivity, solvents are selected from different selectivity groups that lie close to the triangle apexes. The selected solvents must also be miscible with the strength adjusting solvent. For example, for reversed-phase chromatography a suitable selection might be methanol (group II), acetonitrile (group VI) and tetrahydrofuran (group III) mixed with water to control solvent strength, and for liquid-solid chromatography methyl t-butyl ether (group I), chloroform (group VIII) and methylene chloride (group VII) mixed with hexane to control solvent strength. The composition of isoeluotropic solvents is simply calculated using the additive nature of the volume fractions and their respective solvent weighting factors, or by using empirical transfer rules [487,510,520,521]. If, as in the methanol-water (60:40) example used above, the solvent strength was about optimum for a particular separation and one wished to change solvent selectivity to adjust resolution, then the volume fraction of the new solvent required to give an isoeluotropic mixture could be calculated from equation (4.46). Using acetonitrile as an example and a value of $S_T = 1.56$, we have

$$1.56 = S_a\phi_a + S_w\phi_w$$

$$1.56 = (3.2)\phi_a + (0)\phi_w$$

and

$$\phi_a = 1.56/3.2 = 0.49 \text{ or } 49\%$$

where the subscript a refers to the properties of acetonitrile. Thus a mixture of acetonitrile-water (49:51) has similar solvent strength to a mixture of methanol-water (60:40). In a similar manner a mixture of tetrahydrofuran-water (35:65) is isoeluotropic with the two previous binary solvent mixtures. Alternatively, the following empirical equations can be used to estimate the equivalent solvent composition of isoeluotropic binary solvent mixtures given that experimental conditions have been established using methanol as the organic modifier in reversed-phase liquid chromatography

$$\phi_a = -0.49\phi_m^3 + 0.953\phi_m^2 + 0.447\phi_m$$

$$\phi_t = -0.42\phi_m^3 + 0.702\phi_m^2 + 0.433\phi_m$$

where ϕ_a and ϕ_t are the volume fractions of acetonitrile and tetrahydrofuran required to prepare a binary solvent mixture with water equivalent in strength to a given volume fraction of methanol, ϕ_m. Since the separation time for a mixture is dependent on the elution time for the last component to leave the column, any exceptional behavior of this compound will dramatically influence the calculation of the necessary solvent strength for elution of the sample in an acceptable time. For this reason the above approaches are only approximate and empirical corrections may have to be applied, as needed, to maintain an acceptable separation time for the sample with mobile phases of different composition.

Binary solvent mixtures provide a simple means for controlling solvent strength but only limited opportunities for controlling solvent selectivity. With ternary and quaternary solvent mixtures it is possible to fine tune solvent selectivity while maintaining a constant solvent strength [522-524]. In addition, there are only a small number of organic modifiers that can be used as binary mixtures with water. Many more mobile phases of the same solvent strength can be prepared if ternary and quaternary solvent systems are used. An example of mobile phase selectivity optimization using a ternary solvent mixture is shown in Figure 4.29 [522]. The separation in Figure 4.29A was obtained with the binary mobile phase methanol-water (50:50). The solvent strength of this mixture is about correct for the separation based on the sample capacity factor values, but components 1 and 2 are not separated. This indicates that a change in solvent selectivity is required. The separation obtained with the isoeluotropic binary solvent mixture tetrahydrofuran-water (32:68), Figure 4.29B, shows a change in component resolution with a separation time similar to that of Figure 4.29A. Components 1 and 2 are now well separated but components 2 and 3 are not. Mobile phase strength and selectivity are optimized in the ternary solvent system methanol-tetrahydrofuran-water (10:25:65), Figure 4.29C, in which all the sample components are adequately separated.

4.6.2 Selectivity Optimization for Isocratic Separations

The limited separation capacity of conventional packed columns in liquid chromatography makes selectivity optimization the most powerful approach to achieving the desired resolution in a chromatogram. Selectivity can be optimized following a wide range of protocols varying from an intuitive trial and error approach to automated computer-aided methods [478-481,484, 485,525,526]. However, no approach has emerged to date as a superior method compared to all others, and in practice, the approaches taken in an individual laboratory are usually dictated by the degree of expertise of the chromatographer and the availability of equipment and software to implement some of the more recent computer-aided approaches. This latter group of techniques will be emphasized in this section using reversed-phase separations to illustrate applications.

The first step in developing an optimization strategy is to define the parameter space to be searched for an acceptable

470

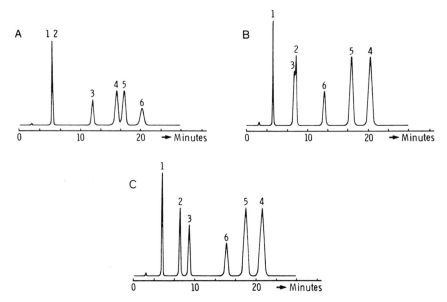

Figure 4.29 An example of the use of ternary solvents to control
mobile phase strength and selectivity in reversed-phase liquid
chromatography. A, methanol-water (50:50); B, tetrahydrofuran-
water (32:68); C, methanol-tetrahydrofuran-water (35:10:55). Peak
identification: 1 = benzyl alcohol; 2 = phenol; 3 = 3-
phenylpropanol; 4 = 2,4-dimethylphenol; 5 = benzene; and 6 =
diethylphthalate. (Reproduced with permission from ref. 522.
Copyright Elsevier Scientific Publishing Co.)

separation. The parameter space is that combination of
experimental variables and their limiting values that define the
search area. An acceptable separation must be within the parameter
space otherwise the search techniques will only result in failure.
On the other hand the parameter space should not be set
artificially large, since the search techniques will become
inefficient. The parameter space is reduced as much as possible by
including only those variables which have a significant effect on
the selectivity and by assigning sensible limits to each variable.
The number of significant variables that affect selectivity is a
matter of judgment, although factorial designs can be used to
identify those variables contributing most to selectivity. In most
cases a few preliminary scouting experiments are used to define
the limits for the experimental variables.

The parameter space is next searched for a global optimum by
performing a series of separations according to a formal

experimental design. The number of separations required will depend on the specifics of the experimental design. Three broad strategies are commonly employed. Simultaneous or grid search methods are used to model the quality of the separation throughout the parameter space involving the collection of data at predefined points followed by inspection of the chromatograms to define the optimum separation conditions. In sequential methods a small number of initial experimental results are used to direct the chromatographer to conditions where better results are expected, for example, simplex optimization. The third type of search technique is based on iterative and interpretive methods, in which the object is to model the retention surface of the individual components in the sample on the basis of experimental retention data. These methods are perhaps the most popular because of the small number of required experiments, typically 7-10. Solute recognition, however, is essential and the accuracy of any predictions dependens on the model used to describe the response surface.

Interpretive methods require the recognition of individual peaks in each chromatogram so that regression equations can be fitted to the model used to define the response surface [527-529]. Peak recognition can be achieved by injecting individual standards for each solute in each chromatogram, from a comparison of peak areas or peak ratios, or from complete UV spectra resulting from multiwavelength detection. The commonly used strategy of injecting all solutes separately is not only slow but obviously not feasible for samples of unknown composition. The accuracy of peak area and peak ratio methods is limited where there are large differences in component concentrations and by changes in detector response and baseline noise associated with solvent changes. Also, peak assignments based on relative areas will not reveal the real number of components present. The problem of recognizing the individual solutes in the chromatograms during the optimization process is complicated because of the large amount of overlap that can be expected to occur between the various peaks in the chromatograms. An efficient method to determine whether all sample peaks are eluted from the column is also required. The recording of multiwavelength spectra with a photodiode array detector and application of chemometric techniques to the data, is the preferred approach to this problem. The success of even these methods, however, depends on the resolution, the number of components, detector noise, the spectral similarity and the relative amounts of the components. Clearly, optimization is simplified when the exact number of components (or relevant components) in the mixture is known and standards are available for each component. For samples with an unknown composition, determining the total number of detectable components is very difficult, particularly if individual components are present in very different concentrations, but is also very relevant to the optimization process, as otherwise it is not possible to identify with absolute certainty when a separation is complete.

If the selectivity optimization is carried out manually, then the quality of the separation may be compared by visual inspection of the chromatograms. This approach, however, is inadequate for automated optimization procedures, and the quality of a chromatogram must be expressed by a single-valued and easily calculated mathematical function [478,479,484,525,530-533]. This function is variously referred to as an objective function, chromatographic response function (CRF) or a chromatographic optimization function (COF). The variation of the magnitude of this function over the parameter space is called the response surface. In the usual case, the response surface resembles a mountain area with several peaks and valleys. The highest feature of the response surface represents the global optimum and defines that combination of experimental parameters producing the best possible separation as determined by the objective function. Smaller features represent local optima expressing the best separation, as defined above, in a section of the parameter space, but an inferior separation with respect to the global optimum. The purpose of the search technique is to locate the global optimum and to distinguish it from inferior local optima.

The purpose of the objective function is to express the desires of the separation by a single number that can be used to rank experimental chromatograms [478,479,484]. The problem is that individual chromatograms normally contain several peaks, each pair with its own resolution. There is no straightforward manner to uniquely define the resolution of all peaks simultaneously by a single number. More likely, for any given value of the objective function there will be a large number of separations that could produce the same numerical value, not all of which will agree with the stated goals of the separation. In addition, the objective function may have to consider the number of peaks identified in

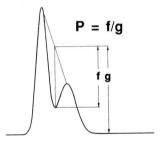

Figure 4.30 Definition of the peak separation function P = f/g.
(Reproduced with permission from ref. 479. Copyright Elsevier
Scientific Publishing Co.)

the chromatogram and the separation time in expressing the goals
of the separation. The ideal objective function has not been
developed so far and a large number of less than perfect solutions
have been described in the literature. Rather than detail all
possible approaches a few representative objective functions will
be outlined briefly and their relative merits discussed.

For manual optimization methods the peak separation
function, P, is easy to determine and can be calculated as shown
in Figure 4.30 [479]. The chromatographic response function for
the chromatogram is then simply the sum of the ln P values for the
n adjacent peak pairs.

$$CRF = \Sigma_{i=1}^{n} \ln(P_i) \qquad\qquad (4.47)$$

The P criterion is quite robust but difficult to calculate
automatically and its performance begins to decline for P < 0.75,
making it less satisfactory for difficult separations. A major
disadvantage of the P criterion is that separations can give rise
to equal values for the separation function as peak order changes
or as resolution between different peak pairs change. In addition
there is no consideration given to the importance of the
separation time.

Rather than the separation function, resolution between
individual peak pairs is used in most automated optimization
procedures because it is easier to calculate, although non-
Gaussian peaks and overlapping peaks can present problems due to
the difficulty of estimating peak widths. A simple objective
function would be to consider only the separation between the
worst separated peak pair, ignoring all others. If a set of

chromatograms is to be compared, then this is a reasonable approach, but it does not provide a suitable criterion for locating a single global optimum, since very different peaks may show the lowest separation in adjacent chromatograms and many optima will be indicated, rendering it impossible for a searching or modeling procedure to make effective progress. The sum of all the resolutions will reflect gradual improvements in different separations, but on its own is of little value. Two peaks that are well resolved and easy to separate will dominate the sum and optimization may result in this peak pair being overseparated, while the most difficult pair to separate is ignored. One way of avoiding this problem is to sum the resolutions but to limit the maximum resolution that can be assigned to any peak pair. Taking the separation time into account the chromatographic optimization function (COF) can be defined as

$$COF = \Sigma_{i=1}^{n} A_i \ln(R_i/R_{id}) + B(t_x - t_n) \qquad (4.48)$$

where R_i is the resolution of the ith pair, R_{id} the desired resolution for the ith pair, t_x the maximum acceptable retention time for the last eluted peak, t_n the observed separation time for the last eluted peak, and A_i and B are arbitrary weighting factors used to indicate which peaks are more important to separate than others, and to allow flexibility in setting the acceptable separation time. In use, the COF gives a single number which tends towards zero as the optimum separation is reached while poor chromatograms produce large, negative values. The COF, however, makes no allowance for peak crossovers and identical values of the COF can result from chromatograms with different numbers of separated peaks. Rather than add the individual resolution values the product of those values can be used. In this case the aim is to space the peaks evenly throughout the chromatogram, since the lowest value of the resolution has a dominant effect. A simple resolution product may still give a higher assessment to an inferior chromatogram but this can be overcome by using the relative resolution product, where the denominator defines the maximum possible value for the resolution product in the given chromatogram. Also, the relative resolution product can be modified to include a term that incorporates the importance of the separation time in obtaining the desired separation.

None of the functions considered so far specifically take into account the number of peaks found in the chromatogram. If the object of the analysis is to detect the maximum number of peaks, even if the resolution between individual peak pairs was poorer, then equations (4.47) and (4.48) would be inferior. A chromatographic response function, CRF, that takes into account the simultaneous importance of resolution, separation time, and the total number of detectable peaks can be expressed as follows

$$CRF = \Sigma_i R_i + n^a - b[t_x - t_n] - c[t_o - t_1] \tag{4.49}$$

where n is the number of peaks observed, t_o the minimum desired retention time for the first detected peak, t_1 the observed retention time for the first detected peak, and a, b and c are weighting factors that can be adjusted to change the emphasis of the various contributions to the CRF. Usually, the exponent a is chosen so large (e.g., a = 2) that the appearance of a new peak raises the criterion significantly. Most multi-term functions share the problem that the numerical value for the CRF can become dominated by one of the terms in the expression and fail to represent the intended mix of terms.

From the above discussion it should be obvious that the selection of an appropriate objective function is a difficult task. The choice of the objective function is a critical factor in automated methods development, since it is used to define the response surface. It is highly likely that different objective functions will result in the production of different response surfaces and the location of different optimum experimental conditions for the separation. Yet, it is not possible to set hard guidelines for the selection of the objective function, which must be chosen by practical experience keeping the objectives of the separation in mind.

Solvent optimization in reversed-phase liquid chromatography is commenced by selecting a binary mobile phase of the correct solvent strength to elute the sample with an acceptable range of capacity factor values (1 < k <10 in general or 1 < k < 20 when a larger separation capacity is required). Transfer rules (section 4.6.1) are then used to calculate the composition of other isoeluotropic binary solvents with complementary selectivity. In practice, methanol, acetonitrile and tetrahydrofuran are chosen as the selectivity adjusting solvents blended in different proportions with water as the strength adjusting solvent. The three isoeluotropic binary solvents can be placed at the apexes of a triangle whose boundaries define the parameter space. The parameter space contains all those ternary and quaternary mobile phases that can be produced by combining the three binary mobile phases. Ideally, the solvent strength is constant throughout the parameter space but the selectivity is continuously variable, representing all possible combinations between the extremes defined by the properties of the initial three binary solvents.

The solvent strength of the binary solvents providing adequate retention can be easily estimated from a separation by gradient elution or from a series of sequential isocratic chromatograms of decreasing solvent strength [478-480,484,487, 520,525,534-537]. The steps involved in identifying the composition of a mixture of acetonitrile-water with the correct solvent strength for the elution of a five component mixture by a series of sequential isocratic chromatograms are illustrated in Figure 4.31. Scouting starts with the strongest solvent first (acetonitrile), followed by subsequent separations using binary solvents containing increasing amounts of the strength adjusting solvent (water), initially added in 20% (v/v) increments. The column holdup time in the example is 1 min and the desired retention time for the last eluting peak is 10 min (k = 10), corresponding to a binary mobile phase containing 40% (v/v) acetonitrile. The original step size of 20% (v/v) may have to be reduced in the region of the desired retention range to obtain the optimum value for the solvent strength.

The gradient elution method is a more efficient approach for defining the solvent strength for a separation and can also indicate whether the separation can be achieved isocratically in an acceptable separation time. The approach is based on the theory of linear solvent strength gradients (see section 4.6.3) and will only be briefly commented on here. The calculations are not difficult but rather tedious in the absence of appropriate software [520,534-537]. An initial gradient of 0-100% methanol at a linear rate of increase that depends on the column holdup time is employed for the trial separation. If the ratio of the difference in retention time between the first detected peak and the last detected peak during the gradient exceeds about 25% of the gradient time, it is unlikely that a suitable isocratic method can be developed for this sample due to the wide capacity factor

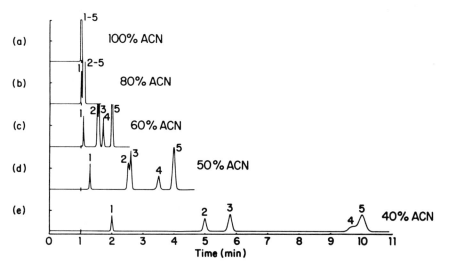

Figure 4.31 Sequential, isocratic elution using a stepwise reduction in solvent strength to identify a binary solvent of acceptable strength for elution of a five component mixture. In this example the column holdup time was 1 min.

range of the sample components. The retention time at the midpoint of the separated components in the gradient, together with the properties of the gradient, are used to calculate the approximate binary solvent composition for the isocratic separation. Alternatively, the binary solvent mixture used to elute the first detected peak with k = 1 and the last detected peak with k = 10 can be calculated. The optimum isocratic mobile phase will lie between these two values. If the difference in solvent composition is too large then this indicates that an isocratic separation will not be possible. If two gradient runs are used, in which only the gradient time is varied (e.g., 20 and 60 min), the resulting retention data can be used to predict retention as a function of mobile phase composition in the corresponding isocratic separations [535-540]. Plots of log k against mobile phase composition based on equation (4.16) can be constructed. In those cases where the individual slopes are different, or if the slopes are the same but the intercepts (k_w) are different, solvent strength optimization can be employed to change band spacing. Method development, aided by computer simulation, will usually be faster with this approach than solvent selectivity optimization with different mobile phases. The overall approach, though, is much less powerful because changes in band spacing accompanying solvent strength changes are not as large as those observed by changing mobile phase solvents. The data can be plotted as relative resolution maps; a plot of the resolution between adjacent bands exhibiting the lowest resolution for a column with a constant number of theoretical plates as a function of the binary mobile phase composition. These plots resemble window diagrams (section 2.5.5) and are interpreted in a similar manner. This approach appears particularly promising for separating simple mixtures containing components that differ in molecular weight or functionality.

The PESOS (Perkin-Elmer solvent optimization system) is an example of a grid search solvent optimization system in which the parameter space is searched in small steps according to a fixed experimental design, Figure 4.32 [485,525,541]. Chromatograms are recorded at each of the measurement points and the best separation is selected either visually or by maximizing an objective function. The method is easily automated and provides a complete picture of the response surface. The global optimum should be obtained but the number of experiments and, therefore, time can be excessive, particularly for simple separations. Although peak tracking is not built in, small stepwise changes in the conditions usually result in small and regular peak movements, making manual peak tracking relatively easy.

Sequential optimization strategies like simplex designs rely on the result from previous experiments to provide guidance and direction for future experiments [478,479,490,526,528,541-543]. A simplex is defined as a geometric figure having one more vertex than the number of variables being optimized. Thus, a two variable simplex is a triangle and a three variable simplex a tetrahedron, etc. The optimization proceeds by rejecting the vertex which has the worst experimental response and the location of the next data point is found by reflecting the simplex in the opposite direction. The direction of advance of the simplex is dependent solely on the ranking of responses. Modern versions allow extension and contraction of the simplex as shown in Figure 4.32. These operations increase the efficiency of the simplex by allowing it to expand and thus accelerate towards the optimum region when, having located it approximately, the simplexes contract and reduce the search region until the optimum is precisely located. The great advantages of the simplex procedure

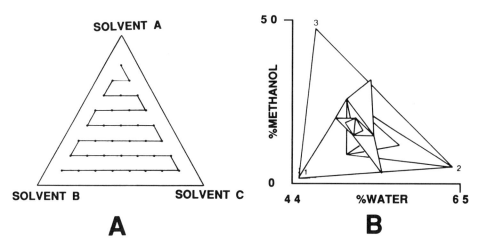

Figure 4.32 Experimental design showing the grid search solvent optimization system employed by PESOS (A) and an example of a simplex search for a global optimum (B).

are that it is able to optimize many interdependent variables with no prior knowledge about the mode of separation or the complexity of the sample. Nor does it require any preconceived model of the retention behavior of solutes, and so does not require that the solutes be identified, or recognized, in individual separations. The method has the further advantages of permitting the introduction of new variables during the optimization process for the price of just one additional experiment per variable, as well as allowing the progress of the optimization to be assessed at all times rather than at the end of the experimental sequence. The simplex approach has several disadvantages as well. The success of the optimization process depends entirely on the quality of the objective function used to rank each experiment. Since each vertex is ranked comparatively, the objective function must clearly define the final goal of the separation, which as we saw earlier, is not always easy to achieve. Also, the simplex is poor at dealing with irregular response surfaces and may become stalled on a local optimum instead of proceeding to the global optimum. The simplex searches the response surface for the global optimum rather than mapping the response surface, so that at the end of the optimization procedure little insight into the shape of the response surface is obtained. Although in favorable cases the simplex will probably locate the global optimum with fewer

experiments than for grid search methods, it still requires a
relatively large number of experiments. A typical grid search
might require on the order of 100 experiments while for the
simplex approach perhaps 25-40 would be required [526]. Although
both methods can be fully automated, a typical chromatographic
experiment is likely to require about 30 min or so for the
separation and column re-equilibration steps and, therefore,
neither method is particularly fast as an approach for identifying
the global optimum.

Interpretive methods involve modeling the retention surface
(as opposed to the response surface) on the basis of experimental
retention time data [478-480,485,525,541]. The model for the
retention surface may be graphical or algebraic and based on
mathematical or statistical theories. The retention surface is
generally much simpler than the response surface and can be
described by an accurate model on the basis of a small number of
experiments, typically 7 to 10. Solute recognition in all
chromatograms is essential, however, and the accuracy of any
predictions is dependent on the quality of the model.

The mixture-design statistical technique is an example of an
interpretive approach that has been widely used for solvent
optimization in reversed-phase liquid chromatography [544-548].
After reduction of the parameter space to three binary solvents of
equal eluotropic strength with capacity factor values in the
optimum range, a fixed experimental design is applied. This design
consists of seven experiments, separations in each of the binary
solvents which can be taken as the apexes of a triangle,
separations in each of the three ternary solvent mixtures obtained
by alternately mixing equal volumes of any two of the binary
solvents (corresponds to the midpoint on each side of the
triangle), and finally the quaternary solvent mixture obtained by
mixing equal volumes of all three binary solvents (corresponds to
the midpoint of the triangle). The retention time data obtained
can be treated in several ways. A visual comparison of the seven
chromatograms may indicate that a successful separation has been
achieved already, for example Figure 4.33, or indicate a region of
the parameter space where a few more trial experiments could be
expected to yield the desired result. This restricted area of the
parameter space could be searched again using a smaller triangle
with new apexes selected around the optimum region, for example
from Figure 4.33, by choosing points 3, 5 and 6 or 4, 6 and 7 as

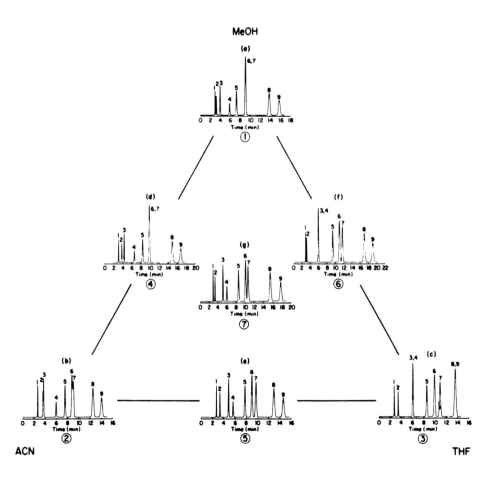

Figure 4.33 Separation of nine substituted naphthalenes using the mixture-design statistical technique to optimize the composition of the mobile phase (Reproduced with permission from ref. 480. Copyright John Wiley and Sons).

the new apexes of the triangle. Although in the particular case illustrated by Figure 4.33, the separation obtained at point 5 would likely be deemed acceptable and further optimization would be superfluous provided that it was known that the mixture contained only nine components. A second approach would be to construct a retention surface from the experimental data by fitting the data to a suitable mathematical model, for example, equation (4.50) and (4.51)

$$\ln k = a_1 X_1 + a_2 X_2 + a_3 X_3 + a_4 X_1 X_3 + a_5 X_2 X_3 + a_6 X_1 X_2 X_3 \qquad (4.50)$$

$$\ln k = a_1 X_1^2 + a_2 X_2^2 + a_3 X_3^2 + a_4 X_1 X_3 + a_5 X_2 X_3 + a_6 X_1 X_2 \qquad (4.51)$$

where X_1, X_2 and X_3 are the fractions of the pseudocomponents (the binary isoeluotropic mixtures of water and the respective modifiers) and the a terms are coefficients evaluated for each solute using the observed values for the capacity factors in the seven experiments with known values of X_1, X_2 and X_3 defined by the experimental design [478,490,523,526,549]. Likewise, the response surface can be fitted to equation (4.50) or (4.51) by replacing the capacity factor term with a suitable objective function. However, the general limitations of objective functions for expressing unambiguously the desired features required in the chromatogram and inherent problems with peak crossovers as the mobile phase composition is changed, have resulted in this approach being largely replaced by the use of overlapping resolution maps. The capacity factors for each solute are predicted at all mobile phase compositions within the parameter space. A grid is constructed for which the pseudocomponents (X) are changed by one percent at a time and at each mobile phase composition the resolution, R_s, is predicted for all peak pairs from

$$R_s = n^{1/2} (k_2 - k_1) / 2 (k_2 + k_1 + 2) \qquad (4.52)$$

where n is the number of theoretical plates for the column (assumed constant for all solutes) and k_1 and k_2 are the capacity factors of adjacent peaks. A resolution contour plot is prepared for each pair of components depicting the resolution for the pair in all solvent compositions. A desired minimum resolution (e.g., R_s = 1.5) between peak pairs is then specified and this enables areas on the contour plot to be eliminated as being of no further interest since they do not meet the desired resolution criterion. Figure 4.34 shows the peak-pair resolution maps for the nine substituted naphthalenes separated in Figure 4.33 with the solvent composition regions for each adjacent peak pair that fails to meet

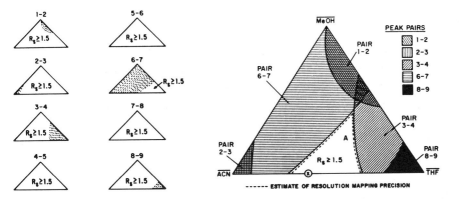

Figure 4.34 Peak-pair resolution maps and an overlapping resolution map for the separation of nine substituted naphthalene compounds by reversed-phase liquid chromatography illustrated in Figure 4.33. (Reproduced with permission from ref. 545. Copyright Elsevier Scientific Publishing Co.)

the resolution criterion shaded in [545]. By superimposing all the specific resolution maps of the various peak pairs a single figure, called an overlapping resolution map (ORM), is generated. Areas in the parameter space where the desired separation can be achieved for all solutes in the mixture are easily identified. In agreement with the results from Figure 4.33, the optimum region for the separation predicted by the ORM (Figure 4.34) is the same. The ORM does not define a unique solution but indicates a region or regions where successful separations can be obtained. Also, the ORM does not explicitly use the separation time as a criterion for the quality of the separation but this is handled by the experimental design. The solvent strength throughout the parameter space is approximately constant and the separation time for all chromatograms within the parameters space should, therefore, also be approximately constant.

Iterative methods are a further example of the interpretive approach to solvent optimization [486,525,526,541,550]. The overall scheme involves selection of the parameter space, data collection at model points, peak labelling, modeling of the retention surface, and calculation of response surfaces followed by the prediction of an optimum value. The initial steps for solvent optimization are similar to those discussed for the mixture-design statistical approach. Linear or non-linear plots of ln k against mobile phase composition for each solute are constructed for the entire range of ternary mixtures derived from the pairs of limiting binary pseudocomponents. A response surface is then constructed from the modeled retention surface using a suitable objective function and used to predict an optimum composition. A sample is then chromatographed with this mobile phase composition and the quality of the chromatogram assessed. In the event that it is found to be inadequate the new retention data for each solute are then used with the original data to refine the response surface plot. The new plot is used to define a second predicted optimum composition and a separation is conducted with this mobile phase. The above process is repeated, as required, until sufficient iterations have been conducted to produce a model that is sufficiently close to the true behavior to enable an optimum separation to be predicted. The characteristic features of the iterative regression design are that the response surface is derived indirectly from the retention surface of individual solutes, which are easier to model. The number of data points and the retention model are not fixed in advance but updated continuously by verifying the predicted optimum. The retention surfaces are described by a succession of linear or quadratic segments until the true curves have been found with sufficient accuracy. Among the advantages of the iterative approach are that an impression of the complete parameter space is maintained throughout the procedure and the number of chromatographic separations required is adapted to the degree of non-linearity of the retention surfaces. Among the disadvantages are that the procedure is computationally intensive, particularly if the number of parameters to be optimized is fairly large. Since it relies on an objective function to determine the quality of the chromatogram, it suffers from all the problems associated with matching the selected function to the goals of the separation.

Also, since it relies on accurate retention data for each solute in all chromatograms, peak labelling is required.

In spite of considerable progress in developing automated systems for method development in liquid chromatography the final objective of a "smart box" that can solve a wide range of problems is still a long way off. There are several stumbling blocks. The selection of the separation model and the parameter space has a profound impact on the success of the optimization step but these decisions are still largely a matter of judgment. The objective function remains a weak link of the optimization procedures as the quality of a chromatogram is not easily encoded by a single numerical value. The objective function is not unique and several different chromatograms will produce the same value, even though some separations would be judged poorer than others. Few laboratories have a wide range of software to choose from; many of the procedures discussed in this section require software restricted to the purchase of a particular instrument or software that is not commercially available. When the time required for a separation, column re-equilibration, peak tracking, number of separations required to predict an optimum separation are considered, the approaches discussed are not particularly rapid and an experienced chromatographer might well develop an acceptable separation in a shorter time. Clearly, the current developments in automated methods development, computer simulations and expert systems have to be considered as an aid to the chromatographer and not a replacement for that strange cocktail of experience, intuition, knowledge and at times luck, that are of importance in solving separation problems in a timely manner.

4.6.3 Gradient Elution Liquid Chromatography

So far we have discussed elution in liquid chromatography under isocratic conditions: that is, with a mobile phase whose solvent strength remains constant throughout the separation. This is generally the preferred method of separation based on convenience, simplicity and reproducibility, but is not appropriate for those samples containing components of widely different relative retention. For samples of this type isocratic elution produces poor resolution of early eluting peaks and inconveniently long retention times and broad peaks for late eluting components. Gradient elution is frequently used to

486

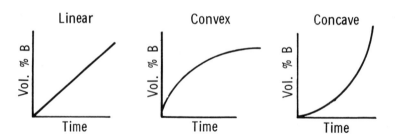

Figure 4.35   Idealized shapes for mobile phase gradients.

overcome this problem, often referred to as "the general elution problem". An alternative approach is the use of coupled columns for multidimensional liquid chromatography discussed in section (8.7.4). In gradient elution liquid chromatography the composition of the mobile phase is varied throughout the separation so as to provide a continual increase in solvent strength and thereby a more convenient elution time and sharper peaks for all sample components [551]. Improvements are seen in peak shape, resolution, sample detectability and separation time.

In this section a phenomenalogical approach to the description of solvent gradients will be adopted based largely on the linear solvent strength model proposed by Snyder [552-555]. This is the least complicated of the models available and provides a reasonable approximation for typical experimental conditions. Mathematically more rigorous approaches have been developed by Jandera and Churacek [551,556], Schoenmakers et al. [520,534,557] and Tomellini et al. [558].

Most practical separation problems can be solved using binary gradients in which the volume fraction of the stronger eluting solvent is progressively increased with time. Ternary and quaternary solvent gradients may be useful for better control of selectivity but they are more difficult to program and interpret [556,559,560]. The shape of the solvent gradients used tend to conform to one of the idealized shapes shown in Figure 4.35. A convex gradient provides a faster increase in the concentration of the stronger eluent in a binary mixture at the beginning of the separation than a linear gradient. Where a linear gradient was the optimum profile, a convex gradient leads to the elution of bands with a lower average capacity factor and a shorter total separation time. As a result, early eluting bands appear sharpened and bunched together with a loss in resolution and later eluting bands appear wider and better resolved. Convex gradients would be favored for the separation of poorly resolved bands eluting at the end of a linear gradient. A concave gradient produces an effect opposite to a convex gradient with wider, better resolved bands at the start of the chromatogram and sharper, less well resolved bands appearing at the end. Complex gradients can be constructed by combining several gradient segments represented by the shapes in Figure 4.35 and/or by incorporating step increases in the stronger solvent, or by combining isocratic segments with gradient segments. Multiple segmented gradient programs are favored for more efficient optimization when the distribution of bands within the chromatogram are uneven, suggesting steeper gradients or step gradients in regions that are relatively empty of bands, or in chromatograms which exhibit pronounced changes in band spacing as the gradient steepness is varied [551,555, 561,562]. A potential problem with the design of multiple segmented gradient programs in which band spacing changes with the gradient steepness, is that the steepness of an earlier segment can affect the separation of bands eluted in a later segment. In practice, this means that the design of an optimal gradient program incorporating several different segments may require a large number of trial-and-error experiments to define the preferred experimental conditions. In essence, the more complex the type of program employed, the more time consuming will be the method development process, so that complex gradients should be reserved for difficult samples that failed to yield an acceptable result with continuous gradient forms.

Linear solvent strength gradients are convenient for optimization studies and the linear solvent strength gradient model sufficiently accurate to predict retention and resolution in gradient separations by computer simulations based on a small number of experiments [480,535,536,539,552-555,562-565]. An example of an optimal linear solvent strength gradient is shown in Figure 4.36 [557]. In an ideal linear solvent strength gradient the effective average value of the capacity factor is roughly equal for all components eluting at different times during the separation, all bands are eluted with approximately constant band widths, and the resolution between adjacent band pairs with

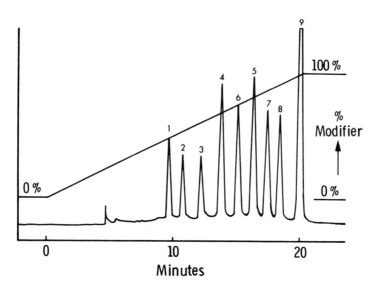

Figure 4.36 Example of an optimum linear-solvent-strength gradient. Peak identification: 1 = benzyl alcohol; 2 = 2-phenylethanol; 3 = o-cresol; 4 = nitrobenzene; 5 = diethyl o-phthalate; 6 = benzophenone; 7 = naphthalene; 8 = biphenyl; and 9 = anthracene. (Reproduced with permission from ref. 557. Copyright Elsevier Scientific Publishing Co.)

similar values of the separation factor are equal. In a linear solvent strength gradient the capacity factor for each solute at the column inlet, k_i, decrease during gradient elution according to

$$\log k_i = \log k_o - b(t/t_m) \qquad (4.53)$$

where k_o is the value of k_i determined isocratically in the starting solvent composition for the gradient, b the gradient steepness parameter, t the time after the start of the gradient and sample injection and t_m the column holdup time. In practice equation (4.53) is a reasonable approximation to experimental data, at least for the optimum capacity factor range ($1 < k_i < 10$). Ideally the gradient steepness parameter, b, should remain constant throughout the solvent program and hence have the same value for all sample components, a condition that is rarely exactly true under normal operating conditions. The gradient steepness parameter is defined by equation (4.54)

$$b = \Delta\phi S t_m / t_G \qquad (4.54)$$

where $\Delta\phi$ is the change in the volume fraction of the stronger eluting solvent during the gradient, S the change in log k for unit change in the volume fraction of stronger eluting solvent [see equation (4.16)], and t_G the gradient time (the time from the start to the end of the gradient). The gradient steepness parameter will clearly vary with the value of S, which is only roughly constant for similar compounds [535,539,562,566,567]. Systematic variations in S are generally observed among homologs, oligomers and samples comprising a parent skeleton with varying numbers of functional group substituents. The separation of these substances by a linear solvent strength gradient usually results in the tendency for later eluting bands to bunch together with a resulting decrease in resolution. Better results may be obtained with a convex gradient. High molecular weight polymers and biopolymers also exhibit a systematic variation of S as a function of molecular weight. An average value of S can be approximated by $S = 0.48(MW)^{0.44}$, but if the sample molecular weight range is large, so will be the variation of S, and therefore b. The value of S for macromolecules can be 30- to 100-fold greater than that for small molecules, indicating that for these substances retention is extremely sensitive to changes in mobile phase composition and typical gradients will require different experimental conditions than those used for small molecules [562,566,567].

A linear solvent strength gradient closely parallels a corresponding isocratic separation, as far as the effect of solvent strength on the separation is concerned. Specifically, a change in the quantity 1/b or t_G in gradient elution has a similar effect on a separation as does a change in the capacity factor for isocratic separations. Thus, analogous to isocratic separations, it should be relatively straightforward to predict changes in band width, resolution and separation time accompanying changes in the gradient steepness parameter in gradient elution chromatography. The retention time of a solute during gradient elution, t_g, is given by

$$t_g = (t_m/b)\log[2.3k_0b(t_s/t_m) + 1] + t_s + t_D \tag{4.55}$$

where t_s is the retention time of a solute under nonretaining conditions and t_D the dwell time for the chromatograph (the time it takes a change in mobile phase composition to pass from the gradient mixer to the column inlet). For small molecules $t_s = t_m$

and equation (4.55) is simplified. The following discussion will be explicit for small molecules only. The band width in gradient elution separations is the result of three more or less independent processes: the normal broadening of a sample band as it moves through the column, the band compression phenomenon arising from the faster migration of the tail of bands in gradient elution compared to the equal migration of all parts of a band in isocratic elution, and to the instantaneous capacity factor of the band, k_f, as it leaves the column [552,563-565]. The band width, w, is given reasonably well by

$$w = 1.1t_m(1 + k_f)/n^{1/2} \tag{4.56}$$

and k_f by

$$k_f = 1/[2.3b + (1/k_o)] \tag{4.57}$$

The plate number in equation (4.56) corresponds to the value when the effective value of the capacity factor (equal to k when the band is at the column midpoint) is equal to the capacity factor in isocratic elution for the same column. The effective value of the capacity factor, k_e, is simply $1/1.15b$. In most cases k_o will be large and equation (4.57) is simplified by equating $1/k_o$ to zero. The resolution between two adjacent bands in a gradient program, again analogous to isocratic elution, is expressed by equation (4.58)

$$R_s = [n^{1/2}/4][(\alpha - 1)/\alpha][k_e/(1 + k_e)] \tag{4.58}$$

noting in this case than n is defined as above and that the resolution, R_s, is defined with respect to the effective capacity factor. To obtain optimum resolution in the gradient elution mode the effective capacity factor should fall between 2 and 10, approximately. The effective capacity factor is controlled by the gradient steepness parameter, which in turn is related to the experimental variables by equation (4.54). An optimum value for b is set by choosing an effective capacity factor in the desired range, for example $k_e = 3$, and using this value to calculate b from the relationship between b and k_e. The value for b is then used to fix the range of experimental variables defined in equation (4.54), such as the gradient time and range of solvent compositions to be used. Note that the optimum gradient time for a given value of b depends on the column holdup time and should be adjusted accordingly when columns of different dimensions are used.

Linear solvent strength gradients can be used to predict isocratic retention in reversed-phase chromatography based on the experimental data for a single [552,553] or two gradient runs [534-536,539]. The latter approach provides the more accurate prediction of experimental isocratic parameters provided that the gradient times used for the two separations differ by a factor of about 3 or 4 and other parameters remain the same. The gradient retention time for each solute in the two gradients is entered into equation (4.55), which can then be solved numerically to give values of k_0 and b from which values of k_w and S in equation (4.16) are determined. The isocratic capacity factor for the solute is then defined as a function of the volume fraction of the stronger eluting solvent within the limits of equation (4.16) (see section 4.5.4). Also, the retention data from two gradient runs can be used to optimize gradient separations in which changes in selectivity accompanying changes in solvent strength occur using relative resolution maps (see section 4.6.2).

A suitable gradient for a separation can be described by three characteristics; the initial and final mobile phase compositions, the gradient shape, and the gradient steepness. Solvent selection rules parallel those for isocratic separations (section 4.6.2), since the solvent composition largely controls selectivity, while the gradient parameters affect solvent strength. The choice of solvent composition at the start of the gradient has an influence on the separation of the initial bands in the chromatogram but very little influence on the separation of the later eluting bands. Its composition should be adjusted to provide optimum resolution for the early eluting bands in the chromatogram with minimum time delay before the appearance of the first peak. The strength of the terminal solvent composition at the end of the gradient influences the selectivity of the separation (i.e., relative peak position) and the retention time and peak shape of later eluting bands. When bands elute long after the end of the gradient program is complete, a stronger terminal solvent is required. If the terminal solvent is too weak, the separation time may become inconveniently long and later eluting bands broadened and difficult to detect. Within these solvent constraints, the gradient program is selected and optimized. Some general guidelines for predicting the optimum gradient profile for

TABLE 4.25

SELECTION OF OPTIMAL LINEAR SOLVENT STRENGTH GRADIENTS IN LIQUID
CHROMATOGRAPHY

Column Type	Gradient Shape	Range or Rate
Reversed-Phase BPC	Linear increase in %B (Organic modifier)	5-7% B/t_m
Normal-Phase BPC	Vary composition of binary mobile phase linearly to give change in polarity equal to 0.5 P'/t_m	12-15% B/t_m
Liquid-Solid	Vary composition of binary mobile phase in a concave fashion to give a change in solvent strength equal to 0.02 $\epsilon°/t_m$	(10-18)t_m for a 0-100% gradient(a)
Ion-Exchange	Increase salt concentration by 1.6 fold per t_m in a concave gradient	
	Increase pH by 0.2 units/t_m, linear gradient (cation exchange)	pH varied from 2-6 in 20t_m
	Decrease pH by 0.2 units/t_m, linear gradient (anion exchange)	pH varied from 8-2 in 30t_m

(a) Gradients from 5-95% B are generally preferred from a practical
point of view

linear solvent strength gradients are summarized in Table 4.25
[553]. This table provides a starting point for gradient selection
but some fine tuning will probably be necessary for optimization.

Computer aided methods for the optimization of gradient
elution separations are very similar to the approaches described
for isocratic separations. Computer simulations using the linear
solvent strength gradient model enable gradient profiles and
initial and final gradient conditions to be predicted from the
experimental input from two gradient runs [536,539,555,563]. The
advantage of the simulation approach is that several changes to
the gradient conditions can be made and evaluated without the need
for additional experimental data. The simplex optimization
procedure and the mixture-design statistical approach have been
used to optimize the initial and final mobile phase compositions
and the gradient shape for gradient elution separations
[556,559,568,569]. The approaches are very similar to those used
to optimize mobile phase compositions in isocratic liquid
chromatography (section 4.6.2). They retain the same strengths and

weakness discussed for optimizing isocratic separations and have not been widely used.

4.7 PREPARATIVE-SCALE LIQUID CHROMATOGRAPHY

The purpose of preparative-scale liquid chromatography is the isolation of materials conforming to a specified purity in amounts that depend on the intended use of the product [8,570-572]. Possible uses include the isolation of materials for structural elucidation, for biological or sensory evaluation, for organic synthesis or commercial applications. The scale of the operation includes laboratory, pilot plant and process-scale systems. Process-scale separations will not be treated specifically in this section, since they represent a specialized area of chemical manufacturing and economic forecasting that the analytical chemist is infrequently exposed to [8,573-577]. However, process-scale methods do not vary in intent or fundamental approach to those used in the laboratory, except that the economics of the operation of the plant are considered of primary importance in the design and selection of columns and equipment, etc. Process-scale liquid chromatography is of considerable importance in the manufacture of high-value added products, such as pharmaceuticals, flavors and biotechnology products, where the rather high unit costs of purification are not considered prohibitive.

From its inception liquid chromatography has been used as a preparative technique. Most chemists are familiar with the gravity-feed glass column systems containing coarse adsorbent packings that are the main stay of laboratory practice. These low cost and easy to prepare and operate columns have many virtues. Not among them, however, are high resolution, short separation times and easy automation. When any of these factors are considered crucial then more sophisticated systems are required based on smaller particle sorbents with a narrow particle size distribution, that in turn are operated at above atmospheric pressure in the optimum mobile phase velocity range. These goals are best achieved using medium pressure and high pressure liquid chromatography. Before proceeding to these methods we will review some of the lower cost alternatives that may be suitable for the solution of many problems demanding only a modest increase in separation capacity compared to classical gravity-feed columns.

One of the earliest attempts to improve the efficiency of classical column chromatography was dry column chromatography [570]. In this case the column was dry packed to the desired length with TLC grade sorbent, and the separation achieved by developing the column with sufficient solvent to reach the lower end of the bed. Bands on the column are removed by extrusion, slicing (if a Nylon column is used) or by digging out, and the products freed from the sorbent by solvent extraction. The method is fast, requires very little solvent, and provides higher resolution than classical column techniques due to the use of sorbents of a smaller average particle size. It is suitable for the recovery of small quantities of materials (loading capacities being about 1% w/w), amounts similar to those obtained by preparative TLC (section 7.8). In order to generate reasonable flow rates for the recovery of sample components by elution when fine particle sorbents are used, either a vacuum at the column base or overpressure at the column head is required. For simple separations a vacuum filtration apparatus with a sintered-glass filter funnel containing a bed of sorbent may suffice [570,578,579]. The columns are dry packed with 10-40 micrometer diameter sorbent and consolidated by applying solvent to the packing and then sucking it dry. The sample, in a suitable weak solvent, is applied directly to the packing or onto a layer of preadsorbent (e.g., celite) to minimize uneven sample application. The column is then eluted with the appropriate solvent mixtures, pulling the column dry between each fraction collected. A vacuum manifold enables the column end to remain under vacuum (1-10 mm Hg) at all times while collecting fractions sequentially. Solvent changes can be made very simply as the column head is maintained at atmospheric pressure. Vacuum liquid chromatography provides resolution comparable to TLC separations while using a simple apparatus. Instead of vacuum, a slight gas overpressure can be used to increase the sample throughput. This method is often called flash chromatography [570,580,581]. The only modification required to the classical technique is a flow control valve at the head of the column that is connected to a pressure-regulated gas supply. A glass column of suitable length is either dry packed or slurry packed with suitable material (usually 40-63 micrometers in diameter). As the gas overpressure is only a few atmospheres, the bed height is restricted to less than 30 cm, typically 15-20 cm. The choice of column dimensions for a particular separation can be

TABLE 4.26

GENERAL CONDITIONS FOR PREPARATIVE-SCALE SEPARATIONS BY FLASH
CHROMATOGRAPHY
Mobile phase selection: (1) Choose a low viscosity solvent which
separates the mixture and moves the desired component to an R_f of
ca. 0.35; (2) If several compounds are to be separated which run
close together on TLC, adjust the solvent strength to put their
midpoint at an R_f value of ca. 0.35; (3) If compounds are well
separated, choose a mobile phase which provides an R_f value of ca.
0.35 for the least retained component.

Column Diameter (cm)	Volume of eluent(a) (ml)	Sample loading for a particular TLC resolution (mg)		Typical fraction size (ml)
		$\Delta R_f \geq 0.2$	$\Delta R_f \geq 0.1$	
1	100	100	40	5
2	200	400	160	10
3	400	900	360	20
4	1600	1600	600	30
5	2500	2500	1000	50

(a)Typical volume of solvent required for column packing and
sample separation.

deduced from the considerations given in Table 4.26. Typical
solvent flow rates are on the order of 5 cm/min. Flash
chromatography provides a rapid and inexpensive general method for
the preparative-scale separation of mixtures requiring only
moderate resolution in amounts of 0.01 to 10 g depending on the
difficulty of the separation and the column diameter (typically,
1-5 cm).

4.7.1 Medium Pressure Preparative Liquid Chromatography

A popular variation on the classical column techniques is
medium pressure liquid chromatography [570,582-589]. This
represents a midpoint in the transition between the classical and
high pressure chromatographic methods. By using glass and
polymeric materials for construction of the apparatus and column
packings with average particle diameters between 15 and 25
micrometers the cost of the system is kept comparatively low while
efficiency and sample throughput are much higher than for the
classical technique. The construction materials usually limit
column pressures to less than 500 p.s.i., and in some cases to
less than 200 p.s.i. The bursting pressure of a glass or plastic
tube depends on the wall thickness and internal diameter. For
heavy walled glass tubing, for example, the maximum safe operating

pressure is about 1200 p.s.i. for 3 mm I. D. columns, 600 p.s.i. for 10 mm I. D. columns and 150 p.s.i. for 25 mm I. D. columns. These pressures and the high flow rates used with large diameter columns can be generated by inexpensive flow-metering pumps. All connections between the pump, injector, column and detector are generally made with Teflon tubes, or similar materials, having specially-designed plastic end fitting that make the assembly of the equipment straightforward. The operation of a medium pressure liquid chromatograph is similar to that of a high pressure liquid chromatograph and requires no special description. A fraction collector is normally used instead of, or in conjunction with, a flow-through detector of low sensitivity (a short path length UV detector or refractive index detector) for sample collection. In the absence of on-line monitoring TLC can be used to regroup similar fractions. For small diameter columns, which require smaller solvent reservoirs for sample elution, a gas-pressurized holding coil may suffice in place of a pump for chromatographic operation.

The column length in medium pressure liquid chromatography is limited by the available operating pressure and is not usually greater than one meter, with 25-60 cm being most common. The column diameter is established by the sample size and is usually 1 to 10 cm. The wider bore columns contain several kilograms of packing and can separate charges of 15-100 g depending on the ease of obtaining the separation under concentration overload conditions. Column packings are usually 15-25, 25-40 or 40-65 micrometers in diameter and include bonded phase packings, inorganic adsorbent and polymeric materials. Prepacked columns are available commercially, or columns can be dry packed or slurry packed in the laboratory without the need for specialized packing equipment [582,585,588,590]. The sample is usually injected on-stream using a valve injector or, for very large sample volumes, through the pump. For wide bore columns the design of the column inlet and end fittings is crucial to ensure an even distribution of the sample across the column diameter. This can be achieved with a fritted disc and plunger assembly (adjustment of the plunger position allow column voids to be filled) or by drawing the column inlet into a conical constriction which is then packed with glass beads.

4.7.2 High Pressure Preparative Liquid Chromatography

When selecting the appropriate preparative-scale liquid chromatographic method for conditions requiring higher resolution or shorter processing times than those discussed previously, it is necessary to consider the intended use of the purified material. For example, 1-10 mg of a pure sample would suffice for spectroscopic identification, while 1-10 g might be the minimum amount of an intermediate useful for an organic synthesis. The amount of sample required dictates the size of the column and the operating conditions necessary for the separation, as indicated in Table 4.27 [590]. The simplest approach to increasing the amount of sample recovered is to scale up an analytical separation using the same column packing, column length and linear mobile phase velocity while increasing the column diameter. The analytical separation should be optimized to maximize the separation factor between critical peak pairs at the expense of a longer separation time to permit the use of higher column loadings before the bands overlap. If the sample loading does not exceed the linear region of the sorption isotherm the results obtained are easily extrapolated from the analytical separation. With most analytical instruments the largest column size that can be used is about 25 cm long and 2.5 cm internal diameter containing approximately 65 g of silica-based packing and requiring an optimum flow rate of around 10-15 ml/min. For larger columns special purpose preparative-scale instruments are usually needed to provide adequate flow and pressure capacity. Also, for the isolation of the maximum amount of sample a different approach to analytical separations is required. Success will be judged by the production rate and recovery yield obtained [591]. The production rate is the amount of the purified fraction containing the corresponding component at the required degree of purity per unit time. The recovery yield is the ratio between the amount of the component of interest that is collected in the product fraction and the amount injected in the column with the feed. The object is to maximize the production rate and the recovery yield, which invariably results in operating the column in an overloaded condition, such that the conditions predicted from an analytical separation have no direct meaning in establishing the optimum conditions for the separation. In fact, column operation under nonlinear or non-ideal conditions is very complex and it is not as easy to predict

TABLE 4.27

MAXIMUM ALLOWABLE COMPOUND SIZE WITH RESPECT TO STATIONARY PHASE
LOADING AND COLUMN DIMENSIONS

Preparative column type	Application	Required Amount of Pure Sample (mg)	Stationary Phase (g)	Column I.D. (mm)	Peak Capacity (mg)(a)
Analytical	Mass spectrometry	0.001	0.2-3.2	1-5	0.2-3.2
Analytical	IR	0.1	0.2-3.2	1-5	0.2-3.2
Wide bore analytical	NMR	0.1-10	3-12	6-11	3-12
Wide bore analytical	Elemental analysis	1-25	7-25	6-11	7-25
Long narrow	Synthesis	100-1000	25-100	10-30	20-1000
Short thick	Large scale	10^3-10^5	10^2-10^4	20-100	10^3-10^5
Industrial	Commercial	10^4-10^6	10^3-10^5	10^2-10^3	10^4-10^6

(a) Rough estimate of the amount of pure material in a single peak
that could be isolated from a single injection. Will vary with the
separation factor between peaks and whether concentration overload
conditions are used.

optimum operating conditions in this case as it is for the less
demanding, although less powerful, scale-up approach to
preparative liquid chromatography. Several general reviews of high
pressure preparative-scale liquid chromatography are available
[8,570-573,590,592,593] as well as application reviews for
specific compound classes [570-573,593-595].

The scale-up approach to preparative HPLC is quite
straightforward and is the approach likely to be taken in an
analytical laboratory that requires sufficient material for
identification purposes or to purify an analytical standard on an
occasional basis. Separations are carried out in the linear region
of the sorbent isotherm, which provides an upper sample mass limit
of about 0.1-1.0 mg/g of sorbent. Increasing the column diameter
at constant column length increases the weight of packing, and
therefore the sample capacity of the column, without dramatically
influencing the resolution obtained if the mobile phase linear
velocity remains constant. The loading capacity of a column and
the required mobile phase flow rate to maintain a constant linear

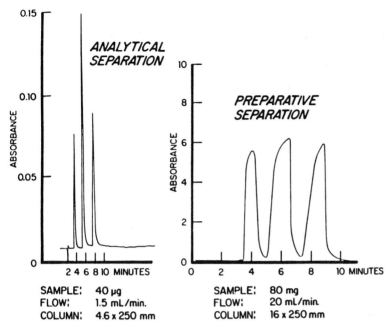

SAMPLE: 40 µg
FLOW: 1.5 mL/min.
COLUMN: 4.6 x 250 mm

SAMPLE: 80 mg
FLOW: 20 mL/min.
COLUMN: 16 x 250 mm

Figure 4.37 Preparative-scale separation of bilirubin isomers by high pressure liquid chromatography. The analytical separation was optimized to maximize the separation factor at the smallest practical value for the capacity factor and then the sample size scaled-up to that allowed by the larger amount of packing in the preparative column. (Reproduced with permission from Perkin-Elmer Corporation).

velocity are directly proportional to the column cross-sectional area. Thus, to scale up an analytical separation a larger diameter column packed with the same material and operated at the same linear velocity as the analytical column is used. A practical example is shown in Figure 4.37 for the analytical and preparative separation of a mixture of bilirubin isomers. The irregular shaped peaks in the preparative separation are due to detector saturation and not column overload.

At this point it is worth considering the demands made on the instrumentation for operation with wide bore columns and, in particular, the adaptation of analytical instruments for this purpose [596,597]. The pump requirements for preparative separations differ from those in analytical HPLC as the ability to generate high flow rates at moderate backpressures is crucial to the efficient operation of wide bore columns. A flow rate maximum of 100 ml/min with a pressure limit of 3000 p.s.i. is considered adequate. Few analytical reciprocating piston-type pumps are capable of reaching this volume delivery rate; some are capable of operating at 30-60 ml/min, which suffices in many instances; still others have a maximum delivery rate below 10 ml/min, which is barely adequate for all but semipreparative use.

The mode of injection differs appreciably from analytical methods where the object is to create discrete narrow bands by point or narrow zone injection of low sample amounts. Point injection of a large sample (mass or volume) in preparative liquid chromatography would result in a local overloading of the column packing with a deleterious effect on column performance. To minimize this effect the sample should be applied evenly over the entire column cross-sectional area of the column inlet. If the sample moving along the wall takes more time to enter the column than the sample moving in the center, because the path length is longer and/or the average velocity slower, the separated bands may be broadened considerably. To avoid this phenomenon, the column should be covered with a thin metal frit having a permeability lower than that of the column, and the mobile phase evenly distributed over the frit using several methods, such as distributor plates with sectors cut out, nozzles or a thin empty chamber above the frit. Injection of the feed through a sample loop is reserved for small samples, such as those used with semipreparative columns. The uneven flow velocity distribution obtained with large sample loops is undesirable. For sample volumes of several milliliters, it is easier to pump the sample onto the column with the eluent pump or, preferably, a dedicated injection pump.

Detection requirements in preparative-scale chromatography also differ from analytical operations where detectors are selected for their sensitivity. Sensitivity is not of overriding importance in preparative-scale chromatography; the ability to accommodate large column flow rates and a wide linear response range are more useful. The sensitivity of the refractive index detector is usually quite adequate for preparative work but the flow capacity of many analytical detectors of this type is limited. The common analytical UV detectors have a limited dynamic range and are easily overloaded when separating preparative-size samples. Variable wavelength detectors can be detuned from the absorption maxima for the sample components to increase the dynamic range of the detector at the higher concentration end.

Some analytical detectors have interchangeable flow cells which allow replacement of the analytical cell with one of shorter path length to reduce sensitivity. Preparative flow cells usually have wider bore inlet and outlet connections to accommodate high mobile phase flow rates. For similar reasons wide bore capillary tubing of 0.4-0.6 mm I. D. should be used to make column connections instead of the narrow bore tubing used with analytical columns. To avoid damage to the flow cell it may be advisable to operate analytical detectors with a low-dead volume flow splitter, so that only a few percent of the total column eluent passes through the detector. In this case it may be necessary to provide a flow restrictor in the line that does not lead to the detector to balance the flow resistance. Finally, if only one or two fractions are to be collected and the number of runs is low, then the sample can be collected manually in suitably-sized containers; a fraction collector is otherwise required.

With columns of large internal diameters wall strength and the connection of end fittings becomes increasingly important [574,592,598]. Large-scale preparative liquid chromatographic columns must be very heavily walled compared to analytical columns if they are to be used with small diameter particles, or alternatively, they must be operated at significantly lower inlet pressures, reflecting their lower bursting pressure. Some commercially available stainless steel columns with an internal diameter of 50 mm, for example, have a maximum operating pressure of about 1,400 p.s.i. Compression fittings may not seal adequately to wide diameter columns at the required operating pressures and flanged end fittings are generally used. They are either bolted or fastened with special collars that can withstand the very high pressures to be used.

Dry packing or slurry packing of large diameter columns is more difficult than conventional diameter columns due to the scale of the packing process and the high volume flow rates and pressures required to consolidate the column bed [573,592,593, 598-601]. In fact very little information has been published on the techniques employed but it is an established fact that columns of large internal diameter are less stable than analytical-sized columns. This might arise from several sources, such as from a lower packing density resulting from the problems of obtaining the necessary flow velocity and packing pressure to reach equivalent packing densities to analytical-sized columns. Pressure surges

502

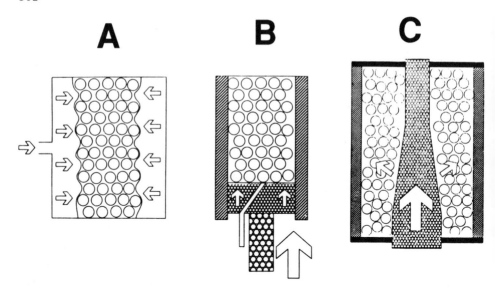

Figure 4.38 Schematic diagrams of continuous compression preparative liquid chromatographic columns. The arrows represent the force directions. A = radial compression, B = axial compression and C = annular expansion.

from the eluent pumps are not as well dampened as in analytical instruments, adding to the instability of the column bed. For columns with diameters larger than about 25 mm, packed with small diameter particles, inefficient dissipation of frictional heat generated by the flow of mobile phase can be a problem, since the stainless steel mantle and packed bed do not have the same thermal expansion properties. Through repeated thermal expansion and contraction the packed bed structure changes, and a void is produced at the top and along the walls of the column. The formation of a high permeability zone near the wall can give rise to peak doubling [602].

The solution to the above problem is to adapt the column volume continuously to the changing packed-bed volume by radial or axial compression [592,598]. Axial compression can be exerted from above with an adjustable column head, or from the bottom with a fixed or floating piston. Radial compression can be achieved with a gas or fluid between a flexible mantle and rigid container, or with prongs. A combination of axial and radial compression is obtained by inserting a wedge or plunger in the center of the column top, which is screwed down as the column conditions require. The three methods of compression in terms of the direction of applied forces are summarized in Figure 4.38.

Radial compression columns are used by Waters Associates (see section 4.3.1) [144-146,603]. A flexible-walled plastic cartridge, 30 x 5.7 cm, is dry packed with a rather coarse sorbent of about 75 micrometers average particle diameter. This cartridge is placed into a stainless steel pressure vessel and the ends of the cartridge are mechanically sealed to the inlet and outlet ends of the pressure vessel. Nitrogen at 400 p.s.i. is introduced into the space between the inside wall of the pressure vessel and the outside wall of the cartridge to consolidate the packed bed. Two cartridges can be operated in series to improve the efficiency and increase the sample capacity. The instrument also allows the use of recycle and peak shaving techniques. Typical eluent flow rates are on the order of 50-500 ml/min with sample charges of 1 to 50 g.

Axxial and Prochrom offer preparative liquid chromatographs for the laboratory employing axial compression technology [592, 598,604-606]. The column design consists of a cylinder which is approximately three times longer than the column length desired and a sliding piston with appropriate fittings to prevent leaks between the piston and the barrel. In order to prepare a column, the cylinder is first filled with a slurry of the packing material, the injection head bolted to the cylinder, and the piston slowly raised forcing the solvent out of the cylinder through a frit in the injection head fitting, which retains the packing. When the process of compression is complete, the cylinder can be locked into position to prevent it from being pushed backwards if the mobile phase inlet pressure exceeds the piston pressure used to consolidate the bed. An alternative design employs a floating piston, which uses the mobile phase inlet pressure to continuously consolidate the column packing. After the separation, the injection head is unbolted and the packing pushed out by the piston. Thus, it is not necessary to elute all components through the column; well-retained components can be recovered or disposed of by sectioning the column bed. With the larger column, samples of 50 to 100 g can be separated per injection. Column length, capacity, selectivity and resolution are controlled by the type and amount of column packing material added to the cylinder. This approach is obviously very flexible and cost effective as the column dimensions and column packing can be easily varied to meet the needs of each individual separation. These advantages would seem to make axial compression preferable to radial compression.

So far we have considered preparative-scale chromatography when the column is not overloaded and the desired amount of sample has been isolated by simply scaling-up the dimensions of the column and adjusting the operating conditions accordingly. The theory pertaining to analytical columns is applicable and only minor operating modifications are required to ensure success. Since there is a limit to the column size that can be prepared with the same efficiency as analytical columns, the only practical way to further increase sample throughput is to overload the column. There is no adequate theory for columns operated in a mass overload condition and the success of this approach must be judged on empirical grounds. As it represents the most cost effective means of obtaining bulk sample quantities and is widely used in many laboratories, there is no need to fear or avoid operating above the column capacity limit for linear chromatography. A column is considered to be overloaded when solute capacity factor values change by more than 10% as the sample size is increased. A column might be overloaded either by increasing the sample concentration while maintaining a constant injection volume (concentration overload) or by increasing the injection volume while maintaining a constant sample concentration (volume overload) [607-611]. Under volume overload conditions, the sample concentration is kept constant and is confined to the linear range of the adsorption isotherm, but the volume injected is very large. For rectangular injection pulses of constant height (concentration) and increasing width (volume), the elution band becomes higher and wider. Ultimately, it becomes flat topped but remains symmetrical, Figure 4.39 [608]. In concentration overload a small sample volume is injected, but its concentration exceeds the linear range of the adsorption isotherm. Accordingly, the band profile broadens and becomes assymetric; in the case of Langmuir type isotherms, the profile becomes close to a triangle, with an almost vertical front and a slanted tail. The production rate in the elution mode increases with increasing volume and/or concentration of the sample injected on the column. Since the band width of all components increases in both cases, there is a limit to the sample size that can be used effectively. When the sample size increases from the very low levels typically used in analytical applications, the recovery yield remains constant and the production rate increases linearly with increasing sample size. When the band of the compound of interest touches its neighbor, the recovery yield starts to decline with increasing sample size, since the wings of the elution band must be clipped

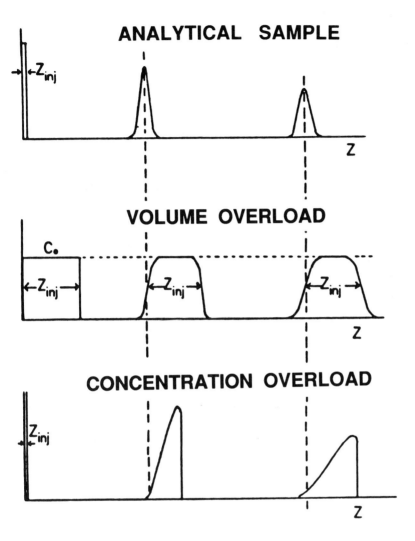

Figure 4.39 Development of peak profiles during migration along the column for analytical and overload samples. (Reproduced with permission from ref. 608. Copyright Elsevier Scientific Publishing Co.)

to eliminate contamination. Eventually, a maximum value for the production rate is reached, but since the separation is carried out under non-linear conditions, optimum separation conditions are no longer predictable from data obtained under analytical conditions. Figure 4.40 is an example of a separation performed under concentration overload conditions. The dotted line

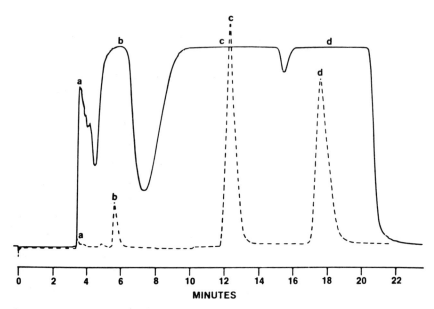

Figure 4.40   Separation of alkylbenzyl alcohol isomers by
reversed-phase semipreparative HPLC in a mass overload condition.
The dotted line is the equivalent analytical separation on the
same column. (Copyright Whatman Inc.)

represents the analytical separation used to optimize (maximize)
the separation factor. Trial and error was then used to establish
the maximum allowable sample size to optimize the production rate.

The properties of chromatographic band profiles under non-
linear conditions are much more difficult to predict and to model
mathematically than is the case for linear chromatography
[591,608,612-622]. Exact models that accurately relate the
concentration profile of the band at the column outlet to the
equilibrium isotherm, the kinetics of mass transfer, and the
profile of the injection band are available for a limited set of
boundary conditions. At high sample concentrations equilibrium
isotherms are non-linear, the concentration in the stationary
phase at equilibrium increasing either faster or more slowly than
the concentration in the mobile phase. Accordingly, solutes at
different concentrations tend to move along the column at
different velocities, and either the front or rear of the band
will become steeper. The general solution to this problem is to
write the mass balance equations for the chemical species
involved, combine these with the equations describing the mass
transfer kinetics, and to solve the system of partial differential

equations obtained. The thermodynamic properties of the system involved are easier to solve than the kinetic properties. Prominent models, therefore, carry out a differentiation of the equilibrium isotherm and make approximations for the mass transfer contribution. The semi-ideal model, for example, assumes that mass transfer is fast enough for the two phases to be always close to equilibrium, the effect of axial diffusion and the finite rate of radial mass transfer merely combining and resulting in an apparent diffusion larger than the true axial diffusion, but having the same consequences. Predictions from this model seem reasonable for columns having more than 1000 theoretical plates with fairly low loading factors. The related ideal model further assumes that the column efficiency is infinite. This model is poor at describing band profiles with low loading factors, but for efficient columns with high loading factors the results are quite good. With large sample sizes the contribution of the non-linear behavior of the equilibrium isotherm to the width of the band profile becomes more important. This is the contribution accounted for by the ideal model.

Since the velocity associated with a concentration of one component is a function of the concentration of all the mixture components, the band profiles of the mixture are no longer independent of each other [591,619,620,623-627]. Compared to the results obtained for a single peak, displacement and tag-along effects are observed, which control the shape of the profiles of the individual component bands of a mixture when these bands remain unresolved for at least part of their time on the column. The intensities of the displacement and tag-along effects depend essentially on the sample size, the composition of the feed and the parameters of the competitive equilibrium isotherm of the components involved, as determined by the loading factors (ratio of the actual amount injected to the column saturation capacity for each compound). The intensity of the tag-along effect is measured by the length of the concentration plateau of the second component left behind by the first component. This plateau results from the fact that the velocity associated with a certain concentration of the second component is a decreasing function of the local concentration of the first component (because of the competition for access to sorption sites). An increase in the ratio of the column loading factor for the second component relative to the first will lead to an increase in the intensity of the tag-along effect. The opposite is true for the displacement effect, which affects the first eluting peak of the pair, and results in its elution as a characteristic concentration plateau (see section 4.7.3).

It is very difficult to provide an optimum set of conditions for operation of a liquid chromatograph under overload conditions due to the complex interactions among a large number of parameters [591,592,595,608,620,625]. The following general observations seem to be applicable in most cases. The column efficiency should be as high as possible and separations should be carried out using concentration overload conditions. The production rate of a purified fraction from the same feed increases with (α^2 - 1), where α is the relative retention of the desired compound and its neighboring impurity (i.e., the ratio of the origin slopes of their equilibrium isotherms). Preparative columns should be operated at higher reduced linear velocities than is typical of analytical columns. There is an optimum value of d_p^2/L (d_p is the particle diameter and L the column length) for which the production rate is a maximum. This ratio is not easy to calculate for a given set of separation conditions, but the ratio d_p^2/L is more important than specifying values of the individual parameters themselves. The production rate increases rapidly with the available inlet pressure when the optimum column is specified.

4.7.3 Preparative Liquid Chromatography Using the Displacement Mode

Displacement is an alternative non-linear separation mechanism to elution, where the sample components are displaced from the stationary phase by a solution of a compound which has a higher affinity for the stationary phase than for all of the sample components [8,595,628-635]. Under optimum conditions the components of the sample are eluted as a train of rectangular bands with concentrations larger than they were in the sample applied to the column. The advantages of the displacement mode for preparative chromatography are that the separation products are obtained in much higher concentration than in the elution mode, solvent consumption is lower, and tailing is reduced as a result of the self-sharpening boundaries obtained. Although separation times are usually shorter than in the elution mode, the sample processing rate is lower than expected due to the necessity of having to regenerate the column after each separation. Regeneration is essential to wash out all of the displacer substance from the column. This is often quite complicated and time consuming. Displacement chromatography usually requires some system for on-line analysis, since detectors commonly used in elution chromatography do not provide sufficient information on the boundary regions, which are usually both sharp and represent high sample concentrations. Perhaps the major disadvantage of displacement chromatography is a general lack of familiarity with this separation mode and its rather difficult experimental optimization coupled with a general lack of published applications to use as a guide for selecting separation conditions.

In displacement chromatography the column is first equilibrated with a carrier solvent, chosen for its low solvent strength in the separation system, and its ability to dissolve the feed components at sufficiently high concentration to allow the introduction of large samples. After equilibrium has been achieved, the feed mixture is introduced into the carrier stream and preconcentrated at the column inlet. Thus, it is important that the stationary phase selected exhibits strong retention of the feed components with the carrier solvent as eluent. The next step consists in pumping a solution of the displacer at high concentration through the column. The displacer is a substance that has a greater affinity for the stationary phase than any of the sample components and causes the sample components to move down the column at speeds determined by the displacer front velocity. Stronger adsorbing components of the sample displace from the surface of the stationary phase those having weaker retention until the separation is achieved. The mixture separates into a displacement train, and once fully developed, the train is composed of adjacent square bands of near uniform concentration, all moving at the same velocity. After all the feed components are eluted, the column has to be regenerated and re-equilibrated with the carrier solvent before another injection can be made.

The critical parameters for separation by displacement are the displacer concentration, the loading factor (ratio of the sample size to the column saturation capacity) and the column efficiency. The choice of displacer is probably the most critical step. For correct development to occur the adsorption isotherm of the displacer must overlie those of the feed components. The concentration of the displacer controls the separation time and the concentration of the sample zones (the length of the separated zones is proportional to the amount of substance injected but not the height). The choice of carrier solvent is less critical. It must be a good solvent for the displacer and the sample, as well as providing high capacity factor values for the sample in the chromatographic system. The optimum flow rate is usually low compared to elution chromatography to minimize the effects of axial dispersion. The column must be long enough for the development train to be fully formed but need not be longer than this minimum length for a given injection size. An unusual feature of displacement chromatography is that as the amount of sample injected is increased the column length must be increased accordingly, since the length of the zones in the displacement train increase with increasing sample amounts.

4.8   CAPILLARY ELECTROPHORETIC SEPARATION METHODS

Capillary electrophoretic methods are a family of separation techniques in which separations are achieved by the differences in migration rates for sample components in an electric field applied to a capillary tube filled with an electrolyte [636-641]. The presence of an electric field causes the movement of sample ions by electrophoretic migration and bulk flow of the electrolyte solution by electroosmosis. In capillary zone electrophoresis (CZE) ions are separated according to their relative mobility (approximately equivalent to their size-to-charge ratio) in free solution. In capillary gel electrophoresis (CGE) ions are separated by their electrophoretic mobility through a porous gel network. In general, ion mobility through a gel network decreases as the ion size increases. CGE, therefore, is ideally suited to the separation of oligomers (e.g, oligonucleotides) which have a constant charge-to-mass ratio. Electrokinetic capillary chromatography (EKC) is a variant of CZE in which a pseudostationary phase is added to the electrolyte solution and allows a separation to be achieved based on the distribution of the sample between the bulk electrolyte and the pseudostationary phase. The pseudostationary phase is usually a surfactant above its critical micelle concentration, a complexing agent, such as cyclodextrins or a polymeric ion exchanger. The stationary phase is not immobilized in this instance but migrates with a different velocity to the bulk electrolyte and it is this combination of differential migration rates for the two phases combined with the

selective distribution mechanism between phases, that results in
the separation. A unique feature of EKC is that it can separate
neutral molecules as well as ions. Another variation of the CZE
method is isoelectric focusing, which is suitable for the
separation of amphoteric compounds, such as proteins. The
separation capillary is filled with sample and a mixture of
ampholytes used to create a pH gradient. When an electric field is
applied the ampholytes migrate to their respective isoelectric
points and, in so doing, form a relatively stable pH gradient. The
sample components migrate electrophoretically through this
gradient until they reach a pH at which their net charge is zero
(isoelectric point). They remain focused and immobile at this
position for as long as the electric field is applied. The
separated zones are detected by displacing the electrolyte
solution passed a suitable detector by applying pressure to one
end of the capillary.

In isotachophoresis (displacement electrophoresis) the
sample solution is sandwiched between a leading and a terminating
electrolyte [641-646]. The leading electrolyte contains ions with
a mobility higher than that of any ion in the sample, and
conversely, the terminating electrolyte contains ions with a
mobility lower than that of any of the sample ions. When an
electric field is applied the leading electrolyte will attempt to
"pull away" from the sample ions. This results in a gap where the
conductivity is dropping and the electric field is rising. The
increased field will "pull" the analyte ions along until their
velocity reaches that of the leading electrolyte. Eventually, a
steady state condition is attained in which all sample ions (of
the same charge type) are migrating as single bands arranged in
order of their relative mobility and with an identical velocity.
The steady state concentration of the analyte is largely
determined by the leading electrolyte concentration. Accordingly,
if the analyte is more dilute than the concentration of the
leading electrolyte, the analytes may become more concentrated as
they are being separated. A feature useful in trace analysis and
multidimensional and micropreparative applications of capillary
electrophoresis [647,648]. Detection in isotachophoresis is
sometimes achieved by UV absorption, but more commonly electrical
methods are used. In isotachophoresis each zone travels at the
same velocity with the electric field in each successive zone
higher than the one that preceded it. A pair of electrodes placed

at the end of the capillary tube can be used to measure the voltage for each zone. Each zone passing the detector causes a successive step change in the detector signal. The identity of the analyte can be obtained from the position of the step and the quantity from the length of the step. Although isotachophoresis has been in use for over 20 years, it was conventionally performed in plastic tubes of 0.2 to 0.5 cm internal diameter using electrolyte pumps to generate buffer flow. Its use in a capillary tube format is quite recent and very limited data are available for analytical applications [647-649].

The capillary electrophoretic methods represent a new approach to electrophoretic separations rather than the application of a new separation principle. Conventional, mainly planar electrophoretic methods, have been routinely used in the life sciences for decades [640,650-652]. These methods have contributed substantially to our knowledge of modern biochemistry, but are generally considered to be too slow, labor intensive, prone to poor reproducibility and of limited quantitative capability. The attractiveness of the capillary methods is the possibility of achieving faster separations, higher efficiency, reliable quantitative detection and ease of automation. In many cases these goals remain the aims of the capillary methods, which have not reached a mature state of development so far. Consequently, in the following sections we will concentrate on the theoretical background to the techniques with limited coverage of practical applications or method development.

In electrophoresis ions or charged particles move in a liquid under the influence of an electric field E, at a rate given by

$$u_e = \mu_e E \tag{4.59}$$

where u_e is the electrophoretic migration velocity of an ion and μ_e its electrophoretic mobility [636-639,653-656]. The electrophoretic mobility of an ion depends on its charge density, and thus on its overall valence and size, as well as the viscosity and dielectric constant of the medium. The electrophoretic mobility of an ion is also a strong function of temperature. At a constant temperature ions move at different rates, according to their charge density, and so can be separated. However, for large colloidal particles and micelles, etc., the charge density is more or less independent of size and these particles migrate at approximately the same velocities. Advantage is taken of this phenomenon in electrokinetic capillary chromatography discussed later.

Bulk flow of the mobile phase in electrophoresis can also occur by electroosmosis. When an insulator is immersed in a solution of an electrolyte, an electrical double layer results at the interface. This is caused by the generation of a charged surface arising from the adsorption of ions from solution, or by the dissociation of surface functional groups. In the case of silica, the surface will usually be negatively charged by virtue of the partially ionized acidic silanol groups. When such a surface is in contact with a solution there is a slight excess of positive charges in solution to balance the fixed negative charges on the silica surface. Some of these excess ions are free to move within the liquid while others are immobilized at the surface. The excess charge density associated with the free ions falls off rapidly with distance from the surface, along with the associated electrical potential, which is proportional to the charge density. The thickness of this layer, the electrical double layer, is small, typically about 10 nm. The potential at the boundary between the charged surface and the start of the diffuse part of the double layer is called the zeta potential, denoted by ξ, and is typically between 0 and 100 mV. The excess free cations in the liquid form a kind of charged sheath which encloses the main bulk of the electrolyte. When a potential gradient, E, is applied across the length of the column the excess ions within the double layer move towards the appropriately charged electrode carrying any enclosed liquid along with them. Shearing of the solution occurs only within the diffuse part of the electrical double layer resulting in a plug flow profile provided that the tube diameter is much larger than the thickness of the electrical double layer, which it generally is. The velocity with which the bulk liquid migrates is given by

$$u_{eo} = \epsilon \xi E / \eta \qquad (4.60)$$

where u_{eo} is the electroosmotic flow velocity, ϵ the dielectric constant of the solution and η the solution viscosity.

In capillary zone electrophoresis the individual ions move at a velocity u_e given by equation (4.59) in a plug of liquid which is itself moving at a velocity u_{eo} given by equation (4.60). The actual velocity with which an ion is moving is the sum of u_e and u_{eo}, which do not necessarily have to have the same sign (migrate in the same direction). In general, when a single, fixed position detector is used, it is convenient to arrange for the electroosmotic velocity of the bulk solution to exceed the electrophoretic velocity of any of the sample ions. In this way both negatively and positively charged ions will migrate in the same direction with the bulk flow passed the detector while being separated by differences in their electrophoretic velocity. It should be noted that neutral molecules will migrate through the capillary at the electroosmotic velocity but remain unseparated in the absence of a distribution mechanism, as is employed in micellar electrokinetic chromatography, for example. In capillary gel electrophoresis there is no electroosmotic flow and the only migratory process is electrophoresis. In this case ions of opposite charge will migrate towards the oppositely charged electrodes at either end of the separation capillary.

Figure 4.41 provides an illustration of the flow profile of the mobile phase through a capillary tube using pressure and electroosmotic forces as the driving mechanism [653]. With the pressure driven system, which is typical of high pressure liquid chromatography, the flow profile is parabolic, with the flow velocity being zero at the wall and twice the mean velocity at the center. For the electroosmotic flow system the flow profile is essentially flat, the shear forces responsible for solvent migration originating in the electrical double layer close to the column wall, and as there is no charge imbalance within the core, the flow profile within the column as a whole is near perfect plug flow. These differences in flow profile have important consequences for separations in open tubes (or packed columns). In pressure driven systems the flow variation across the column diameter contributes significantly to the zone broadening mechanism, while for plug flow, at least in the ideal case, only axial diffusion should contribute to zone broadening. Consequently, the ultimate efficiency for plug flow in capillary zone electrophoresis, or zero bulk flow as in capillary gel electrophoresis, is given by

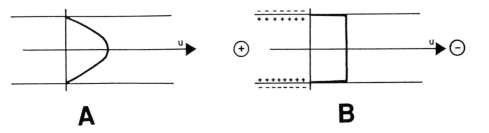

Figure 4.41 Flow profiles in a capillary column for a pressure
driven system (A) and an electroosmotically driven system (B).
(Reproduced with permission from ref. 653. Copyright Friedr.
Vieweg & Sohn).

$$n = Lu/2D_m \qquad\qquad (4.61)$$

where n is the number of theoretical plates, u the appropriate
expression for the mobile phase velocity, L the effective column
length for the separation and D_m the diffusion coefficient for the
analytes in the mobile phase. This expression is very different
from its more familiar pressure-driven chromatography analog
(section 1.4). For example, it tells us that higher mobile phase
velocities and higher potential fields will provide larger numbers
of theoretical plates. There is no dependence of the column
efficiency on the column diameter and higher molecular weight
solutes with smaller values of D_m, should increase column
efficiency. The column efficiency observed in practice, however,
may be substantially smaller than predicted by equation (4.61) due
to a number of experimental considerations [636,639,653-655,
657-662]. The column length, internal diameter and the electric
field are not independent variables, but are related through the
Joule heating of the electrolyte and its effect on the mobile
phase flow profile. Heat is generated homogeneously throughout the
electrolyte but the temperature variation across the bore of the
capillary is parabolic. The heat generated is approximately
linearly dependent on the field strength squared and the molar
conductivity and concentration of the electrolyte. Thus higher
electric fields cause increased radial temperature gradients
between the center of the tube and the column wall, leading to
sample diffusion and solvent density and viscosity differences in
the flow direction which result in zone broadening. Heat

production will be less for electrolytes of low conductivity, such as zwitterionic buffers, and for low electrolyte concentrations, although to avoid zone distortion due to conductivity changes, the sample must be effectively diluted in the buffer at the time of injection, that is, the electrolyte concentration must remain substantially higher than the sample concentration. The properties of the sample then are more relevant in determining the concentration of electrolyte in the separation capillary than the need to minimize Joule heating. The most effective means of minimizing temperature gradients across the column diameter is to reduce the column diameter to promote rapid heat transfer to the column walls. At the field strengths typically used in capillary electrophoretic methods radial temperature gradients are generally not significant for columns with diameters below about 100 micrometers. Thus, although there is no direct relationship predicted between the column diameter and column efficiency by equation (4.61), for high efficiency small diameter columns must be employed to maintain a plug flow profile. The temperature rise also depends on the removal of heat from the outside of the capillary wall. Forced air convection, flowing liquid baths or Peltier coolers are often employed with small diameter capillaries to improve the efficiency of heat removal and the reproducibility of migration times.

Miniaturization of the separation system is required to enable high field strengths to be used to minimize zone broadening resulting from inefficient heat dissipation. However, miniaturization causes other experimental factors to become important in determining the observed system efficiency [663]. Extracolumn injection and detector void volumes are of critical concern. According to Zare, in almost all published reports of capillary zone electrophoresis the observed peak profiles were dominated by the contribution from the sample injection length [660]. A second major contribution to band broadening is sorption of the analyte ions at the column wall or on the gel matrix. If wall retention occurs, mass transfer across the column becomes necessary for equilibrium, and a large contribution to the plate height arises. The control of the injection band length and sorption interactions within the column are important practical problems to be solved before the true potential efficiency of capillary electrophoretic methods can be realized. Zare has also pointed out that peak widths measured in capillary electrophoresis should be corrected to allow for the finite length of the detector slits with respect to the peak widths, and for the fact that the separated zones move passed the detector with different velocities, if true chromatographic peak widths are to be obtained [660].

Electrokinetic chromatography is characterized by the incorporation of a pseudostationary phase component in the bulk electrolyte. If the sample components distribute themselves selectively between the pseudostationary phase and the bulk aqueous electrolyte and provided that the bulk electrolyte solution and the pseudostationary phase migrate at different velocities, then high efficiency separations can be achieved [664-671]. A representative example of the separation process is illustrated in Figure 4.42. In this case the pseudostationary phase is a surfactant, sodium dodecyl sulfate, at a concentration above its critical micelle concentration. Whereas the bulk aqueous phase migrates at a velocity strictly determined by electroosmotic flow, the migration of the micelles is retarded due to the additional opposing electrophoretic forces exerted on the charged micelles. The micellar pseudostationary phase is neither immobilized nor does it consist of a distinct phase in the conventional sense, since micelles are distributed homogeneously throughout the bulk electrolyte. When a sample (X in Figure 4.42) is added to the electrolyte, it forms an equilibrium between the micelles and bulk electrolyte and is separated into individual components due to the differential migration velocities of bulk electrolyte and micelles towards the detector. The range of possible retentions is defined by the window created between the retention time of a solute which is not incorporated into the micelles and moves with the electroosmotic velocity, t_o, and the retention time of a solute which is totally incorporated into the micelles and moves with the velocity of the micelles, t_c. A solute, which, for example, is half incorporated into the micellar phase, will move with the average velocity of the two solutes described above. The capacity factor of a neutral solute in the above system is calculated from the retention times using equation (4.62)

$$k_{(EKC)} = (t_R - t_o)/(1 - t_R/t_c)t_o \qquad (4.62)$$

where $k_{(EKC)}$ is the capacity factor for electrokinetic chromatography (and is not numerically identical to the capacity factor in conventional chromatography) and t_R the retention time of

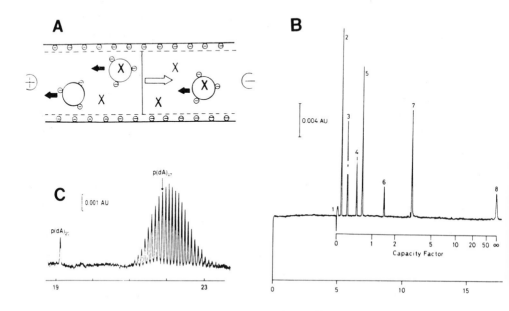

Figure 4.42 Representation of the mechanism of capillary electrokinetic chromatography: large open circle = micelle, x = analyte, black arrow = electrophoretic migration of micelle, and open arrow = electroosmotic flow, (A). Electrokinetic separation of some test solutes: 1 = methanol, 2 = resorcinol, 3 = phenol, 4 = 4-nitroaniline, 5 = nitrobenzene, 6 = toluene, 7 = 2-naphthol, and 8 = Sudan III; capillary 650 x 0.05 mm (migration length 500 mm), voltage ca 15 KV, temperature 35°C, and absorption detection at 210 nm, (B). Capillary gel electrophoretic separation of a poly(deoxyadenylic acid) test mixture with oligomer numbers 40 to 60. Capillary 800 x 0.075 mm (migration length 600 mm) filled with poly(acrylamide) gel and eluted with a 0.1 M Tris-0.25 M borate-7 M urea, pH 7.6 buffer at an applied potential of ca. 25 KV, (C). (A and B are reproduced with permission from ref. 665 and C from ref. 691).

a solute eluting between t_o and t_c. The resolution, R_s, for a neutral pair of solutes is given by

$$R_s = [n^{1/2}/4][(\alpha - 1)/\alpha][k_2/(1 + k_2)][(1 - (t_o/t_c))/(1 + (t_o/t_c)k_1)]$$

$$(4.63)$$

where n is the number of theoretical plates, α the separation factor (k_2/k_1), and k_2 and k_1 the capacity factors defined in equation (4.62) for the earlier and later migrating of the two peaks, respectively. Large values of the capacity factor are unfavorable for obtaining high resolution since the last term in equation (4.63) becomes approximately zero. The optimum value of the capacity factor for maximum resolution depends on the ratio of t_o/t_c and is around 0.8-5 [666] or equivalent to $(t_o/t_c)^{1/2}$ [668]. The magnitude of the capacity factor usually increases with increasing concentration of the micelles and k, α, and t_o/t_c are changed by selecting a different micellar substance for the pseudostationary phase or by using a mixed micellar pseudostationary phase [665,666,672-674]. Most published applications deal with the separation of neutral, water soluble molecules. More hydrophobic solutes bind too strongly to micellar surfactants like sodium dodecyl sulfate to be effectively separated. Bile salts have a smaller solubilizing effect compared to sodium dodecyl sulfate and are more useful for hydrophobic analytes [671,673,675,676]. The fact that the bile salt monomers are chiral allows for enantiomer separations of certain compounds. The addition of cyclodextrins to the micellar solution allows the separation of electrically neutral, hydrophobic compounds and peptides of similar net charge [664,677,678]. For hydrophobic compounds the sample distributes itself between the cyclodextrin and micelles. When the solute is included in the cyclodextrin cavity it migrates with the electroosmotic velocity and when it is incorporated into the micelles it migrates with the micellar velocity. Accordingly, the differential distribution of the sample between the cyclodextrin and the micelles enables a separation to be achieved. Ion-exchange electrokinetic chromatography was developed for the separation of ions having identical electrophoretic mobility [679,680]. A polymer ion having a charge opposite to that of the analyte is employed as a modifier of the electrophoretic mobility. Both the analyte and the polymer ions are subject to electrophoresis, but the migration directions differ. Accordingly, analyte ions bonded to the polymer through ion-pair formation migrate in the opposite direction to the free analyte ions. In this case the separation principle is based on differences in the ion-pair formation constants, and not on differences in electrophoretic mobility. The above few examples serve to illustrate the versatility of electrokinetic chromatography and its applicability to a wide range of general analytical problems.

An example of a modular apparatus for capillary electrophoretic separation methods is shown in Figure 4.43 [637-639,681-684]. It offers a choice of automated sample introduction methods with on-column detection and has a thermostating system for the separation capillary for efficient heat dissipation and temperature control. The electrophoretic module of the instrument is fabricated from non-metallic insulating materials, such as perspex or PTFE for safety reasons. The high voltage power supply, sampling procedures and data acquisition are controlled by computer. The separation capillary can be rinsed and refilled with buffer after each analysis by applying vacuum to the capillary outlet. In operation, two high voltage electrodes are dipped into buffer reservoirs at either end of the capillary. These are moved into position pneumatically. In typical experiments a voltage of 10-50 KV is applied across a capillary column of 0.5-1.5 m in length and 25-100 micrometers internal diameter. The column contains an electrolyte whose concentration is in the range 0.001 to 0.1 M. The sample is introduced as a narrow band at the upstream end of the column and are detected, usually on-column, with a suitable miniaturized detector near the column outlet. Typical electroosmotic flow velocities are on the order of 1 mm/s.

Nearly all columns presently used for capillary electrophoretic separations are made from polyimide-protected fused silica tubes of small internal diameter. For capillary zone electrophoresis and capillary electrokinetic chromatography an immobilized stationary phase is not required, and theoretically at least, no wall coating is necessary. Practically, even the slightest wall adsorption of the sample could dramatically reduce the separation efficiency due to slow sorption kinetics. For proteins and peptides in particular, binding of the charged macromolecules to the capillary surface as a result of electrostatic interactions has often been a problem. One solution involves the deactivation of the fused silica surface by chemical modification, for example, the surface attachment and polymerization over the fused silica surface of a thin film of poly(acrylamide), poly(ethylene glycol), poly(ethyleneimine), etc., [636,639,685-687]. Although these approaches have been

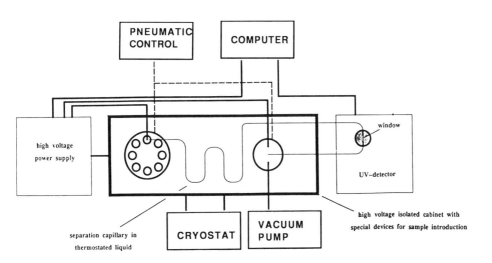

Figure 4.43 Schematic diagram of a modular apparatus for capillary electrophoretic separation methods. (Reproduced with permission from ref. 681. Copyright Friedr. Vieweg & Sohn).

successful in individual cases, problems associated with film stabilities over a wide pH range, column lifetime and column to column reproducibility remain to be solved and standardized coating procedures developed. Other methods of reducing protein-wall interactions include operating at low pH, so that there is little charge on the capillary surface and therefore little protein binding, operating under conditions at which all proteins are negatively charged and repelled from the capillary wall, and adding modifiers to the buffer to block the charged sites on the capillary wall [639,661,688,689]. For capillary gel electrophoresis, poly(acrylamide) gels are prepared in the capillary and in some cases also chemically bonded to the column wall [690-694]. The pore size is controlled by varying the ratio of acrylamide monomer and the crosslinking reagent. Problems that can arise in column fabrication include bubble formation due to the natural shrinkage of the gels that occurs during polymerization and crosslinking of surface-bonded gels and slow extrusion of free gels from the column induced by the electroosmotic flow. Gel extrusion can be minimized by eliminating the electroosmotic flow by covering the column wall with a thin film of poly(acrylamide) prior to preparing the gel. Sodium dodecyl sulfate poly(acrylamide) gel electrophoresis (SDS-PAGE) provides high resolution separations of proteins, as well as molecular weight information for the separated zones. It is also well suited to the separation of high molecular weight oligomeric mixtures, such as poly(deoxynucleotides) as illustrated in Figure 4.42 [691].

Preparative applications of capillary electrophoretic methods are limited by the need to maintain small column diameters for efficient heat dissipation, small injection bands for acceptable efficiency, and low sample concentrations with respect to the buffer concentration, to maintain constant field gradients over the length of the capillary. Using longer columns to increase sample throughput is limited by the need for higher voltages to maintain the optimum sample velocity and leads to longer separation times. Possible methods of increasing sample throughput include the use of rectangular columns with a narrow height for efficient heat dissipation and a wide base to increase the injection volume [695], or by using multiple columns operated in parallel [696]. These methods should allow the recovery of a few hundred nanograms of sample per injection with high column efficiency.

The general acceptance of capillary electrophoretic methods, particularly for quantitative analysis, outside of the research laboratory will require advances in column technology and instrumentation. The former should be easier to achieve by comparison to instrumental problems. Typical columns used in capillary electrophoretic methods often have a total volume less than 1 microliter. Operation at high efficiency requires at a minimum that injection and detection volumes should be no more than a few nanoliters. In principle, injection is easy to achieve using electromigration or hydrodynamic flow [637-639,655, 681,697-700]. For injection by electromigration a low voltage is applied for a brief period of time while one end of the capillary resides in the sample solution. A portion of the sample enters the capillary by the collective contributions from both electrophoretic migration of the analytes and electroosmotic flow of the sample solution. Because of the electrophoretic component of electromigration injection, discrimination occurs for ionic species, with the more mobile components being injected in larger quantities than the less mobile species. The electrical resistance of the sample solution alters both the electrophoretic and electroosmotic flow rates, such that for sample solutions of different conductivity the absolute amount of sample injected will vary. Whenever possible the samples should be prepared in the operating buffer to avoid this problem. Hydrodynamic injection is accomplished by sucking a small volume of sample into the capillary by applying vacuum to the opposite end of the capillary, by overpressure of the sample reservoir, or by hydrostatic pressure, created by raising the sample container and column end to a certain height above the opposite end of the capillary for a period of time. Density differences between the sample solution and the buffer can result in changes in the injection volume as the capillary is inserted and withdrawn from the sample solution. It was noted earlier that the injection zone length is often one of the major contributors to a reduction in column performance in capillary electrophoresis. Typical injection volumes generally exceed those desirable from a column efficiency point of view.

Absorption, fluorescence, electrochemical, conductivity, radioactivity and mass spectrometry have all been applied as detection techniques for capillary electrophoresis with varying degrees of success [636-639]. Only absorption detection is available as standard equipment on commercial instruments at present. The very small absolute sample amounts injected (100 pg to 100 fg) and the small absolute detection volumes required to maintain column efficiency place stringent demands on the detection system. Detection problems are perhaps the weak element in capillary electrophoretic systems at present [655]. To minimize the detection volume in the absorption mode, on-column detection is the most popular method using special slits and optics to reduce stray light and sensitivity to refractive index changes [701-704]. The limited cell path length (equivalent to the column diameter), source instability and optical noise limit sample detection limits to about 20-100 fmol in favorable cases, corresponding to a concentration limit of 10^{-7} to 10^{-5} M. This is barely adequate for many applications, since low mass sensitivity is always required because of the small sample volumes that are injected. Laser-induced fluorescence offers higher sensitivity in the femtomole to attomole range using detectors under development in research laboratories, but is not a universal detection technique [703,704]. Pre- and postcolumn derivatization techniques can be used to extend the range of application of fluorescence detection. Perhaps a more general approach is indirect fluorescence using a buffer with fluorescent properties, such as sodium salicylate [705-707]. The fluorescent salicylate can produces a high level of laser-induced fluorescence as it elutes continuously passed the detection zone. The analyte signal is based upon the charge-induced physical displacement of a coeluting, fluorescent buffer ion by a nonfluorescent, nonabsorbing analyte ion to maintain charge neutrality. A decreased level of fluorescence for the analyte zone is monitored by the detector. Detection limits in the attomole and femtomole range have been obtained for typical analytes. Amperometric detection is not as easily applied to capillary electrophoresis as are spectrophotometric methods [637,638,708]. Typical electric currents under normal operating conditions can be up to six orders of magnitude greater than electrochemical currents obtained at a suitable amperometric detector and tend to overwhelm the detector signal. Using a single carbon fiber as a detector located very close to the end of the capillary, but electrically insulated from it, it has proven possible to detect easily oxidized species in ultrasmall samples from biological environments [708]. Electrospray ionization and co-axial flow fast atom bombardment mass spectrometry have been demonstrated recently to provide useful mass spectra when interfaced with capillary electrophoretic methods [647,709,710]. Typical capillary flow rates of less than a microliter/min are compatible with the mass spectrometer vacuum systems, but the small absolute sample masses and their often high molecular weights present sensitivity problems.

4.9 REFERENCES

1. Cs. Horvath, B. Preiss, and S. Lipsky, Anal. Chem., 39 (1967) 1422.
2. J. Kirkland, J. Chromatogr. Sci., 7 (1969) 361.
3. R. Majors, J. Chromatogr. Sci., 11 (1973) 88.
4. R. E. Majors, J. Assoc. Off. Anal. Chem., 60 (1977) 186.
5. H. Giesche, K, K, Unger, U. Esser, B. Eray, U. Trudinger, and J. N. Kinkel, J. Chromatogr., 465 (1989) 39.

6. L. F. Colwell and R. A. Hartwick, J. Liq. Chromatogr., 10 (1987) 2721.
7. K. K. Unger and B. Anspach, Trends Anal. Chem., 6 (1987) 121.
8. P. R. Brown and R. A. Hartwick (Eds.), "High Performance Liquid Chromatography", Wiley, New York, NY, 1989.
9. K. Unger (Ed.), "Packings and Stationary Phases in Chromatographic Techniques", Dekker, New York, NY, 1989.
10. D. Berek and I. Novak, Chromatographia, 30 (1990) 582.
11. J. H. Knox, B. Kaur, and G. R. Millward, J. Chromatogr., 352 (1986) 3.
12. K. Makino and B.-H. Kim, J. Chromatogr. Sci., 27 (1989) 659.
13. K. K. Unger, R. Janzen, and G. Jilge, Chromatographia, 24 (1987) 144.
14. D. Josic, W. Hofmann, and W. Reutter, J. Chromatogr., 371 (1986) 43.
15. K. M. Gooding and F. E. Regnier, "HPLC of Biological Macromolecules. Methods and Applications", Dekker, New York, NY, 1990.
16. W. S. Hancock (Ed.), "High Performance Liquid Chromatography in Biotechnology", Wiley, New York, NY, 1990.
17. M. Kawahara, H. Nakamura, and T. Nakajimu, J. Chromatogr., 515 (1990) 149.
18. T. Kawasaki, M. Niikura, and Y. Kobayashi, J. Chromatogr., 515 (1990) 91 and 125.
19. R. K. Iler, "The Chemistry of Silica", Wiley, New York, NY, 1979.
20. K. K. Unger, "Porous Silica", Elsevier, Amsterdam, 1979.
21. R. P. W. Scott, Adv. Chromatogr., 20 (1982) 167.
22. C. Horvath (Ed.), "High-Performance Liquid Chromatogrphy, Advances Perspectives", Academic Press, New York, NY, Vol. 2, 1980.
23. C. J. Brinker and G. W. Scherer, "Sol-Gel Science. The Physics and Chemistry of Sol-Gel Processing", Academic Press, San Diego, CA, 1990.
24. K. K. Unger, J. N. Kinkel, B. Anspach, and H. Giesche, J. Chromatogr., 296 (1984) 3.
25. A. Wojcik and L. Kwietniewski, J. Chromatogr., 435 (1988) 55.
26. N. D. Danielson and J. J. Kirkland, Anal. Chem., 59 (1987) 2501.
27. R. W. Stout, G. B. Cox, and T. J. Odiorne, Chromatographia, 24 (1987) 602.
28. J. Nawrocki and B. Buszewski, J. Chromatogr., 449 (1988) 1.
29. M. Verzele, C. Dewaele and D. Duquet, J. Chromatogr., 329 (1985) 351.
30. C. Dewaele and M. Verzele, J. Chromatogr., 282 (1983) 341.
31. M. Verzele, J. Lammens, and M. Van Roelenbosch, J. Chromatogr., 178 (1979) 463.
32. D. A. Hanggi and P. W. Carr, J. Liq. Chromatogr., 7 (1984) 2323.
33. M. Verzele, J. Chromatogr., 295 (1984) 81.
34. H. Engelhardt and H. Muller, J. Chromatogr., 219 (1981) 395.
35. F. V. Warren and B. A. Bidlingmeyer, Anal. Chem., 56 (1984) 950.
36. L. C. Sander and S. A. Wise, J. Chromatogr., 316 (1984) 163.
37. J. Kohler, D. B. Chase, R. D. Farlee, A. J. Vega, and J. J. Kirkland, J. Chromatogr., 352 (1986) 275.
38. J. Kohler and J. J. Kirkland, J. Chromatogr., 385 (1987) 125.
39. M. Mauss and H. Engelhardt, J. Chromatogr., 371 (1986) 235.
40. P. C. Sadek. C. J. Koester, and L. D. Bowers, J. Chromatogr. Sci., 25 (1987) 489.

41. J. Nawrocki, D. L. Moir, and W. Szczepaniak, Chromatographia, 28 (1989) 143.
42. S. P. Boudreau and W. T. Cooper, Anal. Chem., 61 (1989) 41.
43. S. H. Hansen, P. Helboe, and M. Thomsen, J. Chromatogr., 368 (1986) 39.
44. P. Helboe, S. H. Hansen, and M. Thomsen, Adv. Chromatogr., 28 (1989) 196.
45. R. Schwarzenbach, J. Chromatogr., 334 (1985) 35.
46. B. Law and P. C. Chan, J. Chromatogr., 467 (1989) 267.
47. R. M. Smith and J. O. Rabuor, J. Chromatogr., 464 (1989) 117.
48. T. Cserhati, Chromatographia, 29 (1990) 593.
49. C. Laurent, H. A. H. Billiet, and L. de Galan, Chromatographia, 17 (1983) 253.
50. C. F. Simpson (Ed.), "Techniques in Liquid Chromatography", Wiley, New York, NY, 1982.
51. J. G. Dorsey and K. A. Dill, Chem. Rev., 89 (1989) 331.
52. L. C. Sander and S. A. Wise, CRC Crit. Revs. Anal. Chem., 18 (1987) 299.
53. H. Engelhart, H. Low, W. Eberhardt, and M. Mauss, Chromatographia, 27 (1989) 535.
54. H. Figge, A. Deege, J. Kohler, and G. Schomburg, J. Chromatogr., 351 (1986) 393.
55. M. Hanso, K. K. Unger, and G. Schomburg, J. Chromatogr., 517 (1990) 269.
56. A. Kurganov, O. Kuzmenko, V. A. Davankov, B. Eray, K. K. Unger, and U. Trudinger, J. Chromatogr., 506 (1990) 391.
57. J. H. Knox and A. Pryde, J. Chromatogr., 112 (1975) 171.
58. J. J. Pesek and H.-D. Lin, Chromatographia, 28 (1989) 565.
59. M. P. Rigney, T. P. Weber, and P. W. Carr, J. Chromatogr., 484 (1989) 273.
60. R. M. Chicz, Z. Shi, and F. E. Regnier, J. Chromatogr., 359 (1986) 121.
61. R. W. Stout and J. J. Destefano, J. Chromatogr., 326 (1985) 63.
62. R. W. Stout, S. I. Sivakoff, R. D. Ricker, H.-C. Palmer, M. A. Jackson, and T. J. Odiorne, J. Chromatogr., 352 (1986) 381.
63. U. Bien-Vogelsang, A. Deege, H. Figge, J. Kohler, and G. Schomburg, Chromatographia, 19 (1985) 170.
64. V. Rehak and E. Smolkova, Chromatographia, 9 (1976) 219.
65. E. Grushka (Ed.), "Bonded Stationary Phases in Chromatography", Ann Arbor Science, Ann Arbor, 1974.
66. H. Engelhart and P. Orth, J. Liq. Chromatogr., 10 (1987) 1999.
67. L. C. Sander and S. A. Wise, Adv. Chromatogr., 25 (1986) 139.
68. T. L. Ascah and B. Feibush, J. Chromatogr., 506 (1990) 357.
69. L. C. Sander and S. A. Wise, Anal. Chem., 56 (1984) 504.
70. B. Buszewski, Chromatographia, 28 (1989) 574.
71. N. H. C. Cooke and K. Olsen, J. Chromatogr. Sci., 18 (1980) 512.
72. K. Jones, J. Chromatogr., 392 (1987) 1 and 11.
73. G. Schomburg, A. Deege, J. Kohler, and U. Bien-Vogelsang, J. Chromatogr., 282 (1983) 27.
74. J. N. Kinkel and K. K. Unger, J. Chromatogr., 316 (1984) 193.
75. K. D. Lork, K. K. Unger, and J. N. Kinkel, J. Chromatogr., 352 (1986) 199.
76. D. E. Layden (Ed.), "Silanes, Surfaces, and Interfaces", Gordon and Breach Science Publishers, New York, NY, 1986.
77. W. W. Yau, J. J. Kirkland, and D. D. Bly, "Modern Size-Exclusion Liquid Chromatography", Wiley, New York, NY, 1979.
78. S. H. Chang, K. M. Gooding, and F. E. Regnier, J. Chromatogr., 120 (1976) 321.

79. W. Muller, J. Chromatogr., 510 (1990) 133.
80. J. J. Kirkland, J. L. Glach, and R. D. Farlee, Anal., Chem., 61 (1989) 2.
81. M. J. J. Hetem, J. W. de Haan, H. A. Claessens, L. J. M. van de Ven, and C. A. Cramers, Anal. Chem., 62 (1990) 2288 and 2296.
82. L. W. Yu and R. A. Hartwick, J. Chromatogr. Sci., 27 (1989) 176.
83. A. Nomura, J. Yamada, and K.-I. Tsunoda, Anal. Chem., 60 (1988) 2509.
84. D. Farin and D. Avnir, J. Chromatogr., 406 (1987) 317.
85. G. Foti, C. Martinez, and E. Sz. Kovats, J. Chromatogr., 461 (1989) 243.
86. G. E. Berendsen, K. A. Pikaart, and L. de Galan, J. Liq. Chromatogr., 3 (1980) 1437.
87. A. Yu Fadeev and S. M. Starover, J. Chromatogr., 447 (1988) 103.
88. B. Buszewski, D. Berek, J. Garaj, I. Novak, and Z. Suprynowicz, J. Chromatogr., 446 (1988) 191.
89. B. W. Sands, Y. S. Kim, and J. L. Bass, J. Chromatogr., 360 (1986) 353.
90. B. I. Diamondstone, S. A. Wise, and L. C. Sander, J. Chromatogr., 321 (1985) 319.
91. S. C. Antaki and J. Serpinet, Chromatographia, 23 (1987) 767.
92. J. Goworek, F. Nooitgedacht, M. Rijkhof, and H. Poppe, J. Chromatogr., 352 (1986) 399.
93. C. Lullmann, H.-G. Genieser, and B. Jastorff, J. Chromatogr., 323 (1985) 273.
94. S. D. Fazio, S. A. Tomellini, H. Shih-Hsien, J. B. Crowther, T. V. Raglione, T. R. Floyd, and R. A. Hartwick, Anal. Chem., 57 (1985) 1559.
95. T. R. Floyd, N. Sagliano, and R. A. Hartwick, J. Chromatogr., 452 (1988) 43.
96. R. K. Gilpin, Anal. Chem., 57 (1985) 1465A.
97. R. K. Gilpin, J. Chromatogr. Sci., 22 (1984) 371.
98. E. Bayer, K. Albert, J. Reiners, M. Nieder, and D. Muller, J. Chromatogr., 264 (1983) 197.
99. B. Pfleiderer, K. Albert, and E. Bayer, J. Chromatogr., 506 (1990) 343.
100. H. A. Claessens, J. W. De Haan, L. J. M. van de Ven, P. C. de Bruyn, and C. A. Cramers, J. Chromatogr., 436 (1988) 345.
101. K. Jinno, J. Chromatogr. Sci., 27 (1989) 729.
102. P. Kolla, J. Kohler, and G. Schomburg, Chromatographia, 23 (1987) 465.
103. N. Tanaka and M. Araki, Adv. Chromatogr., 30 (1989) 81.
104. J. R. Benson and D. J. Woo, J. Chromatogr. Sci., 22 (1984) 386.
105. H. Hatano, J. Chromatogr., 332 (1985) 227.
106. D. T. Gjerde and J. S. Fritz, "Ion Chromatography, 2nd. Edn., Huethig, Heidelberg, 1987.
107. D. P. Lee, J. Chromatogr., 443 (1988) 143.
108. K. A. Tweeten and T. N. Tweeten, J. Chromatogr., 359 (1986) 111.
109. L. D. Bowers and S. Pedigo, J. Chromatogr., 371 (1986) 243.
110. F. Nevejans and M. Verzele, J. Chromatogr., 350 (1985) 145.
111. H. W. Stuurman, J. Kohler, S. O. Jansson, and A. Litzen, Chromatographia, 23 (1987) 341.
112. R. M. Smith and D. R. Garside, J. Chromatogr., 407 (1987) 19.
113. N. Tanaka, K. Hashizume, M. Araki, H. Tsuchiya, A. Okume, K. Iwaguchi, S. Ohnishi, and N. Takai, J. Chromatogr., 448 (1988) 95.

528

114. F. Nevejans and M. Verzele, J. Chromatogr., 406 (1987) 325.
115. F. Nevejans and M. Verzele, Chromatographia, 20 (1985) 173.
116. J. V. Dawkins, L. L. Lloyd, and F. P. Warner, J. Chromatogr., 352 (1986) 157.
117. H. Wada, J. Chromatogr., 322 (1985) 255.
118. N. Tanaka, K. Hashizume, and M. Araki, J. Chromatogr., 400 (1987) 33.
119. S. Yamamoto, K. Nakanishi, and R. Matsuno, "Ion-Exchange Chromatogrpahy of Proteins", Dekker, New York, NY, 1988.
120. R. E. Barron and J. S. Fritz, J. Chromatogr., 284 (1984) 13.
121. P. R. Haddad and P. E. Jackson, "Ion Chromatography. Principles and Applications", Elsevier, Amsterdam, 1990.
122. D. J. Burke, J. K. Duncan, L. C. Dunn, L. Cummings, C. J. Siebert, and G. S. Ott, J. Chromatogr., 353 (1986) 425.
123. O. A. Shpigun and Yu. Zolotov, "Ion Chromatography in Water Analysis", Ellis Horwood, Chichester, Great Britain, 1988.
124. H. Small, "Ion Chromatography", Plenum Press, London, 1989.
125. T. S. Stevens and M. A. Langhorst, Anal. Chem., 54 (1982) 950.
126. R. W. Stingsby and C. A. Pohl, J. Chromatogr., 458 (1988) 241.
127. J. R. Stillian and C. A. Pohl, J. Chromatogr., 499 (1990) 249.
128. L. M. Warth, J. S. Fritz, and J. O. Naples, J. Chromatogr., 462 (1989) 165.
129. B. J. Bassler and R. A. Hartwick, J. Chromatogr. Sci., 27 (1989) 162.
130. F. J. Yang (Ed.), Microbore Column Chromatography. A Unified Approach to Chromatography", Dekker, New York, NY, 1989.
131. M. Verzele, C. Dewaele, and D. Duquet, J. Chromatogr., 391 (1987) 111.
132. M. Martin and G. Guiochon, Chromatographia, 10 (1977) 194.
133. R. A. Mowery, J. Chromatogr. Sci., 23 (1985) 22.
134. P. R. Haddad and R. C. L. Foley, J. Chromatogr., 407 (1987) 133.
135. E. Von Arx, J. Chromatogr., 209 (1981) 310.
136. Y. Hirata and K. Jinno, J. High Resolut. Chromatogr., 6 (1983) 196.
137. K. Jinno, Chromatographia, 25 (1988) 1004.
138. R. T. Kennedy and J. W. Jorgenson, Anal. Chem., 61 (1989) 1128.
139. G. Vigh, E. Gemes, and I. Inczedy, J. Chromatogr., 147 (1978) 59.
140. R. Gill, J. Chromatogr., 354 (1986) 169.
141. D. Ishii, K. Watanabe, H. A. Sal, Y. Hashimoto, and T. Takeuchi, J. Chromatogr., 332 (1985) 3.
142. R. P. W. Scott (Ed.), "Small Bore Liquid Chromatography Columns. Their Properties and Uses", Wiley, New York, NY, 1984.
143. M. Konishi, Y. Mori, and T. Amano, Anal. Chem., 57 (1985) 2235.
144. J. N. Little, R. L. Cotter, J. A. Pendergast, and P. D. McDonald, J. Chromatogr., 126 (1976) 439.
145. J. S. Landy, J. L. Ward, and J. G. Dorsey, J. Chromatogr. Sci., 21 (1983) 49.
146. P. A. Hyldburg, C. M. Sparacion, J. W. Hines, and C. D. Keller, J. Liq. Chromatogr., 10 (1987) 2639.
147. P. Welling, H. Poppe, and J. C. Kraak, J. Chromatogr., 321 (1985) 450.
148. T. Takeuchi and D. Ishii, J. Chromatogr., 190 (1980) 150.

149. L. R. Snyder and J. J. Kirkland, "Introduction to Modern Liquid Chromatography", 2nd Edn., Wiley, New York, NY, 1979.
150. R. D. Davies, J. High Resolut. Chromatogr., 4 (1981) 270.
151. J. H. Knox, J. Chromatogr. Sci., 15 (1977) 353.
152. D. C. Shelly and T. J. Edkins, J. Chromatogr., 411 (1987) 185.
153. D. C. Shelly, V. L. Antonucci, T. J. Edkins, T. J. Dalton, J. Chromatogr., 458 (1989) 267.
154. S. A. Karapetyan, L. M. Yakushina, G. G. Vasijarov, and V. V. Brazhnikov, J. High Resolut. Chromatogr., 8 (1985) 148.
155. J. A. Anderson, J. Chromatogr. Sci., 22 (1984) 332.
156. Sh. A. Karapetyan, L. M. Yakushina, G. G. Vasijarov, and V. V. Brazhnikov, J. High Resolut. Chromatogr., 6 (1983) 440.
157. J. Bowermaster and H. McNair, J. Chromatogr., 279 (1983) 431.
158. R. F. Meyer and R. A. Hartwick, Anal. Chem., 56 (1984) 2211.
159. I. Halasz and G. Maldener, Anal. Chem., 55 (1983) 1842.
160. R. E. Majors, Anal. Chem., 44 (1972) 1722.
161. H. R. Linder, H. P. Keller, and R. W. Frei, J. Chromatogr. Sci., 14 (1976) 234.
162. H. Menet, P. Gareil, M. Caude, and R. Rosset, Chromatographia, 18 (1984) 73.
163. R. P. W. Scott and P. Kucera, J. Chromatogr., 169 (1979) 51.
164. Y. Yamauchi and J. Kumanotani, J. Chromatogr., 210 (1981) 512.
165. J. Asshauer and I. Halasz, J. Chromatogr. Sci., 12 (1974) 139.
166. P. H. Shetty, P. J. Youngberg, B. R. Kersten, and C. F. Poole, J. Chromatogr., 411 (1987) 61.
167. N. K. Vadukul and C. R. Loscombe, J. High Resolut. Chromatogr., 6 (1983) 488.
168. T. J. N. Webber and E. H. McKerrel, J. Chromatogr., 122 (1976) 243.
169. P. A. Bristow, P. N. Brittain, C. M. Riley, and B. F. Williamson, J. Chromatogr., 131 (1977) 57.
170. M. Broquaire, J. Chromatogr., 170 (1979) 43.
171. M. De Weerdt, C. Dewaele, and M. Verzele, J. High Resolut. Chromatogr., 10 (1987) 553.
172. K.-K. Karlsson and M. Novotny, Anal. Chem., 60 (1988) 1662.
173. M. Verzele, C. Dewaele, M. De, Weerdt, and S. Abbot, J. High Resolut. Chromatogr., 12 (1989) 164.
174. D. C. Shelly, J. C. Gluckman, and M. V. Novotny, Anal. Chem., 56 (1984) 2990.
175. H. J. Cortes, C. D. Pfeiffer, B. E. Richter, and T. S. Stevens, J. High Resolut. Chromatogr., 10 (1987) 446.
176. J. C. Gluckman, A. Hirose, V. L. McGuffin, and M. Novotny, Chromatographia, 17 (1983) 303.
177. C. Borra, S. N. Han, and M. Novotny, J. Chromatogr., 385 (1987) 75.
178. F. Andreotini, C. Borra, and M. Novotny, Anal. Chem., 59 (1987) 2428.
179. D. Ishii (Ed.), "Introduction to Microscale High-Performance Liquid Chromatography", VCH Publishers, New York, NY, 1988.
180. R. E. Pauls and R. W. McCoy, J. Chromatogr. Sci., 24 (1986) 66.
181. H. P. Keller, F. Erni, H. R. Lindner, and R. W. Frei, Anal. Chem., 49 (1977) 1958.
182. K. Kawata, M. Uebori, and Y. Yamazaki, J. Chromatogr., 211 (1981) 378.
183. D. Ishii and T. Takeuchi, J. Chromatogr. Sci., 22 (1984) 400.
184. T. Tsuda and M. Novotny, Anal. Chem., 50 (1978) 271.
185. V. L. McGuffin and M. Novotny, J. Chromatogr., 255 (1983) 381.

186. T. Tseuda, I. Tanaka, and G. Nakagawa, Anal. Chem., 56 (1984) 1249.
187. S. Folestad, B. Josefsson, and M. Larsson, J. Chromatogr., 391 (1987) 347.
188. P. R. Dluzneski and J. W. Jorgenson, J. High Resolut. Chromatogr., 11 (1988) 332.
189. D. Ishii and T. Takeuchi, Adv. Chromatogr., 21 (1983) 131.
190. T. Tsuda, T. Tsuboi, and G. Nakagawa, J. Chromatogr., 214 (1981) 283.
191. R. Tijssen, J. P. A. Bleumer, A. L. C. Smit, and M. E. van Kreveld, J. Chromatogr., 218 (1981) 135.
192. J. D. Vargo, J. Chromatogr., 385 (1987) 33.
193. J. W. Jorgenson and E. J. Guthrie, J. Chromatogr., 255 (1983) 335.
194. J. G. Atwood, G. L. Schmidt, and W. Slavin, J. Chromatogr., 171 (1989) 109.
195. F. M. Rabel, J. Chromatogr. Sci., 18 (1980) 394.
196. G. D. Reed and C. R. Loscombe, Chromatographia, 15 (1982) 15.
197. E. Lundanes, J. Dohl, and T. Greibrokk, J. Chromatogr. Sci., 21 (1983) 235.
198. M. J. Walters, J. Assoc. Off. Anal. Chem., 70 (1987) 465.
199. L. C. Sander, J. Chromatogr. Sci., 26 (1988) 380.
200. K. Kimata, K. Iwaguchi, S. Onishi, K. Jinno, R. Eksteen, K. Hosoya, M. Araki, and N. Tanaka, J. Chromatogr. Sci., 27 (1989) 721.
201. P. A. Bristow and J. H. Knox, Chromatographia, 10 (1977) 279.
202. C. F. Poole and S. K. Poole, Anal. Chim. Acta, 216 (1989) 109.
203. A. Berthod, J. Liq. Chromatogr., 12 (1989) 1169 and 1187.
204. B. A. Bidlingmeyer and F. V. Warren, Anal. Chem., 56 (1984) 1583A.
205. H. Engelhart and M. Jungheim, Chromatographia, 29 (1990) 59.
206. A. P. Goldberg, Anal. Chem., 54 (1982) 342.
207. C. T. Nant and S. Hodges, Chromatographia, 24 (1987) 805.
208. B. Walczak, L. Morin-Allory, M. Lafosse, M. Dreux, and J. R. Chretien, J. Chromatogr., 395 (1987) 183.
209. D. E. Berendsen, P. Schoenmakers, and L. de Galan, J. Liq. Chromatogr., 3 (1980) 1669.
210. L. C. Sander and S. A. Wise, J. High Resolut. Chromatogr., 11 (1988) 383.
211. R. P. W. Scott and P. Kucera, J. Chromatogr., 142 (1977) 213.
212. L. Nondek, B. Buszewski, and D. Berek, J. Chromatogr., 360 (1986) 241.
213. H. Englehart, D. Dreyer, and H. Schmidt, Chromatographia, 16 (1982) 11.
214. M. Verzele and C. Dewaele, Chromatographia, 18 (1984) 84.
215. P. C. Sadek and P. W. Carr, J. Chromatogr. Sci., 21 (1983) 314.
216. S. G. Weber and W. G. Tramposch, Anal. Chem., 55 (1983) 1771.
217. C. B. Cox and R. W. Stout, J. Chromatogr., 384 (1987) 315.
218. R. Ohmacht and I. Halasz, Chromatographia, 14 (1981) 155 and 216.
219. A. Alhedai, D. E. Martire, and R. P. W. Scott, Analyst, 114 (1989) 869.
220. G. E. Berendsen, P. J. Schoenmakers, L. de. Galan, G. Vigh, Z. Varga-Puchoy, and J. Inczedy, J. Liq. Chromatogr., 3 (1980) 1669.
221. R. J. Smith, C. S. Nieass, and M. S. Wainwright, J. Liq. Chromatogr., 9 (1986) 1387.
222. J. H. Knox and R. Kaliszan, J. Chromatogr., 349 (1985) 211.
223. M. J. Wells and C. R. Clark, Anal. Chem., 53 (1981) 1341.

224. K. Jinno, N. Ozaki, and T. Sato, Chromatographia, 17 (1983) 341.
225. R. A. Djerki and R. J. Laub, J. Liq. Chromatogr., 10 (1987) 1749.
226. M. C. Hennion and R. Rosset, Chromatographia, 25 (1988) 43.
227. M. Shibukawa and N. Ohta, Chromatographia, 25 (1988) 288.
228. J. H. Knox, R. Kaliszan and G. Kennedy, J. Chem. Soc., Faraday Symposium, 15 (1980) 113.
229. H. A. H. Billiet, J. P. J. van Dalen, P. J. Schoenmakers, and L. de Galan, Anal. Chem., 55 (1983) 847.
230. L. R. Snyder, "Principles of Adsorption Chromatography", Dekker, New York, NY, 1968.
231. L. R. Snyder and H. Poppe, J. Chromatogr., 184 (1980) 363.
232. Cs. Horvath (Ed.), "High Performance Liquid Chromatography. Advances and Perspective", Academic Press, New York, NY, Vol. 3, 1983.
233. L. R. Snyder, J. Chromatogr., 255 (1983) 3.
234. R. P. W. Scott and P. Kucera, J. Chromatogr., 171 (1979) 37.
235. L. R. Snyder, J. L. Glajch, and J. J. Kirkland, 218 (1981) 299.
236. L. R. Snyder and J. L. Glajch, J. Chromatogr., 248 (1982) 165.
237. L. R. Snyder and T. C. Schunk, Anal. Chem., 54 (1982) 1764.
238. L. R. Snyder, J. Chromatogr., 255 (1983) 3.
239. L. R. Snyder, M. D. Palamareva, B. J. Kurtev, L. Z. Viteva, and J. N. Stefanovsky, J. Chromatogr., 354 (1986) 107.
240. W. T. Cooper and P. L. Smith. J. Chromatogr., 355 (1986) 57.
241. P. L. Smith and W. T. Cooper, J. Chromatogr., 410 (1987) 249.
242. L. D. Olsen and R. J. Hurtubise, J. Chromatogr., 474 (1989) 347.
243. L. R. Snyder and J. Glajch, J. Chromatogr., 214 (1981) 1 and 21.
244. M. D. Palamareva and H. E. Palamarev, J. Chromatogr., 477 (1989) 235.
245. E. Soczewinski, J. Chromatogr., 388 (1987) 91 and 99.
246. B. R. Suffolk and R. K. Gilpin, J. Chromatogr. Sci., 24 (1986) 423.
247. L. D. Olsen and R. J. Hurtubise, J. Chromatogr., 479 (1989) 5.
248. M. Verzele, F. Van Damme, C. Dewaele, and M. Ghijs, Chromatographia, 24 (1987) 302.
249. S. M. Petrovic, S. M. Lomic, and I. Sefer, Chromatogaphia, 23 (1987) 915.
250. P. D. Rice and B. R. Bobbitt, J. Chromatogr., 437 (1988) 3.
251. J. H. Park and P. W. Carr, J. Chromatogr., 465 (1989) 123.
252. T. M. Krygowski, J. P. Radomski, A. Rzeszowiak, P. K. Wrona, and C. Reichardt, Tetrahedron, 37 (1981) 119.
253. D. L. Saunders, J. Chromatogr. Sci., 15 (1977) 372.
254. S. R. Abbott, J. Chromatogr. Sci., 18 (1980) 540.
255. H. Engelhardt, J. Chromatogr. Sci., 15 (1977) 380.
256. J.-P. Thomas, A. P. Brun, and J. P. Bounine, J. Chromatogr., 172 (1979) 107.
257. E. Van den Eeckhout, P. Jenjintanant, G. Bens, and P. De Moerloose, Chromatographia, 23 (1987) 893.
258. H. Engelhardt and W. Bohme, J. Chromatogr., 133 (1977) 67.
259. R. A. Bredeweg, L. D. Rothman, and C. D. Pfeiffer, Anal. Chem., 51 (1979) 2061.
260. E. L. Weiser, A. W. Salotto, S. M. Flach, and L. R. Snyder, J. Chromatogr., 303 (1984) 1.
261. M. De Smet and D. L. Massart, Trends Anal. Chem., 8 (1989) 96.
262. S. A. Wise, S. N. Chesler, H. S. Hertz, L. P. Hilpert, and W. E. May, Anal. Chem., 49 (1977) 2306.

532

263. W. A. Dark. J. Liq. Chromatogr., 9 (1982) 1645.
264. D. Karlesky, D. C. Shelley, and I. Warner, Anal. Chem., 53 (1981) 2146.
265. A. M. Siouffi, M. Righezza, and G. Guiochon, J. Chromatogr., 368 (1986) 189.
266. A. W. Salotto, E. L. Weiser, K. P. Caffey, R. L. Carty, S. C. Racine, and L. R. Snyder, J. Chromatogr., 498 (1990) 55.
267. B. A. Bidlingmeyer, J. K. Del Rios, and J. Korpi, Anal. Chem., 54 (1982) 442.
268. G. L. Schmitt and D. J. Pietrzyk, Anal. Chem., 57 (1985) 2247.
269. H. Lingeman, H. A. Van Munster, J. H. Beynen, W. J. M. Underberg, and A. Hulshoff, J. Chromatogr., 352 (1986) 261.
270. H. Lingeman and W. J. M. Underberg, Trends Anal. Chem., 7 (1988) 346.
271. B. Law, Trends Anal. Chem., 9 (1990) 31.
272. M. T. Kelly, M. R. Smyth, and D. Dadgar, J. Chromatogr., 473 (1989) 53.
273. D. M. Brown and D. J. Pietrzyk, J. Chromatogr., 466 (1989) 291.
274. R. J. Flanagan and I. Jane, J. Chromatogr., 323 (1985) 173.
275. J. B. Green, J. Chromatogr., 358 (1986) 53.
276. D. J. Mazzo and P. A. Snyder, J. Chromatogr., 438 (1988) 85.
277. S. H. Hansen, P. Helboe, and M. Thomsen, J. Chromatogr., 409 (1987) 71.
278. S. H. Hansen, P. Helboe, and M. Thomsen, J. Chromatogr., 360 (1986) 53.
279. S. H. Hansen, P. Helboe, and M. Thomsen, J. Pharm. Biomed. Anal., 2 (1984) 165.
280. J. F. K. Huber, M. Pawlowska, and P. Markl, Chromatographia, 17 (1983) 653.
281. J. F. K. Huber, M. Pawlowska, and P. Markl, Chromatographia, 19 (1985) 19.
282. J. P. Crombeen, S. Heemstra, and J. C. Kraak, Chromatographia, 19 (1985) 219.
283. J. P. Crombeen, S. Heemstra, and J. C. Kraak, J. Chromatogr., 286 (1984) 119.
284. P. H. Shetty, S. K. Poole, and C. F. Poole, Anal. Chim. Acta, 236 (1990) 51.
285. J. F. K. Huber, M. Pawlowska, and P. Markl, Chromatogr., 500 (1990) 257.
286. H. W. Stuurman, C. Pettersson, and E. Heldin, Chromatographia, 25 (1988) 685.
287. J. P. Crombeen, H. Poppe, and J. C. Kraak, Chromatographia, 22 (1986) 319.
288. O. Mikes, "HPLC of Biopolymers and Biooligomers. PART A: Principles, Materials and Techniques", Elsevier, Amsterdam, 1988.
289. A. M. Krstulovic and P. R. Brown, "Reversed-Phase High Performance Liquid Chromatograhy: Theory, Practice and Biomedical Applications", Wiley, New York, NY, 1982.
290. B. L. Karger and R. W. Giese, Anal. Chem., 50 (1978) 1048A.
291. P. Oroszlan, R. Blanco, X.-M. Lu, D. Yarmush, and B. L. Karger, J. Chromatogr., 500 (1990) 481.
292. G. Thevenon and F. E. Regnier, J. Chromatogr., 476 (1989) 499.
293. J. C. Gesquiere, E. Diesis, M. J. Cung, and A. Tartar, J. Chromatogr., 478 (1989) 121.
294. B. L. Karger and R. Blanco, Talanta, 36 (1989) 243.
295. R. M. Smith, Adv. Chromatogr., 26 (1987) 277.

296. A. Tchapla, H. Colin, and G. Guiochon, Anal. Chem., 56 (1984) 621.
297. L. R. Snyder, J. W. Dolan, and J. R. Gant, J. Chromatogr., 165 (1979) 3.
298. T. L. Hafkenscheid and E. Tomlinson, Adv. Chromatogr., 25 (1986) 1.
299. T. Braumann, J. Chromatogr., 373 (1986) 191.
300. D. Reymand, G. N. Chung, J. M. Mayer, and B. Testa, J. Chromatogr., 391 (1987) 97.
301. P. Jandera and J. Kubat, J. Chromatogr., 500 (1990) 281.
302. A. Kaibara, C. Hohda, N. Hirata, M. Hirose, and T. Nakagawa, Chromatographia, 29 (1990) 275.
303. R. Kaliszan, "Quantitative Structure-Chromatographic Retention Relationships", Wiley, New York, NY, 1987.
304. D. J. Minick, B. A. Brent, and J. Frenz, J. Chromatogr., 461 (1989) 177.
305. J. G. Dorsey and B. P. Johnson, J. Liq. Chromatogr., 10 (1987) 2695.
306. B. P. Johnson, M. G. Khaledi, and J. G. Dorsey, Anal. Chem., 58 (1986) 2354.
307. P. C. Sadek, P. W. Carr, R. M. Doherty, M. J. Kamlet, R. W. Taft, and M. H. Abraham, Anal. Chem., 57 (1985) 2971.
308. J. H. Park, P. W. Carr, M. H. Abraham, R. W. Taft, R. M. Doherty, and M. J. Kamlet, Chromatographia, 25 (1988) 373.
309. J. H. Park, M. D. Jang, D. S. Kim, and P. W. Carr, J. Chromatogr., 513 (1990) 107.
310. Cs. Horvath, W. R. Melander, and I. Molnar, J. Chromatogr., 125 (1976) 129.
311. Cs. Horvath and W. R. Melander, J. Chromatogr. Sci., 15 (1977) 393.
312. W. R. Melander and Cs. Horvath, Chromatographia, 18 (1984) 353.
313. H. J. Mockel, G. Welter, and H. Melzer, J. Chromatogr., 388 (1987) 255.
314. D. E. Martire and R. E. Boehm, J. Phys. Chem., 87 (1983) 1045.
315. K. A. Dill, J. Phys. Chem., 91 (1987) 1980.
316. P. T. Ying, J. G. Dorsey, and K. A. Dill, Anal. Chem., 61 (1989) 2540.
317. K. D. Sentell and J. G. Dorsey, J. Chromatogr., 461 (1989) 193.
318. W. J. Cheong and P. W. Carr, J. Chromatogr., 499 (1990) 373.
319. S. Hjerten, Adv. Chromatogr., 19 (1981) 111.
320. Y. Kato, Adv. Chromatogr., 26 (1987) 97.
321. N. T. Miller and B. L. Karger, J. Chromatogr., 326 (1985) 45.
322. K. Benedek, J. Chromatogr., 458 (1988) 93.
323. M. L. Heinitz, L. Kennedy, W. Kopaciewicz, and F. E. Regnier, J. Chromatogr., 443 (1988) 173.
324. A. J. Alpert, J. Chromatogr., 359 (1986) 85.
325. M. N. Schmuck, M. P. Nowlan, and K. M. Gooding, J. Chromatogr., 371 (1986) 55.
326. T. Ueda, Y. Yasui, and Y. Ishida, Chromatographia, 24 (1987) 427.
327. J. L. Fausnaugh, L. A. Kennedy, and F. E. Regnier, J. Chromatogr., 317 (1984) 141.
328. Z. El Rassi, L. F. De Ocampo, and M. D. Bacolod, J. Chromatogr., 499 (1990) 141.
329. Y. Kato, T. Kitamura, and T. Hashimoto, J. Chromatogr., 333 (1985) 202.
330. S. L. Wu, K. Benedek, and B. L. Karger, J. Chromatogr., 359 (1986) 3.

534

331. J. L. Fausnaugh and F. E. Regnier, J. Chromatogr., 359 (1986) 131.
332. S. C. Goheen and S. C. Engelhorn, J. Chromatogr., 317 (1984) 55.
333. B. A. Barford, T. F. Kumosinki, N. Parris, and A. E. While, J. Chromatogr., 458 (1988) 57.
334. J. A. Smith and M. O'Hare, J. Chromatogr., 496 (1989) 79.
335. X. Geng, L. Guo, and J. Chang, J. Chromatogr., 507 (1990) 1.
336. W. R. Melander, D. Corradini, and Cs. Horvath, J. Chromatogr., 317 (1984) 67.
337. A. Katti, Y.-F. Maa, and Cs. Horvath, Chromatographia, 24 (1987) 646.
338. J. Gehas and D. B. Wetlaufer, J. Chromatogr., 477 (1989) 249.
339. Cs. Horvath, W. Melander, and I. Molnar, Anal. Chem., 49 (1977) 142.
340. J. P. Foley and W. E. May, Anal. Chem., 59 (1987) 102.
341. J. P. Foley, Anal. Chim. Acta. 231 (1990) 237.
342. B. A. Bidlingmeyer, J. Chromatogr. Sci., 18 (1980) 525.
343. J. P. Foley and W. E. May, Anal. Chem., 59 (1987) 110.
344. N. Tanaka, T. Yoshimura, and M. Araki, J. Chromatogr., 406 (1987) 247.
345. D. W. Armostrong, Sepn. Purificn. Methods, 14 (1985) 213.
346. J. G. Dorsey, Adv. Chromatogr., 27 (1987) 167.
347. M. G. Khaledi, Trends Anal. Chem., 7 (1988) 293.
348. M. J. Koenigbauer, J. Chromatogr., 531 (1990) 79.
349. J. K. Strasters, E. D. Breyer, A. H. Rodgers, and M. G. Khaledi, J. Chromatogr., 511 (1990) 17.
350. A. Berthod and A. Roussel, J. Chromatogr., 449 (1988) 349.
351. M. F. Borgerding, R. L. Williams, W. L. Hinze, and F. H. Quina, J. Liq. Chromatogr., 12 (1989) 1367.
352. R. W. Williams, Z. Fu, and W. L. Hinze, J. Chromatogr., Sci., 28 (1990) 292.
353. Cs. Horvath, W. Melander, and A. Nahum, J. Chromatogr., 186 (1979) 371.
354. R. M. Smith, S. J. Bale, S. G. Westcoft, and M. Martin-Smith, Analyst, 114 (1989) 771.
355. M. L. Marina, J. C. Diez-Masa, and M. V. Dabrio, J. Liq. Chromatogr., 12 (1989) 1973.
356. R. S. Ramsey, V. W. Chan, B. M. Dittmar, and K. H. Row, J. Chromatogr., 468 (1989) 167.
357. G. Schill, in J. A. Mirnsky and Y. Marcus (Eds.), "Ion-Exchange and Solvent Extraction", Dekker, New York, NY, 6 (1974) 1.
358. G. Schill, H. Ehrsson, J. Vessman, and D. Westerlund, "Separation Methods for Drugs and Related Organic Compounds", Swedish Pharmaceutical Press, Stockholm, Sweden, 1983.
359. J. Crommen, B. Fransson, and G. Schill, J. Chromatogr., 125 (1976) 327.
360. E. Tonlimson, T. M. Jefferies, and C. M. Riley, J. Chromatogr., 159 (1978) 315.
361. M. T. W. Hearn, Adv. Chromatogr., 18 (1980) 59.
362. R. H. A. Sorel and A. Hulshoff, Adv. Chromatogr., 21 (1983) 87.
363. M. T. W. Hearn (Ed.), "Ion-Pair Chromatography. Theory and Biological and Pharmaceutical Applications", Dekker, New York, NY, 1985.
364. R. K. Gilpin, S. S. Yang, and G. Werner, J. Chromatogr. Sci., 26 (1988) 388.
365. R. Gloor and E. L. Johnson, J. Chromatogr. Sci., 15 (1977) 413.

366. Q. Xianren and W. Baeyens, J. Chromatogr., 456 (1988) 267.
367. A. Zein and M. Baerns, J. Chromatogr. Sci., 27 (1989) 249.
368. G. K.-C. Low, A. Bratha, H. A. H. Billiet, and L. De Galan, J. Chromatogr., 478 (1989) 21.
369. A. Bartha, G. Vigh, and J. Stahlberg, J. Chromatogr., 506 (1990) 85.
370. A. Bartha, G. Vigh , H. A. H. Billiet, and L. De Galan, J. Chromatogr., 303 (1984) 29.
371. M. C. Gennaro, J. Chromatogr., 449 (1988) 103.
372. A. Bartha, G. Vigh, and Z. Varga-Puchony, J. Chromatogr., 499 (1990) 423.
373. A. Bartha, H. A. H. Billiet, L. De Galan, and G. Vigh, J. Chromatogr., 29 (1984) 91.
374. A. P. Goldberg, E. Nowakowska, P. E. Antle, and L. R. Snyder, J. Chromatogr., 316 (1984) 241.
375. H. A. H. Billiet, J. Vuik, J. K. Strasters, and L. De Galan, J. Chromatogr., 384 (1987) 153.
376. P. M. J. Coenegracht, N. Van Tuyen, H. J. Metting, and P. J. M. Coenegracht-Lamers, J. Chromatogr., 389 (1987) 351.
377. G. Winkler, P. Briza, and C. Kunz, J. Chromatogr., 361 (1986) 191.
378. U. Juergens, J. Liq. Chromatogr., 11 (1988) 1925.
379. A. Bartha and G. Vigh, J. Chromatogr., 485 (1989) 383.
380. Cs. Horvath, W. Melander, I. Molnar, and P. Molnar, Anal. Chem., 49 (1977) 2295.
381. B. A. Bidlingmeyer, S. N. Deming, W. P. Price, B. Sachok, and M. Petrusek, J. Chromatogr., 186 (1979) 419.
382. J. Stahlberg, J. Chromatogr., 356 (1986) 231.
383. J. Stahlberg, Chromatographia, 24 (1987) 820.
384. J. Stahlberg and A. Bartha, J. Chromatogr., 456 (1988) 253.
385. J. Stahlerg and I. Hagglund, Anal. Chem., 60 (1988) 1958.
386. A. Bartha, G. Vigh, and J. Stahlberg, J. Chromatogr., 506 (1990) 85.
387. G. Schill and J. Crommen, Trends Anal. Chem., 6 (1987) 111.
388. E. S. Yeung, Acc. Chem. Res., 22 (1989) 125.
389. G. Schill and E. Arvidsson, J. Chromatogr., 492 (1989) 299.
390. W. E. Barber and P. W. Carr, J. Chromatogr., 316 (1984) 211.
391. S. Levin and E. Grushka, Anal. Chem., 58 (1986) 1602.
392. S. Levin and E. Grushka, Anal. Chem., 59 (1987) 1157.
393. J. Stahlberg and M. Almgren, Anal. Chem., 61 (1989) 1109.
394. S. Golshan-Shirazi and G. Guiochon, Anal. Chem., 62 (1990) 923.
395. H. Poppe, J. Chromatogr., 506 (1990) 45.
396. Y. Yokoyama and H. Sato, J. Chromatogr. Sci., 26 (1988) 561.
397. R. E. Smith, "Ion Chromatography Applications", CRC Press, Boca Raton, FL, 1988.
398. H. F. Walton and R. D. Rocklin, "Ion Exchange in Analytical Chemistry", CRC Press, Boca Raton, FL, 1990.
399. H. Heftmann (Ed.), "Chromatography. Part B: Applications", Elsevier, Amsterdam, 5th. Edn., 1990.
400. G. Schmuckler, J. Liq. Chromatogr., 10 (1987) 1887.
401. W. T. Frankenberger, H. C. Mehra, and D. T. Gjerde, J. Chromatogr., 504 (1990) 211.
402. D. P. Lee and M. T. Bunker, J. Chromatogr. Sci., 27 (1989) 496.
403. S. C. Churms, J. Chromatogr., 500 (1990) 555.
404. M. Verzele, G. Simoens and F. van Damme, Chromatographia, 23 (1987) 292.
405. K. B. Hicks, Adv. Carbohydr. Chem. Biochem., 46 (1988) 17.
406. W. Voelter and H. Bauer, J. Chromatogr., 126 (1976) 693.

536

407. J. G. Tarter (Ed.), "Ion Chromatography", Dekker, New York, NY, 1987.
408. C. A. Pohl and E. L. Johnson, J. Chromatogr. Sci., 18 (1980) 442.
409. P. R. Haddad and A. L. Heckenberg, J. Chromatogr., 300 (1984) 357.
410. D. A. Colenutt and P. J. Trenchard, Environ. Pollution (series B), 10 (1985) 77.
411. J. S. Fritz, Anal. Chem., 59 (1987) 335A.
412. P. K. Dasgupta, J. Chromatogr. Sci., 27 (1989) 422.
413. K. Tanaka and J. S. Fritz, J. Chromatogr., 409 (1987) 271.
414. B. K. Glod, A. Piasecki, and J. Stafiej, J. Chromatogr., 457 (1988) 43.
415. P. Walser, J. Chromatogr., 439 (1988) 71.
416. E. Papp and P. Keresztes, J. Chromatogr., 506 (1990) 157.
417. D. R. Jenke and G. R. Pagenkopf, in J. A. Jonsson (Ed.), "Chromatographic Theory and Basic Principles", Dekker, New York, NY, 1987, p 313.
418. P. R. Haddad and A. D. Sosimenko, J. Chromatogr. Sci., 27 (1989) 456.
419. A. Rahman and N. E. Hoffman, J. Chromatogr. Sci., 28 (1990) 157.
420. T. Watanabe and M. Kubota, Anal. Chim. Acta, 228 (1990) 61.
421. P. R. Haddad and R. C. Foley, J. Chromatogr., 500 (1990) 301.
422. D. T. Gjerde, J. Chromatogr., 439 (1988) 49.
423. W. Kopaciewicz, M. A. Rounds, J. Fausnaugh, and F. E. Regnier, J. Chromatogr., 266 (1983) 3.
424. M. A. Rounds and F. E. Regneir, J. Chromatogr., 283 (1984) 37.
425. D. A. Bergman and D. J. Winzor, J. Chromatogr., 391 (1987) 67.
426. T. W. L. Burke, C. T. Mant, J. A. Black, and R. S. Hodges, J. Chromatogr., 476 (1989) 377.
427. M. Dizdaroglu, J. Chromatogr., 334 (1985) 49.
428. A. N. Hodder, M. I. Aguilar, and M. T. W. Hearn, J. Chromatogr., 476 (1989) 391.
429. W. R. Melander, Z. El Rassi, and Cs. Horvath, J. Chromatogr., 469 (1989) 3.
430. F. M. Rabel, Adv. Chromatogr., 17 (1979) 53.
431. Y. Baba, J. Chromatogr., 485 (1989) 143.
432. P. R. Haddad, Chromatographia, 24 (1987) 217.
433. D. T. Gjerde, Intern. J. Environ. Anal. Chem., 27 (1986) 289.
434. H. Small, T. S. Stevens, and W. C. Bauman, Anal. Chem., 47 (1975) 1801.
435. R. D. Rocklin, C. A. Pohl, and J. A. Schibler, J. Chromatogr., 411 (1987) 107.
436. H. Shintani and P. K. Dasgupta, Anal. Chem., 59 (1987) 802.
437. K. Irgum, Anal. Chem., 59 (1987) 358 and 363.
438. R. D. Rocklin, M. A. Rey, J. R. Stillian, and D. L. Campbell, J. Chromatogr. Sci., 27 (1989) 474.
439. J. Janca (Ed.), Steric Exclusion Liquid Chromatography of Polymers", Dekker, New York, NY, 1984.
440. P. L. Dubin (Ed.), "Aqueous Size-Exclusion Chromatography", Elsevier, Amsterdam, 1988.
441. B. J. Hunt and S. R. Holding (Eds.), Size Exclusion Chromatography, Blackie, Glasgow, UK, 1989.
442. S. T. Balke, P. Cheung, R. Lew, and T. H. Mourey, J. Liq. Chromatogr., 13 (1990) 2929.
443. L. Fischer, "Gel Filtration Chromatography", Elsevier, Amsterdam, 2nd. Edn., 1980.

444. D. W. Gruenwedel and J. R. Whitaker (Eds.), "Food Analysis Principles and Techniques", Dekker, New York, NY, Vol. 4, 1987.
445. T. Provder (Ed.), "Size-Exclusion Chromatography. Methodology and Characterization of Polymers and Related Materials", American Chemical Society, Washington, DC, 1984.
446. A. L. Lafleur and M. J. Wornat, Anal. Chem., 60 (1988) 1096.
447. H. Waldmann-Meyer, J. Chromatogr., 410 (1987) 233.
448. A. A. Gorbunov, L. Ya Solovyova and V. A. Pasechnik, J. Chromatogr., 448 (1988) 307.
449. H. Engelhart and G. Ahr, J. Chromatogr., 282 (1983) 385.
450. H. Engelhart and U. M. Schon, Chromatographia, 22 (1986) 388.
451. B. Anspach, H. U. Gierlich, and K. K. Unger, J. Chromatogr., 443 (1988) 45.
452. Annual Book of ASTM Standards, ASTM. Philadelphia, PA, 1990, Vol.8.03, D.3536-76, p.104.
453. H. G. Barth, J. Chromatogr. Sci., 18 (1980) 409.
454. E. Pfannkock, K. C. Lu, F. E. Regnier, and J. G. Barth, J. Chromatogr. Sci., 18 (1980) 430.
455. J. Janca, Adv. Chromatogr., 19 (1981) 37.
456. O. Chiantore and M. Guaita, J. Chromatogr., 353 (1986) 285.
457. T. Provder (Ed.), "Detection and Data Analysis in Size-Exclusion Chromatography", American Chemical Society, Washington, DC, 1987.
458. M. Potschka, J. Chromatogr., 441 (1988) 239.
459. A. R. Cooper (Ed.), "Determination of Molecular Weight", Wiley, New York, NY, 1989, p.263-299.
460. Yu. A. Eltekov, J. Chromatogr., 365 (1986) 191.
461. K. Miyazaki, Y. Tanaka, and M. Saito, J. Chromatogr., 454 (1988) 357.
462. L. Andersson, J. Chromatogr., 325 (1985) 37.
463. N. Chikazumi, Y. Mukoyama, and H. Sugitani, J. Chromatogr., 479 (1989) 85.
464. W. Cheng and D. Hollis, J. Chromatogr., 408 (1987) 9.
465. W. W. Yau, C. R. Ginnard, and J. J. Kirkland, J. Chromatogr., 149 (1978) 465.
466. Y. Kato, T. Matsuda, and T. Hashimoto, J. Chromatogr., 332 (1985) 39.
467. M. Potschka, Anal. Biochem., 162 (1987) 47.
468. S. Coulombe, J. Chromatogr., Sci., 26 (1988) 1.
469. T. Takagi, J. Chromatogr., 506 (1990) 409.
470. I. S. Krull, R. Mhatre, and H. H. Stuting, Trends Anal. Chem., 8 (1989) 260.
471. G. Glockner, "Polymer Characterization by Liquid Chromatography", Elsevier, Amsterdam, 1987.
472. G. Glockner and J. H. M. van der Berg, Chromatographia, 19 (1984) 55.
473. G. Glockner and J. H. M. van der Berg, J. Chromatogr., 352 (1986) 511.
474. R. Schultz and H. Engelhart, Chromatographia, 29 (1990) 325.
475. G. Glockner, Trends Anal. Chem., 4 (1985) 214.
476. M. A. Quarry, M. A. Stadalius, T. H. Mourey, and L. R. Snyder, J. Chromatogr., 358 (1986) 1.
477. R. Schultz and H. Engelhart, Chromatographia, 29 (1990) 205.
478. J. C. Berridge, "Techniques for the Automated Optimization of HPLC Separations", Wiley, New York, NY, 1985.
479. P. J. Schoenmakers, "Optimization of Chromatographic Selectivity. A guide to Method Development", Elsevier, Amsterdam, 1986.
480. L. R. Snyder, J. L. Glajch, and J. J. Kirkland, "Practical HPLC Method Development", Wiley, New York, NY, 1988.

538

481. P. J. Schoenmakers and M. Mulholland, Chromatographia, 25 (1988) 737.
482. P. J. Schoenmakers, A. Peeters, and R. J. Lynch, J. Chromatogr., 506 (1990) 169.
483. S. S. Williams, J. F. Karnicky, J.-L. Excoffier, and S. R. Abbott, J. Chromatogr., 485 (1989) 267.
484. J. C. Berridge, Chemomet. Intel. Labor. Syst., 3 (1988) 175.
485. J. W. Dolan and L. R. Snyder, J. Chromatogr. Sci., 28 (1990) 379.
486. H. A. H. Billiet and L. de Galan, J. Chromatogr., 485 (1989) 27.
487. L. de Galan, D. P. Herman, and H. A. H. Billiet, Chromatographia, 24 (1987) 108.
488. G. D'Agostino, L. Castagnetta, F. Mitchell, and M. J. O' Hare, J. Chromatogr., 338 (1985) 1.
489. C. E. Goewie, J. Liq Chromatogr., 9 (1986) 1431.
490. P. M. J. Coenegracht, A. K. Smilde, H. J. Metting, and D. A. Doornbos, J. Chromatogr.,485 (1989) 195.
491. J. W. Dolan, L. R. Snyder, and M. A. Quarry, Chromatographia, 24 (1987) 261.
492. P. J. Schoenmakers, N. Dunand, A. C. Cleland, G. Musch, and T. Blaffert, Chromatographia, 26 (1988) 37.
493. P. J. Schoenmakers and N. Dunand, J. Chromatogr., 485 (1989) 219.
494. E. L. Inman, J. K. Frischmann, P. J. Jimenez, G. D. Winkel, M. L. Persinger, and B. S. Rutherford, J. Chromatogr. Sci., 25 (1987) 252.
495. L. M. C. Buydens, J. A. van Leeuwen, M. Mulholland, B. G. M. Vandeginste, and G. Kateman, Trends Anal. Chem., 9 (1990) 58.
496. J. A. van Leeuwen, L. M. C. Buydens, B. G. M. Vandeginste, G. Kateman, and M. Mulholland, Anal. Chim. Acta, 235 (1990) 27.
497. J. A. van Leeuwen, B. G. M. Vanderginste, G. Kateman, M. Mulholland, and A. Cleland, Anal. Chim. Acta, 228 (1990) 145.
498   Sj. van der Wal, Chromatographia, 20 (1985) 274.
499   J. T. Przybytek (Ed.), "High Purity Solvent Guide", Burdick and Jackson Laboratories, Muskegon, MI, 1980.
500. L. R. Snyder in E. S. Perry and A. Weissberger (Eds.), "Separation and Purification", Wiley, New York, NY, 3rd Edn., 1978, p 25.
501. V. Yu. Zelvensky, A. S. Lavrenova, S. I. Samolyuk, L. V. Borodai, and G. A. Egorenko, J. Chromatogr., 364 (1986) 305.
502. F. M. Rabel, J. Chromatogr. Sci., 18 (1980) 394.
503. J. D. Karkas, J. Germershausen, and R. Liou, J. Chromatogr., 214 (1981) 267.
504. D. W. Bristol, J. Chromatogr., 188 (1980) 193.
505. T. R. Griffiths and D. C. Pugh, Coordn. Chem. Revs., 29 (1979) 129.
506. C. Reichardt, "Solvents and Solvent Effects in Organic Chemistry", VCH Publishers, Weinhein, F. R. G., 1988.
507. M. J. Kamlet, J.-L. M. Abboud, and R. W. Taft, Prog. Phys. Org. Chem., 13 (1981) 485.
508. R. W. Taft, J.-L. M. Abboud, M. J. Kamlet, and M. H. Abraham, J. Soln. Chem., 14 (1985) 153.
509. A. F. M. Barton, "Handbook of Solubility Parameters and Other Cohesion Parameters", CRC Press Inc., Boca Raton, FL, 1983.
510. L. R. Snyder, Chemtech., (1979) 750 and (1980) 188.
511. R. Tijssen, H. A. H. Billiet, and P. J. Schoenmakers, J. Chromatogr., 122 (1976) 185.
512. B. L. Karger, L. R. Snyder, and C. Eon, Anal. Chem., 50 (1978) 2126.

513. P. J. Schoenmakers, H. A. H. Billiet, and L. de Galan, Chromatographia, 15 (1982) 205.
514. M. J. Kamlet, R. M. Doherty, J.-L. M. Abboud, M. H. Abraham, and R. W. Taft, Chemtech., (1986) 566.
515. M. H. Abraham, R. M. Doherty, M. J. Kamlet, and R. W. Taft, Chem. Brit., 22 (1986) 551.
516. J. H. Park and P. W. Carr, J. Chromatogr., 465 (1989) 123.
517. L. R. Snyder, J. Chromatogr. Sci., 16 (1978) 223.
518. H. Poppe and E. H. Slaats, Chromatographia, 14 (1981) 89.
519. S. C. Rutan, P. W. Carr, W. J. Choeng, J. H. Park, and L. R. Snyder, J. Chromatogr., 463 (1989) 21.
520. D. P. Herman, H. A. H. Billiet, and L. de Galan, J. Chromatogr., 463 (1989) 1.
521. L. R. Snyder, M. A. Quarry, and L. J. Glajch, Chromatogrphia, 24 (1987) 33.
522. P. J. Schoenmakers, H. A. H. Billiet, and L. de Galan, J. Chromatogr., 218 (1981) 261.
523. P. M. J. Coenegracht, H. J. Metting, A. K. Smilde, and P. J. M. Coenegracht-Lamers, Chromatographia, 27 (1989) 135.
524. P. M. J. Coenegracht, A. K. Smilde, and A. Knevelman, J. Liq. Chromatogr., 12 (1989) 77.
525. L. de Galan and H. A. H. Billiet, Adv. Chromatogr., 25 (1986) 63
526. P. J. Schoenmakers and T. Blaffert, J. Chromatogr., 384 (1987) 117.
527. A. C. J. H. Drouen, H. A. H. Billiet, and L. de Galan, Anal. Chem., 57 (1985) 962.
528. A. G. Wright, A. Fell, and J. C. Berridge, Chromatographia, 24 (1987) 533.
529. J. K. Stasters, H. A. H. Billiet, L. de Galan, B. G. M. Vandeginste and G. Kateman, Anal. Chem., 60 (1988) 2745.
530. S. T. Balke, "Quantitative Column Liquid Chromatography. A Survey of Chemometric Methods", Elsevier, Amsterdam, 1984.
531. P. J. Schoenmakers, J. Liq. Chromatogr., 10 (1987) 1865.
532. A. Peeters, L. Buydens, D. L. Massart, and P. J. Schoenmakers, Chromatographia, 26 (1988) 101.
533. C. P. Cai and N. S. Wu, Chromatographia, 30 (1990) 400.
534. P. J. Schoenmakers, H. A. H. Billiet, and L. de Galan, J. Chromatogr., 205 (1981) 13.
535. M. A. Quarry, R. L. Grob, and L. R. Snyder, Anal. Chem., 58 (1986) 907.
536. L. R. Snyder and M. A. Quarry, J. Liq. Chromatogr., 10 (1987) 1789.
537. P. R. Haddad and S. Sekulic, J. Chromatogr., 392 (1987) 65.
538. L. R. Snyder, J. W. Dolan, and M. A. Quarry, Trends Anal. Chem., 6 (1987) 106.
539. M. A. Quarry, R. L. Grob, L. R. Snyder, J. W. Dolan, and M. P. Rigney, J. Chromatogr., 384 (1987) 163.
540. L. R. Snyder. J. W. Dolan, and D. C. Lommen, J. Chromatogr., 485 (1989) 65.
541. J. C. Berridge, Chemomet. Intell. Labor. Syst., 5 (1989) 195.
542. J. C. Berridge and E. G. Morrissey, J. Chromatogr., 316 (1984) 69.
543. J. C. Berridge, J. Chromatogr., 485 (1989) 3.
544. J. L. Glajch, J. J. Kirkland, Anal. Chem., 55 (1983) 319A
545. J. L. Glajch, J. J. Kirkland, K. M. Squire, and J. M. Minor, J. Chromatogr., 199 (1980) 57.
546. J. L. Glajch, J. J. Kirkland, and L. R. Snyder, J. Chromatogr., 238 (1982) 269.
547. J. L. Glajch, J. J. Kirkland, and J. M. Minor, J. Liq. Chromatogr., 10 (1987) 1727.

540

548. J. L. Glajch and J. J. Kirkland, J. Chromatogr., 485 (1989) 51.
549. R. D. Snee, Chemtech., (1979) 702.
550. P. J. Naish-Chamberlain and R. J. Lynch, Chromatographia, 29 (1990) 79.
551. P. Jandera and J. Churacek, "Gradient Elution in Liquid Chromatography. Theory and Practice", Elsevier, Amsterdam, 1985.
552. L. R. Snyder, J. W. Dolan, and J. R. Gant, J. Chromatogr., 165 (1979) 3 and 31.
553. L. R. Snyder in C. Horvath (Ed.), "High Performance Liquid Chromatography, Advances and Perspectives", Academic Press, New York, NY, Vol. 1, 1980, p. 207.
554. L. R. Snyder, M. A. Stadalius, and M. A. Quarry, Anal. Chem., 55 (1983) 1412A.
555. J. W. Dolan, D. C. Lammen, and L. R. Snyder, J. Chromatogr., 485 (1989) 91.
556. P. Jandera, J. Chromatogr., 485 (1989) 113.
557. P. J. Schoenmakers, H. A. H. Billiet, and L. de Galan, J. Chromatogr., 185 (1979) 179.
558. S. A. Tomellini, R. A. Hartwick, and H. B. Woodruff, Anal. Chem., 57 (1985) 811.
559. Sz. Nyiredy and O. Sticher, J. High Resolut. Chromatogr., 10 (1987) 208.
560. J. J. Kirkland and J. L. Glajch, J. Chromatogr., 255 (1983) 27.
561. W. Markowski and W. Golkiewicz, Chromatographia, 25 (1988) 339.
562. B. F. D. Ghrist, B. S. Cooperman, and L. R. Snyder, J. Chromatogr., 459 (1988) 1, 25, and 43.
563. J. W. Dolan, L. R. Snyder, and M. A. Quary, Chromatographia, 24 (1987) 261.
564. M. Eslami, J. D. Stuart, and K. A. Cohen, J. Chromatogr., 411 (1987) 121.
565. J. D. Stuart, D. D. Lisi, and L. R. Snyder, J. Chromatogr., 485 (1989) 657.
566. J. C. Ford and J. A. Smith, J. Chromatogr., 483 (1989) 131.
567. M. A. Stadalius, H. S. Gold, and L. R. Snyder, J. Chromatogr., 296 (1984) 31.
568. J. C. Berridge, J. Chromatogr., 244 (1982) 1.
569. G. D'Agostino, M. J. O'Hare, F. Mitchell, T. Salomon, and F. Verillon, Chromatographia, 25 (1988) 343.
570. K. Hostettmann, M. Hostettmann, and A. Marston, "Preparative Chromatography Techniques. Applications in Natural Product Isolation", Springer-Verlag, Berlin, FRG, 1986.
571. B. A. Bidlingmeyer (Ed.), "Preparative Liquid Chromatography", Elsevier, Amsterdam, 1987.
572. E. Grushka (Ed.), "Preparative-Scale Chromatography", Dekker, New York, NY, 1989.
573. K. Jones, Chromatographia, 25 (1988) 547.
574. T. J. Gentilucci, S. I. Sivakoff, G. B. Cox, S. D. Stearns, and M. W. Hutchinson, J. Chromatogr., 461 (1989) 63.
575. A. M. Cantwell, R. Calderone, and M. Sienko, J. Chromatogr., 316 (1984) 133.
576. E. P. Kroeff, R. A. Owens, E. L. Campbell, R. D. Johnson, and H. I. Marks, J. Chromatogr., 461 (1989) 45.
577. G. K. Sofer and L.-E. Nystrom, "Process Chromatography. A Practical Guide", Academic Press, New York, NY, 1989.
578. E. J. Leopold, J. Org. Chem., 47 (1982) 4592.
579. N. M. Targett, J. P. Kilcoyne, and B. Green, J. Org. Chem., 44 (1979) 4962.

580. W. C. Still, M. Kahn, and A. Mitra, J. Org. Chem., 43 (1978) 2923.
581. S. R. Moinelo, R. Menendez, and J. Bermejo, Chromatographia, 23 (1987) 179.
582. H. Loibner and G. Seidl, Chromatographia, 12 (1979) 600.
583. M. Radke, H. Willsch, and D. H. Welte, Anal. Chem., 52 (1980) 406.
584. F. Eisenbeib and H. Henke, J. High Resolut. Chromatogr., 2 (1979) 733.
585. T. Leutert and E. von Arx, J. Chromatogr., 292 (1984) 333.
586. D. Schaufelberger and K. Hostettmann, J. Chromatogr., 346 (1985) 396.
587. J. Ru Hwu, J. A. Robl, and K. P. Khoudary, J. Chromatogr. Sci., 25 (1987) 501.
588. G. C. Zogg, Sz. Nyiredy, and O. Sticher, J. Liq. Chromatogr., 12 (1989) 2031 and 2049.
589. Sz. Nyiredy, K. Dallenbach-Toelke, G. C. Zogg, and O. Sticher, J. Chromatogr., 499 (1990) 453.
590. M. Verzele and E. Geereart, J. Chromatogr. Sci., 18 (1980) 559.
591. G. Golshan-Shirazi and G. Guiochon, Anal. Chem., 61 (1989) 1276.
592. G. Guiochon and A. Katti, Chromatographia, 24 (1987) 165.
593. M. Verzele, Anal. Chem., 62 (1990) 265A.
594. J. Lesec, J. Liq. Chromatogr., 8 (1985) 875.
595. J.-X. Huang and G. Guiochon, J. Chromatogr., 492 (1989) 431.
596. K. P. Hupe and H. H. Lauer, J. Chromatogr., 203 (1981) 41.
597. E. J. Kikta, Sepn, Purifn. Methods, 13 (1984) 109.
598. M. Verzele, M. de Coninck, J. Vindevogel, and C. Dewaele, J. Chromatogr., 450 (1988) 47.
599. J. Klawiter, M. Kaminki, and J. S. Kowalczyk, J. Chromatogr., 243 (1982) 207 and 225.
600. K. K. Unger and R. Janzen, J. Chromatogr., 373 (1986) 227.
601. T. Wang, R. A. Hartwick, N. T. Miller, and D. C. Shelly, J. Chromatogr., 523 (1900) 23.
602. E. Geereart and M. Verzele, Chromatographia, 12 (1979) 50.
603. D. J. Pietrzyk and W. J. Cahill, J. Liq. Chromatogr., 5 (1982) 781.
604. E. Godbille and P. Devaux, J. Chromatogr. Sci., 12 (1974) 564.
605. E. Godbille and P. Devaux, J. Chromatogr., 122 (1976) 317.
606. G. Cretier and J. L. Rocca, Chromatographia, 16 (1983) 32.
607. V. R. Meyer, J. Chromatogr., 316 (1984) 113.
608. J. H. Knox and H. M. Pyper, J. Chromatogr., 363 (1986) 1.
609. S. Ghodbane and G. Guiochon, J. Chromatogr., 444 (1988) 275.
610. P. Gareil and R. Rosset, J. Chromatogr.,450 (1988) 13.
611. H. T. Rasmussen and H. M. McNair, J. Liq. Chromatogr., 13 (1990) 3079.
612. L. R. Snyder, J. W. Dolan, and G. B. Cox, J. Chromatogr., 483 (1989) 63, 85, and 95.
613. J. E. Eble, R. L. Grob, P. E. Antle, and L. R. Snyder, J. Chromatogr., 405 (1987) 1, 31, and 51.
614. G. B. Cox, L. R. Snyder, and J. W. Dolan, J. Chromatogr., 484 (1989) 409, 425, and 437.
615. M. J. Gonzalez, A. Jaulmes, P. Valentin, and C. Vidal-Madjar, J. Chromatogr., 386 (1987) 333.
616. F. D. Antia and C. Horvath, J. Chromatogr., 484 (1989) 1.
617. C. A. Lucy, J. L. Wade, and P. W. Carr, J. Chromatogr., 484 (1989) 61.
618. R. Nowakowski, Chromatographia, 28 (1989) 293.

542

619. S. Golshan-Shirazi and G. Guiochon, J. Chromatogr., 461 (1989) 1.
620. S. Golshan-Shirazi and G. Guiochon, Anal. Chem., 61 (1989) 1368.
621. M. Czok and G. Guichon, Anal. Chem., 62 (1990) 189.
622. S. Golshan-Shirazi and G. Guiochon, J. Chromatogr., 506 (1990) 495.
623. S. Golshan-Shirazi and G. Guiochon, Anal. Chem., 62 (1990) 217.
624. Z. Ma and G. Guiochon, Anal. Chem., 62 (1990) 2330.
625. S. Golshan-Shirazi and G. Guiochon, J. Chromatogr., 517 (1990) 229.
626. S. Golshan-Shirazi and G. Guiochon, J. Chromatogr., 523 (1990) 1.
627. J. Newburger and G. Guiochon, J. Chromatogr., 423 (1990) 63.
628. F. Brunner (Ed.), "The Science of Chromatography", Elsevier, Amsterdam, 1985, p. 179.
629. A. W. Liao, Z. El Rassi, D. M. LeMaster, and Cs. Horvath, Chromatographia, 24 (1987) 881.
630. J. Frenz, Ph. van der Schrieck, and Cs. Horvath, J. Chromatogr., 330 (1985) 1.
631. H. Kalasz and C. Horvath, J. Chromatogr., 239 (1982) 423.
632. S. Golshan-Shirazi, B. Lin, and G. Guiochon, Anal. Chem., 61 (1989) 1960.
633. K. Valko, P. Slege, and Bati, J. Chromatogr., 386 (1987) 345.
634. G. Vigh, G. Quintero, and G. Farkas, J. Chromatogr., 506 (1990) 481.
635. F. Cardinali, A. Ziggiotti, and G. C. Viscomi, J. Chromatogr., 499 (1990) 37.
636. A. S. Cohen, A. Paulus, and B. L. Karger, Chromatographia, 24 (1987) 15.
637. A. G. Ewing, R. A. Wallingford, and T. M. Olefirowicz, Anal. Chem., 61 (1989) 292A.
638. R. A. Wallingford and A. G. Ewing, Adv. Chromatogr., 29 (1989) 1.
639. B. L. Karger, A. S. Cohen, and A. Guttman, J. Chromatogr., 492 (1989) 585.
640. J. W. Jorgenson, Anal. Chem., 58 (1986) 743A.
641. Ch. Schwer and E. Kenndler, Chromatographia, 30 (1990) 546.
642. F. M. Everaerts, J. L. Beckers, and Th. P. E. M. Verheggen, "Isotachophoresis: Theory Instrumentation and Applications", Elsevier, Amsterdam, 1976.
643. P. Bocek, M. Deml, P. Gebauer, and V. Dolnik, "Analytical Isotachophoresis", VCH Publishers, FRG, Weinheim, 1988.
644. J. L. Beckers and F. M. Everaerts, J. Chromatogr., 480 (1989) 69.
645. P. Gebauer, L. Krivankova, and P. Bocek, J. Chromatogr., 470 (1989) 3.
646. T. Hirokawa, K. Nakahara, and Y. Kiso, J. Chromatogr., 463 (1989) 51.
647. C. G. Edmands, J. A. Loo, C. J. Barinaga, H. R. Udseth, and R. D. Smith, J. Chromatogr., 474 (1989) 21.
648. D. Kaniansky and J. Marak, J. Chromatogr., 498 (1990) 191.
649. W. Thormann, J. Chromatogr., 516 (1990) 211.
650. P. G. Righetti, J. Chromatogr., 516 (1990) 3.
651. O. Vesterberg, J. Chromatogr., 480 (1989) 3.
652. Z. Deyl, (Ed.), "Separation Methods", Elsevier, Amsterdam, 1984.
653. J. H. Knox and I. H. Grant, Chromatographia, 24 (1987) 135.
654. J. H. Knox, Chromatographia, 26 (1988) 329.

655. J. H. Konx and K. A. McCormack, J. Liq. Chromatogr., 12 (1989) 2435.
656. A. A. A. M. Van de Goor, B. J. Wanders, and F. M. Everaerts, J. Chromatogr., 470 (1989) 95.
657. F. Foret, M. Deml, and P. Bocek, J. Chromatogr., 452 (1988) 601.
658. E. Grushka, R. M. McCormick, and J. J. Kirland, Anal. Chem., 61 (1989) 241.
659. G. O. Roberts, P. H. Rhodes, and R. C. Snyder, J. Chromatogr., 480 (1989) 35.
660. X. Huang, W. F. Coleman, and R. N. Zare, J. Chromatogr., 480 (1989) 95.
661. M. J. Gordon, K.-J. Lee, A. A. Arias, and R. N. Zare, Anal. Chem., 63 (1990) 69.
662. J. M. Davis, J. Chromatogr., 517 (1990) 521.
663. H. K. Jones, N. T. Nguyen, and R. D. Smith, J. Chromatogr., 504 (1990) 1.
664. J. Snopek, I. Jelinek, and E. Smolkova-Keulemansova, J. Chromatogr., 452 (1988) 571.
665. S. Terabe, Trends Anal. Chem., 8 (1989) 129.
666. S. Terabe, K. Otsuka, and T. Ando, Anal. Chem., 57 (1985) 834.
667. S. Terabe, K. Otsuka, and T. Ando, Anal. Chem., 61 (1989) 251.
668. J. P. Foley, Anal. Chem., 62 (1990) 1302.
669. K. Ghowsi, J. P. Foley, and R. J. Gale, Anal. Chem., 62 (1990) 2714.
670. P. Gareil, Chromatographia, 30 (1990) 195.
671. R. O. Cole, M. J. Sepaniak, and W. L. Hinze, J. High Resolut. Chromatogr., 13 (1990) 579.
672. H. T. Rasmussen, L. K. Goebel, and H. M. McNair, J. Chromatogr., 517 (1990) 549.
673. H. Nishi, T. Fukuyama, M. Matsuo, and S. Terabe, J. Chromatogr., 513 (1990) 279.
674. K. Otsuka and S. Terabe, J. Chromatogr., 515 (1990) 221.
675. H. Nishi, T. Fukuyama, M. Matsuo, and S. Terabe, J. Chromatogr., 498 (1990) 313.
676. H. Nishi, T. Fukuyama, M. Matsuo, and S. Terabe, J. Chromatogr., 515 (1990) 233.
677. S. Terabe, Y. Miyashita, O. Shibala, E. R. Barnhardt, L. R. Alexander, D. G. Pallerson, B. L. Karger, K. Hosoya, and N. Tanaka, J. Chromatogr., 516 (1990) 23.
678. J. Liu, K. A. Cobb, and M. Novotny, J. Chromatogr., 519 (1990) 189.
669. S. Terabe and T. Isemura, J. Chromatogr., 515 (1990) 667.
680. S. Terabe and T. Isemura, J. High Resolut. Chromatogr., 515 (1990) 667.
681. J. A. Lux, H.-F. Yin, and G. Schomburg, Chromatographia, 30 (1990) 7.
682. G. Schomburg, Chromatographia, 30 (1990) 500.
683. M. J. Sepaniak, D. F. Swaile, and A. C. Powell, J. Chromatogr., 480 (1989) 185.
684. V. Rohlicek and Z. Deyl, J. Chromatogr., 494 (1989) 87.
685. K. A. Cobb, V. Dolnik, and M. Novotny, Anal. Chem., 62 (1990) 2478.
686. J. A. Lux, H. Yin, and G. Schomburg, J. High Resolut. Chromatogr., 13 (1990) 145.
687. S. A. Swedberg, Anal. Biochem., 185 (1990) 51.
688. H. H. Lauer and D. McManigill, Anal. Chem., 58 (1986) 166.

544

689. M. M. Bushey and J. W. Jorgenson, J. Chromatogr.,480 (1989) 301.
690. H.-F. Yin, J. A. Lux, and G. Schomberg, J. High Resolut. Chromatogr., 13 (1990) 624.
691. A. Guttman, A. S. Cohen, D. N. Heiger, and B. L. Karger, Anal. Chem., 62 (1990) 137.
692. A. Guttman, A. Paulus, A. S. Cohen, N. Grinberg, and B. L. Karger, J. Chromatogr., 448 (1988) 41.
693. D. N. Heiger, A. S. Cohen, and B. L. Karger, J. Chromatogr., 516 (1990) 33 and 49.
694. A. Paulus and J. I. Ohms, J. Chromatogr., 507 (1990) 113.
695. T. Tsuda, J. V. Sweedler, and R. N. Zare, Anal. Chem., 62 (1990) 2149.
696. T. Hanai, H. Hatana, N. Nimura, and T. Kinoshita, J. High Resolut. Chromatogr., 13 (1990) 573.
697. H. E. Schwartz, M. Melera, and R. G. Brownlee, J. Chromatogr., 480 (1989) 129.
698. X. Huang, M. J. Gordon, and R. N. Zare, Anal. Chem., 60 (1988) 375.
699. E. Grushka and R. M. McCormick, J. Chromatogr., 471 (1989) 421.
700. D. J. Rose and J. W. Jorgenson, Anal. Chem., 60 (1988) 642.
701. T. Wang, R. A. Hartwick, and P. B. Champlin, J. Chromatogr., 462 (1989) 147.
702. J. S. Green and J. W. Jorgenson, J. Liq. Chromatogr., 12 (1989) 2527.
703. H. Jwerdlow, S. Wu, H. Harke, and N. J. Dovichi, J. Chromatogr., 516 (1990) 61.
704. B. Nickerson and J. W. Jorgenson, J. Chromatogr., 480 (1989) 157.
705. W. G. Kuhr and E. S. Yeung, Anal. Chem., 60 (1988) 2642.
706. B. L. Hogan and E. S. Yeung, J. Chromatogr. Sci., 28 (1990) 15.
707. T. W. Garner and E. S. Yeung, J. Chromatogr., 515 (1990) 639.
708. T. M. Olefirowicz and A. G. Ewing, Anal. Chem., 62 (1990) 1872.
709. R. D. Smith, J. A. Loo, C. G. Edmands, C. J. Barinaga, and H. R. Udseth, J. Chromatogr., 516 (1990) 157.
710. M. A. Moseley, L. J. Deterding, K. B. Tomer and J. W. Jorgenson, J. Chromatogr., 516 (1990) 167.

CHAPTER 5

INSTRUMENTAL ASPECTS OF HIGH PRESSURE LIQUID CHROMATOGRAPHY

5.1   INTRODUCTION

Modern liquid chromatography features columns with small diameter particles of high packing density requiring high pressures for operation at their optimum mobile phase velocities (section 1.7.7). A block diagram of a suitable instrument for high pressure liquid chromatography is shown in Figure 5.1. For isocratic operation a single pump will suffice; for gradient elution a single pump and electronically operated proportioning valves allow continuous variation in the mobile phase composition or, alternatively, independent pumps in parallel are used to pump different solvents into a mixing chamber to vary the mobile phase composition. Between the pump and the injector there may be a series of devices which correct or monitor the pump output. Such devices include pulse dampeners, mixing chambers, flow controllers and pressure transducers. Their function is to ensure that a homogeneous, pulseless liquid flow is delivered to the column at a known pressure and volumetric flow rate. They may be operated either independently of the pump or in a feedback network which continuously updates the pump output. An injection valve is connected to the head of the column for loading the sample solution under ambient conditions and then inserting a known

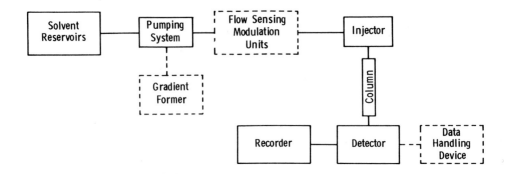

Figure 5.1 Block diagram of a high pressure liquid chromatograph. Dotted lines refer to optional components.

volume of the sample into the fully pressurized mobile phase flow. Injection valves must be capable of routine high pressure operation and resistant to corrosion by a wide variety of mobile phases. Only the column or the whole solvent delivery system may be theromstated to improve reproducibility of retention times and/or column performance, although in most cases operation under ambient conditions prevails. The separation is recorded continuously, on-line, and on the low pressure side of the column. High pressure stability, therefore, is not a general requirement for the detector, but a small operating volume, high sensitivity and a fast response are. A wide range of bulk and physical property detectors are available including refractive index, spectrophotometric and electrochemical devices.

Commercially available instruments for high pressure liquid chromatography cover a wide range of sophistication, features and cost [1-3]. Two general philosophies have emerged. The modular approach allows the chromatograph to be assembled from components that operate independently of each other. This allows flexibility but usually only a limited degree of automation unless a system controller, computer or microprocessor, is independently wired to all components and used to sequence their operation. The integrated approach is represented by a complete instrument assembly, usually with a single control module at which all (most) operating parameters are entered. The main advantage of this approach is that the purchaser acquires an instrument of (claimed) optimized performance. On the other hand flexibility is lost, since it may not be possible to operate modules independently of each other or the central processing unit, and although modules, such as detectors, may be exchangeable, this may only apply to modules purchased from the original manufacturer. Another modern trend has been the development of autosamplers for unattended operation [1-4]. The most advanced of these are more than programmable autoinjectors; they are automated chemistry stations (sometimes referred to as robotic autosamplers) and can perform such routine operations as dilutions, standard additions, derivatization, liquid-liquid extraction, variable size injections for calibration and out-of-order sample sequencing. Standard protocols for evaluating autosampler operation as well as other instrument components have been proposed and should be useful for comparing performance specifications of different instruments [5]. Another trend has been the development of biocompatible and corrosion resistant instruments for the separation of biopolymers and for ion chromatography. The most commonly used material for instrument fabrication is stainless steel due to its favorable mechanical properties and excellent corrosion resistance. However, corrosion of stainless steel has been observed for a wide range of aqueous, halide-ion containing and halogenated solvents, often caused by erosion of the protective oxide layer that covers the steel surface [6,7]. Certain proteins tend to get denatured by contact with stainless steel. Oxidizing or complexing eluents used for cation exchange separations can be contaminated with detectable levels of iron, chromium and nickel from stainless steel. For samples affected in the above way instruments machined from titanium and/or fluoropolymers for the key components are available.

A continuing trend in the scientific literature has been towards the miniaturization of chromatographic instruments for use with narrow bore columns [8-12]. Columns of small internal diameter lead to a number of advantages, such as lower solvent consumption, greater mass sensitivity, and easy interfacing to other separation and detection techniques (see section 1.7.7). Miniaturization requires new or adapted pumping systems, injectors, gradient formers, detectors, etc. The academic interest in miniaturized column systems has not been matched by that of the instrument manufacturers, with some justification. At present suitable instruments for the operation of conventional packed columns and small bore columns with internal diameters between 1 and 2 mm are commercially available. It may be necessary to buy a specialized instrument constructed to minimize extracolumn dispersion for the operation of small bore columns of 1 mm internal diameter, as shown in Figure 5.2 [14-18]. In the case of packed capillary columns and open tubular columns there are very few commercially available systems or modules that have been optimized to meet the particular demands these columns make on instrument design. The reasons for this are explained in the next section. At present, these column types are used mainly in research laboratories with home made, or extensively modified commercial instruments, frequently employing compromised operating conditions in which only a fraction of the quoted benefits for miniaturization are realized.

5.2 REQUIREMENTS WITH RESPECT TO EXTRACOLUMN DISPERSION

Under ideal conditions, the peak profile recorded during a separation should depend only on the operating characteristics of the column and should be independent of the instrument in which the column resides. Under less than ideal conditions, the peak profile will be broader than the column profile by an amount equivalent to the extracolumn band broadening. This broadening arises from dispersion and mixing phenomena which occur in the injector, column connecting tubes and detector cell, as well as from electronic constraints which govern the response speed of the detector and data system. The various contributions to extracolumn band broadening can be treated as independent factors additive in their variances (σ^2), according to the relationship [17-26]:

$$\sigma^2_T = \sigma^2_{col} + \sigma^2_{inj} + \sigma^2_{con} + \sigma^2_{det} + \sigma^2_{tc} \tag{5.1}$$

where σ^2_T is the total peak variance observed in the chromatogram, σ^2_{col} the peak variance due to the column, σ^2_{inj} the variance due to the volume and geometry of the injector, σ^2_{con} the variance due to connecting tubes, unions, frits, etc., σ^2_{det} the variance due to the volume and geometry of the detector, and σ^2_{tc} the variance due to the finite response time of the electronic circuits of the detector and data system. For experimental evaluation equation (5.1) can be simplified to

$$\sigma^2_T = \sigma^2_{col} + \sigma^2_{ext} \tag{5.2}$$

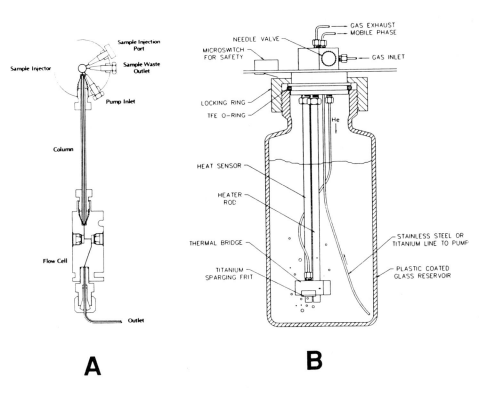

Figure 5.2   A, valve, column and flow cell assembly of a
miniaturized liquid chromatograph for use with small bore columns
(Reproduced with permission from ref. 14) and B, mobile phase
reservoir designed for solvent degassing by heat and helium
sparging (Reproduced with permission from ref. 34. Copyright
Elsevier Scientific Publishing Co.)

where σ^2_{ext} is the sum of all extracolumn contributions to the peak
variance and represents the instrumental contribution to band
broadening. The column contribution to the peak variance can be
written in time or volume units, equation (5.3)

$$\sigma^2_{col} = t_R^2/n = V_R^2/n = V_m^2(1 + k)^2/n \qquad (5.3)$$

where t_R and V_R are the retention time and volume, respectively,
for a peak, V_m the column holdup volume, k the peak capacity factor

and n the true column efficiency in the absence of extracolumn band broadening. If σ^2_{col} from equation (5.3) is inserted into equation (5.2) then a plot of σ^2_T against t_R^2, V_R^2 or $(1 + k)^2$ will be linear. The true column efficiency can be obtained from the slope of the line and σ^2_{ext} from the intercept value on the vertical axis. A commonly accepted criterion for the instrumental contribution to zone broadening is that this should not exceed 10% of the column variance. This corresponds to a loss of 10% of the column efficiency and about 5% in the column resolution. A peak eluting from a column will occupy a volume equivalent to 4 σ units and the above criterion can then be used to establish working limits for the acceptable extracolumn variance and volume for typical separation conditions using the properties of an unretained peak as the worst case or limiting condition. Some representative values are summarized in Table 5.1. A more extensive range of values for different column dimensions and particle sizes are given in section 1.7.9. Extracolumn volumes, which include the detector and injector volumes, should be less than about 20 microliters for conventional packed columns, 1 microliter for small bore packed columns, 100 nanoliters for packed capillary columns and 100 picoliters for open tubular columns. From an engineering perspective it should be feasible to manufacture instrument components with volumes around 0.1 microliters and above, but very difficult to miniaturize these components further. Consequently, instrumentation that allows the full potential of open tubular columns, and to a lesser extent packed capillary columns, to be realized, would seem to be unlikely to emerge in the near future. The generation of low column flow rates and the need for very sensitive detection techniques are also substantial instrument problems that have to be addressed for these columns.

Experimentally there are two methods of determining the extracolumn band broadening of a chromatographic instrument. The linear extrapolation method, discussed above, is relatively straightforward to perform and interpret but rests on the validity of equation (5.1) and (5.3). The assumption that the individual contributions to the extracolumn variance are independent, may not be true in practice, and it may be necessary to couple some of the individual contributions to obtain the most accurate values for the extracolumn variance [20]. It is assumed in equation (5.3)

TABLE 5.1

TYPICAL RANGES FOR VALUES OF COLUMN VARIANCE AND PEAK VOLUMES OF
AN UNRETAINED PEAK

Column Type	Column Internal Diameter (mm)	Peak Variance $(\mu l)^2$	Peak Volume (μl)
Conventional Packed	4-5	25-2500	20-200
Small Bore Packed	1-2	1-25	1-12
Packed Capillary	0.2-0.5	$10^{-3}-10^{-1}$	0.2-0.8
Open Tubular	0.005-0.02	$10^{-7}-10^{-3}$	$10^{-5}-10^{-2}$

that solute diffusion coefficients in the mobile and stationary phases are identical and that any variation in the column plate height as a function of the capacity factor can be neglected [25]. These assumptions are unlikely to be true in all cases diminishing the absolute accuracy with which extracolumn variance can be calculated using the linear extrapolation method. The zero length column method uses a liquid chromatograph without a column, so that an injected peak has a total dispersion with contributions from the external dispersion only [17]. This approach is experimentally demanding and ignores the contribution that column frits and connectors make to the extracolumn variance.

Band broadening due to injection arises because the sample is introduced into the column as a finite volume. A solute zone is formed at the column head which reflects the degree of sample axial displacement during the injection time. This solute zone is generally less than the injection volume due to retention of the solute by the column packing. The elution strength of the injection solvent and the effect of solvent dilution with the column mobile phase will determine the extent of the zone displacement. This situation is too complex to be described by a simple mathematical model. Two extreme views, those of plug or exponential injection, can be used to define the limiting cases. Valve injection occurs largely by displacement (plug injection) accompanied by various contributions from exponential dilution (mixing). The variance due to the injection profile is described by equation (5.4)

$$\sigma^2_{inj} = V^2_{inj}/K \qquad (5.4)$$

where V_{inj} is the injection volume and K is a constant with values between 1-12 depending on the characteristics of the injector [20,27-29]. For plug injection K = 12; more typical values for K under conditions of valve injection are 2-8.

Band dispersion in open tubes is due to poor radial mass transfer of the solute resulting from the laminar nature of the solvent flow. Laminar flow in a cylindrical tube is accompanied by a parabolic velocity profile, the flow velocity in the center of the tube being at a maximum, whereas the velocity at the tube wall is zero. This range of solvent velocity from the wall of the tube to the center causes a significant increased dispersion of any solute band passing through it. The variance due to connecting tubes is given by

$$\sigma^2_{con} = \pi r_c^4 L / 24 D_m F \qquad (5.5)$$

where r_c is the tube radius, L the tube length, D_m the solute diffusion coefficient in the mobile phase and F the column flow rate. The dispersion in connecting tubes can be minimized by using short lengths of tubing with small internal diameters. The volume per unit length for open tubes of different diameters is summarized in Table 5.2. Tubing with too small a diameter, however, increases the risk of plugging, and tubing of too short a length limits the flexibility for spatial arrangement of the instrument modules. An alternative approach is the use of serpentine tubing which introduces radial convection to break up the parabolic flow [30]. Dispersion in a serpentine tube is about 20% of that for a straight tube and allows longer connecting tubes to be used. Strictly speaking equation (5.5) requires a certain minimum efficiency to be an accurate description of the band broadening process. In many cases typical connecting tubes may not reach this minimum requirement, and the use of equation (5.5) leads to an over estimate of the tube contribution to band broadening [31].

With modern instruments the contribution of the detector cell volume to extracolumn variance is often the most significant of the extracolumn terms. Depending on the design of the detector cell, it can either have the properties of a tube with plug flow or act as a mixing volume. In practice, most detector cells behave in a manner somewhere between those represented by the two extreme models. The variance can be calculated using equation (5.4)

TABLE 5.2

VOLUME PER UNIT LENGTH FOR CONNECTING TUBES OF DIFFERENT DIAMETERS

Tube Internal Diameter		Volume per Centimeter
in	mm	μl
0.005	0.13	0.13
0.007	0.18	0.25
0.010	0.25	0.51
0.020	0.50	2.03
0.030	0.75	4.56
0.040	1.00	8.11
0.046	1.20	10.72

substituting the product of the cell volume over the flow rate for the injection volume when plug flow is dominant. If the cell volume is approximately 10% of the peak volume detected, then extracolumn band broadening from the detector will be insignificant. Increasing the flow rate through the detector cell results in a decrease in the peak residence time in the cell, and hence in a decrease of its extracolumn band broadening contribution, at the expense of sensitivity [32]. Detection and data systems can also cause band broadening due to their speed of response which is primarily a function of the time constant associated with the filter network used to diminish high frequency noise [32,33]. The variance contribution due to the detector response time is the product of the detector time constant and the flow rate squared. For high efficiency analytical columns a value of 0.1 s or smaller for the time constant is desirable. The maximum acceptable detector time constant for a particular separation is approximately given by $0.1t_R/n^{1/2}$ where t_R is the peak retention time and n the column efficiency.

5.3 SOLVENT RESERVOIRS AND SOLVENT DEGASSING

The solvent reservoir is a storage container made of a material resistant to chemical attack by the mobile phase. In its simplest form a glass jug, solvent bottle or Erlenmeyer flask with a cap and a flexible hose connection to the pump is adequate. The PTFE connecting hose is terminated on the solvent side with a 2 micrometer pore size filter to prevent suspended particle matter from reaching the pump. In more sophisticated instruments the solvent reservoir may also be equipped for solvent degassing, as shown in Figure 5.2 [34]. Degassing may be required for efficient pump and detector operation, particularly for aqueous organic mobile phases and for gradient forming devices using low pressure mixing. Degassing, in this cases, is used to prevent gas bubble formation when different solvents are mixed or the mobile phase is depressurized. A further problem associated with oxygen in the mobile phase is the oxidative degradation of samples and phases and a reduction in the sensitivity and baseline stability of ultraviolet, fluorescence and electrochemical detectors [2,34-37]. Solvated oxygen complexes absorb significantly in the low wavelength UV region, 190-260 nm, and changes in the mobile phase oxygen concentration lead to detector baseline drift, random noise, a loss of sensitivity, and a large baseline offset at high sensitivity. Oxygen can cause intersystem energy transfer from the excited state of fluorescent compounds resulting in "quenching" of the fluorescence signal through this nonradiative decay pathway. Oxygen removal to very low levels is an absolute requirement for electrochemical detection in the reduction mode, since the oxygen reduction current contributes deleteriously to the residual current and detector noise. In the above cases temperature fluctuations and oxygen concentration may be coupled together in reducing detector baseline stability at high sensitivity. Degassing can be carried out by applying a vacuum above the solvent, heating with vigorous stirring, ultrasonic treatment, or sparging with helium. All the above methods are fairly efficient at preventing bubble formation. Refluxing solvents was found to be most efficient method of reducing oxygen concentrations while sparging with helium was less efficient but more convenient. Vacuum degassing and ultrasonic treatment are not very effective at reducing oxygen levels. Helium sparging removes about 80-90% of the dissolved air within 10 min by drawing the unwanted dissolved gases from solution as they equilibrate with the helium bubbles. Since helium has a very low solubility in common solvents, after sparging the solvents are nearly gas-free. Redissolution of unwanted gases is prevented by continued sparging with helium or by maintaining the mobile phase in a helium atmosphere. Particular care is needed when sparging with helium to prevent back diffusion of air into the solvent reservoir against the much lighter helium gas. This includes the exclusion of Teflon (or related plastic) tubing, permeable to air, from the solvent delivery system. An alternative method for oxygen removal is catalytic reduction over a platinum-on-alumina catalyst in the presence of methanol [38]. A short column packed with the catalyst is inserted between the pump and injector. Provided that the mobile phase contains on the order of about 1% methanol (or more) oxygen is very efficiently reduced producing formaldehyde and formic acid. It is necessary to ensure that these products do not interact with or degrade the analyte. On-line solvent degassers based on gas membrane technology are now commercially available. These are very efficient and convenient for solvent degassing without consuming expensive helium. Their efficiency at removing oxygen to low levels is not indicated.

5.4 SOLVENT DELIVERY SYSTEMS

The solvent delivery system is one of the most important components of the liquid chromatograph since its performance directly affects retention time reproducibility and detector baseline stability [1,2,39,40]. As well as the pump, other components such as check valves, flow controllers, mixing chambers, pulse dampeners, and pressure transducers make up the solvent delivery system. Some of these components may be electronically linked to the pump to control its output or switch it off if a default value is exceeded.

The types of pumps used in modern liquid chromatographs can be divided into two categories: constant pressure pumps, such as gas displacement and pneumatic amplifier pumps, and constant volume models, such as reciprocating-piston and syringe pumps, Figure 5.3. The gas displacement pump is the simplest and least expensive of all pumps but is limited to low pressure operation, has a limited reservoir volume, and limited possibilities for gradient elution. A gas cylinder with regulator is used to pressurize a liquid in a cylinder or holding coil and then force it out through a check valve. The solvent flow is pulse free but the volume delivery depends on the column back pressure. Gas displacement pumps are sometimes used as mobile phase pumps in open tubular liquid chromatography and for reagent addition in postcolumn reaction detectors. The pneumatic amplifier pump uses a large bore gas piston to drive a liquid piston of smaller size in a dual chamber configuration. The pressure on the eluent piston is proportional to the gas pressure and the area ratio of the two pistons. The pneumatic amplifier pump is equipped with a power-return stroke that permits very rapid refilling of the empty piston chamber with mobile phase. This pump can provide high

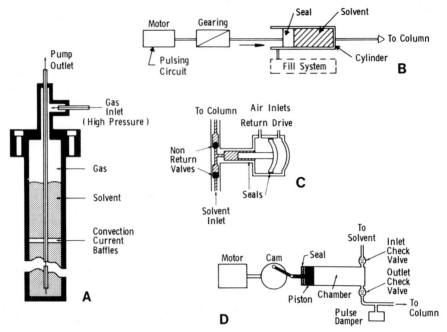

Figure 5.3 Different types of pumps used in high pressure liquid chromatography. A, gas displacement pump; B, syringe pump; C, pneumatic amplifier pump; D, reciprocating pump.

pressures and relatively high volumetric flow rates. It is used primarily for slurry packing columns and in some preparative-scale liquid chromatographs. Syringe pumps employ solvent displacement by a mechanically controlled piston advancing at a constant speed in a fixed volume chamber. The pump output is relatively pulse free, very high pressures can be obtained, and gradient and flow programming are quite straightforward, at least in principle. Syringe pumps were widely used in the early development of instrumental liquid chromatography but high cost, limited solvent reservoir capacity, and problems with solvent compressibility resulted in their decline for use with columns of conventional bore. Since solvents are compressible, a finite time is required before the volumetric flow rate is constant at high pressures [41,42]. The time required to reach steady-state flow conditions is proportional to the volume of the piston reservoir, the compressibility of the solvent (typically ca. 0.5-1.5 % per 100 bar), and the steady state inlet pressure, and is inversely proportional to the flow rate. Most modern pumps are equipped with a fast pump option to rapidly pressurize the solvent in the reservoir and minimize the delay time when the pump is first started. Solvent compressibility problems are less significant for pumps with small reservoir volumes operated at low flow rates. Thus, syringe pumps with cylinder volumes of 10-50 ml are widely used in instruments designed specifically for small bore and packed capillary column operation, where typical flow rates are less than 100 microliters/min [14-16]. Syringe pumps are also commonly used in instruments designed for supercritical fluid chromatography.

Most general-purpose pumps used today are of the single- or multi-head reciprocating-piston type. Each stroke of the piston displaces a small volume (typically 40-400 microliters) from a pump chamber equipped with inlet and outlet check valves. During the intake (fill) stroke, the piston is withdrawn from the cylinder, lowering the pressure upstream from the inlet check valve. The inlet check valve opens, and liquid flows into the cylinder. The outlet check valve closes, because the system pressure exceeds the cylinder pressure. On the delivery stroke, the piston moves into the cylinder, the inlet check valve closes, the outlet check valve opens, and liquid flows to the column. A special seal of filled (usually carbon) fluoropolymer around the piston prevents leakage of the solvent during piston motion. Pistons and check valve balls and ball seats are usually made from sapphire or ruby, and the pump head from stainless steel. Recently, titanium, Hastelloy, and fluoropolymers have been used for pump heads, so the device can be operated with corrosive solvents, or used to separate metal-sensitive samples [43]. Early single-piston pumps delivered a pulsating flow due to their simple sinusoidal stroke and required that various pulse dampening devices (discussed later) were used to reduce the pump pulsations observed at the detector. Modern single-piston pumps use an asymmetrical cam to drive the piston, which speeds up the piston movement during the refill stroke, and produces a uniform delivery over most of the cycle regardless of flow rate. With a dual-head reciprocating-piston pump both chambers are usually driven by the same motor through a common cam, such that the two pistons are 180° out of phase. As one chamber is emptying the other is refilling; thus the two flow profiles overlap, leading to partial cancellation of the peaks and troughs in the total flow output. Triple-head pumps work on the same principle, this time each piston being 120° out of phase with its neighbor, and theoretically provides a greater cancellation of flow pulsations. An alternative design for a dual-head reciprocating-piston pump uses the second piston as a pulse compensator. It operates in a valveless chamber connected in series to the main chamber. The compensating piston withdraws at a slower rate during the delivery stroke of the main piston, accumulating liquid to be dispensed when the main chamber is refilling.

Nearly all reciprocating-piston pumps use a pulse dampener in series with the pump head(s), and most also use some sort of electronic feedback control to vary the pump motor speed within each cycle to further reduce pulsations, as well as to maintain accurate flow as the column backpressure varies. Mechanical pulse dampeners store energy during the pressurizing stroke and release it during the refill stroke. Pulse dampening has been achieved using a variety of devices, such as bellows or coiled tubes, often in conjunction with a fluid or gas ballast reservoir. An inexpensive pulse dampener can be prepared from a length of flattened stainless steel tubing of significantly smaller cross section than the pressure peak profile of the mobile phase and a Bourdon tube pressure regulator. A commercially available pulse dampener uses a flattened length of PTFE tubing immersed in a degassed compressible liquid. The flexibility of the tubing and the compressibility of the fluid thus absorb any pressure fluctuations in the solvent stream. Alternatively, the flow output can be passed over a thin, flexible, steel membrane, one side of which is in contact with a compressible fluid. All instruments should contain a pressure sensing device to prevent overpressurization which can result in damage to the system. A strain-sensitive semiconductor crystal, whose resistance varies with the applied pressure, is generally used.

Reciprocating-piston pumps deliver a constant flow at a fixed backpressure. At high pressures some minor flow variability may arise due to the compressibility of the mobile phase. Some instruments incorporate a flow controller which provides a fixed backpressure for the pump to work against, independent of the column backpressure. The influence of pressure fluctuations, solvent compressibility, and solvent viscosity on the volumetric output of the pump are thereby eliminated. Reciprocating-piston pumps can provide continuous solvent delivery, fast solvent change over, gradient elution compatibility, and have low maintenance requirements.

For modern high pressure liquid chromatography the solvent delivery system must meet certain general requirements. A typical pump for use with conventional bore columns should have a flow rate range of about 0.1 to 10 ml/min and a maximum operating pressure of about 6,000 p.s.i. (conversion factors: 1 bar = 14.5 p.s.i. = 0.9868 atm = 1.02 kg/cm^2 = 0.1 MPa). Different pumps are required for microcolumn and/or preparative-scale operation due to the large flow rate differences incurred in operating columns with different internal diameters. The volumetric flow rate should be stable and accurate to better than a few percent over the normal operating range for the flow delivery system [5]. A set of criteria for measuring the performance of the solvent delivery systems is summarized in Table 5.3. In general terms, a high degree of accuracy in pump resetability and flow rate control with a minimum of pump pulsation and drift are the hallmarks of a good pump. Flow reproducibility can be evaluated from the changes in the elution time of an unretained solute over time. The solvent delivery system proportioning accuracy can be determined by replacing the column in the chromatograph by a length of capillary tubing and measuring the step height increases on the detector trace for mixing two solvents (or the same solvent) in which a small amount (< 1%) of a UV absorbing substance (e.g., acetone) is added to one of the solvent reservoirs. Suitable solvent increments are 10% for 0-100% solvent B and 1% for 1-10% solvent B for a binary solvent mixture AB, Figure 5.4. Most pumps have difficulty in mixing a few percent of one solvent in another and the largest variations in composition accuracy can be expected for solvent mixtures containing only a few percent of one of the solvents. The contribution of the solvent delivery system to baseline noise can be determined by the difference in baseline disturbance for measurements made with and without flow through the detector cell (see section 1.8.1). Other considerations for pump selection might be price, chemical resistance to corrosive solvents, serviceability, ease of operation, and the time required for solvent change over.

5.4.1 Formation of Solvent Gradients

Sample components having a wide range of capacity factor values are not conveniently separated by isocratic elution.

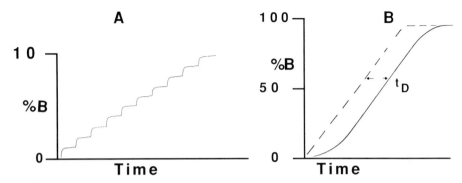

Figure 5.4 A, an example of the proportioning accuracy of a reciprocating-piston pump in the composition range 0-10% B in 1% increments. B, gradient shape as a function of the design of the solvent delivery system. The Broken line describes the set gradient entered at the gradient controller and the solid line the gradient delivered to the column inlet. The dwell time is marked as t_D.

TABLE 5.3

PUMP SPECIFICATIONS

Specification	Description
Pump Resetability	The ability to reset the pump to the same flow rate repeatedly.
Flow Rate Accuracy	The ability of the pump to deliver exactly the flow rate indicated by a particular setting.
Pump Pulsation or Noise	Flow changes sensed by the detector as a result of pump operations, such as piston movement and check valve operation.
Short Term Precision	The accuracy of the volume output of the pump over a few minutes.
Drift	Measure of the generally continuous increase or decrease in the pump output over relatively long periods (e.g., hours).

Gradient elution conditions are frequently used for these samples (section 4.6.3). A mobile phase gradient is formed by mixing two or more solvents, either incrementally or continuously, according to some predetermined gradient shape. The most frequently used gradients are binary solvent systems with a linear, concave or convex increase in the percent volume fraction of the stronger solvent. The basic requirement of a gradient device can be stated as follows: it should have the ability to produce a reasonable

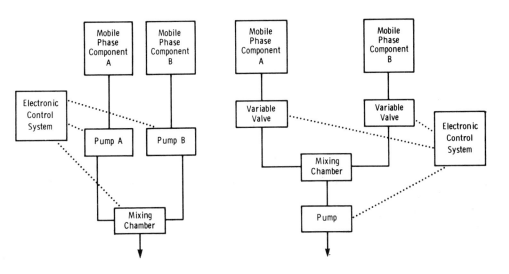

Figure 5.5   Two methods for generating binary solvent gradients.

choice of useful gradient profiles; it must provide homogeneous mixing of the mobile phase before it reaches the column; it should be able to create reproducible and accurate gradients over the full solvent mixing range (0-100% B); and it should minimize the time delay between the point at which the solvents are combined and delivered to the column. Gradient devices are usually classified into two types, Figure 5.5, depending on whether the solvents are mixed on the low or high pressure side of the pump. Low pressure mixing devices use time-proportioning electrovalves, regulated by a microprocessor or similar device, to control solvent delivery to the pump [1,44]. To obtain the highest possible accuracy the operation of the electrovalves should be synchronized with the movement of the piston. This gradient architecture is little influenced by solvent compressibility effects and can completely eliminate errors connected with thermodynamic volume changes due to mixing of the solvents.

In high pressure mixing devices each solvent is pumped separately in the proportions required by the gradient into a mixing chamber before being delivered to the column. Solvent compressibility and thermodynamic volume changes on mixing may influence the accuracy of the composition delivered to the column.

A significant disadvantage of the high pressure mixing architecture is the need for a separate pump for each solvent used.

The mixer in a gradient system is required to dampen fluctuations in eluent composition caused by the fact that the solvent delivery from reciprocating-piston pumps occurs in discrete steps. Effective reduction of detector baseline noise caused by eluent variations requires a mixer that has a large volume relative to the piston volume of the pump. Typically, a compromise must be made between the desire to have a fast response to solvent composition changes at the column inlet and detector noise caused by short-term composition variations. Generally, homogeneous mixing can be achieved either by static flow mixing devices (open or packed tubes) or by actively stirred dynamic mixing chambers. Mixing chambers used in instruments designed to accommodate conventional bore columns have volumes of about 1-4 ml. Miniaturization of gradient systems that can produce accurate and precise gradients for small bore and packed capillary columns is not a simple matter [11,14,16,45-49]. The simplest approach is to use two syringe pumps operating in the microliter flow rate range and a static mixer of 5-10 microliter volume. A single or series of microexponential dilution vessels with a single pump is another popular approach for generating solvent gradients with microliter/min flow rates. The gradient is formed in a stirred mixing vessel. After the initial solvent is introduced via a switching valve, the final solvent is allowed to dilute the initial solvent exponentially. A disadvantage of these systems is that the geometrical configuration of the mixers must be altered to change the gradient shape. A more versatile system based on storing and mixing the gradient in an injection loop was described recently [49].

Differences in gradient elution separations that arise from the use of different equipment are due mainly to differences in the dwell volume, mixing volume, and the use of malfunctioning or poorly designed equipment [5,50-52]. Blank runs with a UV-absorbing compound added to the mobile phase can be used to measure the actual gradient that will be delivered to the column compared with the gradient selected. With the sample injector connected directly to the detector (column removed) and methanol containing about 0.01% acetone in solvent reservoir A and methanol containing 0.1% acetone in solvent reservoir B the UV detector wavelength and sensitivity is adjusted to obtain about 90% full-scale absorbance for solvent B. The blank gradient is then run from 0 to 100% B (or other desired range) over an appropriate time (e.g., 20 min) with the normal column flow rate. A typical example for a linear gradient is shown in Figure 5.4, the broken line representing the set gradient and the solid line the gradient delivered to the column inlet. Gradient delay and rounding are observed to varying degrees in all gradient systems. The gradient delay time, marked as t_D in Figure 5.4, is determined by the volume of the mixer plus all volume elements between the mixer and column inlet (including the pump for low pressure gradient mixing). For contemporary instruments designed for conventional packed column use, typical dwell volumes are 0.5-6.0 ml, corresponding to a dwell time of 0.5-6.0 min at a flow rate of 1 ml/min. These time differences are substantial and influence the time basis on which a change in composition is registered at the column inlet for different instruments. The rounding of the ends of the gradient is related to the volume of the mixer. The programmed gradient and the experimental gradient will be most alike when the dwell volume and mixing volume are as small as possible but the physical layout of the solvent delivery system and the need to dampen fluctuations in the mobile phase composition dictates that a compromise size is used in practice. It is important that the column holdup volume and the gradient dwell volume are known to accurately relate solvent composition to the elution position of peaks in gradient elution chromatography (section 4.6.2).

The variability in retention volumes and peak widths in gradient elution chromatography is due primarily to the limits of precision and reproducibility of the mobile phase composition, the flow rate, and the column temperature. Random or systematic deviations from the preset mobile phase volume composition and flow rate are caused by imperfect functioning of the mechanical parts of pumps (plungers, valves, seals, etc.,) or the electronic parts of the system. Deviations from the instrument gradient can also occur due to solvent demixing, resulting from preferential adsorption of the stronger solvent by the stationary phase.

5.5 INJECTION DEVICES

The ideal sample introduction system should reproducibly and conveniently insert a range of selectable sample volumes into the column as a sharp plug without adversely affecting the efficiency of the column. No injection device for liquid chromatography is ideal in this respect, but the superiority of valve injection has been adequately demonstrated and it is now universally used in virtually all commercial instruments as the basis for both manual and autoinjection products [1,2,53,54]. Earlier approaches using septum type injectors, analogous to those used in gas-liquid chromatography, have passed into disuse for a variety of reasons, such as limited pressure capability, poor resealability, contamination of the mobile phase, disruption of the column packing, etc., [39,55]. Many of these problems could be solved by different ingenious injection port designs, but in the end the loop-injection valve approach emerged as the preferred method of sample introduction because of its convenience, ease of use, reliability, and ease of automation. A typical injection valve is illustrated in Figure 5.6. The valve consists of a rotating seal (or rotor) and a fixed body with a number of external ports for attachment of sample loops, fill ports, waste lines, and column and pump connections. External sample loops are constructed from stainless steel tubing of various diameters, and are connected to the valve body using standard compression fittings. Rotation of the valve core connects different ports in series allowing the sample loading operation to be separated from the flow path of the mobile phase and then for the sample-containing loop to be rapidly inserted into the mobile phase flow path. All of the materials in the valve that come into contact with the sample or mobile phase must be inert and nonsorptive of the sample to avoid memory effects. Most manufacturers offer both electric and pneumatic actuation of their valves, proprietary polymer seals that are resistant to most liquid chromatographic solvents, and valve-bodies of stainless steel or more chemically resistant Hastelloy or tantalum. Injection valves can be used in systems operating up to 10,000 p.s.i. at ambient temperature and can be used at temperatures well above ambient, although usually with lower operating pressure limits.

The sample is loaded at atmospheric pressure into an external or internal loop, or groove in the valve core and introduced into the mobile phase stream by a short rotation of the valve. The volume of sample injected is normally varied by changing the volume of the sample loop or by partially filling a sample loop with a fraction of its nominal volume. External sample loops have volumes from about 5 microliters up to about 1 ml,

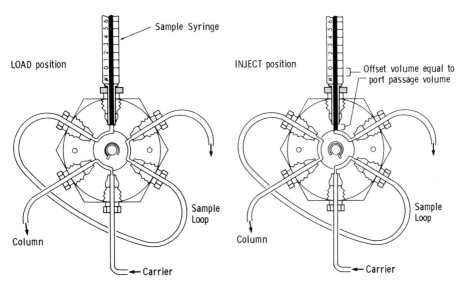

Figure 5.6 Microvolume injection valve showing the valve configuration in the load and inject position. (Reproduced with permission from Valco Instruments, Inc.)

although typical injection volumes for conventional diameter columns are 10-50 microliters. The lower limit is dictated by the diameter of the tube used for the sample loop and the distance between adjacent sample ports. The upper limit by the characteristics of the injection using normal flow rates, essentially poor injection profiles generated by laminar flow through the sample loop [28]. Using internal sample loops or grooves cut in the surface of the core to contain the sample, injection volumes can be reduced while maintaining the simplicity of valve injection [56]. Miniaturization of the injection volume is important for applications using small bore and packed capillary columns. The lowest nominal volume presently available for the groove-type injection valve is 60 nanoliters with 100, 200, 500, or 1000 nanoliter volumes possible with valves of this type, or using exchangeable fixed internal sample loops. The use of these valves and split injection, heart cut, or moving injection techniques allows still smaller sample volumes to be injected which are absolutely essential for some of the separation techniques employing microcolumns [11,12,56-59]. The moving-injection technique uses an actuator to move the valve from the load to the inject position and back to the load position very rapidly, so that only part of the sample loop is injected into the column. In this way samples from 100 to 3 nanoliters can be injected with RSD 2-5% depending on the flow rate and the time the valve stays in the inject position (typically 20-300 ms). With split injection the major part of both the sample and mobile phase are vented to waste with typical split ratios of 50:1 through 500:1 giving on-column injection volumes of 20-200 nanoliters. The split may be operated continuously (this is convenient when conventional pumps are used with small bore columns) or only during injection when the heart cut technique is used. A number of splitting devices have been described varying from a controlled leak from a column end fitting to T-type splitters incorporating a flow restrictor in the split line (microflow valve, packed column, fused silica capillary, etc.) to control the split ratio. Splitting can contribute to band broadening in low dispersion systems. The split ratio can vary with changes in the viscosity of the sample injected and drift in the column permeability. Berry has described the construction of a variable sample splitter based on a standard injection valve of the groove type [56,58]. One of the ports of the valve was equipped with two outlets and using the moving injection technique sample contained in the groove in the rotor was split between the column and a resistor (a short packed column). With a low resistor, the groove injector acts like a fixed volume loop injector, injecting part of the groove volume onto the column, down to about 30 nanoliters. With a high resistor, the groove injector acts like a split-injector with various fractions of the zone above one port being swept onto the column, depending on the flow rate, restrictor size, etc.

Some practical considerations when using a sample loop injector are important [2,60-63]. As a sample is introduced into the loop, it pushes the mobile phase ahead of it out of the loop. In this process, the leading edge of the sample becomes diluted because the fluid traveling through the tube has a parabolic flow profile as a result of laminar flow. This causes the core of the fluid stream to be moving faster than the edges in the vicinity of the tube wall. As a consequence, to fill the loop homogeneously, the sample injected must be at least twice the volume of the loop to displace all of the diluted sample formed initially as the sample is introduced. Likewise, if a partial loop injection is contemplated, the volume injected should be less than half the sample loop volume, to prevent the diluted front of the sample band from leaving the loop during injection. Laminar flow also causes the sample band to be diluted as it is pushed from the loop when inserted into the mobile phase stream. About 5-10 loop volumes are required to completely displace the sample from the loop. Thus the entire sample does not enter the column as a plug, but with an exponential decay injection profile that contributes to the zone dispersion of the separated sample bands in an adverse manner. The moving injection technique can be used to cut off the tailing portion of the sample injection band to retain maximum column efficiency. However, provided that the loop volume is not more than about 20% of the peak volume of the first peak of interest in the chromatogram, the shape of the injection profile can be ignored for practical purposes. This is the normal case for typical injection volumes used with conventional bore columns, but it becomes a more significant problem as the column bore is decreased.

It is generally recommended that whenever possible samples are prepared for injection in the same solvent mixture as used for the mobile phase, or in a solvent that is weaker than the mobile phase but miscible with it [60-62]. Injection solvents stronger than the mobile phase can cause variations in retention time and peak widths as a function of the injection solvent composition, and in the case of fairly large injection volumes, split peaks may also be observed. These effects are caused by variations in the penetration length of the sample zone at the head of the column due to displacement of the sample from the stationary phase by the injection solvent plug during the time it is diluted by the weaker mobile phase. The method of coupling the injection valve to the column can also cause a decrease in column efficiency. The length of the connecting tube should be short, 5 cm is an acceptable and convenient length for conventional bore columns, and of small internal diameter, 0.15 or 0.50 mm [63].

5.6 DETECTORS

Separations in liquid chromatography occur in a dynamic manner and, therefore, require detection systems which work on-line and produce an instantaneous record of the column events. A prototypic detector must have good sensitivity to deal with low concentrations of analytes typical of analytical liquid chromatography, a small detector volume to avoid additional band broadening due to extracolumn dispersion, and a fast response

TABLE 5.4

TYPICAL SPECIFICATIONS FOR HPLC DETECTORS

Property	Detector Type				
	UV/visble Absorption	Fluorescence	Refractive Index	Electrochemical	Conductivity
Limit of Determination					
Commercial (mass, ng)	0.1-1	10^{-2}-10^{-3}	10^{2}-10^{3}	0.01-1	0.5-1
Commercial (concentration, g/ml)	10^{-8}	10^{-11}	10^{-7}	10^{-10}	10^{-8}
State of the Art (mass, pg)	1	0.01	10^{4}	0.1	500
Linear Range (max)	10^{5}	10^{3}	10^{4}	10^{6}	10^{4}
Gradient Compatibility	Yes	Yes	No	No	No

compatible with the rapidly changing analyte concentrations. Detectors in common use belong to one of two classes: bulk property or solute property detectors [2,64-68]. Bulk property detectors, such as the refractive index detector and the conductivity detector, measure the difference in some physical property of the solute in the mobile phase compared to the mobile phase alone. Bulk physical property detectors are fairly universal in application but generally have low sensitivity and a limited dynamic range, Table 5.4. This arises primarily from the fact that the magnitude of the detector signal depends on the difference in properties between the solute and the mobile phase and not on the properties of the solute alone. Thus bulk property detectors are usually adversely affected by small changes in the mobile phase composition and temperature, which usually precludes the use of such techniques as flow programming or gradient elution.

Solute property detectors, such as spectroscopic and electrochemical detectors, respond to a physical or chemical property characteristic of the solute which, ideally, is independent of the mobile phase. Although this criterion is rarely met in practice, the signal discrimination is usually sufficient to permit operation with solvent changes (e.g., flow programming, gradient elution, etc.) and to provide high sensitivity with a wide linear response range, Table 5.4. Solute-specific detectors complement bulk property detectors as they provide high sensitivity and selectivity. Detector selectivity is much more important in liquid chromatography compared with gas chromatography since, in general, separations must be performed with a much smaller number of theoretical plates, and for complex mixture both column separation and detector discrimination may be equally significant in obtaining an acceptable result. Sensitivity is important, not only for trace analysis, but also for compatibility with the small sizes and miniaturized detector volumes associated with microcolumns in liquid chromatography.

The detection process in liquid chromatography has presented more problems than in gas chromatography. There is no equivalent to the sensitive, universal flame ionization detector for use in liquid chromatography. Indeed, the search for a liquid chromatographic detector with high sensitivity, a wide linear response range, predictable response, and the capability of detecting all solutes equally seems to have run full course, and this goal may never be reached. Detector technology has been the principle problem preventing the widespread use of microcolumns in liquid chromatography [8-13]. Current trends are directed towards the development of specific detectors for particular problems. The primary detectors, refractive index, spectroscopic and electrochemical will be described in the following sections.

5.6.1 Refractive Index Detectors

The refractive index detector was one of the first on-line detectors to be developed for liquid chromatography and it remains widely used today, probably second in popularity only to the UV/visible absorption detector [64-69]. It is a universal detector but is limited for some applications by poor sensitivity. Because of the temperature dependence and to a lesser degree the pressure dependence of refractive indices, all commercially available refractive index detectors are differential and provide a signal which is in some way related to the difference in refractive indices of the liquid in a sample and a reference cell. A solute will be detected only if its refractive index differs from that of the mobile phase. In the same chromatogram both positive and negative peaks may be observed depending on whether the analyte has a higher or lower index of refraction with respect to that of the mobile phase. The coelution of two peaks with opposite responses might also result in signal cancellation. The ultimate limit of the performance of the refractive index detector is caused by fluctuations in the refractive index of the eluent. The detector is sensitive to small changes in solvent composition, and the majority of problems associated with its practical operation can be traced to this cause. Changes in composition can occur from incompletely mixed mobile phases, leaching of prior samples or solvents from the column, or changes in the amount of dissolved gases in the solvent. The temperature and pressure dependence of the refractive index results in an offset in the detector baseline with changes in flow rate. Incomplete equilibrium of the incoming mobile phase temperature with the temperature of the detector flow cell and viscous heating in the inlet to the cell both contribute to this flow sensitivity. For most general applications the sensitivity of the detector to flow and composition changes of the mobile phase preclude the use of flow programming and gradient elution. For operation at high sensitivity in isocratic separations the ability of the solvent delivery system to accurately mix mobile phases and deliver a pulse free flow as well as the thermostating of the whole solvent delivery system, column and detector are important considerations. To be able to detect 10^{-6} to 10^{-8} g of solute, a noise equivalent concentration of 10^{-8} refractive index units is essential, and to maintain this level the temperature of the detection system must be thermostated to within $\pm 10^{-3} {}^{\circ}$C. For these reasons the refractive index detector is rarely operated at close to maximum sensitivity unless absolutely essential to the analysis. Because of the nature of its response, each substance will generally require an individual calibration curve for quantitative analysis. It is most frequently used with conventional diameter packed columns and preparative-scale columns. Its comparatively poor response and sensitivity to environmental factors has limited its use in microcolumn applications, even though small volume flow cells can easily be prepared for use with conventional or laser light sources [70-72].

Commercially available refractive index detectors employ the principles of refraction, reflection or interference of a collimated light beam to measure small differences in refractive index between the mobile phase and the analyte dissolved in the mobile phase. The deflection type refractive index detector (deflection-refractometer), Figure 5.7A, measures the refraction or bending of a light beam when it crosses a dielectric interface separating two media of different refractive index at an angle of incidence other than zero (Snell's law). Light from a tungsten

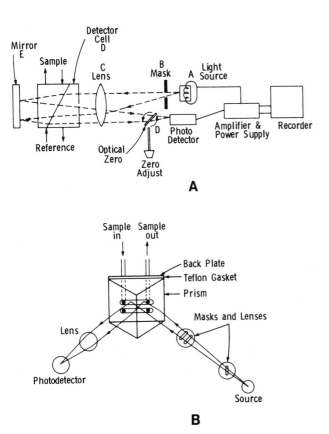

Figure 5.7   Schematic diagram of refractive index detectors
employing the principle of refraction (deflection-refractometer),
A, and the reflection principle (Fresnel-refractometer), B.

lamp passes through a beam mask, collimating lens, sample and
reference cells; is reflected by a mirror; and passes back through
the cells and lens, which focuses the light onto a position
sensitive photocell. A rotatable glass plate between the lens and
photocell is used to zero the output of the detector when the
sample and reference cell contain the same solvent. The sample
cell and reference cell are triangular in shape and located in a
single fused glass assembly. For analytical applications the cell
volume is usually 10 microliters, and the undeflected light beam
leaves the first cell and enters the second cell at 45°. If the
refractive indices are the same in the two cells, a light ray
leaving the first cell will be refracted the same amount but in
the opposite direction to that at which it enters the second cell,

so that the net deflection of the ray is zero after passage through the two cells. If the refractive indices in the two cells are different, there will be a net deflection of the light ray after passage through the two cells, which is doubled after reflection from the mirror and passage back through the cells, and translates into a lateral displacement of the focused beam at the position of the photocell. Deflection-refractometers have the advantage of a wide linear range, the entire refractive index range can be accommodated using a single cell, low volume cells can be fabricated, and the cells are relatively insensitive to air bubbles or the buildup of contaminants on the sample cell windows.

The Fresnel refractive index detector, Figure 5.7B, senses a change in the amount of light transmitted through a dielectric interface between the surface of a glass prism and the liquid to be monitored when the refractive index of the liquid changes. A beam mask and collimating lens is used to produce two parallel beams of light from a tungsten lamp that impinge on the prism-cell interface at slightly less than the critical angle for total internal reflection. The flat sample and reference cells are physically defined by the surface of the prism, the grained surface of the metal cell plate which contains the inlet and outlet ports, and a thin PTFE gasket. A collecting lens focuses the light scattered from the grained metal surface on to a photodetector which is sensitive to changes in intensity. The reference cell compensates for mobile phase variations and for source fluctuations. The difference in intensity of the light reflected in the two beams is related to the refractive index difference between the sample and reference cell by Fresnel's law and has a fairly large linear range when the incident light strikes the cell near its critical angle. A disadvantage of this approach is that two different prisms may be required to cover the full range of refractive index values encountered for typical solvents. On the other hand the advantages of this design are high sensitivity, small cell volume, and ease of cleaning.

The interferometric refractive index detector measures the difference of the speed of light in a sample cell and reference cell (the speed of light in a medium is directly related to refractive index by definition) by means of the interference of the two light beams after passage through the cells. As the relative speed of light in the two cells changes, destructive interference occurs and the energy measured by a photomultiplier detector decreases. An order of magnitude increase in sensitivity compared to conventional detectors is claimed for the interferometer detector. This is because a longer path length can be used to compensate for lower sample concentrations.

5.6.2 Spectrophotometric Detectors

The most important of the spectrophotometric detectors are the UV/visible absorption detector and the fluorescence detector [64-69,73-75]. The UV/visible absorption detector is the most widely used detector in liquid chromatography. Since most organic compounds have some useful absorption in the UV region of the electromagnetic spectrum, these detectors are fairly universal in application, although sensitivity depends on how strongly the sample absorbs light at a particular wavelength and the availability of a transparent mobile phase for the separation corresponding to the wavelength of maximum absorption. Organic compounds absorb most strongly in the wavelength range 180-210 nm, which is also the range in which most solvents have strong absorption bands, making detection difficult in this region. Longer wavelengths are typically used in practice reducing both the sensitivity of the detector and its ability to function as a universal detector. Many organic compounds either do not absorb or exhibit insignificant absorption at wavelengths greater than 210 nm, that is, at those wavelengths most likely to be used for detection. The operation of absorption detectors is based on the measurement of the absorption of monochromatic light in accordance with the well-known Beer-Lambert law. Most detectors provide a signal in absorbance units which is linearly related to sample concentration over a range of 10^4 to 10^5 and detection limits, for the most favorable cases, in the range of 0.1-1.0 ng (about 10^{-8} M in concentration terms).

Fluorescence detection is inherently more sensitive than absorption detection but is further restricted in general application. Only a small percentage of all organic compounds that absorb light are naturally fluorescent, and then only a still smaller fraction are strongly fluorescent. The inherent sensitivity and selectivity of fluorescence detection has generated a great deal of interest in fluorescence labelling (precolumn derivatization) and reaction flow detectors (postcolumn derivatization) for trace analysis of non-fluorescent or weakly fluorescing compounds (see section 8.11.2). This considerably expands the possible applications base for this method of detection. The selectivity of the fluorescence detection process arises because two wavelengths, an excitation wavelength and an emission wavelength, are used in the measuring process, and from the fact that certain structural features are required in a molecule for fluorescence to occur. The molecular requirements for fluorescence cannot be clearly defined, but many fluorescent molecules contain rigid, planar, conjugated systems. Fluorescence is a very sensitive technique since, in contrast to absorption measurements, the emitted radiation is measured against a dark background due to optical separation of excitation (absorption) and emission (signal) processes, permitting the use of higher signal amplification. Under favorable circumstances detection limits of 1-10 pg can be obtained using conventional detectors with a linear range of about 1000. Laser-based fluorescence detectors are capable of still lower mass detection limits and enable very small detector volumes to be used that are essential to preserve the efficiency of microcolumn separations [76,77].

There are several types of absorption detectors commercially available that vary in their flexibility and capacity to provide single or multiple wavelength detection capability [1-3]. The simplest, and among the most sensitive of the absorption detectors, are the fixed single wavelength detectors employing atomic vapor sources, a simple filter, and a photodiode or photomultiplier detector. Typical atomic vapor lamps include mercury with strong emission lines at 254, 313 and 365 nm (among others), cadmium 229 and 326 nm, zinc 308 nm, and magnesium at 206 nm [78]. The source emission lines are narrow and well separated and, therefore, a monochromator is not required, reducing the cost of the detector. The number of wavelengths available for detection can be increased by using a phosphor screen to re-emit the source light at a longer wavelength, for example, using the mercury atomic vapor lamp the emission line at 254 nm can be re-emitted at 280 nm, and some detectors allow switching between these two wavelengths or monitoring the response at both wavelengths simultaneously. The light intensity at the re-emitted wavelength, however, is only a fraction of that at the excitation wavelength produced by the lamp and the signal-to-noise ratio somewhat lower. Using a highly regulated source, dual beam operation, and silicon photodiode detectors, a very high signal-to-noise ratio can be achieved with the fixed single wavelength detector. Their principal limitation is that the sample must have some absorption at the source wavelength to be detected, and the fact that it may not be possible to select the optimum wavelength for maximum sensitivity or selectivity for the separation from the limited number of source lines available. These problems are overcome using continuously variable wavelength detectors or simultaneous multiple wavelength detectors.

The variable wavelength detector is probably the detector most commonly found in analytical laboratories. This detector utilizes a continuum source combined with a monochromator for wavelength selection. Wavelengths are either selected manually in the range 190-800 nm or, in an automated version, the detector may be programmed to change wavelengths during the separation and, with stop flow techniques, scan the complete spectrum of any peak. These detectors use high energy sources, typically a deuterium discharge lamp for the range 190-350 nm and a tungsten lamp for the range 350-800 nm, relatively low optical resolution (i.e., moderate bandpass monochromators) to maximize energy throughput while maintaining an acceptable linear response range and utilizing stable low noise electronics. An example of a variable wavelength detector is shown in Figure 5.8A. Dual beam operation is almost universal to compensate for short-term electrical fluctuations, as well as other time-dependent variations in the source intensity. Lamp noise arises from two principal sources [78]. Eddies developed around the lamp envelope as a result of the heat generated by the lamp and the consequent variation of the refractive index of the surrounding air leads to fluctuations in the amount of light falling on the detector cell. Modern detectors are often fitted with an air circulation fan to improve the dissipation of heat from the source housing. The lamp arc is not homogeneous over its length, or over time, and variations in spectral output and intensity can occur. In practice, electronic noise and lamp noise is almost negligible when using dual beam operation compared to the noise originating at the detector flow cell, discussed later, which effectively limits sensitivity with contemporary detectors. In general, variable wavelength detectors produce a signal-to-noise ratio slightly lower than the fixed wavelength detectors but are not limited in wavelength selection.

Two approaches have been taken for the design of multiple-wavelength detectors [65-68,73-75]. The first approach uses conventional forward optics and a rapidly moving holographic

576

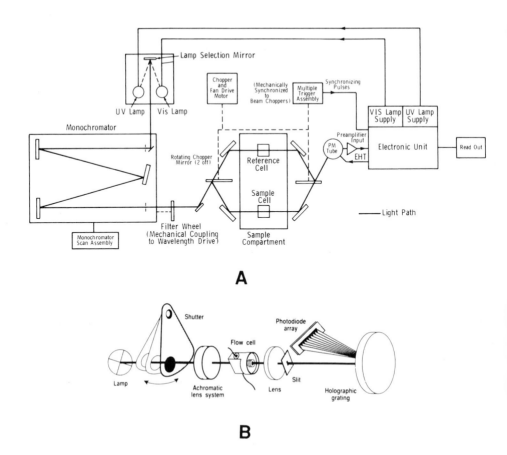

Figure 5.8 Schematic optical diagram of a variable wavelength dual beam absorption detector, A, and a photodiode array detector with reverse optics, B.

grating to disperse the light prior to passage through the flow cell and onto a single photodetector. The full spectra from 190-800 nm can be recorded at a sampling rate of about 10 spectra per second. This should be adequate for conventional columns but may not provide a sufficient number of data points for rapidly eluting peaks from microcolumns. The advantages of this approach arise from the use of conventional forward optics, which provide lower sensitivity to refractive index changes and stray light interference and should result in improved sample detectability. There are few detectors of this type commercially available, and the second option, utilizing reverse optics and a photodiode array sensor is more common [1,65,73,75,79,80]. As shown in Figure 5.8B, the photodiode array multiple wavelength absorbance detectors generally employ a parallel detection approach in which an array of photodiodes is placed across the focal plane of a polychromator. All the light from the source is focused through the flow cell and subsequently dispersed by the grating for simultaneous measurement. The light source is typically a deuterium lamp. The unusual features of the array detector are that the sample is illuminated with white light, and in the general case, the optical system of the detector contains no moving parts. This configuration is made possible by the photodiode array, which is a combined multichannel radiation sensor, storage element, and readout system. The photodiode array consists of several hundred photosensitive diodes (e.g., 1024, 512, 256 or 211) generally configured in a linear pattern on a silicon wafer. Each photodiode is connected in parallel to a storage capacitor. The photocurrent generated by light striking individual photodiodes discharges their respective capacitors. Each photodiode is successively interrogated by a digital shift register, the integrated intensity of the light registered at the diode recorded, and the entire readout process repeated every few microseconds. A microprocessor network is required to manage the data acquisition tasks and process the information in real time. The first generation photodiode arrays were of the self-scanned type with the associated readout electronics located on the same silicon wafer as the diodes. Self-scanned diode arrays provide a wider spectrum coverage and increased wavelength resolution but are relatively noisy. Second generation instruments use discrete (externally-scanned) arrays. These are less expensive but provide reduced spectral coverage (generally UV only) and lower wavelength resolution. On the other hand, their noise characteristics are better, providing improved sample detectability and increased data manipulation speed with the possibility of real time calculation and display of some chromatographic parameters. Diode bunching is another technique used for increasing signal-to-noise performance. In this case, the output from a group of adjacent diodes is electronically summed or averaged to produce one signal output. Practically, this is accomplished by defining a central wavelength for detection and an associated effective bandwidth. These parameters, however, must be set carefully to avoid nonlinearity in the calibration plots [80].

Multiple wavelength detectors generate an enormous amount of data which has to be made available to the analyst in a usable form. From the manufacturer's point of view this is a software problem and from the user's point of view this can be a learning and interpretation problem. Multiple wavelength detectors probably vary most in how they acquire data and enable that data to be manipulated in real time, or in post run calculation and display procedures [73,74,81-85]. One difference compared to variable wavelength detectors is that the complete spectral range is recorded continuously at microsecond time intervals and stored as raw data, or in an abbreviated form. This data matrix can then be interrogated in several ways and at a later date, without having to repeat the separation, should the needs of the analysis change. One common format for viewing the total data matrix is a three-dimensional pseudoisometric plot of absorbance, wavelength and time like the example shown in Figure 5.9A for the separation of two well resolved components [82]. Hidden line removal and false color can be used to minimize loss of information for small peaks obscured by earlier eluting more intense peaks. However, it can still be disconcerting to rotate the plot and see how the peak profiles change with the viewing angle. An alternative method of data display is the contour plot, Figure 5.9B, where concentric isoabsorptive lines are plotted in the wavelength and time plane. The contour plot diagram facilitates the easy identification of optimum wavelengths for detection of each component. Also, serious peak distortions can be seen and used to detect coeluting impurities.

Simpler forms of data display are often adequate for determining peak purity and homogeneity. Spectral overlays using several characteristic wavelengths, as shown in Figure 5.10, can be easier to interpret than the plots in Figure 5.9. The shifting retention times in the second and fourth peaks in the spectral overlay shown in Figure 5.10 suggest the presence of coeluting components. Another common method of determining peak purity is the use of absorbance ratios [82,83]. The computation of the absorbance ratio at two wavelengths for all points across the peak profile is a square-wave function for a single component peak. Coeluting impurities, which absorb at least one of the selected wavelengths, will distort this function. The absorbance ratio depends very little on the experimental parameters but the selection of detection wavelengths and the absorption threshold

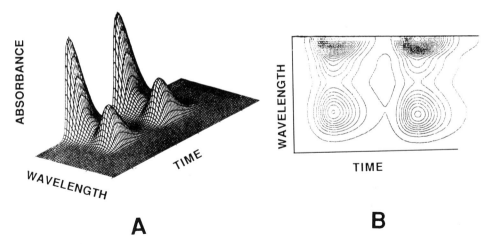

Figure 5.9 An example of display options for the chromatographic and spectroscopic data obtained with a photodiode array detector. A three-dimensional psuedoisometric plot of absorbance, wavelength and time is shown in A. In B the same data as in A is plotted as a contour diagram. (Reproduced with permission from ref. 82. Copyright Elsevier Scientific Publishing Co.)

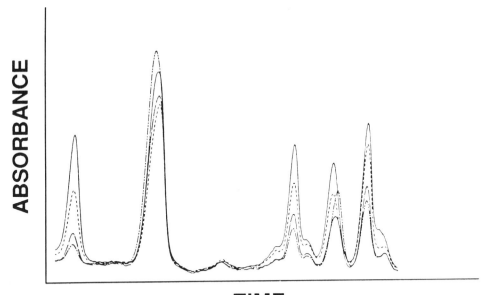

Figure 5.10 An example of the use of spectral overlays to detect coeluting impurities. (Reproduced with permission from ref. 73. Copyright Preston Publications, Inc.)

value can dramatically affect the utility of this technique. If the spectra of all components simultaneously present in a chromatographic peak are known in advance then spectral deconvolution techniques can be used for quantitation based on the sequential application of a least-squares routine to all data points in the chromatogram. Principal component analysis and factor analysis can also be applied to spectral peak profiles to estimate the maximum number of probable components present in a peak cluster.

The photodiode array detector also allows the recording of the complete absorption spectra for each component in the chromatogram and software is available for library searching using the normalized spectra [74,81,82,86]. This approach is far less powerful than is the case for mass spectral or infrared spectral data due to the rather broad and featureless bands that typify absorption spectra. Derivatives of the spectra can go someway to enhancing small differences between individual spectra and improve identification possibilities. However, UV/visible absorption spectra of similar compounds and compounds with a chromophore well separated from the variation in structure are often virtually identical. Also, spectral changes dependent on the experimental conditions (pH, type of organic modifier, temperature, etc.) occur frequently. It seems most likely, therefore, that UV/visible spectral libraries will be created for local use employing specific chromatographic conditions, but the wider creation of spectral libraries for the identification of unknown solutes, as is common practice for infrared and mass spectra, is an unlikely development.

Careful consideration must be given to the design of the detector flow cell as it forms an integral part of both the chromatographic and optical systems. A compromise must be made between the need to miniaturize the cell volume to reduce extracolumn band broadening and the desire to employ long path lengths to increase sample detectability. For columns of conventional internal diameters flow cells of the H-cell, Z-cell (Figure 5.11A), or tapered cell configuration with a cell volume of 10 microliters or less, are commonly used [64,87]. The Z-path is used to minimize stagnant flow regions in the cell and to reduce peak tailing. Typically, the flow is confined by two quartz windows held in place by caps screwed onto the cell body at either end of the tubular cell cavity. The tapered cell is designed so that the cell cavity is conical in shape, the aperture at the entrance for the light beam being narrower than the exit. The purpose of the tapered cavity is to reduce the amount of refracted light that is lost to the cell walls. Miniaturization of the same cell designs for use with small bore columns is possible by

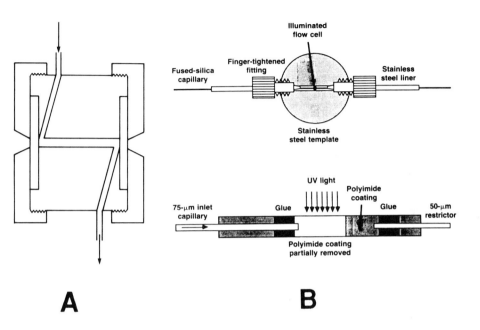

Figure 5.11 Typical flow cells used for absorption detection. A, Z-cell for conventional packed column applications and B, a fused silica microflow cell and holder used with packed capillary columns.

reducing the path length from 10 mm to 3 to 5 mm, providing cell volumes of about 0.2 to 2 microliters [13,88,89]. In order to maintain sensitivity the cell diameter should be reduced while maintaining the longest practical path length dictated by the cell volume requirement. Ultimately, however, reducing the cell diameter decreases the light throughput and increases background noise and susceptibility to refractive index changes. Commercially available flow cells for absorption detection using small bore columns have internal diameters of about 0.5 mm and may be internally coated with a reflective metallic layer to further reduce signal fluctuations caused by refractive index changes in the column eluent. For absorption detection using packed and open tubular capillary columns, detector volumes must be reduced further to the nanoliter range [15,90-93]. The simplest low volume flow cells are prepared by removing a short length of the polyimide protective coating, about 25 mm, from just below the position of the frit, if a packed fused silica capillary column is used. The cell volume and path length will depend on the diameter of the fused silica capillary if the cell is illuminated transversely across the column diameter. The detector volume is also defined by the slit width in the orthogonal direction to the capillary diameter, which cannot be made too small otherwise there is very little light transmitted through the capillary. Cell volumes from about 3 to 200 nanoliters are easily created in this way. A fixed fused silica capillary cell, like the example shown in Figure 5.11B, can be more convenient in practice, since different columns can be attached using a short piece of PTFE tubing without changing the optical alignment for the cell. Also, the cell path length can be made wider than the internal diameter of the separation capillary to increase sample detectability at the possible expense of introducing some additional band broadening due to the larger cell volume obtained.

The principal source of background noise in modern absorption detectors can be attributed to the inhomogeneous changes in refractive index of the eluent flowing through the cell [87,94-97]. This arises from temperature gradients, incomplete mixing of the mobile phase, flow perturbations, and turbulence. Their effect causes some of the incident light which would normally pass directly through the cell, to strike the cell wall instead, and is lost. The photodetector cannot distinguish light lost by refraction to the cell walls from light absorbed by the sample. The continuous variation of the refracted light contribution to the detector signal constitutes the major portion of the noise signal observed when the detector is operated at high sensitivity. Employing an optical design with aperture and field stops external to the cell is generally used to minimize the contribution from variations in the refraction of the flow cell [98]. Standardized procedures have been proposed for evaluating the performance of absorption detectors for liquid chromatography that includes an evaluation of the noise contribution to sample detectability [5,99,100].

Detectors available for fluorescence monitoring in liquid chromatography differ mainly in the method used to generate and isolate the excitation wavelength and to isolate the emission wavelength [64-68,101,102]. Since the signal intensity is directly proportional to the source intensity, high energy line (mercury)

or continuous (deuterium or xenon) arc sources are used. The
mercury source produces line spectra (254, 313 and 365 nm) which
can be isolated with a simple filter. However, the available
emission lines may not overlap with the maximum excitation
wavelength of the sample diminishing the response of the detector.
Deuterium (190-400 nm) and xenon (200-850 nm) arc sources are used
with a grating monochromator for continuously variable selection
of the excitation wavelength. Also, grating monochromators can be
programmed to change either or both excitation and emission
wavelengths during a separation to maintain optimum selectivity
and sensitivity throughout the chromatogram. Emission wavelength
isolation is performed using either filters or a monochromator.
Filters are low cost and simple to use, however, there is
generally a reduction in selectivity relative to wavelength
selection using a monochromator because the filters usually pass a
wider range of wavelengths. Monochromators having a narrow band
pass, provide continuous wavelength selection, and with stop-flow
scanning, can provide an emission spectrum for peaks of interest.
Fluorescence signals are usually of low intensity so that
photomultiplier tubes are commonly used for detection.

For dilute solutions (< 0.05 AU) the measured fluorescence
intensity can be related to sample concentration by equation (5.6)

$$I_f = I_o \, \phi_f (2.3abc) \tag{5.6}$$

where I_f is the fluorescence emission intensity, I_o the excitation
beam intensity, ϕ_f the quantum yield (number of photons
emitted/number of photons absorbed), a the molar absorption
coefficient, b the cell path length, and c the sample
concentration. For fluorescence detectors used in liquid
chromatography equation (5.6) describes the detector response over
about two to three orders of magnitude (linear range of the
detector). Sensitivity depends on the instrument (I_o and the
reduction of scattered and stray light), the sample (quantum
efficiency), and the composition of the mobile phase (solvents,
impurities, etc). The major sources of optical background noise
are: specular scatter, which results from the reflections and
refractions of the excitation wavelength at the various optical
interfaces of the detector; Rayleigh and Raman scattering
occurring in the flow cell material and mobile phase; and

luminescence of impurities present in the flow cell materials and the mobile phase.

It is frequently overlooked that the fluorescence signal from a sample may be dramatically affected in both the position of the wavelength emission maximum and emission intensity by the mobile phase composition and even by the presence of contaminants in the mobile phase. Some of these solvent effects are summarized in Table 5.5. Under less than ideal conditions the constancy with which the pump delivers and mixes the mobile phase and the presence of contaminants in the mobile phase may influence detector sensitivity and reproducibility more than fluctuations in the detector operating system. Fluorescence detection can be used with gradient elution, and unless the mobile phase contains a high level of fluorescent impurities, the detector baseline changes very little during the solvent program.

Flow cells for fluorescence detection with 3-10 microliter volumes are similar in design to those used for absorption measurements. Sample fluorescence is generated in all directions simultaneously and a hemispherical mirror can be used to maximize the fluorescence emission signal collected with an end-on photomultiplier detector. To achieve smaller cell volumes a reduction in path length is required. If high powered lasers are used as the source of excitation energy, a reduction in the cell volume can be achieved while maintaining an acceptable signal level [33,76,77,103-106]. Laser sources provide high intensity levels of collimated monochromatic light easily adapted to illuminating very small volumes. When sample sizes are limited, for example, in some biological applications, laser fluorometry may be the only technique suitable for quantitation by liquid chromatography. When the sample size is not restricted, the sensitivity of laser-based detectors using typical microcolumn chromatographic configurations is less than that obtained using a conventional packed column and a standard detector. The number of suitable wavelengths for excitation is restricted to a few with the range of lasers currently available. Four cell designs have commonly been used with microcolumns. Cells made from fused silica capillaries viewed longitudinally or transversely after removing a section of the polyimide coating are similar to those used for absorption measurements, except that the emission signal is usually measured at right angles to the direction of the excitation beam. Since light scattering and fluorescence

TABLE 5.5

THE INFLUENCE OF MOBILE PHASE COMPOSITION ON FLUORESCENCE EMISSION

Mobile Phase Parameter	Effect on Fluorescence Emission
pH	Both the emission wavelength and fluorescence intensity of ionizable aromatic compounds (i.e., those containing acidic or basic functional groups) are critically dependent upon pH and solvent hydrogen-bonding interactions.
Solvent	Intensity changes of an order of magnitude and large wavelength shifts are found for molecules which can undergo strong solvent interactions. A shift in the fluorescence spectrum to longer wave-lengths is usually observed as the dielectric constant of the solvent is increased. If the solvent absorbs any of the excitation or emission energy, the sensitivity will be reduced.
Temperature	Many compounds show marked temperature dependence with increasing temperature causing a decrease in intensity of 1-2% per °C.
Concentration	At high concentrations fluorescence emission becomes non-linear due to self-absorption by the sample itself or complete absorption of the excitation energy before it reaches the cell center. High fluorescence intensity may overload the photomultiplier tube which returns slowly to its normal operating conditions and misrepresents the actual fluorescence signal until restabilized.
Quenching	Impurities in the mobile phase, particularly oxygen, may entirely quench the signal from low concentrations of fluorescent compounds (see solvent degassing).
Photo-decomposition	High intensity sources may cause sample decomposition, which depends on the residence time of the sample in the detector cell.

impurities in the fused silica tubing contribute significantly to detector noise flow cells without walls have also been used. The flowing drop cell creates a suspended droplet at the end of the fused silica capillary and the liquid ensheathed cell imposes a strong flow of solvent around the eluent flowing out of the column and compresses it into a narrow stream a few micrometers in diameter. Convenience seems to dictate that the fused silica capillary flow cell has been the most widely used detection cell for microcolumn applications. In-column fluorescence detection is also possible using packed fused silica capillary columns by removing a small section of the polyimide coating to create a viewing area before the frit at the end of the column [33,106-109]. The sample zones are now detected in the presence of the stationary phase eliminating the contribution to band broadening from connecting tubes between the column and detector flow cell. Also, the sample zones are focused, such that their concentration is increased by the factor $(1 + k)$, where k is the solute capacity factor, compared to the concentration of the eluted zone. There is also an environmental factor which could result in an enhancement or diminution of the fluorescence emission for the sorbed sample in the stationary phase. On the other hand, the excitation beam probably dose not completely penetrate the packing, so that only a fraction of the adsorbed analyte is excited and the presence of the packing causes additional scattering of the excitation beam manifested as an increase in the background noise. Detection limits do not seem to be very different to those obtained using postcolumn detection, which is generally more convenient [33].

5.6.3 Electrochemical Detectors

Electrochemical detectors based on the principles of capacitance (dielectric constant detector), resistance (conductivity detector), voltage (potentiometric detector), and current (coulometric, polarographic and amperometric detectors) have all been used in liquid chromatography [34,66,67,110-119]. Conductance is a fundamental property of ions in solution and exhibits a simple dependence on ion concentration. Thus, the measurement of conductance is an obvious choice for the continuous and selective monitoring of ionic species in a column eluent. The conductivity detector came to prominence with the rapid development of ion chromatography for the detection of common ions lacking a chromophore and is still predominantly used for this purpose. All voltammetric detectors are based on the principle that when a voltage is applied to an electrode, any electroactive material will undergo electrolysis and produce an electric current. These detection systems afford excellent selectivity, since organic functional groups electrolyse only at specific values of the applied potential. Examples of compounds which can be detected conveniently are phenols, mercaptans, aromatic nitro and halogen compounds, catecholamines, heterocyclic compounds, ketones and aldehydes. When amperometric detection is used, the electrolysis of the electroactive species is not complete, being typically 1-10%. By increasing the electrode surface area the electrolysis may reach 100%, at which point the coulometric limit is reached. The coulometric detector is insensitive to flow rate and temperature changes and responds in an absolute manner, eliminating the need for calibration. However, the coulometric detector is more prone to electrode contamination, presents greater design problems, and requires strict potential control over the entire working electrode surface. Although the conversion of electroactive species is much higher in the coulometric detector, the background noise is also greater than for the amperometric detector; thus both detectors exhibit similar sample detectability. If amperometry is performed with mercury as the electrode material, this is referred to as polarography. Microvolume dropping mercury polarographic detectors can be used in liquid chromatography for the determination of reducible species (the low oxidation potential of mercury precludes its use for oxidizable species). Surface contamination is rarely a problem as the electrode surface is continuously renewed. Disadvantages include high background currents due to current oscillations over the lifetime of the drop, turbulence caused by the liquid flow in the region of the drop, and the need for complex cell designs. Recent innovations in cell design and electronic dampening of the signal have done much to domesticate the polarographic detector, although its general use is far less frequent than that of the other amperometric detectors. Sensitivity is similar to other amperometric detectors but its range of application is more limited. Potentiometric detectors (ion selective electrodes) appear to be far too specific for general applications in liquid chromatography [67,117].

Conductivity detectors have a simple and robust design and are easily miniaturized [67,113,114,117-125]. The detector consists of a small chamber made out of some nonconducting material containing two electrodes, across which a potential is applied. The electrodes are constructed of an inert conducting material, such as stainless steel or platinum, and have a low surface area to reduce the cell's capacitance. Commercially available detectors have cell volumes of 1 to 10 microliters, although detectors with volumes in the nanoliter range have been described in the literature. As a rule, the resistance of the cell is measured in a Wheatstone bridge circuit. During the measurement of the cell resistance, undesirable processes, such as electrolysis or the formation of an electric double layer on the electrodes, may occur. By varying the frequency of the applied potential the electrolytic effect is suppressed or is completely eliminated, and the current flowing through the cell is determined by capacitive impedance. The effect of this impedance can be removed by measuring the instantaneous current (i.e., the current generated at the moment when the potential is applied before the electrical double layer has formed). Some detectors apply a sinusoidal wave potential across the cell electrodes at 1-10 KHz and synchronous detection of the component of the cell current which is in phase with the applied potential frequency. Other detectors use a bipolar pulse conductance technique. Two short-duration, about 100 microsecond, voltage pulses are applied in succession to the cell. The pulses are of opposite polarity but equal in amplitude and duration. At exactly the end of the second pulse the cell current is measured and the cell resistance is determined by applying Ohm's law. Because an instantaneous cell current is measured in the bipolar pulse technique, capacitance does not affect the measurement and an accurate cell resistance measurement is made. Most detectors are equipped with an automatic temperature-compensating circuit, which produces a linear change in resistance as a function of the temperature of the column eluent. The detector cell is insulated to avoid sudden temperature changes. All detectors must have some means of offsetting the background conductivity signal. Conductivity detectors are simple to operate and, in the absence of background electrolyte, provide high sensitivity (10^{-8}-10^{-9} g/ml). Conducting impurities in the mobile phase limit absolute sensitivity.

If a conductivity detector is used to monitor the effluent from an ion-exchange column, the signal observed when a solute ion elutes is proportional to the solute concentration and to the difference in limiting equivalent ion conductances between the eluent and solute ions. The detector response for anion-exchange, for example, is given by

$$\Delta G = (Y_{s^-} - Y_{E^-}) C_{s^-}/K \tag{5.7}$$

where ΔG is the conductance signal, Y_{s^-} and Y_{E^-} the limiting equivalent ionic conductances of the solute and eluent anions, respectively, C_{s^-} the concentration of the solute anion, and K the cell constant which takes into account the physical dimensions of the cell. Sensitive detection can result whenever there exists a considerable difference in the limiting equivalent ionic conductance of the solute and eluent ions.

Many cell designs for solid working electrode amperometric detectors have been devised, but with few exceptions the thin-layer and wall-jet cell with a three-electrode configuration is the most popular [34,66,67,111,115-117,126]. A cross-sectional view of the wall-jet and thin-layer cells is shown in Figure 5.12 along with a complete assembly for a thin-layer amperometric detector showing the electrode positions [34,67]. The column eluent is introduced either parallel to the electrode embedded in the channel wall using the thin-layer cell, or perpendicular to the electrode surface followed by radial dispersion in the wall-jet design. Because amperometric detection is a surface phenomenon, the cell volume is easily miniaturized. Cell volumes of 1-5 microliters are common and by reducing the thickness of the spacers, thin-layer cells with volumes of 100 nanoliters or less have been produced. Using a single carbon fiber as a working electrode detectors of very small volume have been constructed for use in microcolumn liquid chromatography [112,127,128]. The body of the thin-layer cell is usually made out of an insulating material like PTFE and machined to accommodate the eluent channel and various electrodes. The reference and auxiliary electrodes are placed on the downstream side of the working electrode, so that either leakage from the reference electrode or the formation of any electrochemical products at the auxiliary electrode do not interfere with the working electrode. The reference electrode is normally placed close to the working electrode to ensure that the electrical resistance of the cell is kept to a minimum. By inclusion of a reference electrode the potential of the working electrode can be monitored. The auxiliary electrode still serves to carry the current but, if there is any deviation in the potential of the working electrode from the pre-set applied potential, current feedback via the auxiliary electrode can be employed to restore the balance. The types of materials used for the reference electrode (e.g., silver-silver chloride) and auxiliary electrode (e.g., platinum or stainless steel) are not critical, but the choice of working electrode material is very important as it affects the detector performance. A wide variety of materials have been used for the working electrode, but the

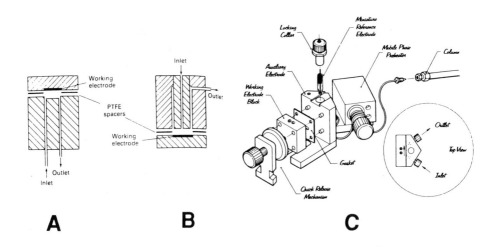

Figure 5.12   Schematic diagram of A, a thin-layer cell, B, a wall-jet cell, and C a complete assembly for a thin-layer amperometric detector.

most popular is glassy carbon. Glassy carbon has a surface that is easily polished, and with a working range of -1.3 to +1.5 V, it is ideal for a wide variety of applications. Metallic electrodes are generally too easily oxidized producing high residual currents for general use in detector flow cells. The chromatogram is recorded by measuring the detector cell current at a fixed potential as the sample is eluted from the column. The background will remain constant as long as the mobile phase velocity and composition do not change and is subtracted from the analytical signal. The resulting detector current is directly proportional to the concentration of electroactive species in accordance with Faraday's law.

Electrochemical detection requires the use of conducting mobile phases containing salts or mixtures of water with water-miscible organic solvents, conditions which are compatible with reversed-phase and ion-exchange chromatography, but are more difficult to achieve with other techniques. These problems can be circumvented, to some extent, if the mobile phase is water miscible by adding a makeup flow of support electrolyte at the column exit. Detector operation is critically dependent on flow rate constancy, solution pH, ionic strength, temperature, cell geometry, the condition of the electrode surface and the presence of electroactive impurities (dissolved oxygen, halides, trace metals, etc.) [34,129,130]. The background detector noise and, thus the ultimate sensitivity of the detector, are controlled by dissolved oxygen, ionic impurities and the contamination of the electrode surface, coupled with transient changes in the pump output. As the response of the detector is critically dependent on the overall solution characteristics of the mobile phase, gradient elution is not normally possible.

Amperometric detectors of the thin-layer cell configuration containing two working electrodes have been developed to improve the overall selectivity or sensitivity of the detection process [34,67,115,117,131,132]. The two working electrodes can be arranged in parallel or series. The peak height ratio at two different potentials for electrodes in parallel provides information on the identity of a peak or an estimate of its purity. The identity of neurotrasmitters and related metabolites in biological samples is commonly confirmed by this method [133]. Using the series configuration compounds eluted from the column can be oxidized (or reduced) on the first electrode and subsequently reduced (or oxidized) on the second. Since the electroactive compounds which irreversibly react on the first electrode are eliminated, determination of a reversible electroactive compound is possible with high selectivity. This technique can be useful for the removal of oxygen interferences in which the upstream electrode reduces oxygen and the analyte, and the downstream electrode oxidatively detects only the reduction product of the analyte [134,135].

Most aliphatic compounds are not considered to be amenable to amperometric detection at a constant potential. Free-radical products from the oxidation of aromatic molecules can be stabilized by π-resonance; hence the activation barrier for the reaction is decreased. A similar mechanism for stabilizing aliphatic free radicals is absent and, consequently, oxidation rates are normally very low, even when anodic reactivity is predicted based on thermodynamic considerations. The activation barrier for oxidation of aliphatic compounds can be decreased at noble-metal electrodes with partially unsaturated surface d-orbitals that can adsorb and thereby stabilize free radical intermediate oxidation products. Using a pulsed amperometric detector with gold or platinum electrodes in alkaline solution certain classes of compounds, such as alcohols, carbohydrates, aliphatic amines, amino acids and sulfur compounds with a nonbonded pair of electrons on the sulfur atom have been determined with moderate sensitivity (0.1-1.0 ppm) after separation by liquid chromatography [136-142]. In pulsed amperometric detection the working electrode is typically cycled to three different potentials at a rate of about 1-2 Hz. A relatively positive potential is used to clean or activate the electrode surface. A more negative potential is then applied to promote adsorption of the analyte on the electrode. Finally, a suitably selected intermediate potential is used for detection. By observing the transient current at the same delay time in the detection cycle a Faradic signal with a reasonable linear dependence on the analyte concentration is obtained. Surface-catalyzed oxidation of the adsorbed analyte is the primary contribution to the analytical signal in the detection mode. The potential is cycled to maintain a clean and active surface for the reaction, which might otherwise be impeded due to fouling of the electrode by accumulated detection products. Organic solvent modifiers can be tolerated in pulsed amperometric detection only if they are not electroactive under the given pulsed-potential waveform. This restricts the use of certain common solvents like methanol and acetonitrile in some circumstances. The pH of the eluent is important for controlling the amplitude of the detector response, with a high pH being required for many reactions. In a recent development a two-step potential waveform at a gold electrode was found to be effective for the detection of carbohydrates [136]. A separate adsorption step was shown to be unnecessary.

5.6.4 Miscellaneous Detectors

Under this heading we will briefly describe the properties of some less well known detectors that are used for specific applications and are not commonly found in most analytical laboratories. The number of detectors that have been described at some point for use in liquid chromatography is enormous and continues to be added to on account of the need for both more sensitive, universal detectors and selective detectors for liquid chromatography. Given the success of the ionization detectors in gas chromatography it comes as no surprise that their adaptation to liquid chromatography should have been an early priority, but their success has only been moderate. The electron-capture detector is easily interfaced to normal phase liquid chromatography using a heated transfer line and nebulizer to generate a stream of vapor, supported by a nitrogen gas flow, which is then throttled through the detector [142]. The main limitations seem to be a decrease in sensitivity of about 10-100 fold compared to detection of the same compounds using gas chromatography and the need for dry solvents containing very low levels of electron-capturing impurities. Also, the fact that most published applications have been for compounds typically analyzed by gas chromatography already, has limited interest in this detector for liquid chromatography. Generally, the column eluent flow rates and nature of the solvents used in liquid chromatography are incompatible with the use of flame-based ionization detectors. The exception is packed capillary columns which can be interfaced directly to element selective flame-based detectors, such as the flame photometric detector and the thermionic ionization detector [101,144-146]. The very low flow rates, in the microliters/min range, make the above combination practical. For conventional diameter columns a solvent elimination interface is required between the column and the detector. A typical example is the moving wire detector [66,147]. The moving wire (or belt) detector transports a thin film of column effluent through an oven where the solvent is selectively evaporated, and into a second zone where the sample is volatilized into the detector, or into a chemical reactor for conversion to an easily detectable product. As an example, the moving wire is passed through a cleaner oven, operated at 850°C, where impurities are removed by oxidation in a stream of air. The wire then moves past the coating block where the column effluent is deposited, and then through an evaporator (180-200°C), which selectively removes solvent in a stream of air. Finally, the wire passes through the oxidizer tube, where sample components are oxidized at 850°C to carbon dioxide and water in a stream of air. The resultant carbon dioxide is mixed with a stream of nitrogen and hydrogen, catalytically reduced to methane in the presence of nickel in the reactor, and finally detected by flame ionization detection. The detector has a universal response (at least for involatile compounds) but is technically complex and limited in sensitivity by the relatively small fraction of the column effluent that can be coated on the wire. The prospects for this detector could change with the development of a more efficient method of applying eluent to the wire, particularly for those compounds lacking a chromophore. Promising results have been demonstrated using spray devices and thermospray deposition onto a moving stainless steel belt with subsequent vaporization or pyrolysis for flame ionization, photoionization or electron-capture detection [148].

An alternative approach to the moving wire detector for the development of a universal detector is the evaporative light scattering detector (ELSD) [149-154]. The column effluent is nebulized via a concentric Venturi nozzle with a fast stream of pressurized gas into a heated tube. The solvent is evaporated in the heated tube and the dry particles formed are drawn through a vent into the detection cell. The concentration of particles is measured by the amount of scattered light from a tungsten lamp or laser at a fixed angle to the emerging beam. The sensitivity of the light scattering detector is about one or two orders of magnitude better than the refractive index detector and the detector is compatible with gradient elution operation. The mobile phase must be free of salts, etc., which would be recorded by the detector limiting its use for some applications, such as ion-exchange and ion-pair chromatography. Since the intensity of the scattered light depends on the particle diameter, the response of the detector is influenced by the changing particle size distribution as a function of the sample size, and is unlikely to be linear. Usually a log-log plot of peak area against sample size is approximately linear over two or three orders of magnitude and is used for calibration purposes. The performance of the detector depends critically on the design of the nebulizer and the flow rate of scavenger gas and tube temperature. This detector may evolve into a useful replacement for the refractive index detector with improved sample detectability and a similar range of applications for involatile samples.

Low-angle laser light scattering (LALLS) detectors are capable of providing unique molecular size information of analytes separated by size-exclusion chromatography [155-158]. A general property of macromolecular species in homogeneous solution is that they scatter light. The light intensity scattered at a certain angle of incidence depends on the relative molecular mass and concentration of the solute in the detector cell. To evaluate the molecular weight term the LALLS detector is coupled in series to a concentration-dependent detector, such as a refractive index or ultraviolet detector, and the response from both detectors is used to calculate molecular weights directly or molecular weight distributions for polymers. The scattered light intensity is highest for small scattering angles and hence scattering is usually measured at very small angles to the incident beam with a photomultiplier. The use of a laser provides high sensitivity and allows small detector volumes to be used. The detector is expensive but unique in the information that it provides. Samples in the microgram range are usually required to obtain molecular weight information.

5.7 REFERENCES

1. V. Berry, CRC Crit. Revs. Anal. Chem., 21 (1989) 115.
2. J. W. Dolan and L. R. Snyder, "Troubleshooting LC Systems", Humana Press, Clifton, NJ, 1989.
3. H. M. McNair, J. Chromatogr. Sci., 25 (1987) 564.
4. Sj. van der Wal, S. J. Bannister, and L. R. Snyder, J. Chromatogr. Sci., 20 (1982) 260.
5. M. J. Rehman, K. P. Evans, A. J. Handley, and P. R. Massey, Chromatographia, 24 (1987) 492.
6. R. A. Mowery, J. Chromatogr. Sci., 23 (1985) 22.
7. P. R. Haddad and R. C. L. Foley, J. Chromatogr., 407 (1987) 133.
8. R. P. W. Scott (Ed.), "Small Bore Liquid Chromatography Columns", Wiley, New York, NY, 1984.
9. P. Kucera (Ed.), "Microcolumn High-Performance Liquid Chromatography", Elsevier, Amsterdam, 1984.
10. M. V. Novotny and D. Ishii (Eds.), "Microcolumn Separations", Elsevier, Amsterdam, 1985.
11. B. G. Belen'kii, E. S. Gankina, and V. G. Mal'tsev, "Capillary Liquid Chromatography", Plenum, New York, NY, 1987.
12. D. Ishii (Ed.), "Introduction to Microscale HPLC", VCH Publishers, New York, NY, 1988.
13. F. J. Yang (Ed.), "Microbore Column Chromatography", Dekker, New York, NY, 1989.

14. H. B. Brooks, C. Thrall, and J. Tehrani, J. Chromatogr., 385 (1987) 55.
15. F. Andreolini and A. Trisciani, J. Chromatogr. Sci., 28 (1990) 54.
16. A. Trisciani and F. Andreolini, J. High Resolut. Chromatogr., 13 (1990) 270.
17. F. W. Freebairn and J. H. Knox, Chromatographia, 19 (1985) 37.
18. P. Roumeliotis, M. Chatziathanassiou, and K. K. Unger, Chromatographia, 19 (1985) 145.
19. J. C. Sternberg, Adv. Chromatogr., 2 (1966) 205.
20. H. H. Lauer and G. P. Rozing, Chromatographia, 14 (1981) 641.
21. K.-P. Hupe, R. J. Jonker, and G. Rozing, J. Chromatogr., 285 (1984) 253.
22. P. J. Naish, D. P. Goulder, and C. V. Perkins, Chromatographia, 20 (1985) 335.
23. W. M. A. Niessen, H. P. M. van Vliet, and H. Poppe, Chromatographia, 20 (1985) 356.
24. K. A. Cohen and J. D. Stuart, J. Chromatogr. Sci., 25 (1987) 381.
25. H. A. Claessens, C. A. Cramers, and M. A. J. Kuyken, Chromatographia, 23 (1987) 189.
26. H. A. Claessens, M. J. J. Hetem, P. A. Leclercq, and C. A. Cramers, J. High Resolut. Chromatogr., 11 (1988) 176.
27. B. L. Karger, M. Martin, and G. Guiochon, Anal. Chem., 46 (1974) 1640.
28. B. Coq, G. Cretier, J. L. Rocca, and M. Porthault, J. Chromatogr. Sci., 19 (1981) 1.
29. H. A. Claessens and M. A. J. Kuyken, Chromatographia, 23 (1987) 331.
30. E. D. Katz and R. P. W. Scott, J. Chromatogr., 268 (1983) 169.
31. J. G. Atwood and M. J. Golay, J. Chromatogr., 218 (1981) 97.
32. P. Gareil and R. Rosset, Chromatographia, 20 (1985) 367.
33. Sj van der Wal, J. Chromatogr., 352 (1986) 351.
34. P. T. Kissinger, J. Chromatogr., 488 (1989) 31.
35. S. R. Bakalyar, M. P. T. Bradley, and R. Honganen, J. Chromatogr., 158 (1978) 277.
36. J. N. Brown, M. Hewins, J. H. M. Ven Der Linden, and R. J. Lynch, J. Chromatogr., 204 (1981) 115.
37. J. Doehl, J. Chromatogr. Sci., 26 (1988) 7.
38. W. A. MacCrehan, S. D. Yang, and B. A. Benner, Anal. Chem., 60 (1988) 284.
39. L. V. Berry and B. L. Karger, Anal. Chem., 45 (1973) 819A.
40. J. P. Foley, J. A. Crow, B. A. Thomas, and M. Zamora, J. Chromatogr., 478 (1989) 287.
41. M. Martin and G. Guiochon, J. Chromatogr., 151 (1978) 267.
42. P. Achener, S. R. Abbott, and R. L. Stevenson, J. Chromatogr., 130 (1977) 29.
43. S. Nilsson, J. High Resolut. Chromatogr., 12 (1989) 56.
44. P. Jandera and J. Churacek, "Gradient Elution in Column Liquid Chromatography", Elsevier, Amsterdam, 1985.
45. K. E. Karlson and M. Novotny, J. High Resolut. Chromatogr., 7 (1984) 411.
46. H. E. Schwartz and G. Brownlee, J. Chromatogr. Sci., 23 (1985) 402.
47. D. Ishii, Y. Hashimoto, H. Asai, K. Watanabe, and T. Takeuchi, J. High Resolut. Chromatogr., 8 (1985) 543.
48. K. Slais and V. Preussler, J. High Resolut. Chromatogr., 10 (1987) 82.
49. K. Slais and R. W. Frei, Anal. Chem., 59 (1987) 376.

50. P. Jandera, J. Churacek, and L. Svoboda, J. Chromatogr., 192 (1980) 37.
51. M. A. Quarry, R. L. Grob, and L. R. Snyder, J. Chromatogr., 285 (1984) 1.
52. J. W. Dolan, D. C. Lommen, and L. R. Snyder, J. Chromatogr., 485 (1989) 91.
53. M. C. Harvey and S. D. Stearns, in J. F. Lawrence (Ed.), "Liquid Chromatography in Environmental Analysis", Humana Press, Clifton, NY, 1983, p. 301.
54. M. C. Harvey and S. D. Stearns, J. Chromatogr. Sci., 21 (1983) 473.
55. A. Schmid, Chromatographia, 12 (1979) 825.
56. V. Berry and K. Lawson, J. Liq. Chromatogr., 10 (1987) 3257.
57. V. McGuffin and M. V. Novotny, Anal. Chem., 55 (1983) 580.
58. V. Berry and K. Lawson, J. High Resolut. Chromatogr., 11 (1988) 121.
59. A. Manz and W. Simon, J. Chromatogr., 387 (1987) 187.
60. T. L. Ng and S. Ng, J. Chromatogr., 329 (1985) 13.
61. N. E. Hoffman, S. Pan, and A. M. Rustum, J. Chromatogr., 465 (1989) 189.
62. D. Vukmanic and M. Chiba, J. Chromatogr., 483 (1989) 189.
63. N. K. Vadukul and C. R. Loscombe, Chromatographia, 14 (1981) 465
64. T. M. Vickrey (Ed.), "Liquid Chromatography Detectors", Dekker, New York, NY, 1983.
65. E. S. Yeung (Ed.), "Detectors for Liquid Chromatography", Wiley, New York, NY, 1986.
66. R. P. W. Scott, "Liquid Chromatography Detectors", Elsevier, Amsterdam, 1986.
67. P. C. White, Analyst, 109 (1984) 677 and 973.
68. E. S. Yeung and R. E. Synovec, Anal. Chem., 58 (1986) 1237A.
69. W. A. Dark, J. Chromatogr. Sci., 24 (1986) 495.
70. C. Fujimoto, T. Morita, and K. Jinno, Chromatographia, 22 (1986) 91.
71. R. E. Synovec, Anal. Chem., 59 (1987) 2877.
72. D. J. Bornhop, T. G. Nolan, and N. J. Dovichi, J. Chromatogr., 384 (1987) 181.
73. T. Alfredson and T. Sheehan, J. Chromatogr. Sci., 24 (1986) 473.
74. B. J. Clark and A. F. Fell, Chem. Brit., (1987) 1069.
75. D. G. Jones, Anal. Chem., 57 (1985) 1057A.
76. E. S. Yeung, in E. H. Piepmeier (Ed.), "Analytical Applications of Lasers", Wiley, New York, NY 1986, p. 557.
77. B. G. Belenkii, J. Chromatogr., 434 (1988) 337.
78. K. P. Jones, Trends Anal. Chem., 9 (1990) 195.
79. S. A. George and A. Maute, Chromatographia, 15 (1982) 419.
80. E. V. Dose and G. Guiochon, Anal. Chem., 61 (1989) 2571.
81. H. Engelhardt and Th. Konig, Chromatographia, 28 (1989) 341.
82. J. G. D. Marr, G. C. R. Seaton, B. J. Clark, and A. F. Fell, J. Chromatogr., 506 (1990) 289.
83. E. L. Inman, M. D. Lantz, and M. M. Strohl, J. Chromatogr. Sci., 28 (1990) 578.
84. P. J. Naish, R. J. Lynch, and T. Blaffert, Chromatographia, 27 (1989) 343.
85. I. Sakuma, N. Takia, T. Dohi, Y. Fukui, and A. Ohkubo, J. Chromatogr., 506 (1990) 223.
86. M. Hayashida, M. Nihira, T. Watanabe, and K. Jinno, J. Chromatogr., 506 (1990) 133.
87. J. N. Little and G. J. Fallick, J. Chromatogr., 112 (1975) 389.
88. J. Doehl and T. Greibrokk, J. Chromatogr. Sci., 25 (1987) 99.

598

89. M. Kamahori, Y. Watanabe, J. Miura, M. Taki, and H. Miyagi, J. Chromatogr., 465 (1989) 227.
90. A. R. Parrott, J. Liq Chromatogr., 10 (1987) 1603.
91. C. Kientz and H. Verweij, J. High Resolut. Chromatogr., 11 (1988) 294.
92. J. P. Chervet and J. P. Salzmann, J. High Resolut. Chromatogr., 12 (1989) 278.
93. M. Verzele, G. Sleenbeke, and J. Vindevogel, J. Chromatogr., 477 (1989) 87.
94. K. Peck and M. D. Morris, J. Chromatogr., 448 (1988) 193.
95. C. E. Evans and V. L. McGuffin, J. Chromatogr., 503 (1990) 127.
96. D. O. Hancock, C. N. Renn, and R. E. Synovec, Anal. Chem., 62 (1990) 2441.
97. G. Torsi, G. Chiavari, C. Laghi, and A. M. Asmudsdottir, J. Chromatogr., 518 (1990) 135.
98. J. E. Stewart, Appl. Optics, 20 (1981) 654.
99. T. Wolf, G. T. Fritz, and L. R. Palmer, J. Chromatogr. Sci., 19 (1981) 387.
100. American Society for Testing Materials, "Standard Practice for Testing Fixed-Wavelength Photometric Detectors used in Liquid Chromatography", Annual Book of ASTM Standards Part 42, E685, Philadelphia, Pennsylvania, 1990.
101. W. Slavin, A. T. Rhys Williams, and R. F. Adams, J. Chromatogr., 134 (1977) 121.
102. E. Johnson, A. Abu-Shumays, and S. R. Abbott, J. Chromatogr., 134 (1977) 107.
103. H. P. M. van Vliet and H. Poppe, J. Chromatogr., 346 (1985) 149.
104. T. J. Edkins and D. C. Shelly, J. Chromatogr., 459 (1988) 109.
105. S. Folestad, B. Galle, and B. Josefsson, J. Chromatogr. Sci., 23 (1985) 273.
106. T. Takeuchi and E. S. Yeung, J. Chromatogr., 389 (1987) 3.
107. M. Verzele and C. Dewaele, J. Chromatogr., 395 (1987) 85.
108. T. Takeuchi and D. Ishii, Chromatographia, 25 (1988) 697.
109. T. Takeuchi and D. Ishii, J. High Resolut. Chromatogr., 11 (1988) 841.
110. D. C. Johnson, S. G. Weber, A. M. Bond, R. M. Wightman, R. E. Shoup, and I. S. Krull, Anal. Chim. Acta, 180 (1986) 187.
111. R. E. Shoup, in Cs. Horvath (Ed.), "High Performance Liquid Chromatography, Advances and Perspectives", Academic Press, New York, NY, 4 (1986) 91.
112. K. Stais, J. Chromatogr. Sci., 24 (1986) 321.
113. P. R. Haddad, Chromatographia, 24 (1987) 217.
114. P. Jandick, P. R. Haddad, and P. E. Sturrock, CRC Crit. Revs. Anal. Chem., 20 (1988) 1.
115. T. Nagatsu and K. Kojima, Trends Anal. Chem., 7 (1988) 21.
116. D. M. Radzik and S. M. Lunte, CRC Crit. Revs. Anal. Chem., 20 (1989) 317.
117. G. Horvai and E. Pungor, CRC Crit. Revs. Anal. Chem., 21 (1989) 1.
118. W. Buchberger, Chromatographia, 30 (1990) 577.
119. H. Takeda, T. Matsumiya, and T. Shibuya, J. Chromatogr., 515 (1990) 265.
120. D. T. Gjerde, Intern. J. Environ. Anal. Chem., 27 (1986) 287.
121. D. T. Gjerde and J. S. Fritz, "Ion Chromatography", Huethig, Heidelberg, 1987.
122. H. Small, "Ion Chromatography", Plenum Press, New York, NY, 1989.

123. D. Qi, T. Okada, and P. K. Dasgupta, Anal. Chem., 61 (1989) 1383.
124. G. J. Schmidt and R. P. W. Scott, Analyst, 110 (1985) 757.
125. P. R. Haddad and P. E. Jackson, "Ion Chromatography. Principles and Applications", Elsevier, Amsterdam, 1990.
126. H. Gunasingham, Trends Anal. Chem., 7 (1988) 217.
127. N. Sagliano and R. A. Hartwick, J. Chromatogr. Sci., 24 (1986) 506.
128. V. F. Ruban, J. High Resolut. Chromatogr., 13 (1990) 112.
129. S. Prabhu and J. L. Anderson, Anal. Chem., 59 (1987) 157.
130. J. T. Bretz and P. R. Brown, J. Chromatogr. Sci., 26 (1988) 310.
131. S. Knapp, M. L. Wardlow, and L. J. Thal, J. Chromatogr., 526 (1990) 97.
132. R. Welpton, P. Dudson, H. Cannel and K. Webster, J. Chromatogr., 526 (1990) 215.
133. G. S. Mayer and R. E. Shoup, J. Chromatogr., 255 (1983) 533.
134. K. Bratin and P. T. Kissinger, J. Liq. Chromatogr., 4 (1981) 321.
135. Y. Haroon, C. A. W. Schubert, and D. V. Hauschka, J. Chromatogr. Sci., 22 (1984) 89.
136. G. G. Neuburger and D. C. Johnson, Anal. Chem., 59 (1987) 150 and 203.
137. D. C. Johnson, Nature, 321 (1986) 451.
138. S. Hughes and D. C. Johnson, Anal. Chim. Acta, 149 (1983) 1.
139. D. C. Johnson and W. R. LaCourse, Anal. Chem., 62 (1990) 589A
140. R. D. Rocklin and C. A. Pohl, J. Liq. Chromatogr., 6 (1983) 1577.
141. W. R. LaCourse, D. C. Johnson, M. A. Ray, and R. W. Slingsby, Anal. Chem., 63 (1991) 134.
142. R. W. Andrews and R. M. King, Anal. Chem., 62 (1990) 2130.
143. U. A. Th. Brinkman and F. A. Maris, Trends Anal. Chem., 4 (1985) 55.
144. V. L. McGuffin and M. Novotny, Anal. Chem., 55 (1983) 2296.
145. Ch. E. Kientz, A. Verweij, G. J. de Jong, and U. A. Th. Brinkman, J. High Resolut. Chromatogr., 12 (1989) 793.
146. Ch. E. Kientz, A. Verweij, H. L. Boter, A. Poppema, R. W. Frei, and U. A. Th. Brinkman, J. Chromatogr., 407 (1989) 385.
147. H. Veening, P. P. H. Tock, J. C. Kraak. and H. Poppe, J. Chromatogr., 352 (1986) 345.
148. L. Yang, G. J. Ferguson, and M. L. Vestal, Anal. Chem., 56 (1984) 2632.
149. R. Schultz and H. Engelhardt, Chromatographia, 29 (1990) 517.
150. M. Righezza and G. Guiochon, J. Liq. Chromatogr., 11 (1988) 2709.
151. G. Guiochon, A. Moysan, and C. Holley, J. Liq. Chromatogr., 11 (1988) 2547.
152. A. Stolyhwo, M. Martin, and G. Guiochon, J. Liq. Chromatogr., 10 (1987) 1237.
153. L. E. Oppenheimer and T. H. Mourey, J. Chromatogr., 323 (1985) 297.
154. J. L. Robinson, M. Tsimidou, and R. Macrae, J. Chromatogr., 324 (1985) 35.
155. M. Martin, Chromatographia, 15 (1982) 426.
156. I. S. Krull, R. Mhatre, and H. H. Stuting, Trends Anal. Chem., 8 (1989) 260.
157. H. H. Stuting and I. S. Krull, Anal. Chem., 62 (1990) 2107.
158. T. Takagi, J. Chromatogr., 506 (1990) 409.

CHAPTER 6

SUPERCRITICAL FLUID CHROMATOGRAPHY

6.1   INTRODUCTION

When a gas or liquid is heated to a temperature above its critical temperature and simultaneously compressed to a pressure exceeding its critical pressure it becomes a supercritical fluid, Figure 6.1. The transition of properties from the gas or liquid phase to the fluid region is quite smooth and not marked by the large changes that generally accompany transitions across phase boundaries. The supercritical fluid region represents an extension of the gas and liquid phase regions in which the properties of the fluid are continuously variable by changing the temperature and pressure above the critical point [1]. The fluid may exhibit gas-like and liquid-like properties at the extremes of its range represented by the dotted lines in Figure 6.1 and some combination of these properties, fluid properties, in between. The viscosity, diffusivity, and solubilizing power of a supercritical fluid, those properties most relevant in defining its use as a mobile phase in chromatography, depend on its density, which in turn is a function of the applied pressure and temperature, Table 6.1 [2]. The densities at 72 atmospheres are representative of conditions close to the critical pressure of carbon dioxide and 400 atmospheres towards the upper pressure generally used in supercritical fluid chromatography (SFC). For carbon dioxide

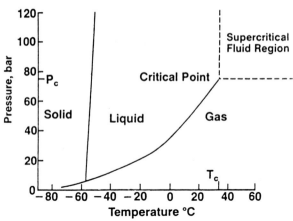

Figure 6.1   Phase diagram for carbon dioxide.

TABLE 6.1

DEPENDENCE OF DENSITY ON TEMPERATURE AND PRESSURE FOR SUPERCRITICAL
CARBON DIOXIDE

Temperature (°C)	Pressure (atms.)	Density (g/ml)
40	72	0.22
	400	0.96
60	72	0.17
	400	0.90
80	72	0.14
	400	0.82
100	72	0.13
	400	0.76
120	72	0.12
	400	0.70
140	72	0.11
	400	0.64

virtually the whole density range from about 0.1-1.0 g/ml is
easily accessible providing a wide range of fluid properties.
Typical properties of a gas (helium) under gas chromatographic
conditions, a liquid (water) under liquid chromatographic
conditions, and carbon dioxide under low and high density
conditions, as might be used in SFC, are summarized in Table 6.2
[3]. Supercritical fluids have densities generally between those
of gases and liquids but closer to those of liquids. Supercritical
fluids have much greater solubilizing power than gases and can,
therefore, be used for the separation of involatile and high
molecular weight samples unsuited to gas chromatography. The

TABLE 6.2

REPRESENTATIVE PHYSICAL PROPERTIES OF TYPICAL CHROMATOGAPHIC MOBILE PHASES

Mobile Phase	Temperature (°C)	Pressure (atms.)	Density (g/ml)	Diffusivity (cm^2/s)	Viscosity (cP)
Helium	200	1.5	2×10^{-4}	0.1-1	0.02
Carbon dioxide					
low density	100	80	0.15	10^{-3}	0.02
high density	35	200	0.8	10^{-4}	0.1
Water	20	*	1.0	10^{-5}	1.0

*liquids are virtually incompressible so the pressure is irrelevant.

solubilizing power of a supercritical fluid is directly related to its density at a constant temperature, which unlike a liquid can be varied over a wide range as shown in Table 6.1. Gases are highly compressible but require inconveniently high pressure to approach liquid densities. Typical solute diffusion coefficients in supercritical fluids are intermediate between those of gases and liquids but generally closer to those of liquids. However, diffusion coefficients may be easily an order of magnitude greater than in liquids, which has important chromatographic implications concerning separation times, column characteristics, and instrument design. Typical minimum plate heights for packed column supercritical fluid chromatography and liquid chromatography are very similar, the most important difference, however, is that the minimum value in SFC is achieved at linear velocities 5 to 10 times greater than for liquid chromatography and separations can be achieved in a much shorter time [4]. Also, the resistance to mass transfer term in SFC is not as large as it is for liquids, which allows a further increase in the separation speed without a significant decrease in efficiency. It should always be possible to obtain faster separations by SFC than by liquid chromatography but these separation speeds do not approach those of gas chromatography, even for low density supercritical fluids. The rate of radial diffusion controls the dimensions of open tubular columns when reasonable efficiency per unit column length is required. Open tubular columns need only have internal diameters of about 0.1-0.5 mm in gas chromatography to obtain reasonable efficiency, conditions that are easily met. For equivalent efficiency in liquid chromatography column internal diameters should be less than about 0.01 mm. These columns are neither easy to prepare nor simple to operate and are rarely used in contemporary practice. Columns packed with particles of small diameters are used predominantly for liquid chromatography. For SFC, increased rates of diffusion allow larger diameter open tubular columns to be used effectively, about 0.05-0.10 mm, that are compatible with reasonable instrument constraints. Thus, both packed and open tubular columns may be used in SFC where each type has distinct advantages. We will come to this later. For now, we should note that since the solute diffusion coefficients vary strongly with fluid density, open tubular columns of typical dimensions will perform more favorably under low density conditions. At high densities, approaching those of liquids, the column dimensions required for optimum performance are far less favorable and approach those of liquid chromatography.

The viscosities of supercritical fluids are intermediate between those of a gas and a liquid, but this time, are much closer to those of a gas than a liquid. For a fixed column pressure drop much longer columns or higher flow rates are possible in SFC compared to liquid chromatography. Liquids, however, are virtually incompressible while gases and supercritical fluids are highly compressible. A large column pressure drop in SFC has the effect of decreasing the density along the length of the column and increasing the local mobile phase velocity leading to additional band broadening and poor reproducibility of retention data. The maximum efficiency achievable and the separation speed in SFC depends ultimately on the allowable column pressure drop which, because of the gas-like viscosities of supercritical fluids, is still likely to favor SFC over liquid chromatography. However, for the same column pressure drop the greater diffusivity of gases means that SFC cannot approach gas chromatography in either the total number of theoretical plates that can be achieved or in separation speed.

The above discussion was clearly weighted towards the kinetic aspects of a separation which are easier to treat fundamentally and do not take into account all of the factors important in selecting a separation technique for a given problem. Gas chromatography is clearly the method of choice for any separation based on the above criteria. However, it cannot be used for involatile samples which prevents its use for many contemporary problems. For these samples SFC should be superior to liquid chromatography all other things being equal. Again, this is not the case, since mobile phases encompassing a wider range of solvent properties are available in liquid chromatography and the number of different separation mechanisms that can be easily exploited to optimize a separation by liquid chromatography exceeds those of SFC. A wider range of detection options is generally possible for SFC, at least when organic solvent modifiers are not used. Maturity and familiarity could be other reasons for selecting liquid chromatography for a particular application. There is a greater understanding of methods development in liquid chromatography compared to SFC and a much larger application data base for liquid chromatography.

From an historical perspective SFC was developed after gas chromatography was well established and high performance liquid chromatography was just beginning. At the early stages of development it was not a simple task to build or operate a supercritical fluid chromatograph which, undoubtedly, contributed to its low popularity. The revived interest in SFC resulted from technological innovations in gas and liquid chromatography that occurred independently of research into SFC but enabled reliable commercial instruments for SFC to be introduced in the 1980s. These included the development of narrow bore, fused silica open tubular columns with immobilized phases; small particles with a narrow size distribution; bonded-phase particles; pulse-free pumps that could deliver stable low flow rates and could be adapted for pressure programming; and high-pressure injection valves with sub-microliter sample volumes. SFC is certainly immature compared to gas and liquid chromatography but has succeeded in identifying application areas for which it is uniquely suited, and has gained acceptance as a separation technique complementary to those of gas and liquid chromatography [2-12].

6.2 COLUMN SELECTION

Virtually all current research in SFC utilizes either small bore packed columns with particles of 5-10 micrometers in diameter optimized for use in liquid chromatography or narrow bore, fused silica open tubular columns with immobilized phases similar to those used in gas chromatography. In the latter case columns of smaller internal diameter, 10-100 micrometers, shorter lengths (generally less than 20 m with 1-10 m being the most common length), and more firmly crosslinked stationary phases are used by comparison with standard columns for gas chromatography. In all other respects column preparation procedures and stationary phase chemistries are identical to those used in gas and liquid chromatography. In specific cases, such as liquid crystalline poly(siloxane) phases [13] and Pirkle-type chiral phases [14,15] improved resolution compared to the use of the same phases in gas and liquid chromatography were claimed. The more ordered structure of the liquid crystalline phase at the lower operating temperatures of SFC afforded improved molecular shape selectivity of polycyclic aromatic compounds compared to gas chromatography. SFC was able to provide faster separations of enantiomers than was observed for liquid chromatography.

Fused silica open tubular columns can be prepared with a very high degree of chemical inertness compared to bonded phase packings. The latter invariably contain unreacted and poorly shielded silanol groups, even after the most exhaustive chemical treatment. These residual silanol groups cause tailing and adsorption of polar solutes, such as organic bases and solutes with hydrogen bonding functional groups [16-27]. Bonded phase packings prepared by surface bonding and crosslinking of poly(siloxane) and poly(butadiene) prepolymers, or dehydrocondensation reactions with polymeric silicon hydride reagents generally exhibit the lowest degree of activity [16,25-27]. Also, wide pore silica packings of low-surface area are generally easier to deactivate than the narrow pore packings favored for analytical liquid chromatography [17-19]. Further, faster separations can be obtained on packings with pore diameters between 15 and 100 nm.

For steric reasons the surface of chemically bonded packings is always heterogeneous containing different concentrations of chemically bonded and free silanol groups (see section 4.2). To a first approximation it can be assumed that the two types of site contribute independently to retention in packed column SFC with nonpolar supercritical fluids, such as carbon dioxide. This will result in poor peak shapes if the two sites exhibit different kinetics or if the sorption isotherms at the two sites are of a different type. Schoenmakers et al. [28] have proposed a model that enables the relative number of bonded ligands and free silanol groups to be estimated from the influence of sample size on retention. In less formal terms it has been established that the interaction of sample proton donor/acceptor and dipolar functional groups with free silanol groups of the column packings causes the characteristic peak tailing and sample adsorption or degradation that occurs in packed column SFC with relatively nonpolar fluids [17-20,22-26,29]. The use of polar modifiers added to the mobile phase, as well as derivatization of polar solute functional groups are common methods of minimizing undesirable solute-stationary phase interactions. There is good evidence that some polar modifiers may react chemically with silanol groups under typical SFC conditions. Ethanol has been shown to form silyl esters and dioxane undergoes ring cleavage to form an unidentified product [30,31]. Steroids containing hydroxyl groups were shown to react chemically with silanol groups under SFC conditions [17]. The mass recovery of the steroids from the column was shown to decline with increasing temperature. Also, secondary and tertiary alcohols have been shown to undergo acid catalyzed elimination reactions on silica packings under SFC conditions [32]. The problems associated with chemical and adsorptive activity of silica- and alumina-based packings in SFC could be circumvented by using materials with a different structure. Graphitized forms of carbon and macroporous polymeric beads have been evaluated for this purpose [4,17,26,27,33-35]. Opinions differ as to the usefulness of these materials. The porous polymer beads developed for reversed-phase liquid chromatography are chemically inert, but their performance seems to deteriorate rapidly under typical SFC conditions. It is thought that this is due to swelling of the beads caused by uptake of the mobile phase. The constant swelling and shrinking of the beads during use eventually results in mechanical disintegration. The retention mechanism on these packings seems to involve partitioning with the polymer backbone as well as selective interactions with solutes that contain π-electron acceptor groups [26,34]. When carbon dioxide is used as the mobile phase retention seems to be high compared to chemically bonded phases, and the efficiency of well retained solutes usually deteriorates compared to earlier eluting peaks. This has tended to limit macroporous polymeric packings to the separation of low molecular weight samples, such as alcohols and carboxylic acids, which are often difficult to separate and elute quantitatively from silica-based packings. The elution of higher molecular weight samples in an acceptable separation time has been demonstrated using mobile phases containing fairly large amounts of organic solvent modifiers [33]. It is not clear whether high molecular weight samples, say with a molecular weight over 1000, can be separated on these columns. Porous graphitic carbon packings, with a unique sponge-like structure of high mechanical stability and small particle size, were recently introduced (see section 4.2.4). These materials are highly retentive under SFC conditions. For low molecular weight solutes they provide highly stereospecific separations of isomers and molecules with only slight structural differences [35]. The strong surface interactions and absence of polar functional groups might make them useful supports for preparing mechanically coated or encapsulated phases [26]. Any oxidation of the packing during manufacture, however, would introduce polar functional groups into the otherwise inert graphite matrix, which could have undesirable effects on their chromatographic properties. Further development of either the macroporous polymeric packings or modified porous graphitic carbon packings could lead to new packing materials for SFC that would solve some of the limitations of silica-based packings, that are used predominantly in contemporary research.

Exhaustive silanization with the most active of silanizing reagents cannot eliminate completely all silanol groups on the silica surface. Encapsulating the deactivated surface by immobilizing a thin film of a polymeric skin over the surface shields the unbound silanol groups from the sample and provides more inert column packings [4,16,23,26,27]. Packings of moderate surface areas based on microparticle silicas with 30 nm and 50 nm pore sizes are now commercially available with a wide range of immobilized ligands, including methyl, octyl, octadecyl, phenyl and cyanoalkyl with a polysiloxane backbone, and a poly(ethylene glycol) phase. Although encapsulation shields silanol groups from the sample, it is unable to prevent all interactions with the sample, because the immobilized film may not cover the whole surface leaving some of the substrate groups exposed and/or because samples soluble in the film must have access to the substrate by diffusion through the film. Since immobilization does not necessarily require chemical bonding to the substrate, the general technique can be applied to other substrates besides silica. Alumina based packings, covered with an immobilized layer of poly(butadiene), are commercially available. These packings provide good separations of nonpolar compounds, particularly hydrocarbons, but like silica, also show undesirable interactions when polar samples are separated. Several research groups are active in the development of immobilized phases on silica and other substrates, and new chromatographic packings can reasonably be expected to emerge from this effort, which will provide improved shielding of substrate properties.

Theoretical approaches to column design in SFC are by no means as straightforward, or as well developed as models used in gas and liquid chromatography. Problems arise from the fact that the mobile phase is compressible and behaves nonideally resulting in additional sources of band broadening that depends on the fluid density and its variation along the length of the column [36]. Mobile phase modification of stationary phase properties, that are also expected to vary with density, are a further complication [37-40]. The supercritical fluid will adsorb onto or absorb into solid and polymeric stationary phases, altering both the chemical and physical properties of the stationary phase. The common use of chemically bonded phases and polar modifiers in SFC serves to obscure even further the exact character of the stationary phase. Pressure, temperature and density relationships for supercritical fluids are normally described by a multi-parameter equation of state, which further complicates any mathematical treatment of the band broadening process [41,42]. It is generally assumed in formulating most models that the mobile phase density varies linearly with column length while more exact calculations indicate that this is not the case. For example, a typical packed column with an inlet pressure of 120 bar (density = 0.6331 g/ml) and outlet pressure of 100 bar (density = 0.4497 g/ml), temperature 320 K, and carbon dioxide as mobile phase, the arithmetic-mean density is reached when typical mobile phase molecules have passed through 63.6% of the column length and have spent 67.5% of their residence time in the column [41]. Based on the numerical solution for a heat balance equation Schoenmakers et al. [43] concluded that for packed columns operated under typical SFC conditions the density profiles along the column are either linear or slightly convex. In most studies instrumental contributions to band broadening and band broadening caused by limited solubility of the sample in the mobile phase were not taken into account, contributing to the spurious nature of the experimental values reported. Experimental values of the plate height for similar columns differ by more than an order of magnitude in the contemporary literature indicating that many of these values, particularly for packed columns, may be unreliable and the conclusions deduced from them, are likely false.

TABLE 6.3

INFLUENCE OF COLUMN INTERNAL DIAMETER ON EFFICIENCY AND SEPARATION TIME FOR OPEN TUBUALR COLUMN SUPERCRITICAL FLUID CHROMATOGRAPHY. D_m = 2x10^{-4} cm^2/s, D_s = 1x10^{-6} cm^2/s, P = 72 atms., T = 40°C, density = 0.22 g/ml and d_f = 0.25 micrometers

Column	Optimum Parameters at Different Capacity Factors(a)			
Diameter (μm)	k = 1		k = 5	
	H_{min} (mm)	u_{opt} (cm/s)	H_{min} (mm)	u_{opt} (cm/s)
25	0.016 (0.080)	0.505	0.021 (0.107)	0.376
50	0.031 0.156)	0.259	0.042 (0.213)	0.190
75	0.046 (0.234)	0.174	0.063 (0.318)	0.126
100	0.061 (0.309)	0.130	0.084 (0.425)	0.095

(a) values in parentheses = 10 u_{opt}

The general approach for kinetic optimization of open tubular columns has been to adopt the familiar Golay equation (equation 1.34) and to assume that the mobile phase can be approximated by an incompressible fluid with ideal gas properties [44-50]. Circumstances that are approximate at best but serve adequately to demonstrate some of the fundamental characteristics of open tubular columns operated at low fluid densities. The column plate height equation can be written in the form given in equation (6.1)

$$H = [2D_m/u] + [f_g(k)d_c^2u/D_m] + [f_s(k)d_f^2u/D_s]$$
(6.1)

where H is the column plate height, u the average mobile phase velocity, D_m the diffusion coefficient of the solute in the mobile phase, D_s the diffusion coefficient of the solute in the stationary phase, d_c the column internal diameter, d_f the stationary phase film thickness, and $f_g(k)$ and $f_s(k)$ are functions of the capacity factor explained elsewhere (section 1.4). Inserting appropriate values for the above variables into equation (6.1) allows a numerical comparison of different column types to be made, Table 6.3 [46-48]. The separation time and column efficiency is strongly dependent on the column internal diameter. The optimum mobile phase velocity is inversely proportional to the column internal diameter, and for the same diameter, decreases as the capacity factor increases. The column efficiency is highest at the optimum linear velocity but the separation times may be too long for practical applications. A linear velocity ten times the optimum velocity is more common in practice, resulting in about a five fold increase in the minimum plate height for the range of capacity factor values indicated in Table 6.3. For fast, high efficiency separations, column internal diameters below 100 micrometers are required and much smaller diameters would be preferred. The optimum column internal diameter is controlled by the rate of radial mobile phase mass transfer and decreases with increasing density due to unfavorable changes in the solute mobile phase diffusion coefficient. Diffusion coefficients increase linearly with the reciprocal of the mobile phase viscosity and at least within the range 0.6-0.9 g/ml for carbon dioxide can be calculated approximately by the Wilke-Chang equation [51,52]. As the diffusion coefficients become more liquid-like column dimensions for reasonable performances start to approach values similar to liquid chromatography and are not easily attained experimentally. For the more favorable case of low density fluids it should be possible to obtain 2,000-9,000 theoretical plates per meter for retained solutes using commercially available columns, Figure 6.2 [46,47]. Also, these values should not be strongly influenced by the film thickness at least up to thicknesses of about 1 micrometer, in contrast to observations made in gas chromatography. Thick films are of interest in capillary SFC because they allow an increase in sample loadability which is of practical importance for trace analysis. Optimum sample sizes for narrow bore open tubular columns of average film thickness are very small (fraction of a microliter). Also demonstrated in Figure 6.2, for mobile phase velocities exceeding about 0.5 cm/s, which in practice they always will be if reasonable separation times are to be obtained, the contribution to the column plate height from longitudinal diffusion in the mobile phase is negligible by comparison to the contribution of resistance to mass transfer in the mobile and stationary phases.

Reduced parameters (section 1.7.10) can be used to compare the potential of open tubular and packed columns to deliver a certain separation potential in SFC [8,43,53-56]. The Golay equation, equation 6.1, can be rewritten as

$$h = (2/\nu) + f_g(k)\nu + f_s(k)\delta_f^2 \nu \tag{6.2}$$

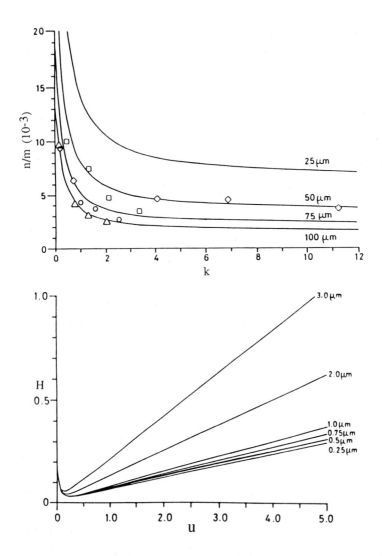

Figure 6.2 Influence of column internal diameter and stationary phase film thickness on efficiency for open tubular columns operated at 10 u_{opt} with supercritical carbon dioxide under low density conditions as the mobile phase. (Reproduced with permission from ref. 46 and 47. Copyright Dr Alfred Huethig Publishers).

where h is the reduced plate height (H/d), ν the reduced velocity (ud/D_m), δ_f^2 the reduced film thickness [$(d_f/d_c)(D_m/d_f)^{1/2}$], and d a parameter characteristic of the column type, either the column internal diameter, d_c, for open tubular columns or the average particle diameter, d_p, for packed columns. There is no exact form of the plate height equation for packed columns under SFC conditions and it has generally been assumed that the Knox equation developed for liquid chromatography can be used in the absence of a more definitive expression, equation (4.2). For high density fluids equation (4.2) may be adequate. At low densities were a substantial column pressure drop exists, a further coefficient with a parabolic variation [$D\nu^2$] has been suggested to improve the agreement with experimental data [56]. Reasonable experimental values of h and ν are h = 3 and ν = 10 for packed columns (at u_{opt}, h = 2 and ν = 3) and for an open tubular column h = 4.5 and ν = 45; the coefficients of the Knox equation are assumed to have their normal values for liquid chromatography (A = 1, B = 2, and C = 0.05). The ultimate performance of a column is assumed to be limited only by the column pressure drop which can be obtained from Darcy's equation (section 1.3) assuming that the fluid is incompressible (which could only be true for fluids not in the immediate vicinity of their critical points).

The separation time expressed in reduced parameters is given by equation (6.3)

$$t_R = (nhd^2/\nu D_m)(1 + k) \tag{6.3}$$

Upon substitution of the reduced parameters given above the separation time for a packed column and an open tubular column would be identical if d_c = 1.73 d_p; given the current limitations of open tubular column technology the column diameter cannot be reduced to the point where these columns can compete with packed columns for fast separations. This is illustrated by the practical example in Figure 6.3 [57]. The separation speed cannot be increased for an open tubular column by increasing the reduced velocity since the reduced plate height is increased proportionately and the ratio (h/ν) in equation (6.3) is approximately constant.

The maximum number of theoretical plates that can be achieved for a particular column depends on the pressure drop per theoretical plate and the maximum pressure drop allowed over the column. Using the same reduced parameters given earlier, the maximum number of theoretical plates obtained for an open tubular

614

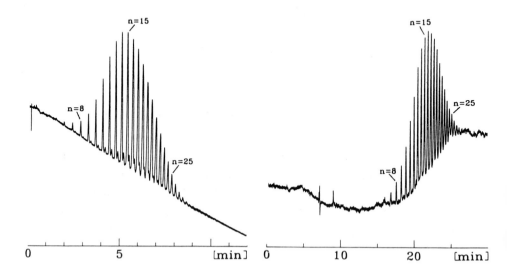

Figure 6.3 Comparison of the separation of the octylphenol poly(ethylene glycol) ether, Triton X-165 on a packed column, left, and an open tubular column, right, using UV detection. For the packed column separation a 10 cm x 2 mm I.D. column packed with Nucleosil C_{18}, d_p = 3 micrometers, temperature = 170°C, and mobile phase carbon dioxide (2 ml/min) and methanol (0.15 ml/min) pressure programmed from 130 to 375 bar in 12 min were used. For the open tubular column separation a 10 m x 50 micrometers I.D., SB-Biphenyl-30, temperature = 175°C, mobile phase carbon dioxide (0.175 ml/min) and 2-propanol (0.0265 ml/min) pressure programmed, 125 bar for 5 min, then ramped from 125 to 380 bar over 19.5 min, and held at 380 bar for 15 min were used. (Reproduced with permission from ref. 57. Copyright Preston Publications, Inc.)

column, n_c, and a packed column, n_p, with the same pressure drop, are related by equation (6.4)

$$n_c = 4.63 \ (d_c/d_p)^2 n_p \qquad (6.4)$$

An open tubular column with an internal diameter of 50 micrometers should be capable of providing about 100-500 times more

TABLE 6.4

SUMMARY OF THE CALCULATED MAXIMUM NUMBER OF THEORETICAL PLATES AND
THE REQUIRED SEPARATION TIME FOR DIFFERENT SFC COLUMNS (k = 4)

Column Type	Column or Particle Diameter (μm)	Maximum Number of Plates	Time Required to Achieve n_{max} (min)	Time Required for 20,000 Plates (min)
Open	8.5	268,000	64.5	4.82
Tubular	17	1,070,000	1030	19.3
Columns	50	8,260,000	77,000	167
	100	37,000,000	1,235,000	667
Packed	3	7,200	0.65	1.8
Columns	5	20,000	5	5
	10	80,000	80	20

theoretical plates than a packed column containing particles with
d_p = 5-10 micrometers. It should be possible to generate about
20,000 theoretical plates in about five minutes using a packed
column and this can be used as a standard by which to judge the
performance of different columns, Table 6.4 [8]. The very large
numbers of theoretical plates that might be available for open
tubular columns in current use can only be obtained with
excessively long and unrealistic separation times. Unless column
internal diameters can be reduced further then more modest numbers
of theoretical plates will have to be accepted with fairly long
separation times, Table 6.5 [8]. When separations can be achieved
with a modest number of theoretical plates it may be possible to
achieve them in an order of magnitude less time with a packed
column than for those open tubular columns currently used. Also,
contemporary open tubular columns will be most useful for
separations requiring comparatively low fluid densities where
diffusion coefficients are more favorable and the greater column
permeability is of importance in lessening the impact of the
column pressure drop. These conditions are typical of those used
for the calculations in Table 6.5. The column plate counts would
be considerably lower and the separation times longer at higher
densities.

6.3 MOBILE PHASE SELECTION

Just about any substance that is stable above its critical
point could be used as a mobile phase in SFC, but in practice only

TABLE 6.5

EFFICIENCY OF DIFFERENT OPEN TUBUALR COLUMNS OPERATED AT 10 u_{opt}
FOR A SOLUTE WITH k = 3, DENSITY = 0.22 g/ml, TEMPERATURE = 40°C
AND MOBILE PHASE CARBON DIOXIDE

Column Diameter (μm)	Column Length (m)	$10u_{opt}$ (cm/s)	h (mm)	n/m(a)	t_R min
100	24	1.1	0.44	2,300	145
75	24	1.4	0.30	3,300	144
50	23	2.0	0.22	4,400	77
25	7	4.3	0.18	5,600	11

(a) correspond to 5.4×10^4, 8.0×10^4, 1.0×10^5 and 3.9×10^4 theoretical
plates per column in order of decreasing column internal diameter.

TABLE 6.6

PHYSICAL PROPERTIES OF POSSIBLE MOBILE PHASES FOR SUPERCRITICAL
FLUID CHROMATOGRAPHY

Fluid	Critical Parameters			Density at 400 atms. (g/ml)	Dipole Moment (D)
	Temperature (°C)	Pressure (atms.)	Density (g/ml)		
Carbon dioxide	31.3	72.9	0.47	0.96	0
Nitrous oxide	36.5	72.5	0.45	0.94	0.51
Sulfur hexafluoride	45.5	37.1	0.74	1.61	0
Xenon	16.6	58.4	1.10	2.30	0
Butane	152.0	37.5	0.23	0.50	0
Pentane	196.6	33.3	0.23	0.51	0
Dichloro-difluoromethane	111.8	40.7	0.56	1.12	0.17
Trifluoromethane	25.9	46.9	0.52	----	1.47
Ammonia	132.5	112.5	0.24	0.40	1.65
Water	374.4	226.8	0.34		
Methanol	240.5	78.9	0.27		

a few substances are commonly used, Table 6.6 [2,12,58,59]. SFC is
generally used for the separation of thermally labile and
involatile substances requiring fluids with low critical constants
and low chemical reactivity. Other favorable properties include
low toxicity, low flammability, and the availability of a high
purity grade of the substance at a reasonable cost. The
solubilizing power of a supercritical fluid depends on its density
and, therefore, those fluids capable of the widest density range
under normal operating conditions will be the most versatile
[60,61]. Compatibility with the properties of detectors commonly

used in SFC, which include flame-based and spectroscopic detectors, is very desirable. Carbon dioxide possesses most of the above properties and is by far the most widely used supercritical fluid. It reacts with some amines to form ureas and carbamates [62]. Not all amines react, and most aromatic amines and N-heterocyclic compounds can usually be separated without difficulty. Most problems are observed for primary and secondary amines with a $pK_b > 9$. Xenon has found use largely because of its spectral transparency extending from the vacuum UV to the NMR region [63]. Sulfur hexafluoride can be used with a conventional flame ionization detectors if the collector electrode is gold-plated in order to prevent corrosion and the hydrogen fluoride generated is vented from the laboratory [64,65]. Ammonia is very corrosive and equipment must be extensively modified for use with ammonia [1,58]. Many stationary phases and solutes are also attacked by ammonia which is little used in practice. The hydrocarbon fluids have unfavorable critical constants and cannot be used with flame-based detectors [12,66].

The solubilizing power of a solvent depends on its density and its capacity for specific intermolecular interactions. The density is related to the experimental parameters of pressure and temperature in a non-linear manner described by an equation of state. Diagrammatically these results are illustrated for carbon dioxide in Figure 6.4 [1]. Close to the critical point the density changes rapidly in a sigmoidal manner with very small changes in pressure. At temperatures further removed from the critical point the isobars are flatter and the change in density with pressure is approximately linear. Also illustrated in Figure 6.4 is the influence of pressure and temperature on the solubility of naphthalene in carbon dioxide [60,67,68]. The complex shape of the curves is explained by two competing effects. As temperature increases the vapor pressure of the solute tends to increase its solubility. At the same time increasing temperature tends to lower the fluid density and to decrease its solubilizing power. At higher pressures the fluid density changes less with temperature and solute solubility depends largely on volatility. At lower pressures naphthalene solubility changes markedly with temperature and decreases significantly with increasing temperature reflecting large changes in fluid density. These results reveal the importance of controlling density in SFC and the rather complex relationship between pressure and temperature that control

618

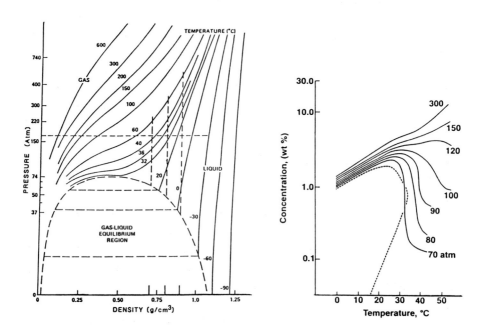

Figure 6.4 On the left is a phase diagram for carbon dioxide. Broken lines indicate isotherm crossing at either constant pressure or density. On the right is illustrated the change in the solubility of naphthalene as a function of temperature and pressure.

density. Increasing density at constant temperature will normally increase solubility providing the appropriate conditions for samples to migrate through the column in SFC. A certain minimum density is required to elute a solute at a given temperature, and if the pressure cannot be adjusted to obtain this density then the sample is not eluted [60,69]. At a constant density, increasing the temperature will normally reduce retention [8,70,71].

As well as density the solvent strength and selectivity of the fluid will affect both the solubilizing power of the fluid for specific solutes and retention in SFC. Solvatochromic indicators have been used to rank fluids commonly used in SFC in terms of their polarity/polarizability characteristics [9,68,72,73]. The π^* scale of Kamlet-Taft is based on spectroscopic measurements of shifts in the absorption maxima of specific test solutes relative to cyclohexane as a base solvent. Values for carbon dioxide as a function of pressure are shown in Figure 6.5. In the region of the critical point relatively small changes in pressure cause large changes in the fluid polarity/polarizability. At higher pressures the change in fluid properties are less marked but do not reach a plateau. These results confirm that the solvent properties of a supercritical fluid are a function of the density of the fluid and can be varied by changing the fluid density. Fluid polarity/polarizability for a series of common supercritical fluids as a function of fluid density are also compared in Figure 6.5. The normal range of values for liquids on the solvent polarity/polarizability scale is 0 to 1 with values close to 1 representing strong dipole-type interactions. With the exception of ammonia the most widely used supercritical fluids provide only very weak polar interactions. Unfortunately, ammonia is a very difficult fluid to use because of its chemical reactivity and is therefore not a viable solvent for general applications. The solvent polarity/polarizability of carbon dioxide and nitrous oxide are very similar so that these two solvents show virtually identical chromatographic behavior. Nitrous oxide might be preferred for samples that react with carbon dioxide [74]. Sulfur hexafluoride is a very weak solvent explaining its use for the separation of nonpolar samples and restricted application to high molecular weight samples. Xenon, ethane and Freon-13 are seen to have properties intermediate between those of sulfur hexafluoride and carbon dioxide. Freon 22 has a reasonable dipole moment and acceptable critical parameters [75]. It was recently recommended for use as a moderately polar supercritical fluid to complement the properties of carbon dioxide. However, it has not been widely evaluated. The solvent polarity/polarizability scale measures only orientation and induction type interactions, but with the possible exception of ammonia, the other fluids in common use are unlikely

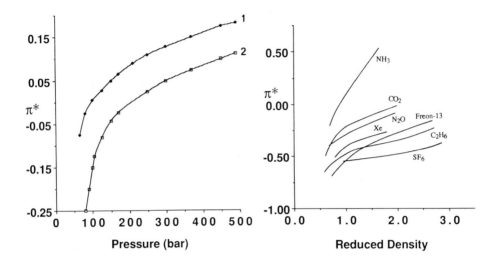

Figure 6.5 Parameters influencing the solvent polarity
/polarizability (π^*) of carbon dioxide and some other supercritical
fluids. On the left π^* is plotted against pressure for carbon
dioxide containing 5.1 mole% of 2-propanol (1) and carbon dioxide
(2). On the right π^* is plotted against the reduced density
(density/critical point density) at a reduced temperature of 1.03
(temperature/critical point temperature) for several common
supercritical fluids. (Reproduced with permission from ref. 9.
Copyright American Chemical Society).

to exhibit significant proton donor/acceptor interactions. These
results indicate the very limited potential for optimizing solvent
selectivity with the supercritical fluids commonly used compared
to the much wider scope available in liquid chromatography. The
most common supercritical fluid, carbon dioxide, has properties
similar to hydrocarbon solvents with the principal intermolecular
interactions being dispersion and induction, since carbon dioxide
has no dipole moment of its own. Similar conclusions to those
based on the solvent polarity/polarizability scale are reached if
the fluids are compared based on the concept of free volume (the
smaller the free volume the closer the molecules are packed
together and the greater will be the strength of intermolecular
interactions) [70,76,77].

One approach to enhancing the range of selective mobile
phases available for SFC is to use binary mixtures of a
supercritical fluid and a polar solvent. In this way the available

range of selective mobile phases can be enhanced, more polar samples separated, peak shapes improved, and resolution and separation time more conveniently optimized. To improve chromatographic performance solvent modifiers are generally used at concentrations below about 2% (v/v). At these concentration levels the modifier may have only small influences on the solubility of the sample in the mobile phase, and is used primarily to mask interactions between the sample and accessible silanol groups on the stationary phase surface [78-80]. If these interactions make a major contribution to retention of the sample then in the presence of a suitable modifier a substantial reduction in retention will occur, but more typically, smaller changes in retention with a decrease in peak asymmetry are observed. When the same modifiers are used with deactivated fused silica open tubular columns at the same concentration, the change in chromatographic behavior is often very small compared to the unmodified mobile phase, indicating that the modifier behaves primarily as a support deactivating agent. Modifiers should be considered as a means of extending the deactivation of well-deactivated chemically bonded packings and not as a means of overcoming the deficiency of poorly deactivated packings. Water and formic acid are the most popular modifiers used with the flame ionization detector to improve the chromatographic properties of proton donor or acceptor solutes and occasionally neutral polar solutes [4,81-83]. Water is only sparingly soluble in supercritical fluid carbon dioxide. Using a saturator column between the pump and injector allows concentrations of about 0.3% (v/v) of water in carbon dioxide to be generated at room temperature for the normal density operating range for carbon dioxide. The water accumulates in the pores of the packing establishing a steady state relationship with the mobile phase composition. Changes in the equilibrium state relationship with changes in pressure and temperature are fairly rapid allowing the experimental conditions to be quickly optimized. Water is a reasonably effective modifier for minimizing silanophilic interactions. Formic acid is often more effective but is limited in general use by its instability and difficulty of purification. Commercially available samples of formic acid have a rather poorly defined composition. Methanol is an effective modifier for masking silanophilic interactions with polar functional groups when spectroscopic detection is used [22,28,84]. In this case deactivation may also involve chemical reaction with the silanol groups, as discussed previously. For basic samples, alkylamines, alone or in combination with other modifiers, are effective deactivating agents [4,85].

At higher solvent modifier concentrations it is possible to adjust the solvent strength and selectivity of the mobile phase in a manner similar to that employed in liquid chromatography, although the range over which the mobile phase properties can be varied may be restricted due to miscibility limitations and unfavorable changes in the critical point parameters for the binary mobile phase [9,70,78,79,86-88]. The minimum temperature and pressure required to ensure that a binary mobile phase remains in the supercritical fluid region can be obtained from phase diagrams, if available, or estimated using various numerical techniques based on an appropriate equation of state and interaction parameters [67,79,86-90]. Typical behavior is shown in Figure 6.6 for binary mixtures of 2-propanol and carbon dioxide. Increasing the mole fraction of 2-propanol causes a continuous increase in the critical temperature while the critical pressure at first rises to a maximum and then falls again at higher mole fractions of modifier. The change in critical temperature is quite sharp and fairly high temperatures must be used to remain in the supercritical state for modest concentrations of modifier. These higher critical constants for binary mobile phases may also limit the range over which density can be changed to control retention. The influence of the addition of 2-propanol on the solvent polarity/polarizability can be seen from Figure 6.5 [9,86,91,92]. The addition of 2-propanol increases the solvent polarity /polarizability of carbon dioxide in a manner that changes with density.

Another approach to increasing the scope of SFC for the separation of polar samples is to modify the sample so that its compatibility with the mobile phase is enhanced. The most common approach has been the derivatization of polar functional groups by chemical reactions similar to those used in gas and liquid chromatography [5,12,93,94]. This serves the purpose of reducing the polarity of the sample which increases its solubility in nonpolar supercritical fluids, provides a means of improving the sensitivity or selectivity of detection using detector-selective derivatizing reagents, and minimizes the importance of adsorptive interactions with the stationary phase, particularly for packed

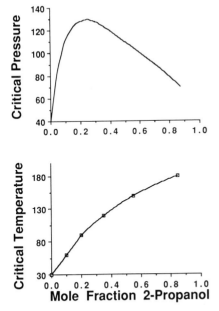

Figure 6.6   Estimated critical temperature and pressure for binary mixtures of 2-propanol in carbon dioxide as a function of the mole fraction of organic modifier.

columns. Secondary chemical equilibria can also be exploited to improve the chromatographic properties of polar solutes using ion-pairing reagents and reverse micelles for certain applications [95-98]. These systems are fairly complicated and not well characterized for the present. Yet, they open up another door by which the limitation of performing separations of polar solutes with virtually nonpolar fluids can be overcome. It has been noted that the principal limitation of ion-pair chromatography in SFC is the low solubility of the ion-pair reagents in the carbon dioxide-modifier mixtures [98].

6.4   PARAMETERS AFFECTING RETENTION

Retention in SFC is a complex function of the experimental parameters and is not as easily rationalized as in the case of gas and liquid chromatography. Retention in SFC is dependent upon temperature, pressure, density, sample concentration, composition of the mobile phase and the composition of the stationary phase. Many of these variables are interactive and do not change in a

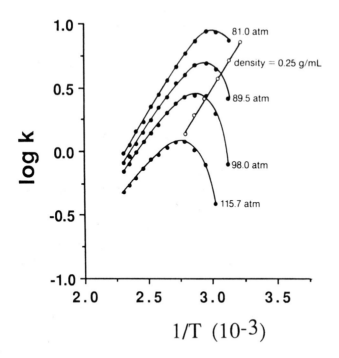

Figure 6.7   Plot of log k against 1/T for hexadecane at constant pressure and constant density with carbon dioxide as the mobile phase.

simple or easily predicted manner. Certain general observations, however, can usually be applied [4,8,70,71,77,99-104]. Changes in retention with temperature at constant density are readily predictable from van't Hoff plots, Figure 6.7. The logarithm of the capacity factor is a linear function of the reciprocal of the column temperature, which persists even down to sub-critical conditions. The change in retention at constant pressure is more complicated, Figure 6.7, and depends on the volatility of the solute and the density of the mobile phase. The shape of the curves can be explained by two competing effects. Under conditions where the solute vapor pressure is not a dominant consideration, increasing temperature will lead to an increase in retention due to a reduction in the density and, therefore, the solubilizing capacity of the mobile phase. This results in a negative slope in the plots shown in Figure 6.7. If the solute has significant vapor pressure then increasing the temperature will result in a reduction in retention giving plots with a positive slope. With increasing pressure (density) the influence of volatility becomes less pronounced. For separations at constant pressure varying the temperature may result in both increases and decreases in retention for some solutes with possible changes in elution order. The thermal stability of the sample is always the overriding consideration when selecting the operating temperature. When operating at constant density the highest practical temperature will normally result in the greatest resolution and shortest separation time due to increasing solute diffusion coefficients resulting in better mass transfer characteristics. When organic modifiers are added to the mobile phase at constant pressure and temperature the retention of analytes will increase or decrease with increasing modifier concentration depending on whether the analytes are more or less soluble in the modifier compared to the supercritical fluid, provided that the column activity level remains the same [79,86].

Temperature, pressure and density may influence retention in otherways than those discussed above. Water solubility in supercritical fluids generally increases with temperature causing a shift in the equilibrium of the number of water deactivated silanol groups to carbon dioxide deactivated groups [4,101]. In this case the solubility of the analyte in the mobile phase increases but so does retention due to increasing stationary phase activity. In SFC, the sorption isotherms of solutes are pressure dependent; at lower pressures the non-linear region of the isotherm is reached at a lower solute concentration than at higher pressures (densities) [28,37,105]. Once in the non-linear region of the sorption isotherm the retention of the concentration maxima of a peak shifts with increasing solute concentration. This can give rise to sample size dependent retention times. The effect should be less noticeable at higher mobile phase densities. Since the capacity factor shows a strong dependence on density, the density drop across a column could cause variations in retention that would depend on the column permeability, the compressibility of the mobile phase, and the magnitude of the density drop [4,54]. Supercritical fluids are most compressible close to the critical point and in this region differences in column permeability could cause changes in retention time.

The elucidation of a formal retention model in SFC is a complex problem. Yonker et al. have proposed a classical thermodynamic model that relates retention as a function of temperature at constant pressure to the volume expansivity of the fluid, the enthalpy of transfer between the mobile and stationary phases, and the changes in the heat capacity of the fluid as a function of temperature [91,106]. Martier has developed an alternative approach based on a statistical thermodynamic treatment using established lattice-gas models to formulate a relationship for the solute partition coefficient [107]. Both approaches are fairly involved mathematically and describe only segments of the experimental data for specific operating conditions [71,108]. Although intuitively useful in relating important aspects of the retention process, their incompleteness and complexity prevents further consideration at this time.

6.4.1 Programming Techniques

The most common separation techniques in SFC employ different gradient or programmed techniques, paralleling trends in gas and liquid chromatography where temperature gradients in gas chromatography and composition gradients in liquid chromatography are routinely used. In SFC pressure, density and velocity gradients are used as well as temperature and composition gradients. These gradients are used individually or in combination to reduce separation times, improve resolution, increase sample detectability, and to extend the molecular weight range or polarity of eluted species [51,109-116]. A typical example is shown in Figure 6.8 for the separation of a mixture of octylphenol poly(ethylene glycol ether) oligomers [57]. At constant density (referred to as isoconfertic or isopycnic conditions by some authors) the first few members of the oligomeric series are sharp and separated to baseline, while later members are retained by ever increasing time increments, broadened, and difficult to detect. Programming an increase in the concentration of the modifier reduces the separation time and sharpens later eluting peaks. For constant composition, programming the pressure achieves similar goals, and in this case enhances resolution as indicated by the separation of the subseries of oligomers in minor concentration. Pressure programming causes an increase in mobile phase density which increases sample solubility. In this mode the type of intermolecular interactions remain largely the same and

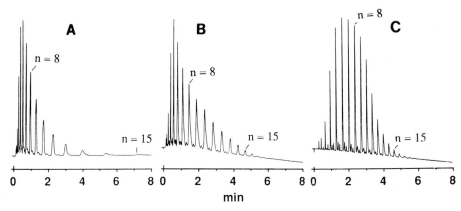

Figure 6.8 Separation of Triton X-114 by SFC using programmed elution on a 10 cm x 2 mm I.D. Nucleosil C_{18} column, 3 micrometer packing, at 170°C with UV detection at 278 nm. The separation on the left was performed under isobaric conditions at 210 bar with a mobile phase of carbon dioxide + methanol (2 + 0.125) ml/min. The separation in the center was obtained using a composition gradient from 0.025 to 0.4 ml/min over 8 min with other conditions as above. The separation on the right was obtained using a pressure program from 130 to 375 bar over 8 min with the same mobile phase used for the isobaric separation. (Reproduced with permission from ref. 57. Copyright Preston Publications, Inc.)

the principal observed change is due to an increase in solvent strength of the mobile phase. Composition gradients generally involve an increase in solvent strength accompanied by a change in selectivity. Density gradients and composition gradients, therefore, do not necessarily produce similar results. Compared to liquid chromatography re-equilibration in both normal and bonded phase chromatography is relatively rapid, increasing sample throughput [117]. Equilibration occurs virtually instantaneously for pressure programming and with a few column volumes of mobile phase for composition programming due to faster diffusion in supercritical fluids compared to liquids. This is very important in normal phase separations.

Pressure or density programming is the most popular of the gradient techniques in SFC. Density is the important parameter with respect to retention but pressure is the physical property which is directly monitored by SFC instruments. If enough experimental density-volume-temperature data are available for the mobile phase then a computer-based algorithm can be used to generate specific density programs. Such data are available for only a few mobile phases, such as carbon dioxide and the n-alkanes. In other cases, and for mixed mobile phases, it may be possible to approximate density changes using an appropriate equation of state. Density programming can be achieved in several ways, for example, by programming the column inlet pressure and fixing the column outlet pressure by a restrictor or backpressure regulator; delivering a constant flow of mobile phase to the column while programming the flow at the column outlet; and by programming the column inlet and outlet pressure independently [109,112,115, 116,118-120]. The most common approach when flame-based detectors are used is to employ a fixed restrictor at the end of the column and to increase the mass flow rate of the mobile phase at the column inlet as a function of time. Volume flow rates of the expanded fluid usually restrict this approach to open tubular columns and microbore packed columns. With larger diameter columns, which are operated at higher volumetric flow rates, optical detectors are obligatory. In this case the restrictor (or backpressure regulator) is placed after the detection cell. Ideally, it is assumed that the pressure drop over the restrictor/regulator is much greater than the pressure drop over the separation column; a situation that is likely to be true for short open tubular columns and packed columns of normal length with coarse particle packings. Changes in linear velocity during pressure programming result from both the pressure-dependent flow through the restrictor at the column exit and the velocity gradient along the length of the column resulting from fluid compressibility. Pressure/density programming, consequently, is performed with a simultaneous mobile phase velocity gradient that can have several effects on the quality of the separation. One problem arises from the fact that under normal circumstances the pressure gradient and velocity gradient are not a simple function of each other, which adds to the complexity of theoretical models attempting to describe retention under different experimental conditions [112,113,116]. Although linear density programming is most commonly used, in certain circumstances nonlinear programming might provide better resolution and shorter separation times [2,109,115]. For example, the retention time of members of a homologous series are a logarithmic function of density. Therefore, asymptotic density programming as shown in Figure 6.9 leads to a more even spacing of components in such samples [115]. Asymptotic density programming is not implemented on all commercially available instruments and is restricted to those

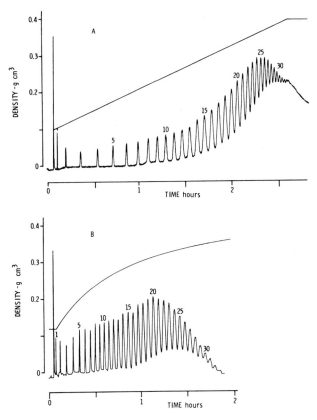

Figure 6.9 Separation of a poly(styrene) oligomer mixture of 2000 average molecular weight by open tubular column SFC using a linear density program (A) and an asymptotic density program in (B) with carbon dioxide as the mobile phase. (Reproduced with permission from ref. 115. Copyright Preston Publication, Inc.)

mobile phases for which extensive density-volume-temperature data are available to generate the required pressure program. The principal purpose of those experiments that employ simultaneous pressure programming of the column inlet and outlet is to maintain a nearly constant linear velocity through the column. These methods are experimental and have not been implemented so far on commercially available instruments.

One of the primary benefits of pressure/density programming is peak compression that results in later eluting peaks having the same width, or an even narrower width, than the earliest peaks in the chromatogram. Qualitatively, this can be ascribed to either positional variations in mobile phase density or velocity along the length of the column. Under programmed conditions the mass flow rate of mobile phase at the column inlet exceeds that at the outlet resulting in a velocity gradient with the velocity of the mobile phase being higher at the column inlet. Consequently, those parts of the sample zone nearest the column inlet are moving faster than the parts further down the column resulting in a compression of the sample zone. In the case of a column with a significant density drop along its length peak compression can occur due to an increase in the capacity ratio of a sample inside the sample zone. The density gradient over the zone results in the front of the zone moving at a slower velocity than the back portion leading to compression of the zone in the direction of migration. The peak compression effect is more pronounced the higher the density and velocity gradients along the column. At some point the benefits of the peak compression mechanism are outweighed by other competing processes that result in decreased resolution. At higher densities diffusion coefficients are decreased and at higher velocities the contribution of the resistance to mass transfer term increases, both leading to a loss in column efficiency. It has also been shown that higher density and velocity gradients lead to a loss in column selectivity [113]. Therefore, in general, to achieve optimal resolution during density programming, both the relative pressure drop along the column and the relative decrease in the mobile phase linear velocity should be kept small. Also, a relatively slow increase in density should lead to improved resolution for difficult separations.

Both positive, negative, and synchronized temperature /density programming have been used in SFC, but less commonly than pressure/density programming [109,114,121-124]. Sample diffusivity increases with temperature which can be used to control retention of volatile solutes. It seems to be generally true in pressure/density programming that high temperatures are preferred for the separation of thermally stable compounds that are soluble in low-density mobile phases. Positive temperature programming at constant pressure is restricted in utility to solutes with significant volatility since density and, therefore, solubility will decline at higher temperatures. Negative temperature programming at constant pressure leads to increasing density and has a similar effect to density programming except that since higher densities are reached at lower temperatures column efficiency declines more noticeably. Since the maximum density that can be reached for a particular mobile phase is controlled by pressure and temperature, the former being fixed by instrumental considerations, this leaves temperature as the only variable that can be changed to obtain the highest densities that might be required to elute certain sample components. Synchronized density and temperature programming should provide the optimal conditions for separating samples covering a wide molecular weight range for which a high maximum final density is required to elute the last few components of a mixture [124]. The slow heat transfer characteristics of certain packed columns precludes their use for temperature programmed separations.

Composition gradients have not been widely used in SFC until comparatively recently [45,110,125]. Some of the reasons for this are the more complex instrument arrangement, continuous change in the critical parameters with mobile phase composition that may not be well understood, and incompatibility with flame-based detectors when significant percentages of organic solvents are used. Density programming is an effective method of changing solvent strength but it has little influence on selectivity. Selectivity is easier to change using composition gradients as commonly employed in liquid chromatography. To generate binary mobile phase gradients two pumps are required. With two syringe pumps the ratio of solvent modifier to primary fluid can be controlled, but simultaneous control of pressure is not straightforward. For packed column applications, and open tubular columns if a precolumn flow splitter is used, reciprocating pumps operated under flow control can be used and the pressure controlled by using a regulator at the detector exit. The compressibility of the fluid and excess molar volumes of the mixture cause significant deviations of the mobile phase composition from the volumetric flow delivered by the pumps.

6.5 INSTRUMENTAL ASPECTS OF SUPERCRITICAL FLUID CHROMATOGRAPHY

A schematic diagram of a chromatograph for SFC is shown in Figure 6.10. In general, the instrument components are a hybrid of components developed for gas and liquid chromatography that have been subsequently modified for use with supercritical fluids. Thus, the fluid delivery system is a pump modified for pressure control and the injection system a rotary valve similar to components used in liquid chromatography. The column oven and

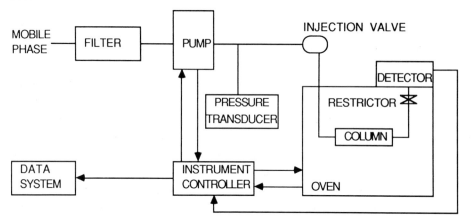

Figure 6.10 Schematic diagram of a supercritical fluid chromatograph.

flame-based detectors are similar to those used in gas chromatography. Alternatively, optical detectors similar to those used in liquid chromatography and modified for high pressure operation can also be used. A unique feature of the chromatograph is a restrictor required to maintain constant density along the column and to control the linear velocity of the fluid through the column. The restrictor is usually placed between the column and detector for flame-based detectors and after the flow cell for optical detectors. The real growth in SFC applications commenced with the introduction of commercial instruments designed for use with packed or open tubular columns in the 1980s. These early instruments are still at an evolutionary stage as improvements continue to be made in all aspects of SFC.

6.5.1 Fluid Delivery Systems

Typical flow rates, as liquid, in SFC correspond to about 1-10 microliters/min for open tubular columns and 0.1-10 ml/min for packed columns [6,8]. Syringe pumps are widely used with open tubular columns and microbore packed columns while reciprocating-piston pumps are more commonly used with packed columns of wider bore, for example, 4-5 mm internal diameter. Reciprocating-piston pumps for liquid chromatography are usually modified with extra check valves and with cooling at the pump heads and at the check valves for SFC with carbon dioxide or other gases at ambient temperature [8,12,126]. Pressurized gases as liquid can be drawn out of cylinders using an eductor (dip) tube. Cylinders with a helium head pressure of over 1000 p.s.i. permit fairly complete filling of syringe pumps with liquid without resorting to external pump cooling [127,128]. Syringe pumps are reliable and easy to use giving a pulseless flow which contributes to their popularity. Reciprocating-piston pumps have the advantage of being able to deliver unlimited volumes with continuous flow of mobile phase but require external cooling to avoid cavitation with some mobile phases and pulse dampeners may also be needed to minimize pulsations in the mobile phase delivered to the column. Particle filters are commonly installed ahead of the mobile phase inlet to the pump to reduce damage to the pump and additional in-line purification cartridges packed with charcoal and/or alumina have been used to purify carbon dioxide.

One difference in the operation of pumps in SFC compared to liquid chromatography is that pressure control is generally more important than flow control. The density of a supercritical fluid varies with pressure and, since the solvent strength of a fluid is a function of its density, pressure must be accurately controlled. A high-precision pressure transducer is installed between the pump and injector for this purpose. Simultaneous measurement of the column temperature and pressure control allows constant density or density programming under computer control if the appropriate isotherms are known or can be approximated. This is usually the case for carbon dioxide and pentane with most commercial instruments but the relevant data and/or programming may not be available for other mobile phases. When using syringe pumps the algorithm controlling pump operation must also accommodate temperature changes resulting from expansion or contraction of the fluid volume in the syringe pump during use. This is relatively simple to allow for with well characterized mobile phases, such as carbon dioxide, but may not be accommodated by the pump controller for less common mobile phases.

Composition gradients are not as well supported by current commercial instruments as are density gradients. The large compressibility of fluids makes the generation of accurate composition gradients more difficult than for liquid chromatography. It is also necessary to ensure that the mixed mobile phase will be homogeneous under the selected operating conditions (see section 6.4). The simplest approach is to purchase cylinders prepared with a certified mixture of components but, in this case, it is not possible to adjust the composition of the mobile phase in a continuous manner unless the mixture is used as feed for one pump in a dual pump system. Binary mixtures of organic solvents with gases at room temperature can be accurately prepared in the laboratory using simple apparatus [129]. Dual syringe pumps are not common for commercial instruments but when operated in the flow control mode they provide a simple means of generating a composition gradient. Simultaneous pressure control, however, is not straightforward. The higher flow rates used with packed columns make the use of reciprocating-piston pumps feasible. These pumps can be operated under flow control with a backpressure regulator positioned after the detector cell of a UV detector to control pressure [110,111,120]. A similar approach can be used with open tubular columns using split flow to overcome the difficulties of adequately mixing the very low volumes of mobile phase required by open tubular columns [125].

6.5.2 Sample Inlets

Injection of the sample into the column in SFC remains a problem for several and varied reasons [6,12,130-132]. The total column volume for typical open tubular columns used in SFC is on the order of a few microliters necessitating the accurate introduction of nanoliter sample volumes in a low-dead volume system. With packed columns large sample volumes are permissible and the higher mobile phase flow rates minimize problems with dead volumes but, in this case, incomplete mixing of the sample solvent with the mobile phase prior to the sample plug entering the column can result in excessive band broadening and even peak splitting.

Nearly all sample introduction systems provided with commercial instruments are based on the use of high pressure rotary valves with internal sample loops of 0.06-0.2 microliters for open tubular columns and 0.2-1.0 microliters for small bore packed columns. The valves may be operated manually but more commonly electronic or pneumatic actuators are used. The sample loop is loaded with a solution of the sample in an organic solvent, the valve switched to the inject position, and the sample displaced into the column by a high pressure liquid that subsequently becomes a supercritical fluid upon entering the column oven. The injection valve is normally mounted on top of the oven at a temperature close to room temperature while the column may be at a much higher temperature. Secondary cooling may be used to maintain the valve temperature close to room temperature by dissipating the heat conducted from the oven. A general requirement of the sample solvent is that it must be able to dissolve the sample at a relatively low temperature. Also, the sample must remain soluble in the liquid mobile phase/solvent mixture as it mixes before becoming supercritical. This is not always a favorable situation as the supercritical fluid mobile phase may have greater solvating capacity than the liquid mobile phase. Failure to maintain the sample in solution can result in memory effects and inaccurate quantitation.

Since the volumes delivered by loop injectors are comparatively large compared to the capacity of standard open tubular columns, split injection techniques are commonly used [6,130,133,134]. A typical arrangement for split/splitless injection using an open tubular column is shown in Figure 6.11. In the dynamic split mode the valve is turned to the inject position and part of the sample enters the column with the remainder exiting via the vent line. For this type of injection a nonlinear relationship between peak areas and the split ratio is found. Also, the split ratio is observed to change with sample viscosity and mobile phase density. Several factors are known to contribute to the above situation, including: (1) sample displacement from the valve has an exponential profile (as opposed to a plug profile) because of the laminar flow of the displacing mobile phase, requiring several loop volumes of mobile phase to displace nearly all of the sample from the loop; (2) slow and possibly inhomogeneous mixing of the injection solvent with the liquid mobile phase; and (3) changes in sample solubility as the mobile phase density changes with temperature. Dynamic split requires rather concentrated samples for a reasonable detector response and is unsuitable for trace analysis.

An alternative form of split injection is the timed split technique, Figure 6.11 [130,131,133]. In this case the column is connected directly to the valve and the valve actuator is controlled electronically to turn the valve to the inject position and back very rapidly with only a portion of the sample in the loop displaced to the column. Timed split allows variable volumes to be injected by changing the valve actuator time and provides more reproducible splitting than the dynamic split technique. However, it suffers from many of the same problems as dynamic split, namely, poor accuracy, split ratios that depend on pressure, and high detection limits.

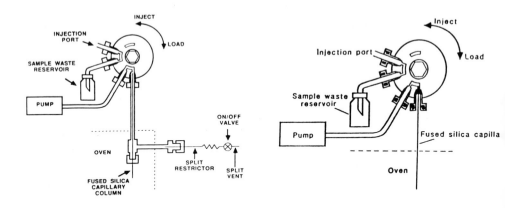

Figure 6.11   Schematic diagram of a split/splitless valve injector (A) and a timed-split injector (B) for open tubular column SFC.

To allow larger sample volumes to be injected into open tubular columns without degrading column performance various direct injection techniques employing retention gaps and pressure trapping have been developed [130,131,133-138]. Direct injection, or splitless injection, of larger volumes is usually accomplished by valve injection at a low mobile phase pressure followed by rapidly increasing the pressure to the desired operating density. This method is effective if a rapid separation of sample and solvent is obtained (usually in a retention gap) and precipitation of the solutes as a narrow band at the head of the column can be achieved. The disadvantages with this method are: (1) solvent elimination times may be very long (and consequently sample separation times); (2) peaks may be broad and/or split due to flooding and incomplete refocusing; and (3) slightly soluble compounds may precipitate in the injector at the initial low density, leading to poor quantitative recoveries and tailing peaks. Several approaches have been described to reduce the solvent elimination time which may require more than 10 min. under normal circumstances. The delayed split method uses the same configuration as the dynamic split except for the addition of an on/off valve in the vent line. With the split vent valve off the sample is injected in the normal way at a selected pressure. After a prescribed delay time the injection valve is returned to the load position and the vent valve opened, which affects an immediate purge of the injection area. The amount of sample transferred into the column depends on the valve actuation time and the delay time. The delayed split injection techniques can be modified for solvent backflushing [130]. Immediately after the vent valve is opened a rapid negative pressure ramp is initiated. This causes a reversal of flow in the first part of the column and precipitation of the sample on the column wall. The solvent, and probably some of the sample, is backflushed out of the column and through the open vent line. In a different approach Greibrokk [135] used a short retention gap coated with a thin-film of stationary phase for solute/solvent separation and a vent valve between the retention gap and column to eliminate the solvent. Solute focusing is favored by injection at low pressures and/or high temperatures corresponding to low mobile phase density. Pressure trapping, as it is called, is the principal mechanism of solute refocusing in SFC [134-137]. This mechanism is effective if the mobile phase density is low enough that the sample is on the verge of being precipitated from solution and the sample is capable of stronger interactions with the stationary phase than mobile phase/stationary phase interactions.

Volume restrictions for valve injection with packed columns are less limiting but band broadening and peak splitting may still be a problem because of solvent effects [132,137]. The sample solution is gradually diluted with mobile phase as it migrates through the connecting tube between the injector and column. In most cases it will also be subjected to a temperature gradient since the column oven temperature generally exceeds the injection valve temperature. The critical constants of organic solvents are generally greater than those of carbon dioxide, for example, and even if the sample solvent and mobile phase were fairly well mixed, it is likely that the sample will be delivered to the column in a mixture of liquid and fluid phases. This has the effect of delivering at least part of the sample to the column in a strong solvent that is not adequately refocused by the column. A solution to this problem is to increase the volume of the connecting tube between the injector and column to provide effective dilution of the sample solvent in the fluid mobile phase. Even under these conditions the sample may still be delivered to the column in a volume of solvent and fluid that is too strong to avoid band broadening entirely. A better solution is represented by the solventless injector shown in Figure 6.12 [132]. Compared to the samples typically separated by SFC most common solvents have substantially higher vapor pressure and can be removed from the sample by evaporation under GC-like conditions in a short precolumn optimized for the purpose. Quite large solvent volumes can be removed enabling dilute solutions to be analyzed. After solvent removal the mobile phase is directed through the precolumn and the sample introduced onto the column without interference from the sample solvent. The advantages of the solventless injector for packed column SFC are: (1) eliminates peak broadening and splitting caused by the injection solvent being too strong; (2) reduces the separation time by eliminating the time wasted while waiting for the solvent to elute; (3) allows the injection of large volumes of dilute samples; (4) allows direct injection of solvents and reagents that might cause column deterioration or have a deleterious effect on the restrictor or detector; (5) enhances sensitivity by allowing for operating conditions that produce narrower analyte peaks; and (6) allows for finer tuning of a separation by controlled post injection density ramps (density compaction) or introduction of samples at higher initial densities.

6.5.3 Restrictors

A restrictor is required to maintain supercritical fluid conditions along the length of the column, to control the linear velocity of the mobile phase through the column, and to aid the process of detection with gas phase detectors. With optical detectors employing a flow through high pressure cell the restrictor is normally positioned after the cell. For flame-based detectors the restrictor is positioned between the column and the detector and must effectively transfer the column eluent from the supercritical fluid phase to the gas phase without introducing excessive dead volume or causing the sample components to condense or form clusters which lead to spiking of the detector signal. The ideal restrictor should be inert, immune from plugging, adjustable, easily replaceable, and effective for all sample types [116,131,139]. Such a restrictor does not exist but commonly used devices include crimped tubing and backpressure regulators for

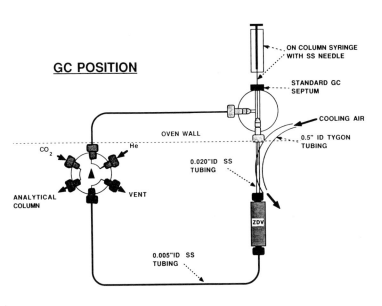

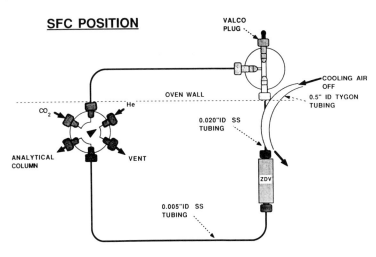

Figure 6.12 Schematic diagram of a solventless injector used for packed column SFC. (Reproduced with permission from ref. 132. Copyright Dr. Alfred Huethig Publishers).

optical detection [110,118,120,125,140] and linear [131,139, 141,142], tapered [139,141-144], converging/diverging [145], integral [131,146,147], and frit [6,148] restrictors for gas phase detectors. Since the restrictor is placed after the detector cell for optical detection relatively simple crimped tubing,

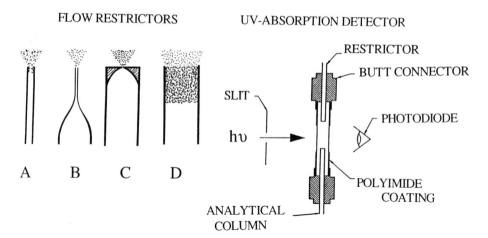

FLOW RESTRICTORS

UV-ABSORPTION DETECTOR

Figure 6.13 Flow restrictors of different design: A, linear; B, tapered; C, integral; and D, frit. On the right side is shown a modified high pressure cell for UV detection using open tubular columns.

mechanically adjustable valves, or pneumatic systems can be used at packed column flow rates with few problems. Density programming with a constant linear velocity is possible by controlling the outlet pressure with an adjustable valve. A sheath-flow nozzle was recently used as a programmable backpressure regulator for open tubular columns [140].

Some typical restrictors, for gas phase detection are shown in Figure 6.13. In early studies linear restrictors were commonly used, consisting of short lengths of narrow bore fused silica tubing (10-15 cm long and 5-15 micrometers internal diameter). Detector spiking often occurs with these restrictors for polar and high molecular weight analytes. This was a result of molecular association and condensation of the analyte during decompression to yield fog particles that entered the detector and created short ion bursts in the flame. A solution to this problem is to carry out the decompression over a very short length and to ensure that the tip of the restrictor is supplied with sufficient heat to avoid condensation of the analyte. Tapered and integral restrictors fit into this category. Tapered restrictors are usually 1-2 cm long with a diameter 50-100 micrometers drawn down to 1-2 micrometers. The taper allows decompression over a short distance, while the thin wall at the tip allows greater heat transfer to the expanding effluent jet, both of which contribute to the elimination of spiking. Reproducible tapered restrictors can be prepared using a robot puller [143]. These restrictors are rather fragile and prone to break unless fitted with a protective fused silica sleeve [144]. Conical integral restrictors are less fragile than tapered restrictors and can be simply prepared (after some practice) by first closing the end of a suitable fused silica capillary in a flame and then sand papering back the tip to produce a pin hole of 1-2 micrometers [131,146]. The integral restrictor provides virtually point restriction but flow rate adjustment is often difficult and the orifice is prone to plugging. The multipath frit restrictor is prepared by bonding an inorganic porous frit inside a short length of fused silica tubing and is more durable than the tapered restrictors. The multiple flow paths and longer residence time of the mobile phase in the frit region allows more efficient heat transfer minimizing spiking [143,148]. No single type of restrictor has emerged as superior to all others, and each design represents a compromise. There is still plenty of incentive to develop a new restrictor that is rugged and adjustable for use over a wide range of selectable flow conditions.

6.5.4 Detectors

Detection in SFC can be achieved in the condensed phase using optical detectors similar to those used in liquid chromatography or in the gas phase using detectors similar to those used in gas chromatography. Spectroscopic detectors, such as mass spectrometry and Fourier transform infrared spectroscopy, are relatively easily interfaced to SFC compared to the problems observed with liquid mobile phases (see Chapter 9). The range of available detectors for SFC is considered one of its strengths.

The flame ionization detector is the most popular of the flame-based detectors. Apart from a reduction in sensitivity compared to expectations based on gas chromatographic response factors [138] and incompatibility with the high flow rates of conventional bore columns (4-5 mm I. D.), the flame ionization detector is every bit as easy to use in SFC as it is in gas chromatography [148,149]. It shows virtually no response to carbon dioxide, nitrous oxide and sulfur hexafluoride mobile phases but is generally incompatible with other mobile phases and mixed mobile phases containing organic modifiers except for water and formic acid. Other gas chromatographic detectors that have been used in SFC include the thermionic ionization detector [148,150], flame photometric detector, [151-153], photoionization detector [12,148], electron-capture detector [148,154], chemiluminescence detector [155,156] and plasma emission detector [157,158]. The bead position, heating current and hydrogen plasma gas flow rate are critical parameters for optimizing the response of the thermionic ionization detector. Sample detection limits and heteroelement selectivity factors are poorer in SFC than for gas chromatography by about one order of magnitude. Sulfur detection limits with the flame photometric detector are only modest due to spectral interference from carbon dioxide and nitrous oxide and also possibly due to quenching of the excited diatomic sulfur emission [151,152]. A rising baseline during density programming is a further unfavorable feature which requires careful choice of restrictor for control [153]. The chemiluminescence detector may be a better choice for the detection of sulfur-containing compounds. In other cases there are too few studies to indicate the general usefulness of these detectors. It is obvious that SFC can be used with a wide range of gas chromatographic detectors but it is equally obvious that few of these detectors have been optimized for operation with supercritical fluids and it is likely that better results can be expected in the future.

After the flame ionization detector the UV absorption detector is the second most popular detector for SFC [159-162]. It is invariably used with wide bore columns, 4-5 mm I. D., and for mobile phases containing organic solvent modifiers, i.e., for those conditions incompatible with flame ionization detection. For narrow bore open tubular columns very small detector cells are required to avoid extracolumn band broadening. For average conditions detector cell volumes below 50 nanoliters are required, which can only be approached by using direct on-column or pseudo on-column detection as shown in Figure 6.13. Either a section of the polymeric coating is removed from the outside of the fused silica capillary or, preferably, the column is inserted into a slightly wider piece of fused silica tubing, 0.25 mm I. D., which acts as the detection cell. The pseudo on-column technique provides improved sensitivity (greater path length) without introducing excessive dead volume. Interfacing a UV detector to packed columns is more straightforward. Dead volume effects are less of a concern and standard cells can be used, although modification to withstand the high operating pressures of SFC may be required [3,159]. One shortcoming of short path length cells, particularly those used for on-column detection, is baseline drift caused by density related retention index changes during pressure programming. These effects can be minimized by cooling the detector cell [161] or by using detectors with forward optics that place the detector cell close to the sensor [160]. A scanning fluorescence detector using a fiber optic design has also been described for use in open tubular column SFC [163,164]. However, little work has been done using fluorescence so far, as is the case with other potential detection principles, such as refractive index, ultrasonic, light scattering and electrochemical detection [12,159,165].

6.6 REFERENCES

1. H. H. Lauer, D. McManigill, and R. D. Board, Anal. Chem., 55 (1983) 1370.
2. F. J. Yang (Ed.), "Microbore Column Chromatography. A Unified Approach to Chromatography", Dekker, New York, NY, 1989.
3. P. J. Schoenmakers and F. C. C. J. G. Verhoeven, Trends Anal. Chem., 6 (1987) 10.
4. H. Engelhardt, A. Gross, R. Mertens, and M. Petersen, J. Chromatogr., 477 (1989) 169.
5. C. M. White and R. K. Houck, J. High Resolut. Chromatogr., 9 (1986) 4.
6. M. L. Lee and K. E. Markides, J. High Resolut. Chromatogr., 9 (1986) 652.
7. C. M. White (Ed.), "Modern Supercritical Fluid Chromatography", Huethig, Heidelberg, 1988.
8. R. M. Smith (Ed.), "Supercritical Fluid Chromatography", Royal Society of Chemistry, London, UK, 1988.
9. R. D. Smith, B. W. Wright, and C. R. Yonker, Anal. Chem., 60 (1988) 1323A.
10. P. J. Schoenmakes and L. G. M. Uunk, Adv. Chromatogr., 30 (1989) 1.
11. W. M. A. Niessen, U. R. Tjaden, and J. van der Greef, J. Chromatogr., 492 (1989) 167.
12. M. L. Lee and K. E. Markides (Eds.), "Analytical Supercritical Fluid Chromatography and Extraction", Chromatography Conferences, Inc., Provo, UT, 1990.
13. H.-C. K. Chang, K. E. Markides, J. S. Bradshaw, and M. L. Lee, J. Chromatogr., 26 (1988) 280.
14. P. Macaudiere, M. Caude, R. Rosset, and A. Tambute, J. Chromatogr. Sci., 27 (1989) 583.
15. A. Dobashi, Y. Dobashi., T. Ono, S. Hara, M. Saito, S. Higashidate, and Y. Yamauchi, J. Chromatogr., 461 (1989) 121.
16. M. Ashraf-Khorassani, L. T. Taylor, and R. A. Henry, Anal. Chem., 60 (1988) 1529.
17. T. A. Dean and C. F. Poole, J. Chromatogr., 468 (1989) 127.
18. A. Nomura, J. Yamada, K.-I. Tsunoda, K. Sakaki, and T. Yokochi., Anal. Chem., 61 (1989) 2076.
19. A. Nomura, J. Yamada, and K.-I. Tsunoda, J. Chromatogr., 448 (1988) 87.
20. J. Doehl, A. Farbrot, T. Greibrokk, and B. Iversen, J. Chromatogr., 392 (1987) 175.

21. M. Petersen, J. Chromatogr., 505 (1990) 3.
22. D. Upnmoor and G. Brunner, Chromatographia, 28 (1989) 449.
23. M. Ashraf-Khorassani, L. T. Taylor, and R. A. Henry, Chromatographia, 28 (1989) 569.
24. M. De Weerdt, C. Dewaele, M. Verzele, P. Sandra, J. High Resolut. Chromatogr., 13 (1990) 40.
25. K. M. Payne, B. J. Tarbet, J. S. Bradshaw, K. E. Markides, and M. L. Lee, Anal. Chem., 62 (1990) 1379.
26. P. J. Schoenmakers, L. G. M. Uunk, and H.-G. Janssen, J. Chromatogr., 506 (1990) 563.
27. L. T. Taylor and H.-G. Chang, J. Chromatogr. Sci., 28 (1990) 357.
28. P. J. Schoenmakers, L. G. M. Uunk, and P. K. De Bokx, J. Chromatogr., 459 (1988) 201.
29. K. Jinno and S. Niimi, J. Chromatogr., 455 (1988) 29.
30. Y. Hirata, J. Chromatogr., 315 (1984) 31.
31. F. P. Schmitz, D. Leyendecker, and D. Leyendecker, J. Chromatogr., 389 (1987) 245.
32. M. B. Evans, M. S. Smith, and J. M. Oxford, J. Chromatogr., 479 (1989) 170.
33. B. Gemmel, B. Lorenschat, and F. P. Schmitz, Chromatographia, 27 (1989) 605.
34. P. Morin, M. Caude, and R. Rosset, J. Chromatogr., 407 (1987) 87.
35. T. M. Engel and S. V. Olesik, Anal. Chem., 62 (1990) 1554.
36. T. A. Berger and J. F. Deye, Chromatographia, 30 (1990) 57.
37. J. R. Strubinger, H. Song, and J. F. Parcher, Anal. Chem., 63 (1991) 98 and 104.
38. C. R. Yonker and R. D. Smith, J. Chromatogr., 505 (1990) 139.
39. S. R. Springston, P. David, J. Steger, and M. V. Novotny, Anal. Chem., 58 (1986) 997.
40. M. V. Novotny and P. David, J. High Resolut. Chromatogr., 9 (1986) 647.
41. D. E. Martire, J. Chromatogr., 461 (1989) 165.
42. D. P. Poe and D. E. Martire, J. Chromatogr., 517 (1990) 3.
43. P. J. Schoenmakers, P. E. Rothfusz, and F. C. C. J. G. Verhoeven, J. Chromatogr., 395 (1987) 91.
44. P. A. Peaden and M. L. Lee, J. Liq. Chromatogr., 5 (1982) 179.
45. P. A. Peaden and M. L. Lee, J. Chromatogr., 259 (1983) 1.
46. S. M. Fields, R. C. Kong, J. C. Fjeldsted, M. L. Lee, and P. A. Peaden, J. High Resolut. Chromatogr., 7 (1984) 312.
47. S. M. Fields, R. C. Kong, M. L. Lee, and P. A. Peaden, J. High Resolut. Chromatogr., 7 (1984) 423.
48. J. A. Crow and J. P. Foley, Anal. Chem., 62 (1990) 378.
49. M. Novotny and S. R. Springston, J. Chromatogr., 279 (1983) 417.
50. H.-G. Janssen and C. A. Cramers, J. Chromatogr., 505 (1990) 19.
51. P. R. Sassiat, P. Mourier, M. H. Caude, and R. H. Rosset, Anal. Chem., 59 (1987) 1164.
52. N. Dahmen, A. Kordikowski, and G. Schneider, J. Chromatogr., 505 (1990) 169.
53. D. R. Gere, R. Board, and D. McManigill, Anal. Chem., 54 (1982) 736.
54. P. J. Schoenmakers and L. G. M. Uunk, Chromatographia, 24 (1987) 51.
55. P. J. Schoenmakers, J. High Resolut. Chromatogr., 11 (1988) 278.
56. P. A. Mourier, M. H. Caude, and R. H. Rosset, Chromatographia, 23 (1987) 21.

57. A. Giorgetti, N. Pericles, H. M. Widmer, K, Anton, and P. Datwyler, J. Chromatogr. Sci., 27 (1989) 318.
58. J. C. Kuei, K. E. Markides and M. L. Lee, J. High Resolut. Chromatogr., 10 (1987) 257.
59. U. Van Wassen, I. Swaid, and G. M. Schneider, Angew. Chem. Int. Ed. Engl., 19 (1980) 575.
60. J. W. King and J. P. Friebrich, J. Chromatogr., 517 (1990) 449.
61. K. D. Bartle, A. A. Clifford, S. A. Jafar, J. P. Kithinji, and G. F. Shilstone, J. Chromatogr., 517 (1990) 459.
62. S. M. Fields and K. Grolimund, J. High Resolut. Chromatogr., 11 (1988) 727.
63. S. B. French and M. Novotny, Anal. Chem., 58 (1986) 164.
64. H. E. Schwartz and R. G. Brownlee, J. Chromatogr., 353 (1986) 77.
65. J. W. Hellgeth, M. G. Fessehaie, and L. T. Taylor, Chromatographia, 25 (1986) 172.
66. C. H. Lochmuller and L. P. Mink. J. Chromatogr., 505 (1990) 119.
67. M. A. McHugh and V. J. Krukonis, "Supercritical Fluid Extraction: Principles and Practice", Butterworth, Boston, 1986.
68. B. A. Charpentier and M. R. Sevenants, "Supercritical Fluid Extraction and Chromatography", ACS Symp. Ser. 366, American Chemical Society, Washington, DC, 1988.
69. T. L. Chester, D. J. Bowling, D. P. Innis, and J. D. Pinkston, Anal. Chem., 62 (1990) 1299.
70. F. P. Schmitz and E. Klesper, J. Chromatogr., 388 (1987) 3.
71. D. R. Luffer, W. Ecknig, and M. Novotny, J. Chromatogr., 505 (1990) 79.
72. C. R. Yonker, R. W. Gale, and R. D. Smith, J. Chromatogr., 371 (1986) 83.
73. J. F. Deye, T.A. Berger, and A. G. Anderson, Anal. Chem., 62 (1990) 615.
74. M. Ashraf-Khorassani, L. T. Taylor, and P. Zimmerman, Anal. Chem., 62 (1990) 1177.
75. C. P. Ong, H. K. Lee, and S. F. Y. Li, Anal. Chem., 62 (1990) 1389.
76. F. P. Schmitz, J. High Resolut. Chromatogr., 10 (1987) 650.
77. A. Hutz, D. Leyendecker, F. P. Schmitz and E. Klesper, J. Chromatogr., 505 (1990) 99.
78. B. W. Wright and R. D. Smith, J. Chromatogr., 355 (1986) 367.
79. S. M. Fields, K. E. Markides, and M. L. Lee, J. Chromatogr., 406 (1987) 233.
80. J. G. M. Janssen, P. J. Schoenmakers, and C. A. Cramers, J. High Resolut. Chromatogr., 12 (1989) 645.
81. A. L. Blilie and T. Greibrokk, Anal. Chem., 57 (1985) 2239.
82. H. E. Schwartz, P. J. Barthel, S. E. Moring, T. L. Yates, and H. H. Lauer, Fresenius Z. Anal. Chem., 330 (1988) 204.
83. F. O. Geiser, S. G. Yocklovich, S. M. Lurcott, J. W. Guthrie, and E. J. Levy, J. Chromatogr., 459 (1988) 173.
84. M. Ashraf-Khorassani and L. T. Taylor, J. High Resolut. Chromatogr., 12 (1989) 40.
85. J. L. Janicot, M. Caude, and R. Rosset, J. Chromatogr., 437 (1988) 351.
86. C. R. Yonker, D. G. McMinn, B. W. Wright, and R. D. Smith, J. Chromatogr., 396 (1987) 19.
87. C. R. Yonker and R. D. Smith, J. Chromatogr., 361 (1986) 25.
88. T. A. Berger and J. F. Deye, Anal. Chem., 62 (1990) 1181.
89. A. W. Francis, J. Phys. Chem., 58 (1954) 1099.

646

90. D. K. Dandge, J. P. Heller, and K. V. Wilson, Ind. Eng. Chem. Prod. Res. Dev., 24 (1985) 162.
91. T. G. Squires and M. E. Paulaitis (Eds.), "Supercritical Fluids", ACS Symp. Ser. 329, American Chemical Society, Washington, DC, 1987.
92. J. M. Levy and W. M. Ritchey, J. High Resolut. Chromatogr., 10 (1987) 493.
93. P. A. David and M. Novotny, J. Chromatogr., 461 (1989) 111.
94. T. L. Chester and D. P. Innis, J. High Resolut. Chromatogr., 9 (1986) 209.
95. T. A. Berger, J. F. Deye, M. Ashraf-Khorassani, and L. T. Taylor, J. Chromatogr. Sci, 27 (1989) 105.
96. W. Steuer, M. Schindler, G. Schill, and F. Erni, J. Chromatogr., 447 (1988) 287.
97. R. D. Smith, J. L. Fulton, H. K. Jones, R. W. Gale, and B. W. Wright, J. Chromatogr. Sci., 27 (1989) 309.
98. W. Steuer, J, Baumann, and F. Erni, J. Chromatogr., 500 (1990) 469.
99. T. L. Chester and D. P. Innis, J. High Resolut. Chromatogr., 8 (1985) 561.
100. U. Van Wassen and G. M. Schneider, Chromatographia, 8 (1975) 274.
101. P. Mourier, P. Sassiat, M. Caude, and R. Rosset, J. Chromatogr., 353 (1986) 61.
102. H. E. Schwartz, R. G. Brownlee, M. M. Boduszynski, and F. Su, Anal. Chem., 59 (1987) 1393.
103. M. Novotny, W. Bertsch, and A. Zlatkis, J. Chromatogr., 61 (1971) 17.
104. F. P. Schmitz and E. Klesper, J. High Resolut. Chromatogr., 10 (1987) 519.
105. C. R. Yonker, R. W. Gale, and R. D. Smith, J. Chromatogr., 389 (1987) 433.
106. C. R. Yonker and R. D. Smith, J. Phys. Chem., 92 (1988) 1664.
107. D. E. Martire, J. Liq. Chromatogr., 10 (1987) 1569.
108. T. A. Berger, J. Chromatogr., 478 (1989) 311.
109. E. Klesper and F. P. Schmitz, J. Chromatogr., 402 (1987) 1.
110. C. R. Yonker and R. D. Smith, Anal. Chem., 59 (1987) 727.
111. A. L. Blilie and T. Greibrokk, J. Chromatogr., 349 (1985) 317.
112. R. D. Smith, E. G. Chapman, and B. W. Wright, Anal. Chem., 57 (1985) 2829.
113. A. Wilsch and G. M. Schneider, J. Chromatogr., 357 (1986) 239.
114. E. Klesper, Fresenius Z. Anal. Chem., 330 (1988) 200.
115. J. C. Fjeldsted, W. P. Jackson, P. A. Peaden, and M. L. Lee, J. Chromatogr. Sci., 21 (1983) 222.
116. S. V. Olesik and L. A. Pekay, Chromatographia, 29 (1990) 69.
117. W. Steuer, M. Schindler, and F. Erni, J. Chromatogr., 454 (1988) 253.
118. Y. Hirata, F. Nakata, and M. Kawasaki, J. High Resolut. Chromatogr., 9 (1986) 633.
119. K. R. Jahn and B. W. Wenclawiak, Anal. Chem., 59 (1987) 382.
120. M. Saito, Y. Yamauchi, H. Kashiwazaki, and M. Sugawara, Chromatographia, 25 (1988) 801.
121. T. Takeuchi, K. Ohta, and D. Ishii, Chromatographia, 25 (1988) 125.
122. Y. Hirta, F. Nakata, and S. Murata, Chromatographia, 23 (1987) 663.
123. B. W. Wenclawiak, Fresenius Z. Anal. Chem., 330 (1988) 218.
124. D. W. Later, E. R. Campbell, and B. E. Richter, J. High Resolut. Chromatogr., 11 (1988) 65.

125. K. Anton, N. Pericles, S. M. Fields, and H. M. Widmer, Chromatographia, 26 (1988) 224.
126. T. Greibrokk, J. Doehl, A. Farbrot, and B. Iversen, J. Chromatogr., 371 (1986) 145.
127. A. C. Rosselli, D. S. Boyer, and R. K. Houck, J. Chromatogr., 465 (1989) 11.
128. T. Gorner, J. Dellacherie, and M. Perrut, J. Chromatogr., 514 (1990) 309.
129. D. E. Raynie, S. M. Fields, N. M. Djordevic, K. E. Markides, and M. L. Lee, J. High Resolut. Chromatogr., 12 (1989) 51.
130. M. L. Lee, B. Xu, E. C. Huang, N. M. Djordjevic, H.-C. K. Chang, and K. E. Markides, J. Microcol. Sepns, 1 (1989) 7.
131. J. Kohler, A. Rose, and G. Schomburg, J. High Resolut. Chromatogr., 11 (1988) 191.
132. T. A. Dean and C. F. Poole, J. High Resolut. Chromatogr., 12 (1989) 773.
133. B. E. Richter, D. E. Knowles, M. R. Anderson, N. L. Porter, E. R. Campbell, and D. W. Later, J. High Resolut. Chromatogr., 11 (1988) 29.
134. G. Schomburg and W. Roeder, J. High Resolut. Chromatogr., 12 (1989) 218.
135. A. Farbrot Bushke, B. E. Berg, O. Gyllenhaal, and T. Greibrokk, J. High Resolut. Chromatogr., 11 (1988) 16.
136. S. B. Hawthorne and D. J. Miller, J. Chromatogr. Sci., 27 (1989) 197.
137. Y. Hirata, M. Tanaka, and K. Inomata, J. Chromatogr. Sci., 27 (1989) 395.
138. A. Munder and S. N. Chesler, J. High Resolut. Chromatogr., 12 (1989) 669.
139. R. D. Smith, J. L. Fulton, R. C. Petersen, A. J. Kopriva, and B. W. Wright, Anal. Chem., 58 (1986) 2057.
140. D. E. Raynie, K. E. Markides, M. L. Lee, and S. R. Goates, Anal. Chem., 61 (1989) 1178.
141. T. A. Berger, Anal. Chem., 61 (1989) 356.
142. R. W. Bally and C. A. Cramers, J. High Resolut. Chromatogr., 9 (1986) 626.
143. T. L. Chester, D. P. Innis, and G. D. Owens, Anal. Chem., 57 (1985) 2243.
144. M. W. Raynor, K. D. Bartle, I. L. Davies, A. A. Clifford, and A. Williams, J. High Resolut. Chromatogr., 11 (1988) 289.
145. C. M. White, D. R. Gere, D. Boyer, F. Pacholec, and L. K. Wong, J. High Resolut. Chromatogr., 11 (1988) 94.
146. E. J. Guthrie and H. E. Schwartz, J. Chromatogr. Sci., 24 (1986) 236.
147. C. K. Huston and R. A. Bernhard, J. Chromatogr. Sci., 27 (1989) 231.
148. B. E. Richter, D. J. Bornhop, J. T. Swanson, J. G. Wangsgaard, and M. R. Andersen, J. Chromatogr. Sci., 27 (1989) 303.
149. B. E. Richter, J. High Resolut. Chromatogr., 8 (1985) 297.
150. P. A. David and M. Novotny, Anal. Chem., 61 (1989) 2082.
151. K. E. Markides, E. D. Lee, R. Bolick, and M. L. Lee, Anal. Chem., 58 (1986) 740.
152. S. V. Olesik, L. A. Pekay, and E. A. Paliwoda, Anal. Chem., 61 (1989) 58.
153. L. A. Pekay and S. V. Olesik, Anal. Chem., 61 (1989) 2616.
154. H.-C. K. Chang and L. T. Taylor, J. Chromatogr. Sci., 28 (1990) 29.
155. W. T. Foreman, C. L. Shellum, J. W. Birks, and R. E. Sievers, J. Chromatogr., 465 (1989) 23.

156. H.-C. K. Chang and L. T. Taylor, J. Chromatogr., 517 (1990) 491.
157. L. J. Galante, M. Selby, D. R. Luffer, G. M. Hieftje, and M. Novotny, Anal. Chem., 60 (1988) 1370.
158. D. R. Luffer and M. Novotny, J. Chromatogr., 517 (1990) 477.
159. D. J. Bornhop and J. G. Wangsgaard, J. Chromatogr. Sci., 27 (1989) 293.
160. S. R. Weinberger and D. J. Bornhop, J. Microcol. Sepns., 1 (1989) 90.
161. S. M. Field, K. E. Markides, and M. L. Lee, Anal. Chem., 60 (1988) 802.
162. D. J. Bornhop, S. Schmidt, and N. L. Porter, J. Chromatogr., 459 (1988) 193.
163. J. C. Fjeldsted, R. E. Richter, W. P. Jackson, and M. L. Lee, J. Chromatogr., 279 (1983) 423.
164. J. C. Glukman, D. C. Shelly, and M. V. Novotny, Anal. Chem., 57 (1985) 1546.
165. M. Lafosse, P. Rollin, C. Elfakir, L. Morin-Allory, M. Martens, and M. Dreux, J. Chromatogr., 505 (1990) 191.

CHAPTER 7

THIN-LAYER CHROMATOGRAPHY

7.1 INTRODUCTION

Thin-layer chromatography (TLC) is a type of liquid chromatography in which the stationary phase is in the form of a thin layer on a flat surface rather than packed into a tube (column). It is a member of a family of techniques that include some types of electrophoresis and paper chromatography, more generally referred to as planar chromatography. Since we will not discuss electrophoresis in this section, and since TLC has virtually superseded paper chromatography in most analytical laboratories, we will confine ourselves to a discussion of TLC. The introduction of new TLC layers prepared from small diameter particles of a narrow size range in the mid-1970s revolutionized the practice of TLC. The techniques employed with these new layers became known as high performance TLC, modern TLC, or instrumental TLC, to distinguish them from earlier practices, generally referred to as conventional TLC. These differences, Table 7.1, are not of a fundamental nature but rather represent a further optimization of all aspects of the separation process in TLC. The tightness of the particle size distribution is just as important as the decrease in actual particle size to an understanding of the properties of modern TLC plates. These new layers require smaller sample sizes and shorter development distances to reveal their separation potential and to provide faster separations; better resolution; and, because spots are more compact and the optical properties of the layer more favorable for in situ detection, much better detection limits. These advantages only become apparent when the appropriate instrumentation is used for sample application, development and scanning densitometry. TLC, because of its methodological simplicity and ease of sample visualization, still finds many applications for the qualitative analysis of simple mixtures, for which little in the way of instrumentation is needed. For this reason, conventional TLC remains popular, and not all laboratories are equipped to perform modern TLC. A number of books provide a good background to conventional TLC practices [1-5], and to modern TLC [4-12], with recent review articles emphasizing modern TLC and its applications [13-18].

Separations by column liquid chromatography (HPLC) and TLC occur by essentially the same physical processes. The two methods have often been considered as competitors when it would be more realistic to consider them as complementary, both having their own strengths and weaknesses. In HPLC each sample component must travel the complete length of the column and the total separation time is determined by the time required for the slowest moving component to reach the detector. While for TLC the total time for the separation is the time required for the solvent front to migrate a predetermined distance, and is independent of the migration distance of the sample components. Excessively retained components result in a considerable loss of time in HPLC while components accumulated at the head of the column are completely eluted, and if this is not possible, permanent alteration of the

Table 7.1

COMPARISION OF CONVENTIONAL AND MODERN TLC

Parameter	Conventional	Modern
Plate size (cm)	20x20	10x10
		10x20
Layer thickness (mm)	0.1-0.25	0.1 or 0.2
Particle size (micrometers)		
Average	20	Between 5 and 15
Distribution	10-60	Narrow
Maximum number of theoretical plates	<600	<5000
Separation number	7-10	10-20
Sample volume (microliters)	1-5	0.1-0.2
Starting spot diameter (mm)	3-6	1-2
Diameter of separated spots (mm)	6-15	2-6
Solvent migration distance (cm)	10-15	3-6
Time for development (min)	30-200	3-20
Detection limits		
Absorption (ng)	1-5	0.1-0.5
Fluorescence (pg)	50-100	5-10
Sample lanes per plate	10	18

properties of the column may occur, eventually making it useless for the analysis. TLC plates are disposed of at the conclusion of each separation and are thus immune from the above problem.

It is simple to demonstrate that it is easier to achieve a greater number of theoretical plates in HPLC than for TLC. For difficult separations HPLC has the greater separation capacity. However, most separations performed by HPLC are done with relatively few theoretical plates, typically < 20,000, with separation numbers between 20 and 40. This is not much greater than values achievable by modern TLC.

In TLC the detection process is static (separations achieved in space rather than time) and free from time constraints, or from interference by the mobile phase, which is removed between the development and detection process. Freedom from time constraints permits the utilization of any variety of techniques to enhance detection sensitivity, which if the methods are nondestructive, may be applied sequentially. Thus, the detection process in TLC is more flexible and variable than for HPLC. For optical detection the minimum detectable quantities are similar for both techniques with, perhaps, a slight advantage for HPLC. Direct comparisons are difficult because of the differences in detection variables and how these are optimized. Detection in TLC, however, is generally limited to optical detection without the equivalent of refractive index and electrochemical detection available for HPLC. The ease of postchromatographic reactions used to enhance optical detection characteristics in TLC largely offsets this disadvantage.

Because of the nature of the method of development, analysis by HPLC is of necessity performed in a sequential manner. Each sample must individually undergo the same sequence of injection, separation, detection and column re-equilibration. The time required to analyze n samples is n times the time required to analyze one sample. TLC allows samples to be separated in parallel with the potential for a large reduction in the analysis time. In contrast to HPLC, the time required to analyze n samples is n divided by the number of sample lanes on the plate, which may vary between 18 and 72 depending on the plate size and method of development. Some of this advantage is lost because it is easier to automate HPLC methods, and unattended overnight operation is possible. Whereas for TLC the individual steps of sample application, development and detection can be automated, but manual intervention is required to move the plate from station to station. In addition, when considering the analysis time for TLC, it is also necessary to consider the time required for sample application and to scan the plate. Even after making suitable allowances for the above points, TLC can generally provide a significant reduction in analysis time when multiple samples of a similar type are to be analyzed.

In the light of the above discussion, TLC methods are most effective for the low-cost analysis of simple mixtures where the number of samples is large, for the rapid analysis of samples requiring minimum sample cleanup, for the analysis of samples containing components that remain sorbed to the stationary phase or contain suspended microparticles, and for the analysis of substances with poor detection characteristics that require postchromatographic treatment for detection [19]. Since only in planar chromatographic techniques is the total sample visible, TLC is the most appropriate technique for screening samples to determine sample recovery and to rapidly select suitable solvent systems for a separation, even if HPLC is selected for the final analysis. Traditionally, TLC has been preferred for large scale screening programs, for example, screening of urine samples to identify drugs of abuse [20-22], and characterizing plant extracts in pharmacology [10,23]; therapeutic monitoring of drugs in biological fluids [16,24]; detection of aflatoxins in agricultural products [20]; and the characterization of lipid extracts [4,5,16,26]. In other cases HPLC methods are preferred, particularly if a large number of theoretical plates are required for a separation and if the sample preparation time per sample is long compared to the separation time. Also, separations by size exclusion and ion-exchange chromatography are usually easier to achieve by HPLC, and HPLC is favored for trace analysis using selective detectors unavailable for TLC.

7.2 THEORETICAL CONSIDERATIONS

For TLC the stationary phase consists of a thin layer of sorbent coated on an inert, backing material. The sample is applied to the layer as a spot or band near to the bottom edge of the plate. The separation is carried out in an enclosed chamber by either contacting the bottom edge of the plate with the mobile phase, which advances through the sorbent layer by capillary forces, or the mobile phase is forced to move through the sorbent layer at a controlled velocity by the application of an external pressure gradient. A separation of the sample results from the different rates of migration of the sample components in the direction traveled by the mobile phase. After development and evaporation of the mobile phase, the sample components are separated in space; their position and quantity being determined by visual evaluation or in situ scanning densitometry.

The fundamental parameter used to characterize the position of a sample zone in a TLC chromatogram is the retardation factor, or R_f value. It represents the ratio of the distance migrated by the sample compared to that traveled by the solvent front. With respect to Figure 7.1, the R_f value for linear development is given by equation (7.1)

$$R_f = Z_x / (Z_f - Z_0) \tag{7.1}$$

where Z_x is the distance traveled by the sample from its origin, $(Z_f - Z_0)$ the distance traveled by the mobile phase from the sample origin, Z_f the distance traveled by the mobile phase measured from its highest position on the plate at the start of the separation, and Z_0 the distance from the sample origin to the position used as the origin for the mobile phase. The boundary conditions for R_f values are $1 \geq R_f \geq 0$. When $R_f = 0$, the spot does not migrate from the origin, and for $R_f = 1$, the spot is unretained by the

654

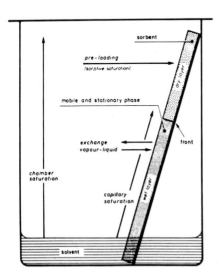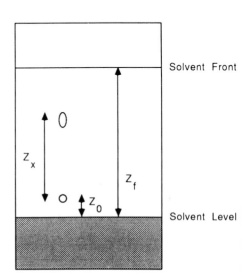

Figure 7.1 Schematic diagram showing the different equilibria for solvent-vapor-sorbent exchange in an unsaturated tank and the nomenclature used to define the R_f value in equation (7.1). (Adapted with permission from ref. 27. Copyright Dr Alfred Huethig Publishers).

stationary phase and migrates with the solvent front. Although R_f values are widely quoted, they are difficult to determine accurately [8,27]. Systematic errors result from the difficulty in locating the exact position of the solvent front. If the adsorbent layer, mobile phase and vapor phase are not in equilibrium then condensation of the vapor phase or evaporation of the mobile phase in the region of the solvent front will give an erroneous R_f value. This value may be either too high or too low, depending on the prevailing conditions.

The capacity factor, k, of a sample zone is defined as the ratio of the time spent by the substance in the stationary phase compared to the time it spends in the mobile phase, and is related to the R_f value by equation (7.2).

$$k = (1- R_f)/R_f \qquad (7.2)$$

Log k, also known as the R_m value, often provides linear relationships between sample and chromatographic properties and is more useful in correlation studies than the R_f value itself.

7.2.1 Solvent Migration Through Porous Layers

In the absence of external forces the mobile phase penetrates the sorbent layer in TLC by capillary action. The flow of solvent at the developing front is generally unsaturated and the speed with which the front moves is dependent on the experimental conditions [8,27-30]. Capillary forces are stronger in the narrow interparticle channels, leading to more rapid advancement of the mobile phase. Larger pores below the solvent front are filled at a slower rate resulting in an increased thickness of the mobile phase layer. If the vapor phase and mobile phase are not in equilibrium, see Figure 7.1, evaporation will cause a loss of mobile phase from the plate surface and a decrease in the solvent front velocity. On the other hand, the dry layer ahead of the solvent front progressively adsorbs vapor, filling some of the pores and interparticle channels, and increasing the apparent velocity with which the solvent front migrates. During the chromatographic process a solvent composition gradient is produced as the mobile phase moves through the sorbent due to selective adsorption by the stationary phase of the solvent component with the higher affinity for the stationary phase. Even single component mobile phases exhibit impurity gradients in the direction of development. Gaining satisfactory control over the above processes in large volume chambers is almost impossible. The use of various kinds of sandwich chambers, which either eliminate or minimize contact of the plate surface with the vapor phase, provide reasonable control over the mobile phase velocity.

It is an empirical fact that in the absence of a significant exchange of solvent flux with the vapor phase the position of the solvent front with respect to time is adequately represented by the simple quadratic equation (7.3), and after differentiation, the velocity of the solvent front by equation (7.4)

$$(Z_f)^2 = \kappa t \qquad (7.3)$$

$$u_f = \kappa/2Z_f \qquad (7.4)$$

where κ is the velocity constant (cm^2/s), t is the time from contacting the sorbent layer with the solvent, and u_f the solvent front velocity. If true equilibrium does not exist during development rather complex correction factors must be applied to equation (7.3) [8,27,28]. Equation (7.4) indicates the well known, undesirable influence of capillary controlled flow in TLC, namely, decreasing solvent velocity with increasing migration distance resulting in longer separation times and a reduced separation potential.

The velocity constant, κ, is related to the experimental conditions by equation (7.5)

$$\kappa = 2K_o d_p (\gamma/\eta) \cos\theta \qquad (7.5)$$

where K_o is the permeability constant of the layer, d_p the average particle diameter, γ the surface tension of the mobile phase, η the mobile phase viscosity, and θ the contact angle. The permeability constant is a dimensionless constant which takes into account the profile of the external pore size distribution, the effect of porosity on the permeability of the layer, and the ratio of the bulk liquid velocity to the solvent front velocity. Experimental values for K_o tend to vary in magnitude for different layers, but typical values for precoated plates fall into the range 0.001 to 0.002 and are not very different from typical values quoted for slurry-packed HPLC columns [31].

Equation (7.5) indicates that the velocity constant should increase linearly with the average particle size. The solvent front velocity should be larger for coarse-particle layers than for fine-particle layers which is in good agreement with experimental observations [27]. Also, from equation (7.5) we see that the velocity constant depends linearly on the ratio of the surface tension of the solvent to its viscosity and that solvents which maximize this ratio (and not just optimize one of the parameters) are preferred for TLC [8,30]. The contact angle for

Table 7.2

VELOCITY CONSTANT AND CONTACT ANGLE FOR WATER-ETHANOL MIXTRUES ON
RP-18 REVERSED-PHASE PLATES.

Water Content (%V/V)	Velocity Constant $(cm^2/s) \times 10^3$	Cos Θ
0	16.6	0.87
4	13.4	0.78
10	12.3	0.76
20	7.8	0.61
30	5.5	0.48
40	3.5	0.34
50	1.5	0.14

most organic solvents on silica gel is generally close to zero
(cos Θ = 1). Also, polar bonded phase layers, such as 3-
aminopropylsiloxane and 3-cyanopropylsiloxane bonded silica, are
completely wet by most common solvents. This is not the case for
reversed-phase layers containing bonded, long-chain alkyl groups,
for which the contact angle of the solvent increases very rapidly
with increasing water content of the mobile phase and, at about
30-40% water, cos Θ becomes less than 0.2-0.3 on highly
hydrophobic layers [32,33]. The mobile phase is no longer able to
ascend the plate and chromatography becomes impossible. Some
typical values for the change in contact angle on reversed-phase
RP-18 plates (Merck) as a function of the composition of binary
water-ethanol mixtures are given in Table 7.2 [32]. To obtain
reasonable solvent compatibility with aqueous mobile phases
hydrophobic layers with a lower degree of silanization and/or
larger particle size are generally used.

The above discussion is applicable to layers unperturbed by
the presence of a vapor phase, such as in a sandwich layer tank.
In practice, most separations are performed in large volume
chambers in the presence of a vapor phase. It is almost impossible
to fully saturate such chambers so that a temporal and spatial
vapor equilibrium is unlikely to exist. Two opposing phenomena can
be expected to influence the rate of solvent migration.
Vaporization of solvent from the wetted layer might reasonably be
expected to depend on the wetted surface area of the plate and the
vapor pressure of the solvent in the tank. The loss of solvent
from the layer will result in a reduction of the mobile phase

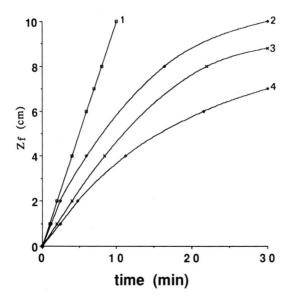

Figure 7.2   Relation between the solvent front position and time for (1) an enclosed layer with forced-flow development, (2) an exposed layer in a saturated chamber with capillary controlled flow, (3) a covered layer (sandwich chamber) with capillary controlled flow, and (4) an exposed layer in an unsaturated atmosphere with capillary controlled flow. (Reproduced with permission from ref. 30. Copyright Dr Alfred Huethig Publishers).

velocity from that indicated by equation (7.3), Figure 7.2 [34]. When the dried plate is placed in the developing chamber it progressively adsorbs solvent vapor. The pores of the unwetted layer ahead of the solvent front fill slowly with adsorbed vapor and the apparent porosity of the layer diminishes. Since the porosity of the layer decreases the velocity constant increases slowly with increasing time. The effect of vaporization is generally small if the atmosphere of the tank is close to saturation while adsorption of solvent vapors by the dry layer will tend to dominate. Thus, the mobile phase velocity in a large volume chamber will tend to be greater than that given by equation (7.3) and should increase continuously with time, if the layer is not conditioned in the chamber atmosphere prior to development.

Forced-flow development enables the mobile phase velocity to be optimized without regard to the deficiencies of a capillary controlled flow system [34,35]. In rotational planar chromatography, centrifugal force, generated by spinning the sorbent layer about a central axis, is used to drive the solvent through the layer [9,36]. The rate of solvent migration is a function of the rotation speed and the rate at which the mobile phase is supplied to the layer. Since the layer is not enclosed, the ultimate velocity of the solvent front is limited by the amount of solvent that can be kept within the layer without floating over the surface. At high rotation speeds the velocity of the solvent front becomes approximately constant in the linear development mode. An alternative approach to forced-flow development is to seal the sorbent layer with a flexible membrane or an optically flat, rigid surface under hydraulic pressure, and to deliver the mobile phase to the layer by a pump [34,37,38]. The mobile phase velocity can be controlled and optimized in this case by adjusting the volume of mobile phase delivered to the layer by the pump. In the linear development mode the mobile phase velocity will be constant and the position of the solvent front at any time t after the start of development is described by equation (7.6)

$$Z_f = u_f t \tag{7.6}$$

where u_f is the linear velocity of the mobile phase under forced-flow development conditions. In the forced-flow mode the mobile phase velocity is no longer dependent on the solvent contact angle and there are no restrictions to solvent selection with reversed-phase layers as was noted for capillary controlled flow systems.

When a liquid is forced though a dry layer of porous particles sealed from the external atmosphere, the air displaced from the layer by the liquid will usually result in the formation of a second front (beta front), moving behind the solvent-air front (alpha front), which is often wavy in character [35,39,40]. During the liquid-air displacement process not all the air is displaced instantaneously escaping ahead of the front. Some is displaced at a slower rate and must leave the sorbent by dissolution in the mobile phase, or as microbubbles moving with the mobile phase. The solubility of the air in the mobile phase depends on the applied pressure, and above some critical pressure, all of the gas will dissolve. The pressure applied to the layer is reduced locally by the resistance to flow of the layer and thus the position of the beta front will depend on both the applied pressure, properties of the sorbent, and flow rate. At pressures higher than the critical pressure the sorbent is completely wetted by the solvent, Figure 7.3. The space between the alpha and beta fronts is often referred to as the disturbing zone and can be

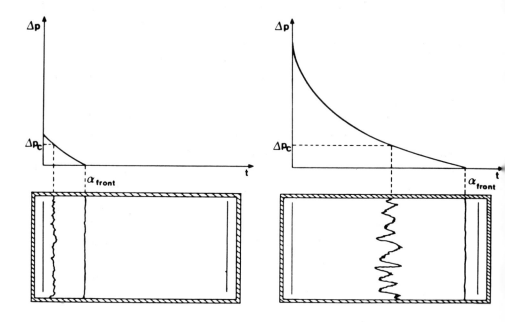

Figure 7.3 Schematic diagram showing the location of the wavy beta front and its relationship to the critical pressure (ΔP_c) at which the gas remaining in the layer is completely soluble in the mobile phase. (Reproduced with permission from ref. 39. Copyright Dr. Alfred Huethig Publishers).

distinguished from the completely wetted region of the layer by its difference in apparent optical density. Samples moving in the disturbing zone, or passed over by it, are often distorted and difficult to quantify by scanning densitometry. The disturbing zone can often be eliminated, or minimized, by predevelopment of the layer with a solvent in which the sample does not migrate to dislodge the trapped air from the layer prior to starting the separation, by using a vacuum pump to reduce the concentration of air prior to development, or by application of a back pressure restrictor to increase the local pressure and thus the solubility of the air in the mobile phase.

Multiple solvent fronts are also observed with mobile phases containing solvents of different strength [8,27,41]. As the mobile phase rises through the layer it becomes depleted in the component with the greatest affinity for the stationary phase. Eventually a secondary front is formed that separates the equilibrium solvent mixture from the solvent mixture now totally depleted in the solvent selectively adsorbed by the stationary phase. For a mobile phase mixture of n solvents as many as n+1 solvent fronts could be formed. In the simplest case, that of a binary mobile phase mixture, the position (R_f value) of the beta front decreases with decreasing concentration of the solvent preferentially adsorbed by the stationary phase. For binary solvents with the same volume composition the R_f values of the beta front will decrease as the affinity of the preferentially adsorbed solvent for the stationary phase decreases. Likewise, the R_f value of the beta front increases as the sorption capacity of the layer is decreased for similar sorbents. Multiple solvent fronts are also observed with chemically bonded stationary phases in which the least polar mobile phase component is selectively adsorbed by the stationary phase; the opposite behavior to normal phase chromatography. Demixing causes sample components to move in regions of different solvent strength and selectivity increasing the complexity of optimizing the total separation.

Typical values for the total porosity (ϵ_T), the interstitial porosity (ϵ_u) and the intraparticle porosity (ϵ_i) of precoated silica gel plates allows some useful comparisons to be made with the properties of slurry-packed HPLC columns [31]. For precoated plates from different manufacturers the various porosity values were fairly similar with average values of ϵ_T = 0.69, ϵ_u = 0.42 and ϵ_i = 0.27. Compared to normal values for slurry-packed HPLC columns the TLC layers exhibit smaller values for the total porosity and intraparticle porosity and comparable values for the interstitial porosity. These results suggest that the packing density for layers and columns is similar in spite of the differences in preparation techniques but a smaller fraction of the intraparticle volume is available to solutes in typical layers compared with columns. Most likely, a substantial amount of the binder must be contained within the pores or, alternatively, blocking the pore entrances and thus reducing the accessible intraparticle pore volume. In all cases the flow resistance parameter (section 1.7.10) for conventional TLC layers was smaller than for high performance layers, but for both types of layers the range of flow resistance parameter values (600-1500) were comparable to typical column values. The flow permeability of columns and layers are similar, the differences observed between the conventional and high performance layers may result from the deleterious effects of the presence of very fine particles in the high performance layers than for the larger particles used to prepare conventional layers. The influence of the fine particles is to block channels through the layer, thus increasing the resistance to solvent flow.

For high performance precoated TLC plates reducing the layer thickness from 0.2 to 0.1 mm increased the mobile phase velocity constant by a factor of 1.1 to 2.5, depending on the solvent [42,43]. The separation performance was virtually independent of the layer thickness but the position of the plate height minimum was shifted to a longer migration distance for the thinner layer. The mobile phase velocity is influenced primarily by the free volume of the layer that has to be filled by the solvent advancing by capillary forces. This results in faster separations for the thinner layers while maintaining similar efficiency. Alternatively, it should be possible to obtain a higher efficiency in an acceptable separation time by simultaneously reducing the average particle size and layer thickness to dimensions less than those used at present to prepare precoated plates.

7.2.2 Band Broadening and the Plate Height Equation

The ultimate chromatographic performance of a TLC plate and, therefore, resolution, is dependent upon the following parameters: the velocity constant of the mobile phase, the diffusion coefficient of the substance in the mobile phase, the mean particle diameter, and the particle size distribution of the stationary phase. There is no doubt that the presence of binder in the layer has an effect but this is not easy to quantify. In all cases performance is improved by using particles of a narrow size distribution. Modern HPTLC plates are prepared from particles of small diameter, and more importantly, of a very narrow size distribution. Separations on these plates are characterized by a series of compact symmetrical spots, with the exception of components eluting close to the solvent front. Zone broadening is dominated only by molecular diffusion. The plate height contribution from resistance to mass transfer can be ignored at normal mobile phase velocities. For conventional TLC plates, elongated and irregularly shaped spots are not uncommon, and in this case the contribution of mass transfer kinetics to spot broadening cannot be ignored.

Layer efficiency can be evaluated in terms of such familiar chromatographic parameters as the number of theoretical plates (n), the height equivalent to a theoretical plate (H), or the separation number (SN). Before establishing the equations to calculate these parameters it is necessary to highlight some of the special features of the TLC process which are different from column chromatographic systems, for which the above parameters have been more widely adopted. In column chromatographic techniques, all substances travel the same migration distance (the length of the column) but have different diffusion times (retention times on the column). This is opposite to TLC, where all substances have the same diffusion time (the plate is developed for a fixed time) but migration distances vary. The chromatographic measures of performance in TLC (e.g., n, H, SN) are all correlated to the migration distance of the substance. Their numerical values are evaluated for a specific R_f value and are thus dependent on their position in the chromatogram.

In modern TLC the distribution of sample within a spot is essentially Gaussian and the number of theoretical plates (n_{obs}) and the plate height (H_{obs}) observed can be conveniently expressed by equation (7.7) and (7.8)

$$n_{obs} = a \ (Z_x/w)^2 \tag{7.7}$$
$$H_{obs} = w^2/aZ_x \tag{7.8}$$

where w is a parameter describing the peak width and a is an appropriate scaling factor (see section 1.4 and Figure 1.1). Peak widths are determined from densitometric recordings. When w is the peak width at the base a has the value 16, and when w is the peak width at half height, a is 5.54. Substituting $Z_x = R_f \ (Z_f - Z_0)$ into either of the above equations demonstrates the general dependency that exists between n_{obs} or H_{obs} and the sample migration distance. The plate height decreases linearly with the reciprocal of the R_f value. Kaiser has proposed an alternative method of determining the plate height based on the linear extrapolation of the peak width at half height (given the symbol b and equivalent to w_h in section 1.4) as a function of $(Z_f - Z_0)$ to determine the expected peak widths at half height corresponding to $R_f = 0$ (b_0) and $R_f = 1$ (b_1) as indicated in Figure 7.4 [6,44]. The real plate number (n_{real}) and the real plate height (H_{real}) can then be defined according to equations (7.9) and (7.10). This approach recognizes the importance of the starting zone dimension on the plate height

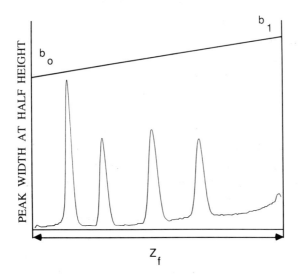

Figure 7.4   Change in peak width at half height as a function of
migration distance for a typical test mixture on a high
performance TLC plate. The values for b_o and b_1 are calculated by
extrapolation using linear regression.

value and provides a simple means to normalize plate height values
using the calculated b values from linear regression for $R_f = 0.5$
or $R_f = 1.0$ without having to have a standard whose spot center is
exactly at the reference value. The value for n_{real} at $R_f = 1$ is
unrealizable in practice, since spots moving close to the solvent
front are generally distorted and flattened in the direction of
migration.

$$n_{real} = 5.54 [Z_x/(b_x - b_o)]^2 \tag{7.9}$$
$$H_{real} = (b_x - b_o)^2/5.54 Z_x \tag{7.10}$$

Ideally to maximize n_{real}, b_o should be small compared to b_1.
In practice the extrapolated value for b_o will generally exceed the
value determined densitometrically for the initial starting zone
due to the very rapid expansion and reshaping of the zone which
occurs as the mobile phase first reaches the starting zone. In
this region the solvent velocity is high, uneven, and the sample
requires a finite time to equilibrate with the mobile phase. The
extrapolated value, then, is a more realistic value of the zone
dimensions at the time of commencing migration than the value
recorded for the spot dimensions prior to contact with the

solvent. Fenimore reports that the eventual spot width is independent of the initial spot size for spots about 1 mm in diameter using high performance TLC plates [7]. Thus, b_o, is determined primarily by the quality of the layer even for starting spots much smaller than 1 mm. Therefore, b_o is always finite compared to the size of developed spots which are generally < 6 mm on high performance TLC plates.

The influence of the layer structure on the plate height can only be interpreted with the aid of a suitable model [8,45-47]. The available models are at best approximate and difficult to test and refine, due to a lack of accurate data for the parameters used. The most recent treatment by Guiochon and Siouffi commences from the assumptions that the TLC plate has the properties of a normal column bed, the local plate height is described by the Knox equation (section 1.7.10), and that the velocity of the eluent is constant at all points in the layer at a given time with the velocity decreasing with time. After further necessary assumptions they arrived at equation (7.11) for the average plate height [45]. The coefficients a, b and c are complex functions of the experimental parameters not identified here for simplicity [44].

$$\overline{H} = [(a/(Z_f - Z_o)][(Z_f)^{2/3} - (Z_o)^{2/3}] + b[Z_f + Z_o] + [c/(Z_f - Z_o)]\log(Z_f/Z_o) \tag{7.11}$$

The last term (c term) is rarely significant and can be neglected in most cases. To a first approximation the first term (a term) can be neglected for fine-particle layers (d_p < 10 micrometers) but must be retained for coarse-particle layers. In a qualitative sense equation (7.11) predicts that under capillary flow controlled conditions with fine-particle layers the plate height first passes through a minimum and then increases sharply for longer migration distances, Figure 7.5 [1]. For coarse-particle layers the plate height is less dependent on the migration distance and eventually the two curves crossover, indicating that a greater number of theoretical plates can be obtained using coarse-particle layers and long migration distances. This contrary finding is easily explained by the relative permeability of the layers. The mobile phase velocity for the fine-particle layer declines rapidly with the migration distance until eventually the zone broadening becomes diffusion controlled [b term in equation (7.11)]. The coarse-particle layer is more permeable than the fine-particle layer and the solvent velocity is higher for longer

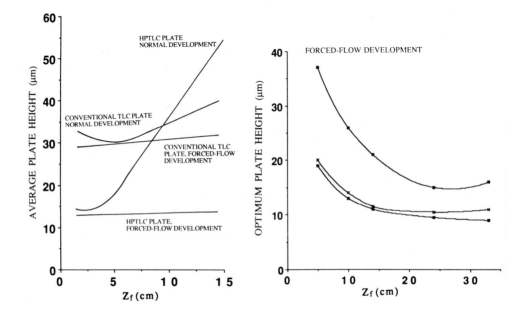

Figure 7.5 Left, variation of the average plate height of fine- and coarse-particle layers as a function of the solvent migration distance and method of development. Right, relationship between the optimum plate height and solvent migration distance for forced-flow development.

plate lengths and, so is the efficiency. For fine-particle layers with a development length of about 5 cm it should be possible to obtain 2,500-5,000 theoretical plates but it will be very difficult to exceed this number using capillary flow controlled development. For coarse-particle layers (d_p about 15 micrometers) a development length of about 15 cm will be required to obtain around 5,000 theoretical plates, and although it is possible to exceed this number, it is not easy, and will lead to long separation times. In practice, conventional TLC plates are often prepared from sorbents of a wide particle size distribution and may provide only a fraction of the number of theoretical plates calculated theoretically.

In forced-flow TLC the average plate height is largely independent of the migration distance and is more favorable for fine-particle than coarse-particle layers, Figure 7.5. This arises from the optimization of the mobile phase velocity by external force overcoming the limitations of capillary flow controlled systems, namely, the quadratic decrease in mobile phase velocity with time. In the forced-flow mode there is an optimum linear velocity for a fixed development length corresponding to the minimum in the plot of the average plate height as a function of the mobile phase velocity [37,48]. This optimum value decreases with the migration distance and only becomes approximately constant for solvent migration distances exceeding about 25 cm, Figure 7.5. Compared to capillary flow controlled systems zone broadening by diffusion is now restricted to a minor role even for long migration distances, since the optimum mobile phase velocity is always higher than that observed for ascending development by capillary flow. The average plate height in forced-flow TLC is approximately constant, so that the number of theoretical plates increases linearly with the solvent migration distance. Plate height values as low as 8 micrometers have been observed corresponding to a limit of 31,000 theoretical plates for a substance migrating 25 cm. The ultimate efficiency of the forced-flow system is limited only by the particle size, the available plate length, and the pressure required to maintain the optimum mobile phase velocity.

7.2.3 Resolution and Separation Capacity

The resolution, R_s, between two sample zones is defined as the ratio between the separation of the two zone centers and the average width of the zones expressed by equation (7.12) [49].

$$R_s = 2(Z_{x2} - Z_{x1})/(w_{b1} + w_{b2}) \tag{7.12}$$

where Z_x is the migration distance of the zone center, w_b the width of the zone at its base, and the subscripts 1 and 2 refer to the individual zones numbered such that the larger number corresponds to the zone with highest R_f value. To optimize the resolution of individual sample zones it is necessary to know how R_s varies with the experimental parameters, such as the layer efficiency, the ratio of the equilibrium constants governing the separation process, and the position of the zones within the chromatogram. Equation (7.12) can be expressed in terms of the R_f values of the two zones by substituting $Z_x = R_f(Z_f - Z_o)$, and for two zones in close proximity, can be further simplified by assuming $w_{b1} = w_{b2} = w$, to give equation (7.13)

$$R_s = (Z_f - Z_o)(R_{f2} - R_{f1})/w \tag{7.13}$$

A zone of width w has a corresponding plate height given by equation (7.7), which after substitution into equation (7.13), gives

$$R_s = (n^{1/2}/4)(R_{f2} - R_{f1})/R_{f2} \tag{7.14}$$

Equation (7.14) is the analog of the classical resolution equation used for column chromatography (section 1.6). However, in TLC neither n nor $(R_{f2} - R_{f1})$ are independent of the ratio of the equilibrium constants controlling the separation process or the characteristics of the layer. The value of n is strongly dependent on R_f for capillary flow controlled conditions, and to a crude approximation can be replaced by $n = n_1 R_{f2}$ where n_1 is the number of theoretical plates passed over by the zones if they moved to the solvent front [50]. The capacity factor, k, in terms of the R_f value is given by equation (7.2), which can be substituted along with $n = n_1 R_{f2}$ into equation (7.14), and after rearranging gives

$$R_s = [(n_1 R_{f2})^{1/2}/4][(k_1/k_2) - 1][(1 - R_{f2})] \tag{7.15}$$

Qualitatively equation (7.15) is adequate to describe the influence of layer quality, selectivity, and zone position in the chromatogram upon resolution for a single unidimensional development under capillary flow controlled conditions. The variation of R_s with R_f is not a simple function as can be seen from Figure 7.6. The resolution increases with the layer efficiency in a manner that depends linearly on the R_f value. Relatively small changes in selectivity have an enormous impact on

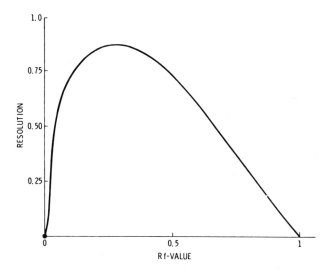

Figure 7.6 The variation of resolution of two closely migrating zones as a function of the R_f value of the faster moving zone.

the ease of obtaining a given separation in TLC, since the total number of theoretical plates that can be made available for a separation is not very large, or large enough to make kinetic optimization the most worthwhile approach to improve resolution. In TLC, separations are fairly easy when $(R_{f2} - R_{f1}) > 0.1$ and very difficult or impossible for $(R_{f2} - R_{f1}) \leq 0.05$ in the region of the optimum R_f value for the separation. The influence of zone location on resolution shows the opposite behavior to that of the layer quality. At large values of R_{f2}, the term $(1 - R_{f2})$ will decrease and the resolution will become zero at $R_{f2} = 1$. Differentiation of equation (7.15) indicates that the maximum resolution of two difficult to separate substances will occur at an R_f value of about 0.3. It can be seen from Figure 7.6 that the resolution does not change significantly for R_f values between 0.2 and 0.5; within this range, the resolution is greater than 92% of the maximum value (75% between $R_f = 0.1$ and 0.6).

Resolution in forced-flow development is not restricted by the same limitations that apply to capillary flow controlled systems. The maximum resolution achieved usually corresponds to the optimum mobile phase velocity and R_s increases approximately linearly with the solvent migration distance [48]. Thus there is no theoretical limit to R_s with forced-flow development; the upper bounds are established by practical constraints (plate length, separation time and inlet pressure). For the same solvent/sorbent combination, sample resolution should always be greater in the forced-flow mode compared to capillary flow controlled conditions, at least in the absence of solvent demixing.

The potential of a chromatographic system to provide a certain separation can be estimated from its separation number, SN, also referred to as the spot capacity in TLC. The separation number in TLC is defined as the number of spots that can be completely separated ($R_s = 1$) between $R_f = 0$ and $R_f = 1$ [6]. It is calculated in an approximate form by equation (7.16) and more exactly by equation (7.17) with b_0 and b_1 as defined in Figure 7.4.

$$SN = [(Z_f - Z_0)/(b_0 + b_1)] - 1 \tag{7.16}$$
$$SN = \log(b_0/b_1)/[\log(1 - b_1 + b_0)/(1 + b_1 - b_0)] \tag{7.17}$$

Equation (7.16) provides values that are about 25% smaller that those calculated by equation (7.17) [8]. Models used to interpret the separation number for different TLC conditions are far from straightforward [51,52]. The separation number is a complex function of the characteristics of the solvent and the layer as well as the development length. Approximate solutions are usually achieved by iteration procedures, and consequently in this section we will consider only the phenomena and the conclusions reached, without delving deeply into the mathematical relationships. The results show that with capillary controlled flow it should be easy to achieve a separation number of 10 to 20, but it is extremely difficult to reach 25 and practically impossible to exceed 30 except in extreme circumstances. To obtain separation numbers greater than 25, very long plates and prohibitively long separation times are required. The separation number for forced-flow TLC should be identical to those of column chromatography, where values of 150-200 are possible but difficult to achieve. In practice, shorter bed lengths and limited inlet pressures compared to those used in column chromatography set a practical limit of about 80 for the separation number with commercially available equipment.

As an analogy, the separation number envisages the chromatogram as being similar to a string of beads, each bead touching its neighbor with no unoccupied space between the beads. The separation number is thus an inflated estimate of the real separation capacity, since real chromatograms do not usually consist of an array of equally spaced peaks. In general, unless the separation number exceeds the number of components in the sample by a significant amount the separation will normally be difficult to achieve.

7.3 STATIONARY PHASES FOR TLC

TLC plates may be prepared in the laboratory by standardized procedures [1-3], although the exacting experimental conditions required for their reproducible preparation are more easily obtained in a manufacturing setting. Consequently, most laboratories, today, use commercially available precoated plates. Precoated plates for high performance, conventional and preparative TLC are available in thicknesses from 0.1 to 2.0 mm supported by either glass, aluminum or plastic backing sheets. Most plates also contain a binder such as gypsum, starch or salts of poly(acrylic acid), in amounts of 0.1 to 10% (w/w), to impart the desired mechanical strength, durability and abrasion resistance to the sorbent layer. A UV-indicator, such as manganese-activated zinc silicate of a similar particle size to the sorbent, may be added to the layer for visual evaluation of separated samples by fluorescence quenching. TLC plates with a binary layer of two different, separated sorbents, forming a narrow interface parallel to one edge, are available for two-dimensional TLC. If one of the layers is a form of silica with very weak retention properties, it can be used as a concentrating zone, to aid sample application. Recent commercial developments include the introduction of new materials for TLC in the form of flexible sheets prepared from silica or bonded phase silica homogeneously mixed and entrapped in a matrix of poly(tetrafluoroethylene) micro fibrils [53,54]. These sheets contain about 90% by weight of sorbent with the balance being PTFE. The absence of a backing support imparts unusual flexibility to the sheets facilitating isolation of samples by cutting out sections of the chromatogram of interest. Porous glass sheets have been found suitable for TLC [55]. The glass sheets are mechanically strong and can be washed with strong acids for repeated use. These porous glass plates are different from the sintered glass plates prepared by sintering a low melting point glass with a high melting point sorbent as a thin layer supported by a glass backing plate that have been commercially available for a number of years [56]. The sintered glass layers are highly porous and allow chromatographic development to proceed normally, but compound retention is considerably reduced compared to standard TLC plates. As well as plates, glass rods with a sintered layer of adsorbent are also available, and are used primarily in conjunction with a scanning flame ionization detector [56-58].

Sorbents generally used in TLC include silica, alumina, chemically bonded silica, cellulose, polyamide and polymeric ion-exchange resins. Polyamide [23,59] and polymeric ion-exchange resins are only available in a low performance grade and are used primarily for qualitative analysis [16]. Sorption on polyamide layers is mainly due to the reversible formation of hydrogen bonds between the functional groups of the sample and the carbonyl oxygen of the amide group, as well as contributions from a more general reversed-phase type mechanism. Polyamide TLC plates have been used mainly for the separation of phenols, amino acid derivatives, heterocyclic nitrogen compounds and carboxylic and sulfonic acids. Ion-exchange layers are usually based on poly(ethyleneimine), poly(styrene)-divinylbenzene and diethyl-aminoethyl cellulose resins and powders and are used primarily for the separation of inorganic ions and biological polymers. Several types of gels, suitable for use in thin-layer gel chromatography (TLG), are commercially available [60,61]. These include the Sephadex series of crosslinked dextrans, the Bio-Gel P series of crosslinked poly(acrylamide) gels, and the Enzacryl crosslinked poly(acryloylmorpholine) gels. These gels must be swollen in an appropriate buffer/solvent prior to spreading as a layer, and consequently, the plates must be prepared in the laboratory. The published applications of TLG are confined almost entirely to the separation of macromolecules of biochemical interest, and to proteins in particular. TLG is used to make comparisons of different protein fractions, to estimate the relative concentrations of components in a mixture, and to determine molecular weights. The mechanism of separation is primarily size-exclusion.

7.3.1 Inorganic Oxide Adsorbents

The most common inorganic adsorbents used in TLC are silica gel, alumina and kieselguhr; silica gel being by far the most important. Kieselguhrs are made from natural diatomites and are an impure form of silica, being about 90% or so silica with the

TABLE 7.3

TYPICAL PROPERTIES OF INORGANIC OXIDE ADSORBENTS USED FOR TLC

Parameter	Silica Gel	Alumina	Kieselguhr
Specific surface area (m^2/g)	200-800	50-350	1-4
Specific pore volume (ml/g)	0.5-2	0.1-0.4	1-3
Average pore diameter (nm)	4-20	2-35	10^3-10^4
Concentration of active sites(a) ($micromol/m^2$)	8	13	-

(a)Silanol groups for silica and oxide ions for alumina

balance as alumina, ferric oxide and other metal oxides. The chromatographic properties of the inorganic adsorbents depends on their surface chemistry and composition, specific surface area, specific pore volume, and average pore diameter. Relevant properties of the inorganic oxide sorbents used for TLC are summarized in Table 7.3 [23]. Because of the presence of oxide ions, the surface of alumina is quite basic (estimated as approximately pH 12). Acids with a pK_a lower than about 13 transfer protons to this surface, producing charged conjugate bases that are strongly adsorbed. The surface of silica is mildly acidic due to the presence of silanol groups, pH about 5, although this value can be adjusted over a wide range by buffering with appropriate solutions [1,3,62]. Also, the activity of an adsorbent will depend on the concentration of adsorbed impurities, particularly water. Deactivation by water occurs through the gas phase and depends on the relative humidity at which the plates are stored and used. Adjustment of the relative humidity in a closed vessel is relatively easy to achieve using aqueous solutions of sulfuric acid or saturated salt solutions [1,8]. Upon removing the TLC plate from one environment to another of a different relative humidity, re-equilibration occurs very rapidly making the absolute control of retention by adjusting the activity of the layer difficult. In practice, the best method of control is to perform the equilibration and development process in the same chamber without removing the plate. With all other parameters held constant, an increase in layer activity will lead to lower R_f values and a decrease to higher R_f values. The R_f value also decreases with increasing surface area for similar activity levels. Retention on silica gel is controlled by the number and type of functional groups present in the sample and their spatial

674

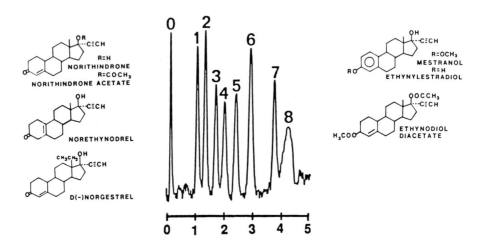

Figure 7.7 Separation of ethynyl estrogens on silica gel 60 HPTLC plates using two 15-min developments in the solvent system hexane-chloroform-carbon tetrachloride-ethanol (7:18:22:1 v/v). Identification: 0 = methyl green (lane marker); 1 = 17α-ethynylestradiol; 2 = norethindrone; 3 = norgestrel; 4 = norethynodrel; 5 = norethindrone acetate; 6 = mestranol; 7 = ethynodiol diacetate; 8 = solvent front (Reproduced with permission from ref. 63. Copyright Dr. Alfred Huethig Publishers).

location. Proton donor/acceptor functional groups show the greatest retention, followed by dipolar molecules, and finally nonpolar groups. The influence of functional group type and spatial position is illustrated in Figure 7.7 for the separation of ethynyl estrogens, which are the active ingredients in oral contraceptives [63]. The estrogens with phenolic groups are the most strongly retained, followed by hydroxyl, and ketone and ester groups. Subtle differences are also seen due to crowding of a functional group and conformation of the rings, which allow the separation of components of very similar chemical properties.

TABLE 7.4

PHYSICAL PROPERTIES OF SILICA GELS USED TO PREPARE HPTLC PLATES

Property	Merck Si 60	Whatman HP-K
Average pore diameter (nm)	6	8
Pore volume (ml/g)	0.82	0.70
Specific surface area (m^2/g)	550	300
Mean particle size (micrometers)	5	5
Layer thickness (mm)	0.2	0.2
pH of 10% aqueous suspension	7.0	7.0-7.2

The physical properties of two commercial silica gels used to prepare HPTLC plates are summarized in Table 7.4 [64,65]. They differ primarily in pore diameter and pore volume which results in a concomitantly larger specific surface area for the Merck plate, and all other things being equal, a lower R_f value when developed with identical solvents. More specific differences in the type and distribution of silanol groups for the two silicas may result in selectivity differences for difficult to separate compounds.

There are several ways of adjusting the selectivity of silica gel layers by impregnation with a modifier, generally achieved by simply immersing the layer into a solution of the modifier in a volatile solvent and allowing the solvent to evaporate, leaving an even coating of the modifier throughout the layer [1,3,10,66-68]. For example, hydrophilic layers can be prepared by impregnation with dimethylformamide, dimethyl sulfoxide, poly(ethylene glycol), etc., while lipophilic layers are prepared with liquid paraffin, undecane, silicone oils, etc. The main disadvantages of these impregnated layers are that the mobile phase must be selected so as not to destroy the homogeneity of the modified layer and, if the impregnating reagent is a liquid of low viscosity, the separated zones will broaden on standing due to diffusion in the liquid film. These reasons have led to the preferential use of chemically bonded silica layers, although their chromatographic properties and separation mechanisms are not necessarily identical to those of physically impregnated layers. An exception to this observation is the use of selective reagents that can form reversible complexes with the sample. Some common examples are silver nitrate used for the separation of saturated and unsaturated compounds, boric acid to differentiate between isomers with vicinal hydrogen bonding functional groups, and caffeine or picric acids to improve the separation of polycyclic aromatic compounds. Sometimes, stronger complexes and better resolution are achieved at subambient temperatures, allowing resolution to be controlled by temperature optimization.

TLC plates with concentrating zones are prepared from two layers of silica gel having different retention properties [69]. The two layers abut each other, parallel to one edge, forming a very narrow interface. The concentrating zone is usually a strip about 25 mm high in the direction of development and possibly also slightly thinner than the separating zone, which is prepared from the same silica gel used to coat normal TLC plates and occupies most of the plate area. One manufacturer uses a synthetic porous silica of medium pore volume, extremely large pore diameter (ca. 5000 nm) and extremely small surface area (ca. 0.5 m^2/g). The same silica gel is used to prepare precoated layers for the separation of very polar compounds, such as carbohydrates, nucleic acid derivatives, phosphates and sulfonates, etc., that is, substances not easily separated on regular layers of higher activity [70]. The concentrating zone simplifies the process of sample application, since microliter volumes can be applied either as spots or bands to any position on the concentrating zone. Alternatively, the entire zone can be immersed in a dilute solution of the sample. During development, the sample migrates out of the concentrating zone and is focused at the interface as a narrow band resulting in excellent chromatographic efficiency. However, since the distribution of the sample may not be even within the band, the quantitative accuracy of densitometric measurements may be lowered. TLC plates with concentrating zones are most useful when large sample amounts are applied, when very dilute sample solutions are used, or when crude samples (e.g., biological fluids) are applied directly to the plate. HPTLC plates with concentrating zones can also be used in laboratories lacking the sophisticated equipment necessary for spotting nanoliter volumes onto regular HPTLC plates.

7.3.2 Chemically Bonded Layers

Chemically bonded layers are prepared by reacting silica gel with various functionalized organosilane reagents forming siloxane bonds with some of the silanol groups originally present on the silica surface [10,23,71-77]. Chemically bonded siloxane layers

TABLE 7.5

CHARACTERISTC PROPERTIES OF PRECOATED CHEMICALLY BONDED PLATES

Manufacturer	Derivatizing Reagent	Percent Silanol Groups Reacted	Carbon Loading (%)	Average Particle Size (μm)
Merck				
RP-2	Bifunctional	50		5-7
RP-8	Bifunctional	25	8.9	11-13
RP-8	Bifunctional	37		5-7
RP-18	Bifunctional	22	15.4	11-13
RP-18	Bifunctional	35		5-7
Amino	Trifunctional	50	5.8	5-7
Cyano	Bifunctional	27		5-7
Whatman				
KC-2	Monofunctional		4.5	10-14, or 20
KC-8	Monofunctional		8.5	10-14
KC-18	Trifunctional	16	12.5	10-14
Diphenyl	Difunctional		8.5	10-14
Macherey-Nagel				
Sil C18-100	Trifunctional	45		5-10
Sil C18-50	Trifunctional	30		5-10

with dimethyl, diphenyl, ethyl, octyl, octadecyl, 3-aminopropyl, 3-cyanopropyl, and a spacer bonded propanediol group [$Si(CH_2)_3OCH_2CH(OH)CH_2OH$] are commercially available. The choice of original silica substrate and method of preparation result in substantial differences in products which might otherwise be considered similar, Table 7.5 [10,73,74,78,79]. The reagents used to prepare TLC plates are not necessarily the same as those used by the same manufacturers to produce products for column liquid chromatography. The reversed-phase layers produced by Merck are prepared from dichloroalkylmethylsilanes with the alkyl groups methyl (RP-2), octyl (RP-8), and octadecyl (RP-18). The reagent is attached to the surface by a combination of one or two surface bonds. The amino phase is prepared with 3-aminopropyl-triethoxysilane and has a much higher surface coverage than the cyano phase which is prepared from 3-cyanopropyldichloro-methylsilane. The bonded 3-cyanopropylsiloxane phase is anchored to the silica substrate almost entirely by bidentate linkages and the 3-aminopropylsiloxane phase by a combination of bidentate and tridentate links. The latter are most likely due to crosslinking of the reagent during reaction with the surface. The extent of reaction with silanol groups is much greater for the 3-

aminopropylsiloxane bonded phase, which is a different type of
bonded phase to the 3-cyanopropylsiloxane phase. The two Macherey-
Nagel reversed-phase octadecylsiloxane bonded layers are of the
polymeric type, prepared from a trifunctional reagent, that differ
primarily in the extent of silanization. The series of Whatman
products are prepared using different chemistries. The
dimethylethylsiloxane and the dimethyloctylsiloxane bonded layers
are prepared from monofunctional reagents and should have a well
defined "brush" type surface. The octadecylsiloxane bonded phase
is prepared from a trifunctional reagent and has a polymeric,
crosslinked, surface bonded film. The octyl-, octadecyl- and
diphenylsiloxane bonded phases are also endcapped by reaction with
an undisclosed silane to minimize the concentration of free
silanol groups. Even so, the number of unreacted silanol groups is
the highest of the octadecylsiloxane bonded phases in Table 7.5
[79].

The water content of the mobile phase has a dramatic
influence on the mobile phase velocity for revered-phase plates
(see also section 7.2.1) [10,72-74,80-83]. The concentration of
silanol groups on the silica surface is about 8 micromol/m^2 but for
steric reasons only about 50% of these groups can be reacted with
bulky silanizing reagents. Those reversed-phase layers prepared
with silica silanized close to completion can no longer be
developed using highly aqueous solvents. In this case the
hydrophobic repulsive forces are stronger than the capillary
forces moving the solvent through the layer (this does not apply
to forced-flow development). With binary mobile phases containing
more than about 40% (v/v) water the mobile phase velocity becomes
very slow and/or the precoated layers swell and flake off from the
glass backing plate. This drawback for the Whatman KC layers can
be overcome by adding inorganic salts to the mobile phase (e.g.,
3% (w/v) sodium chloride, lithium chloride, or ammonium acetate
solution, etc., rather than water). The polarity of the mobile
phase can then be varied up to about 80-90% water in aqueous
solution without greatly changing the mobile phase velocity. The
Whatman plates are prepared with an amphoteric binder to enhance
water compatibility. The addition of a salt to the mobile phase is
used primarily as a "mass ion" effect to salt out the binder and
prevent the layer from detaching from its support [84]. This
approach cannot be expected to work with other reversed-phase
layers. Both Merck and Macherey-Nagel produce reversed-phase

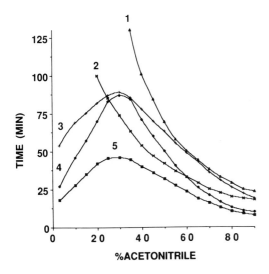

Figure 7.8 Dependence of the migration time for a 7 cm development in an unsaturated chamber for different Merck reversed-phase layers using water-acetonitrile as the mobile phase. Identification: 1 = RP-18 HPTLC plate; 2 = RP-2 HPTLC plate; 3 = RP-18W HPTLC plate; 4 = RP-18 TLC plate; and 5 = RP-8 TLC plate.

layers that can be completely wetted by water. These plates are given the additional prefix W and are similar products to those described in Table 7.5, except that they have a defined and reproducible lower degree of modification. The Merck plates show a maximum in the development time (minimum for the mobile phase velocity) corresponding to about 60% water, Figure 7.8 [74]. It is claimed that the occurrence of the maximum can be explained primarily by the particular swelling properties of the layer-stabilizing binder in this region of the mobile phase composition. Increasing the particle size and/or reducing the degree of surface modification are general ways of increasing the mobile phase velocity and water compatibility of revered-phase layers. Layers prepared from reagents with short alkyl chains also show greater compatibility with aqueous mobile phases.

The hydrophobic character of the reversed-phase layers depends on the degree of modification and the chain length of the organic groups bonded to the surface. When nonpolar solvents are used as the mobile phase retention is likely to be governed by an adsorption mechanism involving the unreacted silanol groups as active sites. For polar organic solvents and aqueous mobile phases a reversed-phase partition mechanism is probably dominant. Reversed-phase layers have been used to separate a wide variety of compounds [71,74]. Ionic and ionizable compounds, generally difficult to separate on silica gel, can be separated by using ion-pair reagents [10,85-87]. A rather unique application of reversed-phase chromatography is the fractionation of polymers of high molecular weight in a mobile phase composition gradient [23,88]. The least polar component of a binary mobile phase must be a good solvent for the polymer and selectively adsorbed by the layer, while the second solvent must be a poor solvent for the polymer. The polymer fractions are then separated by precipitation in the mobile phase gradient with an apparent R_f value that can be simply related to its molecular weight.

Polar bonded phases can be used in either the normal or reversed-phase mode [10,74-76]. They are completely compatible with aqueous mobile phases including pure water. A minimum in the mobile phase velocity as a function of the volume percent of organic solvent in aqueous mixtures is generally observed. The composition of the mobile phase at the minimum depends on the choice of modifier and is usually between 50-70% (v/v) modifier. The 3-aminopropylsiloxane bonded layer can act as a weak, basic ion-exchanger for the separation of polyanions [89-91]. When the mobile phase is an organic solvent mixture the 3-aminopropylsiloxane bonded layers behave as weak adsorbents and affect separations predominantly by the type and number of polar functional groups present in each component. The 3-aminopropylsiloxane bonded plates have been used to separate sugars, which after separation can be induced to fluoresce by heating for about 3 min at 170°C while adsorbed on the layer [92]. The 3-cyanopropylsiloxane bonded silica layers show different selectivity to the 3-aminopropylsiloxane bonded silica layers when used in the normal or reversed-phase mode [75,76,81,93,94]. The polarity of the 3-cyanopropysiloxane bonded silica layers lying between that of the amino modification and the dimethylsiloxane reversed-phase plates. Lithium chloride and other salts have been added to the mobile phase for the separation of some polar compounds to suppress secondary interactions and achieve more compact spots, for example, in the separation of amino acid derivatives, alkaloids, pharmaceutical compounds and preservatives. Siloxane bonded layers containing spacer bonded propanediol groups are the most recent addition to the range of polar chemically bonded layers and have quickly found a number of useful applications based on their ability to function as adsorbents with selectivity somewhat similar to silica gel but with much weaker interactions [75,76,95,96].

7.3.3 Cellulose Layers

Precoated TLC plates with both native and microcrystalline cellulose layers have been used for several years, primarily for the separation of very polar compounds (e.g., carbohydrates, carboxylic acids, amino acids, nucleic acid derivatives, phosphates, etc). Celluloses are naturally occurring polysaccharides of general formula $(C_6H_{10}O_5)$n occurring in the form of long fibers (native cellulose) or as rods (microcrystalline cellulose). It is generally assumed that a partition mechanism is responsible for retention on these low surface area sorbents. HPTLC plates coated with a special grade of microcrystalline cellulose are commercially available [70]. High performance layers prepared from wide pore (5000 nm) and very low surface area (ca. 0.5 m^2/g) silica gel has been recommended for separating similar substances to those normally separated on cellulose [70]. Compared to cellulose the silica plates are non-swelling in organic solvents and can be used with aggressive visualizing reagents without darkening or reaction.

7.3.4 Chiral Stationary Phases

The resolution of enantiomers by liquid chromatography using chiral stationary phases is based on the formation of reversible diastereomeric complexes of different stability between the sample and stationary phase. Since the formation of the complexes is strongly dependent on the structure of the sample, there are no universal chiral stationary phases. The specific advantages of TLC for enantiomeric separations result from its low cost, convenience and speed [10,97,98]. The main limitation, particularly with respect to column liquid chromatography, is the small number of phases currently available.

The oldest method of resolving enantiomers by TLC takes advantage of the natural chiral properties of cellulose and triacetylcellulose resulting from the helical structure of the polymers [98]. Amino acid derivatives have been resolved on silica gel layers impregnated with chiral acids or bases, for example,

(+)-tartaric acid or (+)-ascorbic acid [99]. The separation mechanism, in this case, is based on the formation of diastereomeric salts as intermediate complexes. Chemically bonded chiral phases have been prepared by the ionic or covalent bonding of the chiral selectors N-(3,5-dinitrobenzoyl)-R-(-)-α-phenyl-glycine or N-(3,5-dinitrobenzoyl)-L-leucine [100] and (R)-(-)-1-(1-naphthyl)ethylisocyanate [101] to precoated aminopropyl-siloxane bonded silica plates. The dinitrobenzoyl amide phases are identical to the Pirkle phases used in column liquid chromatography (section 8.15). They have been used to separate β-amino alcohols, benzodiazepines and hexobarbital. These plates, however, have two disadvantages. The plates discolor on standing and the strong absorption of UV radiation by the selector makes quantitation by scanning densitometry difficult. The naphthylurea chiral phase has been used to separate racemic α-methylarylacetic acid 3,5-dinitroanilides and amines as 3,5-dinitrobenzoyl amides. The weaker UV absorption of the selector compared to the derivatives, in this case, permits in situ detection.

Cyclodextrins as chemically bonded layers [102] or mobile phase additives [103-105] have been used successfully to resolve a wide variety of alkaloids, steroids and dansyl- and naphthylamide-amino acid derivatives. The low solubility in aqueous solution and high cost of cyclodextrins restricted the use of these additives initially. These limitations were overcome by the availability of synthetically prepared β-cyclodextrins incorporating hydroxy-propyl, hydroxyethyl or maltosyl substituents to a defined degree of modification. These modified cyclodextrins have much higher solubility in aqueous organic solvent mixtures and can be used in reversed-phase TLC. It was found that a minimum concentration of additive was required to achieve a given separation of enantiomers and that the resolution of the enantiomers increased with increasing concentration of the additive until a plateau region was reached or the solubility limit of the additive was achieved. At very high concentrations of additive the viscosity of the mobile phase may become too great to allow reasonable separation times on reversed-phase layers. Tivert and Backman have demonstrated the separation of amino alcohol enantiomers by TLC using the chiral counterion N-carbobenzoxy glycyl-L-proline as a mobile phase additive [106].

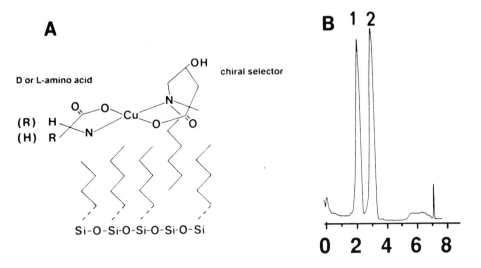

Figure 7.9 A, Schematic representation of complex formation
between an amino acid and copper in the presence of a physically
immobilized chiral selector. B, separation of D-leu-L-leu(1) and
L-leu-D-leu(2) on a CHIR plate with methanol-propan-1-ol-water
(5:1:4) as the mobile phase with detection at 410 nm. (Reproduced
with permission from ref. 109. Copyright Dr. Alfred Huethig
Publishers).

The most widely used approach for the separation of
enantiomers by TLC is based on a ligand exchange mechanism using
commercially available reversed-phase plates impregnated with a
solution of copper acetate and (2S,4R,2'RS)-4-hydroxy-1-(2-
hydroxydodecyl)proline in optimized amounts, Figure 7.9
[10,97,98,107-109]. Enantiomers are separated based on the
differences in the stability of the diastereomeric complexes
formed between the sample, copper, and the proline selector. As a
consequence, a prime requirement for separation is that the sample
must be able to form complexes with copper. Such compounds include
the amino acids and their derivatives, α-substituted carboxylic
acids, peptides, some lactones and certain thiazolidine
derivatives [98].

7.4 DEVELOPMENT TECHNIQUES

Development in TLC is the process by which the mobile phase moves through the sorbent layer, thereby inducing differential migration of the sample components. The principal development modes used in TLC are linear, circular and anticircular with the velocity of the mobile phase controlled by capillary forces or forced-flow conditions. In any of these modes the development process can be extended by using continuous development or multiple development.

7.4.1 Linear and Radial Development

The linear development mode represents the simplest situation and is more widely used than radial development. The samples are applied along one edge of the plate and the separation performed by allowing the solvent to migrate to the opposite edge. Viewed in the direction of development, the chromatogram consists of a series of compact symmetrical spots of increasing diameter, or if samples are applied as bands, in rectangular zones of increasing width. Spots eluting close to the solvent front may be distorted and elliptical or rod-like in shape. The usable separation area of the plate is defined by an R_f range of approximately 0.1 to 0.8.

If the position of sample application and the point of entry of the mobile phase are at the center of the plate and the flow of mobile phase is towards the periphery of the plate, then this mode of development is called circular chromatography [6,110]. Samples can be injected into the mobile phase, in which case they will be separated as a series of concentric rings. If the samples are applied as a cluster of spots in a radial pattern around the solvent entry position, after development, spots near the origin remain symmetrical and compact while those near the solvent front are compressed in the direction of development and elongated at right angles to this direction, Figure 7.10(A).

In anticircular development the sample is applied along the circumference of an outer circle and developed toward the center of the plate [6,110,111]. Spots near the origin remain compact while those toward the solvent front are considerably elongated in the direction of migration, but changed very little in width when viewed at right angles to the direction of development, Figure 7.10(B). Elongation of the sample spots at high R_f is unavoidable

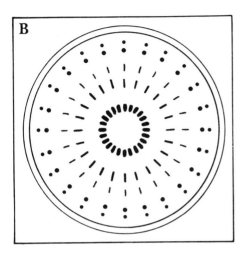 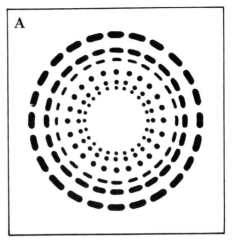

Figure 7.10 Circular development with the point of solvent entry at the plate center (A). Anticircular development from the outer circle towards the center (B). (Reproduced with permission from ref. 13. Copyright American Chemical Society).

and arises as a consequence of the lateral compression induced by the mobile phase flow through a continuously decreasing plate area. The unique features of anticircular development are its high speed and large sample capacity. The mobile phase velocity remains approximately constant under capillary controlled flow conditions on account of the equivalence between the quadratic decrease in mobile phase velocity and a similar reduction in the plate area to be wetted.

The R_f values for a substance measured in linear, circular and anticircular chromatograms, for which the flow conditions vary, can be related to each other by equation 7.18

$$R_f = [R_{f(C)}]^2 = 1 - [1 - R_{f(AC)}]^2 \qquad (7.18)$$

where R_f is the R_f value obtained by linear development, $R_{f(C)}$ the R_f value for the same component determined by circular development, and $R_{f(AC)}$ the R_f value of the same component determined by anticircular development. If the point of solvent entry and sample application are not identical then a more complex relationship is required to relate R_f to $R_{f(C)}$ that takes into account the distance between the position at which the sample was applied to the layer and the mobile phase entry position [91]. The circular development mode is advantageous for the separation of components of low R_f values and anticircular development for the separation of components with high R_f values. Both circular and anticircular chromatograms require a scanning densitometer capable of radial or peripheral scanning, an option available with some instruments.

7.4.2 Continuous Development

Continuous development is used in capillary controlled flow TLC to enhance the resolution of difficult samples using an optimized solvent and a shorter bed length [112,113]. In continuous development the mobile phase is allowed to traverse the layer until it reaches some predetermined position on the plate, at which point it is continuously evaporated. Evaporation of the mobile phase usually occurs at the plate atmospheric boundary using either natural or forced evaporation. The movement of the mobile phase to the air boundary occurs by conventional development, but once it reaches the boundary, additional forces are applied by the evaporation of the solvent. Eventually a steady state (constant velocity) is reached, where the mass of solvent evaporating at the boundary is equivalent to the amount of new solvent entering the layer. Sandwich-type chambers for continuous development have been reviewed by Soczewinski [9] and Perry [112] has outlined the use of the short-bed continuous development chamber for optimized continuous development with variable selection of the plate length.

A weaker solvent must be used for separations by continuous development compared to normal development to prevent the more mobile spots from accumulating at the solvent/air boundary. Under these conditions the resolution of the spots is enhanced by the greater selectivity afforded by using weaker solvents. Samples are then resolved by very short migrations, the increase in selectivity results in a decrease in the number of theoretical plates (bed length) required for the separation, and the spots remain compact and easier to detect. A theoretical model for predicting separations by continuous development under capillary flow controlled conditions has been proposed, and verified by Nurok [114-116]. Nurok has shown that for a given spot separation the separation time will always be shorter by continuous development than by conventional development, when the experimental parameters are correctly optimized.

7.4.3 Multiple Development

In unidimensional multiple development, the TLC plate is developed for some selected distance, then withdrawn from the developing chamber, and adsorbed solvent evaporated before returning the plate to the developing chamber and repeating the development process [7,113,117-121]. This is a very versatile strategy for separating complex mixtures since the primary experimental variables of plate length, time of development if continuous development is used, and composition of the mobile phase, can be changed at each development step, and the number of steps varied to obtain the desired separation. Quantitative measurements by scanning densitometry can be made at several steps in the sequence and, therefore, it is unnecessary for all components to be separated at one time provided they can be resolved (chromatographically or spectroscopically) at different segments in the development sequence. Multiple development provides higher resolution of complex mixtures than normal or continuous development, can easily handle samples of a wide polarity range (stepwise gradient development), and because the separated zones are usually more compact, leads to lower detection limits. Equipment for automated multiple development is commercially available [15,117,122].

A unique feature of multiple development TLC, that leads to an increase in the efficiency of the chromatographic system, is the spot reconcentration mechanism [14,117,123,124]. Every time the solvent front traverses the stationary sample it compresses the spot in the direction of development. Initially the spot will be symmetrical and at each development step, it will become more oval shaped until, if a sufficiently large number of developments are used, it will be compressed to a thin band, Figure 7.11. The compression occurs because the mobile phase first contacts the bottom edge of the spot where the sample molecules start to move forward before those molecules still ahead of the solvent front. Once the solvent front has reached beyond the spot, the reconcentrated spot migrates and is broadened by diffusion in the usual way. If a balance, or equivalence, can be struck between the reconcentration and zone broadening mechanisms, it is possible to move a spot a considerable distance without significant zone broadening compared with the size of the zone after the first development.

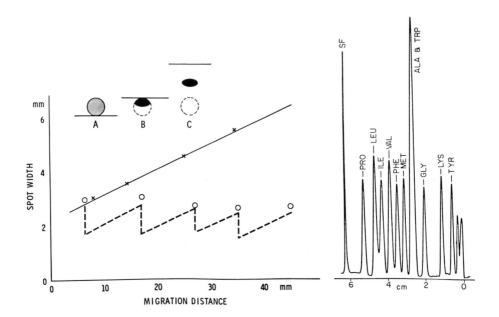

Figure 7.11 Separation of a mixture of PTH-amino acid derivatives by unidimensional multiple development (right) illustrating the use of the spot reconcentration mechanism to control zone broadening (left). A = spot size after the first development and solvent front of the second development (line). B = spot area (black oval) after reconcentration by the second development. C = spot size after the second development. (Reproduced with permission from ref. 117. Copyright Dr. Alfred Huethig Publishers).

Multiple development techniques using stepwise solvent gradients enable a subset of optimal separation conditions to be used to separate a mixture of wide polarity that cannot be separated using a single mobile phase [117,119,120,125]. As an example of this approach the separation of 20 common protein amino acid PTH-derivatives is shown in Figure 7.12 [126]. Five developments with four changes in mobile phase composition in a short-bed continuous development chamber were used for the separation. The first development, Figure 7.12(A), was made with methylene chloride for five minutes using a plate length of 3.5 cm. This step was performed to order the PTH-amino acids at the origin to enhance their resolution in later segments. Although PTH-proline is baseline resolved from other PTH-amino acids in this segment, there is no need to make any determinations at this

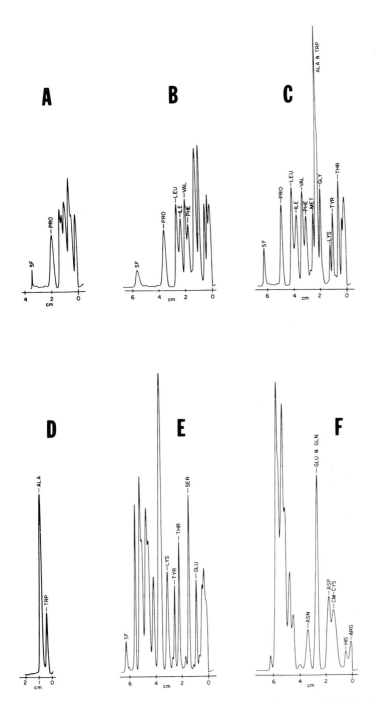

Figure 7.12   Separation of protein amino acid PTH-derivatives
using multiple development with sequential changes in mobile phase
composition (see text for details). (Adapted with permission from
ref. 126. Copyright Elsevier Scientific Publishing Co.)

point since it remains baseline resolved in the next four development steps. After the plate was dried, it was redeveloped for 10 minutes in methylene chloride-2-propanol (99:1) using a plate length of 7.5 cm. In all subsequent segments the plate length was unchanged. At this development step, Figure 7.12(B), the PTH-derivatives of proline, leucine, isoleucine, valine and pheylalanine are separated. The third development segment, Figure 7.12(C), was a repeat of the second and provides a better separation of the peaks resolved in step 2, as well as enabling the PTH-derivatives of methionine, alanine/tryptophan, glycine, lysine, tyrosine and threonine to be identified. In this segment of the development sequence, tryptophan frequently appears as a shoulder on the alanine peak, but is not adequately resolved for identification purposes. The PTH-derivatives of alanine and tryptophan can be separated almost to baseline by a 10 minute development in hexane-tetrahydrofuran (9:1) with a plate length of 3.5 cm (see Figure 7.12(D)), although not as part of the development sequence now being considered. The mobile phase was changed to methylene chloride-2-propanol (97:3) for the fourth, 10 minute development, segment. The separation of the PTH-derivatives of lysine, tyrosine and threonine is improved over that obtained in the third segment and the derivatives of serine and glutamine are now baseline resolved as well, Figure 7.12(E). A much more polar solvent was required for the final 10 minute development segment, ethyl acetate-acetonitrile-glacial acetic acid (74.3: 25:0.7) to separate the four remaining polar PTH-amino acid derivatives, Figure 7.12(F). The whole development sequence requires less than 1 h, only segments 3,4, and 5 need be scanned to indicate individual amino acid derivatives, and standards can be run simultaneously with samples (up to 32 on a 20x10 cm plate) to improve sample identification from position locations.

One disadvantage of the multiple development technique concerns the separation of samples of similar polarity. These tend to migrate together from the origin and become separated higher up the plate. For difficult separations fairly long plate lengths are needed and at each subsequent development sequence, a substantial amount of time is wasted while the mobile phase reaches the level of the lowest spot on the plate. For HPTLC layers with capillary controlled flow conditions the mobile phase velocity may no longer be adequate to maintain optimum separation conditions. A solution to this problem is to move the position of solvent introduction to higher positions on the plate for each successive development [127], or to remove a portion of the lower edge of the plate at some, or each subsequent development step [8,118]. Provided that the correct mobile phase velocity, number of development segments, and time for each development are selected then the sequential repositioning of the solvent entry position allows a spot to traverse virtually the whole length of the plate without experiencing significant zone broadening. In a specific example, by normal development an average value of approximately 2,000-3,000 theoretical plates was obtained; for normal multiple development this increased to 5,000-10,000; while, for multiple development with repositioning of the solvent entry position at each segment a value of 15,000-25,000 was obtained [118]. In these experiments the mobile phase composition, stationary phase, and total development time were identical and ten segments were used in the multiple development experiments.

Mixtures containing components spanning a wide polarity range cannot be separated using a single mobile phase since in any single mobile phase some components will be unretained or resident at the sample origin and, therefore, unresolved. Exponential, stepwise and continuous solvent gradients generated externally and applied by gravity to the TLC layer [9,125,128,129], or controlled by a pump in forced-flow TLC [130], have been used. However, the results from gradient elution are less predictable than those associated with column liquid chromatography due to the formation of multiple zones of different solvent composition resulting from solvent demixing. At the present state of instrumental development in TLC the multiple development technique provides a more certain approach to optimizing a gradient separation, particularly in the form of automated multiple development [9,15,131-134]. The distinctive feature of this method is that development starts with the most polar solvent (for the shortest development distance) and concludes with the least polar solvent (for the longest migration distance). The use of solvents of gradually decreasing solvent strength for the individual chromatographic steps leads to a step gradient whose effect corresponds to a linear solvent gradient, provided that a sufficient number of development segments are used. The separation potential of automated multiple development is illustrated by the chromatogram in Figure 7.13 [9]. A spot capacity of about 80 is claimed for this technique using a 10 cm HPTLC plate.

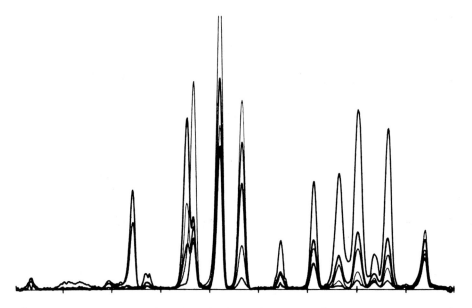

Figure 7.13 Separation of a test mixture using automated multiple development with a universal mobile phase gradient from acetonitrile through dichloromethane to carbon disulfide on a silica gel HPTLC plate. The chromatogram was scanned at different wavelengths to enhance the chromatographic information. (Reproduced with permission from ref. 9. Copyright Dr. Alfred Huethig Publishers).

It has proven difficult to develop a theoretical model for multiple development under capillary flow controlled conditions and too little work has been done using forced-flow conditions for any concrete conclusions to be reached [135]. A model has been proposed for the simplest of all cases; that of p developments with the same solvent for the same migration distance [135-137]. Under these specific conditions the zone width and the position of the zone in the chromatogram are given, to a first approximation, by

$$w_p = w_i (1 - R_{f2})^p \qquad (7.19)$$
$$R_{fa} = 1 - (1 - R_{f2})^p \qquad (7.20)$$

where R_{f2} refers to the R_f value of the faster moving of two zones, R_{fa} the apparent R_f value after p developments, w_p the zone width after p developments, and w_i the initial zone width. The condition for maximum resolution in p developments is given by

$$R_{f2} = 1 - [2/(p + 2)]^{1/2} \qquad (7.21)$$

TABLE 7.6

MAXIMUM RESOLUTION IN MULTIPLE DEVELOPMENT THIN-LAYER
CHROMATOGRAPHY

Number of Developments (p)	Optimum R_f		Relative Resolution(a)
	Single Development (R_f)	Position of band after multiple development (R_{fa})	
1	0.33	0.33	1.00
2	0.29	0.50	1.31
3	0.26	0.61	1.52
5	0.22	0.71	1.73
10	0.16	0.82	2.00

(a) R_s measured divided by R_s for a single development (R_f = 0.33)

Numerical values for maximum resolution using equation (7.28) are
given in Table 7.6. A fifty percent increase in resolution is
easily achieved by multiple development (p = 3). Further
significant increases in resolution are possible, but only when a
large number of developments are used. At maximum zone separation,
the average position of the zones will be constant, 0.632 (Z_f -
Z_0), and independent of the single development R_f value. The number
of developments required for this condition is given by

$$p_{max} = -1/\ln(1 - R_{f2}) \qquad (7.22)$$

where p_{max} is the number of developments producing maximum
separation. Since the maximum extent of separation increases as R_f
decreases, difficult to separate components should be repeatedly
irrigated with solvents that provide low R_f values (equivalent to
using the most selective solvent system for the separation).

7.4.4 Two-Dimensional Development

In two-dimensional TLC the sample is spotted at the corner
of the layer and developed along one edge of the plate
[113,117,138,139]. The solvent is then evaporated, the plate
rotated through 90°, and redeveloped in the orthogonal direction.
If the same solvent is used for both developments then the sample
will be redistributed along a line from the corner at which the
plate was spotted to the corner diagonally opposite. In this case
only a very slight increase in resolution can be anticipated (a
factor of the square root of two) corresponding to the increased

migration distance for the sample. The realization of a more efficient separation system implies that the resolved sample should be distributed over the entire plate surface. This can be achieved only if the selectivity of the separation mechanism is different in the orthogonal directions. Under these conditions the spot capacity for two-dimensional TLC can be very large, on the order of several thousand at the upper limit, and easily exceeds that of column systems, such as used in modern HPLC. The upper limit of separation performance, however, will depend on the mode of separation, whether by development or elution, and whether by capillary or forced-flow dominated conditions.

For two-dimensional TLC under capillary flow controlled conditions it should be possible to achieve a spot capacity, in theory, on the order of 100 to 250, but difficult to reach 400 and nearly impossible to exceed 500 [52,140]. Theoretical calculations indicate that by forced-flow development it should be relatively easy to generate spot capacities well in excess of 500 with an upper bound of several thousand, depending on the choice of operating conditions.

The spot capacity for two-dimensional TLC is less than the product of two unidimensional developments but is still considerably greater than the value for column chromatography. Two reasons contribute to the loss in separation potential in two-dimensional development. At the start of the second development the separated spots have increased in size due to the first development and are thus always larger than the initial starting size, this decreases the spot capacity in the second development compared to the first. Also, during the second development the spots spread laterally, and consequently, they must be separated with a resolution greater than unity at the beginning of the second development if they are to have a resolution of unity at the end.

Such high spot capacities can only be achieved if the separation mechanisms in the orthogonal directions are complementary to one another. In practice, several methods have been used to reach this goal and are briefly summarized in Table 7.7. The first approach listed, using a single sorbent and two solvent systems of different selectivity, is the easiest to implement in practice, particularly using chemically bonded sorbents with functional groups that allow normal, reversed-phase, and ion-exchange mechanisms to be exploited [9,74,140,142].

TABLE 7.7

POTENTIAL METHODS FOR GENERATING TWO DIFFERENT RETENTION
MECHANISMS IN ORTHOGONAL DIRECTIONS

---

(1) A single sorbent layer is used and alternately developed in
     orthogonal directions with two solvent systems exhibiting
     different selectivities for the sample components.

(2) A bilayer plate prepared from two sorbents with different
     selectivities can be used. The sorbent layer for the first
     development is a narrow strip that abuts the much larger area
     used for the second development. Commercially available plates
     have silica gel and revered-phase layers as adjacent zones.

(3) A layer containing an intimate mixture of two different
     sorbents can be used with different mobile phases such that
     the retention mechanism for the two, orthogonal, developments
     is governed by the properties of one of the sorbents in each
     direction.

(4) Between the first and second development the characteristics
     of the layer are modified by impregnation with a chemical
     reagent or immiscible solvent prior to the second development.

(5) Between the first and second development the properties of the
     sample are modified by chemical reaction or derivatization
     prior to the second development.

However, it is not a simple task to find appropriate solvent
systems that are able to distribute the sample over the whole
sorbent layer. It is more likely that sample components will
congregate along the diagonal and only a fraction of the potential
separation capacity will be reached. An alternative to mobile
phase optimization is stationary phase optimization using a
bilayer plate [139,142,143]. The larger area of the bilayer plate
used in the second development may be either a normal or reversed-
phase layer. To avoid distortion of the chromatogram it may be
necessary to transfer the sample from the narrow to the large
layer by using an intermediate development with a strong solvent
and refocusing of the sample at the interface between the two
layers. An alternative approach to using a bilayer plate is to
make a normal phase separation on silica gel with simultaneous
paraffin impregnation for a subsequent reversed-phase separation
on the paraffin impregnated layer in the second development [144].
Impregnating the sorbent layer with a selective reagent between
developments is also relatively easy to carry out. For example,
Grinberg and Weinstein separated a mixture of dansyl amino acid
derivatives into their enantiomeric forms by first separating the

derivatives in a reversed-phase system prior to development in the orthogonal direction after impregnating the sorbent with the chiral copper complex of N,N-di-n-propyl-L-alanine to resolve the enantiomers [145]. Similarly, silver nitrate impregnated silica has been used to affect a separation of saturated and unsaturated lipids after reversed-phase separation on a bilayer plate [146].

Several methods of computer simulation of two-dimensional chromatograms derived from unidimensional chromatographic data have been devised using either visual inspection or a mathematical separation function to judge the quality of the simulated chromatograms [113,142,143,147-149]. Since any spot in a two-dimensional chromatogram can be defined by a pair of x and y coordinates, the quality of a separation can be established by comparing the separation distance between all pairs of components in the chromatogram for any given combination of solvent systems and/or stationary phase layers. Separation distances are computed and not resolution, due to the difficulty of adequately describing the zone dimensions by a reliable mathematical relationship. To ensure that a certain minimum resolution is obtained for all spots the minimum useful separation distance can be defined as a default value and included as a component of the separation function. No single chromatographic function has been found ideal for evaluating all chromatograms but they serve a useful purpose in minimizing the number of simulated chromatograms that need be submitted to visual inspection [113,148,150,151]. In practice, plate to plate variations in the solvent velocity constant, layer activity, and solvent demixing and layer modification may adversely affect the agreement between simulated and experimental results. However, since computer simulation requires only a fraction of the time and materials required for performing actual separations this approach is very promising for aiding methods development by two-dimensional TLC.

The Achilles heel of two-dimensional TLC is the difficulty of recording and quantifying chromatograms. Slit-scanning densitometers designed for recording unidimensional chromatograms are not easily adapted to this problem [152,153]. Electronic scanning using image analysis technology is still under development and lacks the general sensitivity, wavelength range, and affordability of slit scanning densitometers [20,154-156]. It seems reasonable that in the future these problems will be solved, but for the present time the use of electronic scanning of TLC chromatograms is restricted to a few research laboratories. An apparatus for two-dimensional TLC under forced-flow conditions with elution chromatography for the second direction has been described [140,157,158]. The separated spots are eluted through a pair of quartz plates running along the complete length of the plate and opposite the solvent entry position. The quartz plates form the detection cell which is evenly illuminated with monochromatic UV light and focused onto a photodiode array detector. The separated spots can be reconstructed from the signal of the array as a three-dimensional plot of concentration against time with the plate length as abscissa. The detection and instrumental problems with this technique remain formidable and commercial instrumentation is unavailable at present. It seems unlikely that two-dimensional TLC will be more widely used, except in qualitative analysis, until the problems of detection can be solved by routine, low-cost instruments comparable in performance to slit-scanning densitometers in current use for recording unidimensional chromatograms.

7.5 MOBILE PHASE OPTIMIZATION

The selection of a mobile phase for the separation of simple mixtures may not be a particularly difficult problem and can be arrived at quite quickly by trial and error. Solvent systems can be screened in parallel using either several development chambers or a device like the Camag Vario KS chamber, which allows the simultaneous evaluation of a number of solvents by allowing each of these to migrate along parallel channels scored on a single TLC plate [8]. However, whenever the number of components in a mixture exceeds all but a small fraction of the spot capacity for the TLC system, a more systematic method of solvent optimization is required.

Given the similarity in the retention mechanisms it is hardly surprising that the principal methods of solvent selection in TLC are identical to those followed in column liquid chromatography. Since the solvent used for the separation is evaporated prior to detection, a wider range of UV absorbing solvents are commonly used in TLC than is the case for HPLC. Solvents must be of high purity, since involatile impurities remain sorbed to the layer causing sloping and unstable baselines in scanning densitometry [19,159]. Stabilizers and antioxidants commonly added to certain kinds of solvents (e.g., ethers, tetrahydrofuran, dioxane) are the most frequent problem. The most significant difference between column and planar liquid chromatographic methods relates to the non-equilibrium state that occurs during separations by TLC. In TLC the mobile phase encounters a dry layer that selectively adsorbs the solvent of highest affinity, denuding the advancing mobile phase of this component and producing a solvent strength gradient in the direction of development (see section 7.2.1 for more details). Demixing, as it is called, may be complete, resulting in the formation of zones with sharp boundaries separating the chromatogram into sections of different solvent composition and, therefore, selectivity. These considerations hinder optimization strategies based on the composition of the solvent added to the developing chamber.

Solvents used in liquid chromatography are characterized by their strength and selectivity based on experimental organizational schemes such as those proposed by Snyder. Thus solvents can be assigned to one of several selectivity groups based on their characteristic intermolecular interactions and ranked in terms of their ability to competitively migrate samples of different types in a chromatographic system, described by a solvent strength parameter (see section 4.6 for details). In TLC, a solvent of the correct strength for a unidimensional development will migrate the sample into the R_f range 0.2-0.8, or thereabouts, and if of the correct selectivity, will distribute the sample evenly throughout this range, or meet some other arbitrary resolution criterion established for the particular separation. Solvents of the same strength but different selectivity are prepared by choosing solvents from different selectivity groups and diluting them to the correct volume fraction with a weak solvent according to their characteristic solvent strength parameter, as explained for column liquid chromatography (section 4.6). The further mixing of these solvents to produce ternary and quaternary solvent mixtures of the same strength provides additional fine tuning of the selectivity. The use of these guided trial and error procedures are reviewed elsewhere [8,112], and shortly we will look at one of these models, the PRISMA model, in some detail, as representative of the general approach most often been used in TLC. Computer-aided strategies for mobile phase optimization in TLC using window diagrams [142,160], overlapping resolution maps [161,162], simplex methods [9,163-165], and pattern recognition procedures [166,167] have been rarely used in TLC compared to HPLC. In these procedures some form of statistical design is used to select a group of solvents for evaluation, or alternatively, the results obtained from the arbitrary selection of a series of solvents are compared to indicate the best separation obtained. Comparisons are made based on the selection of a mathematical function to define the quality of a separation by a numerical index, or a technique, such as overlapping resolution maps, is used for visual evaluation. To use these methods effectively a fairly large number of data points (mobile phase compositions) are required and the choice of the separation function can lead to the prediction of different optimum mobile phases, which may be confusing. However, for the optimization of complex mixtures such approaches seem warranted, and are likely to be more fruitful than the trial and error procedures more frequently adopted. Prior to such methods becoming more widely used it will be necessary to demonstrate their greater utility in TLC, and to provide a greater degree of standardization for routine application.

An important difference between the statistical mixture design techniques popular in HPLC and the PRISMA model is that the former yields a computed optimum solvent composition while the latter relies on a structured trial and error approach, which is readily adaptable to TLC. Solvent changes and re-equilibration in HPLC can be quite time consuming, so that it becomes attractive to minimize the number of experiments, while for TLC, experiments can be performed in parallel and time constraints are less significant. Changes in solvent strength are also more rapidly adjusted empirically within the PRISMA model when theoretical considerations are found inadequate or require modification due to differences in the experimental approach.

The PRISMA model was developed by Nyiredy for solvent optimization in TLC and HPLC [142,168-171]. The PRISMA model consists of three parts; the selection of the chromatographic system, optimization of the selected mobile phases, and the selection of the development method. Since silica is the most widely used stationary phase in TLC, the optimization procedure always starts with this phase, although the method is equally applicable to all chemically bonded phases in the normal or reversed-phase mode. For the selection of suitable solvents the first experiments are carried out on TLC plates in unsaturated

TABLE 7.8

SOLVENT STRENGTH PARAMETERS FOR SOLVENTS COMMONLY USED IN NORMAL
PHASE TLC

Solvent	Selectivity Group	Solvent Strength
n-Butyl Ether	I	2.1
Diisopropyl Ether		2.4
Methyl tert.-Butyl Ether		2.7
Diethyl Ether		2.8
n-Butanol	II	3.9
2-Propanol		3.9
1-Propanol		4.0
Ethanol		4.3
Methanol		5.1
Tetrahydrofuran	III	4.0
Pyridine		5.3
Methoxyethanol		5.5
Dimethylformamide		6.4
Acetic Acid	IV	6.0
Formamide		9.6
Dichloromethane	V	3.1
1,1-Dichloroethane		3.5
Ethyl Acetate	VI	4.4
Methyl Ethyl Ketone		4.7
Dioxane		4.8
Acetone		5.1
Acetonitrile		5.8
Toluene	VII	2.4
Benzene		2.7
Chloroform	VIII	4.1
Water		10.2

chambers with 10 solvents, chosen from the different selectivity
groups of Snyder, and underlined in Table 7.8 [171]. After these
first TLC experiments with single solvents, the solvent strength
has either to be reduced or increased so that the substance zones
are distributed in the R_f range 0.2-0.8. If the substances migrate
into the upper third of the plate the solvent strength has to be
reduced by dilution with hexane, the strength adjusting solvent
(solvent strength = 0). If the substances remain in the lower
third of the plate with the single solvents their solvent strength
has to be increased by the addition of water or acetic acid. A
similar procedure is followed in the reversed-phase mode except

that solvent selection is limited to water miscible solvents, the empirically derived solvent weighting factors are used in place of Snyder's values, and water is used as the strength adjusting solvent (solvent strength = 0). Other solvents than those indicated as preferred solvents can be tested at this point to obtain the best possible separation. From these trial experiments, those solvents showing the best separation are selected for further optimization in the second part of the model. If it is anticipated that forced-flow development will be used for the separation then it is advantageous to include one solvent in the selected solvent combinations in which the sample does not migrate [169]. This solvent can be used in a prerun to eliminate the disturbing zone caused by undissolved gases in the sealed layer, which have a detrimental effect on the separation.

Between two and five solvents can be selected for construction of the PRISMA model. Modifiers such as acids, ion-pair reagents, etc., can be added to improve the separation and reduce tailing. Modifiers are generally used in low and constant concentration so that their influence on solvent strength can be neglected. The actual PRISMA model, Figure 7.14, is a three dimensional geometrical design which correlates the solvent strength with the selectivity of the mobile phase [170]. The model consists of three parts: the base or platform representing the modifier; the regular part of the prism with congruent base and top surfaces; and the irregular truncated top prism (frustum). The lengths of the edges of the prism (S_A, S_B, S_C) correspond to the solvent strengths of the neat solvents (A, B and C). Since different solvents usually have different solvent strengths, the lengths of the edges of the prism are generally unequal and the top plane of the prism will not be parallel and congruous with its base. If the prism is cut parallel to its base at the height of the lowest edge (determined by the solvent strength of the weakest solvent, solvent C in Figure 7.14), the lower part gives a regular prism, where the top and planes are parallel equilateral triangles. The upper frustum of the model is used for mobile phase optimization of polar compounds in normal phase chromatography, while the regular part is used for the separation of nonpolar and moderately polar substances. For reversed-phase chromatography, the regular part of the prism is used to optimize the separation of both polar and nonpolar substances.

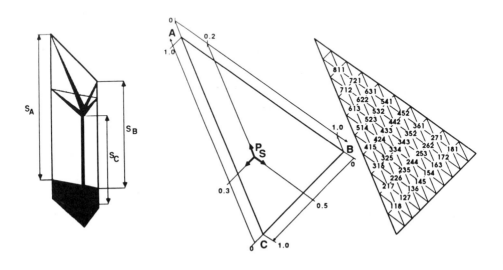

Figure 7.14 The PRISMA mobile phase optimization model showing the construction of the prism and the selection of selectivity points (Reproduced with permission from ref. 170. Copyright Marcel Dekker, Inc.)

For polar compounds optimization is always started on the top irregular triangle of the model, either within the triangle, when three solvents are selected, or along one side, when two solvents are selected. In the triangle shown in Figure 7.14, a certain solvent composition (P_s) can be characterized by the volume fractions corresponding to its axes, in the case shown for P_s, this corresponds to 30% solvent A, 20% solvent B, and 50% solvent C, which can be described by the three digit number 325. P_s is a selectivity point, and all solvent compositions on the surface of the triangle can be described by three-coordinate selectivity points. Optimization is commenced by selecting solvent combinations corresponding to the center point P_s = 333 and three other points close to the apexes of the triangle P_s = 811, 181 and 118. If the separation obtained is insufficient other selectivity points are tested around the solvent combination that gave the best separation. On changing the selectivity points on the top triangle the solvent strength changes as well, especially when the solvent strengths of the solvents used to construct the prism differ considerably. The strength of the solvent should be adjusted with the strength adjusting solvent to maintain the separation in the optimum R_f range. It may also be advisable to change the selectivity points by small increments if regular step sizes cause large changes in resolution.

The regular center portion of the prism is used to optimize the mobile phase composition for the separation of nonpolar and moderately polar samples. The initial solvent composition corresponds to the center of the triangular top face of the regular prism; this composition is then diluted to bring all sample components into the R_f range 0.2-0.8. At this solvent strength three more chromatograms are run corresponding to the selectivity points close to the apexes of the triangle. These initial runs are then used to choose selectivity points for further chromatograms until the best solvent composition is located. Failure to obtain the beginning of a separation requires that a new prism be constructed, using a different solvent for at least one of the edges.

The optimization of the solvent strength by varying the selectivity points has to be carried out until at least a beginning separation is obtained. At this point the third part of the PRISMA model can be used to select the appropriate development mode. If an increase in efficiency is required to improve the overall resolution of the sample then forced-flow linear development should be used. If the separation problem exists in the upper R_f range then anticircular development may be the best choice, if in the lower R_f range, then circular development is favored. If at this point the separation is still inadequate then a new stationary phase or additional solvents must be selected and the PRISMA model utilized again to optimize the new system. Nyiredy et al. [171] have summarized the above procedure in the form of decision flow charts which are useful guides for implementing all steps of the model.

7.6 QUANTITATIVE EVALUATION OF THIN-LAYER CHROMATOGRAMS

At best, inspection by eye of a TLC plate is capable of detecting about 1-10 micrograms of colored components with a reproducibility rarely better than 10-30%. Excising the separated spots, eluting the substance from the sorbent material, and measurement by solution photometry are time consuming and fairly inaccurate. Difficulties in accurately locating the spot boundary by eye, incomplete elution of the sample from the sorbent, and non-specific background absorbance due to colloidal sorbent particles in the analytical solution add to the problem. The above process is too tedious to be an acceptable detection method for a high speed chromatographic technique. Instrumentation for in situ measurements of TLC chromatograms first appeared in the mid-1960s and is now considered essential for the accurate determination of both spot size and location, for a true measure of resolution, and for rapid, accurate quantitation. Modern versions of these early densitometers have evolved into sophisticated, automated, computer-controlled devices.

In situ measurements of substances on TLC plates can be made by a variety of methods: reflectance, transmission, simultaneous reflectance and transmission, fluorescence quenching and fluorescence [7,13,19,20,23,172-174]. All optical methods for the quantitative evaluation of TLC chromatograms are based upon measuring the difference in optical response between a blank portion of the medium and regions where a separated substance is present. For a given compound, measurements in the transmission mode provide greater peak heights, up to a factor of two, accompanied by significantly higher baseline noise when compared to reflectance [175]. In the reflectance mode most of the scattered light arises from layers close to the surface and is influenced less by changes in the thickness of the TLC medium, which generates much of the background noise in transmission measurements. Transmission measurements are limited to those wavelengths greater than 320 nm due to strong absorption by the glass backing plate and by silica gel itself at shorter wavelengths. Reflectance measurements can be made at any wavelength from the UV to the near infrared (185-2500 nm).

UV-absorbing compounds can be measured by fluorescence quenching as well as by reflectance. The fluorescence quenching technique provides a means of visualizing spots absorbing UV-light on special TLC plates incorporating a fluorescent indicator. When such a plate is exposed to UV light of short wavelength, the UV-absorbing spots appear dark against the brightly fluorescing background of a lighter color. The UV-absorbing spots behave in a similar way to an optical filter, absorbing a portion of the fluorescence excitation radiation and thus diminishing the fluorescence emission intensity. The fluorescence quenching method can only be applied to those substances whose absorption spectra overlap the excitation spectrum of the fluorescent indicator. The fluorescent indicators in common use have a maximum absorption around 280 nm and virtually no absorption below 240 nm. Fluorescence quenching measurements are thus less specific and less sensitive than absorption measurements. Fluorescence quenching is mainly used as a visualization technique in qualitative analysis.

When monochromatic light falls on an opaque medium some light is reflected from the surface, some absorbed by the medium and dissipated in some way, such as by conversion to heat, and the remainder is diffusely reflected or transmitted by the medium. For quantitation it is the diffusely reflected/transmitted component that is of importance and it must be assumed that the specularly reflected component from the sorbent surface is very small. The specularly reflected component contributes to the background signal at the detector (noise) but carries no indication of the properties of the sample. The propagation of light within an opaque medium is a very complex process that can only be solved mathematically if certain simplifying assumptions are made [20,23,172,176,177]. The most generally accepted theory is due to P. Kubelka and F. Munk, which is often expressed in the following form

$$(1 - R^*)^2/2R^* = 2.303aC/S \tag{7.23}$$

where R^* is the reflectance for an infinitely thick opaque layer, a the molar absorption coefficient of the sample, C the sample concentration, and S the scatter coefficient for the layer. Equation (7.23) is only an approximation for thin layer plates, since it is derived explicitly for a layer of infinite thickness. However, it serves to illustrate the general properties of a solid-sorbent matrix on the observed sample response and explains why calibration curves obtained on TLC plates do not obey Beer-Lambert's Law. It predicts, for example, a nonlinear relationship between signal and sample concentration in the absorption mode, an increase in response with larger molar absorption coefficients, and an increase in response for sorbents having a low scatter coefficient. Since equation (7.23) is only approximate it is not used directly for quantitation in TLC. The principal method of quantitation is by calibration using a series of standards spanning the concentration range of the sample to be determined. As illustrated in Figure 7.15, calibration curves are individual in shape, generally comprise a pseudolinear region at low sample concentration curving towards the concentration axis at higher concentration, and eventually reach some asymptotic value where signal and sample concentration are no longer correlated. The extent of the individual ranges, or sections of the calibration curve, are frequently very different for different substances and, while in some instances the pseudolinear range may be adequate for most analytical purposes, in others no suitable linear range may exist. The lack of general linearity of calibration curves has promoted the use of several mathematical transformation techniques to linearize the nonlinear portions of the calibration curves [20,173,178,179]. Typical examples include the conversion of the sample and/or signal into reciprocals, logarithms, squared terms, or use of the Michaelis-Menten function. None of these methods has a sound theoretical basis for treating calibration curves in TLC and can lead to the propagation of significant errors in the transformed data. Their general use, therefore, cannot be recommended without adequate testing. Alternatively, nonlinear polynomials described by equations (7.24) and (7.25) can be used without disturbing the error distribution.

$$\ln R = A_o + A_1(\ln C) + A_2(\ln C)^2 \qquad (7.24)$$

$$R = A_o + A_1C + A_2C^2 \qquad (7.25)$$

Fluorescence measurements are fundamentally different to absorption measurements [20,173,180]. The fluorescence intensity depends only on the population of sample molecules and can be calculated in several ways. Independent of the method chosen at low sample concentrations the fluorescence signal, F, is adequately described by equation (7.26)

$$F = \phi I_o abC \qquad (7.26)$$

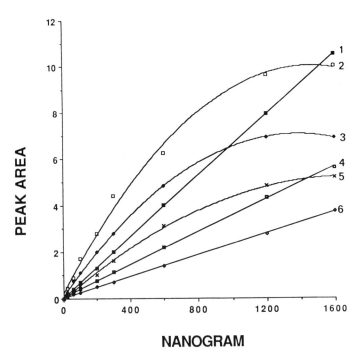

NANOGRAM

Figure 7.15 Some typical calibration curves for several substances measured by absorption in the reflectance mode. Substance
identification: 1 = practolol; 2 = azobenzene; 3 = diphenyl-
acetylene; 4 = alprenolol; 5 = estrone; and 6 = pamatolol.
(Reproduced with permission from ref. 20. Copyright Elsevier
Scientific Publishing Co.)

where ϕ is the quantum yield, I_0 the intensity of the excitation
source, a the absorption coefficient, b the thickness of the TLC
layer, and C the sample amount. Since all terms in equation (7.26)
are constant, or fixed by the experiment, the fluorescence
emission is linearly dependent on the sample amount. The
fluorescence signal is also independent of the spot shape provided
that the spot is completely contained within the measuring beam.
Calibration curves in fluorescence are usually linear over two or
three orders of magnitude. At larger sample sizes self-absorption
becomes a problem and calibration lines curve over towards the
weight axis. An alternative method of calibration, the two-point
calibration method, has been suggested for use in screening
studies where several components are to be quantified [121,181].
This method requires only a single standard for each substance to
be determined, and only a single lane for all calibration

standards. It thus preserves the high sample throughput of TLC by minimizing the number of lanes occupied by standards. It is based on the observation that there is a linear relationship between slit width dimensions in fluorescence densitometry [182] and that the slope of the detector response versus slit width curve is proportional to the sample amount [181]. Combining these two observations yields the calibration equation (7.27)

$$M_u = M_s (S_u/S_s) \qquad\qquad (7.27)$$

where M_u is the unknown sample amount, M_s, the known amount of sample (standard), and S_u and S_s the slope of the detector response versus slit width curve for the unknown and standard, respectively. The values for S_u and S_s are determined from two measurements of the signal response at slit widths of 0.4 and 0.8 mm. Thus, in practice, calibration is performed by scanning each sample and the lane occupied by standards twice at the selected values for the optimum slit widths. The two-point calibration method cannot be used to replace normal calibration when the highest degree of accuracy is required. It is used primarily to determine approximate sample amounts in large scale screening programs.

On occasion it will be observed that the fluorescence response for substances in TLC will be far less than expected from measurements made in solution, will occur at different excitation and emission wavelengths than in solution, and may diminish over time at varying rates [10,19,20,173,180]. Adsorption onto the sorbent layer provides additional nonradiative pathways for the dissipation of the excitation energy, which is most likely lost as heat to the surroundings, reducing the observed fluorescence signal. Quenching of the signal by interaction with oxygen or reaction of the solute with oxygen to produce new products with a diminished fluorescence yield are other known mechanisms affecting the magnitude and constancy of the fluorescence response with time of adsorbed fluorescent solutes. There are several practical consequences of these interactions that can influence the reliability of fluorescence densitometry for TLC measurements.

The extent of fluorescence quenching often depends on the sorbent medium and is generally more severe for silica gel than for chemically bonded sorbents [183]. In many cases the emission signal can be enhanced by application of a viscous liquid to the layer before scanning the plate. Common fluorescence enhancing reagents are liquid paraffin, Triton X-100 (an isooctylphenoxy-polyethoxyethanol) and Fomblin (a poly[perfluoroalkyl ether]) [19,20,180,183-185]. The general mechanism of fluorescence enhancement is assumed to be dissolution of the sorbed solute with enhancement in response due to the fraction of solute that is transferred to the liquid phase where fluorescence quenching is less severe. In favorable cases the application of a fluorescence enhancing reagent can increase the signal response by as much as 10- to 200-fold. In those cases where fluorescence quenching is negligible the application of a fluorescence enhancing reagent rarely produces a significant change in response.

Oxygen is a ubiquitous fluorescence quenching reagent and is difficult to exclude from the plate surface during scanning densitometry. Flooding the scanning stage with nitrogen is possible [186] but it is generally more practical to impregnate the sorbent layer with an antioxidant, such as BHT (2,6-di-tert-butyl-4-methylphenol) prior to development, and to add antioxidant to the mobile phase to avoid displacement of antioxidant from the lower portion of the plate by the mobile phase [19,20,183,187]. Some samples undergo rapid photoxidation reactions requiring the use of an antioxidant and precautions to shield the sample on the plate from room light. Fortunately, the above problems are not severe in most cases, and only those substance that are unusually susceptible to catalytic oxidation or oxygen quenching require special handling.

7.7 INSTRUMENTATION FOR THIN-LAYER CHROMATOGRAPHY

TLC measurements are performed according to two divergent philosophies with different goals and expectations in mind. For some quality control applications, reaction monitoring, etc., where only simple mixtures are encountered and only approximate quantitation is expected, the equipment needs are minimal. Samples are applied to the layer with disposable micropipettes, beakers or some similar device can serve as a developing chamber, and detection can be performed by visual inspection after application of a visualizing reagent, if required. The alternative approach attempts to take advantage of the full separation capacity of TLC and requires that quantitation be both accurate and reliable. This form of TLC is completely instrumentalized and forms the basis of this section [7,13,15,172,188,189].

7.7.1 Sample Application

Samples are applied to TLC plates as spots or bands to conform to the demands of minimum size and a homogeneous distribution of sample within the starting zone [7,188,180]. For HPTLC plates, where desirable starting spot sizes are about 1.0 mm, this corresponds to a sample volume of 100 to 200 nl. For conventional TLC plates sample volumes 5 to 10-fold greater are acceptable. The sample solvent must be a good solvent for the sample to promote quantitative transfer from the sample application device to the layer. Also, it must be of low viscosity and sufficiently volatile to be easily evaporated from the plate. Further, it must wet the sorbent layer adequately otherwise penetration of the layer will not occur. In addition it should be a weak chromatographic solvent for the sample on the sorbent selected for the separation. Ideally the least retained sample component should have an R_f < 0.1 if the sample solvent was used as a mobile phase for chromatographic development. With the correct instrumentation few problems are encountered with silica gel, but for reversed-phase layers, which are not wet completely by some aqueous solvent mixtures, it may be necessary to use polar organic solvents for sample application, and to modify the general application procedure to minimize predevelopment problems.

Manual devices for sample application are inappropriate if scanning densitometry is to be used for detection. First of all, the starting position of each spot must be known precisely. This is most easily achieved with mechanical devices operating to a precise grid mechanism. The sample must be applied to the layer without disturbing the surface, something that is near impossible to achieve using manual application. When using dosimeters, samples and standards must be applied in the same solvent and identical volumes to ensure that identical starting zones are formed [10]. Calibration curves should not be prepared by spotting incremental volumes from a single standard for the above reason. Poor linearity and non-zero intercepts are commonly observed when calibration standards are spotted in variable volumes from a fixed concentration standard.

For most quantitative work in TLC the sample is applied to the layer using a fixed-volume dosimeter comprising a platinum-iridium capillary of 100 or 200 nl volume, sealed into a glass support capillary of larger bore [6]. The capillary tip is polished to provide a smooth planar surface of small area (ca. 0.05 mm^2), which is brought into contact with the plate surface using a mechanical device to discharge its volume. These dosimeters are unsuitable for hand spotting, since this invariably causes damage to the layer and frequent clogging of the dosimeter with sorbent particles from the layer. Mechanical application of the sample is made possible by attaching a metal collar to the glass support capillary so that it can be held by a magnet and lowered to the plate surface under controlled conditions. An example of this type of sample applicator is the Nanomat III shown in Figure 7.16. A spring mechanism allows the applicator head to be lowered and lifted from the plate surface while the frictional forces holding the dosimeter to the applicator head control the pressure with which the dosimeter engages the layer. A click-stop grid mechanism is used to aid in the even spacing of the samples on the plate and to provide a frame of reference for sample location during scanning densitometry. An automated sample spotting device has been developed which uses a flexible fused silica capillary tube as the applicator and a motor driven syringe to suck up and deposit on the layer sample volumes in the range of 100 nl to 20 microliters as spots or bands [15]. Controlled by a microprocessor, it can be programmed to select samples from a rack of vials and deposit fixed volumes of the sample, at a controlled rate, to selected positions on the plate. The applicator automatically rinses itself between sample applications and can spot a whole plate with different samples and standards without operator intervention. Automated multi-spotters are also available for simultaneous sample application.

As an alternative to the dosimeter, samples can be spotted with a microsyringe [7]. For accurate dispension of nanoliter volumes, preselected in the range 50-230 nl, the syringe is controlled by a micrometer screw gauge. A fixed lever mechanism is provided for the repetitive application of a constant sample volume. The microsyringe delivers the sample volume by displacement as opposed to capillary action and, therefore, does not deform the layer surface. The microsyringe needle is brought only close enough to the plate surface for the convex sample drop of the ejected liquid to touch the layer surface.

Samples may be applied as bands to TLC plates using either specially prepared plates with a concentrating zone (see section 7.3.1) or by using a band applicator such as the Linomat IV,

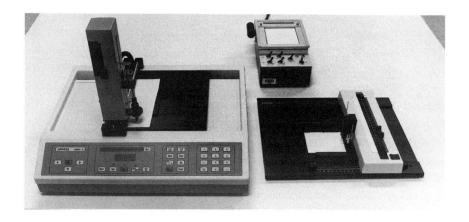

Figure 7.16 Apparatus for sample application in modern TLC. In the foreground is shown the Nanomat III spot applicator and the Linomat IV band applicator, and in the background, the Transpot contact spotter.

Figure 7.16 [15]. Band applicators work by mechanically moving the plate on a stage beneath a fixed syringe. A controlled nitrogen atomizer sprays the sample from the syringe, forming narrow, homogeneous bands on the plate surface. The plate is moved back and forth under the atomizer to apply bands of any length between 0 (spots) and 190 mm and sample volumes between 1-500 microliters. The rate of sample deposition is also controllable. The advantages of band application are that for quantitative scanning densitometry the sample band can be made longer than the slit length of the light source, which minimizes errors due to positioning of the sample within the light beam, and different volumes of a single standard concentration can be applied for calibration. Also, the standard addition method of quantitation is easily carried out by overspraying the sample already applied to the layer with a solution of the standard. The primary disadvantages of band application relate to the time required to apply a sample compared to other techniques and the additional time consumed in cleaning of the syringe between applications.

Any of the methods discussed so far are suitable for spotting solutions of low viscosity, or viscous solutions by repetitive application after dilution. Samples from environmental and biological sources, however, invariably yield viscous residues; dilution and repetitive sample application can be tedious and time consuming, or impossible for these samples. The contact spotter, Figure 7.16, provides a simple means of spotting viscous samples and samples of large volume, up to 100 microliters, onto HPTLC plates. This apparatus is designed for the solventless sample application of evaporated residues of several samples simultaneously at precise locations on the thin-layer plate [191,192]. The transfer medium is a fluorinated polymer film, coated with perfluorokerosene, positioned over a series of depressions in a metal platform. The film is forced to conform to the shape of the platform surface by applying slight vacuum through small holes in each depression. The sample solutions are pipetted into these depressions, and the solvent is evaporated by gentle heat and a flow of nitrogen. The HPTLC plate is then positioned over the film, sorbent side down, and with slight pressure replacing the vacuum, the spots are transferred simultaneously to the plate. Under conditions of low humidity, static charges may develop on the fluorocarbon film, causing movement of the samples out of the concave depressions on the surface of the spotter. This results in misalignment of the spots on the surface of the plate. The static charge can be eliminated by wiping both surfaces of the film with antistatic paper moistened with ethanol before applying the samples, or by using a variety of readily available antistatic guns. Crystalline samples do not penetrate the sorbent layer, and they must be transferred in a small volume of a nonvolatile carrier solvent, such as octanol, dodecane, acetophenone, etc., which may be added to the sample solution prior to evaporation to give a final volume of about 5-20 nl, when the residue is transferred. A small amount of a colored dye, that does not migrate in the chromatographic system, can be added to each sample prior to evaporation to mark the exact position of the sample lanes for scanning densitometry.

7.7.2 Development Chambers

Development chambers commonly used in TLC can be divided

into two categories [8]. Normal chambers, abbreviated to N-type, have a depth of gas phase in front of the layer greater than about 3 mm. Chambers with a depth of gas space in front of the layer less that 3 mm are called small chambers or sandwich chambers, abbreviated to S-type. Either type may be further characterized by whether the gas phase is saturated or unsaturated with vapors of the mobile phase, indicated by adding a subscript s or u to the chamber designation. For conventional TLC, large volume chambers lined with paper that has been saturated with mobile phase, are routinely used. Such chambers are adequate for qualitative work but fail to provide the controlled environment necessary to obtain reproducible R_f values. Equilibration of the vapor phase with the sorbent layer of the plate is easily achieved in the compact sandwich chamber [9]. A sandwich chamber is formed by clamping the TLC plate and a glass plate together, separated by a U-shaped spacer a few millimeters thick. Development is carried out by placing the open end of the chamber in a solvent trough containing the mobile phase.

The twin-trough chamber, Figure 7.17, is the simplest of the HPTLC developing chambers [193]. It consists of a standard developing tank with a raised, wedge-shaped bottom to minimize solvent consumption. The wedged bottom divides the tank into two compartments, so that it is possible to either develop two plates simultaneously, or to use one compartment to precondition the sorbent layer prior to development [79]. The plate is placed in one compartment and the mobile phase or the preconditioning solvent in the other. Examples of sorbent preconditioning procedures that are easily carried out in the twin-trough chamber are control of humidity with water-sulfuric acid mixtures and plate deactivation with aqueous ammonia solutions for the separation of basic components. The mobile phase can be added directly to the plate compartment or, when the mobile phase is first used for preconditioning, the chamber can be tilted to allow the transfer of solvent into the compartment for development. The twin-trough chamber is widely used for routine quality control applications.

The horizontal developing chamber is a sandwich-type chamber for the simultaneous development of samples from both ends toward the plate center, or for conventional edge-to-edge development [194]. The mobile phase is transported from the reservoir to the

Figure 7.17   Developing chambers used for HPTLC. Right, short-bed continuous development chamber and left, twin trough chamber.

sorbent layer by two glass slides; the liquid rises by surface tension and capillary forces. The mobile phase then travels through the sorbent layer by capillary action. When a plate is developed from both edges simultaneously, the chamber must be leveled to allow the two solvent fronts to migrate at the same speed and meet precisely in the middle. At this point the capillary forces balance out and the chromatographic development ceases. The sandwich configuration of the horizontal developing chamber is not suitable for all solvents. Those that contain volatile acids or bases, and for mixtures with a comparatively larger concentration of a volatile polar solvent, such as methanol or acetonitrile, should not be used [15]. This is because of the restricted access of the more or less saturated gaseous phase with the unwetted region of the layer in the sandwich configuration; the tank configuration will normally be superior when the vapor phase modification of the sorbent layer is important for obtaining the desired separation.

Related to the linear development chamber is the Vario-KS chamber [8,195]. The plate is also developed in the horizontal position in a sandwich configuration. However, this chamber is supplied with a variety of conditioning trays, divided into lanes or square segments, and up to six mobile phase reservoirs. Thus, the plate may be segregated into six environmentally separated lanes that can be developed with six different mobile phases simultaneously. The activity of the sorbent layer can be adjusted as desired for each of the six sample lanes independently, or any desired activity gradient can be formed in the direction of mobile phase migration for each lane. The Vario-KS chamber is used primarily for scouting the conditions necessary for optimum separation of a mixture and also to assess the influence of preloading of the sorbent layer with different solvent vapors.

The short-bed continuous development chamber (SB/CD chamber) is designed specifically for use with continuous and multiple development techniques, Figure 7.17. The chamber has a low profile and a wide base to permit development close to the horizontal position and to minimize the chamber volume. The base contains four glass ridges running nearly the entire length of the chamber. The four ridges, along with the back wall of the chamber, provide five positions which permit a choice of the plate length used for a particular separation. One end of the HPTLC plate protrudes from the chamber, enabling continuous evaporation of the mobile phase at the junction formed by the plate surface and the cover lid of the chamber.

The U-chamber is designed for optimal circular HPTLC development [6,110]. The mobile phase velocity is electronically controlled by a stepping motor that drives a captive syringe. The syringe feeds the mobile phase directly onto the horizontal plate via a platinum-iridium capillary. The chamber volume is miniaturized to provide optimum conditions for the saturation of the vapor phase in contact with the plate. The sorbent layer can be equilibrated prior to, and also during development, with any desired volatile vapor. An injection valve is provided to introduce the sample onto the plate surface via the mobile phase. In this mode, the chromatogram consists of a series of concentric rings of varying circumference. A circular heating block is available as an accessory to evaporate the mobile phase at the solvent front for use in the continuous development mode.

The anticircular development chamber contains many of the features found in the U-chamber. The plate is developed horizontally, the chamber volume is small, and the vapor phase can be controlled by external means [111]. An outer circle is cut from the adsorbent layer to allow the solvent to migrate in only the desired direction. This adsorbent-free annulus provides a barrier that prevents solvent migration towards the outer edge of the plate. A very narrow circular channel in direct contact with the solvent reservoir is contacted with the sorbent layer sealing the chamber to start the development process. The mobile phase moves by capillary forces through the bed and cannot be controlled by external means. Mobile phase consumption in both the circular and anticircular development modes is very low.

The AMD device for automated multiple development is shown in Figure 7.18 [9,15,117,131]. The separation sequence generally involves 1 to 25 individual developments with each development being longer than the previous one by about 3-5 mm. Between development, the solvent is completely removed from the developing chamber and the layer is dried under vacuum. Also, provision is made for conditioning the layer with an atmosphere of controlled composition prior to each development. The distinctive feature of this method is that development starts with the most polar solvent (for the shorter development distance) and concludes with the least polar solvent (for the longest development distance). The solvent composition can be changed for every segment in the development program and is generally altered for a large number of the development segments. The use of solvents of gradually decreasing solvent strength for the individual chromatographic steps leads to a step gradient whose effect corresponds to a linear solvent gradient, provided that a sufficient number of development segments are used. Thus the AMD technique is suitable for the separation of mixtures with components differing widely in their polarity and for the identification of suitable solvent systems for those samples poorly resolved using gradient techniques.

Although there are theoretical advantages to forced flow development in TLC (see section 7.2.1), the design of instrumentation to take advantage of this operating mode is more complicated. Two general approaches have been used; the mobile phase velocity can be controlled in an open system using centrifugal forces or the layer can be sealed and the mobile phase pumped through the layer using a mechanical pump. The Rotachrom device (Petazon, Zug, Switzerland) uses centrifugal forces created by spinning the plate around a central axis to move the solvent through the layer [9,10,36]. The rotation speed can be varied from

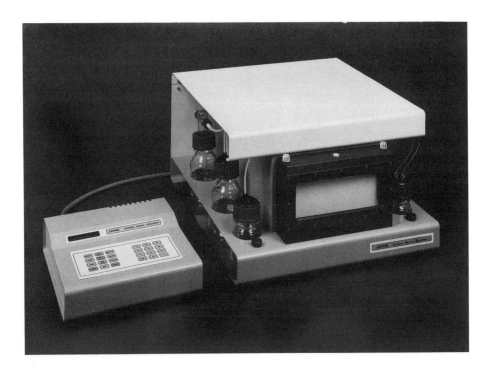

Figure 7.18   Apparatus for automated multiple development.
(Courtesy of Camag Scientific, Inc.)

80 to 1,500 rpm. By scrapping out zones in the layer, either
circular, anticircular or linear development is possible. Both
analytical and preparative separations are possible, with thicker
layers being used to increase sample throughput. A collection
system fixed to the rotor allows separated fractions to be
conveniently collected as they leave the layer in the circular
development mode. The Chromatotron device (Harrison Research, Palo
Alto, CA, USA) is a simplified version of the apparatus described
above for preparative, circular, centrifugal TLC [9,196].

Devices for chromatographic development with the vapor space
above the layer eliminated by using a pressurized sandwich
configuration include the Chrompres 25 (Labor Mim, Budapest,
Hungary) overpressure development chamber, Figure 7.19, and the
HPPLC 3000 (Institute fur Chromatographie, Bad Durkheim, FRG) high
pressure planar liquid chromatograph. Both devices represent
different philosophies. The HPPLC 3000 is designed for circular or

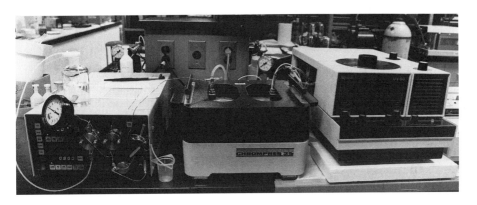

Figure 7.19   Chrompes 25 overpressure development chamber for
forced-flow TLC. The configuration shown is for on-line detection.

anticircular development of 10 x 10 cm HPTLC plates under very
high pressure, created by a hydraulic press, with the mobile phase
applied to the layer by a mechanical pump [9,35,38,197].
Separations are very fast, perhaps only requiring a few seconds
for the separation of simple mixtures. Samples may be applied to
the layer as a series of spots surrounding the solvent entry
position or injected by a valve, similar to HPLC, in which case a
single sample is separated as a number of concentric rings. The
Chrompres 25 operates at much lower pressures, up to about 25
atms., in which the layer is sealed by an elastic membrane
pressurized on one side by pumping water into a chamber directly
above it [10,35,37]. A second eluent pump, operating at pressures
below that created by the water cushion, delivers a constant
volume of mobile phase to the layer. A slit in the plastic sheet
in contact with the layer at the entrance and exit position of the
mobile phase on the layer provides an even flow of mobile phase
across the layer and allows operation in the continuous
development mode with on-line detection using a standard HPLC
detector, if desired [198]. Since only a single sample can be
separated with on-line detection, this is normally used to
identify fractions for collection in preparative chromatography.
The plates used for linear development must be scraped at their

edges and coated with a sealant, such as liquid paraffin or latex glue, to avoid leakage of the mobile phase from the sides during development.

7.7.3 Scanning Densitometry

Instruments for scanning densitometry using absorbance or fluorescence measurements in the reflectance or transmission mode first appeared in the middle 1960s and since then have undergone continuous change with the introduction of new technology [6-8,10,13,15,20,23,172-174]. The most obvious technical change has been the greater use of computers and microprocessors, which have revolutionized data handling and permitted a greater degree of automation of the scanning process.

Commercial instruments for scanning densitometry share many features in common, Figure 7.20. Different lamps must be used as light sources in order to cover the entire UV-visible range from 200 to 800 nm. Halogen or tungsten lamps are used for the visible and deuterium lamps for the UV region. High-intensity mercury or xenon arc sources are preferred for fluorescence measurements. To select the measuring wavelength either a monochromator or filter is used. Filter densitometers usually employ a mercury line source and filters to pass only light corresponding to individual wavelengths available from the source. Their main advantage is their low cost, otherwise broad spectral sources and grating monochromators offer greater versatility for optimizing sample absorption wavelengths, and are generally preferred. For fluorescence measurements a monochromator or filter is used to select the excitation wavelength. A filter, which transmits the emission wavelength envelope but attenuates the excitation wavelength, is placed between the detector and the plate. Photomultipliers or photodiodes are generally used for signal measurements.

Two optical geometries are predominantly used in contemporary scanning densitometers. The single-beam mode is the simplest optical arrangement and is capable of producing excellent quantitative results, but spurious background noise resulting from fluctuations in the source output, inhomogeneity in the distribution of extraneous adsorbed impurities contaminating the layer, and irregularities in the plate surface can be troublesome. These problems can be minimized by good analytical practices and through electronic correction. Background disturbances can be

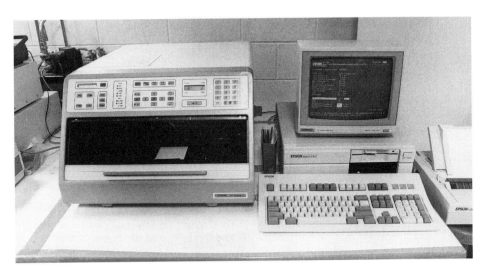

Figure 7.20 Computer-controlled automated scanning densitometer for quantitative TLC.

compensated for, to some extent, by double-beam operation. The two beams can be either separated in time at the same point on the plate or separated in space and recorded simultaneously by two detectors. Densitometers employing the latter principle are no longer in production.

In the single-beam dual-wavelength mode, fluctuations caused by scattering at a light absorbing wavelength are compensated for by subtracting the fluctuations at a different wavelength at which the spot exhibits no absorption but experiences the same scatter [7,199,200]. The two beams are altered by a chopper and combined into a single beam to provide the difference signal at the detector. As the scatter coefficient is to some extent wavelength-dependent the background correction is better when the measuring and reference wavelengths are as nearly identical as possible. This requirement is often difficult or impossible to meet, since absorption spectra are usually broad, and it may be impossible to find two similar wavelengths at which absorption occurs for one wavelength and not for the other.

For technical reasons the sample beam is fixed and the plate is scanned by mounting it on a movable stage controlled by

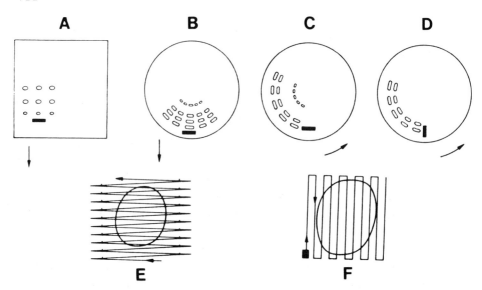

Figure 7.21 Methods of scanning thin-layer chromatograms. A =
linear slit scanning; B = radial scanning; C and D = peripheral
scanning; E = zig-zag scanning of an individual spot; F = meander
scanning of an individual spot. The arrows indicate the direction
of plate movement.

stepping motors. The common method of scanning is slit scanning in
which the sample beam is shaped into a rectangular area on the
plate surface through which the plate is transported in the
direction of development, Figure 7.21 [20]. Each scan, therefore,
represents a lane whose length is defined by the sample migration
distance and whose width is determined by the slit dimensions.
Some instruments have a turntable-type scanning stage for
peripheral and radial scanning used for chromatograms developed in
the circular or anticircular mode. As well as slit scanning, point
scanning can also be used. The measuring beam is shaped into a
spot or rectangle of dimensions much smaller than the
chromatographic zones to be scanned. By moving the scanning stage
in the x and y direction a zig-zag or meander scan is possible.
Zig-zag and meander scanning allow zones of any shape to be
accurately quantified using computer algorithms to perform the
integration. For very small spots errors may arise because the
scanning beam dimensions cannot be made sufficiently small to
permit a sufficient number of sampling points for the spot. The
method employed for baseline correction is also important.

For linear slit scanning densitometers it is well
established that chromatographic resolution and sample
detectability can be impaired by the choice of slit dimensions
defining the area of the measuring beam on the plate surface, the
scan rate, and the total electronic time constant of the

densitometer and recording device [10,173,175,182,183,201]. In the absorption mode the ratio of the slit height to spot diameter has a large influence on sample detectability. Slit heights less than the diameter of the spot produce the highest sensitivity but may lead to unacceptable errors due to incorrect alignment of the sample beam and spot centers throughout the lane. Generally, a slit height equivalent to the diameter of the largest spot to be scanned is selected as a compromise. In the fluorescence mode the signal increases linearly with increasing slit width and shows a general increase with increasing slit height until a slit height as large as the spot diameter is reached. For slit heights greater than the spot diameter there is little further change in response. At high scan rates resolution can be degraded and the signal attenuated if the time constant of the signal processing device is inadequate. A standard protocol has been suggested for comparing the sensitivity of slit scanning densitometers based on the use of standard substances, chromatographic conditions, and instrument parameters [10,202,203].

Most modern densitometers are designed for automatic scanning of a complete plate. In simple instruments a microprocessor is used for this purpose. Normally, the first lane to be scanned is manually positioned in the measuring beam and then the values for lane length, distance between lanes, and the number of lanes to be scanned is entered. The densitometer then scans the plate following the established geometric pattern without further intervention by the operator. The disadvantage of this method is that if the samples migrate irregularly then the spot may become misaligned with respect to the beam position and erroneous data generated. More sophisticated programs may recenter the measuring beam for the first spot of each lane prior to scanning in a linear manner, execute a meander or zig-zag scan function, or optimize the beam/spot co-ordinates for each spot in the chromatogram.

Automatic scanning routines combined with video integration allow the correct determination of the baseline under each peak. During the scan mode the background signal before and after each spot is acquired and used to compute the baseline position. The baseline in TLC may vary at different positions in the chromatogram due to matrix interferents and impurity gradients in the sorbent layer. These can be eliminated from the signal by software control or with the intervention of the operator by video integration of the chromatographic data displayed on a monitor. A further advantage of computers is that they simplify report-writing by combining graphical, statistical, and word processing capabilities as well as simplifying the archival process of data storage. Digital data stored on floppy disks can be reassessed and recalculated numerous times as the goals and interests of the analysis change without having to rerun the chromatogram. These, and other considerations, have been reviewed in the application of computers to TLC [7,23,174,204].

The principal sources of error in scanning densitometry have been identified as the reproducibility of sample application, the reproducibility of chromatographic conditions, the reproducibility of positioning the spot in the center of the measuring beam, and the reproducibility of the measurement [203,205,206]. The measurement error can be determined by repeatedly scanning a single lane of the TLC plate without changing any experimental variables between scans. It is composed of errors due to the optical measurement, electronic amplification, and the recording device. The measurement error is dependent on the signal-to-noise ratio but for a properly adjusted instrument typical values fall into the range 0.2-0.7% [7,205]. The error in positioning the spot in the measuring beam and the sample application error should be controlled by good analytical practices. The chromatographic error is generally the most significant error and is only reduced by minimizing the variability in the development process. The data pair technique can be used to minimize errors due to migration differences as a result of edge effects, deviations in layer thickness, nonlinear solvent fronts, etc. [207]. In this technique an internal compensation approach is used by pairing up the measurements of two spots on the same plate. In modern scanning densitometry with HPTLC plates the relative standard deviation from all errors can be maintained below 2% making it a very reliable quantitative tool.

Most scanning densitometers make some provision for either manually or automatically recording the in situ spectra of any desired number of spots [203]. For automatic spectrum recording a motor-driven monochromator controlled by a computer is used. For manual recording of spectra the spot is scanned repetitively while the monochromator position, or fluorescence emission filter for recording the fluorescence spectrum, is changed by fixed-wavelength increments between scans. A line connecting the individual peak maxima gives the substance-characteristic absorption or fluorescence emission envelope. The band pass of most monochromators used in scanning densitometers is in the range 10-30 nm, so that the resultant low-resolution absorption spectra obtained are rarely sufficient for substance identification in themselves, and differences between the sorbent spectrum and solution spectrum, which is more likely to be available for comparison, can be significant. In the case of the fluorescence spectra, the solution and in situ spectra may show little correspondence due to the concert of energy loss and energy conversion mechanisms available to the adsorbed sample. It is less time-consuming to scan a separation sequentially at several characteristic wavelengths than to record the full spectra of each spot [120,121,173,209]. The ratios of the response values obtained at these characteristic wavelengths can be used to confirm the similarity between samples and standards or to indicate contamination of a sample spot with other components. If standards are run on the same plate with the samples then the reproducibility of absorbance ratios is reasonable, R.S.D. = 1-6% [120]. Combining the information from coincidence of migration properties of sample and standards in the same chromatographic system and acceptable agreement between the absorbance or fluorescence emission response ratios is the most widely used technique for in situ substance identification in modern TLC.

The detection process in TLC can be considered static as the sample zones are stationary when the development process is terminated. All information concerning the chromatographic experiment can be made available as a three dimensional array in which the x and y coordinates define the spot position and the z direction the sample amount using image analysis techniques for optical detection [9,20,154-156,210] and radiation-sensitive detectors for radioisotopically labeled substances [10,153,188, 211,212]. For optical detection scanning takes place electronically using a combination of video digitizer, light source, monochromators and appropriate optics to homogeneously illuminate the plate and focus the image onto a vidicon tube or charged-coupled video camera. The main advantages of electronic scanning are fast acquisition of data, absence of moving parts, simple instrument design, and compatibility with data analysis of two-dimensional chromatograms that are difficult to scan using conventional slit-scanning densitometers. However, given the limitation of todays technology, electronic scanners cannot compete with mechanical scanners in terms of cost, sensitivity, available wavelength measuring range, and dynamic response range. Modern imaging detectors for radioisotopically labeled compounds employ windowless gas-flow proportional counters as detectors and are very sensitive. The proportional counter is filled with a mixture of argon-methane gas, which is ionized locally by collision with beta or gamma rays produced by decay of the radioisotopic sample zones. The local bursts of ionized gas molecules are sensed by a position-sensitive detector and stored in computer memory. These signals are accumulated for quantitative measurements.

Flame ionization detection has been used for the detection of samples of low volatility that lack a chromophore for optical detection. The separation is performed on specially prepared thin, quartz rods coated with sorbent sintered onto the surface of the rod [55,56,213,214]. The rods are developed in the normal way, usually held in a support frame that also serves as the scan stage after the rods have been removed from the development chamber and dried. Several rods can be held in the support frame and automatically scanned in order. The rods are moved at a controlled speed through a hydrogen flame and the signal processed in a similar manner to the flame ionization detector used in gas chromatography. The linear working range of the detector is about 3-30 micrograms for most substances and the response nearly universal. Careful standardization of the chromatographic and detection process is required to generate reproducible and reliable data.

7.8 PREPARATIVE THIN-LAYER CHROMATOGRAPHY

The principal difference between analytical TLC and preparative TLC is one of scale and not of procedure or method. Scale up is achieved by increasing the thickness of the layer and the length of the edge of the plate to which the sample is applied. Preparative TLC plates range in size from 20 x 20 cm to 20 x 100 cm and are coated with a sorbent layer 0.5 to 10.0 mm thick. The most commonly used layer thicknesses are 1.0 and 2.0 mm. Analytical layers are suitable for micropreparative applications and when high resolution is required. In general, the loading capacity increases with the square root of the layer thickness but the resolution is usually less for thicker layers.

Tapered plates, prepared with a gradual increase in thickness of the layer from 0.3 mm to 1.7 mm, can be used to improve resolution of the sample [215]. On the tapered layer the solvent front velocity decreases as the thickness of the layer increases. This results in the formation of a negative velocity gradient in the direction of solvent migration. As a result the lower portion of a zone moves faster than the top portion, keeping each component focused as a narrow band. Plates with concentrating zones are useful for optimizing sample application.

Sample application is one of the most critical steps in preparative TLC [215,216]. The sample, usually as a 5-10% solution in a volatile organic solvent, is applied as a band along one edge of the plate. The maximum sample load for a silica layer 1.0 mm thick is about 5 mg/cm (lower for cellulose and reversed-phase layers). Any of the band applicators used for analytical TLC are equally applicable to preparative TLC (section 7.7.1). Manual sample application by syringe or pipette must be performed with care to avoid damaging the layer and producing irregularly shaped developed zones. The sample zones can be refocused to some extent, by a short, about 1 cm, predevelopment with a strong solvent. When applying the sample to the sorbent layer, a margin of 2-3 cm is frequently left at either vertical edge of the plate to avoid uneven development. Plates with an inert concentrating zone solve many of the problems of manual sample application. The sample can be applied to the concentrating zone without much skill and the sample pushed to the boundary between the concentrating zone and separation layer with an appropriate solvent, where it is automatically refocused into a narrow line (see section 7.3.1) [69]. It is important for all sample application methods described above that the sample solvent is completely evaporated prior to separation otherwise the band spacing may be irreproducible.

Most of the changes that have taken place in preparative TLC in the last decade have occurred in the method of development. Conventionally, large volume tanks holding a number of preparative plates in a rack were commonly used. By ascending development run times of 1-2 h were not unusual. Nowadays, much higher efficiency and shorter separation times can be obtained using forced-flow development, such as rotation planar chromatography [9,10,196,215, 217,218] and overpressure layer chromatography [10,37,215,219] (see section 7.7.2). These methods allow conventional development and elution with on-line detection to be used. The elution mode allows samples to be automatically collected as they leave the layer. Fine-particle layers can also be used with forced-flow development to improve resolution. However, for the most complex mixtures preparative HPLC is still generally superior, although column costs are much greater than those of plates, and greater care is required to avoid column contamination [5,215].

In conventional preparative TLC, after the separation the bands are visualized and the separated zones carefully marked for removal. The zones are scraped off the plate with a spatula or similar tool onto glassine weighing paper. A number of devices based on the vacuum suction principle are commercially available for removing the marked zones from the plate. These devices work like a vacuum cleaner and collect the adsorbent material in a Soxhlet thimble or a glass chamber with a fritted base. The sample is separated from the adsorbent by Soxhlet extraction, elution or solvent extraction. For extraction, a small quantity of water is usually added to dampen the silica gel prior to shaking for several minutes with several portions of an organic solvent. In an alternative method, enough water is added to completely cover the adsorbent material and the aqueous suspension is extracted several times with an immiscible organic solvent. Prior to solvent evaporation, colloidal silica can be removed by filtration through a membrane filter. The possibility that recovery will be non-quantitative by any of the above methods must always be kept in mind.

7.9 REFERENCES

1. E. Stahl (Ed.), "Thin-Layer Chromatography", Springer-Verlag, New York, NY, 1969.
2. J. C. Touchstone and M. F. Dobbins, "Practice of Thin-Layer Chromatography", Wiley, New York, NY, 1978.
3. J. G. Kirchner, "Thin-Layer Chromatography", 2nd Edn., Wiley, New York, NY, 1978.
4. B. Fried and J. Sherma, "Thin-Layer Chromatography. Techniques and Applications", 2nd. Edn., Dekker, New York, NY, 1986.
5. J. Sherma and B. Fried (Eds.), " Handbook of Thin-Layer Chromatography", Dekker, New York, NY, (1990).
6. A. Zlatkis and R. E. Kaiser (Eds.), "HPTLC: High Performance Thin-Layer Chromatography", Elsevier, Amsterdam, 1977.
7. W. Bertsch, S. Hara, R. E. Kaiser, and A. Zlatkis (Eds.), "Instrumental HPTLC", Huethig, Heidelberg, 1980.
8. F. Geiss, "Fundamentals of Thin Layer Chromatography", Huethig, Heidelberg, 1987.
9. R. E. Kaiser (Ed.), "Planar Chromatography", Vol. 1, Huethig, Heidelberg, 1986.

10. F. A. A. Dallas, H. Read, R. J. Ruane, and I. D. Wilson (Eds.), "Recent Advances in Thin-Layer Chromatography", Plenum, London, 1988.
11. H. Jork, W. Funk, W. Fischer, and H. Wimmer, "Thin-Layer Chromatography, Reagents and Detection Methods", VCH Publishers, New York, NY, vol. 1a, 1990.
12. N. Grinberg (Ed.), "Modern Thin-Layer Chromatography", Dekker, New York, NY, 1990.
13. D. C. Fenimore and C. M. Davis, Anal. Chem., 53 (1981) 252A.
14. C. F. Poole, Trends Anal. Chem., 4 (1985) 209.
15. D. E. Jaenchen and H. J. Issaq, J. Liq. Chromatogr., 11 (1988) 1941.
16. A. M. Siouffi, E. Mincsovics, and E. Tyihak, J. Chromatogr., 492 (1989) 471.
17. C. F. Poole and S. K. Poole, Anal. Chem., 61 (1989) 1257A.
18. M. L. Langhorst, J. Planar Chromatogr., 2 (1989) 346.
19. C. F. Poole, S. K. Poole, T. A. Dean, and N. M. Chirco, J. Planar Chromatogr., 2 (1989) 180.
20. C. F. Poole and S. K. Poole, J. Chromatogr., 492 (1989) 539.
21. B. R. Hepler, C. A. Sutheimer, and I. Sunshine, Clin. Toxicol., 22 (1985) 503.
22. A. K. Singh, K. Granley, M. Ashraf and U. Mishra, J. Planar Chromatogr., 2 (1989) 41.
23. L. R. Treiber (Ed.), "Quantitative Thin-Layer Chromatography and Its Industrial Applications", Dekker, New York, NY, 1987.
24. D. C. Fenimore, C. M. Davis, and C. J. Meyer, Clin. Chem., 24 (1978) 1386.
25. V. Betina, J. Chromatogr., 477 (1989) 187.
26. V. M. Dembitsky, J. Chromatogr., 436 (1988) 467.
27. F. Geiss, J. Planar Chromatogr., 1 (1988) 102.
28. G. Guiochon and A. Siouffi, J. Chromatogr. Sci., 16 (1978) 598.
29. V. G. Gimpelson and V. G. Berezkin, J. Liq. Chromatogr., 11 (1988) 2199.
30. C. F. Poole, J. Planar Chromatogr., 2 (1989) 95.
31. W. P. N. Fernando and C. F. Poole, J. Planar Chromatogr., 3 (1990) 389.
32. G. Guiochon, G. Korosi, and A. Siouffi, J. Chromatogr Sci., 18 (1980) 324.
33. H. Halpaap, K.-F. Krebs, and H. E. Hauck, J. High Resolut. Chromatogr., 3 (1980) 215.
34. H. Kalasz, Chromatographia, 18 (1984) 628.
35. E. Tyihak and E. Mincsovics, J. Planar Chromatogr., 1 (1988) 6.
36. Sz. Nyiredy, L. Botz, and O. Sticher, J. Planar Chromatogr., 2 (1989) 53.
37. Z. Witkiewicz and J. Bladek, J. Chromatogr., 373 (1986) 111.
38. R. E. Kaiser and R. I. Rieder, J. Amer. Oil Chem. Soc., 66 (1989) 79.
39. Sz. Nyiredy, S. Y. Meszaros, K. Dallenbach-Tolke, K. Nyiredy-Mikita, and O. Sticher, J. High Resolut. Chromatogr., 10 (1987) 352.
40. A. Velayudhan, B. Lillig, and C. Horvath, J. Chromatogr., 435 (1988) 397.
41. L. Zlatanov, C. Gonnet, and M. Marichy, Chromatographia, 21 (1986) 331.
42. S. K. Poole, H. D. Ahmed, M. T. Belay, W. P. N. Fernando, and C. F. Poole, J. Planar Chromatogr., 3 (1990) 133.
43. S. K. Poole, H. D. Ahmed, and C. F. Poole, J. Planar Chromatogr., 3 (1990) 277.
44. C. F. Poole, J. Planar Chromatogr., 1 (1988) 373.

730

45. G. Guiochon and A. Siouffi, J. Chromatogr. Sci., 16 (1978) 470.
46. A. M. Siouffi, F. Bressolle, and G. Guiochon, J. Chromatogr., 209 (1981) 129.
47. B. Belenkii, O. Kurenbin, L. Litvinova, and E. Gankina, J. Planar Chromatogr., 3 (1990) 340.
48. H.-E. Hauck and W. Jost, J. Chromatogr., 262 (1983) 113.
49. C. F. Poole and S. K. Poole, J. Planar Chromatogr., 2 (1989) 165.
50. G. Guiochon, F. Bressolle, and A. Siouffi, J. Chromatogr. Sci., 17 (1979) 368.
51. G. Guiochon and A. M. Siouffi, J. Chromatogr.,245 (1982) 1.
52. G. Guiochon, M. F. Gonnord, A. Siouffi, and M. Zakaria, J. Chromatogr., 250 (1982) 1.
53. S. K. Poole, W. P. N. Fernando, and C. F. Poole, J. Planar Chromatogr., 3 (1990) 331.
54. L. Botz, Sz. Nyiredy, E. Wehrli, and O. Sticher, J. Liq. Chromatogr., 13 (1990) 2809.
55. M. Yoshioka, H. Araki, M. Kobayashi, F. Kaneuchi, M. Seki, T. Miyazaki, T. Utsuki, T. Yaginuma, and M. Kakano, J. Chromatogr., 507 (1990) 221.
56. T. Okumura, J. Chromatogr., 184 (1980) 37.
57. M. Ranny, "Thin-Layer Chromatography with Flame Ionization Detection", D. Reidel Publishing Co., Dordrecht, The Netherlands, 1987.
58. R. G. Ackman, C. A. Macleod, and A. K. Banerjee, J. Planar Chromatogr., 3 (1990) 452.
59. K. T. Wang, Y.-T. Lin, and I. S. Y. Wang, Adv. Chromatogr., 11 (1974) 73.
60. J. N. Miller, in R. Epton (Ed.), "Chromatography of Synthetic Biological Polymers", Ellis Horwood, Chichester, 1978, p. 181.
61. L. Fisher, "Gel Filtration Chromatography", 2nd. Edn., Elsevier, Amsterdam, 1980.
62. Z. Illes and T. Cserhati, J. Planar Chromatogr., 1 (1988) 231.
63. J. A. Berndt and C. F. Poole, J. Planar Chromatogr., 1 (1988) 174.
64. J. Ripphahn and H. Halpaap, J. Chromatogr., 112 (1975) 81.
65. H. Halpaap and J. Ripphahn, Chromatographia, 10 (1977) 613.
66. H. Halpaap and J. Ripphahn, Chromatographia, 10 (1977) 643.
67. G. P. Tomkinson, I. D. Wilson, and R. J. Ruane, J. Planar Chromatogr., 2 (1989) 224.
68. W. Funk, G. Donnevert, B. Schuch, V. Gluck, and J. Becker, J. Planar Chromatogr., 2 (1989) 317.
69 H. Halpaap and K.-F. Krebs, J. Chromatogr., 142 (1977) 823.
70. H. E. Hauck and H. Halpaap, Chromatographia, 13 (1980) 538.
71. U. A. Th. Brinkman and G. De Vries, J. High Resolut. Chromatogr., 5 (1982) 476.
72. U. A. Th. Brinkman, Trends Anal. Chem., 5 (1986) 178.
73. E. Heilweil and F. Rabel, J. Chromatogr. Sci., 23 (1985) 101.
74. W. Jost and H. E. Hauck, Adv. Chromatogr., 27 (1987) 129.
75. H. E. Hauck, M. Mack, S. Reuke, and H. Herbert, J. Planar Chromatogr., 2 (1989) 268.
76. T. Cserhati and H. E. Hauck, J. Chromatogr., 514 (1990) 45.
77. F. Galli, F. Gasparrini, D. Misiti, G. Natile, and G. Palmieri, J. Chromatogr., 409 (1987) 377.
78. M. Marichy, C. Gonnet, and L. Zlatanov, Chromatographia, 21 (1986) 105.
79. J. W. de Haan, L. J. M. van de Ven, G. de Vries, and U. A. Th. Brinkman, Chromatographia, 21 (1986) 687.
80. G. Grassini-Setrazza, I. Nicoletti, C. M. Polcaro, A. M. Girelli, and A. Sanci, J. Chromatogr., 367 (1986) 323.

81. J. Dingenen and A. Pluym, J. Chromatogr., 475 (1989) 95.
82. H. Halpaap, K.-F. Krebs, and H. E. Hauck, J. High Resolut. Chromatogr., 3 (1980) 215.
83. U. A. Th. Brinkman and G. De Vries, J. Chromatogr., 265 (1983) 105.
84. T. E. Beesley, in J. C. Touchstone (Ed.), "Thin-Layer Chromatography. Clinical and Environmental Applications", Wiley, New York, NY, 1982, p. 1.
85. W. Jost, H. E. Hauck, and H. Herbert, Chromatographia, 18 (1984) 512.
86. W. Jost and H. E. Hauck, J. Chromatogr., 264 (1983) 91.
87. A. Junker-Buchheit and H. Jork, J. Planar Chromatogr., 1 (1988) 214.
88. D. W. Armstrong, K. H. Bui, and R. E. Boehm, J. Liq. Chromatogr., 6 (1983) 1.
89. W. Jost and H. E. Hauck, J. Chromatogr., 261 (1983) 235.
90. G. Grassini-Strazza and I. Nicoletti, J. Chromatogr., 322 (1985) 149.
91. W. Jost and H. E. Hauck, Anal. Biochem., 135 (1983) 120.
92. R. Klaus, W. Fischer, and H. E. Hauck, Chromatographia, 28 (1989) 364.
93. W. Jost, H. E. Hauck, and W. Fischer, Chromatographia, 21 (1986) 375.
94. J. S. Kang and S. Ebel, J. Planar Chromatogr., 2 (1989) 434.
95. T. J. Good and A. G. Taketomo, J. Planar Chromatogr., 2 (1989) 336.
96. L. Witherow, R. J. Thorp, I. D. Wilson, and A. Warrander, J. Planar Chromatogr., 3 (1990) 169.
97. M. Mack and H. E. Hauck, J. Planar Chromatogr., 2 (1989) 190.
98. M. Mack and H. E. Hauck, Chromatographia, 26 (1988) 197.
99. R. Bhushan and I. Ali, J. Chromatogr., 392 (1987) 460.
100. P. E. Wall, J. Planar Chromatogr., 2 (1989) 228.
101. C. A. Brunner and I. Wainer, J. Chromatogr., 472 (1989) 277.
102. A. Alak and D. W. Armstrong, Anal. Chem., 58 (1988) 582.
103. D. W. Armstrong, J. R. Faulkner, and S. M. Han, J. Chromatogr., 452 (1988) 323.
104. J. D. Duncan and D. W. Armstrong, J. Planar Chromatogr., 3 (1990) 65.
105. L. Lepri, V. Coas, P. G. Desideri, and L. Checchini, J. Planar Chromatogr., 3 (1990) 311.
106. A.-M. Tivert and A. Backman, J. Planar Chromatogr., 2 (1989) 472.
107. U. A. Th. Brinkman and D. Kamminga, J. Chromatogr., 330 (1985) 375.
108. Sz. Nyiredy, K. Dallenbach-Toelke, and O. Sticher, J. Chromatogr., 450 (1988) 241.
109. M. Mack, H. E. Hauck, and H. Herbert, J. Planar Chromatogr., 1 (1988) 304.
110. R. E. Kaiser, J. Planar Chromatogr., 1 (1988) 265.
111. R. E. Kaiser, J. High Resolut. Chromatogr., 1 (1978) 164.
112. J. A. Perry, J. Chromatogr., 165 (1979) 117.
113. H. Cortes (Ed.), "Multidimensional Chromatography. Techniques and Applications", Dekker, New York, NY, 1990.
114. D. Nurok, R. M. Becker, and K. A. Sassic, Anal. Chem., 54 (1982) 1955.
115. R. E. Tecklenburg, G. H. Fricke, and D. Nurok, J. Chromatogr., 290 (1984) 75.
116. R. E. Tecklenburg and D. Nurok, Chromatographia, 18 (1984) 249.

117. C. F. Poole, S. K. Poole, W. P. N. Fernando, T. A. Dean, H. D. Ahmed, and J. A. Berndt, J. Planar Chromatogr., 2 (1989) 336.

118. C. F. Poole, H. T. Butler, M. E. Coddens, S. Khatib, and R. Vandervennt, J. Chromatogr., 302 (1984) 149.

119. L. Zhou, C. F. Poole, J. Triska, and A. Zlatkis, J. High Resolut. Chromatogr., 3 (1980) 440.

120. K. Y. Lee, C. F. Poole, and A. Zlatkis, Anal. Chem., 52 (1980) 837.

121. H. T. Butler, M. E. Coddens, S. Khatib, and C. F. Poole, J. Chromatogr. Sci., 23 (1985) 200.

122. K. Burger, Fresenius' Z. Anal. Chem., 318 (1984) 228.

123. T. H. Jupille and J. A. Perry, J. Amer. Oil Chem. Soc., 54 (1976) 179.

124. J. A. Perry, J. Chromatogr., 113 (1975) 267.

125. W. Markowski, J. Chromatogr., 485 (1989) 517.

126. S. A. Schuette and C. F. Poole, J. Chromatogr., 239 (1982) 251.

127. P. Buncak, Fresenius' Z. Anal. Chem., 318 (1984) 289.

128. E. Soczewinski and W. Markowski, J. Chromatogr., 370 (1986) 63.

129. E. Soczewinski and G. Matysik, J. Planar Chromatogr., 1 (1988) 354.

130. J. Vajda, L. Leisztner, J. Pick, and N. Anh-Tuna, Chromatographia, 21 (1986) 152.

131. E. Zietz and I. Ricker, J. Planar Chromatogr., 2 (1989) 262.

132. M. F. M. Trypsteen, R. G. E. Van Severen, and B. M. J. De Spiegeleer, Analyst, 114 (1989) 1021.

133. E. Menziani, B. Tosi, A. Bonora, P. Reschiglian, and G. Lodi, J. Chromatogr., 511 (1990) 396.

134. U. de la Vigne and D. Janchen, J. Planar Chromatogr., 3 (1990) 6.

135. T. H. Jupille and J. A. Perry, J. Chromatogr., 99 (1974) 231.

136. J. A. Thoma, Anal. Chem., 35 (1963) 214.

137. G. Goldstein, Anal. Chem., 42 (1970) 140.

138. M. Zakaria, M.-F. Gonnord, and G. Guiochon, J. Chromatogr., 271 (1983) 127.

139. H. J. Issaq, Trends Anal. Chem., 9 (1990) 36.

140. G. Guiochon, L. A. Beaver, M.-F. Gonnord, A. M. Siouffi, and M. Zakaria, J. Chromatogr., 255 (1983) 415.

141. L. van Poucke, D. Rousseau, C. van Peteghem, and B. M. J. de Spiegeleer, J. Planar Chromatogr., 2 (1989) 395.

142. D. Nurok, Chem. Revs., 89 (1989) 363.

143. S. Habibi-Goudarzi, K. J. Ruterbories, J. E. Steinbrunner, and D. Nurok, J. Planar Chromatogr., 1 (1988) 161.

144. I. D. Wilson, J. Chromatogr., 287 (1984) 183.

145. N. Grinberg and S. Weinstein, J. Chromatogr., 303 (1984) 251.

146. M. H. Jee and A. S. Ritchie, J. Chromatogr., 299 (1984) 460.

147. E. K. Johnson and D. Nurok, J. Chromatogr., 302 (1984) 135.

148. D. S. Risley, R. Kleyle, S. Habibi-Goudarzi, and D. Nurok, J. Planar Chromatogr., 3 (1990) 216.

149. B. de Spiegeleer, W. van den Bossche, P. de Moerloose, and D. Massart, Chromatographia, 23 (1987) 407.

150. J. E. Steinbrunner, D. J. Malik, and D. Nurok, J. High Resolut. Chromatogr., 10 (1987) 560.

151. D. Nurok, S. Habibi-Goudarzi, and R. Kleye, Anal. Chem., 59 (1987) 2424.

152. M. Prosek, M. Pukl, A. Golc-Wondra, and D. Fercej-Temeljotov, J. Planar Chromatogr., 2 (1989) 464.

153. J. C. Touchstone (Ed.), "Planar Chromatography in the Life Sciences", Wiley , New York, NY, 1990, p. 185.

154. J. A. Cosgrove and R. B. Bilhorn, J. Planar Chromatogr., 2 (1989) 362.
155. D. H. Burns, J. B. Callis, and G. B. Christian, Trends Anal. Chem., 5 (1986) 50.
156. V. A. Pollak and J. Schulze-Clewing, J. Planar Chromatogr., 3 (1990) 104.
157. G. Guiochon, M.-F. Gonnord, L. A. Beaver, and A. M. Siouffi, Chromatographia, 17 (1983) 121.
158. M.-F. Gonnord and A.-M. Siouffi, J. Planar Chromatogr., 3 (1990) 206.
159. J. A. de Schutter, G. van der Weken, W. van den Bossche, and P. de Moerloose, Chromatographia, 20 (1985) 739.
160. D. Nurok, R. M. Becker, M. J. Richard, P. D. Cunningham, W. B. Gorman, and C. L. Bush, J. High Resolut. Chromatogr., 5 (1982) 373.
161. W. Qin-Sun and Y. Bing-Wen, Chromatographia, 28 (1989) 473.
162. W. Qin-Sun and W. Heng-Yan, J. Planar Chromatogr., 3 (1990) 15.
163. C. K. Bayne and C. Y. Ma, J. Liq. Chromatogr., 10 (1987) 3529.
164. C. K. Bayne and C. Y. Ma, J. Liq. Chromatogr., 12 (1989) 235.
165. A. G. Howard and I. A. Bonicke, Anal. Chim. Acta, 223 (1989) 411.
166. B. M. J. de Spiegeleer, P. H. M. de Moerloose, and G. A. S. Slegers, Anal. Chem., 59 (1987) 62.
167. B. M. J. de Spiegeleer and P. de Moerloose, J. Planar Chromatogr., 1 (1988) 61.
168. Sz. Nyiredy, B. Meier, C. A. J. Erdelmeier, and O. Sticher, J. High Resolut. Chromatogr., 8 (1985) 186.
169. K. Dallenbach-Toelke, Sz. Nyiredy, B. Meier and O. Sticher, J. Chromatogr., 365 (1986) 63.
170. Sz. Nyiredy, K. Dallenbach-Toelke, and O. Sticher, J. Liq. Chromatogr., 12 (1989) 95.
171. Sz. Nyiredy, K. Dallenbach-Toelke, and O. Sticher, J. Planar Chromatogr., 1 (1988) 336.
172. R. J. Hurtubise, "Solid Surface Luminescence Analysis", Dekker, New York, NY, 1981.
173. E. D. Katz (Ed.), "Quantitative Analysis Using Chromatographic Techniques", Wiley, New York, NY, 1987.
174. I. M. Bohrer, Topics Curr. Chem., 126 (1984) 95.
175. M. E. Coddens, S. Khatib, H. T. Butler, and C. F. Poole, J. Chromatogr., 280 (1983) 15.
176. F. A. Huf, J. Planar Chromatogr., 1 (1988) 46.
177. I. E. Bush and H. P. Greeley, Anal. Chem., 56 (1984) 91.
178. S. Ebel, Topics Curr. Chem., 126 (1984) 71.
179. B. de Spiegeleer, W. Vanden Bossche, P. de. Moerloose and H. Stevens, Chromatographia, 20 (1985) 249.
180. W. R. G. Baeyens and B. L. Ling, J. Planar Chromatogr., 1 (1988) 198.
181. H. T. Butler and C. F. Poole, J. Chromatogr. Sci., 21 (1983) 385.
182. H. T. Butler and C. F. Poole, J. High Resolut. Chromatogr., 6 (1983) 77.
183. C. F. Poole, M. E. Coddens, H. T. Butler, S. A. Schuette, S. S. J. Ho, S. Khatib, L. Piet, and K. K. Brown, J. Liq. Chromatogr., 8 (1985) 2875.
184. D. Woolbeck, E. V. Kleist, I. E. Elmadfa, and W. Funk, J. High Resolut. Chromatogr., 7 (1984) 473.
185. S. S. J. Ho, H. T. Butler, and C. F. Poole, J. Chromatogr., 281 (1983) 330.
186. B. Lin Ling, W. R. G. Baeyens, H. Marysael, and K. Stragier, J. High Resolut. Chromatogr., 12 (1989) 345.

734

187. T. A. Dean and C. F. Poole, J. Planar Chromatogr., 1 (1988) 70.
188. D. M. Wieland, T. J. Manger, and M. C. Tobes, (Eds.), "Analytical and Chromatographic Techniques in Radiopharmaceutical Chemistry", Springer-Verlag, New York, NY, 1985.
189. J. C. Touchstone, J. Chromatogr. Sci., 26 (1988) 645.
190. R. E. Kaiser, J. Planar Chromatogr., 1 (1988) 182.
191. D. C. Fenimore and C. J. Meyer, J. Chromatogr., 186 (1979) 555.
192. G. Malikin, S. Lam and A. Karmen, Chromatographia, 18 (1984) 253.
193. P. Petrin, J. Chromatogr., 123 (1976) 65.
194. T. H. Dzido, J. Planar Chromatogr., 3 (1990) 199.
195. F. Geiss and H. Schlitt, Chromatographia, 1 (1968) 392.
196. E. Stahl and J. Muller, Chromatographia, 15 (1982) 493.
197. R. E. Kaiser, "Einfuhrung in die HPPLC", Huethig, Heidelberg, 1987.
198. E. Mincsovics, E. Tyihak, and A. M. Siouffi, J. Planar Chromatogr., 1 (1988) 141.
199. H. Yamamoto, T. Kurita, J. Suzuki, R. Hira, K. Nakano, M. Makabe, and K. Shibata, J. Chromatogr., 116 (1976) 29.
200. M.-L. Cheng and C. F. Poole, J. Chromatogr., 257 (1983) 140.
201. H. T. Butler, S. A. Schuette, F. Pacholec, and C. F. Poole, J. Chromatogr., 261 (1983) 55.
202. M. E. Coddens and C. F. Poole, Anal. Chem., 55 (1983) 2429.
203. J. Allwohn and S. Ebel, J. Planar Chromatogr., 2 (1989) 71.
204. R. E. Kaiser, J. Planar Chromatogr., 2 (1989) 323.
205. S. Ebel and E. Glaser, J. High Resolut. Chromatogr., 2 (1979) 36.
206. V. A. Pollak, Adv. Chromatogr., 30 (1989) 201.
207. H. Bethke, W. Santi, and R. W. Frei, J. Chromatogr. Sci., 12 (1974) 392.
208. S. Ebel and J. S. Kang, J. Planar Chromatogr., 3 (1990) 42.
209. H. T. Butler, M. E. Coddens, and C. F. Poole, J. Chromatogr., 290 (1984) 113.
210. V. A. Pollak, Adv. Chromatogr., 30 (1989) 221.
211. H. Filthuth, J. Planar Chromatogr., 2 (1989) 198.
212. I. Smith and V. Furst, J. Planar Chromatogr., 2 (1989) 233.
213. M. Ranny, M. Zbirovsky, and V. Konecny, J. Planar Chromatogr., 3 (1990) 111.
214. R. G. Ackman and W. M. N. Ratnayake, J. Planar Chromatogr., 2 (1989) 219.
215. Sz. Nyiredy, Anal. Chim. Acta, 236 (1990) 83.
216. L. Botz, Sz. Nyiredy, and O. Sticher, J. Planar Chromatogr., 3 (1990) 10.
217. Sz. Nyiredy, S. Y. Meszaros, K. Dallenbach-Tolke, K. Nyiredy-Mikita, and O. Sticher, J. Planar Chromatogr., 1 (1988) 54.
218. L. Botz, Sz. Nyiredy, and O. Sticher, J. Planar Chromatogr., 3 (1990) 401.
219. G. C. Zogg, Sz. Nyiredy,, and O. Sticher, J. Planar Chromatogr., 1 (1988) 261.

CHAPTER 8

SAMPLE PREPARATION FOR CHROMATOGRAPHIC ANALYSIS

8.1   INTRODUCTION

In most instances, mixtures obtained from biological or

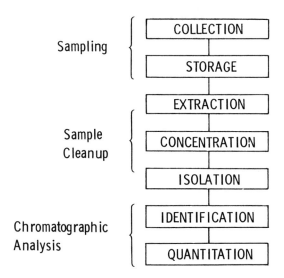

Figure 8.1 Flow diagram of the procedures used to prepare samples for chromatographic analysis.

environmental sources are too complex, too dilute, or are incompatible with the chromatographic system to permit analysis by direct injection. Preliminary fractionation, isolation, and concentration of the sample components of interest are needed prior to analysis. Figure 8.1 illustrates a flow diagram for a typical organic analysis. A representative sample is obtained and then fractionated and concentrated by extraction and cleanup procedures to ensure effective separation, detection, and system compatibility in the final chromatographic determination. A wide range of physical, chemical, and chromatographic methods have been employed for sample preparation and references [1-14] provide a good overview of this topic. Many specialized publications exist describing sample preparation procedures for individual analytes or sample types that are too numerous to be referenced here. Our aim, instead, is to discuss the characteristics of the most widely used sample preparation techniques in their broadest sense.

8.2 PROBLEMS IN OBTAINING A REPRESENTATIVE SAMPLE

Obtaining a representative sample is an important part of any analysis, since errors or faults in the sampling protocol cannot be corrected at any later point in the analysis [15-21].

Sampling may be a problem because of the size or inhomogeneity of the sample. If one desired to establish the concentration of a chemical in a lake, then size would preclude total analysis, and subsamples would have to be taken. Any single subsample might be representative of water at a specific location, depth and time but may not be representative of all water in the lake. The subsample may be biased by being in close proximity to a point source of the analyte. Also, the sample would be inhomogeneous containing water and suspended matter. The analyte might be distributed differently between the two phases. The concentration of suspended matter may vary with location and depth. To obtain a representative value for the analyte of interest multiple subsamples would be needed whose location and time of sampling must be in accordance with some statistical plan that can be validated [16-19]. Also, it may be necessary to sample the two phases separately; adsorption techniques would be convenient for the soluble organics and filtration for the particulates. For studies in occupational hygiene the sampling plan must generally allow for the peak transient exposure of a worker to a toxic chemical, as well as a weighted average for exposure over a regular work day. The frequency of the sampling will be very different for both analyses, one being short and repetitive, the other accumulative. Large samples such as vegetable matter are invariably inhomogeneous and in a form not suited to chromatographic analysis. Sampling in this case may involve such operations as drying, cutting, grinding, mincing, pressing or sieving to reduce the size and nonuniformity of the sample. As a dried, fine, free-flowing powder the sample could be remixed to provide a homogeneous subsample.

The above examples are only meant as illustrations of typical problems encountered in obtaining representative subsamples for analysis. For the analytical data to be meaningful a workable plan for acquiring samples must be implemented and if possible validated by statistical techniques [15,17,20,21]. Sampling plans are now frequently incorporated in material submitted to regulatory agencies, such as the United States Food and Drug Administration and Environmental Protection Agency, or are established as part of the protocol to establish compliance with regulations policed by these and similar agencies [2,18,20-23].

Additional problems may arise in the analysis due to contamination, careless collection and handling, or alteration of the sample during storage. At trace analyte concentrations almost any surface which the sample contacts becomes a possible source of contamination. Chemical processes that occur in the sample between the time of sample collection and analysis can invalidate the analytical results. Examples of such processes are photo-decomposition, adsorption, vaporization loss, thermal decomposition, microbial action and chemical reaction. Samples retained for analysis at a later date should be stored in the dark, in glass containers, and maintained at subzero temperatures, since these conditions tend to inhibit the processes listed above. Preservatives, antioxidants and adjustment to the sample pH may also be called for. Tissue and food samples when macerated may release enzymes capable of changing the sample composition upon storage. It may be preferable to store these samples whole, in the frozen state under nitrogen, and to macerate them just prior to analysis [5,24,25]. Both the properties of the analyte and the matrix dictate the level of precautions needed to maintain the integrity of the subsample up to the time of analysis.

8.3 ISOLATION AND CONCENTRATION TECHNIQUES USING PHYSICAL METHODS

The most frequently used methods of analyte isolation and concentration for organic compounds involve distillation, extraction and adsorption techniques. Some typical applications of these techniques and their attendant advantages and disadvantages for the analysis of trace organic solutes in water are summarized in Table 8.1 [4,26]. These methods will be elaborated on below and in subsequent sections of this chapter.

8.3.1 Distillation

Distillation is a suitable technique for the isolation of volatile organic compounds from liquid samples or the soluble portion of solid samples [24,27-30]. The physical basis of separation depends on the distribution of constituents between the liquid mixture and the vapor in equilibrium with that mixture. The more volatile constituents are concentrated in the vapor phase, which is collected after condensation. The effectiveness of the separation is dependent on the physical properties of the

TABLE 8.1

COMPARISON OF REPRESENTATIVE METHODS FOR THE CONCENTRATION OF TRACE ORGANIC COMPOUNDS IN WATER

Method	Principle	Scope	Comments
Freeze Concentration	Water sample is partially frozen, concentrating the dissolved substances in the unfrozen portion.	All sample types	Minimizes sample losses due to volatilization or chemical modification. Principal losses occur due to occlusion, adsorption, evaporation and channelling in the ice layer. Limited sample size.
Lyophilization	Water sample is frozen and pure water is removed by sublimation under vacuum.	Nonvolatile organics	Can handle large sample volumes. Selective loss of volatile organics. Inorganic constituents concentrated simultaneously.
Vacuum Distillation	Water is evaporated at reduced pressure and at or near ambient temperature.	Nonvolatile organics	Slow process when sample volumes are large. Inorganics also concentrated. Sample contamination is low but sample may be modified due to thermal degradation or chemical and microbial reactions.
Reverse Osmosis	Water is forced through a membrane by application of pressure thereby enriching the water sample in constituents which ordinarily cannot pass through the membrane.	Molecular weight > 200	Compounds of small size are not concentrated. Inorganic materials may contaminate sample. Membranes may either adsorb constituents or release impurities into the sample.

Method	Sample types	Description
Ultra-filtration	Large molecules	Water sample is filtered under pressure through a membrane that will pass molecular constituents below a certain size and retain those above that size. Porosity of membrane determines the size of molecules concentrated. Usually used for compounds > 1000 molecular weight. Can concentrate large sample volumes at low temperatures.
Solvent Extraction	All sample types	Aqueous sample is partitioned with an immiscible organic solvent. Extraction efficiency depends on the affinity of the solute for the organic solvent. Samples with a high affinity for water are not extracted. Extractions can be performed by a simple single equilibration or by multiple equilibrations with fresh solvent. Solvent impurities concentrated along with sample.
Surface Adsorption	All sample types	Water sample is passed through a column of the adsorbent and the adsorbed organic constituents subsequently eluted with a smaller volume of organic solvent. Adsorbents include charcoal, macroreticular resins, polyurethane foams, bonded phases and ion-exchangers. Generally have high capacity but sample discrimination may be a problem. Sample modification and incomplete recovery are further possible problems.
Gas-stripping	Volatile samples	An inert purge gas sweeps over or sparges through the aqueous sample thereby transporting volatile constituents from the liquid phase to the gas phase, which permits them to be subsequently trapped cryogenically or on adsorbents. Examples include closed-loop stripping, purge-and-trap and headspace techniques. Limited to small sample volumes. Extraction efficiency depends on the physical and chemical properties of the solute.

components in the mixture, the equipment used, and the method of
distillation. The latter includes simple distillation, fractional
distillation, flash distillation, steam distillation, codistill-
ation and molecular distillation. Further, the above processes may
be performed at atmospheric pressure or at reduced pressure for
compounds too unstable to be distilled at their boiling point.
Simple distillation refers to the process whereby the vapor
leaving the liquid is not subject to partial condensation or
contact with condensed liquid prior to reaching the condenser. The
composition of the vapor leaving the liquid does not change as it
moves from the surface of the liquid to the condenser. Equipment
requirements are minimal usually consisting of a flask fitted with
a condenser and a product receiver (a fractionating column is not
used). It is used when an efficient separation process is not
required, such as for stripping a liquid from a soluble involatile
solid, separating simple mixtures were the components differ
widely in boiling point, and in obtaining product cuts with
defined boiling point ranges, for example, petroleum products can
be characterized by the percent distilled in different boiling
point ranges. Individual fractions are then suitable for analysis
by gas chromatography. Fractional distillation is used when a more
efficient separation process is required. Both a fractionating
column and a reflux-ratio-controlling device are incorporated into
the simple distillation apparatus discussed above. This type of
distillation is an equilibrium process in which the composition of
the distillate and the boiling liquid is constantly changing as
the distillation proceeds. For simple mixtures pure components may
be isolated for further examination. A wide range of fractionating
columns can be used, such as a simple empty tube, a tube with
indentations (Vigreux column), a tube packed with spirals,
helixes, spheres, etc., a tube packed with glass or a wire gauze,
or a tube with a stationary or moving element, such as a spinning
band column. The purpose of the fractionating column is to provide
adequate contact between the rising liquid vapors and the
returning condensed liquid for equilibrium to be established. The
single-stage concentric tube, rotary film, climbing-film and
falling-film evaporators, operated at atmospheric or reduced
pressure, are the most widely used for isolating organic
volatiles, particularly for fragrance and flavor evaluation. The
fast-action climbing and falling-film evaporators are more
efficient and less likely to produce artifacts than the standard

distillation equipment found in most organic synthesis laboratories. A spinning-band column, for example, uses a rotor with a wire wrapped around it in a tightly fitting column or a spiral band which is rotated on a vertical shaft in the column. Rotation of the band enhances the vapor-liquid contact, and its centrifugal motion throws the liquid and vapor onto the column wall. The wall is contacted by the edges of the spinning band which combs the liquid from the wall enhancing the contact of liquid with the vapor. Spinning-band columns can generate about 30-300 theoretical plates depending on their design and dimensions, can be operated at quite high throughputs with a low pressure drop, and small sample sizes can be handled (at least relative to conventional fractional distillation apparatus). Fractional distillation procedures are more commonly employed for the large scale isolation of relatively pure fractions of organic compounds than for sample preparation, since gas chromatography is frequently able to separate the same samples directly and simple distillation (or low efficiency fractional distillation) is more convenient and rapid for obtaining fractions with a defined boiling point range.

Probably the most common distillation method used as a form of sample preparation for chromatographic analysis is steam distillation [31,32]. Solvent extraction and gas phase stripping methods are generally inefficient procedures for isolating polar, acidic, or basic compounds in an aqueous matrix due to the low efficiency of water immiscible solvents for the extraction of these compounds and their low volatility and high water affinity which results in a very slow transfer to the gas phase using sparging techniques. Steam distillation is a simple distillation procedure in which vaporization of the mixture is achieved by either continuously blowing live steam through the mixture or by boiling water and the mixture together. In either case the volatile organic components are entrained and carried along with the steam at a rate proportional to their relative partial pressure and molecular weight at the temperature of the distillation. For a binary mixture of steam and an organic component, A, the composition of the condensing vapors, and consequently the rate of distillation of A, will be given by equation (8.1)

$$m_A = m_S \ (P_A M_A)/(P_S M_S) \qquad (8.1)$$

where m_A is the mass flow rate of A in the vapor, m_S the mass flow rate of steam, P_A the partial pressure of A, P_S the partial pressure of steam, M_A the molecular weight of A, and M_S the molecular weight of water. If isolation and enrichment are to be achieved simultaneously, the conditions most desirable for the determination of trace sample components, then the ratio of $(P_A M_A)/(P_S M_S)$ must be favorable otherwise a large volume of steam condensate will be needed for efficient recovery of A, and the final volume of condensate will be too large for direct analysis. Low molecular weight and reasonably volatile components are the sample types likely to be efficiently isolated by steam distillation. Since many substances form azeotropes with water of lower boiling point than the parent compound, this helps to extend the molecular weight range of the sample types that can be steam distilled. The recovery of organic volatiles may be changed by simultaneous gas sparging, addition of a large amount of an inorganic salt, by pH adjustment, or by the addition of a codistiller, such as benzene or toluene [31-34]. In individual circumstances any of these techniques may lead to an increase or decrease in the recovery of the desired analyte and specific recommendations cannot be given. Collecting the distillate below the surface of a cold solvent is also frequently used to improve the recovery of the analytes.

Various types of apparatus have been used for steam distillation. Figure 8.2(A) shows a steam distillation assembly used for the isolation of halogenated fumigants in cereal products [35]. The aqueous homogenate of the sample is placed in compartment B and steam generated externally in boiling flask A which also serves to heat the sample container providing good thermostating of the sample compartment. An overpressure releasing device attached to the side arm of the steam generator prevents pressure build up should the steam tube leading to the sample chamber become blocked. The volatile components are carried over with the steam and the aqueous condensate collected in a separatory funnel. Figure 8.2(B) shows an example of a microdistillation apparatus used for the isolation of essential oils from small plant samples [36]. The plant material (0.2-0.3 g) was suspended in 50 ml of distilled water in flask A, electrically heated to boiling and distilled for 30 min. The essential oil was collected in bubble B containing 100 microliters of a solvent less dense than water (avoids losses through emulsion formation) and

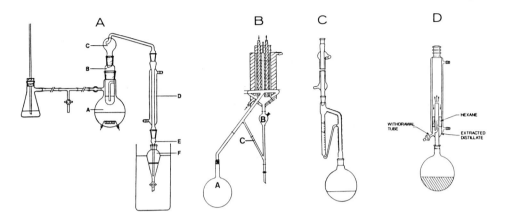

Figure 8.2 Different apparatus used for steam distillation. A = modified Garman steam distillation apparatus: A, steam generation flask; B, sample chamber; C, splash head; D, condenser; E, delivery tube; F, separatory funnel. B = micro steam distillation apparatus: A, boiling flask; B, collection bulb; C, water return arm; D, Condenser. C = steam distillation apparatus using a Dean and Stark type trap. D = Nielsen-Kryger steam distillation apparatus.

excess water continuously recycled to the boiling flask through arm C. A refrigeration cooler was used to maintain the condenser at 2-3°C. The simple steam distillation apparatus shown in Figure 8.2(C) consisting of a boiling flask, condenser, and Dean and Stark type trap was used to determine petroleum hydrocarbons in water and homogenized mussels [37]. Alkaline saponification was used to increase the recovery of hydrocarbons from the mussels and a small volume of organic solvent was added as a codistiller to improve the collection of the analytes. A distillation time of one hour was sufficient to recover most volatile hydrocarbons (chrysene, eicosane, tetracosane needed a longer time). Figure 8.2(D) illustrates a further design of a simple steam distillation apparatus used to determine chlorinated benzenes in sediments [38]. Ten grams of sediment and 250 ml of water were placed in the boiling flask and 10 ml of water and 10 ml of hexane added to the condenser. A Tenax trap attached to the top of the condenser was

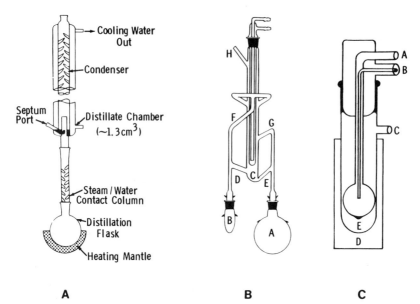

A   **B**   **C**

Figure 8.3 Apparatus used for sample preparation involving distillation or sublimation. A, distillation apparatus for organic volatiles in water; B, micro apparatus for continuous steam distillation with continuous liquid-liquid extraction of the distillate; C, micro sublimation apparatus.

used to retain analytes lost from the condenser. Recoveries of chlorobenzenes from a spiked sediment sample averaged 81 ± 12% for the steam distillation method compared to 55 ± 11% for Soxhlet extraction.

Peters has described a small, all-glass distillation-concentration apparatus that is particularly well-suited for the isolation of volatile, polar organic solutes (alcohols, nitriles, aldehydes, ketones, etc.) from water frequently yielding a clean extract that can be analyzed by gas chromatography without further treatment, Figure 8.3(A) [39]. The apparatus consists of a round-bottomed flask, condenser, condensate collection (distillate) chamber, steam/water contact column and an overflow return tube for returning a portion of the condensate to the distillation flask. The volatile polar organics azeotropically distill into the distillation chamber and are preferentially retained there. The overflow of condensate back to the distillation flask is stripped by the rising steam and the volatile polar organics are recycled back into the distillate chamber. Two factors are important in judging the efficiency of the apparatus: the boiling point difference between water and the compound or its azeotrope and the time required to establish equilibrium between the vapor phase and condensate. At least a 1°C boiling point difference is required and an equilibration time less than one hour is preferred. The absolute recovery of volatile organics in the distillate chamber is typically 80%, with the remaining 20% of the compounds entrained in the reflux condenser. By continually withdrawing the organic concentrate from the distillate chamber after equilibrium has been established, virtually 100% recovery can be obtained. Concentration factors of 300- to 400-fold are possible with this apparatus.

For large sample sizes phase separation occurs after condensation in steam distillation but for small samples, as we have already seen, solvent extraction is required for adequate recovery of the analyte. The ability of the solvent to extract the analytes from the condensed water will, therefore, also influence the recovery of the analyte. Verzele has designed a micro steam distillation continuous liquid-liquid extraction apparatus based on an earlier macro design by Likens and Nickerson, which is well suited to chromatographic applications [40-42]. The apparatus can handle 10-100 ml of aqueous solution or 1-20 g of solid material blended with water and provides an extraction of the distillate into 1.0 ml of pentane or methylene chloride, Figure 8.3(B). The sample is placed in flask A and the extraction solvent in flask B. To commence operation, the phase separator, C, is charged with a mixture of pentane and water, ice water is circulated through the cold-finger condenser, and the pentane reflux is established by immersing flask B in a hot water bath. After 5 min., steam is generated by applying heat to flask A. The vapor channels F and G are insulated. The construction of the apparatus shown in Figure 8.3(B) is such that the low-density solvent (pentane) returns through arm D to flask B and the high-density solvent (water) returns through arm E to flask A. Reversing the positions of flask A and B enables the extraction to be performed with a heavier-than-water solvent, such as methylene chloride. After enrichment is complete, normally one hour is sufficient, steam generation is stopped while solvent extraction is continued for another 20 min to ensure that all the steam-distillable material is collected. An internal standard is added to the extraction solvent to allow for evaporative entrainment losses during operation. For aroma constituents and pesticide compounds recoveries of 90-100% were typically obtained. For polar and high-boiling compounds recoveries were much lower and longer processing times were needed due to the low concentration of the analytes codistilled with the steam [43,44]. This problem can be avoided by direct introduction of aqueous sample onto the top of the condenser, the falling film of aqueous sample is then efficiently mixed with condensing solvent vapors and the analytes of low volatility stripped by liquid extraction. The efficiency of the stripping of the analytes of low volatility depends on the existence of a favorable distribution constant for the analyte between water and the extracting solvent and the flow rate of the aqueous sample through the condenser.

Assisted distillation (also called sweep co-distillation) has been used for the isolation of low concentrations of volatile pesticides in animal and vegetable fats and oils [45-49]. A fluid fat sample is injected by syringe through a septum injector into an all glass fractionating tube designed to fit in a multi-port heating block which can process several samples simultaneously (currently 10 with commercially available equipment). The fractionating tube, which is packed with silanized glass beads in the annular space between the center tube and glass wall, Figure 8.4(A), or is fabricated to provide a close fit between the inner and outer tubes, is designed so that the sample passes down the center tube, with the carrier gas, then out and up onto the glass beads/tube wall. The high temperature and high velocity flow of nitrogen through the tube disperses the fat as a thin film over the support facilitating the stripping of the more volatile pesticides which are collected in a trap containing sodium sulfate and partially deactivated Florisil. Elution of the trap with solvent usually provides an extract sufficiently fat free to be directly injected into a gas chromatograph. Typical operating conditions for organochlorine pesticides are 1 g of fat, a distillation temperature of 235°C, a nitrogen flow rate of 230 ml/min., and a processing time of 30 min. Under these conditions the recovery of thermally stable pesticides is generally 80-100% although, of course, individual recoveries depend on the volatility of the pesticides and their resistance to thermal and/or catalytic decomposition.

8.3.2 Sublimation and Freeze Concentration

Sublimation, the direct vaporization and condensation of a

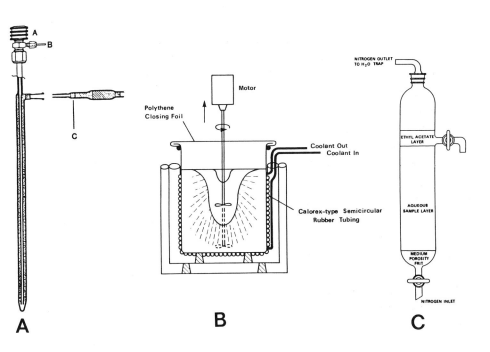

Figure 8.4   Apparatus for sample preparation using physical separation. A = fractionation tube and trap for assisted distillation (A = septum injector, B = carrier gas inlet, C = Florisil trap for collecting volatile pesticides);   B, Shapiro-type freeze concentrator; and C, apparatus for solvent sublation.

solid without passing through a liquid phase, is a useful method of sample cleanup for those compounds, such as the polycyclic aromatic hydrocarbons, that can be sublimed at reasonable temperatures [50,51]. A simple apparatus for vacuum sublimation is shown in Figure 8.3(C), which is designed to provide a large surface for vaporization and condensation with only a short pathlength in between. The extraction yield with vacuum sublimation is dependent on the vapor pressure of the components at a given temperature and pressure and also on the nature of the sample matrix. Thus, different times are required to reach constant extraction yields for different compounds, and for the same compounds in different sample matrices. Optimization of the method for quantitative trace analysis can be very difficult and the method may be slow, although under favorable circumstances a relatively clean extract is obtained that can be analyzed usually without further treatment.

Lyophilization (freeze-drying), the removal of water by vacuum sublimation of ice, is a convenient method for concentrating involatile organic compounds in aqueous solution and for preserving biological samples [52-54]. Most of the bulk weight of tissues is mainly water so that its selective removal considerably reduces the space required for storage. After freeze-drying the tissue sample can usually be ground and sieved to a powder providing a homogeneous sample for subsequent analysis. Commercial apparatus for freeze-drying is widely available in many designs and different degrees of sophistication, particularly in biochemistry facilities. The essential steps are that the sample is first frozen and then placed under high vacuum. The heat absorbed by the sample from its surroundings causes the ice to sublime, which is then recondensed on a large surface condenser held at cryogenic temperatures. Sample losses, particularly of volatile organic compounds, occur when the bulk of the water has been removed and the temperature rises while the sample remains under a relatively high vacuum. Acidic and basic compounds can be converted to their salts prior to freeze-drying to minimize vaporization losses. High concentration factors can be obtained as large volumes of aqueous solutions containing traces of organic substances can be concentrated. Inorganic substances are concentrated simultaneously with the organic substances and the subsequent desalting of the extract can be a problem in some analyses.

Freeze concentration has been used for the concentration of aqueous solutions of organic volatiles and substances that are heat labile using an apparatus similar to that shown in Figure 8.4(B) [24,55,56]. For successful results, the contact layers between the liquid and solid phases should be continuously disturbed by stirring or shaking and part of the solution should remain unfrozen at the end of the concentrating procedure.

Artifact formation is uncommon by this procedure, but sample losses due to occlusion, adsorption, evaporation and channeling in the ice layer may occur. For concentrating natural waters precipitate formation at high salt concentrations limits recoveries and concentration factors to about 10 to 100 depending on the starting concentration of inorganic salts. Vaporization is the principal source of poor recovery of organic volatiles, particularly if the apparatus is not enclosed. On the laboratory scale, a 2 liter sample volume can be concentrated to a 40-50 ml volume in about nine hours with a 90-100% recovery of the organic species.

8.3.3 Foam Fractionation and Solvent Sublation

Adsorptive bubble separation techniques are generally used for the concentration and isolation of detergents and other surface active materials (as well as some inactive materials that form adducts with an added collector surfactant) from water [13]. Foam fractionation is the removal of dissolved material by foaming, the efficiency of which increases with decreasing bubble size, increasing gas flow rate, column height and decreasing surfactant concentration. Increasing efficiency with decreasing surfactant concentrations arises from interference caused by micelle formation at high concentration which are not as efficiently sorbed at the gas-liquid interface as monodispersed surfactant molecules in dilute solution. The recovery of surfactant is often very sensitive to pH which must be optimized empirically. The most convenient form of foam fractionation for analytical purposes is solvent sublation, Figure 8.4(C). Solvent sublation combines foam fractionation with solvent extraction of the surface concentrated surfactant into a water immiscible organic solvent on top of the water sample.

8.3.4 Membrane Processes

Separations in membrane processes result from differences in the transport rates of analytes or solvent molecules through a membrane interface. The transport rate is usually determined by the existence of a driving force, such as a concentration, pressure or temperature gradient and the mobility and concentration of analytes within the interface. The most useful membrane processes for sample preparation are dialysis, ultrafiltration and reverse osmosis [30,57-59]. In dialysis analytes are separated by their ability to diffuse through a membrane as a result of a concentration gradient. The overall efficiency of the process is controlled by the ratio of the flow rates of the sample feed and dialysate solutions and the rate constant for solute transport between the two solutions, which is determined by the properties of the membrane, the fluid channel geometry, and the local fluid velocities. The relative size of different molecules largely determines the permeation selectivity of a membrane in the absence of strong specific interactions. Consequently, dialysis has been most effective in separating small molecules from large molecules in biological fluids, such as serum and urine, etc, and for desalting protein extracts. Dialysers are used commonly in many automated analyzers for isolating drugs or other substance from biological fluids. A recent example of this method includes the combination of a low volume dialyser and trace enrichment column for the automated analysis of drugs in biological fluids by liquid chromatography [60-63]. The automated sample preparation procedure eliminates the gross effects of the sample matrix on column operation and permits reliable quantitative trace analysis by analyte preconcentration. Sample preparation takes place concurrently with the separation of the previous sample.

Ultrafiltration is a separation process in which large molecules are separated from solution by filtration through membranes. The filtration process is aided by hydrostatic pressure, the use of modern microporous membranes with a skin of porous polymer (0.1-0.2 micrometers) on a highly porous support (100-200 micrometers) which provide fast filtration rates and resistance to plugging, and a mechanism for agitation of the sample close to the surface of the membrane. Membranes are commercially available for separating molecules in the molecular weight range of about 1,000 to 1 million. Ultrafiltration is primarily used for the isolation of low or high molecular weight analytes in the presence of each other, as a method of concentrating high molecular weight analytes, to desalt organic concentrates, and to fractionate samples into different molecular size ranges. Another application is the determination of the fraction of unbound drug in biological fluids for therapeutic monitoring, since the free rather than the total drug concentration more accurately reflects the amount of circulating drug available for binding to receptors [64]. Since the volume of ultrafiltrate for these analyses is very small, cartridge units that fit into a laboratory centrifuge are frequently used [65].

Reverse osmosis is a similar separation technique to ultrafiltration except that membranes of a much smaller pore diameter are employed and the operating pressure is much higher. The operating pressure must exceed the natural osmotic pressure for the system resulting in the movement of solvent, usually water, from the solution of high analyte concentration to that of low analyte concentration. Reverse osmosis is thus suitable for preconcentrating relatively large volumes of dilute solutions, such as river or drinking water. For semivolatile organic solutes, however, permeation of the membrane by some analytes limits recoveries. The rate of permeation of organic solutes through the membrane depends on the chemical compatibility of the membrane and analytes, not sieving, as was the case for ultrafiltration. Reverse osmosis, therefore, can be used to separate solutes of similar size.

8.4   ISOLATION AND CONCENTRATION TECHNIQUES USING SOLVENT EXTRACTION

Solubilizing all or part of a sample matrix by contacting with liquids is one of the most widely used sample preparation techniques for gases, vapors, liquids or solids. Additional selectivity is possible by distributing the sample between pairs of immiscible liquids in which the analyte and its matrix have different solubilities. Equipment requirements are generally very simple for solvent extraction techniques, Table 8.2 [4,10], and solutions are easy to manipulate, convenient to inject into chromatographic instruments, and even small volumes of liquids can be measured accurately. Solids can be recovered from volatile solvents by evaporation. Since relatively large solvent volumes are used in most extraction procedures, solvent impurities, contaminants, etc., are always a common cause for concern [65,66].

Some guidelines for predicting the results from distributing a sample between two immiscible solvents are summarized in Table 8.3 [67,68]. The efficiency of an extracting solvent, E, depends primarily on the affinity of the solute for the extracting solvent, K_D, the phase ratio, V, and the number of extractions, n. For simple batchwise extractions K_D should be large, as there is a practical limit to the volume of the extracting solvent   and   the

TABLE 8.2

THE ISOLATION OF ORGANIC COMPOUNDS BY LIQUID EXTRACTION

Sample Type	Equipment Requirement	Comments
Organic vapors in air or other gaseous mixtures	Impinger, gas wash bottle, or similar device	*Solvent acts as a selective extractant, retaining the sample because of its higher affinity for the solvent compared to the gas. *Solutions of chemical or physical complexing agents may be used to improve the extraction efficiency. *Solubility of the extractants may be adjusted by changes in the temperature of the extracting solvent.
Aqueous solutions	Separatory funnel Continuous extractor Counter-current distribu-tion apparatus	*In its simplest form an aliquot of the aqueous solution is shaken with an equal volume of an immiscible organic solvent. *Limited to small sample volumes and solutes with large distribution constants. *Several extractions are required when the distribution constant is small. The addition of salts, pH adjustment, ion-pairing reagents, etc., can be used to improve the distribution of organic solutes into the extracting solvent. *When the distribution constant is very small continuous liquid-liquid extraction or countercurrent distribution apparatus is required. *Large sample volumes may be extracted using continuous and countercurrent distribution methods.
Organic liquids	Mixing device	*Selective extraction by mixing with an organic solvent. Trituration of semi-liquid samples. *Precipitation or freezing used to remove coextractants.
Solid samples	Shaker Homogenizer Soxhlet extractor	*Solid samples such as soil are usually mechanically shaken with an appropriate solvent for a set period of time. *Extraction efficiency may be improved by warming the solvent or by heating to reflux. *Bulky samples such as plant materials are dried, cut, ground, pressed or milled and sieved prior to extraction to promote even and efficient extraction. *Tissue samples are homogenized in the presence of a water miscible organic solvent to promote efficient extraction. *Samples which are difficult to extract efficiently with a few solvent exchanges can be extracted continuously at room temperature or at the boiling point of the solvent in a Soxhlet apparatus.

TABLE 8.3

BASIC RELATIONSHIPS FOR PREDICTING SOLUTE DISTRIBUTION IN LIQUID-
LIQUID PARTITION

Nernst Distribution Law: Any neutral species will distribute
between two immiscible solvents such that the ratio of the
concentrations remains a constant.

$$K_D = [A]_o/[A]_{aq}$$

where K_D is the distribution constant, $[A]_o$ the concentration
of A in the organic phase, and $[A]_{aq}$ the concentration of A in the
aqueous phase.

The fraction of A extracted is given by E,

$$E = [A]_o V_o/([A]_o V_o + [A]_{aq} V_{aq}) = K_D V/(1 + K_D V)$$

where V is the phase ratio V_o/V_{aq}, V_o the volume of organic
phase and V_{aq} the volume of aqueous phase.

The fraction extracted in n successive extractions is:

$$E = 1 - [1/(1 + K_D V)]^n$$

when $K_D V$ is 10, 99% of the solute is extracted with n = 2, when
$K_D V$ is 1, 99% of solute is extracted with n = 7, and when $K_D V$
is 0.1, 50% of solute is extracted with n = 7.

Countercurrent Distribution: The relationship of the distribution
constant K_D of the solute in a CCD process to the concentration
in the various separatory funnels or stages is given by:

$$[E + (1 - E)]^n = 1$$

The fraction $T_{n,r}$ of the solute present in the rth stage for
n transfers is given by:

$$T_{n,r} = n!(K_D V)/[r! (n-r)! (1 + K_D V)^n]$$

---

number of extractions that can be performed before the method
becomes tedious and results in a very dilute sample extract. For
extractions in which K_D is small, automated countercurrent
distribution apparatus should be used, so that n can be made very
large. For the intermediate case a higher extraction efficiency is
obtained by sequential extraction with several portions of solvent
rather than performing a single extraction with the same total
volume of solvent.

For some systems the value of K_D may be made more favorable
by adjusting pH to prevent ionization of acids or bases, by
forming ion pairs with ionizable solutes, by forming hydrophobic
complexes with metal ions, or adding neutral salts to the aqueous

phase to diminish the solubility of the analyte [68-70]. The ion-pair extraction principle can be used for all kinds of ionizable organic substances, but it offers particular advantages for compounds that are difficult to extract in an uncharged form, such as strong acids and bases and aprotic ions, such as quaternary ammonium ions. The extraction efficiency of an ion pair depends on the concentration of the counterion in the aqueous phase, the properties of the counterion and the characteristics of the extraction solvent. Thus some considerable versatility in choosing the experimental conditions is possible.

Although solvent extraction is simple and does not require complex equipment, it is not entirely free of practical problems. The formation of emulsions can be a problem if they cannot be readily broken up by conventional techniques, such as filtration through a glass wool plug or phase separation filter papers, centrifugation, refrigeration, salting out, or addition of a small volume of organic solvent. The rate and extent of extraction may be different for a solute in a test system and in the presence of a sample matrix. Partial association of drug substances with proteins in plasma samples has been recognized as one instance where the extraction efficiency may vary substantially from that obtained using water as a model system. Dilution of the plasma sample with water or the formation of a homogeneous extract prior to forming the two phase system will often solve this problem. For analytes at trace concentrations contamination from solvents and glassware or losses due to adsorption is a frequently encountered problem. Glassware should be thoroughly cleaned and rinsed with pure solvent and, if stored, maintained in a dust-free area. Heating in a muffle furnace or high temperature vacuum oven is often a convenient method of decontaminating glassware. Solution or vapor phase silanization can be used to deactivate clean glass surfaces [71]. Whenever possible the sample should contact only those materials that can be easily cleaned, such as stainless steel, Teflon and glass.

8.4.1 Extraction of Gases and Organic Vapors

Impingers and bubblers, Figure 8.5, containing liquids have been used extensively for collecting high boiling, reactive or polar substances that cannot be quantitatively recovered from solid sorbents [8,72]. They are most frequently used as personal sampling systems and in combination with filters and various

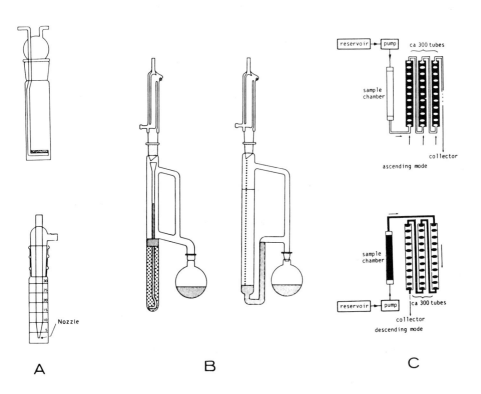

Figure 8.5 Apparatus for liquid extraction. A, bubblers and impingers; B, lighter-than and heavy-than water continuous liquid-liquid extractors; C, droplet countercurrent chromatograph.

sorbents for sampling particles, vapors and gases in industrial stacks, engine exhausts, etc. Bubblers and impingers have several disadvantages for personal monitoring. They are clumsy to wear, there is a large dilution factor due to the large solution volumes generally required, high air sampling rates and, hence, large pumps are needed, and evaporative losses must be controlled or accounted for in some way.

The efficiency with which analytes are collected depends on several factors of which the most important are: solubility of the analyte in the collecting liquid; rate of diffusion of the analyte into the liquid; vapor pressure of the analyte at the sampling temperature; contact area (bubble size) and time of contact (air flow rate); vapor pressure of collecting solvent; and chemical reactivity of the analyte with the reagent. The impinger design, which does not insure intimate mixing of the gas as it passes through the solvent, is suitable for gases and vapors that are strongly retained by some mechanism, particularly those that react chemically. Bubblers provide a more efficient contacting of gas and solvent but require more careful flow calibration and are more difficult to clean. Typical applications of bubblers and impingers include the determination of reactive gases, such as hydrogen sulfide and sulfur dioxide; acidic or basic organic vapors of low volatility, such as amines, carboxylic acids and phenols; and of chemically unstable compounds, such as aldehydes, chloroacetyl chloride, etc., after conversion to a more stable product with favorable detection characteristics.

8.4.2 Extraction of Aqueous Solutions

The most common method of isolation and sample cleanup involves contacting a filtered aqueous solution with an appropriate immiscible organic solvent in a laboratory separatory funnel of appropriate size. Some specific examples are discussed later. With multicomponent samples a single solvent or solvent mixture is unlikely to extract all components equally causing discrimination. This discrimination may be useful if the solvent discriminates against the extraction of solutes that are not of interest in the analysis.

Continuous liquid extraction techniques are used when the sample volume is large, the distribution constant is small, or the rate of extraction is slow. The efficiency of extraction depends on many factors including the viscosity of the phases, the magnitude of the distribution constant, the relative phase volumes, the interfacial surface area, and the relative velocity of the phases. Numerous continuous extractors using lighter-than-water and heavier-than-water solvents have been described [3,27,42,73,74]. Generally, either the lighter or heavier density

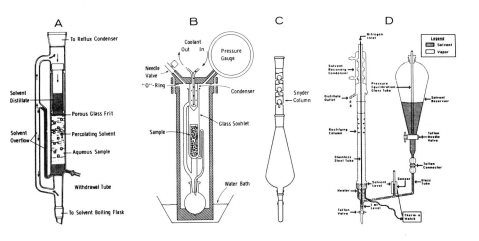

Figure 8.6 Apparatus used for sample preparation involving solvent extraction. A, heavier-than-water continuous liquid-liquid extractor; B, pressurized Soxhlet extractor for use with supercritical fluids; C, Kuderna-Danish evaporative concentrator; D, automated evaporative concentrator.

liquid is boiled, condensed, and allowed to percolate repetitively through an immiscible companion solvent in which the sample is dissolved. In this way, the total extraction of the solute in the companion solvent can be carried out continuously and progressively with a fixed volume of organic solvent. Some examples of laboratory scale continuous extractors are shown in Figures 8.5 and 8.6. Other designs are capable of handling large sample volumes by coupling several units in series [75], can be used for on-site sampling [76,77], or can be easily modified to work with variable sample volumes [78,79].

An additional method of increasing the efficiency of liquid-liquid extraction is based on the countercurrent distribution principle [80-86]. Early countercurrent distribution apparatus were really automated versions of the batchwise extraction process. Countercurrent distribution apparatuses allow one thousand or more equilibration steps between two immiscible liquids of fixed volume and different densities. Mixing, separation, and phase transfer between stages are automated using an electrical timer and motor-driven shaft, on which the individual stages are mounted in series. These apparatuses were complex and of large size, separations were time consuming, and the volume of solvent required (fixed by the size of the apparatus) were often excessive. Increasing interest in countercurrent separations in recent years is due to the development of small laboratory-scale units which work in a continuous manner, such as the droplet countercurrent chromatograph and a variety of flow-through coil planet centrifuge devices. The droplet countercurrent apparatus, Figure 8.5(C), consists of a number of narrow bore silanized glass tubes held in the vertical position. The tubes are connected head to toe by capillary Teflon tubes and the whole assembly filled with stationary phase. The sample dissolved in either the stationary or mobile phase (or a mixture of both) is then pumped into the apparatus where it enters the head or bottom of the first tube, depending on the relative density of the mobile phase. A steady stream of droplets is formed and maintained by the continuous delivery of mobile phase. When a droplet reaches the end of the tube it is delivered by the Teflon capillary to the next tube where droplets are formed again, and so on, until all tubes have been traversed and the separated components in the mobile phase are collected. The large number of droplets provides for an efficient partition process but the method is time consuming (typically 1 to 4 days) and the necessity of droplet formation limits the choice of solvent systems. Shorter separation times and greater efficiency may be obtained using centrifugal coil planet centrifuge devices. In this case the coiled separation column is subjected to a planetary motion: revolution around the central axis of the centrifuge produces a strong centrifugal force field to ensure the retention of stationary phase while rotation about the column holder axis introduces efficient mixing of the two solvent phases to promote the distribution process. The mobile phase is pumped continuously through the coiled column and the separated components collected at the exit of the column.

Countercurrent chromatography is particularly attractive for the preliminary fractionation of complex mixtures prior to further analysis or for the isolation of compounds which are unsuitable for separation by liquid chromatography due to undesirable interactions with solid stationary phases (e.g., irreversible adsorption, denaturization, etc.). For these reasons most of the proven applications in the literature involve the isolation of natural products or biochemical compounds from crude plant or animal extracts and biological fluids.

8.4.3 Extraction of Solid Samples

There is neither a single extraction technique nor a single solvent that is effective for the efficient extraction of all organic compounds from all sample matrices. In the absence of a universal extraction procedure efficient recovery of a specific analyte from a particular matrix requires tailoring of both the extraction technique and solvent to the problem at hand. The most common extraction techniques include Soxhlet extraction, shake flask methods, sonication and various homogenization techniques [87-92]. Less well characterized methods include microwave extraction [93] and solvent leaching at high temperature with forced solvent flow [94]. Using microwave irradiation it is claimed that the degradative effects of high temperatures can be avoided and extractions performed more rapidly than by classical Soxhlet extraction. The forced flow solvent leaching method is suitable for particle samples which can be packed into a heated stainless steel tube through which a solvent close to its boiling point is forced to flow by nitrogen overpressure. For certain samples this method is faster than classical Soxhlet extraction and requires a much smaller volume of solvent. Extraction yields seem to be similar to Soxhlet extraction.

Shake flask methods work well when the analyte is very soluble in the extracting solvent and the sample matrix is a liquid or porous solid. Equilibrium of the analyte between the solid and solvent should strongly favor the solvent and the solid and solvent should be easily separated by centrifugation or filtration. Shake flask methods are probably the least efficient of the solid extraction methods and generally produce the lowest yields for difficult to extract matrices. Sonication and homogenization methods provide a more efficient contact between the solid and solvent usually resulting in greater recovery of the analyte. Tissue samples are usually homogenized with a water miscible organic solvent to simultaneously commute and extract the sample. The sample can be homogenized alone or after mixing the sample with another agent, such as dry ice or diatomaceous earth. Samples containing a great deal of fat are more easily handled in this way. Sonic treatment, either by immersion in a sonic bath or by using a probe device, is commonly used for coarse granular samples, such as soil, sediments and air particulates. When a sonic bath is used several samples can be extracted simultaneously in an unattended fashion. The gentle heating that occurs with the sonic bath also seems desirable to increase the rate at which equilibrium is attained. It seems to be a general fact that whereas homogenization and sonication can yield similar recoveries of many analytes from diverse matrices, they do not generally yield recoveries exceeding those obtained by Soxhlet extraction, and were discrepancies exist, higher recoveries are usually obtained by Soxhlet extraction. Although, in any particular case, the analyte recovery is affected by the physical and chemical nature of the sample matrix and the structure and concentration of the analyte, which precludes the development of any hard and fast rules. Even though a lower recovery of analyte may be obtained by sonication it may still be preferable in screening studies because it is rapid, employs less solvent than Soxhlet extraction, and frequently extracts fewer interferences.

Soxhlet extraction is probably the most widely used method for the extraction of organic analytes from solid samples. The method works best for samples which can be commuted to a free flowing powder and for analytes of high solubility in the extracting solvent. Ideally the solvent or solvent mixture used for the extraction should have a high affinity for the analyte, a low affinity for the analyte matrix, and a high volatility and low viscosity to aid its removal from the extracted sample. The basis of the Soxhlet method is that a suitable solvent is vaporized, condensed, and allowed to percolate through the solid sample contained in an extraction thimble. The return of the solvent to the boiling flask is discontinuous; working on the siphon principle, the solvent is returned only when a certain volume of solvent is accumulated in the extraction chamber. Soxhlet extractors are available for milligram to kilogram sample sizes and for extractions at either room temperature or near the boiling point of the solvent [27,95,96]. Unattended operation is possible which is fortunate as extraction times are usually fairly long, for example, 24 hours. The main disadvantage is that the compounds extracted must be stable at the boiling point of the extracting solvent, as eventually they are accumulated in the boiling flask. Jennings has described a pressurized system that permits the use of liquid carbon dioxide with a standard glass Soxhlet extractor, Figure 8.6(B), for the extraction of solid samples and for the recovery of organic volatiles from adsorbents [3,97]. The extractor, which can utilize dry ice as a source of carbon dioxide, has the advantage of achieving high extraction efficiencies under an inert atmosphere and at low temperatures. The carbon dioxide is allowed to dissipate at subambient temperatures, yielding a solvent-free extract for analysis. Supercritical fluids, and in particular carbon dioxide, are excellent solvents for many organic compounds [98].

8.4.4 Solvent Reduction Methods

Liquid extraction uses large solvent volumes compared to the sample volumes that can be utilized for analysis by chromatographic instruments. This dilution of the sample is often too great for direct determination without a preconcentration step. Large sample volumes may be evaporated using a rotary evaporator [99,100], Kuderna-Danish evaporative concentrator, Figure 8.6(C), automated evaporative concentrator (EVACS), Figure 8.6(D) [101], or the gas blow-down method. Rotary evaporators are available in most laboratories and provide a convenient means of solvent evaporation under reduced pressure. Volatile compounds are largely lost during concentration with this apparatus and even the recovery of less volatile materials may be lower than expected due to entrainment of the sample in the solvent vapors, adsorption on the glass walls of the flask and apparatus, as well as from uncontrolled expulsion from the flak due to uneven evaporation. Slow evaporation and ice bath temperatures are recommended to improve the recovery of semi-volatile analytes [100]. The Kuderna-Danish evaporative concentrator is generally operated at atmospheric pressure under partial reflux conditions using a three-ball Snyder column with an efficiency of about 2.7 theoretical plates. Condensed vapors in the Snyder column are returned to the boiling flask, washing down organics from the sides of the glassware; the returning condensate also contacts the rising vapors and helps to recondense volatile organics. Although the Kuderna-Danish concentrator provides a slower rate of evaporation than rotary evaporators, it generally provides higher recoveries of trace organic analytes [99,100]. It is generally not possible to reduce sample volumes of several hundred milliliters to less than one milliliter in a single apparatus. Samples are usually concentrated to 5-10 ml and then transferred to a micro Kuderna-Danish evaporator or to a controlled-rate evaporative concentrator [102,103]. The Kuderna-Danish technique is wasteful of solvent which is not recovered, time consuming, and requires constant attention to control any violent boiling of the solvent in the first stage and to avoid evaporation to dryness in the second stage. The automated evaporative concentrator, EVACS, is designed for the unattended solvent evaporation of organic extracts for which there is a boiling point difference of about 50°C between the solvent and analytes. The flow of solvent from a pressure equalized conical reservoir is fed at a controlled rate into a concentration chamber where the solvent is vaporized through a short distillation column with about 3.5 theoretical plates. The distillate is withdrawn at the head of the column while maintaining the reflux ratio as high as possible. Nitrogen gas in the absence of heat is used to complete the concentration of the sample from about 10 ml to 1.0 ml. The use of liquid level monitors enables the concentrator to operate unattended without fear of evaporating the solvent to dryness.

The gas blow-down method is suitable for the evaporation of volatile solvents of less than 25 ml volume. A gentle stream of pure gas is passed over the surface of the extract contained in a conical-tipped vessel or culture tube partially immersed in a water bath or heating block [100,104]. The solvent evaporation rate is a function of the gas flow rate, the position of the gas inlet tube relative to the refluxing solvent, the water bath temperature, and the solvent surface area. High gas flow rates must be avoided to prevent loss of sample by nebulization. The choice of gas is important to avoid contaminating the sample. Cylinder gases such as prepurified nitrogen or helium should be used and further purified if contamination is a problem. Two convenient high capacity methods of gas decontamination are passage through a U-tube filled with a mixture of molecular sieve and carbosieve immersed in a dry ice-acetone coolant bath [105] or by using a femtogas purification train [106].

8.4.5 Micromethods of Solvent Extraction

Since the solvent evaporation process is slow and the

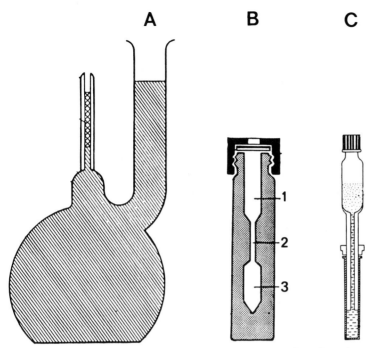

A   B   C

Figure 8.7 Apparatus for micromethods of solvent extraction. A, micro extraction apparatus; B, Keele micro-reactor; C, Mixor phase separation device.

possibility of contamination from the solvent and glassware high, micromethods using small solvent and sample volumes are attractive and compatible with the sample size requirements of chromatographic techniques. Figure 8.7(A) is an example of a micro-extraction apparatus which enables up to 980 ml of an aqueous sample to be extracted with 200 microliters of hexane [107]. The sample is extracted by shaking, and after allowing sufficient time for phase separation, additional water is added through the side arm, forcing the lighter organic phase into a capillary tube where it can be sampled by syringe. Typical recoveries of hydrocarbons, chlorinated pesticides, and phthalate esters at trace levels average 40-70% for one extraction and 89-99% for three consecutive extractions. Similarly, glass syringes with a luer-lok fitting can be used to extract sample volumes of 1 to 50 ml with solvent volumes of 20 microliters and greater [108]. Extraction takes place in the syringe barrel and, after phase separation, the organic layer is easily dispensed through a narrow-bore glass transfer tube attached by a PTFE sleeve to the luer-lok fitting. Morgan et al [109,110] have developed a micro-reactor and extractor for very small sample volumes, Figure 8.7(B). The micro-reactor consists of a bottom conical tip chamber of 75 microliter volume connected to a larger upper chamber of about 600 microliters volume by a capillary neck with a volume of about 5 microliters. After mixing the phases by pumping in and out with a standard syringe, either the heavier or lighter phase can be removed by forcing it into the neck region. Figure 8.7(C) is a diagram of a commercially available piston device that can be used to extract from 2 to 20 ml volumes of solution with a few hundred microliters of solvent [111]. The device consists of a glass mixing reservoir and glass piston with a channel to connect the upper and lower reservoirs. A pumping motion of the piston forces the liquid back and forth through the narrow channel between the upper and lower reservoirs creating efficient mixing of the two liquid phases. The piston can be adjusted to allow phase separation to occur and is then gently depressed to force the lower density liquid into the upper reservoir for convenient isolation of the two phase compartments. The books by Ma and Horak [27] and Dunges [112] contain other examples of micro-apparatus useful for extractions. When handling microvolume samples in relatively large volume vessels the influence of the headspace volume on sample composition needs to be considered [113] and special attention may also have to be given to methods of further concentrating the sample volume if low recoveries are to be avoided [114].

8.4.6 Some Practical Examples

Figure 8.8 is an example of the use of solvent extraction to isolate and concentrate the insect molting hormone ecdysterone $(2\beta,3\beta,14\alpha,20,22,25$-hexahydroxy-5$\beta$-cholest-7-en-6-one) from the desert locust [115,116]. The hormone is very soluble in methanol which was selected for the extraction. The anesthetized insects were macerated in a blender with methanol and the insoluble residue discarded. Hexane was used to remove fats and lipids from the polar fraction and selective partition with butanol isolates ecdysterone from water-soluble coextractants.

Figure 8.9 is an example of the use of solvent extraction to isolate polychlorinated biphenyls from a fat sample [6]. In this example the matrix is chemically modified to improve the selectivity of the extraction. The fat is first hydrolyzed by refluxing in 1 N ethanolic potassium hydroxide prior to the

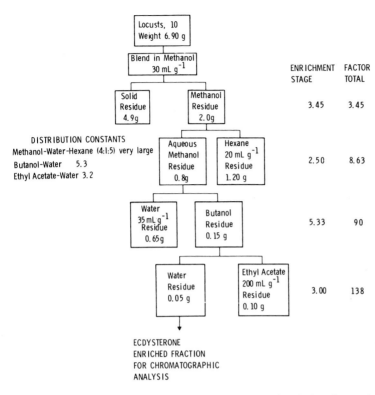

Figure 8.8   Extraction scheme used to isolate insect molting hormone (ecdysterone) from the desert locust.

addition of hexane and water. The hydrolyzed fats are removed in the ethanol-water phase concentrating the moderately polar polychlorinated biphenyls into the hexane layer.

Figure 8.10 provides a general outline for the isolation of drugs from whole blood, urine and faeces [117]. Basic drugs are generally extracted at pH > 7.0 from whole blood while acidic drugs are best extracted below pH 5.0, preferably from plasma or serum (which already has the bulk of the haemprotein removed). The separation of serum requires natural clotting of fibrinogen, which precipitates out the bulk of the haem- and lipoprotein. The separation of plasma requires the addition of a anticoagulant (oxalate, citrate, heparin, etc.) to prevent clotting followed by centrifugation to separate the red cells from the lipids and lipoprotein fractions. Lipids can be removed in a subsequent step by extraction with a nonpolar solvent, if desired. Proteins can be precipitated by several methods, such as by addition of an acid (perchloric acid, trichloroacetic acid, hydrochloric acid, etc.),

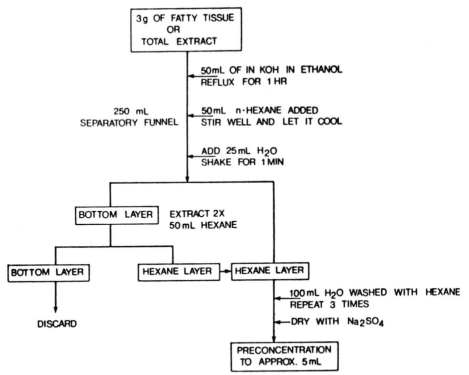

Figure 8.9 Solvent partition scheme for the isolation of polychlorinated biphenyls from fat samples.

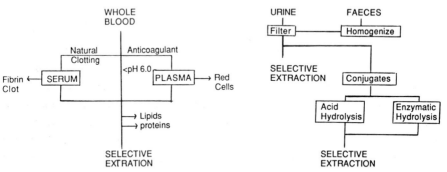

Figure 8.10 General scheme for the isolation of drugs from biological fluids.

salting-out with ammonium sulfate, zinc sulfate/barium hydroxide, etc., dilution with an organic solvent, such as ethanol or acetonitrile, or by hydrolysis with a proteolytic enzyme [117-121]. The lipid/protein free extract of whole blood is then suitable for the selective isolation of drugs using different solvents, changes in pH, ion-pair formation, etc., based on the physical characteristics of the substance to be isolated. The lipid/protein concentrations in urine is usually much lower than in plasma and direct extraction of the filtered sample is possible for many drugs. This provides information concerning the free drug concentration. Urine contains substantial amounts of bound and conjugated drugs which are generally not directly extractable without hydrolysis. This is usually performed by acid or enzymatic hydrolysis (e.g., incubation with β-glucuronidase/sulfatase at 37°C for 2-12 hours).

Figure 8.11 is an example of a general solvent extraction scheme for isolating neutral, base, weak acid and strong acid fractions from a sample that is soluble in methylene chloride [122]. Note that the acid and base fractions are extracted by changing the pH of the aqueous phase to manipulate the selectivity (K_D values) of the solvent system. Neutral solutes are not affected by changes in pH but further fractionation is possible using selective chemical reagents. Sodium bisulfite or Girard T and P reagents can be used to extract aldehydes and ketones as water-soluble derivatives. The chemistry and properties of Girard T (trimethylethylammonium acetyl hydrazide chloride) and Girard P (pyridinium acetyl hydrazide chloride) reagents are reviewed in reference [123]. These derivatives can be easily hydrolyzed back to the parent compound at high pH.

Dimethyl sulfoxide provides a high extraction efficiency for compounds that contain hydrogen bonding or π-electron rich sites [124]. The solvent system pentane-dimethyl sulfoxide is useful for the fractionation of polar and unsaturated solutes, notably the separation of aliphatic and polycyclic aromatic hydrocarbons, Figure 8.12. Partitioning between pentane and dimethyl sulfoxide provides a separation of polar and nonpolar solutes. The dilution of the dimethyl sulfoxide phase with water diminishes the extent of π-electron attraction with the solutes without dramatically diminishing hydrogen bonding interactions. Back extraction of the aqueous dimethyl sulfoxide phase with pentane provides a separation between hydrogen bonding and neutral polar solutes.

8.5 SAMPLE CLEANUP USING LIQUID CHROMATOGRAPHY

Liquid chromatography is frequently used to further fractionate a sample after solvent extraction on the basis of differences in polarity, size or ion-exchange capacity. Either

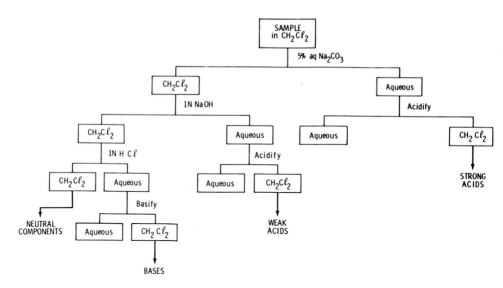

Figure 8.11 Solvent fractionation scheme for isolating neutral, base, and acid fractions from a sample soluble in methylene chloride.

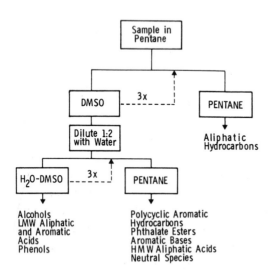

Figure 8.12 Selective solvent partition scheme for isolating polar and unsaturated solutes.

column or thin-layer techniques may be used, but column methods are generally preferred, since sample recovery is more

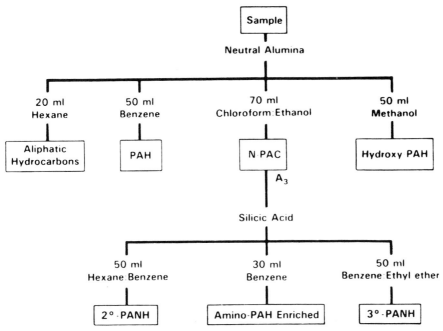

Figure 8.13   Class fractionation of polycyclic compounds on columns of alumina and silica gel (Reproduced with permission from reference 127. Copyright American Chemical Society).

straightforward. Large-scale gravity fed columns, short-bed gravity fed or forced-flow columns, and medium- and high-pressure analytical and semipreparative columns are all used for specific applications depending on the amount and complexity of the sample. Large-scale columns are used to obtain bulk fractions for chemical identifications or biological testing [125] and the smaller columns for minimizing interferences during the final chromatographic analysis.

The most widely employed adsorbents for sample cleanup are silica gel, alumina, Florisil, carbon and diatomaceous earths, Table 8.4. Silica, alumina and Florisil are used to provide class separations by dividing the sample into fractions containing a similar number and type of functional groups. Typical examples are the isolation of lipids by chromatography over silica gel using chloroform to elute simple lipids, acetone to elute glycolipids, and methanol to elute phospholipids [126]. A further example is the group fractionation of polycyclic aromatic compounds in synthetic fuels and similar samples over neutral alumina and silica gel with a step solvent gradient, Figure 8.13 [127]. The sample is first separated into aliphatic hydrocarbons, polycyclic

TABLE 8.4

ADSORBENTS FREQUENTLY USED FOR SAMPLE CLEANUP

Type	Composition	Comments
Silica Gel	$SiO_2 \cdot xH_2O$	Prepared from sodium silicate using the sol-gel procedure. Silica gel is the most widely used general purpose adsorbent for sample cleanup although it may irreversibly bind some strongly basic substances. Generally considered to be slightly acidic in character but this may vary with the method of preparation. Activated by heating at about 180 °C for 8-12 hours.
Alumina	$Al_2O_3 \cdot xH_2O$	Prepared by the low temperature dehydration (<700° C) of alumina trihydrate and is a mixture of α alumina with smaller amounts of α alumina (less active form) and sodium carbonate. Neutral alumina (pH 6.9-7.1) is the most widely used form and is suitable for the separation of hydrocarbons, esters, aldehydes, ketones, lactones, quinones, alcohols and weak organic acids and bases. Basic alumina (pH 10-10.5) is used to separate acid-labile substances. Basic alumina in aqueous solution has strong cation-exchange properties. Acid alumina (pH 3.5- 4.5), prepared by acid washing neutral alumina, acts as an anion exchanger and is used to separate organic acids and inorganic compounds. Alumina is activated by heating at 400°C for 8-12 Hours.

Florisil	Magnesium Silicate	Prepared by precipitation from a mixture of magnesium sulfate and sodium silicate solutions and calcined at about 1200 °C. Very porous with a surface area of about 200 -250 m²/g. Batch to batch variation in properties associated with the presence of variable amounts of sodium sulfate. Widely used for the cleanup of pesticide residues in environmental extracts and fats and oils. Basic compounds may be irreversibly adsorbed. Activated by heating at about 130 °C for 8-12 hours.
Celite	Diatomaceous Earth	A form of flux calcined silica of low surface area used mainly as a filter aid and inert support in gas and liquid-liquid chromatography.
Carbon	Activated Charcoal	Prepared by the low temperature pyrolysis of vegetable matter followed by thermal conditioning in air or steam to extend the pore structure. Produces a heterogeneous surface of large area containing variable amounts of acidic functional groups. General purpose adsorbent used to remove organic species from aqueous solution.

aromatic hydrocarbons, nitrogen-containing polycyclic aromatic compounds and hydroxyl-containing polycyclic aromatic hydrocarbons on alumina and the nitrogen-containing polycyclic aromatic hydrocarbons further subdivided into primary, secondary and tertiary amines on silica gel. Florisil has adsorption characteristics complementary to alumina and silica gel [128], and is widely used to isolate pesticides from environmental extracts [129,130].

The column dimensions used are dictated by the sample size and the resolution required. Sample loads are usually about three percent of the weight of stationary phase with wide columns used for large samples and long columns for difficult separations. Careful control of the activity level of the adsorbent is important to obtain well defined fractions. This is achieved by accurately controlling the water content of the adsorbents prior to the separation and predrying extracts with anhydrous sodium sulfate, or a similar drying agent, prior to applying the extract to the column. As an additional precaution a layer of anhydrous sodium sulfate can be placed on top of the adsorbent bed. The adsorbent activity is controlled by the purposeful addition of a known amount of water to the dry adsorbent or by conditioning the adsorbent at a controlled temperature until the adsorbed water reaches equilibrium. Absorbent activity can be estimated by the Brockmann scale, Table 8.5, which, although based on the elution characteristics of a series of test dyes, provides a reasonable general indication of sample retention [129]. Adsorbents may also be standardized using the equivalent retention volume for a standard solute, such as naphthalene, eluted under standard conditions [127]. A calibration curve for the equivalent retention volume as a function of added water enables the water content of an unknown adsorbent sample to be determined. Curves of this kind show a steep change in retention for low water levels eventually flattening out at higher levels. This emphasizes the importance of controlling adsorbent activity, particularly at low deactivation levels, to obtain reproducible sample fractionation on different columns. The activity of different batches of Florisil is usually determined by its adsorptive capacity for lauric acid in hexane. After equilibration with the adsorbent the excess lauric acid is determined by titration [131]. Irreversible adsorption and catalytic degradation of sensitive solutes can occur on all adsorbents and is a source of low recovery for some analytes.

TABLE 8.5

STANDARDIZATION OF ADSORBENT ACTIVITY

Brockmann Activity Grade	Percentage of Water (w/w)		
	Alumina	Silica Gel	Florisil
I	0	0	0
II	3	5	7
III	6	15	15
IV	10	25	25
V	15	38	35

Silica or alumina coated with chemical reagents, such as concentrated sulfuric acid, sodium hydroxide, alkaline potassium permanganate, silver nitrate, etc., can be used to improve the selectivity of the isolation of the analytes from their matrix [50,132-134]. Silver modified silica gel selectively retains olefins by formation of charge-transfer complexes and enhances their separation from alkanes. Acid and base modified silica enhances the retention of acidic and basic compounds from neutrals and alkaline permanganate oxidizes easily oxidized species to more polar solutes that can be separated from neutral analytes.

Whenever the degree of purification of the analyte is inadequate using gravity fed columns, improved results can be obtained using medium pressure and high pressure liquid chromatography [135-139]. Other advantages include saving of time and solvents, reduced solvent contamination problems and ease of automation and monitoring of fraction cuts. Repeated injections of crude extracts onto the column results in an accumulation of coextracted material and loss of column efficiency, which generally means that it is necessary to employ a wash cycle between injections, and if gradient elution is used, to re-equilibrate the column to the appropriate starting conditions. These requirements will usually be reflected in a reduced sample throughput. The capacity of the column depends on the amount of stationary phase available but sample extracts containing up to about 25 mg of material in 200 microliters of mobile phase can be processed using a semipreparative column. Simple matrices, such as animal fats, can often be injected in much larger amounts [136]. Figure 8.14 is an example of a class fractionation of the soluble organic fraction obtained from a diesel engine exhaust particulate sample using a semipreparative column containing silica gel [140].

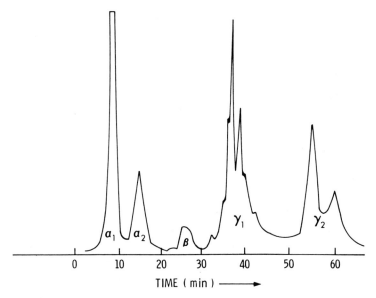

Figure 8.14 Semipreparative class separation of a diesel engine exhaust sample. Column 25 cm x 7.9 mm, 10 micrometer Porasil. Solvent gradient hexane to 5% methylene chloride over 5 min., linear gradient to 100% methylene chloride over 25 min., isocratic for 10 min., linear gradient to 100% acetonitrile over 10 min., isocratic for 5 min., step change to tetrahydrofuran for 10 min., step reverse to acetonitrile for 5 min. The α_1 and α_2 fractions contain 2-4 and 4-6 ring PAHs and saturated aliphatics; β fraction contains 6-8 ring PAHs, hydroxyl- and nitro-PAHs with 2 rings; the γ_1 fraction contains hydroxyl-PAHs (3-4 rings), nitro-PAHs (2-6 rings), PAH quinones (3-5 rings) and PAH ketones (2-3 rings); and the γ_2 fraction contains PAH quinones (3-5 rings), hydroxyl-PAHs (5-7 rings), PAH ketones (3-4 rings) and dinitro-PAHs (3 rings).

Each peak contains components of a similar type, which were collected and subsequently analyzed by gas chromatography and mass spectrometry. Similar techniques have been used to isolate organochlorine pesticides from animal fats and oils [136,138]. Normal phase chromatography on a semipreparative column packed with a 3-aminopropylsiloxane bonded silica packing has been widely used for separating mixtures of polycyclic aromatic compounds into fractions containing isomeric polycyclic aromatic hydrocarbons and their alkyl-substituted homologs with similar ring numbers suitable for subsequent profiling by gas chromatography [141]. Size-exclusion liquid chromatography has become a standard technique for the isolation of pesticides, herbicides, fungicides, aflatoxins, steroids, etc., in samples containing high molecular weight polymers, lipids, fats and oils [49,142-144]. Separations are based on size differences with the high molecular weight components eluting ahead of the low molecular weight analytes. Both polar and nonpolar analytes are isolated in the same fraction in a manner complementary to the polarity separations obtained using inorganic oxide adsorbents. Since the sample rarely contaminates the column (total elution is achieved), the method lends itself to automation and unattended operation.

8.5.1 Solid-Phase Extraction

Solid-phase extraction using small cartridges filled with sorbents of a small particle size has rapidly established itself as an important sample preparation technique for either matrix simplification or trace enrichment [145,146]. It has prospered at the expense of liquid-liquid extraction which is considered labor intensive and frequently plagued by practical problems, such as emulsion formation. Liquid-liquid extraction tends to consume large volumes of high purity solvents which may have significant health hazards and disposal costs associated with their use. Solid-phase extraction has the advantages of lower costs, reduced processing times, substantial solvent savings, and simpler processing procedures. Solid-phase extraction procedures are easily automated using robotics, or special purpose flow processing units that simultaneously extract samples and prepare them for automatic injection [135,145-148] or centrifugal analyzers, which can batchwise process multiple samples [149,150]. This increases sample throughput and reduces labor costs. Solid-phase extraction is convenient for field sampling applications as it minimizes the transport and storage problems associated with bulk samples which have to be returned to the laboratory for processing. Solid-phase extraction techniques are not without problems, but these are different problems to those of liquid-liquid extraction techniques, and the two approaches can be considered complementary. The batch to batch reproducibility of sorbent packings is poor, particularly for chemically bonded phases, which can affect the possibility of implementing standards protocols in critical cases. Since the packing materials are derivatives of products manufactured for HPLC, they are subject to all the same problems observed in the preparation of reproducible column packings (see section 4.2.2). For example, the recovery of basic drugs from silica-based chemically bonded packings can be a precarious process [151]. The chemical background from impurities, contaminants, antioxidants, etc., commonly observed for solid-phase extraction products may interfere in the subsequent analysis of the sample [152]. Cleaning of the cartridges and running blanks to establish background contamination levels diminishes sample throughput and adds significantly to solvent consumption and processing costs. It is also possible for sample processing problems, such as column overloading, displacement and blocking of pores, that is processes related to the competition of the analyte and sample matrix for sorbent sites, to go unnoticed, leading to changes in the recovery of the analyte. Notwithstanding the above problems solid-phase extraction techniques have established themselves as routine techniques in many laboratories, particularly for the analysis of clinical samples [151,153-155], water samples [156-159] and environmental extracts [160-164]. Some further details of applications are summarized in Table 8.6.

The solid-phase sorbents available include the common inorganic adsorbents (silica, alumina, Florisil), siloxane-bonded silica materials (octadecyl, octyl, ethyl, cyclohexyl, phenyl, 3-cyanopropyl, diol, 3-aminopropyl, N-propylethylenediamine, benzenesulfonylpropyl, sulfonylpropyl, carboxymethyl, diethylaminopropyl, trimethylaminopropyl, etc.), nonpolar and ion-exchange macroreticular resins, and some specialty products, such as resin-bonded phenylboronic acids [165] for the isolation of steroids, catecholamines and nucleotides containing a vicinal diol group. Since the materials are similar to those currently used in HPLC except for particle size, the whole range of chemistries used in modern packing technology can be used to prepare any LC packing in a form suitable for solid-phase extraction. The packing materials for solid-phase extraction have particle sizes in the range 30 to 60 micrometers packed into cartridges formed from purified polyethylene, polypropylene or glass tubes. The packing material is sandwiched between two porous frits with a pore diameter of about 20 micrometers, that is, small enough to contain the packing but not too small to impede the flow of sample and solvents through the cartridge. The bottom end of most cartridges are terminated in a Leur-lok fitting for easy connection to a sampling manifold or to connect a syringe needle to direct the effluent into a collection vial. The flow rate of solution through the cartridge is usually controlled by vacuum suction using a vacuum box manifold that can process several samples simultaneously or

TABLE 8.6

CHARACTERISTICS OF SILICA BASED BONDED-PHASE SORBENTS

Sorbent Type	Sample Type	Typical Applications
Octadecyl	Reversed-phase extraction of nonpolar compounds.	Drugs, essential oils, food preservatives, vitamins, plasticizers, pesticides, steroids, hydrocarbons.
Octyl	Reversed-phase extraction of moderately polar compounds. Compounds bound too tightly by octadecyl silica.	Priority pollutants, pesticides.
Phenyl	Reversed-phase extraction of nonpolar compounds. Provides less retention of hydrophobic compounds.	Does not seem to be widely used.
Cyanopropyl	Normal phase extraction of polar compounds.	Amines, alcohols, dyes, vitamins, phenols.
Silica Gel	Adsorption of polar compounds.	Drugs, alkaloids, mycotoxins, amino acids, flavinoids, heterocyclic compounds, lipids, steroids, organic acids, terpenes, vitamins.
Diol Functionality	Normal phase extraction of polar compounds (similar to silica gel).	Proteins, peptides, surfactants.
Aminopropyl	Normal phase extraction.	Carbohydrates, peptides nucleotides, steroids, vitamins.
Dimethyl-aminopropyl	Weak anion-exchange extraction.	Amino acids.
Aromatic Sulfonic Acid Functionality	Strong cationic-exchange extraction and reversed-phase extraction (eliminates ion pairing when used in place of octadecyl silica.	Amino acids, catecholamies, nucleosides, nucleic acid bases.
Quaternary Amines	Strong anion-exchange extraction.	Antibiotics, nucleotides, nucleic acids.

the cartridge is designed to allow a syringe to be connected at the top to force liquid through the cartridge. The cartridges are available in several sizes containing from 35 mg to 5 g of sorbent with the 100 mg and 500 mg sorbent cartridges being the most commonly used. As a rough approximation the total sample capacity of the cartridges is about 1-5% of the sorbent mass or 0.4-0.6 mequiv./ml for ion-exchange sorbents. The sample volume that can be processed depends primarily on the breakthrough volume of the analyte, the concentration of the analyte matrix, and the sample flow rate [146,154,166]. Normally, sample volumes are limited to less than 2 liters, but because of the small volume of solvent needed to elute the extracts from the cartridge, concentration factors up to 1000 fold are easily achieved in favorable circumstances. For clinical samples only small sample volumes are generally available, except for urine, and the concentration factors achieved are much lower.

A new approach to solid-phase extraction takes advantage of the rapid processing possibilities achieved by enmeshing the sorbent particles in a web of PTFE microfibrils to create a membrane, which can be used with a simple vacuum-assisted filtration apparatus [167,168]. The membranes are 25 or 47 mm in diameter and 0.5 mm thick containing about 90% by weight of sorbent particles of a smaller average particle size, 8 micrometers, than used in the conventional solid-phase extraction cartridges. The 47 mm membrane disk contains about 500 mg of sorbent and has a similar collection efficiency for standard compounds as a typical cartridge containing the same amount of sorbent. Also, membrane extraction disks have been used successfully for the on-line preconcentration of pesticides and herbicides by HPLC [168]. Further studies are required for a wider range of problems to establish the general utility and preference of the membrane approach over the use of packed cartridges.

Whether membranes or cartridges are used for solid-phase extraction, sample processing involves four distinct steps. The sorbent bed is conditioned with an organic solvent to extend the bonded-phase ligands and increase the surface area available for interaction with the sample and to remove impurities from the sorbent. Failure to carry out this stage effectively will result in poor recoveries of the analyte due to reduced retention and the appearance of interference peaks in the chromatogram. Before applying the sample to the sorbent it is returned to the ready condition by washing with 3 or 4 bed volumes of a solvent as similar to the sample solvent in polarity, ionic strength and pH as possible. The sample is then sorbed onto the cartridge. Some common problems encountered in sample processing and possible solutions are summarized in Table 8.7. After the sample is loaded onto the sorbent, the cartridge is rinsed with a weak solvent to remove undesired matrix components. The choice of solvent for this step is critical if matrix simplification is to be achieved. The

TABLE 8.7

SAMPLE PROCESSING PROBLEMS IN SOLID-PHASE EXTRACTION

Sample Type	Remedy
Viscous Sample	Dilute to reduce processing time
Particle Matter	Remove by prefiltering or centrifugation. Addition of concentrated HCl is effective in dissolving inorganic particles when processing water samples.
Large Volume Aqueous Samples	Add 1-3% (v/v) organic solvent, such as methanol, to insure that the sorbent remains wetted by the sample.
Easily Ionized Substances	Adjust pH to enhance interaction with the sorbent.
Ionic Strength	Maintain approximately constant for samples and standards.
Biological Fluids	Deproteination often required to give acceptable recoveries

solvent must effectively desorb the matrix components without affecting the analytes of interest. In the final step, the analytes are eluted by a solvent of sufficient strength to desorb the analytes in a small volume without displacing more strongly sorbed matrix components.

Solid-phase extraction is a convenient sample preparation technique for gas, liquid and thin-layer chromatography. In liquid chromatography the degree of sample cleanup is generally higher when the retention mechanisms employed for the sorbent extraction and analytical separation are complementary. When identical sorbents are used for extraction and separation the extraction column functions as an additional guard column to protect the analytical column from contamination by either particles or strongly retained matrix components. Figure 8.15 provides an example of the use of an ion-exchange sorbent cartridge to isolate N,N-dimethylaminopyrene from an urban air particulate extract prior to analysis by reversed-phase liquid chromatography [13]. N,N-Dimethylaminopyrene was formed by reductive alkylation of 1-nitropyrene, a potent bacterial mutagen and suspect human carcinogen. The nitro group quenches the fluorescence of the pyrene ring, and 1-nitropyrene cannot be detected at trace levels using fluorescence. The dimethylamino derivative is strongly

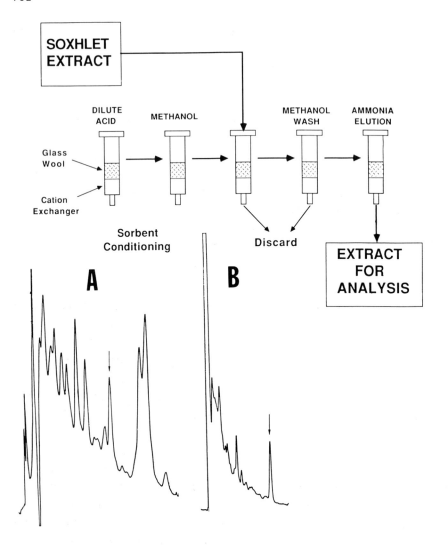

Figure 8.15   Isolation of N,N-dimethylaminopyrene by ion-exchange extraction using a solid-phase extraction cartridge prior to analysis by HPLC. The glass wool plug was used to filter out the reduction catalyst. Chromatogram A was obtained by derivatizing 1-nitropyrene in an urban air particulate sample and chromatogram B the same sample after ion-exchange cleanup. The arrow marks the position of the derivative. Reversed-phase chromatography on an octadecylsiloxane bonded silica column with acetonitrile/water (75:25) as the mobile phase was used for the separation. The derivative was detected using fluorescence excitation at 240 nm and emission at 370 nm.

fluorescent and simple to isolate by ion-exchange chromatography. The ion-exchange cartridge is first activated with phosphoric acid and then washed with methanol. The sample is applied to the cartridge in a mixture of methanol and benzene, which are the solvents for the derivatization reaction. Non-sorbed matrix components are then washed through the ion-exchange cartridge with methanol. The derivative is subsequently recovered by elution with a small volume of ammonia. After solid phase extraction interfering coextractants are largely eliminated from the chromatogram and a more reliable baseline is obtained for quantitation. For more difficult problems it may be necessary to combine solid-phase extraction with semipreparative liquid chromatography for adequate sample cleanup [135].

8.6 SORPTION TECHNIQUES FOR TRACE ENRICHMENT OF ORANICS IN WATER

Solid-phase extraction is widely used for trace enrichment of very dilute solutions, such as natural waters, where large sample volumes may have to be processed to yield concentrations of analyte sufficient for convenient detection. By selection of the correct sorbent semivolatile and involatile neutral and ionic species can be recovered in high yield from water, although because of the nature of the manipulation and sampling steps, the recovery of organic volatiles is generally poor and gas phase sparging and headspace techniques are considered more suitable for these analytes. The efficiency with which organic solutes are recovered from water depends on the relative importance of the analyte-water, analyte-sorbent and sorbent-water interactions. For effective sorption the analyte-water and sorbent-water interactions should be weak and the analyte-sorbent interactions strong. With the exception of ion-exchange sorbents this essentially dictates that those sorbents of general use will be nonpolar, since otherwise the volume of sample which can be processed before breakthrough of the analytes will be small due to strong water-sorbent interactions. For this reason the most commonly employed sorbents are carbon, macroreticular polymeric resins, polyurethane foams and siloxane-bonded silica materials; the latter were discussed in section 8.5.

8.6.1 Granular Activated Carbon

Carbon adsorbents are available in many forms differing in surface area, pore structure and functionality depending on the starting material for their preparation and subsequent activation. Graphitized carbon blacks have low surface areas, usually less than 100 m^2/g and typically 5-30 m^2/g. They have a low capacity for low molecular weight analytes but have been used successfully for the isolation of heavier nonpolar analytes that might be difficult to recover from more active forms of carbon [169,170]. The most common form of carbon used for trace enrichment is granular activated carbon, generally prepared by the low temperature oxidation of vegetable charcoals. These materials have a large surface area (300-2000 m^2/g), a wide pore diameter distribution, and a heterogeneous surface containing active functional groups [171-173]. Protocols based on the use of large filter cartridges containing granular activated carbon have been standardized for the determination of dissolved organic compounds in drinking and surface water where analyte concentrations are generally less than 50 micrograms per liter. Precleaned carbon of about 0.5 mm particle diameter packed into a 18 x 3 inch diameter glass tube and a sampling flow rate of 120 ml/min are fairly typical conditions to process a volume of about 1,200 liters of water requiring about one week for completion [171-175]. The carbon adsorbent is then removed from the cartridge, spread on a glass tray, and dried at 35-40°C until free flowing. The adsorbed organics are removed by Soxhlet extraction with chloroform and/or ethanol to yield about 20-200 mg of material. Such large scale sampling is used primarily for health effects research and chemical identification.

Analytical sampling using smaller sample volumes is less common, since other martials, such as the macroreticular porous polymers, are often more efficient [176,177]. Problems in using carbon for analytical work are related to large variations in the physical and chemical properties of different sources of carbon, the greater difficulty in obtaining low background contamination levels compared to other sorbents, and the fact that many organic compounds are adsorbed so strongly that desorption is accomplished in very low yield. Activated carbon has a very complex surface structure containing a wide range of functional groups including phenolic, carboxylic, quinone and lactone groups. The principal binding mechanisms include hydrophobic interactions, charge-transfer complexation, hydrogen bonding and cation exchange. Low recovery of many organic solutes is often associated with the strength of these multiple binding interactions, which provide an efficient adsorption mechanism accompanied by inefficient solvent elution, leading to a low overall recovery. Irreversible adsorption can occur in some cases as well as catalytic transformations to different products.

8.6.2 Macroreticular Porous Polymer Adsorbents

Macroreticular resins are prepared by suspension polymerization of, for example, styrene-divinylbenzene copolymers in the presence of a substance which is a good solvent for the monomer but a poor swelling agent for the polymer [178-180]. Each resin bead is formed from many microbeads joined together during the polymerization process to create a network of holes and channels. This results in greater mechanical stability, permeability and surface area than is found for styrene-divinylbenzene gels. Characteristic properties of the Amberlite XAD series of macroreticular resins manufactured by Rohm and Haas are summarized in Table 8.8 [178,180,181]. The XAD-1, -2 and -4 resins are aromatic in character, very hydrophobic, and possess no ion-exchange capacity. Amberlite XAD-7 and -8 are acrylic ester resins, non-aromatic in character, and possess a weak ion-exchange capacity (approximately 0.01 mequiv./g for XAD-8). The acrylic ester resins are more hydrophilic than the styrene-divinylbenzene copolymers; accordingly they show a higher adsorptive capacity for polar solutes (but lower for nonpolar solutes), are easily wetted by and absorb more water and, in general, have lower breakthrough volumes for many analytes due to the more favorable water-sorbent interactions. Compared to carbon the overall analyte recovery for macroreticular resins is often better and irreversible adsorption and catalytic activity is greatly diminished.

To obtain accurate analytical data at trace concentrations careful attention to procedural details is essential [9,176, 179-187]. Prior to use the polymeric beads must be prewet to promote complete permeation of the water into all of the pores of the polymer, and the adsorbent column must be operated so that the sample passes through these pores. The column is first washed with several bed volumes of a water miscible solvent, such as methanol, followed by several bed volumes of pure water. The volume of water should be sufficient to displace all of the conditioning solvent from the pores. The column bed should be long and narrow to avoid poor recovery resulting from channeling of the water flow and packed with coarse beads (0.85-0.35 mm) to facilitate rapid

TABLE 8.8

PHYSICAL PROPERTIES OF AMBERLITE XAD RESINS
STY-DVB = styrene-divinylbenzene copolymer

Type	Composition	Character	Average Pore Diameter (nm)	Specific Surface Area (m^2/g)	Specific Pore Volume (ml/g)	Solvent Uptake (g/g) of dry resin
XAD-1	STY-DVB	hydrophobic	20	100	0.69	– – – –
XAD-2	STY-DVB	hydrophobic	8.5-9	290-300	0.69	0.65-0.70
XAD-4	STY-DVB	hydrophobic	5	750	0.99	0.99-1.10
XAD-7	Methyl methacrylate polymer	moderately hydrophilic	8	450	1.08	1.89-2.13
XAD-8	Methyl methacrylate polymer	moderately hydrophilic	25	140	0.82	1.31-1.36

sampling by gravity flow or forced flow using a peristaltic pump. Relatively large sample volumes, 1-100 liters, can be pumped through a short sorbent column, 6 x 1.5 cm, at high flow rates, up to 20 column bed volumes per minute, although 4-7 column bed volumes per minute is more typical. The sorbed sample is recovered by elution with a small volume of organic solvent such as methanol, diethyl ether, ethyl acetate, acetone, etc. For efficient desorption penetration of the eluent into the pores is essential. For water immiscible eluents most authors recommend that the water be drained from the column and replaced by the eluting solvent followed by an equilibration period. The eluting solvent is then displaced, replaced, and the process repeated as required to elute all the analytes. Ideally, a strong solvent should be used to minimize further sample processing. The solvent evaporation step after elution has frequently been cited as a procedure were substantial losses of semivolatile analytes can occur (see section 8.4.4). After evaporating the eluting solvent to a small volume enrichment factors of 1000 to 10 million can be obtained. Thermal desorption and Soxhlet extraction have also been used as alternative methods for sample recovery [178,180]. Thermal desorption avoids the use of elution solvents and, therefore, provides the highest degree of sample enrichment. Unfortunately, the thermal stability of the Amberlite-type macroreticular polymers is not very high (< 250-275°C) and, therefore, limits the technique to easily desorbed analytes. Tenax is more thermally stable than the Amberlite polymers and produces relatively little chromatographic background at usual desorption temperatures. Tenax, however, has a low surface area, 19-30 m^2/g, limiting its sorption capacity which results in lower recoveries and restricts the sample volume that can be processed. In spite of these limitations its use in studies of water pollution involving trace concentrations of organic compounds has been demonstrated [184,185]. The analytical methodology involves column sampling in the usual way, centrifugation and vacuum desiccation to remove entrapped water, and thermal desorption at 280°C for gas chromatography.

Breakthrough volumes for macroreticular polymers can be calculated theoretically from a knowledge of the analyte solubility [181] or determined experimentally [180,182,188]. Model systems are usually prepared by adding the test sample in a small volume of a miscible organic solvent to water. The water must be free of suspended particles that can irreversibly adsorb the solutes, all glassware must be clean, and the spike level should not exceed the water solubility of the analyte. Volatile compounds with a lower density than water can collect at the water surface and evaporate from the open sample reservoir unless the air space above the reservoir is minimized [182]. An alternative method of estimating the breakthrough volume is to apply a small amount of the analyte to the top of the sorbent bed and then to pass a large volume of water through the column. If all of the applied model compound can be recovered by using the eluting solvent, 100 % efficiency is assumed. In either case, the values obtained for model systems may be inaccurate for real samples, where solution conditions and the presence of coextractants can influence the efficiency of the adsorption and desorption steps [180,183].

Neutral polymers, such as the styrene-divinylbenzene type, are efficient for the sorption of neutral hydrophobic solutes from water but have little affinity for polar and ionic solutes. Partially dissociated organic analytes (e.g., phenols, carboxylic acids, etc.,) are generally weakly sorbed by the resins and the percentage retained diminishes with dilution due to a higher degree of dissociation. Adjusting the pH of the sample to inhibit dissociation leads to increased sorption for these compounds but average recoveries at low concentrations are often not comparable to those of neutral analytes. Surprisingly, long-chain hydrocarbons are only poorly retained by macroreticular polymers in spite of their low water solubility [183]. This has been explained as resulting from the inability of this kind of solute to fit into the polymer lattice. For similar reasons high molecular weight analytes are poorly retained by high surface area (small pore diameter) sorbents due to size exclusion effects. Otherwise, the high affinity of the macroreticular polymers for neutral organic compounds (typical recoveries of 80-100 % for alcohols, aldehydes, ketones, esters, ethers, halogen compounds, aromatic compounds, pesticides, plasticizers and industrial chemicals at part per billion and part per trillion levels can be obtained; applications are reviewed in references 9, 178 and 180), high sample capacity, low water retention, chemical inertness (except towards chlorine and condensate from stack gases), and ease of sample recovery are features that have contributed to their wide acceptance for sampling fresh water, waste water, salt water and biological fluids. Their principal disadvantage is the slow and careful purification required prior to use in trace analysis to obtain reasonable blank values [178,180,183]. Fines and residual sodium chloride and sodium carbonate used in the production process are removed by slurrying in water or methanol and monomers, contaminants and anti-bacterial agents by sequential Soxhlet extraction for 4-12 hours each with methanol, acetonitrile, diethyl ether, for example. Many polymerization impurities originate from the use of impure technical grade starting materials, e.g., benzoic acid, alkylbenzenes, styrene, naphthalene, biphenyl, alkylacetophenones, methylbenzoate and alkyl derivatives of styrene and naphthalene. The solvent-purified resins are stored in glass-stoppered bottles under methanol until used. Purified resins do not store well in the dry state. It is hypothesized that alternating expansion and contraction of the resin beads while stored in the dry state or during changes in solvent, release interstitially trapped contaminants incompletely removed during the cleaning process.

8.6.3 Polyurethane Foam Adsorbents

Polyurethane foams have been used as sorbents for neutral dissolved organic compounds in water [189,190]. Extraction columns are prepared from sections of cellular (foamed) polyurethanes inserted as a plug or the column bed can be prepared in situ by reacting a solution of an isocyanate with a polyol. The open-pore polyurethanes are composed of agglomerated spherical particles, 1-10 micrometers in diameter, bonded to one another in a rigid and highly permeable structure. Their chief attribute is their favorable hydrodynamic and aerodynamic properties which permit sampling to occur at very high flow rates. The polyurethane foams have demonstrated efficient retention of polynuclear aromatic hydrocarbons and pesticides, although no significant analytical advantages over the widely-used macroreticular porous polymers have been shown.

8.6.4.   Ion-Exchange Resins

Ion-exchange resins can be used for the efficient isolation of ionic substances in the presence of complex non-ionizable matrices or for the efficient collection and concentration of low levels of ionic analytes in water and biological fluids. In this respect their properties are complementary to those of the macroreticular porous polymers which, generally, yield poor recoveries for hydrophilic ionizable analytes. Silica-based (see section 8.5) and macroreticular polymeric ion-exchangers are mainly used for the isolation of low molecular weight carboxylic, sulfonic and phosphoric acids, phenols, amides, amines and inorganic ions from water and biological fluids [136,191-194]. This includes profiling of amino acids, organic acids, nucleosides and nucleotides in clinical samples used for the recognition of various metabolic diseases. Also, ion-exchange chromatography is widely used for the isolation of organic acids and bases from coal-derived fluids and synthetic fuels [195]. The hydrophobic nature of the matrix, small pore size and denaturing effect of the porous polymer ion-exchangers generally limit their usefulness for the isolation of large molecules, such as proteins, polypeptides, nucleic acids, polynucleotides and polysaccharides which are usually isolated using hydrophilic ion exchangers based on cellulose, dextran or agarose supports [30].

Ion exchange is particularly attractive for the isolation of ionizable substances since neutral molecules, which may interfere in the final chromatographic analysis, are easily washed from the ion exchanger without affecting the recovery of the ionized components. Adjusting either the pH or ionic strength of the sample solution or eluting solvent enables the selectivity of the isolation step to be increased and the sample fractionated by its ability to bind to the ion exchange groups in competition with the ions in the mobile phase. For the recovery of organic acids from DEAE-Sephadex pyridinium acetate is frequently used as the competing ion in the mobile phase, since it is easily removed from the extract before analysis by lyophilization [194]. Barium hydroxide is sometimes added to biological fluids prior to sampling to precipitate inorganic acids (e.g., sulfate, phosphate, etc.) which might interfere in the recovery of organic acids [194]. A combination of macroreticular porous polymer and ion-exchange resins have been used for the comprehensive isolation of semivolatile and involatile dissolved organic solutes from water [180,196].

8.7 MULTIDIMENSIONAL AND MULTIMODAL SAMPLE FRACTIONATION AND TRACE ENRICHMENT TECHNIQUES

Multidimensional chromatography involves the separation of a sample using two or more columns in series were the individual columns differ in their capacity or selectivity. Multimodal separations employ two or more chromatographic methods in series, such as the on-line coupling of liquid chromatography/gas chromatography or supercritical fluid chromatography/gas chromatography. Both methods employ the transfer of one part of the chromatogram from the first column to another via some suitable interface as a means of affecting trace enrichment of selected analytes, improving the resolution of parts of a complex mixture, or to increase sample throughput employing such techniques as heartcutting, backflushing, foreflushing, cold trapping and recycle chromatography. Some typical applications for multidimensional gas chromatography, by way of illustration, are summarized in Table 8.9.

The function of the interface is to ensure compatibility in terms of flow, solvent strength, and amount of transferred sample from one column to the next. The design requirements of the interface obviously differ for the different chromatographic modes and will be treated separately in the following sections. However, the unit operations of heartcutting, backflushing, etc., have a common definition and can be described at this point in general terms. Heartcutting allows selected sections of the chromatogram from the first column to be transferred to the second. It is the

TABLE 8.9

APPLICATIONS OF MULTIDIMENSIONAL GAS CHROMATOGRAPHY FOR SAMPLE
PREPARATION

Application	Comments
Solvent removal	Large amounts of solvents and excess derivatizing reagents, etc., can be excluded from the main separation column. Prevents column deterioration and improves the performance of sensitive, selective detectors.
Enrichment of trace components	This can be achieved by multiple injections into a precolumn with selective trapping and storage of the fraction of interest. Often performed with packed precolumns because of their high sample capacity and because maximum resolution is not required in the pre-separation step.
Separation of trace components buried under major peaks	It is possible to vent from the system the major portion of large peaks and transfer the trace components of interest into the trap.
Heartcutting of a limited section at the beginning of the chromatogram	In a sample having a very wide boiling point range, only the components of interest are transferred into the main separation column. With a dual oven instrument high boiling components can be backflushed at an elevated temperature, thus maintaining a high sample throughput.
Switching of single peaks or selected cuts throughout a chromatographic run	Single peaks or whole areas of a chromatogram can be switched to a second column to provide better separation. With intermediate trapping and columns of different selectivity two sets of retention index data can be obtained.

most common procedure in multidimensional/multimodal chromatography. Backflushing allows sample components with large capacity factor values to be eluted from the first column by reversing the direction of mobile phase flow threw the column, usually after heartcutting. These components can be backflushed through a detector to permit analysis or simply vented to waste; in both cases the primary objective is to minimize the separation time for the analytes of interest, which are not necessarily those backflushed. Foreflushing is usually used as a method of solvent stripping where this would have a deleterious effect on the operation of the second column or the detector used to monitor the separation. This might happen when a large volume of sample is injected on the first column for trace enrichment purposes or the sample contains derivatizing reagents, etc., which would mask the response of the detector to the analytes derivatized. This is achieved by ensuring that the majority of the solvent, etc., is eluted from the first column before selectively transferring the analytes of interest to the second column. Cold trapping (cryogenic refocusing) is a term used in multidimensional gas chromatography to describe the refocusing of the analytes at the head of the second column after transfer from the first. It reconcentrates the transferred fraction into a narrow band prior to reinjection on the second column to maintain resolution and is virtually always required if the interface has substantial dead volume or the flow rate of the first column exceeds that of the second, for example, in the coupling of a packed column to an open tubualr capillary column, if total analyte transfer is required. Cold trapping has a parallel in liquid chromatography were a large volume of eluent from the first column can be transferred to the second and the analytes refocused provided that the mobile phase used for the first column is a weak solvent with respect to the retention mechanism of the analytes on the second column. In gas chromatography cold trapping is achieved by temporarily cooling an intermediate trap or a small section of the second column to a temperature sufficiently low to preclude migration of the transferred analytes at that temperature into or along the column. In recycle chromatography the entire sample or some fraction of it is returned to the head of the column and reinjected sequentially until a desired separation is achieved. It is frequently used in preparative liquid chromatography to increase the available number of theoretical plates for a separation without resorting to the use of long columns. It is rarely used for sample cleanup.

Multidimensional or multimodal separations can be performed on-line or off-line, particularly in liquid chromatography. Trapping of a component for reinjection in gas chromatography is more difficult and likely to result in sample losses. The advantages of off-line techniques in liquid chromatography are that the effluent is easy to collect, reconcentrate and exchange solvents if required. Trace solutes are easily concentrated from large sample volumes and it is possible to combine two LC modes that use incompatible mobile phases. The disadvantages are that the off-line approach is difficult to automate, is time consuming, presents a greater opportunity for sample loss (such as through adsorption onto glassware, oxidation by exposure to the atmosphere, entrainment during solvent evaporation, etc.,) and is less likely to yield accurate quantitative data. Because of these problems automated multidimensional and multimodal systems are to be preferred and have grown substantially in reliability and ruggedness in the last few years.

8.7.1 Multidimensional Gas Chromatography

The interface in multidimensional gas chromatography must provide for the quantitative transfer of the effluent from one column to the next without altering the composition of the transferred sample or degrading the resolving power of the second column [197-203]. Two types of interfaces are commonly used for this purpose, those employing microvolume, multiport, switching valves and those using pneumatic switches. Pneumatic switches, also commonly referred to as Deans' switches, are based on the balance of flow at different positions in the chromatographic system using in-line restrictors and precision pressure regulators [204]. The direction of flow between columns can be changed by opening and closing the valves located external to the column oven. Thus, no moving parts are located inside the oven and dead volume effects are not important since there are no unswept volumes. Problems can arise from back diffusion of sample into the switching lines causing memory effects and the stability of the pressure controllers, which must remain reproducible under all conditions. Since switching is controlled by pressure, widely varying flows, as might exist in interfacing a packed column to an open tubular capillary column, are easily handled [199,200, 205-207].

Intuitively, the simplest approach to switching effluent from one column to another is by means of a mechanical valve [208-212]. Since the valve forms part of the sample passageway, the chosen valve must be chemically inert, free from outgassing products, gas tight at all temperatures, have a low heat capacity and small internal dimensions, and operate without lubricants. Modern miniaturized multifunctional valves meet most of these requirements with some exceptions. During passage through the valve the sample comes into contact with the stainless steel valve body and elastomer material (fluorocarbon-filled crosslinked polyimide) used to machine the rotor. Dead volumes are typically very small (ca. 0.2 microliters) and typical sample residence times only a few microseconds. Partial sorption of very polar solutes, however, has been observed to occur in some cases. Separate heating of the valve may be necessary if rapid temperature programming is used due to the thermal lag of the valve, which can cause peak asymmetry due to cold trapping. These limitations do not apply to all sample types and multifunctional valves are acceptable for many less demanding chromatographic problems. They are generally more straightforward to operate than a pneumatic switch and can be coupled to pneumatic or electric actuators for automated operation.

An intermediate cryogenic trap is essential whenever a band refocusing mechanism is required as part of the switching process [198,199,213-216]. Examples include sample transfer from packed to open tubullar capillary columns, preconcentration by multiple injections, and accurate determinations of retention index values on the second column. The trap is usually a short length of platinum/iridium tubing, fused silica tubing or a short portion of the second column. In some cases the trap may be packed or coated with liquid phase to enhance the recovery of volatile analytes. It is necessary to provide a temperature gradient within the trap to avoid breakthrough caused by microfog formation. A trap should not only effectively retain the substances that are directed into it but should also be of low mass so that it can be rapidly heated to instantaneously introduce the trapped fraction into the second column. The trap is usually cooled by circulating the vapors from dry ice or liquid nitrogen through a shroud surrounding the trap. The trap is heated either by forced convection with preheated nitrogen, by ballistically temperature programing the column oven containing the trap, or in the case of metal traps and metal coated fused silica traps, by resistive heating using a high amperage circuit.

The simplest type of two-dimensional gas chromatography for heart cutting or trace enrichment using a packed precolumn and a capillary column is shown in Figure 8.16 [205]. Almost any modern gas chromatograph could be converted into a similar unit with the addition of a few auxiliary components. Preliminary separation takes place on the packed column, the effluent from which is directed either to a vent or to the capillary inlet by the Deans' switch. The effluent reaching the capillary inlet is split three ways. One portion passes to a detector used to monitor the preseparation, a second portion enters the capillary column and is

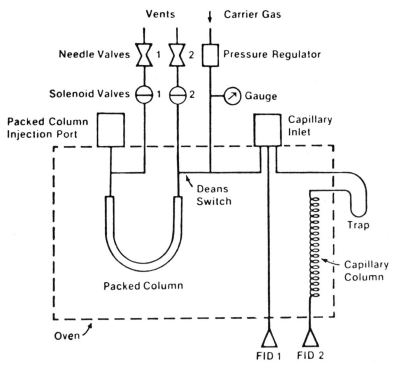

Figure 8.16 A simple two-dimensional gas chromatograph for packed column fractionation or enrichment using a Deans' switch and intermediate trap for transfer to a capillary column. (Reproduced with permission from ref. 205. Copyright Dr. Alfred Huethig Publishers).

reconcentrated in the cold trap located outside the oven, and the remaining portion passes out the split vent. Provision is made for backflushing sample components of low volatility from the precolumn by placing a tee leading to a second solenoid valve between the packed column and the injection port. A liquid nitrogen Dewar is used to manually cool the trap and resistive heating provides rapid reinjection onto the capillary column.

A number of conversion kits are now commercially available that enable any single- or two-oven gas chromatograph to be converted into a multidimensional chromatograph, for example, Figure 8.17 [200,217-219]. To simplify installation all the pneumatic controllers and programing components are located in a single unit outside the chromatograph. Inside the chromatograph are positioned the cold trap, column bracket module which contains the column connecting fittings and connecting T-piece (midpoint-restrictor). Changing the dimensions of the midpoint restrictor

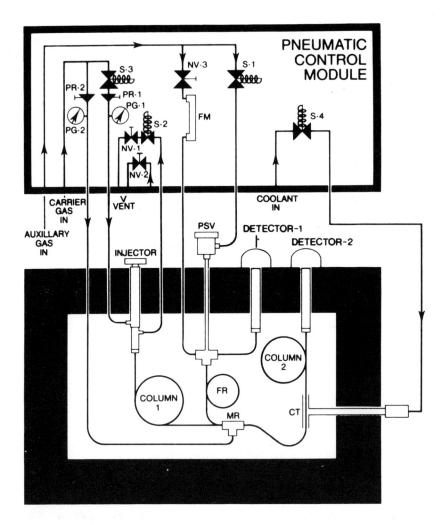

SOLENOID VALVES
S-1 Controls pneumatic shut-off valve.
S-2 Controls injection splitter and back-flush
     operation.
S-3 Controls pre-column carrier gas input.
S-4 Controls CO_2 or optional liquid nitrogen cold
     trap.

PRESSURE REGULATORS
PR-1 Sets pre-column carrier gas input pressure.
PR-2 Sets mid-point carrier gas pressure.

PRESSURE GAUGES
PG-1 Monitors pre-column pressure.
PG-2 Monitors mid-point pressure.

FIXED RESTRICTOR
FR   Deactivated transfer line from mid-point
     restrictor to monitor detector.

PERIPHERAL HARDWARE
FM   Flow Meter
MR   Mid-point restrictor
CT   Cold Trap
PSV Low dead volume pneumatically controlled
     shut-off valve.

NEEDLE VALVES
NV-1 Controls injector split flow.
NV-2 Controls mid-point split flow for packed to
     capillary column operation.
NV-3 Controls make-up gas flowrate.

Figure 8.17 Conversion of a standard gas chromatograph to a dual
capillary column multidimensional chromatograph using a commercially available conversion kit. (Reproduced with permission from
Scientific Glass Engineering).

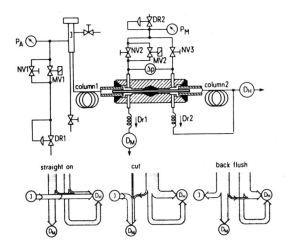

DR precision pressure regulator

NV needle valve

P_A prepressure gauge

P_M middle pressure

Dr flow restrictor

D_M monitor detector

D_H main detector

Figure 8.18 Schematic diagram of a dual oven two-dimensional gas chromatograph employing live switching between two capillary columns. (Reproduced with permission from Siemens AG).

enables either a packed or open tubular column to be used as the first column. The midpoint restrictor enables a split flow to the monitor detector for the first column and then by changing the relationship between the reference midpoint pressure and the column pressure the effluent from column one flows into the second column. Reversing the ratio prevents flow of effluent from column one to column two.

A more sophisticated instrument employing two independent temperature controlled ovens with sample transfer by live switching is shown in Figure 8.18 [198,213,214,219-222]. The critical component of the system is the double T-piece with tube connections for the adjustment of pressures and a platinum/iridium or fused silica capillary for the connection of the two capillary columns. The direction of flow through the double T-piece is controlled by slight pressure differences applied between its ends, ΔP in Figure 8.18. The flows through the makeup gas lines are adjusted with the two needle valves NV2 and NV3 with the pressure difference ΔP indicated by the manometer P_m. The T-piece is constructed in such a way that any gas flowing from one end to the other must pass through the connecting capillary, which is inserted loosely into both the first and second column. Effluent gas flow from the first column can be directed to a detector via a restrictor, D_r1, or through the second column and then to a detector. Moreover, the first column can be backflushed by activating solenoid valves MV1 and MV2. An example of the live switching technique for the separation of the components of a coal derived gasoline fraction is shown in Figure 8.19. The overlapping of components of different polarity but similar boiling point is circumvented by using a polar column as the first column. The selectivity of this column for the resolution of the hydrocarbon isomers is inadequate, however, and therefore, the hydrocarbon fraction was heartcut to a nonpolar column for separation. The analysis time remains acceptable as both separations are performed simultaneously.

8.7.2 Multimodal Liquid Chromatography/Gas Chromatography

The separation mechanisms in liquid chromatography and gas chromatography are complementary providing a powerful combined tool for the separation of complex mixtures [203,223-227]. The combined technique is well suited to the analysis of a single component or of components of similar characteristics, forming a single peak or narrow fraction in the liquid chromatogram. This fraction can then be transferred directly to the gas chromatograph avoiding manual manipulations and reducing the possibility of contamination or losses during solvent removal. Since the final determination of the analyte is by well-established gas chromatographic detectors, greater sensitivity and selectivity as well as more straightforward interfacing to a mass spectrometer is possible than is the case for liquid chromatography. A more detailed discussion of the advantages of LC/GC has been given by Grob and is summarized in Table 8.10 [226,228]. However, certain general incompatibilities of the direct coupling of LC to GC resulted in very little interest in this technique until recently. Separations by liquid chromatography occur with concurrent

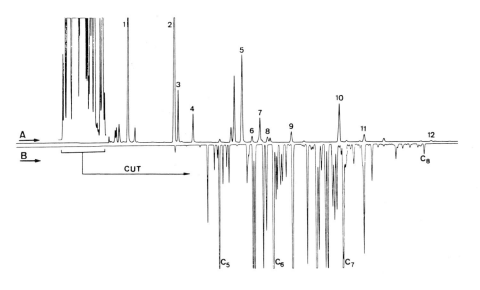

Figure 8.19   Two-dimensional separation of the components of a coal derived gasoline fraction using live switching. Column A was 121 m open tubular column coated with poly(ethelene glycol) and column B a 64 m poly(dimethylsiloxane) thick film column. Both columns were temperature programmed independently taking advantage of the two oven configuration. Peak identification: 1 = acetone, 2 = 2-butanone, 3 = benzene, 4 = isopropylmethylketone, 5 = isopropanol, 6 = ethanol, 7 = toluene, 8 = propionitrile, 9 = acetonitrile, 10 = isobutanol, 11 = 1-propanol, and 12 = 1-butanol. (Reproduced with permission from Siemens AG).

dilution of the sample requiring that a fairly large volume of solvent be transferred to the gas chromatograph that may easily exceed the conventional injection volume by several hundred fold. This problem was only solved by the development of miniaturized liquid chromatographic systems and immobilized stationary phases and retention gaps in gas chromatography. The earliest attempts at interfacing LC to GC generally employed a splitter interface with a fraction of the total LC effluent routed through a syringe of an autosampler which injected a fixed volume of the eluent, usually one microliter or less, onto an open tubular column [229]. There are two fundamental problems with this kind of interface. The GC sample volume represents only a minor fraction of the LC peak volume so that the sensitivity of the method is very poor and

TABLE 8.10

ADVANTAGES OF DIRECTLY COUPLED LC/GC

(a) HPLC is an extremely powerful technique for preseparating complex samples.

(b) Direct transfer assures complete transfer of the HPLC fraction, producing maximum sensitivity and circumventing problems in quantitative analysis.

(c) On-line detection in the HPLC eluent enables cutting of very narrow windows, enhancing separation of by-products from the fraction of interest.

(d) The technique is easily automated. However, even if carried out manually, coupled LC/GC is less time consuming than most other alternative methods.

(e) The possibility of introducing large volumes of solvent into the GC system rules out losses of volatile solute material during intermediate solvent evaporation steps.

(f) GC detectors are generally more selective and sensitive than current LC detectors and provide greater scope for trace analysis of complex mixtures.

---

trace analysis is not possible. As described the sampling process is discriminatory in that each peak eluting from the LC contains several components inhomogeneously distributed along the peak x-axis, such that sampling the eluted peak at different positions will result in different chromatograms being obtained. Reliable sample quantitation will not be possible. Also, during the time the sample occupies the open tubular column no further samples can be injected leading to low productivity. The productivity problem can be partially solved by incorporating a multifunctional valve in the interface. For example, Philip and Anthony used a 34 port multifunctional valve with sixteen 100 microliter loops to collect a series of 100 microliter volume fractions from a separation of synthetic coal fluid by size-exclusion chromatography with automated injection into the GC of 0.1 microliter subfractions from each loop [230]. Because of the small GC injection volume only major components could be determined. The peak discrimination problem can be overcome using an isokinetic eluent splitter fabricated from a bundle of 25 micrometer fused silica capillaries epoxyed together and randomly selecting the desired number of capillaries on the outlet side to give the desired split ratio, the remainder being used to direct the bulk of the eluent to waste

[231]. In this way the whole eluted peak and not just a fraction of it is sampled but sensitivity is sacrificed, since only a small fraction of the available eluent is injected into the GC.

One approach to quantitative eluent transfer is to miniaturize the LC system by using a packed capillary column and injecting a larger volume of sample into the GC than is conventionally practiced, [203,227]. In this case a multifunctional valve and a short retention gap was used as the interface with LC eluent flow rates generally less than 50 microliters/min. Using concurrent evaporation of the solvent a fairly large section of the chromatogram can be transferred to the retention gap while maintaining a well defined solute band for reinjection. The major limitations of this approach are associated with the specialized technology of the LC column and instrument, the need to maintain low dead volumes in all regions of the chromatograph and interface, and the small sample size tolerated by the packed capillary column which is a problem in trace analysis.

Several groups have extended the use of the valve-retention gap interface to microbore LC columns with internal diameters up to about 3 mm [232-241]. The use of short retention gaps with complete or partial concurrent solvent evaporation has been established as the preferred transfer technique, Figure 8.20 [224]. During concurrent solvent evaporation, the eluent is introduced into the retention gap at a speed allowing immediate evaporation and discharge of the solvent vapor through the column or interface between the retention gap and column. The latter approach offers some specific advantages. A vapor exit, located behind 2-3 m of uncoated precolumn and a retaining precolumn, accelerates eluent evaporation due to shortening of the flow path and prevents passage of large amounts of vapor through the column and detector. The length of the flooded zone is generally less than one meter and short retention gaps of 1-2 meters are adequate independent of the volume of eluent introduced. Optimum conditions for concurrent solvent evaporation are high carrier gas flow rates and an injection (oven) temperature close to the solvent boiling point at the prevailing inlet pressure. Under these conditions solvent trapping and phase soaking are inefficient band reconcentration mechanisms and cold trapping must be relied upon for refocusing of the initial solute bands. Consequently, sharp peaks will only be obtained for solutes with a boiling point at

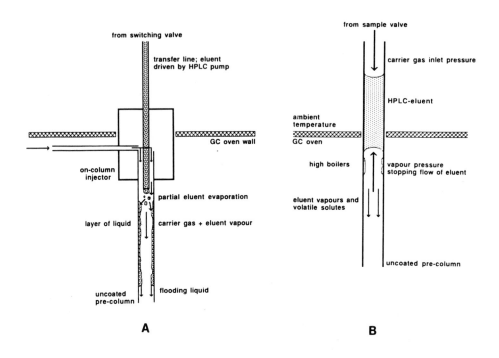

Figure 8.20 Schematic representation of process of (A) partial
concurrent evaporation and (B) concurrent evaporation using a
loop-type interface for the transfer of eluent from an LC column
to a GC column. (Reproduced with permission from ref. 224.
Copyright Elsevier Scientific Publishers).

least 50-100°C higher than the boiling point of the solvent. The
use of concurrent solvent evaporation, therefore, is limited to
the analysis of solutes of moderate to low volatility.

For partial concurrent solvent evaporation the eluent is
introduced at a rate exceeding the amount of simultaneously
evaporating solvent and, hence, there is a steady expanding
flooded zone during eluent introduction. However, the rates of
eluent introduction and solvent evaporation are adjusted to each
other to cause substantial concurrent solvent evaporation and a
corresponding shortening of the flooded zone. Concurrent solvent
evaporation is simplified if the carrier gas is introduced behind
the plug of liquid such that the carrier gas pushes the liquid
into the thermostated retention gap. This is easily achieved using
a sample loop and switching valve to isolate the column fraction

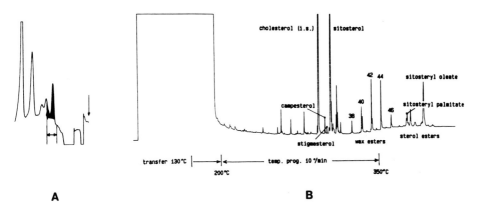

Figure 8.21 Separation of an olive oil sample after esterification
with pivalic anhydride on a 10 cm x 2 mm I. D. silica gel column
with hexane-dichloromethane (4 + 1) containing 0.05% acetonitrile
as the mobile phase (A). The fraction shaded in (A), 750 μl, was
transferred to the GC column for separation and detection using a
loop-type interface and concurrent solvent evaporation. The GC
separation (B) was carried out as indicated using a 15 m x 0.32 mm
I. D. glass capillary column coated with PS-225, film thickness
0.15 μm. (Reproduced with permission from ref. 242. Copyright
Elsevier Scientific Publishers).

of interest from the LC and then transfer it as a plug to the
retention gap. The column temperature must be high enough to cause
the solvent vapor pressure to exceed the carrier gas inlet
pressure, with the effect that the solvent vapor pressure at the
front of the plug prevents the liquid from flowing far into the
thermostated column. Without loop-type introduction carrier gas
and eluent are mixed in the retention gap which requires a careful
balance between carrier gas flow rate, sample introduction rate,
and temperature to establish a steady state situation. An example
of the use of the loop-type interface and concurrent evaporation
for solvent removal is shown in Figure 8.21 for the determination
of esterified sterols and wax esters in oils and fats by LC/GC
[242].

The above methods work well for LC fractions in volatile and
relatively nonpolar organic solvents. It is much more difficult to
apply the same techniques to reversed-phase eluents containing an

appreciable amount of water [239,243-245]. Water is an unfavorable solvent for the precolumn evaporation techniques due to its high boiling point, relatively large volume of vapor produced per unit volume of liquid, poor surface wetting characteristics due to its high surface tension, and poor solvent trapping efficiency for organic solutes [239,243-245]. Water also hydrolyzes all types of deactivated glass surfaces rendering the flooded section of the uncoated precolumn highly adsorptive. Using the concurrent solvent evaporation technique it was shown that the organic solvent must evaporate more slowly or at the same rate as water, otherwise water is left behind the evaporating solvent mixture and, as water does not wet these surfaces, it floods the precolumn and separation capillary resulting in unsatisfactory chromatographic separations. Most common organic modifiers used in reversed-phase chromatography are considerably more volatile than water and thus do not promote the required conditions for the successful operation of evaporative interfaces employing solute refocusing mechanism. This problem can be circumvented by adding a co-solvent of low volatility, such as butoxyethanol at 5-20% (v/v), to the mobile phase. The co-solvent does not completely evaporate during the transfer process and retains the solute material in the uncoated precolumn until evaporated at some time after the end of the sample transfer process. Other approaches include evaporation and selective recondensation of the solute in a cold trap maintained at a temperature high enough to prevent recondensation of the mobile phase. This approach is limited to analytes of low volatility that can be quantitatively cold trapped at temperatures close to the boiling point of water. Alternatively, the organic solutes can be stripped from water using a microcolumn packed with bonded phase sorbent. After displacing excess water with nitrogen the analytes are eluted with hexane directly into the GC using the partial concurrent solvent evaporation interface. Many problems remain to be solved, however, before the routine use of LC/GC with aqueous solvents will become common practice.

8.7.3 Multimodal and Related Techniques Involving Supercritical Fluid Extraction

Current interest in supercritical fluid extraction as a sample preparation technique for chromatographic analysis is intense, in spite of it receiving very little attention until the mid-1980s. Although neglected by analytical chemists, during the same period supercritical fluid extraction gained acceptance and found many applications as an industrial scale process technique, used for example, to decaffeinate coffee and to extract nicotine from tobacco [98,246]. Supercritical fluid extraction has been used off-line for sample preparation using a modified Soxhlet extractor (see section 8.4.3) [97,247]. In contemporary practice, however, the main use has been in combined techniques in which supercritical fluid extraction using a micro-extractor is coupled on-line to gas, liquid, thin-layer or supercritical fluid chromatography [223,248-257].

Supercritical fluids have a number of advantages over liquids for selective extraction [257,258]. The low viscosity and the absence of surface tension in supercritical fluids increases the speed of percolation of the fluid into the interstices of the matrix. The high diffusion coefficients of solutes in supercritical fluids permits rapid transfer out of the sample matrix. Both processes combine to enhance the rate of extraction of solutes compared with liquids. Extractions are usually complete in less than 30 min, while several hours may be required using Soxhlet extraction. The efficiency of extraction can be varied over a fairly wide range by manipulating temperature and pressure (density) and by the addition of small quantities of organic solvent modifiers [250]. Modifiers can be introduced as mixed fluids in the pumping system, with the aid of a second pump, or by simply injecting the modifier as a liquid onto the sample before beginning the extraction. The selection of modifiers and their concentrations is largely empirical because very little analyte solubility data exists for modified supercritical fluids, in addition, the competitive interactions between the modified supercritical fluid and the target analytes for the sorptive sites of the bulk matrix are poorly understood. The fractionation range for a particular analyte can be defined as the density range between which the solute initially dissolves and the density at which it attains maximum solubility in the compressed fluid. Within this range the density can be manipulated to obtain maximum discrimination for the analyte over its matrix. High selectivity, however, is only found under conditions where a significant difference in physical properties (e.g., molecular weight, polarity) exists. The recovery or transfer of analytes to a chromatographic system is easily achieved for fluids that are gases at atmospheric pressure by releasing the pressure on the fluid and allowing any solute to precipitate. Decompression through a micro-orifice occurs with cooling, enhancing the efficiency of sample recovery. Sample losses are likely to be less than for solvent stripping and simple interfaces should sufffice for total sample transfer.

A disadvantage of supercritical fluids for extraction is that most common fluids used for extraction (carbon dioxide, nitrous oxide, sulfur hexafluoride, etc.) are weak solvents, limiting the polarity and molecular weight range of analytes that can be efficiently extracted. Also, for trace analysis the availability of fluids of adequate purity may be a problem.

Supercritical fluid extraction can be performed in a static system with the attainment of a steady-state equilibrium or in a continuous leaching mode (dynamic mode) for which equilibrium is unlikely to be obtained [257,260]. In most instances the dynamic approach has been preferred, although the selection of the method probably depends just as much on the properties of the matrix as those of the analyte. The potential for saturation of a component with limited solubility in a static solvent pool may hinder complete recovery of the analyte. In a dynamic system, the analyte is continuously exposed to a fresh stream of solvent, increasing the rate of extraction from the matrix. In a static system movement of the analytes is controlled entirely by diffusion.

The instrumental requirements for supercritical fluid extraction are quite simple. A pump is essential to generate the extraction pressure in a thermostated extraction vessel. The soluble sample components are then swept from the vessel through a flow restrictor into a collection device that is normally at ambient pressure. The fluid used for supercritical fluid extraction are usually gases at ambient conditions and are vented from the collection device while the extracted sample components are retained. Analytical supercritical fluid extraction is usually performed with syringe pumps designed for, or part of, a supercritical fluid chromatograph, although simpler pumps would probably suffice, since complex density gradients are generally not required. Contemporary syringe pumps also lack the solvent capacity required for extracting large sample sizes. Gas compressors are simple and have been used for the extraction of large scale samples (e.g., > 50 g). Temperature is usually controlled by placing the extraction cell in a chromatographic oven or in a simple tube heater. Either fused silica capillary tubing of 10-50 micrometers internal diameter or micrometering valves are used to maintain the desired extraction pressure and to discharge the fluid from the extraction vessel. Off-line supercritical fluid extraction is inherently simpler to perform because only the extraction step need be understood, and the extract can be analyzed by any appropriate method. Most attention has been devoted to on-line techniques, however, which eliminate sample handling between extraction and separation of the sample, and will be discussed subsequently. In the off-line mode the sample extracts are often collected in a few milliliters of liquid solvent, or collected on sorbents and recovered by solvent elution. Direct depressurization into an empty vessel is often less successful due to aerosol formation. The cooling of the solvent caused by the expanding supercritical fluid prevents rapid evaporation of the collection solvent that would be expected because of the high gas flow rate, and even relatively volatile analytes are quantitatively recovered [257,259,261,262]. In some cases it may be necessary to heat the solvent to avoid blocking the restrictor by formation of crystalline material at the restrictor outlet. After collection, degassing the solvent by ultrasonication may improve the precision of quantitative analysis by minimizing bubble formation during syringe manipulation.

Supercritical fluid extraction can be coupled directly to GC by depressurizing the supercritical fluid extract inside a conventional split/splitless injector or by inserting the extraction cell restrictor through an on-column injector into the capillary column itself [250-252,255,257,263-266]. The sample extracts are cryogenically focused at the head of the GC column during the extraction step, and the subsequent GC separation is performed in the usual way. On-column injection gives the highest sensitivity for trace components since all of the extract is collected on the column while split injection can use larger samples (up to 15 g) and is simpler to perform with wet samples (restrictor less likely to plug). Peak shapes obtained are essentially identical to those obtained using conventional solution injection for both interfaces. As far as the criteria for the design of a suitable interface are concerned, the rate of analyte extraction and the sample size to be extracted are important considerations. The limited sample capacity of open tubular columns in GC dictates that extractors with a volume generally less than 0.5 ml and a sample size of 1-100 mg be used.

These extractors can be conveniently fabricated from a modified column end fitting or from guard column cartridges used for LC [252,259]. Alternatively, a number of manufacturers of supercritical fluid chromatographic instruments offer a variety of extraction vessels in different sizes and configurations to meet most demands. Supercritical fluid extraction coupled to GC is also suitable for the recovery of samples adsorbed on Tenax or polyurethane foam samplers [254,267]. The advantage of supercritical fluid extraction in this case is the enhanced recovery of high molecular weight and thermally labile analytes compared with thermal desorption.

Supercritical fluid extraction can be coupled to supercritical fluid chromatography using a series of switching valves and either a loop interface or an accumulator trap [257,260,268-273]. In the loop interface the fluid from the extraction cell passes continuously through an injection loop and into a collection vessel. Injection of a predetermined fraction of the extract onto a packed column is made by switching the loop so that it is in-line with the flow of fluid to the analytical column. A similar interface was used for coupling supercritical fluid extraction to LC except that a packed loop was used for sample collection and subsequent expansion of the supercritical fluid. The sample residue left behind in the loop was then redissolved by passing the mobile phase for the LC separation through it [248].

The accumulator interface allows the extract and fluid to be separated in a cryogenic trap which can be subsequently rapidly heated for injection of the sample extract into the separation column. The operation of a commercially available instrument utilizing this principle is shown in Figure 8.22 [271,272]. In the load position, the mobile phase from the syringe pump enters the tee and is directed via the multiport valve, A, to the extraction vessel, B, and then to the selection valve, C. The fluid from the extraction vessel is routed back to the multiport valve, A, and via a fused silica capillary restrictor of 50- or 25-micrometers internal diameter into the cryogenic collection trap, which is cooled as low as -50°C using an auxiliary supply of dry carbon dioxide. All of the extracted materials are accumulated in the cryogenic trap and the expanded fluid vented through valve A to the atmosphere. In the inject position, the pump is initially equilibrated at the density for the start of the separation (if

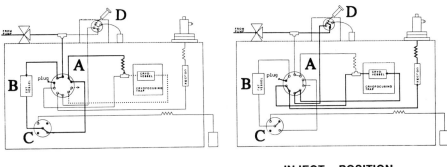

LOAD   POSITION   INJECT   POSITION

Figure 8.22 Schematic diagram of the Suprex MPS/225 integrated
supercritical fluid extractor, cryogenically focused interface and
supercritical fluid chromatograph. The bold lines represent the
direction of fluid flow in the load and inject positions.
(Reproduced with permission from Suprex Corporation).

different from the conditions used for the extraction) and then
the multiport valves B and C are simultaneously switched to the
inject position. In this configuration, the mobile phase passes
through the tee, the injection valve D, and the multiport valve A
into the cryogenic trap, which is then ballistically heated to
200°C. The mobile phase transports the extracted components from
the trap back to valve A and onto the column for separation.
Additional fluid flow from the other outlet of the tee to the
cryogenic collection trap prevents backflushing of the extracted
material into the restrictor. Good quantitative extraction and
recovery of a range of different analytes at a variety of
concentrations was demonstrated using the above integrated system.
Cambell and Lee have described a supercritical fluid extraction
apparatus that permits on-line extraction, packed-column
fractionation, and fraction collection in cooled pressurized

collection vessels [274]. In this system extraction and fractionation occur simultaneously. The approach is most useful for group type separations for which a well defined critical density for extraction and fractionation is likely to exist. An instrument for on-line two-dimensional supercritical fluid chromatography using a packed capillary column and open tubular capillary column connected by a rotary valve and cryogenic trap interface has also been described [275].

To date most of the work which has been done with supercritical fluid extraction has concentrated on the extraction of analytes from solid matrices or liquids supported on an inert solid carrier matrix. The extraction of aqueous matrices presents particular problems [276-278]. The co-extraction of water causes problems with restrictor plugging, column deterioration, and phase separation if a nonpolar solvent is used for sample collection. Also, carbon dioxide may have limited extraction efficiency for many water soluble compounds.

8.7.4 Multidimensional Liquid Chromatography

Multidimensional liquid chromatography encompasses a variety of techniques used for sample separation, cleanup and trace enrichment [12,279-289]. A characteristic feature of these methods is the use of two or more columns for the separation with either manual or automatic switching by a valve interface of fractions between columns. These techniques require only minor modification to existing equipment, and of equal importance, enable the sample preparation and separation procedures to be completely automated.

For multidimensional chromatography a standard high pressure liquid chromatograph is used with the addition of a switching valve. This valve may be a simple manually-operated six- or ten-port valve or it may be automatically controlled either via a fixed interval timer or coupled to the operation of an in-line detector and controlled by a microprocessor [287,290-292]. Depending on the complexity of the column switching network, several switching valves may be needed to permit the controlled operation of column conditioning, sample loading, flushing of interferents from the sorbed sample, backflushing or foreflushing of the sample to the separation column, heartcutting from one column to another, elution from either column, etc. Each valve must be capable of high pressure operation without deterioration, and provide a low-dead volume flow path so as not to significantly

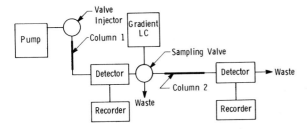

Figure 8.23   Schematic diagram of an on-line two column multidimensional liquid chromatograph. (Reproduced with permission from
reference 293. Copyright Elsevier Scientific Publishing Co.)

broaden the peaks that pass through it. The valve may also contain
a loop of fixed volume for trapping the switched peak prior to
injection on the second column or it may simply be used as a
switch to divert flow between columns. An example of a two-column
multidimensional liquid chromatograph using a six-port switching
valve, intermediate trap and injection valve is shown
schematically in Figure 8.23 [293]. The detector positioned
between the two columns is not an essential component. For complex
chromatograms the precise location of the peak or peaks to be
switched is very important for quantitative analysis, since the
separation of the switched peaks on the second column must not be
compromised by switching additional impurity peaks. The
intermediate detector thus conveniently locates the position of
the peak to be switched in each sample and minimizes sample loss
and contamination. When the detector function is not used to
control the switching process a sequential timer can be used to
automatically switch the peaks of interest. However, interval
timers are unable to account for variability in the elution volume
of the sample from the first column.

       Two general considerations influence the choice of coupled
column systems as discussed so far. To bring about a significant
improvement in resolution the retention mechanisms for the two
columns should be complementary. Secondly, the column flow rate
for the first column must be generally far less or similar to that
of the second column, and the mobile phase for the first column
must be a weak solvent for the sample transfered to the second
column as well as being miscible with the mobile phase for the
second column. By this means the size of the transferred zone is
minimized and reconcentration of the transferred zone on top of

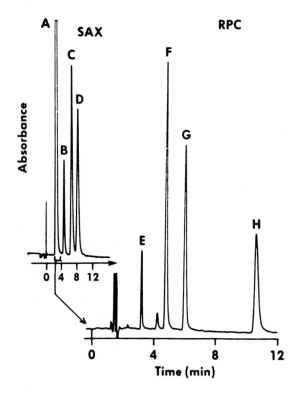

Figure 8.24   Separation of the major deoxyribonucleosides and
their   5'- monophosphate deoxynucleotides on a strong cation
exchange column (column   one) and a reversed-phase column. The
unseparated nucleosides, A, on the ion- exchange column were
switched to the reversed-phase column. Peak identification: A =
nucleosides, B = d-CMP, C = d-AMP, D = d-GMP, E = d-CYD, F = d-
URD, G = THD, and H = d-ADO. (Reproduced with permission from ref.
298. Copyright Preston Publications, Inc.)

the second column is possible. Thus, not all column types are
compatible: typical examples of compatible systems are size-
exclusion   and   normal,   reversed-phase   and   ion-exchange
chromatography; ion-exchange and reversed-phase chromatography;
and polar bonded phase and normal phase chromatography [294-297].
An example of a two-column separation of a mixture of major
deoxyribonucleosides and their 5'-monophosphate deoxynucleotides
on a strong cation exchange column and a reversed-phase column is
shown in figure 8.24 [298]. The neutral deoxyribonucleosides are
switched as a single peak for separation on the reversed-phase
column while their phosphate-containing analogs are resolved by
the ion-exchange column.

The current trend in column switching is to use a very short column, similar to a guard column, for the first column. This provides a faster separation and enables samples which might otherwise damage the analytical column to be analyzed; treating the precolumn as disposable after a certain number of samples have been processed. Longer columns provide higher resolution which might be required for some applications. The properties of short precolumns for sample cleanup and trace enrichment will be discussed subsequently.

One of the primary uses of multidimensional liquid chromatography is to minimize separation time for the analysis of complex mixtures [283]. Consider the following examples illustrated in Figure 8.25. Figure 8.25(A) is an example of a two column and valve arrangement for the separation of a sample made up of components spanning a wide range of capacity factors if a column of high selectivity is used. Components eluting at the start of the chromatogram are well resolved but later eluting peaks with long retention times may be difficult to detect. Conversely, if the column has low selectivity the separation time will be shorter but early eluting components will be poorly resolved. In the multidimensional mode, two columns containing the same stationary phase but of different lengths or two columns of similar length but of different selectivity are used. The first column is short (or of low selectivity) and is used to separate the later eluting components of the chromatogram. The components of low retention are switched to the second column (1) and remain there while the most retained components are separated on the first column (2). When these components have been separated, the least retained components are then separated on the longer (or more selective) second column (3). Compared to the normal chromatogram, the multidimensional chromatogram is reversed; the most retained components are at the front of the chromatogram while the least retained are at the back. The total separation time is reduced, there is less dead space between peaks and, as the peaks elute in a smaller volume of mobile phase, they are easier to detect. The separation may be carried out isocratically provided that the total number of sample components is not too great. This simplifies the instrumentation required, permits the use of nongradient compatible detectors, and eliminates the problem of column equilibration required in gradient elution analysis.

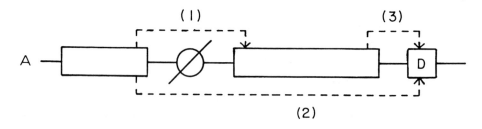

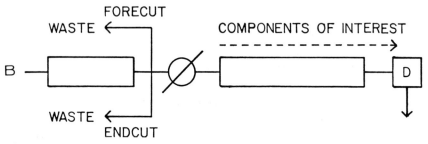

Figure 8.25 Schematic arrangement of a two-column system for separation of a sample comprised of components spanning a wide rage of capacity factors (A) and heartcutting of a group of analytes of similar retention to an analytical column for separation (B).

As a second example consider the problem in which quantitative analysis of a few components of similar retention is required in a complex mixture containing components having a wide rage of capacity factor values. Let us further assume that the peaks of interest are buried in the middle of the normal chromatogram. The time used in separating the components prior to, and after, the peaks of interest is wasted time, as it is not contributing effectively to the separation of the desired components. Using the column configuration of Figure 8.25(B), the short column (or column of low selectivity) is used to quickly isolate the area of the chromatogram of interest, which is then switched to the second column for separation. The frontcut and endcut of the chromatogram are vented to waste. If the components of the endcut are well retained they may be backflushed by reversing the direction of the column flow. The components of interest and any overlapping bands from the first column are meanwhile being efficiently separated on the second column. Snyder has adapted this technique to a process he termed "boxcar chromatography" [299,300]. In boxcar chromatography samples are switched from column one to column two before the previous sample has migrated the complete length of the second column. In this way the separating power of the second column is fully utilized, since it is continuously resolving samples separated only by the minimum time required to ensure that intersample mixing does not occur. This method is capable of very high sample throughput and can be operated automatically for routine analysis.

Since multidimensional chromatography is used to solve the general elution problem, there is obviously some overlap of the technique with that of gradient elution analysis (section 4.6.3). Multidimensional techniques are preferred when a very large number of theoretical plates are required for a separation, when a large number of samples of a similar kind are to be analyzed in the shortest possible time, and for the analysis of complex mixtures which contain only a few adjacent components of interest. For the above separation problems, unidimensional techniques are inherently inefficient as only a small fraction of the column is actually in use at any given time. The multidimensional separation methods provide optimum efficiency for the separation of the components of interest while simultaneously minimizing separation time by decreasing the time spent in separating components of the sample which are not of particular interest.

When sample dilution prevents detection, trace enrichment is required. For on-line trace enrichment techniques, greater accuracy can be expected by minimizing sample losses from adsorption to container walls, degradation upon heating and sample manipulation, as well as minimizing interferences from solvent impurities. Precolumns for trace enrichment are generally very short, bed length 2-20 mm, and packed with small particle diameter, conventional LC sorbents. Nonselective reversed-phase, carbon and porous polymer packings are generally used [301-304], although recently more selective sorbents, such as metal-loaded ion exchangers for the selective isolation of herbicides containing an ethnyl group from water [305-307] and a silver(1)-thiol stationary phase for preconcentration of the AIDS drug AZT from plasma [307,308] have been used. Nonselective sorbents in conjunction with pH adjustment, if required, enable nearly all analytes of low to moderate water solubility to be efficiently collected. A disadvantage of trace enrichment with sorbents of low selectivity is that many sample constituents are concentrated on the precolumn and the final sample complexity may be too great for adequate separation and detection. This problem can sometimes be solved by flushing the precolumn with a wash solvent strong enough

to remove the interfering matrix components but not so strong as
to displace the analytes of interest, which are eluted from the
precolumn and transferred to the analytical column in a subsequent
step. These processes can, of course, be automated using an
additional LC pump to provide solvent to condition the precolumn,
load the sample, wash away matrix interferents, and load the
analytes onto the analytical column, if this is not done with the
eluent pump. Since the precolumns are short, pressure drops are
very low, and the sample solution can be pumped through the
precolumn at high flow rates, typically 5-25 ml/min. Sampling
times are thus short, even for sample volumes of 0.1-10 liters,
which are normal sample sizes for water analysis. Under favorable
circumstances enrichment factors of 10,000 can be obtained.

The physical requirements of the precolumn for trace
enrichment are that its diameter should be comparable or smaller
than that of the analytical column, its length should be as short
as possible, and the sorbent used should be of similar size and
quality to that selected for the analytical column [301,307,
309-311]. The amount and volume of sample solution that can be
preconcentrated on a precolumn depends on the dimensions of the
precolumn, the characteristics of the packing material, the nature
of the solvent (usually water), and the flow rate used during
sampling. The ultimate enrichment factor obtainable is determined
by the breakthrough volume of the solute on the precolumn and the
solubility of the solute in the mobile phase used with the
analytical column. The breakthrough volume can be measured
experimentally or predicted from chromatographic data. The first
approach becomes tedious if multiple solutes are to be determined
and the latter is subject to large errors. The maximum volume that
can be sampled from an aqueous solution onto a sorbent before
breakthrough occurs is given by equation (8.2)

$$V_B = [(n - 2)^{1/2}/n^{1/2}](1 + k_w)V_m \qquad (8.2)$$

where V_B is the breakthrough volume, n the number of theoretical
plates for the precolumn, k_w the solute capacity factor on the
sorbent used to prepare the precolumn with water as the mobile
phase, and V_m the precolumn void volume. For short precolumns the
influence of extracolumn band broadening can be very large, making
the measurement of V_m and n subject to large errors. Capacity
factor measurements are subject to error because many organic
compounds have long retention times and poor peak shapes on

bonded-phase columns when water alone is used as the mobile phase and values obtained by extrapolation from binary mobile phases containing an organic cosolvent are not very accurate (see section 4.5.4).

Trace enrichment and sample cleanup on short precolumns are finding increasing use in the automated determination of drugs in biological fluids by liquid chromatography [12,285,291, 300,312-315]. The main problem in injecting untreated plasma or serum samples, and to a lesser extent urine samples, onto reversed-phases columns is the incompatibility of the sample matrix with the column system. The proteins become adsorbed on the solid phase and denatured or precipitated by components of the eluent, shortening the useful lifetime of the analytical column. Use of a short precolumn serves the dual function of acting as a guard column as well as affecting a preseparation of the analytes from the biological matrix. The precolumn-venting plug technique is one of the more effective means of isolating analytes from biological fluids [285,315]. This technique employs a two-valve arrangement. The sample is introduced by the second valve into a plug of aqueous buffer introduced by the first valve, such that the sample is surrounded by the "plug" liquid during its passage through the precolumn. Using this arrangement protein denaturation will be minimized and precolumn stability increased. The proteins are washed to waste and separated from the analytes which are sorbed onto the precolumn. Typically, up to about 26 ml of plasma corresponding to thousands of 10 microliter injections, or up to 50 injections of 500 microliters can be processed with a single precolumn before a significant loss of performance occurs. Restricted access sorbents represent an alternative approach to the precolumn-venting plug technique for precolumn enrichment or direct column separations [316,317]. A characteristic of these sorbents is that they have an internal surface coated with a retentive bonded phase and an external surface designed to be weakly retentive of biopolymers. The macromolecules found in biological fluids are excluded by their size from exploring the pore volume and are eluted within the void volume of the precolumn avoiding any destructive accumulation on the sorbent. The smaller analyte molecules are not excluded from the pore volume and are retained there by the greater density of bonded phase ligands until subsequently desorbed with an appropriate solvent on to the analytical column for separation. Restricted access sorbents have been shown to be durable in use and long precolumn lifetimes have been obtained.

8.8 HEADSPACE ANALYSIS

Headspace methods provide an indirect method of sample analysis suitable for the determination of organic volatiles [11,318-323]. The gas phase in contact with the sample and not the sample matrix itself is taken for analysis. If the sample is in thermodynamic equilibrium with the gas phase in a closed thermostated vessel, then this method of analysis is referred to as static headspace. If a carrier gas is passed over the sample and the sample volatiles accumulated in a cryogenic or sorbent trap, then the method is generally referred to as dynamic headspace. If the carrier gas is introduced below the surface of the sample and passes through the sample in the form of a stream of small bubbles with the stripped organic volatiles accumulated in a sorbent trap, then this method is generally referred to as dynamic headspace, purge-and-trap, gas phase stripping, or gas phase sparging. The headspace sampling methods are used predominantly for the determination of trace concentrations of volatile substances in samples which are difficult to handle by conventional chromatographic means. Examples include dilute solutions where the matrix would obscure the components of interest, damage the column or require excessively long separation times due to the presence of late eluting peaks, inorganic or high molecular weight polymers which cannot be volatilized or solubilized under normal conditions, and inhomogeneous mixtures, such as blood, sewage, colloids, etc., which require extensive sample cleanup prior to analysis. In the above situations, the advantages of the headspace method are economy of effort and the attainment of a sample which is relatively free from its matrix and the problems associated with the chromatographic properties of the matrix. The main disadvantage of quantitative headspace analysis is the need for careful calibration.

8.8.1 Static Headspace

From the experimental point of view the static headspace sampling technique is very simple. The sample, either solid or liquid, is placed in a glass vial of appropriate size and closed with a Teflon-lined silicone septum. The vial is carefully thermostated until equilibrium is established. The gas phase is sampled by syringe for manual procedures or with an electropneumatic dosing system in automated headspace analyzers. The original analyte concentration (in Henry's law region) is then given by equation (8.3)

$$C_L^0 = [C_G(KV_L + V_G)]/V_L \qquad (8.3)$$

where C_L^0 is the initial analyte concentration in the liquid phase, C_G the concentration of analyte in the gas phase, K the gas-liquid partition coefficient for the analyte at the analysis temperature, V_L the volume of liquid phase, and V_G the volume of gas phase [318-321,324,325]. From equation (8.3) it can be seen that the concentration of the analyte in the headspace above a liquid in equilibrium with a vapor phase will depend on the volume ratio of the gas and liquid phases and the compound-specific partition coefficient which, in turn, is matrix dependent. The sensitivity of the headspace sampling method can be increased in some instances by adjusting the pH, salting out or raising the temperature of the sample. Since concentration of the analyte in the gas phase is proportional to the concentration of the undissociated part of the analyte in solution, adjusting solution pH so that the analyte exists mainly in the undissociated form will improve sensitivity. Salting out by adding an inorganic salt to an aqueous solution or a non-electrolyte such as water to a miscible organic solvent can produce an increase in sensitivity of over 100-fold in favorable cases. Raising the temperature of the sample increases the saturated vapor pressure of the analyte and hence its concentration in the gas phase. Derivatizing reagents forming involatile derivatives with compounds containing particular functional groups can be used to selectively identify analytes containing these groups by making a before addition and after addition comparison of chromatograms.

Analysis of headspace samples can be performed manually using a gas-tight syringe or automatically using one of a large number of commercially available pneumatic headspace analyzers [320,321,325-330]. Since many headspace analyses are carried out at elevated temperatures with pressures exceeding atmospheric, difficulties can arise in obtaining accurate values by syringe sampling due to the undefined expansion of the headspace pressure through the needle into the atmosphere. Selective adsorption of the headspace vapors onto the syringe is another potential problem. This can be estimated, and in some cases corrected for, by the re-injection method [331]. Careful use of syringes, particularly those with a gas lock mechanism, combined with appropriate internal standards, can provide acceptable quantitative data but when the analysis of many samples is contemplated, pneumatic sampling devices are generally superior, Figure 8.26. Pneumatic sampling ensures that both the pressure and volume of the fraction of the headspace sampled are the same and identical for all samples and standards. A constant pressure is provided for by pressurizing the headspace vials with an inert gas to a pressure at least equal to the column inlet pressure. The sample is then either expanded directly into the column or to a sample loop of a thermostated gas sampling valve. Typical headspace sample volumes tend to be 0.5-3.0 ml, which can be quickly transferred to and easily handled by packed columns. For open tubular columns with flow rates of 1-2 ml/min. the complete transfer of the sample is too slow to form a sharp column injection band [332]. The simplest solution to this problem is to use flow splitting to rapidly transfer a fraction of the sample to the column. Sample discrimination does not occur because the vapors and gas are mixed prior to the split but sensitivity is compromised as only a portion of the sample reaches the column. The alternative approach is to transfer the whole of the sample to the column which is then refocused at the head of the column by cold trapping. In some headspace analyzers the whole column oven is cooled [326,329] while in others just a short section of the column is cooled by using an independent cold trap [327,328,330, 333,334].

Figure 8.26(A) is an example of a valve type interface [329]. Helium carrier gas is provided to the headspace sampler and is split into two flow paths. One path is flow-controlled and provides a constant flow of carrier gas which passes from the headspace unit through the heated transfer line to the gas chromatograph. The second flow path is pressure-regulated and, in the standby mode, the sample loop and sampling needle are flushed continuously by the helium flow. At a time determined by the operator, the sampling needle pierces the septum and helium pressurizes the headspace vial to any desired pressure. The headspace gas is then allowed to vent through the sample loop. Once filled, the sample loop is placed in series with the normal carrier gas flow and its contents are driven through the heated

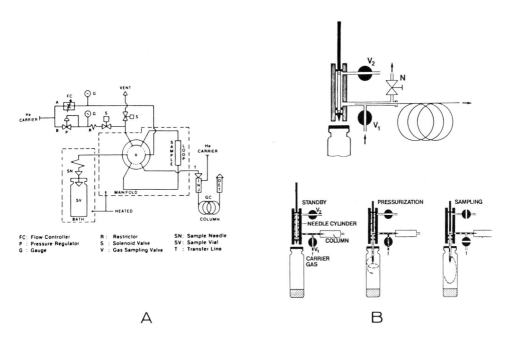

A   B

Figure 8.26   A, valve diagram of the Hewlett-Packard 19395A
Headspace Sampler. B, Headspace sampling with constant headspace
pressure, as carried out with the Perkin-Elmer HS-100 Automatic
Headspace Sampler.

transfer line to the GC injection port. The sampling device shown
in Figure 8.26(B) comprises a movable needle with two vents and a
solenoid valve V_1 in the carrier gas supply line. The needle moves
up and down in a heated cylinder and is sealed by three O-rings.
In the standby position the lower needle vent is placed between
the two lower O-rings and thus sealed from the atmosphere, while
the carrier gas streams through valve V_1 to the column. A small
cross flow purges the needle cylinder to vent any residual sample
vapors through solenoid valve V_2. This sampling head is connected
to the oven of the gas chromatograph by a heated transfer line and
the capillary column is threaded through this line until it ends

shortly before the moving needle. At the end of the selected thermostating period the sampling needle descends, pierces the septum cap, and enters the headspace of the vial. Carrier gas streams into the sample vial and pressurizes it to the column inlet pressure. At the end of the selected pressurization time, valves V_1 and V_2 both close and the pressurized gas in the headspace vial expands through the sampling needle, carrying all volatile compounds onto the column. At the end of the preselected sampling time both valves open again and the carrier gas flow to the column is resumed. Opening the needle valve, N, allows sampling by split flow if required. The addition of an auxiliary carrier gas supply line, operated at a higher pressure than the column carrier gas, and two additional solenoid valves, which are placed before valve V_1, enables the headspace sampler to operate with headspace pressures higher than the column inlet pressure.

A unique feature of the static headspace method is that the information obtained from the experiment, the chromatographic peak area of a substance in the gas phase, is an indirect measure of the concentration of that same substance in the original sample. The liquid and gas phase concentration are related to each other by the partition coefficient and an experimentally-derived proportionality constant. The partition coefficient is matrix dependent and remains unknown for most analyses; it must therefore be accounted for by calibrating the sampling system. Suitable calibration methods are summarized Table 8.11. In principle, all methods used for routine quantitation in GC are applicable except for peak area normalization [318-322,335-337]. If the sample matrix can be obtained in a pure form, then calibration is simply performed by adding known amounts of the analyte to the matrix and analyzing the calibration standards in a manner identical to the original samples. This method is convenient but requires good repeatability as two independent injections are compared. If the sample matrix cannot be duplicated or if the reproducibility of sample injection is inadequate (for example, using a manual syringe method) then the method of standard additions is used. The inherent advantage of the internal standard method is that the influence of the sample matrix on the volatility of the analyte is included in the calibration procedure and no detector response factors need be determined. Each sample must be analyzed twice and it is assumed that the small amount of analyte added to the sample does not change the partition coefficient of the analytes which,

TABLE 8.11

CALIBRATION METHODS FOR QUANTITATIVE HEADSPACE ANALYSIS

Sample Type	Calibration Method	Principle	Examples
Homogeneous Solutions	Model System	Matrix available in pure form. Known amounts of substance added to matrix and determined by the same experimental procedure as samples.	Volatile organic compounds in drinking water, beverages, vegetable oils, mineral oils, etc.
Inhomogeneous Samples (liquid + solid)	Model System Standard Additions	Model systems may be used if a pure sample matrix is available (e.g., blood, milk). In the standard addition method the sample is analyzed twice, the second time after the addition of a known amount of the substance to be determined. The sample must be re-equilibrated after addition of the standard.	Residual monomers in polymer dispersions, alcohol and toxic substances in blood, olfactory substances in milk.
Soluble Solids	Model System Standard Additions	For solid samples forming homogeneous solutions the model system may be used if pure sample matrix materials are available; otherwise, the standard additions method is used.	Inorganic, organic, and polymeric solids and salts. Monomers in polymers.
Insoluble Solids	Multiple Headspace Extraction	MHE can be used for substances of high volatility with a small partition coefficient. Method is based on a stepwise gas extraction at equal time intervals. Normal headspace chromatogram is run, a fraction of the gas phase exhausted, and a second headspace chromatogram is run. The difference in peak areas provides a measure of the total peak area of the analyte.	Volatiles and monomers in insoluble polymers. Aroma volatiles from foodstuffs, fruits, spices, tobacco, etc. Residual solvents in pharmaceuticals and printed films.

in some cases, may be only an approximation [336]. This is the most universal calibration method for liquids and solutions. Solid samples which form homogeneous solutions may be analyzed by either internal or external standard calibration. The solvent must be free of volatile impurities and preferably of low volatility, so that it elutes after the sample components in the chromatogram and does not interfere in their determination. For insoluble solid samples, and for samples for which matrix duplication or the standard additions method are inappropriate, the method of multiple headspace extraction is utilized [335,338-340]. This method is in principle a dynamic gas extraction procedure, but carried out stepwise, in a manner analogous to repeated liquid-liquid extraction in a separatory funnel, except that in this case a single vial is used and the headspace alternately equilibrated, sampled with removal of the remaining headspace (or most of it), followed by replacement of the gas phase. The process can be automated. The concentration of the analyte in the headspace becomes smaller after each extraction, while the partition coefficient remains constant. Repetition of the process would eventually result in the complete stripping of the analyte from the matrix and the original amount of analyte could be obtained from the sum of all the partial chromatographic peak areas for each extraction. Fortunately, exhaustive extraction is not required to calculate the total amount of analyte present. In most cases two measurements suffice to estimate the total area for the analyte using equation (8.4)

$$A_T = A_1^2/(A_1 - A_2) \hspace{4cm} (8.4)$$

where A_T is the sum of all partial peak areas, A_1 the peak area obtained in the first headspace analysis, and A_2 the peak area obtained in the second headspace analysis. The multiple headspace extraction method may fail for solids made up of components of different particle sizes or porosity, for which different rates of migration of the volatiles to the gas phase exist, and for experimental conditions that fail to permit the establishment of an equilibrium [339-341]. Problems of this kind can be recognized by a lack of linearity of semilogarithmic plots of peak area as a function of extraction number.

8.8.2 Dynamic Headspace

Dynamic headspace sampling employs the continuous removal of headspace vapors above a liquid or solid sample by means of a gas flow with subsequent trapping of the sample components by solid-phase extraction or cold trapping. It is used to determine analytes which are too low in concentration or have unfavorable partition coefficients for their determination by static headspace methods [318-322,342]. The efficiency of the dynamic headspace method for solutions can be improved by passing the gas through the solution in the form of small bubbles (purge-and-trap) and by operating the extraction and trapping steps in a closed loop by continuously recirculating a fixed volume of gas through the solution and trap (closed loop stripping analysis). When a gas is used to strip volatile organic substances from a liquid, the rate at which a substance is removed from solution can be expressed by equation (8.5)

$$R = 1 - \exp[-Ft/(KV_L + V_G)] \tag{8.5}$$

where R is the recovery of the analyte (equal to the ratio of the analyte sorbed onto the trap and the initial amount present in the sample), F stripping gas flow rate, t stripping time, K the gas-liquid partition coefficient, V_L sample volume, and V_G volume of gas passed through the liquid in time t [319,343,344]. Equation (8.5) is somewhat idealized as it is assumed in its derivation that: (1) thermodynamic equilibrium exists; (2) no breakthrough occurs in the trap (closed loop stripping analysis); (3) the liquid matrix is involatile; and (4) the partition coefficient is independent of concentration. However, from equation (8.5) it can be seen that the amount of analyte stripped from solution will depend primarily on the substance-specific partition coefficient and the experimental variables, such as flow rate, time and the total volume of stripping gas passed through the solution. In the most favorable case, that of volatile nonpolar analytes, KV_L will be small and the recovery will become independent of the vapor pressure of the analyte and is determined only by the volume of stripping gas. For polar, involatile analytes KV_L can be assumed to be significant and large stripping gas volumes will be needed to achieve reasonable recoveries. Only in those instances where the analyte has a low water solubility (< 2% w/w) and is relatively volatile (b.p. < 200°C) can quantitative extraction be expected, all be it, if the sampling time will be long [345].

A suitable apparatus for the dynamic headspace analysis of urine and similar biological fluids available in large volume, waste and surface waters, tissue homogenates, beverages, and plants and herbs is shown in Figure 8.27(A) [342,346,347]. Taking urine as a general example, an aliquot of a 24-hour urine sample mixed with ammonium sulfate (200 g/l) is charged into a 500 ml sampling bottle and vigorously stirred with a magnetic stirrer. The ground-glass joint at the neck of the sampling bottle is fitted with a condenser connected to a thermostatically controlled fluid circulator. Single or multiple sorbent traps, usually filled with Tenax, are attached to the other end of the condenser. A flow of helium is established through the apparatus at 20 ml/min. The organic volatile fraction is collected by heating the rapidly stirred sample in a water bath set at 90°C; the sampling time usually being one hour. The passage of organic volatiles from urine into the gas phase is influenced primarily by their water solubility, vapor pressure, the possibility that selective adsorption to high molecular weight biological molecules will diminish their vapor pressure, and the area of contact between the gas and liquid phase. The transfer of organic volatiles to the gas phase is generally favored by salting-out, elevated temperatures, and vigorous stirring. Because the equilibration of volatiles between the gas and liquid phases is slow, sampling times are comparatively long; thus it is possible to trap several samples sequentially during the same experiment. These samples may differ quantitatively, a reflection of the fact that the volatiles are removed gradually and incompletely.

The headspace volatiles from biological fluids are comprised of a chemically diverse group of substances of widely different polarity; most are alcohols, ketones, aldehydes, O- and N-hetrocyclic compounds, isocyanates, sulfides, and hydrocarbons containing from 1 to 12 carbon atoms and with boiling points generally less than 300°C [342]. Quantitative differences in headspace profiles are useful for the identification of metabolic disorders.

Gas phase stripping (purge-and-trap) techniques can improve the yield of organic volatiles from water or biological fluids by facilitating the transfer of volatiles from the liquid to the gas phase; it is also more suitable than dynamic headspace sampling when the sample volume is restricted [320-323,347-351]. The technique is used routinely in many laboratories for the analysis

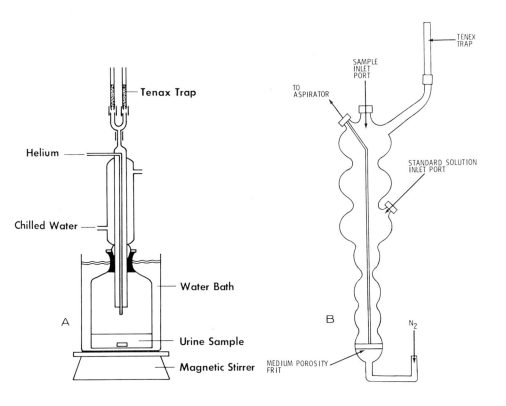

Figure 8.27 A, apparatus for dynamic headspace analysis of urine with sorbent trapping. B, gas phase stripping apparatus (purge-and-trap).

of water suspected of containing low boiling point environmental contaminants, such as organochlorine and aromatic solvents, and automated equipment is available. A suitable apparatus for gas phase stripping is shown in Figure 8.27(B) [352]. Gas phase stripping differs from the dynamic headspace method in that the sampling gas is introduced below the liquid level through a fritted orifice; the finely dispersed bubbles provide maximum surface contact between the gas and liquid phases. As the organic volatiles move into the gas phase they are rapidly and continuously carried away, thus favoring the stripping process. The apparatus shown in Figure 8.27(B) has baffled walls to diminish sample carryover due to foaming, has an inlet port so that standard solutions may be added below the level of the sample by syringe, and contains a center tube which facilitates emptying and cleaning the apparatus. For volatile halocarbons in waste water the stripping process is essentially complete and detection limits below the microgram/liter level can be obtained.

Trapping of the volatiles is again usually performed with a Tenax trap. For the accumulation of substances with boiling points below 30°C a segmented trap containing Tenax backed-up with silica gel, carbon or coated stationary phase is used to improve the trapping efficiency [353,354]. Cold trapping is also used with the purge-and-trap technique for the recovery of organic volatiles particularly when open tubular columns are used for the analysis [349,355-358]. Cold trapping has the advantage that thermally labile and polar compounds are less affected during the trapping and thermal desorption steps and fewer artifact peaks are likely to be produced. Usually, after passage through the sample the purge gas is led through an efficient condenser at a low temperature (ca. - 15°C) in order to remove most of the water vapor without affecting the volatile compounds. These are then collected in a capillary trap consisting of a length of fused-silica tubing cooled by liquid nitrogen. The capillary trap may be empty, packed, or coated with stationary phase depending on the analytes to be determined. The trapped analytes are subsequently thermally desorbed into the analytical column using either split or splitless injection.

Gas phase stripping with adsorption of the volatiles on a small charcoal trap in a closed loop system was popularized by Grob for the determination of very low levels of organic volatiles (nanograms/liters) in drinking water, etc. [322,343, 359-361]. A version of this apparatus is shown in Figure 8.28(A) [360]. A fixed volume of gas is recirculated through the sample and trap which contains 1-5 mg of a specially prepared charcoal. A influent preheater (not shown in Figure 8.28) is generally used to warm the stripping gas to prevent condensation of water in the charcoal trap. The trace organics are then microextracted with a few hundred or less microliters of solvent to avoid the need for a solvent concentration step prior to gas chromatography. Carbon disulfide or methylene chloride have generally been used as the extracting solvent by forcing several volumes of solvent backwards and forwards through the charcoal filters using either a modified syringe or nitrogen pressure [343,359,364]. The sensitivity of the procedure depends very much on maintaining low background levels

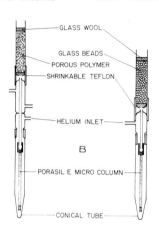

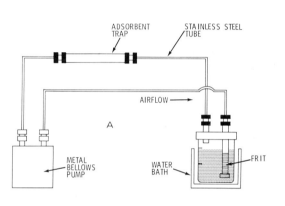

Figure 8.28   Closed loop gas phase stripping apparatus (A).
Transevaporator sampling apparatus for isolating organic volatiles
from small sample volumes (B).

for the apparatus. Poor quality charcoal and contaminated
stripping gas, breakthrough of very polar analytes which deposit
on the apparatus and are only slowly released, and dust particles
escaping from damaged filters that become deposited in the pump
and lines and interact with samples giving ghost peaks have been
recognized as the principal sources of background contamination
[362,365,366]. The nature of the solute and its volatility and
polarity can greatly influence the recovery and processing time.
Heating at 45°C for 30 minutes is reasonable for large survey
studies while from 2-24 hours at 30-35°C will provide higher
recoveries of less volatile analytes. Analytes of high water

solubility and nonpolar analytes with boiling points above 300°C are generally recovered in low yield.

In general terms, the purge-and-trap technique is the method of choice for determining organic volatiles in water because of its ease of operation. If greater sensitivity is required, the closed loop stripping apparatus should be used.

Zlatkis and co-workers have described a transevaporator sampling apparatus for the solvent stripping of organic volatiles from small (5-500 microliter) samples of biological fluids with collection and separation of the organic volatiles from the extracting solvent on an adsorbent column [105,342,367,368]. The transevaporator apparatus is suitable for the semiquantitative analysis of organic volatiles in serum, urine, saliva, cerebrospinal fluid, breast milk, amniotic fluid, sweat and tissue homogenates. With reference to Figure 8.28(B), the sample is injected by syringe into the base of a microcolumn packed with Porasil E. The microcolumn retains most of the water, the high molecular weight polar organics, and the inorganic salts present in biological fluids. It also distributes the sample as a film over a large surface area and thus improves the efficiency with which the organic volatiles are extracted. The stripping gas passes through the microcolumn, removing the organic volatiles which are then collected on a Tenax trap, providing a modified headspace sample. A solvent-extraction profile is obtained using gas pressure to force a volatile solvent such as 2-chloropropane through the microcolumn; this solvent carries the solvent-extractable organic volatiles to a glass bead trap, where the volatiles are collected and the solvent is evaporated. Thermal desorption is used to transfer the organic volatiles from the trap to a capillary column for separation by gas chromatography.

Small solid samples can be analyzed directly by dynamic headspace sampling using a platinum coil and quartz crucible pyrolyzer and cold trap coupled to an open tubular column [341,369,379]. This method has been used primarily for the analysis of mineral samples and of additives, catalysts and byproducts in finished polymers which yield unreliable results using conventional headspace techniques owing to the slow release of the volatiles to the headspace. At the higher temperatures (450-1000°C) available with the pyrolyzer the volatiles are more readily and completely removed from the sample providing for quantitative analysis.

8.9 AIR SAMPLING

In general, three kinds of organic compounds can be distinguished as normal constituents of air or environmental contaminants, requiring different sampling procedures for their analysis: substances whose natural state at atmospheric pressure and ambient temperature is that of a gas; substances having sufficient vapor pressure to sublime or volatilize into the atmosphere and exist as vapors; and compounds of intermediate or restricted volatility attached to solid particles. Liquids and solids may exist in the atmosphere as both a vapor and as a particulate requiring that sampling methods be developed which collect both phases without bias. Gases and very volatile organic compounds (e.g., C_1-C_5 compounds) are usually sampled by grab sampling, condensation or liquid or solid phase extraction. Solid-phase or liquid extraction is commonly used for less volatile organic vapors (C_5-C_{20} compounds) and filtration or impingement for the collection of particles. No single sampling technique is adequate for all problems, which adds to the difficulty of providing general guidelines for ambient air analysis [2,8,9,371-374]. Similar techniques to the above can be adapted for sampling workplace atmospheres (personal monitoring) and for monitoring sources of environmental contamination, such as emission from stack gases, incinerators and automobiles [8,286,375-379]. The high temperatures and the corrosive and reactive nature of the gas plume from combustion processes adds to the difficulty of designing and constructing suitable sampling devices. Bubblers, vapor adsorption tubes and passive samplers are widely used in personal monitoring to measure the concentration of airborne organic volatiles in the region of the mouth. If the monitoring device is to be worn throughout the work period, it is important that the device is lightweight, unobtrusive, silent and permits normal movement. Standards for acceptable levels of exposure can be established by combining analytical data with results from animal toxicity studies and human epidemiological investigations. Standards for individual substances are usually set to reflect ceiling values, a threshold concentration value never to be exceeded even momentarily, and also time-weighted average values, which reflect exposure over a normal working period, for example 8 hours. Threshold limit values are set conservatively to allow for individual susceptibilities resulting from differences in age, weight, health, respiration rate, rates of absorption and personal habits, such as smoking or drinking. They are believed to be safe concentrations below which the worker is assumed to be protected from short term effects, such as chronic toxicity and environmental nuisances, and long term effects, such as cancer.

8.9.1 Volatile Organic Compounds in Air

Volatile hydrocarbons and halocarbons, etc., are usually collected by grab sampling or, in the case of reactive substances, by chemical trapping in bubblers (see section 8.4.1). Solid-phase extraction techniques are less frequently used due to the low breakthrough volumes of many low boiling point organic compounds (see section 8.9.2). Grab samples are taken by pulling air into an evacuated glass bulb, syringe, container or plastic bag. Convenience usually limits the size of the collection vessels to a few liters so that pumping to above atmospheric pressure or cryogenically cooling the collection container are used to increase the sample size for analysis. This can be important when sampling non-urban atmospheres were the mixing volume of volatile organic compounds may be in the parts-per-billion or parts-per-trillion range. Plastic bags made from a variety of materials, such as Teflon, Mylar, Tedlar, etc., and some times laminated with aluminum to reduce their permeability, are available commercially and have the advantages of being lightweight, unbreakable, and are easily filled from a collapsed state by syringe or with a small pump. Rigid containers must be evacuated away from the sampling site to avoid the possibility of contaminating the sample by the vacuum source. However, all such containers, rigid or collapsible, show different rates of adsorption, catalysis, or reaction with various analytes which is generally the overriding consideration for selection. This also controls the permitted duration of storage and acceptable analyte concentration range for analysis. Extensive cleaning and testing of the containers between use, particularly for trace analysis, detract from the apparent simplicity of the grab sampling method [8,380,381].

Grab samples are usually returned to the laboratory for preconcentration, when required, and analysis [380-384]. Cold trapping techniques are used to isolate and concentrate the analytes from air prior to separation. Since air contains relatively large volumes of water and carbon dioxide compared to typical analyte concentrations, these components must be separated from the analytes by using trap to trap fractionation at low temperatures. The rate of vaporization of the volatile organics is much greater than that of the frozen water so that a relatively short evaporation period is sufficient to remove all the analytes leaving most of the water behind. An intermediate drying trap containing a desiccant may also be used if it does not adsorb or react chemically with the analytes. The analytes are then cold trapped for refocusing and injection onto the analytical column for separation by gas chromatography. If insufficient water has been removed from the sample then ice crystals may form and block the cold trap.

8.9.2 Semivolatile Organic Compounds in Air

Organic compounds of intermediate volatility are usually collected by drawing filtered air through a bed of an appropriate sorbent. Porous polymers, various forms of carbon, or inorganic adsorbents, such as silica gel and alumina, are the most frequently used. No single sorbent is useful for all sample types. Tenax, a polymer based on 2,6-diphenyl-p-phenylene oxide, is the most widely used sorbent for trapping volatile compounds at room temperature with a molecular weight above C_7. It is hydrophobic and thus retains little water during sampling, has a high sampling capacity for appropriate analytes, is thermally stable so that high temperatures can be used for rapid desorption, and is reasonably free of impurities that might interfere with the analysis. It is more expensive than the macroreticular porous polymer beads based on polystyrene and poly(methacrylate) polymers but is generally preferred over these because of its greater chemical purity and thermal stability [8,9,180,385-391]. Soxhlet extraction followed by thermal conditioning is generally used to minimize sorbent blanks. Both Tenax and the porous polymers react chemically with ozone, chlorine and nitrating agents which can represent a source of artifact formation when sampling stack gases or during photochemical smog events with high ozone concentrations [392-395].

Various forms of carbon are used to sample those analytes whose breakthrough volume is too low on Tenax for sufficient preconcentration [8,395-399]. Charcoal, graphitized carbon blacks, and carbosieves with surface areas from 5 to 900 m^2/g are commercially available. The high surface area sorbents are used for sampling gases and very volatile organic compounds, such as chloromethane and vinyl chloride, and the lower surface area sorbents for semivolatile organic compounds. The most common method of personal monitoring makes use of a sorbent cartridge filled with charcoal in conjunction with a small pump to maintain a fixed flow of air through the cartridge [377-379,397-401]. Typical sorbent traps are glass tubes (5-6 cm long and 4 mm I.D.) that have been divided with polyethylene plugs; they contain approximately 100 mg of sorbent in the front section and 50 mg in a back section. The two sorbent sections are analyzed separately by eluting the trapped organic volatiles with carbon disulfide followed by gas chromatography. Analysis of the back section confirms whether breakthrough of the analyte of interest on the front section has occurred. Sorbent cartridges containing 600 mg or 1 g are used for long-term sampling of high concentration and/or highly volatile compounds. Charcoal open tubular traps are useful for microanalysis and analyte recovery by thermal desorption [399].

Unfortunately, carbon irreversibly adsorbs some compounds and the chemical and physical properties of various types of sorbent vary tremendously. Batch to batch variations in sorbent efficiency can be pronounced. The recovery of polar compounds from carbon using carbon disulfide may be incomplete. In some cases elution with binary solvent mixtures, e.g., carbon disulfide-water, or polar organic solvents such as dimethylformamide may lead to increased recovery [8,402]. However, careful calibration is necessary when recovery from the sorbent is incomplete. Thermal desorption methods are less commonly used with carbon sorbents, although by no means never used, because of the slow kinetic release of analytes from carbon and the higher catalytic activity of the sorbents compared to polymeric sorbents. Recently, it has been shown that microwave heating can rapidly raise the temperature of carbon to 700°C and permit efficient, kinetically fast, desorption of stable analytes [328,403]. Also, more carefully matching the surface area of the carbon with the analyte sorption/desorption properties increases the success of the thermal desorption method particularly for very volatile analytes that may be troublesome to process in solution without vaporization losses [399].

Silica gel and alumina show high retention of volatile and polar compounds. They can also be coated with chemical reagents for reactive trapping. Silica and alumina take up large volumes of water which can cause breakthrough of the analytes by displacement or due to the heat generated by the adsorption of water. These sorbents tend to be reserved for special purposes or are used in conjunction with other sorbents to prevent breakthrough of analytes with low retention on Tenax and carbon [8,378]. As an understanding of the properties of different sorbents has grown, the use of traps containing several different sorbents, in different optimized amounts, for comprehensive trapping of divergent analytes within a single sorbent cartridge has become more common.

The polyurethane foams are perhaps the least efficient of the common sorbents for trapping volatile compounds. They are limited in application to the sampling of high boiling point compounds, such as airborne pesticides and polychlorinated biphenyls [190,404-408]. They are frequently used for high flow rate sampling of semivolatiles in conjunction with high-volume air filters on account of their low pressure drop compared to standard sorbent cartridges. Analytes are usually recovered by solvent elution or Soxhlet extraction.

A schematic diagram of an apparatus used for trapping organic volatiles in air with Tenax is shown in Figure 8.29(A) [385]. Although sorbent traps of different dimensions are frequently used, a sorbent bed of 1.5 x 6.0 cm is generally sufficient for sampling air volumes of 5 to 200 liters at 10-200 ml/min. The Tenax sorbent is decontaminated before use by Soxhlet extraction with pentane and methanol, dried, and thermally conditioned in a stream of purified helium at 250-350°C until an acceptable analytical blank is obtained. After conditioning, the Tenax traps can be stored in culture tubes with Teflon-line caps until needed. After sample collection the analytes are recovered by thermal desorption into a capillary column or cold trap in an apparatus similar to that shown in Figure 8.29(B). The desorption temperature and time required to strip the sample from the sorbent depends on the properties of the sample, but temperatures of 250-350°C and times from 5-30 min are common. As a general guide, a purge flow of helium at 15 ml/min and a temperature of 270°C for 8 min is sufficient to recover most of the organic compounds trapped from air.

The thermal desorption process is kinetically slow but when packed columns at low temperatures were used very few problems

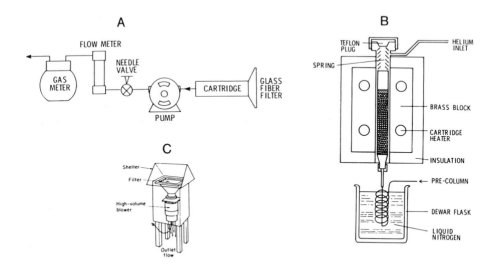

Figure 8.29 Apparatus for sampling airborne organic samples. A, trapping of organic volatiles in air using a sorbent trap; B, thermal desorption chamber; C, high-volume sampler for air particulates.

were experienced. The use of open tubular columns, however, because of their lower capacity for refocusing large volumes of vapor and their susceptibility to blockage by condensed water at cryogenic temperatures, presented problems. Passing a dry stream of gas through the sorbent for a short time at a low temperature, storing the sorbent cartridge in a sealed container with desiccant or vacuum stripping is usually sufficient to avoid the problems associated with ice formation in cold traps [354,385,409,410]. Still a considerable inherent flow mismatch exists between the conditions used for optimum thermal desorption and the required flow rates of narrow bore open tubular columns. Flow splitting was an obvious solution but mitigated against the primary advantage of the thermal desorption method, namely high sensitivity because solvent dilution was avoided. Wide bore columns are less efficient but present less of a flow mismatch problem. All types of open tubular columns can be interfaced to a desorption chamber using one of the following methods: (1) thermally desorb onto a smaller intermediate Tenax trap which can subsequently be more rapidly desorbed into the open tubular column; (2) desorb into a section or loop of capillary tubing cooled to liquid nitrogen temperatures (see Figure 8.29(B)); (3) desorb into a short length of the analytical column or retention gap at cryogenic temperatures; (4) desorb into a cryogenically cooled packed capillary precolumn (used for the quantitative recovery of the most volatile analytes); or (5) desorb into the analytical column with whole column cold trapping [354,384,410-414]. The use of cold trapping allows the desorbed sample to be refocused and the full resolving power of the open tubular column to be maintained. Automated equipment is now available for thermal desorption with cold trapping based on the above principle. The parameter which characterizes the collection efficiency of a particular sorbent for a particular analyte is the breakthrough volume. Since there is no clear consensus on the definition of this parameter, or on the preferred method of measurement, numerous conflicting values can be found in the literature. The breakthrough volume is usually assumed to be that volume of gas, containing analyte, which can be passed through the sorbent bed until its concentration at the outlet reaches some fraction of its inlet concentration. This fraction has been variously defined as 100%, 50%, 5% or 1% of the analyte concentration at the inlet. Several methods have been used to determine the breakthrough volumes of sorbent traps [9,386,388, 389,415-418]. These include passage of a constant concentration of the analyte through a sorbent trap connected to a flame ionization detector and measuring the volume of sample required to give a predetermined detector response, field sampling with dual traps in series, loading traps by liquid or vapor deposition which is then purged with a volume of gas and the amount of sample remaining determined, or by estimation from the measurement of chromatographic retention volume data for the sorbent at different temperatures. The passage of a constant concentration of analyte through a sorbent trap, an example of frontal chromatography, probably best represents practical experimental conditions but suffers from the added difficulty of accurately preparing multiple standards of low concentration in the vapor phase. Depositing the sample as a vapor or as a solution onto one end of the sorbent bed followed by purging with carrier gas is relatively simple to perform but the breakthrough volume is only approached incrementally with this technique. After purging the tube is thermally desorbed to determine the remaining sample. The process is repeated in steps until a significant loss of analyte is determined and may involve several measurements to fix the breakthrough volume. Sampling under field conditions with two sorbent traps in series is simple to do and is included as standard practice in many sampling protocols. A significant finding of analyte on the second sorbent cartridge indicates that the breakthrough volume of the first cartridge was exceeded and the analytical data for that cartridge is unreliable. A value exceeding 10% of the analyte concentration in the second cartridge compared to the value for the first is frequently used as an indication of significant breakthrough from the first cartridge. The chromatographic methods, based on theory that assumes infinite dilution, are perhaps not appropriate to accurately describe the sampling process but they are simple to perform and lead to reasonable values for the breakthrough volume. An approximate linear relationship exists between the logarithm of the specific retention volume of a substance and the reciprocal of the absolute column temperature (van't Hoff equation). Therefore, the retention volume of a substance can be measured at several higher column temperatures and the value at 20°C, or any other temperature, can be obtained by extrapolation. Some representative breakthrough volumes on Tenax measured in this way are given in Table 8.12.

The breakthrough volume depends mainly on the affinity of the analyte for the sorbent, the analyte concentration, humidity, temperature, the efficiency (measured in theoretical plates) of the sorbent trap, and matrix coadsorption. Within reasonable experimental limits, the breakthrough volume for polymer sorbents is independent of normal variations in humidity and the concentration of the analyte (< 100 ppm) in air. Since a wide variety of sampling conditions are employed for collecting organic volatiles on sorbent traps, the idea of a safe sampling volume has been proposed [389,417]. This has been defined as the volume of air containing a particular vapor contaminant that may be sampled under a variety of circumstances without significant breakthrough occurring, and is expressed approximately by equation (8.6); for a more explicit solution see [415,416],

TABLE 8.12

BREAKTHROUGH VOLUMES (liters/g) OF SOME COMMON ATMOSPHERIC
POLLUTANTS ON TENAX-GC
Under field sampling conditions a cartridge containing about 2.2 g
(7.96 ml) of Tenax is commonly used.

Chemical Class	Compound	Boiling Point (°C)	Adsorbent Trap Temperature (°C)		
			21.1	26.7	32.2
Hydro-carbons	hexane	68.7	7.7	5.5	4.1
	heptane	98.4	34.1	25.0	17.7
	1-heptene	93.6	61.4	42.3	29.1
	cyclohexane	80.7	11.8	8.6	6.4
	benzene	80.1	24.5	17.3	12.3
	toluene	110.6	111.4	78.6	55.4
	ethylbenzene	136.2	315.0	221.4	156.4
	biphenyl	256	14381.0	10228.0	7293.6
Halogen-ated hydro-carbons	vinyl chloride	13	0.57	0.45	0.36
	1,2-dichloropropane	95	52.3	36.8	26.4
	1,3-dichloropropane	121	83.6	60.9	44.1
	2,3-dichloropropane	94	69.1	47.3	32.3
	bromobenzene	155	490.0	347.0	246.0
Alcohols	methanol	64.7	0.36	0.27	0.18
	propanol	97.5	6.4	4.5	3.2
	ethylene glycol	197	30.4	21.4	15.0
Amines	dimethylamine	7.4	1.8	1.4	0.91
	pyridine	115	85.9	60.9	43.2
	aniline	184	1724.0	1176.0	802.7
Alde-hyde	acetaldehyde	20	0.91	0.45	0.41
	benzaldehyde	179	1594.0	1083.0	737.3
Ketones	acetone	56	5.4	3.6	2.7
	methyl ethyl ketone	81	17.7	12.3	8.6
	acetophenone	202	1258.0	909.0	654.0
Esters	ethyl acetate	77	32.7	21.8	14.5
	methyl acrylate	80	34.1	22.7	15.4
	methyl methacrylate	100	144.5	95.0	62.3

$$SSV = V_g \, (1 - 2/n^{1/2}) \tag{8.6}$$

where SSV is the safe sampling volume, V_g the specific retention
volume, and n the number of theoretical plates for the sorbent
trap. These equations show clearly that the collection efficiency
of the sorbent increases as n increases and will be
nonquantitative for small values of n. For example, if n = 2 it
would be impossible to collect a sample with an efficiency greater
than 85% because, even at the start of sampling, 15% of the
analyte would pass through the sorbent bed without being
collected. This applies independent of the value for the specific
retention volume, which may be very large. For a collection
efficiency of 99.9%, a sorbent trap with a minimum of 10.7

theoretical plates is required regardless of the sampling volume or analyte being collected. Hence, the sorbent trap should conform to a minimum length and be packed homogeneously with sorbent of a well-defined mesh range compatible with obtaining a minimum pressure drop across the trap at normal sampling flow rates.

8.9.3 Sampling Techniques for Aerosols

Particle organic matter is generally collected by filtration, impingement or electrostatic precipitation [8,57, 372-376,419,420]. The high-volume sampler, Figure 8.29(C), consists of a filter of large surface area made from glass fiber, porous Teflon or some similar inert material, in a covered housing through which air is drawn by a high flow rate blower at about 1.1 to 1.7 m^3/min. The air is drawn in through a free space between the roof and shelter body. In order to reach the filter, the air has to move upwards through the surrounding rectangular aperture, the dimensions of which are such that particles with diameters larger than 100 micrometers have sedimentation speeds that prevents them from entering and being collected. The filter medium must provide high filtering efficiency, be non-hygroscopic and have low head loss properties. This usually precludes filters that effectively collect particles less than 10 micrometers in diameter. Typically, samples are collected over 24 hours, sampling approximately 2000 m^3 of air, to yield about 0.1 g of particles. The particles consist largely of inorganic material from which the organic fraction is isolated by extraction. During sampling, loosely bound analytes may be lost by sublimation and volatilization from the trapped particles and artifact compounds may be produced by reaction with reactive gases in the sampling stream. The adsorption of gas-phase vapors onto the surface of the filter during sampling is another potential source of error.

Cascade impactors make use of a series of sequential jets and collection plates with increasing jet velocities and/or decreasing gaps between the jet and collecting plate to collect and fractionate air particles by mass. As the air stream progresses through the device smaller particles will be collected more efficiently. Cascade impactors are generally used to isolate particles with diameters less than about 10 micrometers, which present the greatest potential for biological activity as they readily pass through the nasal passageway and are adsorbed in the lungs. Cascade impactors can be used to obtain a particle size distribution.

Sampling mobile sources such as stack gases and engine exhausts, Figure 8.30, is more difficult than ambient air because of the wide range of particle sizes likely to be encountered, the unsteady flow of the gases, and the high temperature, humidity, and reactivity of the sampled gases [372,375,376]. The sampling apparatus shown in the top portion of Figure 8.30 is widely used for the isokinetic sampling of stack gases (see Table 8.13 for an explanation of terms) using a heated filter and a cooled impinger train to retain particles. A condenser and sorbent trap can be inserted instead of, or ahead of, the impinger train to retain organic vapors. For the collection of large sample sizes, as might be needed for chemical analysis or biological testing, the source assessment sampling system (SASS) is more appropriate. This apparatus consists of a series of cyclones for particle sizing, a glass or quartz fiber filter for collecting fine particles, a sorbent module for collecting semivolatile organic compounds, and impingers for collecting reactive gases. This apparatus can operate at high flow rates, 110-140 liters/min with a typical sample size of 30 m^3 of flue gas. Figure 8.30, bottom, illustrates a dilution tube used for monitoring vehicle emissions. This apparatus is designed to simulate as closely as possible the dilution of a vehicle exhaust as it would occur during the first few minutes in ambient air from vehicles on the road. The dilution tube receives the exhaust as input, and supplies a known quantity of filtered and humidity-and-temperature-controlled ambient air for rapidly diluting and cooling the exhaust. The particle material is sampled isokinetically and retained on the filters. Other probes shown are used to sample volatile organics by sorbent trapping and reactive gases using impingers.

8.9.4 Passive Samplers

Passive sampling devices for personal monitoring are being developed as a replacement for sorbent sampling using flow-through samplers to reduce the cost of large-scale screening studies and to make the monitor more convenient for the wearer [8,371,400,401,421-424]. Passive samplers contain an adsorbent, usually charcoal, although porous polymers and chemically reactive reagents are also used, in a protective polymeric support and may be worn in the form of a badge. Sampling occurs by means of

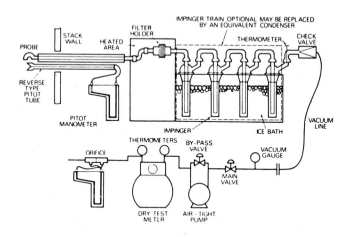

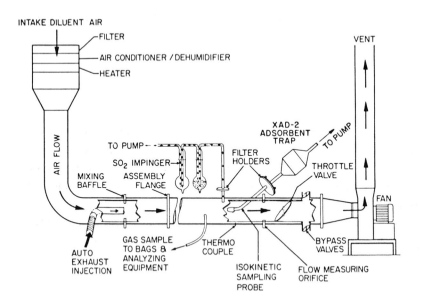

Figure 8.30 Top, sampling train, for collecting particles from stack gases. Bottom, dilution tube used for monitoring and collection of gas- and particle-phase vehicle emissions.

TABLE 8.13

GLOSSARY OF TERMS COMMONLY USED IN AEROSOL SAMPLING

Term	Explanation
Isokinetic Sampling	Collection of samples such that there is no change in the momentum of the particles before they reach the filter. Particularly important to obtain a representative sample when the whole air stream cannot be sampled. Small particles (less than three micrometers) do not require isokinetic sampling as they possess little inertia, but it becomes increasingly important for sampling larger particles.
Pitot Tube	Used to measure the flow rate of the source from which the sample is taken. Consists of two concentric tubes, one with an opening facing upstream and the other with an opening perpendicular to the flow. The pressure difference between the two tubes provides a measure of gas flow rate.
Dry Test Meter	Measures the total integrated volume of air sampled over a given time. The volume flow is measured by mechanical displacement of an internal bellows by the air flow. The displacement is recorded on a mechanical counter via a series of levers (accuracy 2-4%).
Rate Meter	Measures the instantaneous volume flow rate through the sampling systems. An example would be a rotameter or venturi meter. Used to set precise flow rate for flow sensitive sampling devices.

gaseous diffusion through a stagnant air layer or by permeation through a membrane (thus no expensive pumps and associated power supplies are needed) and, provided that the sorbent capacity of the monitor is not exceeded, a measure of the time-weighted average of the concentration of gases and vapors in the breathing zone of a worker is easily obtained.

For the diffusional monitor the steady state mass transport of the vapors follows Fick's Law of diffusion, given by equation (8.7)

$$m = DACt/L \tag{8.7}$$

where m is the total mass of analyte collected, D the molecular diffusion coefficient, A the area of the diffusion channel, L the diffusion path length, C the analyte concentration in the air, and t the sampling time. In deriving equation (8.7) it was assumed that the sorbent is an effective sink for the analyte and, therefore, the analyte concentration at the sorbent surface is zero. By changing the nature of the sorbent and design of the dosimeter (A and L) badges can be produced to sample at any reasonable analyte concentration and over any desired time. The disadvantages of badge dosimeters are that some minimum flow of air at the sampling aperture must be maintained at all times to permit representative sampling and the dosimeter cannot be switched off, so that additional care is needed to avoid contamination of the sampler during storage prior to analysis.

The permeation dosimeter operates on the principle that the rate at which a gas permeates a given membrane is a fixed value, and the total mass of the gas that is transported through the membrane is a function of the concentration of the gas in the ambient atmosphere and the time of exposure. The quantity of analyte collected is given by equation (8.8)

$$m = kCt \qquad\qquad (8.8)$$

where k is a calibration constant that is determined experimentally for each dosimeter and the other terms are the same as given in equation (8.7).

8.9.5 Methods for Preparing Standard Mixtures of Organic Vapors in Air

The preparation of synthetic mixtures of gases or volatile organic liquids in gases is more difficult and potentially less accurate than methods for preparing liquid mixtures [425-428]. This is because gases cannot be easily weighed, volumes may change during handling, and temperature and pressure effects must be considered. Many methods have been used for preparing mixtures with reasonable accuracy, Table 8.14, and these methods can be broadly classified as static or dynamic. Static methods involve preparing and storing the mixture in a closed vessel, for example, a cylinder, flask or plastic bag. The sample volume is thus limited to that of the container. Cylinders must be used to store mixtures at high pressures. Static systems are preferred when comparatively small volumes of mixtures are required at moderately high concentration levels, but losses of components to the vessel walls may occur. Dynamic systems generate a continuous flow of sample mixture and can produce large volumes, with lower surface losses, owing to an equilibrium between the walls and the flowing gas stream. Whether static or dynamic systems are employed, the

TABLE 8.14

METHODS FOR PREPARING STANDARD MIXTURES OF VOLATILE ORGANIC
COMPOUNDS IN A GAS

Method	Principle
Single Rigid Chamber	A known amount of the compound is introduced into a single rigid chamber of known dimensions. Magnetic stirrer or similar device used for homogeneous mixing.
Exponential Dilution Flask	A known amount of pure component or standard mixture is introduced into the vessel, which is stirred for efficient mixing. The vessel is continuously flushed with a steady stream of gas causing the concentration of the vapor to decrease with time.
Gas Stream Mixing	Two or more gas streams flowing at a known rate are mixed to give the desired concentration. Multiple dilution stages may be used to give lower concentrations.
Permeation Tubes	A volatile liquid, when enclosed in an inert plastic tube, may escape by dissolving in and permeating through the walls of the tube at a constant and reproducible rate. The permeation rate depends on the properties of the tube material, its dimensions and on temperature.
Diffusion Systems	The liquid whose vapor is to be the contaminant of the gas phase is contained in a reservoir maintained at a constant temperature. The liquid is allowed to evaporate and the vapor diffuses slowly through the capillary tube into a flowing gas stream. If the rate of diffusion of the vapor and the flow rate of the diluent gas are known, the vapor concentration in the resultant gas mixture can be calculated.

methods used to create homogeneous mixing of the gas and vapor are
important considerations and some provision for creating forced
convection is incorporated into most devices.

The single rigid chamber is the simplest of the mixing
devices. The concentration of the standard mixture produced is
given by equation (8.9).

$$C = C_o \exp (-V_W/V) \tag{8.9}$$

where C is the instantaneous concentration, C_o the initial
concentration, V the container volume, and V_W the volume of sample
withdrawn. The exponential dilution flask, Figure 8.31(A), is a
hybrid static/dynamic system [429-431]. This simple, reliable

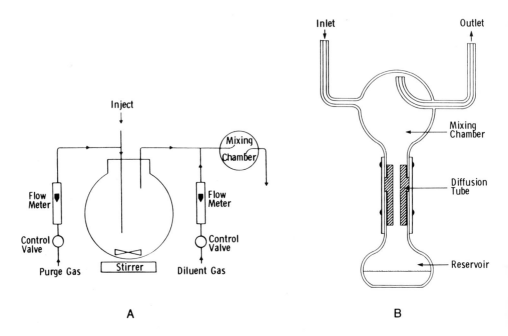

Figure 8.31 Apparatus for preparing standard mixtures of organic vapors. A, exponential dilution flask; B, diffusion tube system.

device is commercially available and has been widely used. It produces an outlet concentration described by equation (8.10).

$$C = C_o \exp(-Qt/V) \qquad (8.10)$$

where Q is the volumetric gas flow rate and t the time after sample introduction. Its principal problems arise from surface adsorption losses, mechanical wear of the mixing device, and the difficulty of accurately measuring the initial sample concentration. However, it is capable of providing adequate accuracy and precision for most analytical applications involving readily volatilized substances. Very low gas phase concentrations can be prepared by mixing the output from the flask with diluent gas.

Permeation tubes are very popular devices for generating standard vapor concentrations. The permeation tube contains a volatile liquid sealed in an inert permeable membrane, usually Teflon or a fluorinated copolymer of ethylene and propylene, through which it diffuses at a fixed and controlled rate. The driving force for the process is the difference in partial

pressures of the analyte between the inner and outer walls of the
tube. This depends on the dissolution of the vapor in the
membrane, the rate of diffusion through the membrane wall, and the
rate at which the vapor is removed from the outer surface of the
membrane [430-432]. The mass permeation rate per unit tube length
can be expressed by equation (8.11).

$$G = 730 \ KMP/\log \ (d_2/d_1) \qquad\qquad (8.11)$$

where G is the mass permeation rate in micrograms/min per cm of
tube length, K the permeation constant for the vapor through the
membrane, M the molecular weight of the vapor, P the gas pressure
inside the tube, d_2 the outside diameter of the tube, and d_1 the
inside diameter of the tube. For samples with low vapor pressures
at room temperature, elevated temperature ovens are used to
increase permeation rates to yield desirable values. Gases or
vapors with high membrane permeability require devices other than
the standard single-walled tubes, for example, multiwalled tubes,
microbottles or permeation wafer devices, to yield reasonable
lifetimes. Commercially available permeation tubes have lifetimes
of several months and provide a simple and inexpensive method of
calibration for laboratories interested in determining only a few
substances or for those who need to perform measurements
infrequently.

The diffusion system, Figure 8.31(B), is a useful and simple
apparatus for preparing mixtures of volatile and moderately
volatile vapors in a gas stream [388]. The method is based on the
constant diffusion of a vapor from a tube of accurately known
dimensions, producing a gas phase concentration described by
equation (8.12).

$$S = (DMPA/RTL) \ \ln \ (P/P-p) \qquad\qquad (8.12)$$

where S is the diffusion rate of vapor out of the tube, D the
diffusion coefficient, M the molecular weight, P the pressure in
the diffusion cell, A the cross-sectional area of the diffusion
tube, R the gas constant, T the temperature, L the length of the
diffusion tube, and p the partial pressure of the diffusing vapor.
Within limits, broad concentration ranges can be prepared by
varying the tube dimensions and/or the flow rate of the diluent
gas. Diffusion tube systems are preferable to permeation tubes
when the latter are not commercially available.

8.10 DERIVATIZATION TECHNIQUES FOR GAS CHROMATOGRAPHY

Gas chromatography is the technique of choice for the separation of thermally stable, volatile organic and organometallic compounds. Unfortunately many compounds of biomedical and environmental interest, particularly those of high molecular weight and/or containing polar functional groups, are thermally labile at the temperatures required for their separation. Derivatization, in principle a microchemical organic synthesis, is used to improve the thermal stability of those compounds that would otherwise be unsuitable for gas chromatographic analysis. In most instances, derivatization reactions are performed to convert protonic functional groups into thermally stable, nonpolar groups. As well as improving the thermal stability of a compound, the derivatized compound often exhibits improved peak shape with a minimization of undesirable column interactions which could lead to irreversible adsorption and asymmetric peak formation. As a wide variety of derivatizing reagents are available the opportunity also exists to purposefully adjust the volatility of a compound to eliminate peak overlaps in the chromatogram. Derivatization can also be used to enhance the detectability of a compound through the introduction of detector-oriented organic groups. In particular, reagents predisposing a substance to detection by the sensitive and selective electron-capture detector are often used.

The anatomy of a typical derivatizing reagent is shown in Figure 8.32. It has two halves: the organic portion which controls volatility and, in the case of reagents used with the electron-capture detector, provides the detector-oriented response; and the reactive group which provides the means by which the organic chain is attached to the substrate. The choice of reactive group controls the application range of the reagent to different functional groups, the selectivity of the reagent towards certain functional groups in the presence of others, and the rate and extent of the reaction. The choice of the organic chain will influence the detection characteristics of the derivative, the rate and completeness of the derivatization reaction (resulting from steric and electronic effects), and the volatility of the derivatized molecule.

Equipment requirements for derivatization reactions are generally simple [27,109,112]; typically, glass tapered reaction

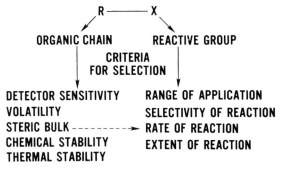

Figure 8.32 Anatomy of a derivatizing reagent.

vials or culture tubes with Teflon-lined plastic screw-caps are used. As the sample requirement for gas chromatography is only a few microliters, reaction vessels tend to be small. Tapered reaction vials are available in sizes from 0.1 to 10.0 ml, with the smaller sizes being the most useful. Reagents and soluble samples are usually measured and handled by syringe. Reaction solutions are mixed by hand agitation, vortex mixing, or with miniature Teflon-coated stirring paddles driven by a magnetic stirrer. Slow reactions are usually accelerated by heating the reaction mixture with a convection oven, hot plate, oil bath or drilled block heater. Screw-capped culture tubes are particularly suitable for reactions occurring under reflux with the air-cooled top section of the tube acting as a condenser. Mininert valves and vacuum hydrolysis tubes are other useful devices. An automated derivatization station for addition of reagents, reaction at elevated temperatures, removal of excess reagents and solvents, and injection into a gas chromatograph is commercially available and should be useful for those laboratories that run routine methods incorporating a derivatization reaction [433,434]

The reagents and conditions used for derivatization reactions are so varied that no attempt at a comprehensive coverage can be presented here. Only general reactions, common reagents, and a few typical applications will be discussed. Extensive compilations of methods and conditions can be found in references [435-441]. Subject reviews covering trialkylsilyl reagents [442,443], cyclizing reagents for bifunctional compounds [444,445], alkylation reactions [446,447], drugs in untreated

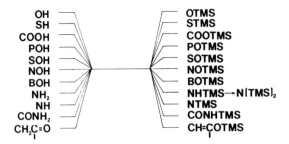

Figure 8.33   Functional groups forming trimethylsilyl derivatives.

biological samples [448], and **reagents** for the preparation of electron-capturing derivatives [449-451] are also available.

8.10.1 Alkylsilyl Derivatives

The most versatile and universally applicable derivatizing reagents for polar molecules containing protonic functional groups are the alkylsilyl reagents. Nearly all functional groups which present a problem in gas chromatography can be converted to alkylsilyl ethers or esters, Figure 8.33. The most common derivatizing reagents are the trimethylsilyl (TMS) reagents. On occasion, higher alkyl homologs or halogen-containing alkyl or aryl substituted analogs of the TMS derivatives are used to impart greater hydrolytic stability to the derivative, improved separation characteristics, increased sensitivity when used with selective detectors, or to provide mass spectra containing greater diagnostic information [443,451]. These newer reagents and their particular areas of application will be discussed later.

The structures of the most widely used trimethylsilylating reagents are given in Figure 8.34. Reactions are usually carried out under anhydrous conditions in glass vials with Teflon-lined screw-caps. Many reactions occur instantaneously at room temperature; slower reactions are accelerated by raising the temperature and/or by adding a catalyst, usually trimethyl-chlorosilane (TMCS). In the absence of a detailed model for the silylation mechanism [442], it is possible to rank the silylating reagents of Figure 8.34 according to their "silyl donor ability" and the functional groups of Figure 8.33 according to their "silyl acceptor ability". For the trimethylsilyl reagents the approximate

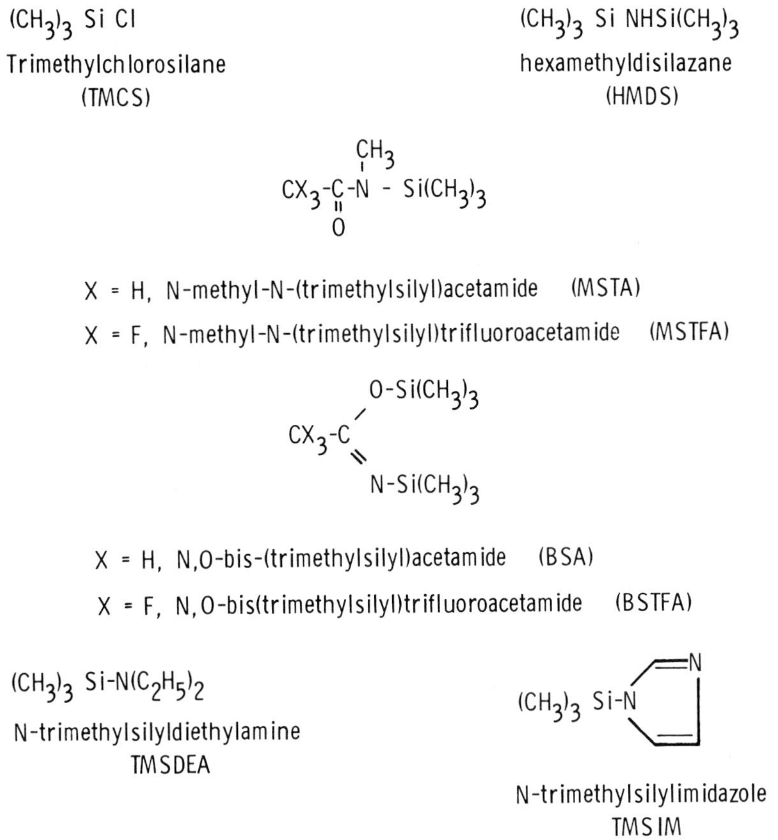

(CH₃)₃ Si Cl
Trimethylchlorosilane
(TMCS)

(CH₃)₃ Si NHSi(CH₃)₃
hexamethyldisilazane
(HMDS)

$$CX_3-\underset{\underset{O}{\|}}{C}-\underset{\underset{}{\overset{CH_3}{|}}}{N} - Si(CH_3)_3$$

X = H, N-methyl-N-(trimethylsilyl)acetamide   (MSTA)

X = F, N-methyl-N-(trimethylsilyl)trifluoroacetamide   (MSTFA)

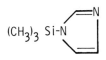

X = H, N,O-bis-(trimethylsilyl)acetamide   (BSA)

X = F, N,O-bis(trimethylsilyl)trifluoroacetamide   (BSTFA)

(CH₃)₃ Si-N(C₂H₅)₂
N-trimethylsilyldiethylamine
TMSDEA

(CH₃)₃ Si-N

N-trimethylsilylimidazole
TMSIM

Figure 8.34 Structures of the most commonly used trimethyl-silylating reagents.

order of "silyl donor ability" is: trimethylsilylimidazole (TMSIM) > N,O-bis(trimethylsilyl)trifluoroacetamide (BSTFA) > N,O-bis(trimethylsilyl)acetamide (BSA) > N-methyl-N-(trimethylsilyl)-trifluoroacetamide (MSTFA) > N-trimethylsilyldiethylamine (TMSDEA) > N-methyl-N-(trimethylsilyl)acetamide (MSTA) > trimethylchloro-silane (TMCS) with base > hexamethyldisilazane (HMDS). For organic functional groups the approximate order of "silyl acceptor ability" is: alcohols > phenols > carboxylic acids > amines > amides. Primary functional groups react faster or more completely than secondary functional groups, which in turn are more reactive than tertiary functional groups. Within the above framework, the reaction between a good "silyl donor" and a good "silyl acceptor" is likely to be facile and quantitative under mild conditions.

TABLE 8.15

OBSERVATIONS ON THE USE OF TRIMETHYLSILYL REAGENTS

Reagent	Property	Comments
TMSIM	Does not react with amino groups. Can be used to form TMS derivatives of carbohydrates in aqueous solution. Does not promote enol-ether formation with unprotected ketone groups.	Most generally useful reagent, preferred for most applications. Exceptions are the formation of N-TMS derivatives and the separation of low molecular weight TMS derivatives
BSTFA or BSA	Reagents of choice for the formation of N-TMS derivatives.	May promote the formation of enol-TMS ethers unless ketone groups are protected.
BSTFA	Produces volatile by-products which do not interfere with the GC analysis of low molecular weight compounds.	Reactivity is similar to BSA and is generally preferred to BSA for most applications.
MSTFA	Most volatile by-products of all.	Particularly useful for the separation of low molecular weight derivatives in the presence of excess reagent.
TMCS	A poor silylating reagent unless used in the presence of base (e.g., pyridine, diethylamine). Causes extensive enol-ether formation with unprotected ketone groups.	Mainly used to catalyze the reaction of other silylating reagents.

There seem to be few applications for which the use of weak "silyl donors" is either necessary or desirable. Other important considerations for the selection of the correct reagents for a particular application are summarized in Table 8.15. The strongest silylating reagent of all is a mixture of TMSIM-BSTFA-TMCS (1:1:1).

The rate of the silylation reaction is also affected by steric factors, the use of catalysts, the choice of solvent and the reaction temperature. The trimethylsilyl group has similar geometry to the tert.-butyl group and is slightly larger in size. Thus, impeded access of reagent to the functional groups to be derivatized can be the rate-determining step. For example, the reaction conditions required for the quantitative derivatization of steroid hydroxyl groups in different steric environments are well known to vary due to the nature of the functional group (primary, secondary, or tertiary) and to changes in the steric environment of the different groups [443,452-454]. Using TMSIM, typical unhindered primary and secondary hydroxyl groups require less than one hour at room temperature for quantitative reaction, the moderately hindered secondary 11β-hydroxyl group is quantitatively reacted in about one hour at 100°C, the tertiary hindered 17α-hydroxyl group in corticosteroids requires from 2 hours at 100°C to 2 hours at 150°C for quantitative reaction depending on the type of functional group on the neighboring carbon atom and the hindered tertiary 14α-hydroxyl group in ecdysones requires 12 hours at 140°C for quantitative reaction. Reaction rates are also influenced to a lesser extent by the choice of solvent. Polar solvents, such as pyridine, dimethylformamide, acetonitrile, dioxane or tetrahydrofuran, promote the reaction and are generally preferred. The silylating reagents themselves have good solubilizing properties for many compounds and can be used without additional solvent. The primary criterion for selecting a solvent is that it must solubilize both substrate and reagents. Increasing the temperature of the reaction will often improve substrate solubility and enhance the rate of reaction. Some difficult-to-derivatize functional groups will only react to completion at elevated temperatures. Normal reaction temperatures are room temperature, 60°C, 100°C, and 150°C; in a few instances, temperatures as high as 200°C have been used.

Besides the wide applicability and ease of use of the TMS reagents, the fact that most reactions occur cleanly without artifact or byproduct formation adds to the attraction of these reagents. Secondary products can be formed under normal conditions for silyl ether formation with compounds containing unprotected ketone groups. These products arise from the formation of enol-TMS ethers [RCH=C(OTMS)R], the yield of the latter being high, and in some cases quantitative in the presence of an acid catalyst, such as TMCS. In general, enol-TMS ethers are hydrolytically and thermally unstable, making them less desirable reaction products even when formed in quantitative yield. To avoid the formation of enol-TMS ethers a reagent which does not promote enol formation, such as TMSIM, or prior protection of the ketone group by conversion to a methoxime derivative is used. Methoxime-trimethylsilyl derivatives are widely used for the analysis of biological extracts and in metabolic profiling studies. Two peaks for each compound may appear in the chromatogram due to the separation of the syn- and anti-methoxime isomers formed upon derivatization. A new reagent, N-methoxy-N,O-bis(trimethylsilyl) carbonate, allows the simultaneous formation of methoxime derivatives of ketone groups and silylation of hydroxyl groups [455].

Multiple products are frequently observed for the separation of TMS-sugar derivatives. At equilibrium reducing sugars can exist in more than one isomeric form known as anomers. Formation of their TMS derivatives followed by gas chromatography will result in multiple peaks corresponding in composition to the equilibrium anomeric mixture [436].

The TMS reagents are generally compatible with other reagents for the formation of mixed derivatives of polyfunctional compounds, provided that the TMS ethers are formed as the last step in the reaction sequence. The TMS ethers exhibit moderate hydrolytic stability and are cleaved under conditions normally used for the formation of alkyl esters, acetates and amides, oximes, cyclic boronic esters, acetonides and most reactions employing acid catalysts. The TMS reagents, with a few exceptions, do not normally cleave any of the common protecting groups used in gas chromatography.

The mass spectra of TMS ethers are characterized by weak or absent molecular ions; the $[M-15]^+$ ion formed by cleavage of a methyl to silicon bond is generally more abundant. This ion can be used to determine the molecular weight provided that it is not mistaken for the molecular ion itself. Dissociation of the molecular ion often results in prominent secondary fragment ions containing the ionized dimethylsiloxy group attached to a hydrocarbon portion of the molecule. In common with alkyl ethers, α-cleavage of the bond adjacent to oxygen is favored. Characteristic ions of hydroxyl TMS ethers are:

$[(CH_3)_3Si]^+$ $\qquad$ $[HO=Si(CH_3)_2]^+$ $\qquad$ $[(CH_3)_2Si=O-Si(CH_3)_3]^+$

$m/z = 73$ $\qquad\qquad$ $m/z = 75$ $\qquad\qquad\quad$ $m/z = 147$

The m/z 73 ion is prominent in virtually all TMS spectra and is often the base peak. The m/z 147 ion is common in polyhydroxyl TMS compounds containing two or more TMS groups in close proximity. The TMS group undergoes a prolific number of intramolecular migrations and rearrangements (including McLafferty rearrangements) to give prominent silicon-containing ions. Perdeuterated silyl reagents are available to aid in elucidating specific fragmentation and rearrangement processes based on a comparison of the TMS and d_9-TMS spectra.

New reagents which are either homologs or analogs of the trimethylsilylating reagents have been developed to improve the hydrolytic stability or detectability of the silyl ether derivatives [443]. The most important are the alkyldimethylsilyl reagents containing a propyl, isopropyl or tert.-butyl substituent [456-459]. The isopropyldimethylsilyl and tert.-butyldimethylsilyl derivatives are one hundred to ten thousand time more hydrolytically stable than the TMS ethers and are widely used as protecting groups in organic synthesis as well as in analytical chemistry. The increase in hydrolytic stability results from the bulkiness of the alkyldimethylsilyl group, which in turn also influences the rate and extent of derivative formation under conditions influenced by stereochemical considerations. The tert.-butyldimethylsilyl chloride, N-methyl-N-(tert.-butyldimethylsilyl) trifluoroacetamide, and tert.-butyldimethylsilylimidazole reagents are now commercially available which has done much to popularize their general use for derivatizing amino acids [460-463], organic acids [464,465], alcohols [457,466] and thiols [467]. The tert.-butyldimethylsilyl ethers have good gas chromatographic properties, are not as volatile as the TMS derivatives (retention times 2 or 3 methylene units higher), are relatively insensitive to moisture, and are stable to adsorption chromatography using columns or thin-layer plates. The N-tert.-butyldimethylsilyl derivatives are more labile than the O-tert.-butyldimethylsilyl ethers but more stable than the O-trimethylsilyl ethers [458]. The mass spectra of the alkyldimethylsilyl ethers have weak molecular ions with an abundant $[M-R]^+$ ion due to cleavage of the bond between silicon and the larger alkyl group. This ion is frequently the base peak of the mass spectrum and is used for trace analysis by GC/MS employing single ion monitoring.

To improve the detectability of silyl ethers, silylation reagents containing an electron-capturing group [443,449-451,468] or cyano group for thermionic detection [469] have been prepared. The 2-cyanoethyldimethylsilyl derivatives are only marginally more sensitive (ca. 5 fold) to the thermionic detector than to the flame ionization detector which limits their usefulness. The

TABLE 8.16

RELATIVE VOLATILITY AND ELECTRON-CAPTURE DETECTOR RESPONSE OF HALOGEN-CONTAINING ALKYSILYL ETHER DERIVATIVES

Derivative	Relative Retention Time (for Cholesterol)	Least Detectable Amount	
		Cholesterol (ng)	Octanol (pg)
TMS	1.00		
CMDMS	2.10	75	
Flophemesyl	3.14	4	4
ISP-flophemesyl	4.57	–	5
BMDMS	5.13	0.5	–
CM-flophemesyl	6.30	–	0.9
t-Buflophemesyl	6.30	–	6
IMDMS	12.82	0.005	–

trimethylsilyl group shows no particular electron-capturing properties of its own, but the necessary detector-oriented response can be conferred on the trimethylsilyl group by introducing a halogen atom (Cl, Br, I) into one of the methyl groups or by replacing a methyl group by a pentafluorophenyl ring [443,470-472]. The structure and common names for the various reagents and derivatives are given below, and their relative volatility and response to the electron-capture detector are summarized in Table 8.16.

$$CH_3$$
$$CH_2X-Si-Y$$
$$CH_3$$

$$CH_3$$
$$C_6F_5-Si-Y$$
$$R$$

X = Cl, Chloromethyl-
dimethylsilyl
derivative (CMDMS)

X = Br, Bromomethyldimethyl-
silyl derivative
(BMDMS)

X = I, Iodomethyldimethyl-
silyl derivative
(IMDMS)

Y = Cl, Chlorosilane

Y = NHSi(CH$_3$)$_2$CH$_2$X
Disilazane

R = CH$_3$, Pentafluorophenyl
dimethylsilyl derivative
(flophemesyl)

R = CH(CH$_3$)$_2$, Pentafluoro-
phenylisopropylmethylsilyl
derivative (ISP-flophemesyl)

R = C(CH$_3$)$_3$, tert. -Butylpenta-
fluorophenylmethylsilyl derivative (tert.-buflophemesyl)

R = CH$_2$Cl, Chloromethylpentafluoro-
phenylmethylsilyl derivative
(CM-flophemesyl)

Y = Cl, Chlorosilane

Y = NH$_2$, Flophemesylamine (only)

Y = N(C$_2$H$_5$)$_2$, Flophemesyldiethylamine (only)

The haloalkyldimethylsilyl reagents have been used to prepare a wide range of derivatives of alcohols, steroids, carboxylic acids, phenols, prostaglandins and carbohydrates. The introduction of the halogen atom into the trimethylsilyl group dramatically increases the retention time of the derivatives, particularly in the case of iodine, however, the detector response to the chloro- and bromo- containing derivatives is only modest. The iodo-containing derivatives are the most useful for trace analysis, but iodomethyl reagents are unstable and do not store well. The IMDMS derivatives are usually prepared from the CMDMS and BMDMS derivatives by halide ion exchange using a saturated solution of sodium iodide in acetone. All the halomethyldimethylsilyl derivatives are usually formed under mild conditions to minimize byproduct formation resulting from secondary reactions with nucleophiles that can attack either at carbon or silicon to displace the halogen from the halomethyl group or eliminate the halomethyl group entirely. The products from these reactions are not usually electron- capturing and may easily be overlooked leading to erroneous quantitation of the substance of interest.

The flophemesyl derivatives have been used to detect trace levels of alcohols, phenols, carboxylic acids, amines, steroids and drugs. Flophemesylamine can be employed to selectively derivatize primary and secondary hydroxyl groups in the presence of tertiary hydroxyl or unprotected ketone groups. Mixtures of flophemesyl chloride and flophemesyldiethylamine are used to derivatize functional groups of low reactivity, but in this case, ketone groups must be protected to avoid formation of labile enol ether derivatives. The flophemesyl derivatives have similar hydrolytic stability to the trimethylsilyl ethers. If a more robust derivative is required to withstand common sample isolation and cleanup procedures for GC, then the tert.-buflophemesyl or ISP-flophemesyl derivatives are recommended. For complete reaction with these bulkier reagents, more forcing conditions may be required than for the flophemesyl reagents, particularly in the case of sterically hindered functional groups.

The halogen-containing alkylsilyl derivatives have never enjoyed extensive popularity for trace analysis. There are several probable reasons for this: modest sensitivity; difficulties in conveniently separating derivatives from excess reagent; interferences in the separation of low molecular weight analytes by reagent impurities and reaction byproducts (particularly the disiloxanes formed by reaction of the reagents with water); and poor hydrolytic stability of some derivatives. In the case of the halomethyldimethylsilyl derivatives, there is the additional problem of potential loss of the halomethyl group during derivative formation as discussed previously.

The mass spectra of the halomethyldimethylsilyl ethers have weak molecular ions and an abundant characteristic daughter ion at $[M-CH_2X]^+$. As the halogen, X, decreases in electronegativity so the bond from the silicon atom to the carbon bearing X becomes stronger and loss of CH_2X less likely. The mass spectra of the flophemesyl ethers often have prominent molecular ions with a higher percentage of the total ion current carried by hydrocarbon fragments than is observed with the TMS ethers. The principal ions of lower m/z are dominated by the presence of fluorosilane ions (m/z 47 $[SiF]^+$, m/z 77 $[Si(CH_3)_2F]^+$, and m/z 81 $[Si(CH_3)F_2]^+$) and fluorocarbon ions originating from fluorocarbon tropylium ions (m/z 181 $[C_7H_2F_5]^+$, m/z 163 $[C_7H_3F_4]^+$, m/z 159 $[C_8H_6F_3]^+$, and m/z 145 $[C_7H_4F_3]^+$) [473].

8.10.2 Haloalkylacyl Derivatives

The haloalkyl acid chlorides and anhydrides are probably the most studied reagents for the introduction of an electron-capturing group into compounds with protonic functional groups (except carboxylic acids), Figure 8.35. Although they contain halogen atoms to provide a response to the electron-capture detector, they are equally suitable for use with the flame ionization detector, and have almost entirely displaced the hydrocarbonacyl derivatives from general use. In particular, the perfluorocarbonacyl reagents (trifluoroacetyl, pentafluoro-propionyl, heptafluorobutyryl) produce stable, volatile derivatives which often elute earlier and with better peak shape than the hydrocarbonacyl derivatives. The choloroacetyl and bromoacetyl derivatives have long retention times often accompanied by poor peak shape and poor thermal stability compared to the perfluoroacyl derivatives.

Derivatives are prepared from the appropriate acid anhydride, or occasionally the acid chloride, usually in the presence of a base such as pyridine, triethylamine, or N,N-dimethyl-4-aminopyridine at elevated temperatures [474-482]. Acylation of amines and phenols (not alcohols) in aqueous solution in the presence of potassium carbonate has been demonstrated [448,483,484], but does not constitute normal practice as reactions are generally performed under anhydrous conditions.

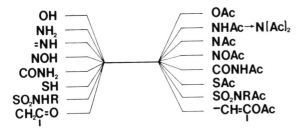

Figure 8.35   Functional groups forming acylated derivatives.

Excess reagents are conveniently removed by either evaporation with a stream of nitrogen or partitioning with an aqueous solution of a weak base. Byproduct formation is generally not a major problem except for analytes which are labile to acidic or basic conditions at elevated temperatures [476,478,479,485,486]. Under conditions where the anhydride can lead to undesirable side reactions (dehydration, enolization, etc.), the haloalkylacyl-imidazole reagent can sometimes be employed.

The relative response of the electron-capture detector to some haloalkylacyl derivatives is summarized in Table 8.17 [451]. In general terms, the monochloroacetyl and chlorodifluoroacetyl derivatives provide a greater response than the trifluoroacetyl derivatives. Increasing the fluorocarbon chain length of the fluorocarbonacyl derivatives increases their electron-capture detector response without inconveniently increasing their retention times. The heptafluorobutyryl and pentafluoropropionyl derivatives are considered to be the best compromise between detector sensitivity and volatility for most applications.

The mass spectra of the halocarbonacyl derivatives frequently have abundant ions at high m/z values. Consequently, they have often found use in studies using selected ion monitoring for detection and quantitation of the derivatives. The abundance of the molecular ion varies and depends primarily on the nature of the compound derivatized. In most cases, ions corresponding to $[C_nF_{2n+1}]^+$ occur abundantly in the mass spectra of the perfluorocarbonacyl derivatives. Also, the $[C_nF_{2n+1}CO]^+$ ion is usually fairly prominent and a loss of $(C_nF_{2n+1}CO_2)$ occurs readily in the mass spectra of alcohols and phenols. Aliphatic amines tend to give abundant ions corresponding to $[C_nF_{2n+1}CONHCH_2]^+$.

TABLE 8.17

RELATIVE RESPONSE OF THE ELECTRON-CAPTURE DETECTOR TO SOME
HALOALKYLACYL DERIVATIVES

Derivative	Amphet-amine	Testos-terone	Thymol	Diethyl-stilbesterol	Benzyl-amine
Acetyl	1.0				1.0
Monofluoroacetyl				0.007	
Monochloroacetyl	1.0	40	0.3	2.7	750
Chlorodifluoro-acetyl		340			
Dichloroacetyl				2.6	
Trichloroacetyl	540			2.1	
Trifluoroacetyl	< 0.1	4		1.7	200
Pentafluoro-propionyl	40	50	1.3	15	5725
Heptafluorobutyryl	90	190	1.0	23	17875
Perfluorooctonyl	230	600		21	
Pentafluoro-benzoyl	770		6.9		

8.10.3 Esterification

Esterification is used to derivatize carboxylic acids and
other acidic functional groups. In a typical reaction the
carboxylic acid is dissolved in an excess of an alcohol which
contains a few percent of an acid catalyst, such as hydrogen
chloride, acetyl chloride, thionyl chloride, boron trifluoride
etherate, boron trichloride or an ion-exchange resin [487-490].
Many esterification reactions are slow and elevated temperatures
are frequently used. Since the esterification reaction is an
equilibrium reaction, it may be necessary to remove the water
formed as a product during the reaction to obtain a quantitative
yield of derivative. A Dean-Stark trap, reflux in a Soxhlet
apparatus containing a desiccant in the thimble, or a solvent
which reacts with water (e.g., 2,2-dimethoxypropane) may be used
for this purpose. For many reactions some water can be tolerated
in the reaction medium if appropriate ratios between the amount of
water and catalyst/reagent are maintained [489]. To derivatize
carboxylic acids with low solubility in higher alcohols the
technique of transesterification is used. For example, cystine and
the basic amino acids are virtually insoluble in n-butanol. To
prepare their butyl esters, the methyl esters are formed first,
the methylating reagents removed by evaporation, and the methyl
esters then converted to the butyl esters by refluxing the residue

with an excess of n-butanol and acid catalyst [491]. The yield of butyl esters is usually quantitative compared to a few percent in the case of direct reaction. Fatty acid esters isolated from biological fluids can be converted to methyl esters for analysis by reaction with trimethylsulfonium hydroxide [492]. Electron-capturing derivatives can be prepared using alcohols such as 2,2,2-trifluoroethanol, 2,2,3,3,3-pentafluoro-1-propanol, 2,2,2-trichloroethyl and hexafluoro-2-propanol [493-495]. The fluorine-containing alcohols are usually used in combination with other electrophoric reagents to improve their sensitivity to detection. As stand alone reagents, they generally lack sufficient sensitivity for use in trace analysis.

8.10.4 Alkylation

The term "alkylation" covers a variety of techniques in which an active hydrogen atom is replaced in a chemical reaction by an alkyl or, sometimes, an aryl group [436,441,446,448,496]. Functional groups which can be alkylated are shown below:

$RCOOH$	--->	$RCOOR_1$
RSO_2OH	--->	RSO_2OR_1
ROH	--->	ROR_1
RSH	--->	RSR_1
RNH_2	--->	$RN(R_1)_2$
R_2NH	--->	R_2NR_1
$RCONH_2$	--->	$RCON(R_1)_2$
RSO_2NH_2	--->	$RSO_2N(R_1)_2$
$RCOCH_2COR$	--->	$RCOCH=C(OR_1)R$

where R_1 represents the position of the alkyl group introduced into the compound derivatized. A variety of reagents and methods are used in their preparation, of which the most important are the reaction with an alkyl halide and catalyst, diazoalkanes, N,N'-dimethylformamide dialkyl acetals, extractive alkylation, pyrolytic alkylation and arylation.

Alkyl halides in the presence of silver oxide will convert any non-hindered carboxylic acid (or its salt) to the corresponding alkyl ester in minutes, and phenolic or thiol groups will also be alkylated rapidly [436]. Hydroxyl groups are alkylated slowly and not always to completion. The alkyl halides most frequently used are the lower molecular weight aliphatic bromides and iodides (e.g., methyl, ethyl, propyl, isopropyl, butyl, etc.) or benzyl and substituted benzyl bromides. The reaction has been extensively investigated for the alkylation of sugars. Oxidative degradation of free sugars and migration of O-acetyl groups can lead to unexpected byproducts. N-acetyl groups are usually stable to the reaction conditions. In a typical reaction, silver oxide is added to a solution of the substrate in an excess of alkyl halide and the mixture is shaken in the dark until the reaction is complete. Slow reactions are usually monitored at intervals by any suitable chromatographic technique and fresh portions of silver oxide are added as necessary.

Barium oxide and sodium hydride are more potent catalysts than silver oxide. With barium oxide catalysis, reactions occur more rapidly but O-acetyl migration is promoted. With sodium hydride, even sterically hindered groups may be quantitatively alkylated but unwanted C-alkylation instead of, or in addition to, O-alkylation is a possibility. Sodium hydroxide is a suitable catalyst for the alkylation of carboxylic acids and alcohols [497].

Diazoalkane alkylating reagents include diazomethane, diazoethane, diazopropane, diazobutane, diazoisobutane and phenyl-diazomethane [435]; diazomethane is most frequently used. The other reagents are used principally to improve chromatographic separation when the methyl derivatives are unsuitable or to avoid loss of volatile methyl derivatives of low molecular weight substances during the removal of excess reagents. Reaction conditions are very simple: a solution of the diazoalkane is added to a solution of the analyte at or below room temperature, until a faint yellow color persists and the evolution of nitrogen gas ceases. Excess reagent may be removed by evaporation or destroyed by addition of acid (e.g., acetic acid).

Diazoalkanes alkylate acidic and enolic groups rapidly and other groups with replaceable hydrogens slowly. Carboxylic and sulfonic acids, phenols and enols are alkylated virtually instantaneously when treated with this reagent. Lewis acid catalysts (e.g., $BF_3.Et_2O$) are used to promote the reaction of substances containing hydrogen atoms of low reactivity (e.g., alcohols). Because of the large difference in reaction rates, many substances containing carboxylic acid and phenolic groups can be selectively alkylated in the presence of less reactive functional groups. Alternatively, methanol can be used to advantage as a solvent for the alkylation of carboxylic acids. When Lewis-acid catalysis is used, acid-labile substances may undergo undesirable transformations induced by the catalyst while O-acylated sugars with free hydroxyl groups are alkylated with O-acyl migration. The diazoalkanes are versatile synthetic reagents in preparative organic chemistry. As well as alkylation, they also undergo a wide range of addition and cyclization reactions which may result in unexpected products with multifunctional compounds [435,498].

The diazoalkanes are toxic substances and may explode on contact with rough surfaces. Consequently, many workers prefer not to make large quantities of these materials when only small quantities are needed for derivatization reactions. Simple micro-diazoalkane generators capable of rapidly preparing small quantities of the reagents, as required, are commercially available [499-501].

Various N,N'-dimethylformamide dialkyl acetals, $(CH_3)_2NCH(OR_1)_2$, are commercially available in which $R_1 = CH_3$, C_2H_5, C_3H_7 and C_4H_9. The reagents are easily hydrolyzed to dimethyl-formamide and the appropriate alcohol so most reactions are performed under scrupulously dry conditions. Carboxylic acids, phenols, amides and thiols are rapidly converted to the corresponding alkyl derivatives while amines are converted to the N-dimethylaminomethylene [$RNH_2 \longrightarrow RN=CHN(CH_3)_2$] derivatives [441]. These reagents also undergo a fairly wide range of unexpected reactions frequently yielding different products from those anticipated.

Extractive alkylation is used to derivatize acids, phenols, alcohols or amides in aqueous solution [435,441,448,502]. The pH of the aqueous phase is adjusted to ensure complete ionization of the acidic substance which is then extracted as an ion pair with a tetraalkylammonium hydroxide into a suitable immiscible organic solvent. In the poorly solvating organic medium, the substrate anion possesses high reactivity and the nucleophilic displacement reaction with an alkyl halide occurs under favorable conditions.

$$RCOO^- + R_4N^+ \longrightarrow [RCOO^-R_4N^+]_{AQ}$$
$$[RCOO^- R_4N^+]_{AQ} \longrightarrow [RCOO^- R_4N^+]_{ORG}$$
$$[RCOO^- R_4N^+]_{ORG} + R_1I \longrightarrow RCOOR_1 + R_4N^+I^-$$

Various tetraalkylammonium ions are used to form extractable ion pairs including tetrabutyl-, tetrapentyl- and tetrahexyl-ammonium. The choice of cation depends on the efficiency with which the ion pair will extract from aqueous solution and then undergo alkylation. Almost any alkyl halide can be used as the alkylating reagent; selection is based on the separation required in the chromatogram and the degree of sensitivity needed for detection purposes. The lower members of the homologous series of alkyl bromides and iodides are used with the flame ionization detector and pentafluorobenzyl bromide with the electron-capture detector. At the end of the alkylation reaction, excess reagent is removed by evaporation, column chromatography or by aqueous extraction after reaction with a hydrophilic compound. When pentafluorobenzyl bromide is used as the alkylating reagent, some additional care is necessary since the reagent is unstable under some conditions used for extractive alkylation, as well as being a potent lachrymatory agent [503]. Extractive alkylation has the advantage of achieving extraction, concentration and derivatization in a single step. The experimental conditions may require extensive optimization but the high selectivity and sensitivity obtainable have made this a popular technique, particularly for pharmaceutical analysis involving biological fluids. Isolation by anion-exchange chromatography with in situ alkylation of the anion attached to the anion exchanger has been suggested as an alternative to extractive alkylation using an ion pair reagent [504].

Extractive alkylation of phenols and carboxylic acids has also been performed using a potassium salt complexed with a crown ether in the organic phase [505,506]. The selective alkylation of the carboxylic acid group in the presence of phenolic substances is possible using a potassium salt of low basicity. Phenols in the presence of carboxylic acids can be alkylated in a biphasic system containing sodium hydroxide [506]. Direct alkylation is also possible under anhydrous conditions employing a strong organic base as a catalyst [507]. Another approach to alkylation is to use solid-supported reagents [508,509]. Amberlite XAD-2 impregnated with pentafluorobenzyl bromide has been used to simplify the isolation and derivatization of acidic substances, such as carboxylic acids, phenols, prostaglandins and cannabinoids in biological fluids. The yield of derivative is a complex function of the pH of the extract, the presence of a cosolvent to increase the wetting of the sorbent with the sample, the method used to impregnate the beads with reagent, temperature and reaction time. These parameters have to be optimized for individual substances with respect to the sample matrix. Under favorable circumstances a quantitative yield of relatively clean extracts can be obtained. Mechanically separating the sorbent from the sample followed by selective elution of the derivatives contribute to the convenience of the above technique when adequate reaction conditions have been developed.

Pyrolytic alkylation is the process whereby a volatile alkyl derivative of an acidic compound is formed by thermal decomposition of a quaternary N-alkylammonium salt of the acid in the heated injection port of a gas chromatograph [446,510]. The alkyl derivative and the other volatile products of the reaction are then swept onto the column by the carrier gas and eluted in the usual manner. Pyrolytic alkylation can be used for the analysis of a variety of organic compounds containing acidic NH and OH functional groups. These include carboxylic acids, phenols, barbiturates, sulfonamides and heterocyclic nitrogen compounds, such as purines, pyrimidines and xanthines. For the preparation of methyl derivatives, aqueous or methanolic solutions of tetramethylammonium hydroxide, phenyltrimethylammonium hydroxide, trimethylanilinium hydroxide, or m-trifluoromethyl-phenyltrimethylammonium hydroxide are generally used. Higher alkyl homologs are prepared with tetraethyl-, tetrapropyl-, or tetrabutylammonium hydroxides, etc.

The derivative-forming process in pyrolytic alkylation involves two sequential reactions: deprotonation of the acidic substrate in aqueous solution by the strongly basic tetra-alkylammonium ion and the thermal decomposition of the quaternary N-alkylammonium salt formed to give a tertiary amine and alkyl derivative. For some weak acids both processes may occur virtually simultaneously in the injector oven of the gas chromatograph.

$$RCOOH \ + \ C_6H_5(CH_3)_3N^+OH^- \ \longrightarrow \ RCOO^-C_6H_5N^+(CH_3)_3 \ + \ H_2O$$

$$RCOO^-C_6H_5N^+(CH_3)_3 \ \longrightarrow \ RCOOCH_3 \ + \ C_6H_5N(CH_3)_2$$

Despite its great utility, pyrolytic alkylation has certain limitations. One major problem is that certain base- and heat-sensitive compounds can be isomerized or degraded under the conditions of high alkalinity and high temperature employed. For example, phenobarbitone undergoes base-catalyzed cleavage and partial degradation to form N-methyl-α-phenylbutyramide as a byproduct [435]. Another problem is that the efficiency of the thermal decomposition reaction is often strongly affected by the rate at which the sample solution is injected. In many instances, slow injection (2-5 seconds per microliter) yields better results in terms of solvent response, peak height and resolution than rapid injection.

By far the most important reagent for arylation is 2,4-dinitrofluorobenzene [435,436]. This reagent will derivatize amine, thiol and phenol groups in buffered aqueous solution or in non-aqueous solvents. Heating is usually required and reactions may be non-quantitative, particularly for sterically-hindered groups. 2,4-Dinitrobenzenesulfonic acid is preferred for some applications due to its higher selectivity for amines and the fact that the acidic derivatizing reagent can be easily separated from the derivative in the reaction medium. The principal limitation of the dinitrophenyl derivatives is their poor gas chromatographic volatility. However, they provide a high response towards the electron-capture detector and can be used for trace analysis.

8.10.5 Oxime Formation

Ketone groups do not normally present difficulties in gas chromatography. However, it may be desirable to protect ketone groups when other functional groups are derivatized to avoid enolization or condensation reactions that might result in byproduct formation. For this purpose oxime formation in either aqueous or organic solvent is a very facile reaction [505,511]. The most common reagents used are methylhdyroxylamine, ethylhydroxylamine and benzylhydroxylamine as their hydrochloride salts. Excess reagent is usually removed by partitioning between water and an immiscible organic solvent. Under some chromatographic conditions two peaks may be observed in the chromatogram for each analyte due to the separation of the syn and anti geometric isomers. For use with the electron-capture detector O-(2,3,4,5,6-pentafluorobenzyl)oxime derivatives are used [512-514]. This derivative is generally superior to the 2,4-dinitrophenylhydrazone and pentafluorophenylhydrazone derivatives which do not store well and frequently produce byproducts [468,515,516].

8.10.6 Pentafluorophenyl-Containing Derivatives

Reagents for the preparation of pentafluorophenyl-containing derivatives are summarized in Table 8.18. These reagents can be used to derivatize a broad spectrum of organic compounds and the derivatives formed are generally stable, have good gas chromatographic properties, and can be determined at trace levels with an electron-capture detector. Some of the reagents in Table 8.17 offer a high degree of specificity, for example, the reaction of pentafluorobenzyl chloroformate with tertiary amines, pentafluorobenzyl alcohol with carboxylic acids, and pentafluorobenzaldehyde with primary amines. The trichloroethyl carbamate derivatives of tertiary amines have a slightly lower detector response than the corresponding pentafluorobenzyl carbamate derivatives, but give fewer interferences and less pronounced solvent fronts [525]. Alkyl chloroformates, in general, are widely used for the derivatization of tertiary amines and are one of the few general reactions applicable for these compounds [526]. Pentafluorobenzaldehyde reacts smoothly with primary amines but may form mixtures of ring-substituted products involving hydrogen fluoride elimination with secondary amines and cyclic derivatives with α-hydroxylamines [527,528]. Pentafluorobenzenesulfonyl chloride reacts smoothly under anhydrous or biphasic conditions to produce stable derivatives of amines and phenols suitable for trace level analysis with the electron-capture detector [529,530].

8.10.7 Reagents for the Selective Derivatization of Bifunctional Compounds

Bifunctional compounds are characterized by the presence of at least two reactive functional groups on a molecular framework that places these groups in close proximity to one another. As such they do not constitute a defined chemical class of substances but are widely distributed among all classes of functionalized molecules (e.g., steroids, lipids, carbohydrates, nucleosides, catecholamines, prostaglandins, amino acids, etc.) In general terms, bifunctional compounds are compounds containing alkyl chains with functional groups on carbon atoms 1,2-, 1,3-, or 1,4- with respect to each other, or aromatic rings with ortho-

TABLE 8.18

REAGENTS FOR THE INTRODUCTION OF THE PENTAFLUOROPHENYL GROUP INTO
ORGANIC COMPOUNDS

Reagent	Functional group type	Reference
Pentafluorobenzoic anhydride	Alcohols	517
Pentafluorobenzoyl chloride	Amines, phenols, alcohols	518, 519
Pentafluorobenzyl bromide	Carboxylic acids, phenols, mercaptans, sulfonamides	section 8.10.4
Pentafluorobenzyl alcohol	Carboxylic acids	520
Pentafluorobenzaldehyde	Primary amines	479, 521
Pentafluorobenzyl chloroformate	Tertiary amines	522, 523
Pentafluorophenacetyl chloride	Alcohols, phenols, amines	524
Pentafluorophenoxyacetyl chloride	Alcohols, phenols, amines	524
Pentafluorophenylhydrazine	Ketones	section 8.10.5
Pentafluorobenzyl-hydroxylamine	Ketones	section 8.10.5

substituted functional groups. Specific reagents can react with
these groups to form stable cyclic derivatives as illustrated by
the examples in Figure 8.36 [436,439,444].

Reagents used to form cyclic derivatives can be divided into
two groups: those which can be used to derivatize a wide range of
functional groups and reagents which are highly selective for
specific functional groups or compounds. Of the general reagents
the most important are the cyclic boronic ester derivatives
introduced by Brooks and co-workers [531-534]. Reference [444]
contains a complete compilation of all published applications on
the use of boronic acids through 1979. Their dominant position as
derivatizing reagents for bifunctional compounds is a consequence
of their broad range of application, ease of reaction, good
thermal and GC properties, and their useful mass spectral
features. Disadvantages include the poor hydrolytic stability
exhibited by many derivatives and the ease of solvolysis observed
in sequential derivatization reactions. Mild conditions are
usually sufficient for derivative formation and a typical reaction
involves mixing the boronic acid and analyte in an anhydrous
solvent at room temperature for a short time (1 to 30 min). In

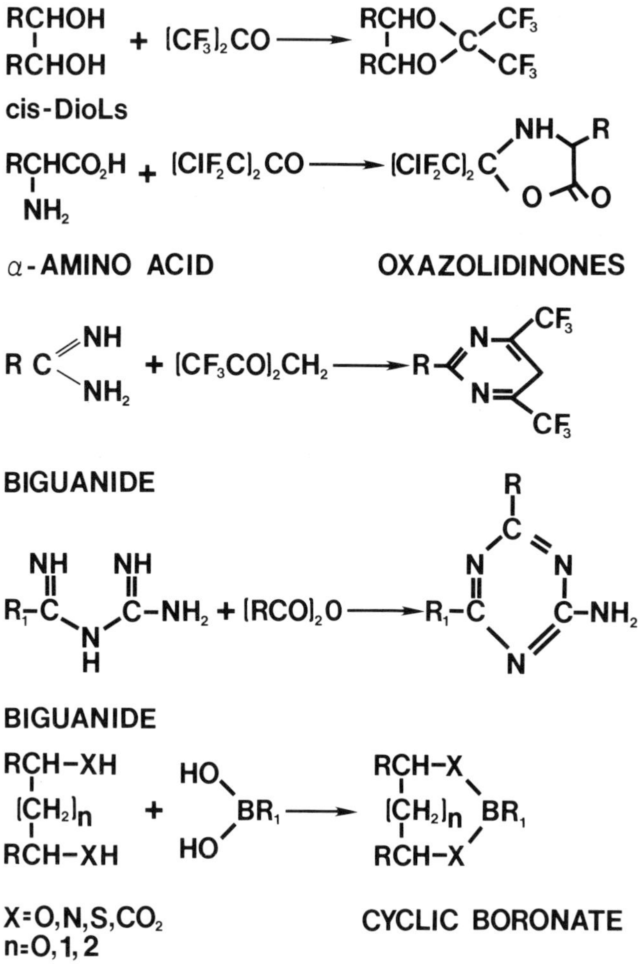

Figure 8.36 Some examples of the reactions used to form cyclic derivatives of bifunctional compounds

some cases excess boronic acid may be required to force the equilibrium reaction to completion and, for those derivatives which are exceptionally moisture sensitive, a means of removing the water produced in the reaction is required (e.g., molecular sieves can be added to the reaction medium, 2,2-dimethoxypropane can be added as a water scavenger, or periodic azeotropic evaporation with benzene or dichloromethane can be used). Direct injection of boronate derivatives with unprotected polar functional groups, without removal of excess boronic acid, invariably results in poor chromatographic performance, exemplified by tailing peaks of reduced peak height. Sequential derivatization of the various functional groups is required in this case and special attention must be paid to the possibility that strong reaction conditions could result in loss of the boronate group.

Methaneboronic acid, butaneboronic acid, tert.-butaneboronic acid, cyclohexaneboronic acid, benzeneboronic acid, and ferroceneboronic acid have all been used to prepare derivatives for gas chromatography. The cyclohexaneboronates and benzeneboronates have long retention times in comparison to the other boronate derivatives. The methaneboronates are very volatile and the small molecular weight increment formed by derivatization is useful in the mass spectrometry of high molecular weight compounds [533]. The butaneboronate derivatives provide a convenient compromise between volatility and stability, making them the most studied derivatives.

The boronate derivatives have useful mass spectral properties with prominent molecular ions. The boronate group is not strongly directing in influencing the mode of fragmentation as charge localization invariably occurs at a center remote from the boronate group due to the electrophilic character of the boron atom. This has the advantage that the abundant ions in the mass spectrum are characteristic of the parent molecule and not the derivatizing reagent. The natural isotopic abundance of boron ($^{10}B:^{11}B = 1:4.2$) aids in the identification of boron-containing fragments in the low resolution mass spectra of the boronate derivatives.

Bronic acids containing electron-capturing subsitituents were developed by Poole and co-workers, Table 8.19 [451,535,536]. In terms of volatility, stability of derivatives, and response to the electron-capture detector the 3,5-bis(trifluoromethyl)benzeneboronic acid, 2,4-dichlorobenzeneboronic acid, and 4-bromobenzeneboronic acid were recommended for general applications. In particular, the 3,5-bis(trifluoromethyl)benzeneboronate derivatives are remarkably volatile, more so than the benzeneboronates, and are suitable for the analysis of bifunctional compounds of low volatility. All the benzeneboronate derivatives are susceptible to solvolysis which is the primary limitation to their general use for trace analysis.

TABLE 8.19

A COMPARISON OF THE VOLATILITY AND   RESPONSE TO THE ELECTRON-
CAPTURE DETECTOR OF BORONIC ESTERS

Boronic Ester	Relative Retention	Minimum Detectable Quantity (pg of pinacol)	Optimum Detector Temperature (°C)
3,5-Bis(trifluoromethyl) benzeneboronate	0.3 ± 0.05	3.0	180
Benzeneboronate	1.0	200.0	150
4-Iodobutaneboronates	1.8 ± 0.5		
4-Bromobenzeneboronates	3.9 ± 0.8	3.0	350
2,6-Dichlorobenzene-boronates	4.3 ± 2.0	18.0	380
2,4-Dichlorobenzene-boronates	4.7 ± 1.7	4.0	380
3,5-Dichlorobenzene-boronates	5.0 ± 1.1	11.0	380
2,4,6-Trichlorobenzene-boronates	6.9 ± 1.8	4.0	380
3-Nitrobenzeneboronates	11.7 ± 3.4	4.0	300
Naphthaleneboronates	18.5 ± 4.6	2550.0	350

Ethylphosphonothioc dichloride (EPTD) or methyldichloro-phosphine and sulfur react with bifunctional compounds containing OH, NH$_2$ or CO$_2$H groups in the presences of triethylamine to form cyclic derivatives similar to the boronate esters [537,438]. EPTD also reacts with ortho-substituted aromatic compounds and some compounds containing enolizable ketone groups. Some derivatives give two peaks on gas chromatography due to the formation of geometric isomers. Low-level detection of the derivatives is possible using phosphorus- or sulfur-selective detectors. The mass spectra provide abundant molecular ions with a characteristic elimination of ethyl sulfide to form a daughter ion which is often the base peak of the mass spectrum. Other ions characteristic of the derivatives are [C$_2$H$_5$PS]$^+$ and [PS]$^+$.

Cyclic diethylsilylene and di-tert.-butylsilylene deriv-atives have been used to protect diol groups in steroids and prostanoids [539-541]. N,O-bis(diethylhydrogensilyl)trifluoro-methylacetamide simultaneously converts isolated hydroxyl groups to the diethylhydrogensilyl ether allowing a single step derivatization to be used for the analysis of corticosteriods.

Phosgene has been used to from a wide range of cyclic derivatives with 1,2- and 1,3-diols, α-hydroxyacids, phenols (as methyl-carbonates), and 2-, 3-, or 4-amino alcohols (as oxazolidinones) [445]. Many of the reactions can be performed in aqueous solution and the derivatives determined selectively using a thermionic detector in the case of the amino alcohols. In acid solution 1,2-diaminobenzene reacts with α-keto acids to form thermally-stable, cyclic quinozalinol derivatives. The phenolic hydroxyl group is usually further derivatized to the TMS ether to improve the chromatographic performance of the quinozalinol derivative [542]. Reagents containing nitro or halogen substituents have been used with the electron-capture detector although the favored method of detection has been by thermionic detection or mass fragmentography [543]. Biguanides undergo two cyclizing reactions that are useful for their gas chromatographic analysis. Condensation with acetylacetone [544] or hexafluoroacetylacetone [545] produces substituted pyrimidine derivatives. The derivatives can be formed in aqueous or physiological solution (biguanides are an important group of pharmacologically-active compounds). With anhydrides, biguanides form cyclic 2,4-disubstituted-2-6-amino-1,3,5-s-triazine derivatives which have good gas chromatographic properties [546].

8.11 DERIVATIZATION TECHNIQUES FOR LIQUID CHROMATOGRAPHY

Derivatives are prepared in liquid chromatography primarily to improve the response of an analyte to a particular detector and less frequently to improve the stability of the analyte in a particular separation system, or to improve the chromatographic separation of a mixture containing overlapping peaks. For many applications in trace analysis liquid chromatography is limited by the availability of suitable sensitive and selective detectors. The refractive index and low wavelength UV detectors provide a fairly universal detection mechanism but are not very sensitive and the fluorescence, chemiluminescence and electrochemical detectors are reasonably sensitive but because of their selectivity may not have a significant response to the analyte of interest. General reviews of derivatization methods and reagents for liquid chromatography are available [1,286,436,437,439, 547,548] as well as specific reviews dealing with the formation of derivatives for fluorescence and chemiluminescence detection

[548-553], and for electrochemical detection [554-556]. Reagents described for fluorescence detection are also suitable for use with the UV detector, although the detection limits will generally be poorer by UV detection.

Derivatization reactions for liquid chromatography can be performed as a precolumn reaction as practiced in gas chromatography or as a postcolumn reaction employing an on-line reaction detector. Both approaches have their advantages and disadvantages. The precolumn approach requires little additional equipment and the reaction conditions are not constrained by time. On the other hand automation of the reaction conditions are less straightforward and the derivatization reaction must be well behaved and generally quantitative. A simple method must exist for separating excess reagent and other products from the derivative or these materials must not interfere in the separation and/or detection of the analytes. Reaction detectors permit easy automation of the derivatization reaction and can employ reactions which are nonquantitative provided that the reaction is reproducible. As all reactions occur postcolumn the separation process and detection process can be optimized independently of each other. Although artifact formation is rarely a problem, both the reagents and byproducts of the reaction must either not respond to the detector under the same conditions used to detect the analyte or must be easily separated from the analyte after derivatization and before detection. Finally, the reaction must be fast enough that column resolution is not destroyed by diffusion in the reaction device. On account of the above limitations not all reactions can be employed for postcolumn detection. The design and applications of reaction detectors are discussed in section 8.11.4.

8.11.1 Derivatives for UV-Visible Detection

The majority of reagents used to introduce UV-chromophores into functionalized molecules contain a reactive group which controls the chemical aspects of reactivity and selectivity and a substituted aromatic moiety that provides the chromophore for detection. Ideally, the chromophoric group should have a large molar absorption coefficient at some convenient measuring wavelength and be small and nonpolar so as not to dominate the chromatographic separation. Table 8.20 summarizes the most widely used chromophoric groups, their wavelength of maximum absorption,

TABLE 8.20

CHROMOPHORES OF INTEREST FOR ENHANCED UV DETECTABILITY

Chromophore	Wavelength for Maximum Absorption (nm)	Molar Absorption Coefficient at 254 nm
Benzyl	254	200
4-Nitrobenzyl	265	6,200
3,5-Dinitrobenzyl		>10,000
Benzoate	230	low
4-Chlorobenzoate	236	6,300
4-Nitrobenzoate	254	>10,000
2,4-Dinitrophenyl		>10,000
Toluoyl	236	5,400
Anisyl	262	16,000
Phenacyl	250	10,000
4-Bromophenacyl	260	18,000
2-Naphthacyl	248	12,000

and their molar absorption measured at 254 nm [557]. Those reagents with a molar absorption coefficient grater than 10,000 can provide detection limits at the low nanogram level for the derivatized analyte.

Suitable reagents for derivatizing specific functional groups are summarized in Table 8.21. Many of the reactions and reagents are the familiar ones used in qualitative analysis for the characterization of organic compounds by physical means. Alcohols are converted to esters by reaction with an acid chloride in the presence of a base catalyst (e.g., pyridine, tertiary amine, etc). If the alcohol is to be recovered after the separation, then a derivative which is fairly easy to hydrolyze, such as p-nitrophenylcarbonate, is convenient. If the sample contains labile groups, phenylurethane derivatives can be prepared under very mild reaction conditions. Alcohols in aqueous solution can be derivatized with 3,5-dinitrobenzoyl chloride.

Amines are usually derivatized via the acid chlorides to form phenyl substituted amides or sulfonamides. Sanger's reagent, 2,4-dinitro-1-fluorobenzene, originally introduced for the identification of N-terminal amino acid residues in proteins, has been widely used for the derivatization of amino-containing compounds in general. N-Succinimidyl-p-nitrophenylacetate and the phenylsuccinimido- and 4-bromophenylcarbamates can be used to derivatize amines under mild conditions without the use of a catalyst. In the case of the succinimidyl reagent the derivative is an amide while the carbamate reagent forms a urea derivative.

TABLE 8.21

REAGENTS FOR THE INTRODUCTION OF CHROMOPHORES INTO FUNCTIONALIZED
MOLECULES

Functional Group Reacted	Derivatizing Reagent	Reference
Alcohols	3,5-Dinitrobenzoyl Chloride	558
	Pyruvoyl Chloride (derivatives)	559
	p-Iodobenzenesulfonyl Chloride	559
	Benzoyl Chloride	560
	p-Nitrobenzoyl Chloride	561
	p-Nitrophenyl Chloroformate	562
	Phenyl Isocyanate	563
	p-Dimethylaminophenyl Isocyanate	564
Amines	3,5-Dinitrobenzoyl Chloride	437
	Pyruvoyl Chloride (derivatives)	559
	p-Methoxybenzoyl Chloride	565
	n-Succinimidyl-p-nitrophenylacetate	557
	Benzoyl Chloride	566
	p-Nitrobenzoyl Chloride	567
	p-Toluenesulfonyl Chloride	568
	2-Naphthacyl Bromide	569
	2,4-Dinitrofluorobenzene	570
	2-Naphthalenesulfonyl chloride	571
	(Dimethylamino)azobenzenesulfonyl chloride	572
	Disuccinimido carbonate reagents	573
Ketones and Aldehydes	2,4-Dinitrophenylhydrazine	574
	p-Nitrobenzylhydroxylamine Hydrochloride	557
	2-Diphenylacetyl-1,3-indandione-1-hydrazone	575
Carboxylic Acids	p-Bromophenacyl Bromide	576
	O-(p-Nitrobenzyl)-N,N'-diisopropylisourea	577,578
	Phenacyl Bromide	579,580
	Benzyl Bromide	576
	N-Chloromethyl-4-nitrophthalimide	581
	Naphthyldiazomethane	582
	Nitrobenzyl Bromide	577
Epoxides	Diethyldithiocarbamate	583
Isocyanates	N-p-Nitrobenzyl-N-n-propylamine	584
Mercaptans	Pyruvoyl Chloride (derivatives)	559
Phenols	Pyruvoyl Chloride (derivatives)	559
	3,5-Dinitrobenzoyl Chloride	557
	p-Iodobenzoylsulfonyl Chloride	559
	p-Nitrobenzenediazonium Tetrafluoroborate	585
	Diazo-4-aminobenzonitrile	439

Phenylisothiocyanate is now commonly used for forming
phenylthiocarbamyl derivatives of amino acids [586]. These

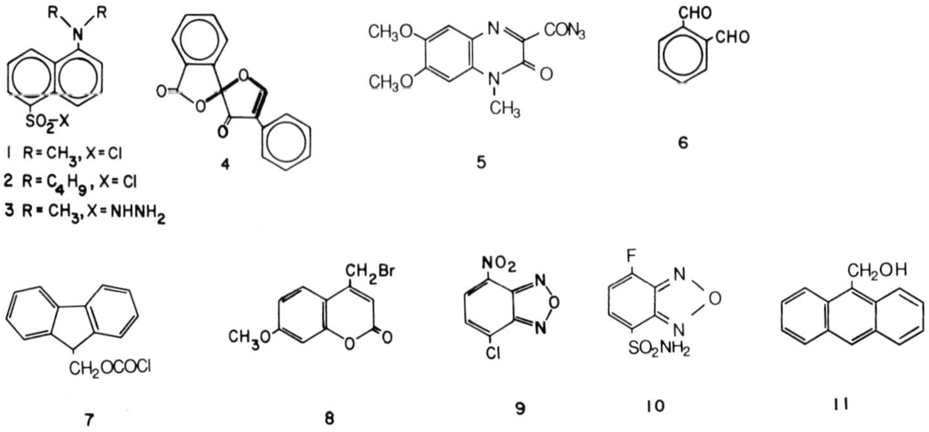

Figure 8.37 Reagents for the formation of fluorescent derivatives.
1 = dansyl chloride; 2 = dabsyl chloride; 3 = dansyl hydrazine; 4
= fluorescamine; 5 = 3,4-dihyro-6,7-dimethoxy-4-methyl-3-oxo-
quinoxaline-2-carbonyl azide; 6 = o-phthaldialdehyde; 7 = fluor-
enylmethyloxycarbonylchloride; 8 = 4-bromomethyl-7-methoxy-
coumarin; 9 = 7-chloro-4-nitrobenzo-2-oxa-1,3-diazole; 10 =
4-(aminosulfonyl)-7-fluorobenzo-2-oxa-1,3-diazole; and 11 =
9-(hydroxymethyl)anthracene.

derivatives are suitable for the separation of the common protein
amino acids by reversed-phase liquid chromatography.

Methods for derivatizing carboxylic acids in anhydrous or
biphasic systems are usually based on well-known alkylation
reactions. Extractive alkylation using a phase transfer catalyst
is convenient for hydrophilic carboxylic acids that would
otherwise be difficult to extract into an organic solvent.
Caboxylic acids in aqueous solution or biological fluids can be
conveniently derivatized using micellar solutions to promote the
reaction [587,588]. O-(p-nitrobenzyl)-N,N'-diisopropylisourea can
be used to derivatize carboxylic acids without the need for a base
catalyst. Aldehydes and ketones can be selectively derivatized
with either 2,4-dinitrophenylhydrazine or p-nitrobenzylhydroxyl-
amine hydrocholride without generally affecting other functional
groups present in the molecule.

8.11.2 Derivatives for Fluorescence Detection

Reagent for introducing fluorescent groups into
functionalized molecules are less numerous than chromophoric

reagents. The physical properties of a molecule required for a significant fluorescent response are rather special and few molecules are naturally fluorescent. When it comes to designing new fluorescent reagents many contain the same fluorogenic group attached to a different reactive group. Adroit choice of the reactive group provides chemical specificity for different functional groups. The structures of some of the most frequently used fluorescent derivatizing reagents are shown in Figure 8.37. The functional groups that can be derivatized with the common fluorescent reagents are summarized in Table 8.22 [547-551]. The response of the derivatives to the fluorescence detector covers quite a wide range depending on the types of fluorophore and the composition of the mobile phase. Those derivatives containing a methoxycoumarin, dimethylaminonaphthalene or benzoxadiazole fluorophore usually provide picogram detection limits while the anthryl- and acridine-containing reagents are usually somewhat less sensitive. Changes in fluorescence intensity and shifts in emission wavelengths with changes in the composition and polarity of the mobile phase occur commonly. Water and other strongly hydrogen-bonding solvents are most likely to have a strong fluorescence quenching effect. Compounds which fluoresce in aqueous solution may be dramatically influenced by solution pH. When developing an assay using fluorescence detection equal consideration should be given to selecting a mobile phase for the separation and to optimize the response of the detector to the separated derivatives.

TABLE 8.22

REAGENTS FOR THE PRECOLUMN PREPARATION OF FLUORESCENT DERIVATIVES
IN LIQUID CHROMTOGRAPHY

Reagent	Abbreviation	Application	Reference
1-Ethoxy-4-(dichloro-s-triazinyl)naphthalene	EDTN	Primary and secondary alcohols and phenols	589
4-Dimethylamino-1-naphthoyl nitrile		Primary and secondary alcohols	590
1-Anthroyl nitrile		Primary and secondary alcohols	591
7-[(Chlorocarbonyl)methoxy]-4-methylcoumarin	CMMC	Primary and secondary alcohols	592

3-Chloroformyl-7-methoxycoumarin		Primary and secondary alcohols	593
3,4-Dihydro-6,7-dimethoxy-4-methyl-3-oxo-quinoxaline-2-carbonyl azide	DMOQ-CON$_3$	Primary, Secondary and tertiary alcohols	594
m-(1-Cyano-2-isoindole)benzoyl azide		Primary and secondary alcohols	595
4-Bromomethyl-7-methoxycoumarin	Br-Mmc	Carboxylic acids and phenols	596-598
4-Bromomethyl-6,7-dimethoxycoumarin		Carboxylic acids	599,600
9-(Chloromethyl)anthracene	9-ClMA	Carboxylic acids	601
9-Anthradiazomethane		Carboxylic acids	602
9-(Hydroxymethyl)anthracene		Carboxylic acids	603,604
9-Bromomethylacridine		Carboxylic acids	605
N-(1-Pyrenyl)bromoacetamide		Carboxylic acids	606
2-(2,3-Naphthalimino)ethyl trifluoromethanesulfonate		Carboxylic acids	607
Dansyl hydrazine	Dns-H	Aldehydes and ketones	608,609
2-Diphenylacetyl-1,3-indandione-1-hydrazine		Aldehydes and ketones	610
7-Hydrazino-4-nitro-benzo-2-oxa-1,3-diazole	Nbd-H	Aldehydes and ketones	611
5-Dimethylaminonaphthalene-1-sulfonylaziridine	Dns-A	Thiols	612
N-(9-Acridinyl)maleimide		Thiols	613
N-[p-(2-Benzoxazolyl)-phenyl]maleimide	BOPM	Thiols	614
4-(Aminosulfonyl)-7-fluoro-benzo-2-oxa-1,3-diazole	ABD-F	Thiols	615
4-(6-methylnaphthalen-2-yl)-4-oxobuten-2-oic acid		Thiols	616
Dansyl chloride	Dns-Cl	Primary and secondary amines, phenols, amino acids and imidazoles	550, 617-619
2,5-Di-n-butylamino-naphthalene-1-sulfonyl chloride	Bns-Cl	as for Dns-Cl	550,617
Fluorescamine	Fluram	Primary amines and amino acids	550
o-Phthalaldehyde	OPA	Primary amines and amino acids	550, 619-623
7-Chloro-4-nitrobenzo-2-oxa-1,3-diazole	Nbd-Cl	Primary and secondary amines, phenols and thiols	550,624
4-Fluoro-7-nitrobenzo-2-oxa-1,3-diazole	Nbd-F	As for Nbd-Cl	625
Fluorenylmethyloxy-carbonylchloride		Amines and alcohols	619, 626-628
3-Benzoyl-2-naphthaldehyde		Amino acids	629
3-(2-Furoyl)quinoline-2-carbaldehyde		Primary amines	630

Dansyl chloride is the most widely used of the derivatizing reagents. It forms derivatives with primary and secondary amines readily, less rapidly with phenols and imidazoles, and very slowly with alcohols. The reaction medium is usually an aqueous-organic mixture (e.g., 1:1 acetone-water) adjusted to a pH of 9.5-10. Dansyl chloride has two major application areas. It is used to determine small amounts of amines, amino acids and phenols, as well as basic drugs and their metabolites in tissues and biological fluids. It finds further use in biochemistry for peptide sequencing and for the fluorogenic labelling of proteins and enzymes. Several analogs of dansyl chloride are also in use. The butyl analog has better storage properties and forms derivatives which are easier to extract into organic solvents. Its reaction and fluorescence properties are very similar to dansyl chloride. Dansyl hydrazine is a selective reagent for the analysis of carbonyl compounds and dansyl aziridine was developed as a selective reagent for the determination of thiols. Several reagents containing a maleimide group that form adducts with sulfhydryl groups are also used to determine thiols [550]. Dansylaminophenylboronic acid and 9-phenanthreneboronic acid have been used for the selective derivatization of bifunctional compounds (see section 8.10.7) [533].

Many reagents have been developed for the determination of amines and for general use in the automated determination of amino acids. Fluorescamine reacts with water, alcohols and primary and secondary amines but only forms stable fluorescent products with primary amines. Primary amines are generally derivatized in aqueous organic solvent mixtures at pH 8-9.5 in less than one minute; hydrolysis of the reagent occurring simultaneously. Fluorescamine has been routinely used as a postcolumn derivatizing reagent for amino acids separated by ion-exchange chromatography. Another widely used reagent for both pre- and postcolumn derivatization of primary amines is o-phthalaldehyde which reacts rapidly with the amino group in an aqueous reducing medium containing 2-mercaptoethanol or ethanethiol in a borate buffer at about pH 10 to yield a fluorescent 1-(alkylthio)-2-alkylisoindole. The reaction time is 1-2 min. at room temperature but the isoindole derivatives are unstable decomposing by a slow spontaneous intermolecular rearrangement. Provided that automated equipment is used for either the pre- or postcolumn reaction this is not a problem [620,623]. Fluorenylmethyloxy-carbonylchloride reacts with both primary and secondary amines in basic solution at room temperature in less than 1 min. The derivatives are stable and should be isolated from excess reagent by extraction, since the reagent and derivatives have identical fluorescence properties [627]. Alternatively, excess reagent can be quenched by the addition of a suitable amine which after reaction is easily separated from the derivatives of interest [628]. 7-Chloro-4-nitrobenzo-2-oxa-1,3-diazole reacts with primary and secondary amines to form intensely fluorescent derivatives but with anilines, phenols and thiols it generally yields only weakly or non-fluorescent products, sometimes in poor yield. Although the reagent is non-fluorescent, it nevertheless interferes in the fluorescence of its products and must be separated from them prior to measurement. 4-Fluoro-7-nitrobenzo-2-oxa-1,3-diazole is more reactive than the chloro analog (50 to 100 times toward amines) and has generally replaced Nbd-Cl in many of its conventional applications. 4-Hydrazine-7-nitrobenzo-2-oxa-1,3-diazole was introduced for the selective derivatization of carbonyl compounds.

Of the reagents used for the derivatization of carboxylic acids only those with a hydroxyl group in the presence of an activating agent such as 2-bromo-1-methylpyridinium iodide offer a high degree of specificity. The other commonly used reagents, which are alkylating reagents, also react to various extents with phenols, thiols and amides. 4-Bromomethyl-7-methoxycoumarin in anhydrous acetone smoothly alkylates carboxylic acids in the presence of potassium carbonate solvolysed by a crown ether catalyst. Coumarinic acid salts resulting from base catalyzed solvolysis of the lactone ring of the reagent are potential interfering fluorescent byproducts [597]. Maintaining anhydrous conditions or forming the potassium salt of the acids with a stoichiometric amount of base catalyst prior to addition of the Br-Mmc reagent minimizes the formation of byproducts. Alternatively, tetra-n-butylammonium salts of carboxylic acids react rapidly with Br-Mmc in acetone under reflux conditions [596].

Reagents for the derivatization of alcohols show variable reactivity and those containing an acid chloride or nitrile group require anhydrous conditions. DMOQ-CON$_3$ is the only reagent that reacts to a significant extent with tertiary alcohols. The product of the reaction is a carbamic acid ester. Detection limits for alcohol derivatives are also frequently no more than modest leaving some scope for the development of new reagents for their determination.

Some fluorescent compounds can be excited chemically enhancing the selectivity and sensitivity with which a particular derivative can be detected [552,553,631]. The limiting factor for the ultimate measurement of fluorescence is the background light that reaches the detector. Since much of this background is from stray light and source instability, a detection method that does not require optical excitation, such as chemiluminescence should be capable of greater sensitivity, perhaps by one to three orders of magnitude in favorable cases. One of the most widely exploited reactions for chemiluminescence employs 1,2-dioxethanedione as the reactive species [631-635]. The unstable reagent is generated in a reaction detector configuration from a mixture of an aryl oxalate and hydrogen peroxide. Interaction of the 1,2-dioxethanedione with the analyte causes its decomposition accompanied by excitation of the analyte to a higher electronic energy level. The excited analyte spontaneously relaxes to a lower energy state by emitting a fluorescence photon characteristic of the analyte. Typical applications include the determination of aldehydes and ketones after reductive amination with 3-aminofluoranthene; carboxylic acids after condensation with 3-aminopyrene; and primary amines after first forming derivatives with either Dns-Cl, Nbd-Cl, or OPA. N-(4-Aminobutyl)-N-ethylisoluminol has been used as a precolumn labeling reagent for amines and carboxylic acids which are subsequently detected after postcolumn mixing with hydrogen peroxide and hexacyanoferrate (III) [636]. Under favorable circumstances detection limits below the picogram level have been obtained.

8.11.3 Derivatives for Electrochemical Detection

Many of the common reagents for introducing nitrophenyl chromophores into a molecule for UV-visible detection are also suitable for use with the electrochemical detector [554,555,637]. Some further examples are the formation of p-aminophenyl derivatives of carboxylic acids [637], N-(4-anilinophenyl)-maleimide derivatives of sulfahydryl compounds [639], and ferrocenoyl derivatives of alcohols [640]. Also, o-phthalaldehyde derivatives of amines and amino acids can be determined at very low levels with electrochemical detection [622]. A comprehensive review through 1984 summarizes all publications to that date employing electrochemically active derivatizing reagents [555]. Reagents for oxidative electrochemical detection are generally preferred over reductive reagents (although the nitrophenyl reagents have been the most frequently used), since reductive electrochemical detection is hindered by operational difficulties based on the need to exclude oxygen from both the mobile phase and sample. The detector response is adversely affected by traces of the derivatizing reagents and/or byproducts which may have to be removed prior to injection. The choice of mobile phase for the separation is limited by the requirements of the detector, for example, in reversed-phase liquid chromatography the amount of organic modifier is limited by the detector preference for an aqueous mobile phase. This may result in poor separations or complete retention of hydrophobic derivatives on the column.

8.11.4 Reaction Detectors

Reaction detectors are a convenient means of performing on-line postcolumn derivatization in HPLC. The derivative reaction is performed after the separation of the sample by the column and prior to detection in a continuous reactor. The mobile phase flow is not interrupted during the analysis and reaction, although it may be augmented by the addition of a secondary solvent to aid the reaction or to conform to the requirements of the detector. Reaction detectors are finding increasing application for the analysis of trace components in complex matrices where both high detection sensitivity and selectivity are needed. Many suitable reaction techniques have been published for this purpose [641-650].

Advantageous features of postcolumn derivatization are that artifact formation is rarely a problem and it is not essential that the reaction employed for derivatization go to completion, as long as it is reproducible. Some general problems with postcolumn reaction detectors are associated with mobile phase incompatibility and reagent selection. Very seldom will the optimum chromatographic eluent also provide an ideal reaction medium. This is particularly true for electrochemical detectors which function properly only within a restricted range of pH, ionic strength and organic modifier concentration [554]. The reagent must not be detectable under the same conditions as the derivative. Finally, the reaction must be fast enough that column resolution is not destroyed by diffusion in the reaction device.

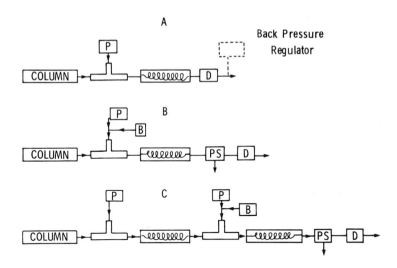

Figure 8.38 Schematic diagrams of some typical reaction detectors used in HPLC. A, non-segmented tubular reactor; B, segmented tubular reactor; C, extraction segmented reaction detector. P = pump, PS = phase separator, B = device for introducing bubbles and D = detector.

This sets an upper limit of about 20 minutes for the slowest reaction that can be easily exploited. The time required for reaction most profoundly influences the design of the reaction detector.

The reaction detector system must provide for the continuous addition of controlled volumes of one or more reagents to the column eluent, followed by mixing of the column eluent-reagent mixture and incubation for some time and temperature governed by the needs of the reaction. Detection of the reaction products is usually by absorption, fluorescence or electrochemical techniques. Some typical reaction detector configurations are shown in Figure 8.38. A piston or membrane pump is used to deliver the reagent to the reactor in a continuous, pulseless and constant flow rate for high reproducibility of the reaction and to permit detector operation at high sensitivity. With segmented flow systems, that generally operate at low pressures, peristaltic pumps are commonly used. Nearly all reaction detectors contain a mixing device, T- or Y-piece, or cyclone, for contacting and homogeneously mixing the column eluent and reagent streams [641,651,652]. The mixing device should be as small as possible to minimize extracolumn band broadening and permit the use of low reagent flow rates to reduce dilution of the chromatographic peaks. Cyclone mixers are superior in this respect as they can be constructed with a smaller dead volume (ca. 0.1 microliters) and work efficiently with reagent flow rates as low as 4% of the column flow rate. T- and Y-pieces with mixing volumes as small as 0.2 microliters have been described, although much larger volumes are more common and efficient mixing requires flow ratios in the range 1:1 to 1:9 depending on the design of the mixer. The reactor itself, usually an open tube or packed bed, should be dimensioned such that it is able to store the combined volumes of the column eluent and the reagent during the reaction time. In an ideal case the reaction should be finished before reaching the detection device. If the segmentation principle is employed, a device for introducing a bubble of liquid or gas into the column eluent is required [651,653]. This results in the separation or segmentation of the column eluent into a series of reaction compartments whose volume is governed by the dimensions of the transfer tube and the frequency of bubble introduction. Prior to detection a phase separator is needed to remove the segmentation agent. Two different principles are used for phase separation. A simple T-piece with one upward opening will remove the air from the liquid flow using density differences. In general the same principle is used for the construction of phase separators for liquid segmented systems. Here the different wettability of glass and organic polymeric tubing is used to separate aqueous and organic segments from each other. Alternatively, differences in the permeability of membranes for organic and aqueous liquids have been used successfully in the construction of phase separators. Dispersion in the phase separator is frequently a considerable source of band broadening. A phase separator is also used when ion-pair formation or extraction is part of the detection system. Reagents or byproducts which could interfere with detector performance are removed in this way. Phase separators are not one hundred percent efficient and a fraction of the detection stream (perhaps 20-40%) goes to waste. For reactions using solid-phase reagents, immobilized enzymes, etc., a packed bed tubular reactor is used [646,647,650,654]. The column eluent flows through the reactor with or without the addition of further reagent, pH buffer, etc. Reaction occurs on the surface of the packing, generating a product that can be detected downstream of the reactor. Reactions that are slow at room temperature may be accelerated by thermostating the reactor to a higher temperature. A cooling coil may be required prior to detection to prevent interference in the operation of the detector or to avoid bubble formation. Models have been devised to predict the influence of postcolumn dispersion on the column separation for the various common reaction detector designs [641,644,648,653,655]. The influence of the reaction rate on detector design will be qualitatively discussed below.

For fast reactions (i.e., < 1 min.), open tubular reactors are commonly used. They simply consist of a mixing device and a coiled stainless steel or Teflon capillary tube of narrow bore enclosed in a thermostat. The length of the capillary tube and the flow rate through it control the reaction time. Reagents such as fluorescamine and o-phthalaldehyde are frequently used in this type of system to determine primary amines, amino acids, indoles, hydrazines, etc., in biological and environmental samples.

The parabolic flow profile of the mixed analyte and reagents through an open tubular reactor restricts the reaction time to less than 1 min. if extensive band broadening and loss of resolution are to be avoided. With optimally deformed open tubes secondary flow is established, even at low flow rates, which breaks up the parabolic profile and minimizes band broadening [656,657]. The easiest way to induce secondary flow is by using knitted or stitched open tubular capillaries. Knitted open tubular (KOT) reactors are generally fabricated from PTFE capillary tubes deformed in such a way that right and left loops with small coiling diameters alternate. The next loop is always bent out of the plane of the preceding loop. Stitched open tubular (SOT) reactors are usually prepared from narrow bore stainless steel capillaries woven through a steel mesh in serpentine fashion with alternate loops displaced to the right and left. The continuous change in the coiling direction and the bend of the coil out of plane leads to a continuous change in the direction of the secondary flow producing a plug like flow profile with almost identical velocities at the wall and in the center of the tube. The effectiveness of the KOT reactor at minimizing extracolumn band broadening compared to a straight capillary tube is illustrated in Figure 8.39. The pressure drop across a KOT reactor is about twice that of an open tube which is the limiting factor in establishing the available reaction time. For Teflon capillaries with a pressure limit of about 10 atmospheres reaction

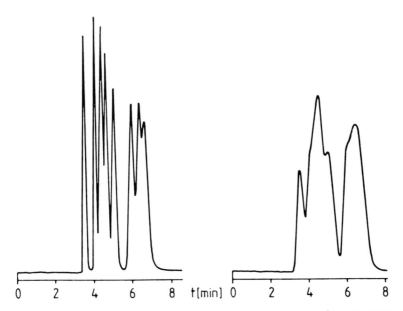

"knitted"capillary   "straight"capillary

Figure 8.39 Comparison of a straight and knitted open tubular reactor of equal length for preserving column resolution. (Reproduced with permission from ref. 656. Copyright Dr Alfred Huethig Publishers).

times approaching 10 min. are possible for typical standard column conditions [641]. However, KOT reactors are usually used for much shorter reaction times, less than 2 min., and segmented systems for the longer reaction times.

For reactions of intermediate kinetics (i.e. those with reaction times from 0.3 to several minutes), and in particular, for reactions employing solid-phase reagents, a packed bed reactor is usually preferred. The packed bed reactor is constructed from a length of column tubing packed with an inert material of small diameter, such as glass beads. Since band broadening is controlled by axial diffusion and convective mixing, packing the reactor bed should be given the same attention as packing the analytical column. The pressure drop across the reactor limits the length and smallest particle size that can be conveniently used. To adjust the reactor volume to the reaction time the inner diameter of the reactor can be increased; typical diameters are 2.5-6.0 mm.

The packed bed reactor is particularly useful for reactions involving solid-phase reagents such as catalysts, immobilized enzymes and metallic reducing agents [641,642,646,647,658]. Since no reagent solutions are added in this case, there are no additional pumps or mixing units required and no dilution or mixing problems associated with postcolumn reagent addition. The solid phase reactor may react with the analyte in a stoichiometric manner, in which case the derivatizing reagent is consumed in the reaction and must be replenished when depleted. Alternatively, solid phase reactors have been reported in which the solid phase reagent catalyzes the derivatization reaction and can be used for long periods of time without regeneration. In the latter case many of the published examples have used immobilized enzymes as the catalyst [646,650].

For slow reactions (reaction times from 5 to 20 minutes), air or liquid segmented reactors are used. These work on the same principle as the auto-analyzers used in many clinical laboratories. The effluent stream is split into segments by the introduction of air or an immiscible liquid bubble at fixed time intervals. The optimal conditions for operation involve small liquid segments with a high frequency of segmentation, short reaction tubes of low internal diameter, high flow rates, and a well-designed phase separation unit just upstream from the detector. Predictions from theory must be tempered with the practical needs of the reaction, which generally dictates the length of the reaction coil. Band broadening occurs mainly by axial diffusion with a small contribution from sample transfer between segments by wetting of the tube wall.

A limitation of postcolumn reaction detectors is that the reagent, always present in large excess, should not interfere with the detection of the derivative. There are several useful reactions, particularly those involving ion-pair formation or complexation, for which reagent interferences are commonly observed. The reaction products are usually of much lower polarity than the reagent, permitting their separation by extraction. Extraction in the continuous flow mode by segmenting with immiscible organic solvent plugs (the extraction solvent) or by air segmentation with an additional flow of extraction solvent and a mixing device can be used. Applications include the ion-pair detection of antibiotics, pharmaceutical compounds possessing a tertiary amine structure not easily derivatized by other techniques, and alkylsulfonate detergents.

The photochemically-induced change of an analyte in a flowing stream can be used as the basis of a reaction detector. Photolysis may be used to convert a substance into more readily detected products, for example, the introduction of fluorescence or electrochemical activity [552,641,649,657,659-663]. Photolysis can also be used to decompose the sample, generating a fragment that can be coupled with an appropriate reagent for detection. Photochemical reaction detectors are fairly simple in design, comprised of a high-powered lamp in a reflective housing with a Teflon or quartz reaction coil wound around the lamp. Teflon has excellent transparency to UV light and is the preferred material for the reaction coil. The reaction time, controlled by the length of the coil and flow rate, is generally optimized by trial and error. Overexposure may result in the decomposition of the product to be detected. The high specificity of the photochemical reaction detector is useful for particular applications but the number of compounds that are photolytically sensitive in a useful manner are not numerous.

8.12   DERIVATIVES FOR THE CHROMATOGRAPHIC SEPARATION OF INORGANIC ANIONS AND CATIONS

Gas chromatography, and more recently liquid chromatography, have been used to separate many inert and stable organometallic compounds [664-667]. Certain metal hydrides and halides have sufficient volatility to be separated by gas chromatography, but only with some difficulty, and generally after certain instrument modifications have been made. Reactive, volatile inorganic compounds, for example SO_2Cl, UF_6, ClO_2F, SF_5Cl, etc., can be separated using perfluorocarbon stationary phases coated on Teflon supports [668]. These columns are used because of their resistance to chemical attack but generally show poor chromatographic efficiency.

Ion-exchange, ion-pair, and ion chromatography are techniques now routinely used for inorganic analysis and were described in detail in Chapter 4. Many ions can be made sufficiently thermally and hydrolytically stable for separation by gas, liquid or thin-layer chromatography after derivatization. It is this aspect of inorganic analysis that will be discussed here. Most of these methods provide the possibility of multi-element or multi-ion determinations. For elemental analysis some overlap with spectroscopic techniques exists with the latter techniques generally being preferred. However, another important area of inorganic analysis is speciation. In this case chromatographic methods are required to separate the individual components prior to detection. Element-tunable spectroscopic detectors are often used in these studies to provide a selective profile of the substances of interest.

The preparation of neutral chelates for the separation of metals by gas chromatography has been studied for many years [436,667,669-672]. The principal limit to the success of this approach is the paucity of suitable reagents which can confer the necessary volatility, thermal stability and chemical inertness on the metal ions. The class of metal chelating reagents that have been most studied are the β-diketones and structurally related monothio-β-diketones, β-dithioketones and ketoenamines. In many instances the complexing enolate anion forms neutral chelates with metals whose preferred co-ordination number is twice their oxidation state; the resultant complexes are effectively co-ordinatively saturated, thus precluding further adduction of solvent or other ligand species. Stable complexes are formed with ions such as Be(II), Al(III) and Cr(III), which conform to the above rule. Other ions such as Ni(II), Co(II), Fe(II) and La(III), which readily adduct additional neutral ligands and assume a higher co-ordination state are difficult to gas chromatograph. Hydrates, which lower the volatility and increase the polarity of the derivative may be formed, resulting in undesirable column behavior, or non-solvated chelates may polymerize or react on-column with active sites to give excessive peak broadening or even irreversible adsorption, particularly at low chelate concentrations. Numerous examples exist of metal chelate derivatives that cannot be quantitatively eluted at low levels by gas chromatography. Non-fluorinated chelates usually have only marginally suitable thermal and chromatographic stability, often requiring elution temperatures that are too high for thermal degradation to be completely absent. The most frequently employed complexing ligands are 1,1,1-trifluoropentane-2,4-dione (HTFA) and 1,1,1,5,5,5-hexafluoropentane-2,4-dione (HHFA). Only in the case of the HTFA derivatives of beryllium, aluminum and chromium have trace level concentrations of the elements in environmental samples been determined [449,670-672].

Liquid chromatographic methods are well suited to the separation and determination of metal chelates that can be extracted into organic solvents. Many chelates also absorb strongly in the UV or visible regions, facilitating detection. The most widely used derivatives are the dithiocarbamates, dithiozonates, ketoenamines, salicylaldimines, dialkyl thiophosphates, 8-hydroxyquinolines and the tetradentate Schiff bases [671,673-676]. It might be expected that many chelates would be more stable to LC conditions than GC but this is not the case. Many complexes are kinetically labile and decompose on the column. For example, the acetylacetonate chelates of Mn(II), Be(II), Co(III), Cr(III), Rh(III), Ir(III), Pd(I) and Pt(II) are stable to reversed-phase liquid chromatography while the derivatives of Fe(III), Co(II), Zn(II) and Pb(II) are degraded on the column [673]. Ligand-substitution reactions may occur between the metal parts of the column and the metal complexes being chromatographed [677]. This is most pronounced for the diethyldithiocarbamate derivatives of Bi(III), Zn(II), Cd(II) and Pb(II), since these complexes are relatively unstable and may readily undergo exchange with nickel from the stainless steel column components. To minimize these interactions PTFE or glass-lined tubing have been used with partial success. Conditioning the column with a concentrated solution of chelates or incorporating disodium ethylenediaminetetraacetic acid into the mobile phase may also be required.

The separation conditions for many metals by ion-exchange chromatography have long been known. On-line detection by conductivity is possible but not particularly sensitive. Postcolumn reaction detectors can be easily interfaced to ion-exchange chromatography with detection limits at the nanogram level [641,674,676,678-681]. Some methods of multi-element detection monitor the decrease in absorbance of the reagent, since formation of different metal chelates usually results in different absorption maxima for each chelate. Typical reagents used are dithizone, arsenazo and eriochrome black T for absorption and oxine for fluorescence.

Selenium(IV) reacts with substituted 1,2-diaminobenzene or 2,3-diaminonaphthalene in acidic solution to form stable cyclic derivatives which can be extracted into an organic solvent and analyzed by gas or liquid chromatography [682,683]. With chloro-, bromo-, or nitro-substituents the piazselenols can be determined with an electron-capture detector at the low picogram level. Se(VI) does not form piazselenol derivatives so the reaction with diaminobenzene can be used to determine the concentration of Se(IV) and Se(VI). Selenium(VI) may be reduced to Se(IV) with boiling hydrochloric acid, providing a value for total Se and a value for Se(VI) by difference. Arylsulfinates, sodium tetraphenylborate in basic or weakly acidic solution, and tetramethyltin, have been used to alkylate inorganic mercury in environmental samples. Alkylmercurial compounds have good GC properties and can be determined at the ppb level with an electron-capture detector [449,670,684].

The most general method for the simultaneous analysis of oxyanions by gas chromatography is the formation of trimethylsilyl derivatives. Trimethylsilyl derivatives of silicate, carbonate, oxalate, borate, phosphite, phosphate, orthophosphate, arsenite, arsenate, sulfate and vanadate, usually as their ammonium salts, are readily prepared by reaction with BSTFA-TMCS (99:1). Fluoride can be derivatized in aqueous solution with triethylchlorosilane and the triethylfluorosilane formed extracted into an immiscible organic solvent for analysis by gas chromatography [685].

Cyanide and thiocyanate anions in aqueous solution can be determined as cyanogen bromide after reaction with bromine [686]. The thiocyanate anion can be quantitatively determined in the presence of cyanide by adding an excess of formaldehyde solution to the sample, which converts the cyanide ion to the unreactive cyanohydrin. The detection limits for the cyanide and thiocyanate anions were less than 0.01 ppm with an electron-capture detector. Iodine in acid solution reacts with acetone to form monoiodoacetone, which can be detected at high sensitivity with an electron-capture detector [687]. The reaction is specific for iodine, iodide being determined after oxidation with iodate. The nitrate anion can be determined in aqueous solution after conversion to nitrobenzene by reaction with benzene in the presence of sulfuric acid [688,689]. The detection limit for the nitrate anion was less than 0.1 ppm. The nitrite anion can be determined after oxidation to nitrate with potassium permanganate. Nitrite can be determined directly by alkylation with an alkaline solution of pentafluorobenzyl bromide [690]. The yield of derivative was about 80% with a detection limit of 0.46 ng in 0.1 ml of aqueous sample. Pentafluorobenzyl p-toluenesulfonate has been used to derivatize carboxylate and phenolate anions and to simultaneously derivatize bromide, iodide, cyanide, thiocyanate, nitrite, nitrate and sulfide in a two-phase system using tetrapentylammonium chloride as a phase transfer catalyst [691]. Detection limits were in the ppm range.

8.13 VISUALIZATION TECHNIQUES FOR THIN-LAYER CHROMATOGRAPHY

Visualization reactions were originally used in thin-layer chromatography (TLC) to enable colorless compounds to be detected by eye and, less frequently, to increase the selectivity of the detection process by reaction of the separated compounds with a reagent having chemical selectivity for a particular functional group or compound class. Several hundred reagents have been described for the above process [692-696]. Many of these reactions are of a qualitative nature which was not a problem when quantitation of sample components was rarely performed by TLC. Some of these reactions have been adapted to the demands of quantitative scanning densitometry as either a pre- or postchromatographic treatment. The advantages of postchromatographic methods are that byproducts of the derivatization reaction do not interfere in the separation and all samples are derivatized simultaneously. The reagent is usually applied to the plate by dipping the plate into a dilute solution of the reagent or by spraying the reagent solution over the plate surface. For quantitative TLC it is absolutely essential that a uniform spraying is obtained. This is not easy to achieve by manual methods and motorized spray units are recommended [697]. Dipping the plate into the reagent solution usually ensures an even application of reagent to the plate and is the preferred technique in most cases.

Prechromatographic derivatization reactions are usually favored when it is desired to modify the properties of the sample to enhance stability during measurement (i.e., minimize oxidative and catalytic degradation, etc.), to improve the extraction efficiency of the substance during sample cleanup, to improve the chromatographic resolution, or to simplify the optimization of the reaction conditions [698-702]. As both pre- and post-chromatographic methods enhance the sensitivity and selectivity of the detection process a choice between the two methods will usually depend on the chemistry involved, ease of optimization, and which method best overcomes matrix and reagent interferences.

Postchromatographic reactions can be classified as reversible or destructive depending on the type of interaction they undergo with the separated compounds, and as selective or universal, based on the specificity of the reaction. The most common reversible methods employ iodine vapor, water, fluorescein, bromine or pH indicators as visualizing reagents [696]. In the iodine vapor method the dried plate is enclosed in a chamber containing a few crystals of iodine; components on the chromatogram are stained more rapidly than the background and appear as yellow-brown spots on a light yellow background. As little as 0.1 to 0.01 micrograms of sample can be visualized in this way and the reaction can be reversed by simply removing the plate from the visualization chamber and allowing the iodine to evaporate. Compounds containing unsaturated hydrocarbon or phenolic groups may irreversibly react with iodine. Mercaptans are oxidized to disulfides. Spraying a TLC plate with water reveals hydrophobic compounds as white spots on a translucent background when the water-moistened plate is held against the light. A TLC plate sprayed with 0.05% aqueous fluorescein, followed by exposure to bromine vapor while still damp, becomes uniformly colored (due to formation of eosin) except at positions where sample components are located. Solutions of pH indicators, for example, bromocresol green and bromophenol blue, are widely used for the detection of acidic and basic compounds. The above methods are all fairly universal and reversible so that they may be used for detection when sample recovery for further studies is required. However, these methods are rarely used for quantitative analysis.

Destructive methods are more common for quantitation. Again, numerous reagents are known [692-694] but among the more generally used are solutions of sulfuric acid, sulfuric acid-acetic anhydride (1:4) and sulfuric acid-sodium dichromate. After spraying, the plate is usually heated in an oven for some time at 110-115°C. Organic compounds are converted to carbon and visualized as black spots on a white-grey background. Alternatively, ammonium bisulfite can be incorporated into the adsorbent layer and, after the development process, sulfuric acid generated in situ by heating the plate in an oven. Using milder conditions and/or more selective reagents converts the sample components into colored or fluorescent materials. When carried out under controlled conditions, these degradative reactions can yield accurate quantitative data. For example, in the determination of lipids by reaction with phosphoric acid/copper sulfate [703] or the determination of monosaccharides by treatment with Hansen's reagent (a mixture of diphenylamine and aniline in phosphoric acid) [704]. Functional group specific or compound class selective reagents can be applied for the determination of low levels of substances in complex matrices such as biological fluids and environmental extracts. Several hundred reactions have been described for this purpose and standard monographs and review articles should be searched for individual examples [696,698-702].

8.14 QUALITATIVE IDENTIFICATION AND MICROREACTION TECHNIQUES FOR GAS CHROMATOGRAPHY

Before the advent of modern hyphenated techniques (GC/MS, GC/FTIR), numerous qualitative physical and chemical tests were devised for the identification of peaks in a gas chromatogram [705]. For the most part these tests were simple to perform, inexpensive, required minimum instrument modification and, in a few instances, provided a simple and easy solution to an otherwise complex problem. They still have some value today as spectroscopic techniques do not solve all problems.

The partition coefficient of a substance between several immiscible solvent pairs can be combined with retention time data to confirm the identity of a substance when a pure standard is available [706]. Devised by Bowman and Beroza, the substance specific partition coefficient ("p-value") was defined as the fractional amount of substance partitioning into the less polar phase of an equal-volume, two-phase system. Only nanogram quantities of sample are required for the measurement and p-values are often sufficiently characteristic to distinguish between closely related substances.

Carbon-skeleton gas chromatography provides a means of identifying the hydrocarbon skeleton of a compound which contains functional groups [707,708]. The basis of the method is the catalytic hydrogenolysis or hydrogenation of the sample prior to separation. Hydrogenolysis involves the removal of heteroatoms such as halides, N, O, or S (usually present as functional groups, e.g., NH_2, OH, SH, etc.), such that the hydrocarbon species so generated retains either the carbon skeleton of the parent or the next lower homolog of the series.

In a typical experimental arrangement, the injection block heater of the gas chromatograph is used to heat a short catalyst bed containing platinum, palladium, copper or nickel coated on a diatomaceous support. The catalyst bed can be the top portion of a packed column or a precolumn connected to a packed or open tubular column. Hydrogen carrier gas flows through the heated catalyst bed (220-350°C) and then into the column. The sample is injected by syringe onto the catalyst bed where a rapid reaction occurs and the products are displaced onto the separation column. Catalyst preparation, activation and conditioning prior to use are important considerations [708,709]. A forked-column arrangement has been described for the quantitative analysis of complex mixtures in which the parent hydrocarbons are also present [710]. The two arms of the fork are identical except that one arm contains support and the other support plus catalyst. Injection of the sample alternately into the two arms enables the natural concentration of the hydrocarbons to be subtracted from the amount found in the reduction mode. Dechlorination prior to capillary column gas chromatography has been performed in solution using nickel boride [$(Ni_2B)_2H_3$] as a reducing agent [711]. With further investigation this reagent may prove to be generally useful for a wide range of reactions.

An important application of carbon-skeleton gas chromatography is the simplification of the analysis of complex samples such as polychlorinated biphenyls, polybrominated biphenyls and polychloroalkanes [709-711]. These complex mixtures of halogenated isomers produce multiple peaks when separated by gas chromatography, making quantitation difficult. The isomers have identical carbon skeletons, resulting in a very simple chromatogram after hydrodechlorination.

The location of the position of double bonds in alkenes or similar compounds is a difficult process when only very small amounts of sample are available [712,713]. Mass spectrometry is often unsuited for this purpose unless the position of the double bond is fixed by derivatization. Oxidation of the double bond to either an ozonide or cis-diol, or formation of a methoxy or epoxide derivative, can be carried out on micrograms to nanograms of sample [713-716]. Single peaks can be trapped in a cooled section of a capillary tube and derivatized within the trap for reinjection. Ozonolysis is simple to carry out and occurs sufficiently rapidly that reaction temperatures of -70°C are common [436,705,707,713-717]. Several micro-ozonolysis apparatuses are commercially available or can be readily assembled in the laboratory using standard equipment and a Tesla coil (vacuum tester) to generate the ozone. Reaction yields of ozonolysis products are typically 70 to 95%, although structures such as triple bonds and β-unsaturated nitriles are quite resistant to the process. After ozonolysis, either pyrolysis or, preferably, reduction with triphenyl phosphine converts the ozonides to aldehydes (or ketones if no hydrogen atoms are attached directly to the carbon atom of the double bond). Identification of the products by gas chromatography enables the position of the original double bond to be established.

Methods have been developed by which compounds containing common functional groups can be subtracted from complex mixtures prior to gas chromatography. Known as reaction or subtraction chromatography, the reactions may take the form of sample preparation techniques (derivatization), precolumn reactions in loops or vessels, on-column or in-syringe reactions, or postcolumn reactions [705,707,713,714,718,719]. The difference between the reacted and unreacted sample chromatograms is used to identify which compounds contain a particular functional group. Some representative reagents and their areas of application are summarized in Table 8.23. In its simplest form the reagent is coated onto a support and packed into a short column that is usually positioned between the injection device and the analytical column or, occasionally, between the column and detector. Such columns expose a large surface area of reagent to the sample, and provided the reaction is rapid, a short column ensures complete subtraction at normal carrier gas flow rates without distorting the peak shape of compounds which do not react. Some reactions are reversible, which affects the operating temperature range for the reaction, others generate secondary byproducts which can confuse the identification, and, of course, all reagents eventually become exhausted or deactivated with use. Sievers used a europium coordination polymer to selectively retain nucleophilic compounds (ketones, aldehydes, alcohols and carboxylic acids) by complexation. The complexation process is reversible and a manifold was constructed that enabled capillary column chromatograms of the total sample, the retained nucleophiles, and the sample without the retained nucleophiles to be obtained [720]. Otherwise, the majority of precolumn reactions for open tubular columns are performed off-line in small glass capillaries that can be flushed with solvent or broken in the injection port to release the sample to the column [713-716].

8.15 SEPARATION OF STEREOISOMERS

Isomers are substances that have the same molecular formula (constitution) but different molecular structure (configuration).

TABLE 8.23

PRECOLUMN SUBTRACTION REAGENTS USED FOR QUALITATIVE SAMPLE
IDENTIFICATION IN GAS CHROMATOGRAPHY

Subtraction Reagent	Compound Functionalities Subtracted
Molecular Sieve 5A	n-Alkanes and other straight-chain molecules in the presence of branched chain molecules.
Salts of Ag, Cu, Hg often mixed with concentrated sulfuric acid	Alkenes, alkynes.
Concentrated sulfuric acid	Aromatics, alkenes, alkynes, basic compounds.
Maleic anhydride	Conjugated dienes.
Boric acid 2-Nitrophthalic anhydride	Primary and secondary alcohols.
Lithium aluminum hydride	Alcohols, aldehydes, ketones, esters, epoxides.
Zinc oxide	Carboxylic acids, partial subtraction of alcohols and phenols.
NaOH-quartz	Phenols.
Phosphoric acid	Epoxides, bases.
Versamid 900	Organic compounds containing active halogens.
Bromine/Carbon tetrachloride	Unsaturated compounds.
Benzidine	Aldehydes, ketones, epoxides.
Sodium borohydride Hydroxylamine Sodium bisulfite Semicarbazide 3,4,5-Trimethoxybenzyl-hydrazine	Ketones, aldehydes.
FFAP o-Dianisidine	Aldehydes.

Structural (configurational) isomers such as n-butanol and diethyl
ether, for example, can differ appreciably in their physical and
chemical properties and, in general, do not present unusually
difficult separation problems. Stereoisomers on the other hand
differ only in the spatial configuration of substituent groups
within a molecule (same atomic bonding order) and are likely to
have very similar physical and chemical properties. Several
different classes of stereoisomers can be distinguished.
Conformational isomers (conformers) can be interconverted by
rotation about single bonds and correspond to different internal
energy minima, for example, the chair and boat conformations of
the cycloalkanes. Stereoisomers that are mirror images of each

other are called enantiomers. Common examples of enantiomers are molecules containing tetrahedral carbon, silicon, sulfur or phosphorus atoms bearing four different substituents, unsymmetrical sulfoxides and substituted aziridines. These molecules lack a plane of symmetry and their mirror images are nonsuperimposable. Asymmetry may also result from the helicity of a macromolecule such as a protein, polymer or helicene. Enantiomers have identical physical and chemical properties except for their ability to rotate the plane of polarized light to equal extents but in opposite directions. Of greater interest is that enantiomers may exhibit different biological activities due to differences in their strength of interaction with the corresponding receptor, have different transport mechanisms, and may be metabolized by different routes. Thus, chiral recognition is an inherent feature of the biological mechanism by which enzymes, chemical agents and drugs perform their prescribed functions. For this reason a knowledge of the enantiomeric purity of natural products and xenobiotics is of considerable importance in biology and medicine.

Stereoisomers that are not mirror images of each other are called diastereomers. Classes of diastereomers include molecules containing more than one asymmetric (chiral) center and geometric isomers that owe their existence to hindered rotation about double bonds, for example, cis and trans isomers. Diastereomers may have different physical and chemical properties that enable them to be separated by normal chromatographic procedures. The general method of separating enantiomers by chromatographic means is after conversion to diastereomers. The diastereomers can be formed by direct interaction with a chiral phase (formation of transient diastereomer association complexes) or after chemical transformation by reaction with an enantiomerically pure (homochiral) derivatizing reagent. Alternative methods of separating enantiomers include the physical sorting of enantiomeric crystals, selective crystallization by formation of diastereomeric salts or by using a chiral solvent, kinetic techniques (asymmetric destruction using enzymes), calorimetric methods, isotope labeling and NMR spectroscopy. For analytical applications the chromatographic methods have been most widely used because of their small sample requirement and favorable detection characteristics, ability to simultaneously resolve mixtures of racemates and to separate mixtures of enantiomers and achiral compounds in the same sample, low cost and high speed coupled with ease of automation.

8.15.1   Separation   of   Enantiomers   as   Their   Diastereomeric Derivatives

Enantiomeric mixtures containing a reactive functional group can be derivatized with a homochiral reagent to produce a mixture of diastereomers [721-723]. Simply forming diastereomeric derivatives does not in itself guarantee resolution, it merely makes resolution possible in a nonchiral separation system. The magnitude of the physical difference between diastereomers and the selectivity and efficiency of the chromatographic system all influence the extent of resolution. Often the resolution of diastereomeric derivatives is enhanced when the chiral centers of the enantiomer and the reagent are in close proximity in the derivative and when the reagent is comprised of conformationally immobile groups or contains bulky groups attached directly to the chiral center. The reaction conditions should be sufficiently mild to avoid racemization or epimerization of the chiral components and the diastereomer transition states should have similar conversion rates for the enantiomers. Also, the homochiral derivatizing reagents should have a reasonable shelf life to be practically useful. The most likely sources of error in establishing the enantiomeric composition of a mixture are the presence of enantiomeric impurities in the derivatizing reagent, racemization during the derivatization reaction, and different rates of reaction for individual enantiomers. If the reaction is to be used for preparative purposes ready conversion of the diastereomers into their respective enantiomeric moieties is also a necessary condition.

Some representative derivatizing reagents for enantiomeric resolution by gas chromatography are given in Table 8.24. These reagents must form derivatives that are thermally stable and free from racemization during the chromatographic process. N-perfluoroacylprolyl chlorides have been widely used to derivatize amines but their usefulness has been questioned due to poor shelf life and possible racemization during reaction [723-725]. The chiral derivatizing reagent α-methoxy-α-(trifluoromethyl)-phenacetyl chloride has been widely used to form diastereomeric derivatives of amines, such as amphetamines, dopamines and

TABLE 8.24

TYPICAL HOMOCHIRAL DERIVATIZING REAGENTS FOR GAS CHROMATOGRAPHIC
SEPARATION OF ENANTIOMERS

Reagent	Functional Group Reacted	Reference
N-Trifluoroacetylprolyl chloride	Amines	725,726
α-chloroisovaleryl chloride	Amines	727
α-methoxy-α-(trifluoromethyl)-	Amines	728,729
phenylacetyl chloride	Alcohols	730
Menthyl chloroformate	Amines	731
	Alcohols	732
Drimanoyl chloride	Amines	733
trans-Chrysanthemoyl chloride	Alcohols	733
1-Phenyethylisocyanate	Alcohols, Amines Carboxylic acids	734
2-Phenylpropionyl chloride	Alcohols	735
2-Butanol	Carboxylic acids	736
3-Methyl-2-butanol	Carboxylic acids	737
Menthyl alcohol	Carboxylic acids	738
2-Octanol	Carboxylic acids	739
1-Phenylethylamine	Carboxylic acids	740
2,2,2-Trifluoro-1-pentylethylhydrazine	Ketones	741

antiarrhythmic drugs [728,729]. It is also widely used for the
resolution of chiral alcohols and hydroxy-fatty acids [730]. 1-
Phenylethylisocyanate has been developed as a versatile
derivatizing reagent for a wide range of enantiomers containing
amine, hydroxyl and carboxylic acid groups [734]. However, the
most frequently used approach to the formation of diastereomeric
derivatives of carboxylic acids is esterification with one of a
variety of enantiomerically pure alcohols or amines.

A selection of reagents for enantiomeric resolution by
liquid chromatography is shown in Figure 8.40 [721-723,742]. Many
of the reagents summarized in Table 8.24 for gas chromatography
are also used in liquid chromatography. Reversed-phase and normal
phase liquid chromatography are used for the separations. The
unique lock-key separation mechanism of normal phase
chromatography is particularly useful for separating
diastereomers. Several of the reagents employed for enantiomeric
resolution by liquid chromatography also contain a chromophore to
improve the detection characteristics of the derivatives. Reagents
containing a N,N'-dimethylaminonaphthyl chromophore such as α-(4-
N,N'-dimethylamino-1-naphthyl)ethylamine can be used with

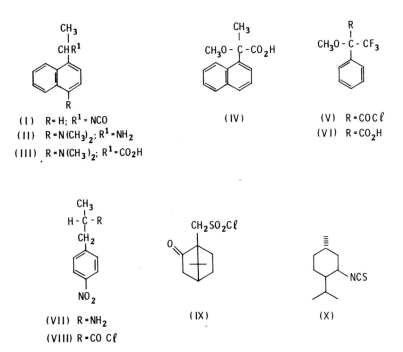

(I) R=H; R^1=NCO
(II) R=N(CH$_3$)$_2$; R^1=NH$_2$
(III) R=N(CH$_3$)$_2$; R^1=CO$_2$H

(IV)

(V) R=COCℓ
(VI) R=CO$_2$H

(VII) R=NH$_2$
(VIII) R=CO Cℓ

(IX)

(X)

Figure 8.40 Representative chiral derivatizing reagents for separating enantiomer by HPLC. (I), α-naphthylethylisocyanate; (II), L-1-(4- dimethylamino-1-naphthyl)ethylamine; (III), L-1-(4-dimethylamino-1-naphthyl)acetic acid; (IV), α-methoxy-α-methyl-1-naphthaleneacetic acid; (V), (+)-α-methoxy-α-trifluoromethyl-phenylacetyl chloride; (VI), carboxylic acid form of (V); (VII), (R)-(+)-α-methoxy-p-nitrobenzylamine; (VIII), acid chloride form of (VII); (IX), (+)-camphor-10-sulfonyl chloride; and (X), (+)-neomenthyl isothiocyanate. Where not explicitly stated the reagent must be in a single enantiomeric form to be of use as a resolving agent.

fluorescence detection for trace analysis [743]. The naphthyl-containing reagents N-1-(2-naphthylsulfonyl)-2-pyrrolidinecarbonyl chloride, α-methoxy-1-naphthaleneacetic acid, 2-methyl-1,1-binaphthalene-2-carbonyl nitrile and 1-naphthylethylisocyanate show little fluorescence and are usually determined with the UV detector. The reactive functional group of the reagent determines its area of application. For example, α-methoxy-α-methyl-1-naphthaleneacetic acid was used to derivatize amines [744] and 1-aminoethyl-4-dimethylaminonathalene the carboxylic acid group [745] of racemic amino acids. 1-(1-Naphthyl)ethylisocyanate and 1-phenylethylisocyanate have been used as chiral reagents for the resolution of amines, amino alcohols and sympathomimetic drugs by liquid chromatography [746,747]. 2,3,4,6-Tetra-O-acetyl-β-D-glucopyranosyl isocyanate is a new chiral derivatizing reagent that has been widely used for the resolution of amino acids, catecholamines, β-adrenergic blocking drugs, amphetamines and amino alcohols [748-750]. Other widely used reagents include 10-camphorsulfonyl chloride for alcohols and amines [751] and 1-methylbenzylamine for carboxylic acids [752]. A series of chiral chloroformate reagents have been introduced for derivatizing amines and alcohols [753]. In a recent development Schill and Petterson have demonstrated that multifunctional ion-pair reagents such as 10-camphorsulfonic acid, di-n-butyltartrate, N-carbobenzoxycarbonylglycine-l-proline and quinine can form diastereomeric ion pairs with protonated amines and carboxylic acids [742,754,755]. Only those analytes that can enter into additional molecular interactions, such as hydrogen bonding, dipole-dipole or steric repulsion simultaneously with the electrostatic interaction holding the ion pair together form diastereomeric ion-pair complexes with different sorption characteristics in the chromatographic system. Formation of the complexes is favored by mobile phases of low polarity.

8.15.2 Separation of Enantiomers Using Chiral Stationary Phases

The separation of enantiomers using chiral stationary phases occurs because the interactions between the enantiomers and the stationary phase result in the formation of transient diastereomeric association complexes having different sorption enthalpies and hence different retention characteristics. Compared with the previous discussion where derivatization techniques were used to form the diastereomers the direct approach has certain advantages. It can be applied to enantiomers lacking reactive functional groups and is not limited by the need for reagents of high enantiomeric purity. Although the extent of resolution on a chiral stationary phase will depend on the enantiomeric purity of the phase, separations are still possible with phases exhibiting a reasonable enantiomeric excess. Also, simple rules permit the prediction of the absolute configuration of enantiomers in the absence of pure standards. However, there is no universal chiral phase for the separation of all enantiomers and, therefore, both the direct and indirect methods of resolution have their uses and particular areas of application.

The first chiral phases introduced for gas chromatography were either amino acid esters, dipeptide, diamide or carbonyl-bis(amino acid ester) phases [721,724,756-758]. In general, these phases exhibited poor thermal stability and are infrequently used today. Real interest and progress in chiral separations resulted from the preparation of diamide phases grafted onto a polysiloxane backbone. These phases were thermally stable and could be used to prepare efficient open tubular columns [734,756,758-762]. These phases are prepared from commercially available poly(cyano-propylmethyldimethylsiloxanes) or poly(cyanopropylmethylphenyl-methylsiloxanes) by hydrolysis of the cyano group and subsequent coupling of the carboxylic acid group with L-valine-tert.-butylamine or L-valine-S-α-phenylethylamide. Separation of each chiral center by several dimethylsiloxane or methylphenylsiloxane units (greater than 7) seems to be essential for good enantiomeric resolving power and thermal stability. The L-valine-tert.-butylamide-containing phase, commercially available as Chirasil-Val, has been used to resolve a wide range of enantiomeric amino acid derivatives, amino alcohols, amines, hydroxy ketones, 2-hydroxy acids and their esters, 3-hydroxy acids, lactones and sulfoxides [724,756,763,764]. It permitted for the first time the resolution of all the enantiomers of the common protein amino acids as their N-pentafluoropropionylamide isopropyl ester derivatives in a single separation, Figure 8.41, and led to the development of the enantiomer labeling technique (using the unnatural D-enantiomer as internal standards) for quantitative amino acid analysis [433,434,756]. The L-valine-S-α-phenylethyl-amide-containing poly(dimethylsiloxane) phase provides a facile method for the configurational analysis of sugars, as well as resolving enantiomers of amino acids and amino alcohols [724,734]. A number of ketones, pharmaceutical compounds, alcohols and hydroxy acids have also been resolved on this phase [724,765-767]. A chiral polysiloxane phase with tartramide substituents has been used for the separation of enantiomers capable of hydrogen bonding interactions with the stationary phase, such as enantiomers containing carboxylic, hydroxyl and amine functional groups [768].

Early attempts to use cyclodextrins and their simple derivatives as chiral stationary phases in gas chromatography met

904

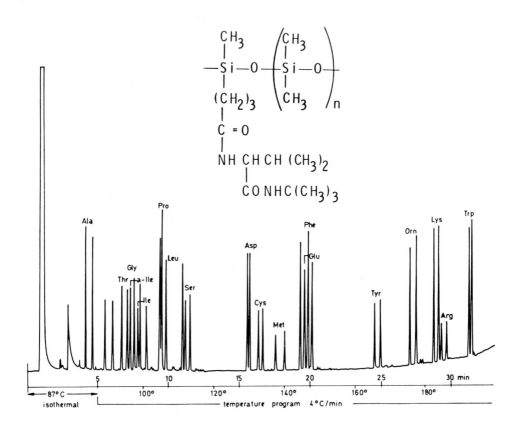

Figure 8.41 Separation of the enantiomers of the common protein amino acids (N-perfluoropropionylamide isopropyl esters) on a 20 m x 0.25 mm I.D. open tubular column coated with Chirasil-Val. (Reproduced with permission from ref. 764. Copyright Elsevier Scientific Publishing Co.)

with limited success due to their unfavorable physical properties [769]. Most compounds were crystalline solids with high melting or decomposition temperatures that made them difficult to exploit as GC stationary phases. Since 1988 interest in the cyclodextrins has intensified with the development of new approaches to overcome their physical limitations and the spectacular success demonstrated in resolving a wide range of low molecular weight enantiomeric compounds including alcohols, amines, amino acids, epoxides, carboxylic acids, esters, lactones, ethers, haloalkanes and hydrocarbons. Many of these compounds are difficult or impossible to resolve by other means and are economically important in evaluating the properties of perfumes, food additives and flavors and as chiral synthetic building blocks for more complex molecules. Some peralkylated cyclodextrins, such as pentyl, 2-hydroxypropyl, and mixed alkylated products, for example, 3-O-butyryl-2,6-di-O-pentyl are viscous liquids that can be coated directly on glass surfaces [769-774]. Many of these products are probably heterogeneous representing variable degrees of derivatization by position and extent, contributing to the lower melting points. The cyclodextrins have three hydroxyl groups not involved in forming the distinctive cavity of these carbohydrates, a primary hydroxyl group located at the narrow base of the torus (6 position) and two secondary hydroxyl groups located at the wider mouth (2 and 3 position). Because of differences in their natural reactivity and steric factors complete and/or selective derivatization is not easy to achieve. These products have good thermal stability, however, and in the absence of oxygen can be used at temperatures from room temperature to around 200°C. Another solution is to dilute the derivatized cyclodextrin in a conventional stationary phase. Melting point and phase transition considerations are no longer important for the dissolved cyclodextrins and columns of higher efficiency can generally be prepared [775-778]. Typical stationary phase mixtures include peralkylated, partially acylated (e.g., 2-O-acetyl-2,6-di-pentyl- or 3-O-heptafluorobutyryl-2,6-di-O-methyl-) or tert.-butyldimethylsilanized cyclodextrins at a concentration of about 5-10% (w/w) in a moderately polar viscous polysiloxane solvent, such as OV-1701. The coating characteristics of the column resemble those of the bulk solvent which can be varied to match the polarity of the samples to be separated. Existing column preparation procedures for the bulk solvent can generally be exploited without significant re-optimization. Recently, polysiloxane-anchored permethyl cyclodextrin phases have been described which can be immobilized on fused silica surfaces [779]. In this case the cyclodextrin is partially alkylated to introduce a 5-pent-1-ene group, the remaining hydroxyl groups permethylated, and the derivatized cyclodextrin incorporated by hydrosilylation into a poly(dimethylsiloxane) polymer containing about 5 percent of Si-H groups.

The mechanism of the enantioselective interactions with derivatized cyclodextrins is not known with certainty. The ability of cyclodextrins to discriminate between the enantiomers of unfunctionalized analytes, such as hydrocarbons implies that the formation of inclusion complexes with the chiral cavity of the cyclodextrins is important. Variations in the separation properties with cavity size tends to support this general assumption. The lack of competing solvent molecules in GC may allow complex formation to occur at relatively high temperatures while in solution it is known that complexation does not occur significantly at elevated temperatures. The high efficiency of open tubular columns allows small selectivity factors to be easily exploited to yield a separation while in LC these would be inadequate. This contributes just as much to the success of the cyclodextrin phases in GC as does their enantioselectivity, which can be poor and yet an acceptable separation obtained. Problems in column technology seem to be solvable. Perhaps a more acute problem is how to define a small number of derivatized cyclodextrins for general use. Before a good understanding of the properties of these phases becomes apparent it is likely that a large and bewildering number of derivatized cyclodextrins will be introduced, found successful for some applications, and only later deleted from use because they simply duplicate the properties of an existing cyclodextrin phase with more favorable synthetic or chromatographic properties.

The last several years have seen an enormous growth in the number and use of chiral stationary phases in liquid chromatography [742,780-791]. Some problems with the gas chromatographic approach are that the analyte must be volatile to be analyzed and larger-scale preparative separations are frequently difficult. For entropic reasons relatively high temperatures tend to minimize the stability differences between the diastereomeric complexes and racemization of the stationary phase over time may also occur. The upper temperature limit for phases such as Chirasil-Val is about 230°C and is established by the rate of racemization of the chiral centers and not by column bleed. Liquid chromatography should be superior in the above respects although gas chromatography would still be preferred for the separation of volatile analytes due to the availability of higher efficiency columns and the absence of competing solvent interactions.

Some general considerations governing the nature of selective enantiomeric interactions for both gas and liquid chromatographic phases (at least of the bonded monomeric ligand type) have been forthcoming [721,742,754,756,781,782,790]. It is generally assumed that three points of simultaneous interaction at least one of which must be stereochemically controlled, are required to distinguish the chirality of a molecule. These interactions usually involve hydrogen bonding, dipole-dipole, $\pi-\pi$ acid/base, and/or steric repulsion. Other researchers have concluded that only one or two interactions are needed to attain sufficient discrimination between diastereomeric molecular complexes. Considering a one-to-one complex viewed at the molecular level this seems rather unlikely. However, in gas or liquid chromatography analytes must interact with clusters or layers of stationary phase (and solvent molecules in the case of liquid chromatography) providing for the possibility of multipoint attachment. These multipoint interactions in a structured solvent environment would be equivalent to the three points of interaction discussed above except that at the molecular level the simultaneous interactions may involve more than two molecules. The final word has not been written on the general nature of chiral recognition mechanisms and gaining a better understanding of this problem will be important to the future development of new chiral stationary phases and the rational selection of a particular phase for a given separation.

These is no formal method of categorizing the types of chiral stationary phases available for liquid chromatography [784]. This diversity is both a strength, in providing a wide range of possibilities to solve a particular problem, and a weakness in that it is often difficult to predict which phase will be most useful for solving a particular problem. Compilations of commercially available chiral stationary phases and their proven applications are available [780-785]. As new phases continue to be introduced at a hectic pace only a brief overview of their properties will be attempted here.

The most widely used stationary phases resemble conventional silica-based column packings in which the chiral ligand is ionically or covalently bonded to the silica surface. Pirkle developed a number of phases containing N-3,5-dinitrobenzamides of α-amino acids, 5-arylhydantoins, 1-aryl-1-alkylamines, N-(2-naphthyl)-α-amino acids and phthalides as the chiral ligand [742,783,792-796]. The ionic or covalently bound N-3,5-dinitrobenzamides of phenylglycine and leucine, and 1-aryl-1-alkylamines, Figure 8.42, found rapid acceptance for the separation of a number of different classes of enantiomeric compounds, including alkyl carbinols, aryl-substituted hydantoins, lactams, succinimides, phthalides, sulfoxides, sulfides, amides and cyclic amides and imides. The Pirkle-type phases are designed to operate using attractive hydrogen bonding, $\pi-\pi$ and dipole stacking interactions between the chiral stationary phase and the enantiomers. Compounds containing amine or carboxylic acid functions are usually derivatized first; amines are converted to amides or carbamates and carboxylic acids to amides or esters. The most commonly used derivatizing reagents contain either a naphthyl moiety or a 3,5-dinitrobenzoyl moiety to ensure maximum $\pi-\pi$ interaction between the analyte and the chiral stationary phase. Separations are usually performed with non-aqueous mobile phases such as hexane with a few percent of a polar modifier. The ionically bonded phases, which can be conveniently prepared by pumping a tetrahydrofuran solution of the N-(3,5-dinitrobenzoyl)-α-amino acid through a prepacked column of 3-aminopropylsiloxane-bonded silica, are less tolerant of polar solvents which can cause elution of the ionic ligand from the column. A similar approach to that of Pirkle lead to the development of the urea-linked chiral stationary phase (R)-N-(1-naphthylethyl-N'-propylsilyl urea) which has many similar applications to the phases discussed above [797].

Microcrystalline cellulose triacetate has been widely used for the analytical and preparative separation of enantiomers in medium pressure liquid chromatography. Their relatively low cost accounts for their widespread use in preparative chromatography but generally microcrystalline cellulose has been replaced by cellulose derivatives (triacetate, tribenzoate, triphenyl-carbamate, and tricinnamate) coated on macroporous silica for analytical applications in high pressure liquid chromatography [742,798-802]. Although the exact enantioselectivity mechanism is unknown the two forms of cellulose probably behave differently. In

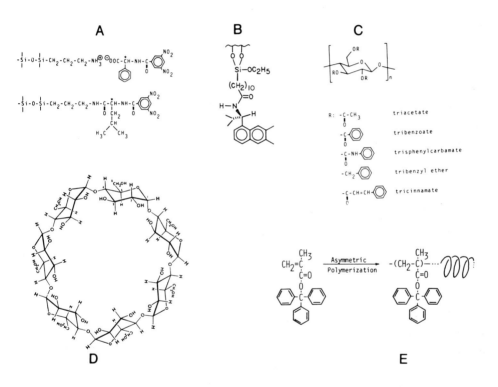

Figure 8.42 Structures of some common chiral stationary phases used in liquid chromatography. A, N-(3,5-dinitrobenzoyl)phenyl-glycine ionically bonded to 3-aminopropylsiloxane-bonded silica and N-(3,5-dinitrobenzoyl)valine covalently bonded to 3-aminopropylsiloxane-bonded silica; B, 1-arylalkylamine-based bonded chiral stationary phase; C, different derivatives of cellulose; D, cyclodextrin with seven glucose rings; and E, formation of triphenylmethyl methacrylate helical polymer.

the case of microcrystalline cellulose phases the helical
structure of the polymer contributes significantly to the

selectivity of the phase by providing a cavity for inclusion complexation which is absent in the phases prepared by coating cellulose on silica. In the latter case hydrogen bonding, π-π and dipole interactions at the polymer surface dominate the chiral recognition mechanism, not unlike that of the bonded phases discussed above. A wide range of alcohols, carbonyl compounds (particularly if the carbonyl group is close to the chiral center), lactones, phosphorus compounds, aromatic compounds and sulfoxides have been successfully resolved on the different celluloses. Nonpolar mobile phases are commonly used, although not exclusively, and the selectivity of the various cellulose polymers changes with the character of the ester or carbamate groups used to derivatize the polymer.

Polymerization of triphenylmethyl methacrylate in the presence of a chiral anion catalyst results in a polymer with a helical structure that can be coated onto macroporous silica [742,804]. Enantioselectivity in this case results from insertion and fitting of the analyte into the helical cavity. Aromatic compounds and molecules with a rigid nonplanar structure are often well resolved on this phase. The triphenylmethyl methacrylate polymers are normally used with eluents containing methanol or mixtures of hexane and 2-propanol. The polymers are soluble in aromatic hydrocarbons, chlorinated hydrocarbons and tetrahydrofuran which, therefore, are not suitable eluents.

Cyclodextrins are natural macrocyclic polymers of glucose containing six (α), seven (β), or eight (γ) glucose units connected via 1,4-linkages [742,781,789,805,806]. The glucose units are joined together to form a torroidal structure with a relatively hydrophobic center and a polar outer surface. The cyclodextrins can be bonded to silica gel via a spacer arm and are generally used with aqueous mobile phases. Only if the less polar (preferably aromatic) part of the solute molecule fits tightly into the cyclodextrin cavity and the substituents interact with the chiral exterior (hydroxyl groups of the glucose units) can enantiomeric separations be expected. The tighter the fit of the guest molecule the better the separation. Small molecules that fit loosely into the cavity are not resolved. Some of the enantiomeric compounds that have been resolved include aromatic amines, acids, sulfoxides, amino acid derivatives, alkaloids metallocenes and several drugs. Alkylation or acylation of the hydroxyl groups around the cyclodextrin cavity provides superior phases for normal phase liquid chromatography [806].

Bovine serum albumin covalently bonded to silica and α-acid glycoprotein immobilized on silica have been used to resolve a wide range of acidic and basic drugs and amino acid derivatives [807-810]. Because of their complex structures, however, the mechanism of chiral recognition is largely unknown. Retention seems to be mainly based on hydrophobic and electrostatic (coulombic) interactions, although hydrogen bonding and charge transfer interaction may also be of significance. The protein phases are generally used with buffered aqueous mobile phases containing a low concentration of organic modifier. The observed resolution depends on temperature, pH, ionic strength and concentration and type of the organic modifier which complicates optimization procedures. The column packings contain very little bound protein and are easily overloaded, even with small sample sizes. Preparative applications have not been possible so far with these packings.

8.15.3 Separation of Stereoisomers by Complexation Chromatography

The rapid and reversible formation of complexes between some metal ions and organic compounds that can function as electron donors can be used to adjust retention and selectivity in gas and liquid chromatography. Such coordinative interactions are very sensitive to subtle differences in the composition or stereochemistry of the donor ligand, owing to the sensitivity of the chemical bond towards electronic, steric and strain effects. A number of difficult to separate mixtures of stereoisomers and isotopomers have been separated by complexation chromatography.

Perhaps the most widely known example of complexation chromatography is the use of silver ions to complex organic compounds containing π-electrons in various kinds of double and triple bonds, or containing heteroatoms, such as N, O, and S with lone pairs of electrons [811,812]. In general, conjugated dienes form weaker complexes with silver than isolated double bonds and double bonds with substituents differing in size and position can frequently be separated. A limitation of silver ion complexation in gas chromatography is the low upper temperature limit of the columns, 65°C or 40°C, as claimed by different authors. In reversed-phase HPLC, the formation of silver complexes with unsaturated ligands results in a decrease in retention due to an increase in the hydrophilic character of the complex compared to the parent ligand [813-815]. Varying the concentration of silver nitrate in the mobile phase enables the retention of the complexed species to be changed over a wide range. For retinyl esters, which contain an unsaturated fatty acid side chain, a linear relationship was found between the logarithm of the capacity factor of the complex and the silver ion concentration in the mobile phase [814]. Silver nitrate added to the mobile phase results in high selectivity for cis/trans isomers in normal phase chromatography [816]. By having silver present in the mobile phase, an equilibrium with respect to silver on the stationary phase is established. The silver ions act as deactivating agents for uncomplexed solutes, which are less retained, and selectively increases the retention of complexed analytes which are more polar than the uncomplexed analytes. The more stable silver complexes will have the higher capacity factors.

Schuring has investigated the use of dicarbonyl rhodium (I)-3-(trifluoroacetyl)-IR-camphorate and the more thermally stable bis-3-(perfluoroacyl)-IR-camphorates of Mn(II), Co(II) and Ni(II) dissolved in a noncoordinating stationary phase, such as squalane or poly(dimethylsiloxane) for the separation of a wide range of stereoisomers of hydrocarbons and oxygen, nitrogen and sulfur-containing electron-donor solutes, such as cyclic ethers, 1-chloroaziridines, thiranes, thiethanes, ketones and aliphatic alcohols by gas chromatography [817-819]. Although the camphorates have been the most widely evaluated other terpeneketonates can be used as the complexing ligand to vary the properties of the coordinating complex [819]. Selective retention results from the fast and reversible chemical interaction of the solute and the metal coordination compound, which gives rise to increased retention that is linearly related to the chemical equilibrium constant for molecular association and the concentration of the metal complex dissolved in the liquid phase. Thermal instability of the coordination complexes (70-110°C) restricts the method to low molecular weight stereoisomers and hydrogen carrier gas causes deterioration of the columns. Metal coordination complexes are widely used for the resolution of volatile insect pheromones, including the determination of enantiomeric composition. Figure 8.43 shows an example of the separation of the enantiomers of

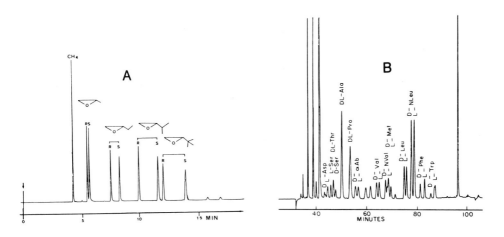

Figure 8.43 Separation of enantiomers using complexation chromatography. A, Separation of alkyloxiranes on a 42 m x 0.25 mm I.D.
open tubular column coated with 0.06 M Mn(II) bis-3-(pentafluoro-
propionyl)-IR-camphorate in OV-101 at 40°C. B, Separation of D,L-
amino acids by reversed-phase liquid chromatography using a mobile
phase containing 0.005 M L-histidine methyl ester and 0.0025 M
copper sulfate in an ammonium acetate buffer at pH 5.5. A stepwise
gradient using increasing amounts of acetonitrile was used for
this separation.

alkyloxiranes using an open tubular column coated with a mixture
of a poly (dimethylsiloxane) phase and an optically active metal
complex.

Ligand-exchange chromatography has been used to resolve
enantiomers of amino acids and their derivatives and some α-
hydroxycarboxylic acids and compounds with an imide structure
based on their ability to form mixed diastereomeric complexes with
transition metal ions such as copper, zinc, nickel and cadmium
[721,723,820-823]. The resolving agent is usually a transition
metal complex containing two molecules of a single

enantiomerically pure amino acid, such as proline, hydroxyproline, or histidine methyl ester, which can rapidly equilibrate with a racemic mixture of another amino acid or suitable ligand to give a mixture of diastereomeric complexes by replacing one of the previously chelated amino acid ligands. The diastereomeric complexes formed in the above process can be separated based on their different stabilities if the copper complex is bonded to a polymer or silica gel support or by their different solubilities and sorption properties in a reversed-phase chromatographic system if the copper complex is added to the mobile phase. Ligand-exchange chromatography is the preferred method of separating amino acid enantiomers by liquid chromatography, Figure 8.43 [824], but its more general application is limited to those substances capable of forming bidentate complexes with transition metal ions. In related studies Grushka used a bonded dithiocarbamate phase on silica to separate nucleotides and nucleosides in the presence of Mg(II) ions [825]. The diphosphate and triphosphate groups of the nucleotides bind Mg(II) very strongly and compete with the dithiocarbamate ligand of the stationary phase for the metal cation. The retention of the nucleotides was found to vary inversely with the magnesium ion concentration in the mobile phase and also exhibited a strong pH dependence.

8.15.4 Separation of Stereoisomers Using Liquid Crystal Stationary Phases

Liquid crystal stationary phases are widely used in gas chromatography for the separation of positional and geometric isomers of rigid molecules, such as substituted aromatic hydrocarbons, polychlorinated biphenyls, dibenzodioxins and steroids [826-829]. The liquid crystalline state represents a specific state of matter intermediate between a crystalline solid and an isotropic liquid. In the liquid crystalline state the phase exhibits the mechanical properties characteristic of a liquid while maintaining some of the anisotropic properties of the solid due to the preservation of a higher degree of order than that associated with isotropic liquids. The liquid crystals used as stationary phases are of the thermotropic type, that is, the liquid crystalline state commences at the melt and is thermally stable until, at some higher temperature (clearing temperature), a transition to an isotropic liquid takes place. The temperature

TABLE 8.25

MOST COMMON STRUCTURES OF LIQUID CRYSTALLINE STATIONARY PHASES
$Y_1(C_6H_4)_nX(C_6H_4)_nY_2$ with n = 1 or more. Y and X are usually
para substituents.

X	Y_1 and Y_2
$(-CH_2-CH_2)_n$	H
$(-CH=CH-)_n$	H
$-OCH_2CH_2O-$	R
$(-OCH_2-)_n$	RO
$-CH=N-N=CH-$	CN
$-C\equiv C-$	Cl, Br, F
$-NC-$	RCOO-
$-NN(O)-$	$RO(CH_2)_nO-$
$-CHN(O)-$	$(CH_3)_2N-$
$-COO-$	R = alkyl or aryl

difference between the melting point and the clearing temperature
determines the mesophase range. Depending on the way the molecules
are ordered, thermotropic liquid crystals are classified into
nematic, cholesteric and smectic types. The mesophase range can be
a single type or several types of liquid crystalline states. The
most useful liquid crystal phases for gas chromatography are
generally of the nematic or smectic type. Smectic phases possess a
two-dimensional structure in which rod-like molecules are arranged
in parallel to give a layered structure whose thickness is
approximately equal to the length of the molecules comprising the
layer. In the nematic phase the parallel orientation is maintained
but the layered structure does not exist and the molecules are
free to move within the limits of the parallel configuration.
Eight types of smectic phases are known but only those possessing
the lowest degree of order are suitable for gas chromatography.

Well over two hundred liquid crystalline phases have been
used in gas chromatography. They are of a variety of chemical
types but all have in common a markedly elongated, rigid, rod-like
structure and generally have polar terminal groups. The most
common types are Schiff bases, esters and azo and azoxy compounds,
as shown in Table 8.25 . In general, their high viscosity and poor
mass transfer characteristics result in column packings of lower
efficiency than those of conventional liquid phases [830].
Retention volumes may vary unpredictably with the amount of liquid
phase and the type of support used [830,831]. Reproducible

retention data can be expected only with high loadings on low surface area supports. At low loadings most of the stationary phase is present as a surface film and solute retention occurs largely by interfacial adsorption. As the amount of liquid phase increases solute partitioning with the bulk liquid phase dominates the retention process. Similar observations have been made with open tubular columns where film thickness and surface preparation procedures strongly influence selectivity and efficiency of the columns [832]. Binary and ternary mixtures of liquid crystal stationary phases often have superior properties in terms of their mesophase range and column efficiencies [833]. The selectivity of liquid crystal stationary phases is often highest at low column temperatures in the supercooled region for single phases or eutectic region for mixed liquid crystals. Column efficiency, however, may show the opposite behavior. New phases with a polysiloxane or hydrocarbon backbone containing liquid crystal functional groups provide higher efficiency and a wider working temperature range, as well as being compatible with modern technology for preparing immobilized phases on fused silica open tubular columns [834-838]. Columns coated with one of these phases, a smectic biphenyl carboxylate ester liquid crystal polysiloxane stationary phase, are now commercially available. This phase seems to be reasonably optimum for the separation of polycyclic aromatic compounds and dibenzodioxin isomers, for example, see Figure 8.44 [839]. Columns are stable to about 280°C and the mesophase range extends from about 100 to 300°C.

In most cases the order of elution for a series of isomers on liquid crystalline stationary phases is generally in accord with the solute length-to-breadth ratios with differences in vapor pressure and solute polarity also being of importance in some cases, leading to an inversion of elution order to that predicted from length-to-breadth ratios [828,829,838]. Long and planar molecules fit better into the ordered structure of the liquid crystal phase whereas nonlinear and nonplanar molecules do not permeate so easily between the liquid crystal molecules of the stationary phase and are more easily eluted from the column.

8.16 PYROLYSIS GAS CHROMATOGRAPHY

Pyrolysis gas chromatography is an indirect method of analysis in which heat is used to transform a sample into a series of volatile products characteristic of the sample and the

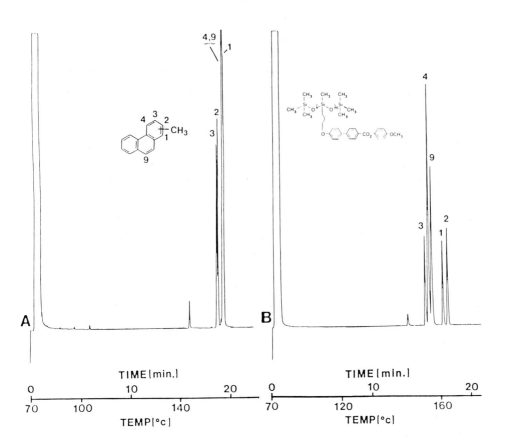

Figure 8.44 Separation of a mixture of methylphenanthrene isomers on open tubular columns coated with the nonselective poly(methyl-phenylsiloxane) SE-54 stationary phase (A) and a smectic biphenyl carboxylate ester liquid crystal polysiloxane stationary phase (B).

experimental conditions [840-843]. The initiation reactions involved in the pyrolysis of most organic materials are accounted for by the generation of free radicals from the cleavage of single bonds or by the unimolecular elimination of simple molecules such as water or carbon dioxide [842,844]. The subsequent reactions of these species via abstraction or combination reactions, or by diffusion, produce products characteristic of the original sample. The products of pyrolysis are identified by gas chromatography, often in combination with mass spectrometry, when more information than a comparative fingerprint (pyrogram) of the sample is required. Samples most commonly analyzed by pyrolysis gas chromatography are complex and polymeric substances, often of variable structure, and are frequently highly polar and nonvolatile. Intractable samples, such as rubbers, polymers, sediments and bacteria, difficult to analyze by conventional gas chromatographic methods, are natural choices for this technique [841,842,845,846]. However, the application of pyrolysis gas chromatography is not limited to such samples alone; small molecules, such as drugs in biological fluids and solvents and monomers entrained in polymers can also be determined by pyrolysis techniques.

Most pyrolysis devices are of the continuous-mode or pulse-mode type with laser disintegration being reserved for special applications [841,842,846-850]. Continuous mode pyrolyzers include a wide range of resistively heated tube furnaces that can be operated on- or off-line; the latter with solvent or cryogenic trapping to collect the pyrolyzate for subsequent injection into the gas chromatograph. Pyrolysis is normally performed by inserting the sample into a continuously-heated zone. The sample may be supported on a boat and dropped or pushed into the heated zone, or is encapsulated in a quartz tube which, after pyrolysis, is broken to release the products. The principal disadvantage of continuous pyrolysis is that the wall temperature of the microfurnace is usually much higher than the temperature of the sample, and heat transfer is relatively slow. The pyrolysis products, therefore, migrate from the sample to hotter regions of the pyrolyzer where additional fragmentation reactions may be initiated to produce secondary products. This possibility is increased by the migration of the products through the residual sample before being released into the pyrolyzer. It will be recognized later that this process differs from that occurring in pulse-mode pyrolyzers and may result in either the formation of different products or the formation of the same products, but with a different distribution between components.

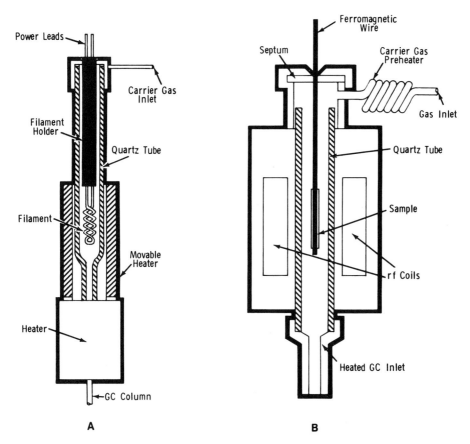

Figure 8.45 Apparatus for pyrolysis gas chromatography. A,
filament or ribbon-type pyrolyzer and B, Curie-point pyrolyzer.
(Reproduced with permission from ref. 848. Copyright American
Chemical Society).

Pulse-mode pyrolyzers include resistively-heated electrical
filaments or ribbons and radio frequency induction-heated wires
[841,842,846,848,849]. The filament or ribbon-type pyrolyzers are
simple to construct, Figure 8.45, and typically consist of an
inert wire or ribbon (Pt or Pt-Rh alloy) connected to a high-
current power supply. Samples soluble in a volatile solvent are
applied to the filament as a thin film. Insoluble materials are
placed in a crucible or quartz tube, heated by a basket-like
shaped or helical wound filament. The coated filament is contained
within a low dead volume chamber through which the carrier gas
flows, sweeping the pyrolysis products onto the column. The
surface temperature of the filament is raised rapidly from ambient
temperature to the equilibrium pyrolysis temperature. This

temperature is maintained for a preset time interval, the power source removed, and the filament allowed to cool. The filament pyrolyzer may also be temperature programmed, either continuously or in steps, to enhance the diagnostic information of the pyrogram. An inductively-heated, Curie-point pyrolyzer is shown in Figure 8.45. The sample, as a thin-film coated on a ferromagnetic wire, a solid contained within a recess forged into the wire, or a solid pressed onto the wire in a hydraulic press (similar to devices used to prepare pellet samples for IR analysis) [851], is heated inductively in a radio frequency field. The eddy currents induced in the wire conductor cause its temperature to rise rapidly until a specific temperature is reached, its Curie point, at which temperature ferromagnetic materials become paramagnetic and the absorption of energy ceases. The temperature then stabilizes. This technique provides a highly reproducible pyrolysis temperature, since the Curie-point temperature of a ferromagnetic material depends only on the composition of the alloy. By selecting wires of different composition temperatures in the range 300 to 1000°C can be obtained.

Since in the pulse-mode pyrolyzers the sample and heat source are in intimate contact, heat transfer is rapid and temperature gradients within the sample should be absent. The primary pyrolysis products are quenched as they rapidly expand away into the cooler regions, diminishing the possibility of forming secondary products. Heating rates are rapid (milliseconds), reproducible, and controllable by adjusting the filament-heating current or alloy composition and radio frequency energy of the Curie-point pyrolyzer. The reproducibility of individual pyrograms depends on the sample size and homogeneity, the temperature profile of the pyrolyzer, and the maintenance of uniform conditions for transferring the pyrolyzates to the column [841,842,846,852]. The smaller the sample the more likely it will pyrolyze in a uniform and reproducible manner. The sample, when possible, should be applied as a uniform, thin film from solution to provide a final sample size of 5 to 100 micrograms after solvent evaporation. The temperature profile, rise time and equilibrium temperature, can be a major cause of poor reproducibility, not only from one type of pyrolyzer to another, but also from laboratory to laboratory using the same type of pyrolyzer [846,853]. Pyrolysis of Kraton 1107 has been suggested as a model molecular thermometer to standardize the equilibrium temperature of pulse-mode pyrolyzers. Kraton 1107 is a copolymer of styrene-isoprene that pyrolyzes to produce isoprene and dipentene, among other products, the ratio of which correlates with the actual pyrolysis temperature experienced by the sample. This calibration curve can be used to adjust the pyrolyzer set point temperature to the actual temperature experienced by a sample. Curie-point pyrolyzers may suffer from small temperature variations due to differences in alloy composition of the wire conductor, sample loading and positioning of the conductor within the induction coil, and fluctuations in the radio frequency energy. As an inlet for a gas chromatograph, the pyrolyzer must conform to the general requirement of introducing the sample to the column as a narrow band. Because of the large swept volume of the pyrolyzer and the fact that the pyrolysis products are produced over a finite time, the direct coupling of the pyrolyzer to open tubular column gas chromatography raises some additional problems [846,854-856]. One solution is a split-type interface in which a flow rate of 20-100 ml/min of gas through the pyrolyzer is used to effectively sweep the sample rapidly to the column but only a small portion (1-5%) of this flow is allowed to enter the column, the remainder being vented. This approach limits the sensitivity of the method. Sorbent trapping or cryogenic trapping of the pyrolyzate for subsequent reinjection by flash vaporization is a more suitable technique. This approach maximizes sensitivity and improves resolution, particularly for the more volatile pyrolysis products eluting early in the chromatogram.

The statistical analysis of pyrograms requires a high reproducibility of the whole analytical method, and for the more complex pyrograms, chemometric techniques are used to establish diagnostic differences between samples [842,857-859]. The use of mass fragmentography to record the pyrogram with a restricted number of m/z values can considerably simplify the features of complex pyrograms without loss of diagnostic information [859,860]. Also, in taxonomic identification of bacteria and fungi, it is of paramount importance that the organism always exhibits the same chemical characteristics. Factors such as sub-culturing, the growth time and the medium upon which the organism is grown will affect the reproducibility of the pyrogram.

922

8.17 REFERENCES

1. J. F. Lawrence, "Organic Trace Analysis By Liquid Chromatography", Academic Press, New York, NY, 1981.
2. R. L. Grob and M. A. Kaiser, "Environmental Problem Solving Using Gas and Liquid Chromatography", Elsevier, Amsterdam, 1982.
3. W. G. Jennings and A. Rapp, "Sample Preparation for Gas Chromatographic Analysis", Huethig, Heidelberg, 1983.
4. C. F. Poole and S. A. Schuette, J. High Resolut. Chromatogr., 6 (1983) 526.
5. K. Beyerman, "Organic Trace Analysis", Ellis Horwood, Chichester, UK, 1984.
6. F. I. Onuska and F. W. Karasek, " Open Tubular Column Gas Chromatography in Environmental Sciences", Plenum Press, New York, NY, 1984.
7. J. Chamberlain, "Analysis of Drugs in Biological Fluids", CRC Press, Boca Raton, FL., 1985.
8. R. G. Metcher, T. L. Peters, and H. W. Emmel, Topics Curr. Chem., 134 (1986) 59.
9. J. Namiesnik, Talanta, 35 (1988) 567.
10. J. Namiesnik, T. Gorecki, and M. Biziuk, Anal. Chim., Acta 237 (1990) 1.
11. F. I. Onuska, J. High Resolut. Chromatogr., 12 (1989) 4.
12. R. D. McDowall, J. Chromatogr., 492 (1989) 3.
13. S. K. Poole, T. A. Dean, J. W. Oudsema, and C. F. Poole, Anal. Chim. Acta, 236 (1990) 3.
14. K. G. Furton and J. Rein, Anal. Chim. Acta, 236 (1990) 99.
15. R. S. Smith and G. V. James, "The Sampling of Bulk Materials", The Royal Society of Chemistry, London, UK, 1981.
16. B. Kratochvil and J. K. Taylor, Anal. Chem., 53 (1981) 924A.
17. G. Kateman and F. W. Pijpers, "Quality Control in Analytical Chemistry", Wiley, New York, NY, 1981.
18. C. J. Kirchner, Environ. Sci. Technol., 17 (1983) 174A.
19. J. K. Taylor, Trends Anal. Chem., 5 (1986) 121.
20. J. K. Taylor, "Quality Assurance of Chemical Measurements", Lewis Publishers, Chelsea, MI, 1987.
21. L. H. Keith (Ed.), "Principles of Environmental Sampling", American Chemical Society, Washington, DC, 1988.
22. D. Dadgar and M. R. Smyth, Trends Anal. Chem., 5 (1986) 115.
23. E. L. Inman, J. K. Frischmann, P. J. Jimenez, G. D. Winkel, M. L. Persinger, and B. S. Rutherford, J. Chromatogr. Sci., 25 (1987) 252.
24. C. Weurman, J. Agric. Food Chem., 17 (1969) 370.
25. R. F. Suckow, J. Liq. Chromatogr., 10 (1987) 293.
26. R. L. Jolley, Environ. Sci. Technol., 15 (1981) 874.
27. T. S. Ma and V. Horak, "Microscale Manipulations in Chemistry", Wiley, New York, NY, 1976.
28. R. Teranishi, R. Flath and H. Sugisawa , "Flavor Research: Recent Advances", Dekker, New York, NY, 1981.
29. P. C. Wankat, "Equilibrium Stage Separations", Elsevier, Amsterdam, 1988.
30. D. W. Gruenwedel and J. R. Whitaker, "Food Analysis. Principals and Techniques", Dekker, New York, NY, Vol. 4, 1987.
31. K. D. Dix and J. S. Fritz, J. Chromatogr., 408 (1987) 201.
32. K. D. Dix and J. S. Fritz, Anal. Chim. Acta, 236 (1990) 43.
33. K. R. Kim, A. Zlatkis, J.-W. Park, and U. C. Lee, Chromatographia, 15 (1982) 559.
34. J. J. Richard and G. A. Junk, Anal. Chem., 56 (1984) 1625.

35. D. Page, W. H. Newsome, and S. B. MacDonald, J. Assoc. Off. Anal. Chem., 70 (1987) 446.
36. C. Bicchi, A. D. Amato, G. M. Nano, and C. Frattini, J. Chromatogr., 279 (1983) 409.
37. P. Donkin and S. V. Evans, Anal. Chim. Acta, 156 (1984) 207.
38. F. I. Onuska and K. A. Terry, Anal. Chem., 57 (1985) 801.
39. T. L. Peters, Anal. Chem., 52 (1980) 211.
40. M. Godefroot, P. Sandra, and M. Verzele, J. Chromatogr., 203 (1981) 325.
41. M. Godefroot, M. Stechele, P. Sandra, and M. Verzele, J. High Resolut. Chromatogr., 5 (1982) 75.
42. G. P. Blanch, G. Reglero, M. Herraiz, and J. Tabera, J. Chromatogr. Sci., 29 (1991) 11.
43. J. Rijks, J. Curvers, T. Noy and C. Cramers, J. Chromatogr., 279 (1983) 395.
44. J. Curvers, T. Noiji, C. Cramers, and J. Rijks, Chromatographia, 19 (1985) 225.
45. A. B. Heath and R. R. Black, J. Assoc. Off. Anal. Chem., 70 (1987) 862.
46. R. L. Brown, C. L. Farmer, and R. G. Millar, J. Assoc. Off. Anal. Chem., 70 (1987) 442.
47. J. Mes and D. J. Davies, Intern. J. Environ. Anal. Chem., 19 (1985) 203.
48. B. G. Luke, J. C. Richards, and E. F. Dawes, J. Assoc. Off. Anal. Chem., 67 (1984) 295.
49. S. M. Walters, Anal. Chim. Acta, 236 (1990) 77.
50. S. K. Poole, T. A. Dean, and C. F. Poole, J. Chromatogr., 400 (1987) 323.
51. U. R. Stenberg and T. E. Alsberg, Anal. Chem., 53 (1981) 2067.
52. R. A. Chalmers and R. W. E. Watts, Analyst, 97 (1972) 224.
53. C. F. Poole, N. J. Evans, and D. G. Wibberley, J. Chromatogr., 136 (1977) 73.
54. S. A. Goldbith, L. Reynold, and W. W. Rothmayr (Eds.), "Freeze Drying and Advanced Food Technology". Academic Press, Orlando, FL., 1975.
55. J. Shapiro, Anal. Chem., 39 (1967) 280.
56. R. E. Kepner, S. Van Straten, and C. Weurman, J. Agric. Food Chem., 17 (1969) 1123.
57. T. D. Brock, "Membrane Filtration: A User's Guide and Reference Manual", Science Tech. Inc., Madison, WI, 1983.
58. T. N. Eisenberg and E. J. Middlebrooks, "Reverse Osmosis Treatment of Drinking Water", Butterworths, London, 1986.
59. J. N. Huckins, M. W. Tubergen, J. A. Lebo, R. W. Gale, and T. R. Schwartz, J. Assoc. Off. Anal. Chem., 73 (1990) 290.
60. D. C. Turnell and J. D. H. Cooper, J. Chromatogr., 395 (1987) 613.
61. M. M. L. Aerts, W. M. J. Beek, and U. A. Th. Brinkman, J. Chromatogr., 500 (1990) 453.
62. L. G. M. Th. Tuinstra, P. G. M. Kienhuis, W. A. Traag, M. M. L. Aerts, and W. M. J. Beek, J. High Resolut. Chromatogr., 12 (1989) 709.
63. D. C. Turnell and J. D. H. Cooper, J. Chromatogr., 492 (1989) 59.
64. A. C. Mehta, Trends Anal. Chem., 8 (1989) 107.
65. J. L. Bock and J. Ben-czra, Clin. Chem., 31 (1985) 1884.
66. B. S. Middelditch, "Analytical Artifacts. GC, MS, HPLC, TLC, and PC", Elsevier, Amsterdam, 1989.
67. W. R. Robins, Anal. Chem., 51 (1979) 1860.
68. G. Schill, H. Ehrsson, J. Vessman, and D. Westerlund, "Separation Methods for Drugs and Related Organic Compounds", Swedish Pharmaceutical Press, Stockholm, Sweden, 2nd. Edn., 1984.

69. A. M. Rustum, J. Chromatogr., 489 (1989) 345.
70. D. C. Leggett, T. F. Jenkins, and P. H. Miyares, Anal. Chem., 62 (1990) 1355.
71. D. C. Fenimore, C. M. Davis, J. H. Whitford, and C. A. Harrington, Anal. Chem., 48 (1976) 2289.
72. P. J. Apps, J. Chromatogr., 504 (1990) 21.
73. T. L. Peters, Anal. Chem., 54 (1982) 1913.
74. J. Czuczwa, C. Leuenberger, J. Tremp, W. Giger, and M. Ahel, J. Chromatogr., 403 (1987) 233.
75. M. C. Goldberg and E. R. Weiner, Anal. Chim. Acta, 115 (1980) 373.
76. M. Ahnoff and B. Josefsson, Anal. Chem., 46 (1974) 658.
77. M. Th. J. Hillebrand and R. F. Nolting, Trends Anal. Chem., 6 (1987) 74.
78. W. A. Hoffman, Anal. Chem., 50 (1978) 2158.
79. V. M. Buchar and A. K. Agrawal, Analyst, 106 (1981) 620.
80. K. Hostettmann, M. Hostettmann, and A. Marston, "Preparative Chromatography Techniques. Applications in Natural Product Isolation", Springer-Verlag, Berlin, 1986.
81. K. Hostettmann, Adv. Chromatogr., 21 (1983) 165.
82. N. B. Mandava and Y. Ito (Eds.), Countercurrent Chromatography. Theory and Practice", Dekker, New York, NY, 1988.
83. Y. Ito, Trends Anal. Chem., 5 (1986) 142.
84. A. Marston, C. Borel, and K. Hostettmann, J. Chromatogr., 450 (1988) 91.
85. W. D. Conway, "Countercurrent Chromatography. Apparatus, Theory and Applications", New York, NY, (1990).
86. K. Hostettmann and A. Marston, Anal. Chim. Acta, 236 (1990) 63.
87. D. M. Hughes, Anal. Chem., 55 (1983) 78.
88. V. Lopez-Avila, R. Northcutt, J. Onstot, M. Wickham, and S. Billets, Anal. Chem., 55 (1983) 881.
89. A. L. Alford-Stevens, W. L. Budde, and T. A. Bellar, Anal. Chem., 57 (1985) 2452.
90. G. A. Junk and J. J. Richard, Anal. Chem., 58 (1986) 962.
91. T. F. Jenkins and C. L. Grant, Anal. Chem., 59 (1987) 1228.
92. M. P. Coover, R. C. Sims, and W. Doucette, J. Assoc. Off. Anal. Chem., 70 (1987) 1018.
93. K. Ganzler, A. Salgo, and K. Valko, J. Chromatogr., 371 (1986) 299.
94. F. Mangani, A. Cappiello, G. Crescentini, F. Bruner, and L. Bonfanti, Anal. Chem., 59 (1987) 2066.
95. H. Matusiewicz, Anal. Chem., 54 (1982) 1909.
96. G. F. Griffin and E. A. DeLwiche, Anal. Chem., 54 (1982) 2616.
97. D. P. Ndiomu and C. F. Simpson, Anal. Chim. Acta, 213 (1988) 237.
98. G. M. Schneider, E. Stahl, and G. Wilke, "Extraction with Supercritical Gases", Verlag Chemie, Weinheim, 1980.
99. F. W. Karasek, R. E. Clement, and J. A. Sweetman, Anal. Chem., 53 (1981) 1050A.
100. D. C. Constable, S. R. Smith, and J. Tanaka, Environ. Sci. Technol., 18 (1984) 975.
101. E. A. Ibrahim, I. H. Suffet, and A. B. Sakla, Anal. Chem, 59 (1987) 2091.
102. G. D. Price and D. A. Carlson, Anal. Chem., 53 (1981) 554.
103. C. E. Higgins and M. R. Guerin, Anal. Chem., 52 (1980) 1984.
104. A. E. Woolaver and D. W. Grund, J. Chromatogr., 388 (1987) 243.
105. A. Zlatkis, C. F. Poole, R. S. Brazell, K. Y. Lee, F. Hsu, and S. Singhawangcha, Analyst, 106 (1981) 352.

106. T. J. Nestrick and L. L. Lamparski, Anal. Chem., 53 (1981) 122.
107. D. A. J. Murray, J. Chromatogr., 177 (1979) 135.
108. J. F. J. van Rensberg and A. J. Hasset, J. High Resolut. Chromatogr., 5 5 (1982) 574.
109. A. B. Attygale and E. D. Morgan, Anal. Chem., 58 (1986) 3054.
110. E. D. Morgan, Anal. Chim. Acta, 236 (1990) 227.
111. I. Peleg and S. Vromen, Chem. Ind., (1983) 615.
112. W. Dunges, "Pre-Chromatographic Micromethods", Huethig Heidelberg, 1987.
113. C. K. Huynh and T. Vu Duc, J. High Reslout. Chromatogr., 8 (1985) 198.
114. S. Blomberg, and J. Roeraade, Chromatographia, 25 (1988) 21.
115. E. D. Morgan and C. F. Poole, J. Insect Physiol., 22 (1976) 885.
116. E. D. Morgan and C. F. Poole, Adv. Insect Physiol., 12 (1976) 17.
117. J. A. F. DeSilva, J. Chromatogr., 273 (1983) 19.
118. K. G. Wahlund and T. Arvidson, J. Chromatogr., 282 (1983) 527.
119. L. L. Ng, J. Chromatogr., 257 (1983) 345.
120. S. Lam and G. Malikin, J. Liq. Chromatogr., 12 (1989) 1851.
121. A. M. Rustum, J. Chromatogr. Sci., 27 (1989) 18.
122. S. G. Colgrove and H. J. Svec, Anal. Chem., 53 (1981) 1737.
123. D. H. Wheeler, Chem. Revs., 62 (1962) 205.
124. D. F. S. Natusch and B. A. Tomkins, Anal. Chem., 50 (1978) 1429.
125. D. Schuetzle and J. Lewtas, Anal. Chem., 58 (1986) 1060A.
126. W. W. Christie, "HPLC and Lipids. A Practical Guide", Pergamon Press, Oxford, UK, 1987.
127. D. W. Later, B. W. Wilson, and M. L. Lee, Anal. Chem., 57 (1985) 2979.
128. D. E. Wells and S. J. Johnstone, J. Chromatogr., 140 (1977) 17.
129. B. McMahon and J. A. Burke, J. Assoc. Off. Anal. Chem., 61 (1978) 640.
130. R. C. Chapman and C. R. Harris, J. Chromatogr., 166 (1978) 513.
131. P. A. Mills, J. Assoc. Off. Anal. Chem., 51 (1968) 29.
132. T. J. Nestrick and L. L. Lamparski, Anal. Chem., 54 (1982) 2292.
133. M. Lienne, P. Gareil, R. Rosset, J. F. Husson, M. Emmelin, and B. Neff, J. Chromatogr., 395 (1987) 255.
134. A. Di Muccio, A. Santilio, R. Dommarco, M. Rizzica, L. Gambetti, A. Ausili, and F. Vergori, J. Chromatogr., 513 (1990) 333.
135. S. Forbes, Anal. Chim. Acta, 196 (1987) 75.
136. A. M. Gillespie and S. M. Walters, J. Liq. Chromatogr., 9 (1986) 2111.
137. S. C. Ruckmick and R. J. Hurtubise, J. Chromatogr., 321 (1985) 343.
138. G. Petrick, D. E. Schulz, and J. C. Duinker, J. Chromatogr., 435 (1988) 241.
139. P. Ciccioli, E. Brancaleoni, A. Cecinato, C. Di Palo, P. Buttini, and A. Liberti, J. Chromatogr., 351 (1986) 451.
140. D. Schuetzle, T. L. Riley, T. J. Prater, T. M. Harver, and D. F. Hunt, Anal. Chem., 54 (1982) 265.
141. S. A. Wise, B. A. Benner, S. N. Chesler, L. R. Hilpert, C. R. Vogt, and W. E. May, Anal. Chem., 58 (1986) 3067.
142. A. H. Roos, A. J. Van Munsteren, F. M. Nab, and L. G. M. Tuinstra, Anal. Chim. Acta, 196 (1987) 95.

926

143. J. F. Lawrence, Intern. J. Environ. Anal. Chem., 29 (1987) 289.
144. S. J. Chamberlain, Analyst, 115 (1990) 1161.
145. R. D. McDowall, J. C. Pearce, and G. S. Murkitt, Trends Anal. Chem., 8 (1989) 134.
146. I. Liska, J. Krupcik, and P. A. Leclerq, J. High Resolut. Chromatogr., 12 (1989) 577.
147. U. Juergens, J. Chromatogr., 371 (1986) 307.
148. H. G. Fouda, J. Chromatogr., 492 (1989) 85.
149. R. C. Williams, Intern. J. Environ. Anal. Chem., 18 (1984) 37.
150. Y. A. Delmotte, J. High Resolut. Chromatogr., 8 (1985) 858.
151. V. Marko, L. Soites, and K. Radova, J. Chromatogr. Sci., 28 (1990) 403.
152. G. A. Junk, M. J. Avery, and J. J. Richard, Anal. Chem., 60 (1988) 1347.
153. M. Tetsuo, H. Ericksson, and J. Sjovall, J. Chromatogr., 239 (1982) 287.
154. L. O. G. Weidolf and J. D. Henion, Anal. Chem. 59 (1987) 1980.
155. S. Scalia and P. Pazzi, Chromatographia, 30 (1990) 377.
156. G. A. Junk and J. J. Richard, Anal. Chem., 60 (1988) 451.
157. M. J. M. Wells and J. Michael, J. Chromatogr. Sci., 25 (1987) 345.
158. M. J. M. Wells, A. J. Rossano, and E. C. Roberts, Anal. Chim. Acta, 236 (1990) 31.
159. P. R. Loconto and A. K. Gaind, J. Chromatogr. Sci., 27 (1989) 569.
160. M. J. Shepherd, M. Holmes, and J. Gilbert, J. Chromatogr., 354 (1986) 305.
161. A. de Kok, M. Hienstra, and C. P. Vreeker, Chromatographia, 24 (1987) 469.
162. J. M. Vinuesa, J. C. M. Cortes, C. I. Canas and G. F. Perez, J. Chromatogr., 472 (1989) 365.
163. W. H. Newsome and P. Collins, J. Chromatogr., 472 (1989) 416.
164. V. Lopez-Avila, J. Milanes, N. S. Dodhiwaia, and W. F. Beckert, J. Chromatogr. Sci., 27 (1989) 209.
165. I. D. Wilson, E. D. Morgan, and S. J. Murphy, Anal. Chim. Acta, 236 (1990) 145.
166. P. Lovkvist and J. A. Jonsson, Anal. Chem., 59 (1987) 818.
167. D. F. Hagen, C. G. Markell, G. A. Schmitt, and D. B. Blevins, Anal. Chim. Acta, 236 (1990) 157.
168. E. R. Brouwer, H. Lingeman, and U. A. Th. Brinkman, Chromatographia, 29 (1990) 415.
169. F. Bruner, G. Furlani, and F. Mangani, J. Chromatogr., 302 (1984) 167.
170. A. Bacaloni, G. Goretti, A. Lagana, B. M. Petronio, and M. Rotatori, Anal. Chem., 52 (1980) 2033.
171. S. D. Faust and O. M. Aly, "Adsorption Processes for Water Treatment", Butterworths, Boston, 1987.
172. M. J. McGuire and I. H. Suffet, "Treatment of Water by Granular Activated Carbon", American Chemical Society, Washington, D.C., 1983.
173. J. W. Neely and E. C. Isacoff, "Carbonaceous Adsorbents for the Treatment of Ground and Surface Waters", Dekker, New York, NY, 1982.
174. R. W. Beulow, J. K. Carswell, and J. M. Symons, J. Amer. Water Works Assoc., 65 (1973) 195.
175. D. K. Chow and M. M. David, J. Amer. Water Works Assoc., 69 (1977) 555.
176. P. van Rossum and R. G. Webb, J. Chromatogr., 150 (1978) 381.
177. A. Tateda and J. S. Fritz, J. Chromatogr., 152 (1978) 329.
178. M. Dressler, J. Chromatogr., 165 (1979) 167.

179. K. A. Kunk and R. Kunin, J. Polymer Sci., Part A1, 6 (1968) 2689.
180. I. H. Suffet and M. Malaiyandi, (Eds.), "Organic Pollutants in Water. Sampling, Analysis, and Toxicity Testing", American Chemical Society, Washington, D.C., 1987.
181. G. R. Aiden, E. M. Thurman, R. L. Malcolm, and H. E. Walton, Anal. Chem., 51 (1979) 1799.
182. G. A. Junk, J. J. Richard, M. D. Grieser, D. Witiak, J. L. Witiak, M. D. Arguello, R. Vick, H. J. Svec, J. S. Fritz, and G. V. Calder, J. Chromatogr., 99 (1974) 745.
183. B. Wigilius, H. Boren, G. E. Carlberg, A. Grimvall, B. V. Lundgren, and R. Savenhed, J. Chromatogr., 391 (1987) 169.
184. J. F. Pankow, L. M. Isabelle, and T. Kristensen, J. Chromatogr., 245 (1982) 31.
185. C. Leuenberger and J. F. Pankow, Anal. Chem., 56 (1984) 2518.
186. M. Ishibashi and M. Suzuki, J. Chromatogr., 456 (1988) 382.
187. M. B. Yunker, F. A. McLaughlin, R. W. Macdonald, W. J. Cretney, B. R. Fowler, and T. A. Smyth, Anal. Chem., 61 (1989) 1333.
188. A. Przyjazny, W. Janicki, W. Chrzanowski, and R. Staszewski, J. Chromatogr., 245 (1982) 256.
189. G. J. Moody and J. D. R. Thomas, Analyst, 104 (1979) 1.
190. G. J. Moody and J. D. R. Thomas, "Chromatographic Separation and Extraction with Foamed Plastic and Rubbers", Dekker, New York, 1982.
191. M. Tetsuo, H. Ericksson, and J. Sjovall, J. Chromatogr., 239 (1982) 287.
192. J. J. Richard and J. S. Fritz, J. Chromatogr. Sci., 18 (1980) 35.
193. J. J. Richard, C. D. Criswell, and J. S. Fritz, J. Chromatogr., 199 (1980) 143.
194. A. Rehman, S. C. Gates, and J. W. Webb, J. Chromatogr., 228 (1982) 103.
195. M. G. Stracham and R. B. Johns, Anal. Chem., 58 (1986) 312.
196. M. Giabbai, L. Roland, M. Ghosal, J. H. Reuter, and E. S. K. Chian, J. Chromatogr., 279 (1983) 373.
197. W. Bertsch, J. High Resolut. Chromatogr., 1 (1978) 85, 187, and 289.
198. G. Schomburg, F. Weeke, F. Muller, and M. Oreans, Chromatographia, 16 (1982) 87.
199. S. Berg and A. Jonsson, J. High Resolut. Chromatogr., 7 (1984) 687.
200. B. M. Gordon, C. E. Rix, and M. F. Borgerding, J. Chromatogr. Sci., 23 (1985) 1.
201. D. E. Willis, Adv. Chromatogr., 28 (1989) 65.
202. G. L. Johnson, A. Tipler, and D. Crowshaw, J. High Resolut. Chromatogr., 13 (1990) 130.
203. H. J. Cortes (Ed.), "Multidimensional Chromatography: Techniques and Applications", Dekker, New York, NY, 1990.
204. D. R. Deans, J. Chromatogr., 203 (1981) 19.
205. R. J. Philips, K. A. Knauss, and R. R. Freeman, J. High Resolut. Chromatogr., 5 (1982) 546
206. S. Wu, W. H. Chatham, and S. O. Farwell, J. High Resolut. Chromatogr., 13 (1990) 229.
207. J. J. Szakasits and R. E. Robinson, Anal. Chem., 63 (1991) 114.
208. J. C. M. Wessels and R. P. M. Dooper, J. Chromatogr., 279 (1983) 349.
209. W. Jennings, J. Chromatogr. Sci., 22 (1984) 129.
210. S. M. Sonchik, J. Chromatogr. Sci., 22 (1986) 22.
211. S. T. Adam, J. High Resolut. Chromatogr., 11 (1988) 85.

928

212. Z. Naizhong and L. E. Green, J. High Resolut. Chromatogr., 9 (1986) 400.
213. H. Neumann and H.-P. Meyer, J. Chromatogr., 391 (1987) 442.
214. M. Oreans, F. Muller, and D. Leonhardt, J. Chromatogr., 279 (1983) 357.
215. S. R. Springston, J. Chromatogr., 517 (1990) 67.
216. W. Frank and H. Frank, Chromatographia, 29 (1990) 571.
217. H. Tani and M. Furuno, J. High Resolut. Chromatogr., 9 (1986) 712.
218. L. G. M. Th. Tunistra, W. A. Traag, A. J. Van Munsteren, and V. Van Hese, J. Chromatogr., 395 (1987) 307.
219. G. Schomburg, H. Husmann, and E. Hubinger, J. High Resolut. Chromatogr., 8 (1985) 395.
220. M. Ahnoff, M. Ervik, and L. Johansson, J. Chromatogr., 394 (1987) 419.
221. H. J. Stan and D. Mrowetz, J. Chromatogr., 279 (1983) 173.
222. F. R. Guenther, S. N. Chesler, and R. M. Parris, J. Chromatogr., 363 (1986) 199.
223. I. L. Davies, M. W. Raynor, J. P. Kithinji, K. D. Bartle, and P. T. Williams, Anal. Chem., 60 (1988) 683A.
224. K. Grob, Trends Anal. Chem., 8 (1989) 162.
225. I. L. Davies, K. E. Markides, M. L. Lee, M. W. Raynor, and K. D. Bartle, J. High Resolut. Chromatogr., 12 (1989) 193.
226. F. Munari and K. Grob, J. Chromatogr. Sci., 28 (1990) 61.
227. H. J. Cortes, E. L. Oliberding, and J. H. Wetters, Anal. Chem. Acta, 236 (1990) 173.
228. K. Grob and B. Schilling, J. High Resolut. Chromatogr., 8 (1985) 726.
229. J. P. Apffel and H. McNair, J. Chromatogr., 279 (1983) 139.
230. C. V. Philip and R. G. Anthony, J. Chromatogr. Sci., 24 (1986) 438.
231. T. V. Raglione, J. Troskosky, and R. Hartwick, J. Chromatogr., 409 (1987) 213.
232. G. Pacciarelli, E. Muller, R. Schneider, K. Grob, W. Steiner, and D. Fronhlich, J. High Resolut. Chromatogr., 11 (1988) 135.
233. F. Munari and K. Grob, J. High Resolut. Chromatogr., 11 (1988) 172.
234. I. L. Davies, M. W. Raynor, P. T. Williams, G. E. Andrews, and K. D. Bartle, Anal. Chem., 59 (1987) 2579.
235. I. L. Davies, K. D. Bartle, G. E. Andrews, and P. T. Williams, J. Chromatogr. Sci., 26 (1988) 125.
236. K. Grob, H. G. Schmarr, and A. Mosandi, J. High Resolut. Chromatogr., 12 (1989) 375.
237. M. Biedermann, K. Grob, and W. Meier, J. High Resolut. Chromatogr., 12 (1989) 591.
238. H.-G. Schmarr, A. Mosandi, and K. Grob, J. High Resolut. Chromatogr., 12 (1989) 721.
239. K. Grob, J. Chromatogr., 477 (1989) 73.
240. F. Munari, G. Dugo, and A. Cotroneo, J. High Resolut. Chromatogr., 13 (1990) 56.
241. E. Dolecka, J. J. Vreuls, G. J. de Jong, U. A. Th. Brinkman, and F. A. Maris, J. High Resolut. Chromatogr., 13 (1990) 405.
242. K. Grob, M. Lanfranchi, and C. Mariani, J. Chromatogr., 471 (1989) 397.
243. Th. Noy, E. Weiss, T. Herps, H. Van Cruchten, and J. Rijks, J. High Resolut. Chromatogr., 11 (1988) 181.
244. E. Noroozian, F. A. Maris, M. W. F. Nielen, R. W. Frei, G. J. de Jong, and U. A. Th. Brinkman, J. High Resolut. Chromatogr., 10 (1987) 17.
245. K. Grob and Z. Li, J. Chromatogr., 437 (1989) 381, 391, 401, 411, and 423.

246. M. McHugh and V. Krukonis, "Supercritical Fluid Extraction: Principles and Practice", Butterworth, Boston, MA, 1986.
247. W. G. Jennings, J. High Resolut. Chromatogr., 2 (1979) 221.
248. K. K. Unger and P. Roumeliotis, J. Chromatogr., 282 (1983) 519.
249. U. Keller and I. Flament, Chromatographia, 28 (1989) 445.
250. S. B. Hawthorne, D. J. Miller, and M. S. Krieger, J. High Resolut. Chromatogr., 12 (1989) 714.
251. J. M. Levy, R. A. Cavalier, T. N. Bosch, A. F. Rynaski, and W. E. Huhak, J. Chromatogr. Sci., 27 (1989) 341.
252. S. B. Hawthorne, D. J. Miller, and M. S. Krieger, J. Chromatogr. Sci., 27 (1989) 347.
253. J. W. King, J. Chromatogr. Sci., 28 (1990) 9.
254. S. B. Hawthorne, Anal. Chem., 62 (1990) 633A.
255. J. M. Levy, A. C. Rosselli, D. S. Boyer, and K. Cross, J. High Resolut. Chromatogr., 13 (1990) 418.
256. M. Lohleit and K. Bachmann, J. Chromatogr., 505 (1990) 227.
257. M. L. Lee and K. E. Markides (Eds.), "Analytical Supercritical Fluid Chromatography and Extraction", Chromatography Conferences, Inc., Provo, UT, 1990.
258. J. W. King, J. Chromatogr. Sci., 27 (1989) 355.
259. F. I. Onuska and K. A. Terry, J. High Resolut. Chromatogr., 12 (1989) 357.
260. M. R. Anderson, J. T. Swanson, N. L. Porter, and B. E. Richter, J. Chromatogr. Sci., 27 (1989) 371.
261. P. Sandra, F. David, and E. Stottmeister, J. High Resolut. Chromatogr., 13 (1990) 284.
262. J. T. Swanson and B. E. Richter, J. High Resolut. Chromatogr., 13 (1990) 385.
263. S. B. Hawthorne, M. S. Krieger, and D. J. Miller, Anal. Chem., 60 (1988) 472.
264. B. W. Wright, S. R. Frye, D. G. McMinn, and R. D. Smith, Anal. Chem., 59 (1987) 640.
265. J. M. Levy and A. C. Rosselli, Chromatographia, 28 (1989) 613.
266. S. B. Hawthorne, D. J. Miller, and J. J. Langenfeld, J. Chromatogr. Sci., 28 (1990) 2.
267. J. H. Raymer and E. D. Pellizzari, Anal. Chem., 59 (1987) 1043.
268. K. Sugiyama, M. Saito, T. Hondo, and M. Senda, J. Chromatogr., 332 (1985) 107.
269. E. Engelhardt and A Gross, J. High Resolut. Chromatogr., 11 (1988) 38.
270. M. E. P. McNally and J. R. Wheeler, J. Chromatogr., 435 (1988) 63.
271. M. Ashraf-Khorassani and J. M. Levy, J. High Resolut. Chromatogr., 13 (1990) 742.
272. M. Ashraf-Khorassani, M. L. Kumar, D. J. Koebler, and G. P. Williams, J. Chromatogr. Sci., 28 (1990) 599.
273. T. A. Dean and C. F. Poole, J. High Resolut. Chromatogr., 12 (1989) 773.
274. R. M. Campbell and M. L. Lee, Anal. Chem., 58 (1986) 2247.
275. Z. Juvancz, K. M. Payne, K. E. Markides and M. L. Lee, Anal. Chem., 62 (1990) 1384.
276. D. Thiebaut, J.-P. Chervet, R. W. Vannoort, G. J. De Jong, U. A. Th. Brinkman and R. W. Frei, J. Chromatogr., 477 (1989) 151.
277. J. L. Hedrick and L. T. Taylor, J. High Resolut. Chromatogr., 13 (1990) 312.
278. J. W. King, J. H. Johnsson, and J. P. Friedrich, J. Agric. Food Chem., 37 (1989) 951.

279. B. Zygmunt, J. Visser, U. A. Th. Brinkman, and R. W. Frei, Intern. J. Environ. Anal. Chem., 15 (1983) 263.
280. J. F. Lawrence (Ed.), "Liquid Chromatography in Environmental Analysis", Humana Press, Clifton, NJ, 1983.
281. C. J. Little, D. J. Tompkins, O. Stahel, R. W. Frei, and C. E. Werkhoven-Goewie, J. Chromatogr., 264 (1983) 183.
282. C. J. Little, O. Stahel, and K. Hales, Intern. J. Environ. Anal. Chem., 18 (1984) 11.
283. C. J. Little and O. Stahel, Chromatographia, 19 (1984) 322.
284. S. A. Berkowitz, J. Liq. Chromatogr., 10 (1987) 2771.
285. D. Westerlund, Chromatographia, 24 (1987) 155.
286. R. W. Frei and K. Zech (Eds.), "Selective Sample Handling and Detection in HPLC. Part A", Elsevier, Amsterdam, 1988.
287. K. A. Ramsteiner, J. Chromatogr., 456 (1988) 3.
288. E. A. Hogendoorn, G. R. van der Hoff, and P. van Zoonen, J. High Resolut. Chromatogr., 12 (1989) 784.
289. P. R. Fielden and A. J. Packham, Anal. Chem., 62 (1990) 2594.
290. D. L. Conley and E. J. Benjamin, J. Chromatogr., 257 (1983) 33.
291. C. E. Werkhoven-Goewie, C. De Ruiter, U. A. Th. Brinkman, R. W. Frei, G. J. De Jong, C. J. Little, and O. Stahel, J. Chromatogr., 255 (1983) 79.
292. F. W. Willmott, I. Mackenzie, and R. J. Dolphin, J. Chromatogr., 167 (1978) 31.
293. E. L. Johnson, R. Gloor, and R. E. Majors, J. Chromatogr., 149 (1978) 571.
294. H. Takahagi, K. Inoue, and M. Horiguchi, J. Chromatogr. 352 (1986) 369.
295. J. V. Alfredson, J. Chromatogr., 218 (1981) 715.
296. J.-B. Lecaillon, N. Febvre, and C. Souppart, J. Chromatogr., 317 (1984) 493.
297. E. J. Benjamin, B. A. Firestone, J. A. Schneider, J. Chromatogr. Sci., 23 (1985) 168.
298. N. Sagliano, S. H. Hsu, T. R. Floyd, T. V. Raglione, and R. A. Hartwick, J. Chromatogr. Sci., 23 (1985) 238.
299. L. R. Snyder, J. W. Dolan, and Sj. Van Der Wal, J. Chromatogr., 203 (1981) 3.
300. A. Nazareth, L. Jaramillo, B. L. Karger, R. W. Giese, and L. R. Snyder, J. Chromatogr., 309 (1984) 357.
301. C. E. Werkhoven-Goewie, U. A. Th. Brinkman, and R. W. Frei, Anal. Chem., 53 (1981) 2072.
302. W. Golkiewicz, C. E. Werkhoven-Goewie, U. A. Th. Brinkman, R. W. Frei, H. Colin, and G. Guiochon, J. Chromatogr. Sci., 21 (1983) 27.
303. R. L. Smith and D. J. Pietrzyk, J. Chromatogr. Sci., 21 (1983) 282.
304. G. E. Goewie and E. A. Hogendoorn, J. Chromatogr., 404 (1987) 352.
305. M. W. F. Nielen, R. E. J. Van Soest, H. E. Van Ingen, A. Frarjam, R. W. Frei, and U. A. Th. Brinkman, J. Chromatogr., 417 (1987) 159.
306. M. W. F. Nielen, H. E. Van Ingen, A. J. Valk, R. W. Frei, and U. A. Th. Brinkman, J. Liq. Chromatogr., 10 (1987) 617.
307. H. Irth and U. A. Th. Brinkman, Trends Anal. Chem., 9 (1990) 235.
308. H. Irth, G. J. De Jong, H. Lingeman, and U. A. Th. Brinkman, Anal. Chim. Acta, 236 (1990) 165.
309. C. E. Goewie, M. W. F. Neilen, R. W. Frei, and U. A. Th. Brinkman, J. Chromatogr., 301 (1984) 325.

310. C. E. Goewie and E. A. Hogendoorn, Sci. Total Environ., 47 (1985) 349.
311. M. W. F. Neilen, R. C. A. Koordes, R. W. Frei, and U. A. Th. Brinkman, J. Chromatogr., 330 (1985) 113.
312. E. M. Bargar, J. Chromatogr., 417 (1987) 143.
313. V. Ascalone and L. Dal Bo, J. Chromatogr., 423 (1987) 239.
314. G. Tamai, H. Yoshida, and H. Imai, J. Chromatogr., 423 (1987) 155 and 163.
315. N. Daoud, T. Arvidsson, and K.-G. Wahlund, J. Chromatogr., 385 (1987) 311.
316. S. E. Cook and T. C. Pinkerton, J. Chromatogr., 368 (1986) 233.
317. J. Haginaka, Trends Anal. Chem., 10 (1991) 17.
318. H. Hachenberg and A. P. Schmidt, "Gas Chromatographic Headspace Analysis", Heyden, London, UK, 1977.
319. J. Drozd and J. Novak, J. Chromatogr., 165 (1979) 141.
320. B. Kolb, "Applied Headspace Gas Chromatography", Heyden, London, UK, 1980.
321. B. V. Ioffe and A. G. Viltenberg, "Headspace Analysis and Related Methods in Gas Chromatography", Wiley, New York, NY, 1984.
322. A. J. Nunez, L. F. Gonzalez, and J. Janak, J. Chromatogr., 300 (1984) 127.
323. P. Schreier, "Analysis of Volatiles. Methods and Applications", Walter de Gruyter, Berlin, 1984.
324. A. G. Vitenberg and N. I. Kalacheva, J. Chromatogr., 368 (1986) 21.
325. R. Guitart, A. Puigdemont, and M. Arboix, J. Chromatogr., 491 (1989) 271.
326. P. Gagliardi and G. R. Verga, J. Chromatogr., 279 (1983) 323.
327. F. Poy and L. Cobelli, J. Chromatogr. Sci., 23 (1985) 114.
328. P. Sandra (Ed.), " Sample Introduction in Capillary Gas Chromatography", Huethig, Heidelberg, Vol. 1, 1985.
329. P. L. Wylie, Chromatographia, 21 (1986) 251.
330. F. Poy, L. Cobelli, S. Banfi, and F. Fossati, J. Chromatogr., 395 (1987) 281.
331. C. Dumas, J. Chromatogr., 329 (1985) 119.
332. L. S. Ettre, J. E. Purcell, J. Widonski, B. Kolb, and P. Pospisil, J. Chromatogr. Sci., 18 (1980) 116.
333. B. Kolb, B. Liebhart, and L. S. Ettre, Chromatographia, 21 (1986) 305.
334. G. Takeoka and W. Jennings, J. Chromatogr. Sci., 22 (1984) 177.
335. B. Kolb, M. Auer, and P. Pospisil, J. Chromatogr., 279 (1983) 341.
336. J. Drozd, Z. Vodakova, and P. Koupil, J. Chromatogr., 518 (1990) 1.
337. A. Hagman and S. Jacobsson, Anal. Chem., 61 (1989) 1202.
338. B. Kolb, Chromatographia, 15 (1982) 587.
339. B. Kolb, P. Pospisil, and M. Auer, Chromatographia 19 (1984) 113.
340. A. G. Vitenberg, J. Chromatogr. Sci., 22 (1984) 122.
341. A. Venema, J. High Resolut. Chromatogr., 11 (1988) 128.
342. A. Zlatkis, R. S. Brazell, and C. F. Poole, Clin. Chem., 27 (1981) 789.
343. J. Curvers, Th. Noy, C. Cramers, and J. Rijks, J. Chromatogr., 289 (1984) 171.
344. A. G. Vitenberg and B. V. Ioffe, J. Chromatogr., 471 (1989) 55.
345. T. A. Bellar and J. J. Lichtenberg, J. Amer. Water Works Assoc., 66 (1974) 739.

932

346. L. C. Michael, M. D. Erickson, S. P. Parks, and E. D.
     Pellizzari, Anal. Chem., 52 (1980) 1836.
347. M, Termonia and G. Alaerts, J. High Resolut. Chromatogr., 8
     (1985) 622.
348. P. Werkhoff and W. Bertschneider, J. Chromatogr., 405 (1987)
     87 and 99.
349. Th. Noy, A. Van Es, C. Cramers, J. Rijks, and R. Dooper, J.
     High Resolut. Chromatogr., 10 (1987) 60.
350. L. C. Michael, E. D. Pellizzari, and R. W. Wiseman, Environ.
     Sci. Technol., 22 (1988) 565.
351. J. W. Eichelberger, T. A. Bellar, J. P. Donnelly, and W. L.
     Budde, J. Chromatogr. Sci., 28 (1990) 460.
352. H. C. Hu and P. H. Weiner, J. Chromatogr. Sci., 18 (1980) 333.
353. R. Otson and C. Chan, Intern. J. Environ. Anal. Chem., 30
     (1987) 275.
354. L. V. Haynes and A. R. Steimie, J. High Resolut. Chromatogr.,
     10 (1987) 441.
355. H. T. Badings, C. de Jong, and R. P. M. Dooper, J. High
     Resolut. Chromatogr., 8 (1985) 755.
356. S. Liu, J. Kang, D. L. Strother, R. J. Carley, and J. D.
     Stuart, J. High Resolut. Chromatogr., 12 (1989) 779.
357. T. N. Barnung and O. Grahl-Nielsen, J. Chromatogr., 466
     (1989) 271.
358. A. Bianchi, M. S. Varney, and J. Phillips, J. Chromatogr.,
     467 (1989) 111.
359. K. Grob and F. Zurcher, J. Chromatogr., 117 (1976) 285.
360. W. E. Coleman, J. W. Munch, R. W. Slater, R. G. Melton, and
     F. C. Kopfler, Environ. Sci. Technol., 17 (1983) 571.
361. M. M. Thomason and W. Bertsch, J. Chromatogr., 279 (1983) 383.
362. J. I. Gomez-Belinchon and J. Albaiges, Intern. J. Environ.
     Anal. Chem., 30 (1987) 183.
363. R. S. Narang and B. Bush, Anal. Chem., 52 (1980) 2076.
364. A. Habich and K. Grob, J. High Resolut. Chromatogr., 7 (1984)
     492.
365. H. Boren, A. Grimvall, and R. Savenhed, J. Chromatogr., 252
     (1982) 139.
366. K. Grob, G. Grob, and H. Habich, J. High Resolut.
     Chromatogr., 7 (1984) 340.
367. A. Zlatkis, C. F. Poole, R. S. Brazell, D. A. Bafus, and P.
     S. Spencer, J. Chromatogr., 182 (1980) 137.
368. A. Zlatkis, C. F. Poole, R. S. Brazell, K. Y. Lee, and S.
     Singhawangcha, J. High Resolut. Chromatogr., 2 (1979) 423.
369. A. Venema, J. High Resolut. Chromatogr., 9 (1986) 637.
370. R. G. Schaefer and W. Puttmann, J. Chromatogr., 395 (1987)
     203.
371. L. H. Keith (Ed.), "Identification and Analysis of Organic
     Pollutants in Air", Butterworth, Woburn, MA, 1984.
372. R. M. Harrion and R. Perry (Eds.), "Handbook of Air Pollution
     Analysis", Chapman and Hall, London, 2nd. Edn., 1986.
373. T. F. Bidleman, Trace Analysis, 4 (1985) 51.
374. J. Rudolph, K. P. Muller, and R. Koppmann, Anal. Chim. Acta,
     236 (1990) 197.
375. L. D. Johnson, Environ. Sci. Technol. 20 (1986) 223.
376. D. Schuetzle, Environ. Hlth. Persp., 47 (1983) 65.
377. L. A. Wallace and W. R. Ott, J. Air Pollut. Control. Assoc.,
     32 (1982) 601.
378. G. Choudhary (Ed.), "Chemical Hazards in the Workplace",
     American Chemical Society, Washington, D. C., 1981.
379. A. L. Linch, "Evaluation of Ambient Air Quality by Personnel
     Monitoring", CRC Press, Clevland, OH, Vol. 1 and 2, 1981.

380. J. Rudolph, F. J. Johnen, and A. Khedim, Intern. J. Environ. Anal. Chem., 27 (1986) 97.
381. S. Muller and M. Oehme, J. High Resolut. Chromatogr., 13 (1990) 34.
382. N. Schmidbauer and M. Oehme, J. High Resolut. Chromatogr., 9 (1986) 502.
383. D. O' Hara, T. Vo, and J. F. Vedder, J. Chromatogr. Sci., 23 (1985) 471.
384. T. Noy, P. Fabian, R. Borchers, F. Janssen, C. Cramers and J. Rijks, J. Chromatogr., 393 (1987) 343.
385. K. J. Krost, E. D. Pellizzari, S. G. Walburn, and S. A. Hubbard, Anal. Chem., 54 (1982) 810.
386. A. I. Clark, A. E. McIntyre, J. N. Lester, and R. Perry, J. Chromatogr., 252 (1982) 147.
387. T. Tanaka, J. Chromatogr., 153 (1978) 7.
388. J. Namiesnik, L. Torres, E. Kozlowski, and J. Mathieu, J. Chromatogr., 208 (1981) 239.
389. R. H. Brown and C. J. Purnell, J. Chromatogr., 178 (1979) 79.
390. M. L. Riba, E. Randrianalimanana, J. Mathieu, L. Torres, and J. Namiesnik, Intern. J. Environ. Anal. Chem., 19 (1985) 133.
391. H. Tang, G. Richards, K. Gunther, J. Crawford, M. L. Lee, E. A. Lewis, and D. J. Eatough, J. High Resolut. Chromatogr., 11 (1988) 775.
392. E. D. Pellizzari and K. J. Krost, Anal. Chem., 56 (1984) 1813.
393. A. Betti, S. Coppi, and C. Bighi, J. Chromatogr., 349 (1985) 181.
394. G. Macleod and J. M. Ames, J. Chromatogr., 355 (1986) 393.
395. P. Ciccioli, B. A. Cecinato, C. Di Palo, A. Brachetti, and A. Liberti, J. Chromatogr., 351 (1986) 433.
396. F. Mangani, A. Mastrogiacomo, and O. Marras, Chromatographia, 15 (1982) 712.
397. J. Rudling, J. Chromatogr., 503 (1990) 33.
398. L. Senf and H. Frank, J. Chromatogr., 520 (1990) 131.
399. K. Grob, A. Artho, C. Frauenfelder, and I. Roth, J. High Resolut. Chromatogr., 13 (1990) 257.
400. D. P. Adley and D. W. Underhill, Anal. Chem., 61 (1989) 843.
401. C. J. Tidwell and D. W. Underhill, Anal. Chem., 61 (1989) 917.
402. I. Johansen and J. F. Wendelboe, J. Chromatogr., 217 (1981) 317.
403. H. J. Neu, W. Merz, and H. Panzel, J. High Resolut. Chromatogr., 5 (1982) 382.
404. R. G. Lewis and M. D. Jackson, Anal. Chem., 54 (1982) 594.
405. W. N. Billings and T. F. Bidleman, Environ. Sci. Technol., 14 (1980) 679.
406. M. P. Ligocki and J. F. Pankow, Anal. Chem., 57 (1985) 1138.
407. C. D. Keller and T. F. Bidleman, Atmos. Environ., 18 (1984) 837.
408. F. R. Guenther. S. N. Chesler, G. E. Gardon, and W. H. Zoller, J. High Resolut. Chromatogr., 11 (1988) 761.
409. J. F. Pankow and L. M. Isabelle, J. Chromatogr., 237 (1982) 25.
410. J. F. Pankow, L. M. Isabelle, and T. J. Kristensen, Anal. Chem., 54 (1982) 1815.
411. J. F. Pankow, J. High Resolut. Chromatogr., 6 (1983) 292.
412. N. H. Mosesman, L. M. Sidisky, and S. D. Corman, J. Chromatogr. Sci., 25 (1987) 351.
413. J. W. Graydon and K. Grob, J. Chromatogr., 254 (1983) 265.
414. J. F. Pankow and M. E. Rosen, J. High Resolut. Chromatogr., 7 (1984) 504.
415. G. I. Senum, Environ. Sci. Technol., 15 (1981) 1073.

934

416. D. Van der Straeten, H. Van Langenhove, and N. Schamp, J. Chromatogr., 331 (1985) 207.
417. S. Coppi, A. Betti, and M. Ascanelli, J. Chromatogr., 390 (1987) 349.
418. M.-L. Riba, N. Tsiropoulos, and L. Torres, J. Chromatogr., 437 (1988) 139.
419. E. T. Peltzer and R. B. Gagosian, Anal. Chim. Acta., 198 (1987) 125.
420. L. Van Vaeck, K. Van Cauwenberghe, and J. Janssens, Atmos. Environ., 18 (1984) 417.
421. R. G. Lewis, J. D. Mulik, R. W. Coutant, G. W. Wooten, and C. R. McMillin, Anal. Chem., 57 (1985) 214 and 219.
422. C. J. Purnell, M. D. Wright, and R. H. Brown, Analyst, 106 (1981) 590.
423. R. W. Coutant and D. R. Scott, Environ. Sci. Technol., 16 (1982) 410.
424. V. E. Rose and J. L. Perkins, Am. Ind. Hyg. Assoc. J., 43 (1982) 605.
425. G. O. Nelson, "Controlled Test Atmospheres, Principles and Techniques" Ann Arbor Science Publishers, Ann Arbor, MI, 1971.
426. R. S. Barratt, Analyst, 106 (1981) 817.
427. W. Muniak, Z. Witkiewicz, E. Woryna, E. Kusinska, A. Twardowski, and B. Goca, J. Chromatogr., 436 (1988) 323.
428. H. Hachenberg, J. High Resolut. Chromatogr., 12 (1989) 742.
429. J. J. Ritter and N. K. Adams, Anal. Chem., 48 (1976) 612.
430. H. Nozoye, Anal. Chem., 50 (1978) 1727.
431. S. Greenhouse and F. Andrawes, Anal. Chim. Acta, 236 (1990) 221.
432. J. Namiesnik, Chromatographia, 17 (1983) 47.
433. E. Bayer, H. Frank, J. Grehardt, and G. Nicholson, J. Assoc. Off. Anal. Chem., 70 (1987) 234.
434. J. Gerhardt, G. Nicholson, H. Frank, and E. Bayer, Chromatographia, 19 (1985) 251.
435. J. D. Nicholson, Analyst, 103 (1978) 1 and 193.
436. K. Blau and G. S. King (Eds.), "Handbook of Derivatives for Chromatography", Heyden, London, UK, 1978.
437. D. R. Knapp, "Handbook of Analytical Derivatization Reactions", Wiley, New York, NY,1979.
438. J. Drozd, "Chemical Derivatization in Gas Chromatography", Elsevier, Amsterdam, 1981.
439. R. W. Frei and J. L. Lawrence (Eds.), "Chemical Derivatization in Analytical Chemistry", Plennum Press, New York, NY, Vol. 1, 1981.
440. J. Drozd, J. Chromatogr., 113 (1975) 303.
441. A. Hulshoff and H. Lingeman, J. Pharm. Biomed. Anal., 2 (1984) 337.
442. A. E. Pierce, "Silylation of Organic Compounds", Pierce Chemical Company, Rockford, IL, 1968.
443. C. F. Poole and A. Zlatkis, J. Chromatogr. Sci., 17 (1979) 115.
444. C. F. Poole and A. Zlatkis, J. Chromatogr., 184 (1980) 99.
445. O. Gyllenhaal and J. Vessman, J. Chromatogr., 435 (1988) 259.
446. W. C. Kossa, J. MacGee, S. Ramachandran, and A. J. Webber, J. Chromatogr. Sci., 17 (1979) 177.
447. A. Hulshoff and A. D. Forch, J. Chromatogr., 220 (1981) 275.
448. J. Vessman, K. E. Karlsson, and O. Gyllenhaal, J. Phar. Biomed. Anal., 4 (1986).
449. A. Zlatkis and C. F. Poole (Eds.), "Electron Capture: Theory and Practice in Chromatography", Elsevier, Amsterdam, 1982.
450. C. F. Poole and A. Zlatkis, Anal. Chem., 52 (1980) 1002A.

451. C. F. Poole and S. K. Poole, J. Chromatogr. Sci., 25 (1987) 434.
452. C. R. Bielby, E. D. Morgan, and I. D. Wilson, J. Chromatogr., 351 (1986) 57.
453. S. H. G. Andersson and J. Sjovall, J. Chromatogr., 289 (1984) 195.
454. E. D. Morgan and C. F. Poole, J. Chromatogr., 116 (1976) 333.
455. D. Knausz, A. Meszticzky, J. Rohonczy, B. Csakvari, F. Szederkenyi, and K. Ujszaszy, J. Chromatogr., 365 (1986) 183.
456. K. Yamashita, K. Watanabe, M. Ishibashi, H. Miyazaki, K. Yokota, K. Horie, and S. Yamamoto, J. Chromatogr., 399 (1987) 223.
457. S. Steffenrud, P. Borgeat, H. Salari, M. J. Evans, and M. J. Bertrand, J. Chromatogr., 416 (1987) 219.
458. M. A. Quilliam and J. B. Westmore, Anal. Chem., 50 (1978) 59.
459. I. A. Blair and G. Phillipou, J. Chromatogr. Sci., 15 (1977) 478.
460. S. L. Mackenzie, D. Tenaschuk, and G. Fortier, J. Chromatogr., 387 (1987) 241.
461. T. P. Mawhinney, R. S. R. Robinett, A. Atalay, and M. A. Madson, J. Chromatogr., 358 (1986) 231.
462. R. J. Early, J. R. Thompson, G. W. Sedgwick, J. M. Kelly, and R. J. Christopherson, J. Chromatgor., 416 (1987) 15.
463. C. J. Biermann, C. M. Kinoshita, J. A. Marlett, and R. D. Steele, J. Chromatogr., 357 (1986) 330.
464. D. L. Schooley, F. M. Kubiak, and J. V. Evans, J. Chromatogr. Sci., 23 (1985) 385.
465. K. R. Kim, M. K. Hahn, A. Zlatkis, E. C. Horning, and B. S. Middleditch, J. Chromatogr., 468 (1989) 289.
466. J. Gilbert, J. R. Startin, I. Parker, M. J. Shepherd, J. C. Mitchell, and M. J. Perkins, J. Chromatogr., 369 (1986) 408.
467. D. C. Landrum and T. P. Mawhinney, J. Chromatogr., 483 (1989) 21.
468. C. F. Poole, A. Zlatkis, W.-F. Sye, S. Singhawangcha, and E. D. Morgan, Lipids, 15 (1980) 734.
469. M. J. Bertrand, S. Stefanidis, and B. Sarrasin, J. Chromatogr., 351 (1986) 47.
470. C. F. Poole, W.-F. Sye, S. Singhawangcha, F. Hsu, A. Zlatkis, A. Arfwidsson, and J. Vessman, J. Chromatogr., 199 (1980) 123.
471. C. F. Poole, S. Singhawangcha, L.-E. Chen Hu, W.-F. Sye, R. S. Brazell, and A. Zlatkis, J. Chromatogr., 187 (1980) 331.
472. A. J. Francis, E. D. Morgan, and C. F. Poole, J. Chromatogr., 161 (1978) 111.
473. C. F. Poole, W.-F. Sye, S. Singhawangcha, and A. Zlatkis, Org. Mass Spectrom., 15 (1980) 486.
474. A. Tavakkol and D. B. Drucker, J. Chromatogr. Sci., 22 (1984) 12.
475. S. Fujihara, T. Nakashima, and Y. Kurogochi, J. Chromatogr., 277 (1983) 53.
476. J. M. Moore, A. C. Allen, and D. A. Cooper, Anal. Chem., 56 (1984) 642.
477. S. L. Mackenzie, D. Tenaschuk, and G. Fortier, J. Chromatogr., 367 (1986) 181.
478. U. A. Th. Brinkman, A. de Kok, and R. B. Geerdink, J. Chromatogr., 283 (1984) 113.
479. T. M. Trainor and P. Vouros, Anal. Chem., 59 (1987) 601.
480. K. Lehtonen and M. Ketola, J. Chromatogr., 370 (1986) 465.
481. H.-B. Lee, J. Chromatogr., 457 (1988) 267.
482. J. Bakthavachalam, R. S. Annan, F. A. Beland, P. Vouros, and R. W. Giese, J. Chromatogr., 500 (1990) 373.

483. H. B. Lee, R. L. Hong-You, and A. S. Y. Chau, J. Assoc. Off. Anal. Chem., 68 (1985) 422.
484. V. Janda and H. van Langenhova, J. Chromatogr., 472 (1989) 327.
485. O. Gyllenhaal, K.-J. Hoffmann, B. Lamm, R. Simonsson, and J. Vessman, J. Chromatogr., 355 (1986) 127.
486. G. Garmerith, J. Chromatogr., 318 (1985) 65.
487. I. Molnar-Perl, M. Pinter-Szakacs, V. Fabian-Vonsik, J. Chromatogr., 390 (1987) 434.
488. I. Molnar-Perl and M. Pinter-Szakacs, J. Chromatogr., 365 (1986) 171.
489. M. Morvai, and I. Molnar-Perl, Chromatographia, 23 (1987) 925.
490. W. Welz, W. Sattler, H. J. Lew, and E. Malle, J. Chromatogr., 526 (1990) 319.
491. C. F. Poole and M. Verzele, J. Chromatogr., 150 (1978) 439.
492. K. D. Muller, H. Husmann, H. P. Nalik, and G. Schomburg, Chromatographia, 30 (1990) 245.
493. D. D. Godse, J. J. Warsh, and H. C. Stancer, Anal. Chem., 49 (1977) 915.
494. D. A. Davis, D. A. Durden, P. Pun-Li, and A. A. Boulton, J. Chromatogr., 142 (1977) 517.
495. P. Pfaffli. H. Savolainen, and H. Keskinen, Chromatographia, 27 (1989) 483.
496. A. Hulshoff and A. F. Forch, J. Chromatogr., 220 (1981) 275.
497. I. Ciucanu and F. Kerek, J. Chromatogr., 284 (1984) 179.
498. J. M. Olias, J. J. Rios, and M. Valle, J. Chromatogr., 467 (1989) 279.
499. H. M. Fales, T. M. Jaouni, and J. F. Babashak, Anal. Chem., 45 (1973) 2302.
500. I. Hazai and G. Alexander, J. High Resolut. Chromatogr., 5 (1982) 583.
501. F. Ngan and M. Toofan, J. Chromatogr. Sci., 29 (1991) 8.
502. M. K. L. Bicking and N. A. Adinolfe, J. Chromatogr. Sci., 23 (1985) 348.
503. O. Gyllenhaal, H. Brotell, and P. Hartvig, J. Chromatogr., 129 (1976) 295.
504. P. Verner and F. Pehal, J. Chromatogr., 382 (1986) 232.
505. B. Schatowitz and G. Gercken, J. Chromatogr., 425 (1988) 257.
506. J. M. Rosenfeld and J. L. Crocco, Anal. Chem., 50 (1987) 701.
507. A. Sonesson, L. Larsson, and J. Jimenez, J. Chromatogr., 417 (1987) 366.
508. J. M. Rosenfeld, M. Mureika-Russel, and S. Yeroushalmi, J. Chromatogr., 358 (1986) 137.
509. J. M. Rosenfeld, S. Yeroushalmi, and E.-Y. Osei-Twum, Anal. Chem., 58 (1986) 3044.
510. S. J. Abraham and W. J. Criddle, J. Anal. Appl. Pyrol., 9 (1985) 53.
511. L. Den Drijver and C. W. Holzapfel, J. Chromatogr., 363 (1986) 345.
512. G. Hoffmann and L. Sweetman, J. Chromatogr., 421 (1987) 336.
513 K. T. Koshy, D. G. Kaiser, and A. L. Vandersilk, J. Chromatogr. Sci., 13 (1975) 97.
514. P. A. Biondi, F. Manca, A. Negri, G. Tedeschi, and C. Secchi, J. Chromatogr., 467 (1989) 315.
515. M. Tomita, T. Okuyama, Y. Hattu, and S. Kawai, J. Chromatogr., 526 (1990) 174.
516. Y. Hoshika and G. Muto, J. Chromatogr., 152 (1978) 224.
517. D. V. Crabtree, A. J. Adler, and G. J. Handelman, J. Chromatogr., 466 (1989) 251.
518. S. B. Margin and M. Worland, J. Pharm. Sci., 61 (1972) 1235.

519. R. T. Coutts, G. A. Torok-both, L. V. Chu, Y. K. Tam, and F. M. Pasutto, J. Chromatogr., 421 (1987) 267.
520. A. J. F. Wickramasinghe and R. S. Shaw, Biochem. J., 141 (1974) 179.
521. J.-C. Lhuguenot and B. F. Maume, J. Chromatogr. Sci., 12 (1974) 411.
522. P. Hartvig and J. Vessman, J. Chromatogr. Sci., 12 (1974) 722.
523. P. Hartvig and J. Vessman, Anal. Lett., 12 (1974) 223.
524. A. Zlatkis and B. C. Pettit, Chromatographia, 2 (1969) 484.
525. P. Hartvig and B. Nashland, J. Chromatogr., 133 (1977) 367.
526. G. Skarping, T. Bellander, and L. Mathiasson, J. Chromatogr., 370 (1986) 245.
527. A. C. Moffat, E. C. Horning, S. B. Martin, and M. Rowland, J. Chromatogr., 66 (1972) 255.
528. P. O. Edlund, J. Chromatogr., 187 (1980) 161.
529. G. B. Baker, T. S. Rao, and R. T. Coutts, J. Chromatogr., 381 (1986) 211.
530. A. Sentissi, M. Joppich, K. O'Connell, A. Nazareth, and R. W. Giese, Anal. Chem., 56 (1984) 2512.
531. G. M. Anthony, C. J. W. Brooks, I. MacLean, and I. Sangster, J. Chromatogr. Sci., 7 (1969) 623.
532. C. J. W. Brooks and I. MacLean, J. Chromatogr. Sci., 9 (1971) 18.
533. N. Ikekawa, Trends Anal. Chem., 9 (1990) 337.
534. C. J. W. Brooks and W. J. Cole, J. Chromatogr., 399 (1987) 207.
735. C. F. Poole, S. Singhawangcha, and A. Zlatkis, J. Chromatogr., 158 (1978) 33.
536. C. F. Poole, S. Singhawangcha, and A. Zlatkis, J. Chromatogr., 186 (1979) 307.
537. C. F. Poole, S. Singhawangcha, L.-E. Chen Hu, and A. Zlatkis, J. Chromatogr., 178 (1979) 495.
538. K. Jacob, G. Schnabl, and W. Vogt, Chromatographia, 19 (1984) 216.
539. C. J. W. Brooks and W. J. Cole, Analyst, 110 (1985) 587.
540. M. Ishibashi, K. Watanabe, K. Yamashita, H. Miyazaki, and S. Krolik, J. Chromatogr., 391 (1987) 183.
541. M. Ishibashi, T. Irie, and H. Miyazaki, J. Chromatogr., 399 (1987) 197.
542. U. Langenbeck, A. Hoinowski, K. Mantel, and U.-H. Mohring, J. Chromatogr., 143 (1977) 39.
543. A. Frigerio, P. Martelli, K. M. Baker, and P. A. Bondi, J. Chromatogr., 81 (1973) 139.
544. M. S. Lennard, J. H. Silas, A. J. Smith, and G. T. Tucker, J. Chromatogr., 133 (1977) 161.
545. S. L. Malcolm and T. R. Marten, Anal. Chem., 48 (1976) 807.
546. S. B. Martin, J. H. Karam, and P. H. Forsham, Anal. Chem., 47 (1975) 545.
547. K. Imai, Adv. Chromatogr., 27 (1987) 215.
548. H. Lingeman and W. L. Underberg, "Detector-Orientated Derivatization Techniques in Liquid Chromatography", Dekker, New York, NY, 1990.
549. Y. Ohkura and H. Nohta, Adv. Chromatogr., 29 (1989) 221.
550. K. Imai, H. Miyano, and T. Toyo'oka, Analyst, 109 (1984) 1365.
551. H. Lingman, W. J. M. Underberg, A. Takadate, and A. Hulshoff, J. Liq. Chromatogr., 8 (1985) 789.
552. J. W. Birks, "Chemiluminescence and Photochemical Reaction Detection in Chromatography", VCH Publishers, New York, NY, 1989.
553. G. J. De Jong and P. J. M. Kwakman, J. Chromatogr., 492 (1989) 319.

938

554. P. T. Kissinger, K. Bratin, G. C. Davis, and L. A. Pachla, J. Chromatogr. Sci., 17 (1979) 137.
555. I. S. Krull, C. M. Selavka, C. Duda, and W. Jacobs, J. Liq. Chromatogr., 8 (1985) 2845.
556. T. Nagatsu and K. Kojima, Trends Anal. Chem., 7 (1988) 21.
557. T. Jupille, J. Chromatogr. Sci., 17 (1979) 160.
558. M. A. Carey and H. W. Persinger, J. Chromatogr. Sci., 10 (1972) 537.
559. R. W. Ross, J. Chromatogr. Sci., 14 (1976) 505.
560. J. Lehrfield, J. Chromatogr., 120 (1976) 141.
561. F. Nachtmann, H. Spitzy, and R. W. Frei, J. Chromatogr., 122 (1976) 293.
562. G. B. Cox, J. Chromatogr., 83 (1973) 471.
563. B. Bjorkuist and H. Tolvonen, J. Chromatogr., 153 (1978) 265.
564. D. G. Musson and L. A. Sternson, J. Chromatogr., 188 (1980) 159.
565. C. R. Clark and M. M. Wells, J. Chromatogr. Sci., 16 (1975) 332.
566. S. Asotra, P. Mladenov, and R. D. Burke, J. Chromatogr., 408 (1987) 227.
567. L. L. Needham and M. M. Kochhar, J. Chromatogr., 114 (1975) 220.
568. T. Sugiura, T. Hayashi, S. Kawai, and T. Ohno, J. Chromatogr., 110 (1975) 385.
569. A. Hulshoff, H. Rosenboom, and J. Renema, J. Chromatogr., 186 (1980) 535.
570. D. M. Barends, J. S. Blauw, C. W. Mijnsbergen, C. J. L. R. Grovers, and A. Hulshoff, J. Chromatogr., 322 (1985) 321.
571. K. Tsuji and K. M. Jenkins, J. Chromatogr., 369 (1986) 105.
572. R. Knecht and J.-Y. Chang, Anal. Chem., 58 (1986) 2375.
573. N. Nimura, K. Iwaki. T. Kinoshita, K. Takeda, and H. Ogura, Anal. Chem., 58 (1986) 2372.
574. I. Miksik, R. Struzinsky, K. Macek, and Z. Deyl, J. Chromatogr., 500 (1990) 607.
575. R. A. Braun and W. A. Mosher, J. Amer. Chem. Soc., 80 (1958) 3048.
576. H. D. Durst, M. Milano. E. J. Kikta, S. A. Connelly, and E. Grushka, Anal. Chem., 47 (1975) 1797.
577. Z. L. Bandi and E. S. Reynolds, J. Chromatogr., 329 (1985) 57.
578. R. Badoud and G. Pratz, J. Chromatogr., 360 (1986) 119.
579. A. Furangen, J. Chromatogr., 353 (1986) 401.
580. P. M. Marce, M. Calull, J. C. Olucha, F. Borrull, and F. X. Rius, Anal. Chim. Acta, 242 (1991) 25.
581. W. Linder, J. Chromatogr., 198 (1980) 367.
582. D. P. Matthees and W. C. Purdy, Anal. Chim. Acta, 109 (1979) 161.
583. D. Munger, L. A. Sternson, A. J. Repta, and T. Higuchi, J. Chromatogr., 143 (1977) 375.
584. K. L. Dunlap, R. L. Sandridge, and J. Keller, Anal. Chem., 48 (1976) 497.
585. K. Kuwata, M. Uebori, and Y. Yamazaki, Anal. Chem., 52 (1980) 857.
586. B. A. Bidlingmeyer, S. A. Cohen, and T. L. Tarvin, J. Chromatogr., 336 (1984) 93.
587. F. A. L. van der Horst, Trends Anal. Chem., 8 (1989) 268.
588. F. A. L. van der Horst, J. M. Reijn, M. H. Post, A. Bult, J. M. Holthuis, and U. A. Th. Brinkman, J. Chromatogr., 507 (1990) 351.
589. M. Schafer and E. Mutscher, J. Chromatogr., 164 (1979) 247.

590. J. Gogo, S. Komatsu, M. Goto, and T. Nambara, Chem. Pharm. Bull. Japan, 29 (1981) 899.
591. J. Goto, N. Goto, F. Shamsa, M. Saito, S. Komatsu, K. Suzaki, and T. Nambara, Anal. Chim. Acta, 147 (1983) 397.
592. K. E. Karlsson, D. Wiesler, M. Alasandro, M. Novotny, Anal. Chem., 57 (1985) 229.
593. C. Hamada, M. Iwasaki, N. Kuroda, and Y. Ohkura, J. Chromatogr., 341 (1985) 426.
594. M. Yamaguchi, T. Iwata, M. Nakamura and Y. Ohkura, Anal. Chim. Acta, 193 (1987) 209.
595. K. Gamoh, H. Sawamoto, and J. Iida, Anal. Chim. Acta, 228 (1990) 307.
596. W. Elbert, S. Breilenbach, A. Neflel, and J. Hahn, J. Chromatogr., 328 (1985) 111.
597. K. D. Ertel, and J. T. Carstensen, J. Chromatogr., 411 (1987) 297.
598. J. H. Wolf, L. Veenma-van Derduin, and J. Korf, J. Chromatogr., 487 (1989) 496.
599. R. Farinotti, Ph. Siard, J. Bourson, S. Kirkiacharian, B. Valeur, and G. Mohuzier, J. Chromatogr., 269 (1983) 81.
600. H. Naganuma and Y. Kawahara, J. Chromatogr., 478 (1989) 149.
601. W. D. Korte, J. Chromatogr., 243 (1982) 153.
602. Y. Yamauchi, T. Tomita, M. Senda, A. Hirai, T. Terano, Y. Tamura, and S. Yoshida, J. Chromatogr., 357 (1986) 199.
603. H. Lingeman, A. Hulshoff, W. J. M. Underberg, and F. B. J. M. Offermann, J. Chromatogr., 290 (1984) 215.
604. J. D. Baty, S. Pazouki, and J. Dolphin, J. Chromatogr., 395 (1987) 403.
605. F. A. L. van der Horst, M. H. Post, J. J. M. Holthuis, and U. A. Th. Brinkman, J. Chromatogr., 500 (1990) 443.
606. S. Allenmark and M. Chelminska-Bertilsson, Chromatographia, 28 (1989) 367.
607. Y. Yasaka, M. Tanaka, T. Shono, T. Tetsumi, and J. Katakawa, J. Chromatogr., 508 (1990) 133.
608. W. F. Alpenfels, R. A. Mathews, D. E. Madden, and A. E. Newson, J. Liq. Chromatogr., 5 (1982) 1711.
609. K. Imai, S. Higashidate, Y. Tsukamoto, S. Uzu, and S. Kanda, Anal. Chim. Acta, 225 (1989) 421.
610. D. J. Pietrzyk and E. P. Chan, Anal. Chem., 42 (1970) 37.
611. G. Gubitz, R. Wintersteiger, and R. W. Frei, J. Liq. Chromatogr., 7 (1984) 839.
612. E. P. Lankmayr, R. W. Budna, K. Muller, F. Nachtmann, and F. Rainer, J. Chromatogr., 222 (1981) 249.
613. B. Kagedal and M. Kallberg, J. Chromatogr., 229 (1982) 409.
614. J. O. Miners, I. Fearnley, K. J. Smith, D. J. Birkett, P. M. Brooks, and M. W. Whitehouse, J. Chromatogr., 275 (1983) 89.
615. T. Toyo'oka, S. Uchiyama and Y. Saito, Anal. Chim. Acta, 205 (1988) 29.
616. V. Cavrini, R. Gatti, P. Roveri, A. Di Pietra, Chromatographia, 27 (1989) 185.
617. M. C. Gennaro, E. Mentasti, C. Sarzanini, and V. Porta, Chromatographia, 25 (1988) 117.
618. H. J. Schneider, Chromatographia, 28 (1989) 45.
619. P. Furst, L. Pollack, T. A. Graser, H. Godel, and P. Stehle, J. Chromatogr., 499 (1990) 557.
620. D. E. Willis, J. Chromatogr., 408 (1987) 217.
621. L. M. Dominguez and R. S. Dunn, J. Chromatogr. Sci., 25 (1987) 468.
622. L. A. Allison, G. S. Mayer, and R. E. Shoup, Anal. Chem., 56 (1984) 1089.

940

623. H. Umagat, P. Kucera, and L.-F. Wen, J. Chromatogr., 239 (1982) 463.
624. M. Ahnoff, I. Grundevik, A. Arfwidsson, J. Fonselius, and B.-A. Persson, Anal. Chem., 53 (1981) 484.
625. H. Miyano, T. Toyo'oka, and K. Imai, Anal. Chim. Acta, 170 (1985) 81.
626. I. Betner and P. Foldi, Chromatographia, 22 (1986) 381.
627. S. Einarsson, S. Folestad, B. Josefsson, and S. Lagerkvist, Anal. Chem., 58 (1986) 1638.
628. B. Gustavsson and I. Betner, J. Chromatogr., 507 (1990) 67.
629. Y.-Z. Hsieh, S. C. Beale, D. Wiesler, and M. Novotny, J. Microcolumn Sepns., 1 (1989) 96.
630. S. C. Beale, Y.-Z. Hsieh, D. Wiesler, and M. Novotny, J. Chromatogr., 499 (1990) 579.
631. U. A. Th. Brinkman, G. J. de Jong, and C. Gooijer, Pure Appl. Chem., 59 (1987) 625.
632. K. Honda, K. Miyaguchi, and K. Imai, Anal. Chim. Acta, 177 (1985) 103 and 111.
633. G. Mellbin and B. E. F. Smith, J. Chromatogr., 312 (1984) 203.
634. K. Miyaguchi, K. Honda, and K. Imai, J. Chromatogr., 316 (1984) 501.
635. B. Mann and M. L. Grayeski, J. Chromatogr., 386 (1987) 149.
636. K. Kawasaki, M. Maeda, and T. Tsuji, J. Chromatogr., 328 (1985) 121.
637. M.-Y. Chang, L.-R. Chen, X.-D. Ding, C. M. Selavka, I. S. Krull, and K. Bratin, J. Chromatogr. Sci., 25 (1987) 460.
638. S. Ikenoya, O. Hiroshima, M. Ohmae, and K. Kawabe, Chem. Pharm. Bull. Japan, 28 (1980) 2941.
639. K. Shimada, M. Tanaka, and T. Nambara, Anal. Chim. Acta, 147 (1983) 375.
640. K. Shimada, S. O'rii, M. Tanaka, and T. Nambara, J. Chromatogr., 352 (1986) 329.
641. I. S. Krull (Ed.), " Reaction Detection in Liquid Chromatography", Dekker, New York, NY, 1986.
642. J. T. Stewart and W. J. Bachman, Trends Anal. Chem., 7 (1988) 106.
643. U. A. Th. Brinkman, Chromatographia, 24 (1987) 190.
644. R. W. Frei, H. Jansen, and U. A. Th. Brinkman, Anal. Chem., 57 (1985) 1529A.
645. U. A. Th. Brinkman, R. W. Frei, and H. Lingeman, J. Chromatogr., 492 (1989) 251.
646. K. Shimada, T. Oe, and T. Nambara, J. Chromatogr., 492 (1989) 345.
647. K.-H. Xie, S. Cilgan, and I. S. Krull, J. Liq. Chromatogr., 6 (1983) 125.
648. H. Jansen, U. A. Th. Brinkman, and R. W. Frei, J. Chromatogr. Sci., 23 (1985) 279.
649. W. R. La Course and I. S. Krull, Trends. Anal. Chem., 4 (1985) 118.
650. L. Dalgaard, Trends Anal. Chem., 5 (1986) 185.
651. A. H. M. Scholten, U. A. Th. Brinkman and R. W. Frei, Anal. Chem., 54 (1982) 1932.
652. H. Engelhart, and U. D. Neue, Chromatographia, 15 (1982) 403.
653. R. S. Deelder, A. T. J. M. Kuijpers, and J. H. M. Van den Berg, J. Chromatogr., 255 (1983) 545.
654. G. Marko-Varga and L. Gorton, Anal. Chim. Acta, 234 (1990) 13.
655. S. Van der Wal, J. Liq Chromatogr., 6 (1983) 37.
656. H. Engelhart and B. Lillig, J. High Resolut. Chromatogr., 8 (1985) 531.
657. J. R. Poulsen, K. S. Birks, M. S. Grandelman, and J. W. Birks, Chromatographia, 22 (1986) 231.

658. R. S. Deelder, M. G. F. Kroll, and J. H. M. Van den Berg, J. Chromatogr., 125 (1976) 307.
659. M. Neider and H. Jaeger, J. Chromatogr., 413 (1987) 207.
660. W. J. Bachman and J. T. Stewart, J. Chromatoge., 481 (1989) 121.
661. J. R. Poulsen and J. W. Birks, Anal. Chem., 61 (1989) 2267.
662. W. J. Bachman and J. T. Stewart, J. Chromatogr. Sci., 28 (1990) 123.
663. L. Dou and I. S. Krull, Anal. Chem., 62 (1990) 2599.
664. G. Schwedt, "Chromatographic Methods in Inorganic Analysis", Huethig, Heidelberg, 1981.
665. T. R. Crompton, "Gas Chromatography of Organometallic Compounds", Plenum Press, New York, NY, 1982.
666. G. Guiochon and C. Pommier, "Gas Chromatography of Inorganics and Organometallics", Ann Arbor Science Publishers, Ann Arbor, 1973.
667. J. C. Macdonald (Ed.), "Inorganic Chromatographic Analysis", Wiley, New York, NY, 1985.
668. R. M. Pomaville and C. F. Poole, Anal. Chim. Acta, 200 (1987) 151.
669. P. C. Uden and D. E. Henderson, Analyst, 102 (1977) 889.
670. P. C. Uden, J. Chromatogr., 313 (1984) 3.
671. K. Robards, E. Patsalides, and S. Dilli, J. Chromatogr., 411 (1987) 1.
672. C. I. Measures and J. M. Edmond, Anal. Chem., 61 (1989) 544.
673. R. C. Gurira and P. W. Carr, J. Chromatogr. Sci., 20 (1982) 461.
674. G. Nickless, J. Chromatogr., 313 (1984) 129.
675. H. Ge and G. G. Wallace, Anal. Chem., 60 (1988) 830.
676. B. D. Karcher and I. S. Krull, J. Chromatogr. Sci., 25 (1987) 472.
677. S. R. Hutchins, P. R. Haddad, and S. Dilli, J. Chromatogr., 252 (1982) 185.
678. P. Jones, P. J. Hobbs, and L. Ebdon, Analyst, 109 (1984) 703.
679. P. Jones, P. J. Hobbs, and L. Ebdon, Anal. Chim. Acta, 149 (1983) 39.
680. S. Elchuk and R. M. Cassidy, Anal. Chem., 51 (1979) 1434.
681. P. K. Dasgupta, J. Chromatogr. Sci., 27 (1989) 422.
682. Y. Shimoishi, J. Chromatogr., 136 (1977) 85.
683. C. F. Poole, N. J. Evans, and D. G. Wibberley, J. Chromatogr., 136 (1977) 73.
684. P. Jones and G. Nickless, J. Chromatogr., 89 (1974) 201.
685. G. Yamamoto, K. Yoshitake, T. Kimura, and T. Ando, Anal. Chim. Acta, 222 (1989) 121.
686. G. Nota and R. Polombari, J. Chromatogr., 84 (1973) 37.
687. St. Grays, J. Chromatogr., 100 (1974) 43.
688. J. C. Hoffsommer, D. J. Glover, and D. Krubose, J. Chromatogr., 103 (1975) 182.
689. R. L. Tanner, R. J. Fajer, and J. Gaffney, Anal. Chem., 51 (1979) 865.
690. H.-L. Wu, S.-H. Chen, K. Funazo, M. Tanaka, and T. Shono, J. Chromatogr., 291 (1984) 409.
691. K. Funazo, M. Tanaka, K. Morita, M. Kamino, T. Shono, and H.-L. Wu, J. Chromatogr., 346 (1985) 215.
692. H. Jork, W. Funk, W. Fischer, and H. Wimmer, "Thin-Layer Chromatography. Physical and Chemical Detection Methods", VCH Publishers, New York, NY, vol. 1a, 1990.
693. J. G. Kirchner, "Thin-Layer Chromatography", Wiley, New York, 2nd. Edn., 1978.
694. J. C. Touchstone and M. F. Dobbins, "Practice of Thin-Layer Chromatography ", Wiley, New York, NY, 1978.

942

695. E Stahl, "Thin-Layer Chromatography", Springer-Verlag, New York, 1969.
696. G. D. Barrett, Adv. Chromatogr., 11 (1974) 145.
697. F. Kreuzig, Chromatographia, 13 (1980) 288.
698. W. Funk, R. Kerler, J.-Th. Schiller, V. Damman, and F. Arndt, J. High Resolut. Chromatogr., 5 (1982) 534.
699. W. Funk, Fresenius Z. Anal. Chem., 318 (1984) 228.
700. C. F. Poole, S. K. Poole, T. A. Dean, and N. M. Chirco, J. Planar Chromatogr., 2 (1989) 180.
701. W. R. G. Baeyens and B. L. Ling, J. Planar Chromatogr., 1 (1988) 198.
702. H. Jork, W. Funk, W. Fischer, and H. Wimmer, J. Planar Chromatogr., 1 (1988) 280.
703. J. C. Touchstone (Ed.), "Advances in Thin-Layer Chromatography: Clinical and Environmental Applications", Wiley, New York, NY, 1982.
704. K. Y. Lee, D. Nurok, and A. Zlatkis, J. Chromatogr., 174 (1979) 187.
705. D. A. Leathard and B. C. Shurlock, "Identification Techniques in Gas Chromatography", Wiley, New York, NY, 1970.
706. M. C. Bowman and M. Beroza, Anal. Chem., 38 (1966) 1544.
707. L. S. Ettre and W. H. McFadden (Eds.), "Ancillary Techniques of Gas Chromatography", Wiley, New York, NY, 1969.
708. M. R. Tirgan and N. Sharifi-Sandjani, Analyst, 105 (1980) 441.
709. M. Cooke, G. Nickless, A. M. Prescott, and D. J. Roberts, J. Chromatogr., 156 (1978) 293.
710. M. Cooke, G. Nickless, and D. J. Roberts, J. Chromatogr., 187 (1980) 47.
711. P. A. Kennedy, D. J. Roberts, and M. Cooke, J. Chromatogr., 249 (1982) 257.
712. L. R. Hogge and J. G. Millar, Adv. Chromatogr., 27 (1987) 299.
713. A. B. Attygalle and E. D. Morgan, Angew. Chem. Int. Ed. Engl. 27 (1988) 460.
714. D. G. Ollett, A. B. Attygale, and E. D. Morgan, J. Chromatogr., 367 (1986) 207.
715. A. B. Attygale and E. D. Morgan, Anal. Chem., 55 (1983) 1379.
716. A. B. Attygale and E. D. Morgan, J. Chromatogr., 290 (1984) 321.
717. L. S. Silbert and T. A. Foglia, Anal. Chem., 57 (1985) 1404.
718. V. G. Berezkin, "Analytical Reaction Gas Chromatography", Plenum Press, New York, NY, 1968.
719. P. Kabo, J. Chromatogr., 205 (1981) 39.
720. J. E. Picker and R. E. Sievers, J. Chromatogr., 217 (1981) 275.
721. R. W. Souter, "Chromatographic Separations of Stereoisomers" CRC Press, Boca Raton, FL, 1985.
722. B. Testa, Xenobiotica, 16 (1986) 265.
723. I. Wainer and D. Drayer (Eds.), "Drug Stereochemistry: Analytical Methods and Pharmacology", Dekker, New York, NY, 1987.
724. W. A. Konig, "The Practice of Enantiomer Separation by Capillary Gas Chromatography", Huethig, Heidelberg, 1987.
725. H. K. Lim, J. W. Hubbard, and K. K. Midha, J. Chromatogr., 378 (1986) 109.
726. R. W. Souter, J. Chromatogr., 108 (1975) 265.
727. W. A. Konig, K. Stoelting, and K. Kruse, Chromatographia, 10 (1977) 444.
728. K. J. Miller, J. Gal, and M. M. Ames, J. Chromatogr., 307 (1984) 335.
729. A. J. Sedman and J. Gal, J. Chromatogr., 306 (1984) 155.

730. J. L. Beneytout, M. Tixier, and M. Rigaud, J. Chromatogr., 351 (1986) 363.
731. J. W. Westley and B. Halpern, J. Org. Chem., 33 (1968) 3978.
732. R. G. Arnett and P. K. Stumpf, Anal. Biochem., 47 (1972) 638.
733. C. J. W. Brooks, M. T. Gilbert, and J. D. Gilbert, Anal. Chem., 45 (1973) 896.
734. W. A. Konig, J. High Resolut. Chromatogr., 5 (1982) 588.
735. S. Hammarstrom and M. Hambert, Anal. Biochem., 52 (1973) 169.
736. J. P. Kamerlin, M. Duran, G. J. Gerwig, D. Ketting, L. Bruinvis, J. F. G. Vliegenthart, and S. K. Wadman, J. Chromatogr., 222 (1981) 276.
737. W. A. Konig and I. Benecke, J. Chromatogr., 195 (1980) 292.
738. J. P. Kamerling, G. J. Gerwig, J. F. G. Vliengenthart, M. Duran, D. Ketting, and S. K. Wadman, J. Chromatog., 143 (1977) 117.
739. D. M. Johnson, A. Reuter, J. M. Collins, and G. F. Thompson, J. Pharm. Sci., 68 (1979) 112.
740. J.-M. Maitre, G. Boss, and B. Testa, J. Chromatogr., 299 (1984) 397.
741. W. E. Pereira, M. Salomon, and B. Halpern, Aust. J. Chem., 24 (1971) 1103.
742. M. Zief and L. J. Crane, "Chromatographic Chiral Separations", Dekker, New York, NY, 1988.
743. H. Nagashima, Y. Tanaka, and R. Hayashi, J. Chromatogr., 345 (1985) 373.
744. J. Goto, M. Hasegawa, S. Nakamura, K. Shimada, and T. Nambara, J. Chromatogr., 152 (1978) 413.
745. J. Hermansson and C. Von Bahr, J. Chromatogr., 221 (1980) 109.
746. A. A. Gulaid, G. W. Houghton, and A. R. Boobis, J. Chromatogr., 318 (1985) 393.
747. M. J. Wilson and T. Walle, J. Chromatogr., 310 (1984) 424.
748. T. Kinoshca, Y. Kasahara, and N. Nimura, J. Chromatogr., 210 (1981) 77.
749. T. Walle, D. D. Christ, U. K. Walle, and M. J. Wilson, J. Chromatogr., 341 (1985) 213.
750. J. Gal, J. Liq. Chromatogr., 9 (1986) 673.
751. D. E. Nichols, A. J. Hoffman, R. A. Oberlender, P. Jacob, and A. T. Shulgin, J. Med. Chem., 29 (1986) 2009.
752. D. Valentine, K. K. Chan, C. G. Scott, K. K. Johnson, K. Troth, and G. Saucy, J. Org. Chem., 41 (1976) 62.
753. M. Ahnoff, S. Chen, A. Green, and I. Grundevik, J. Chromatogr., 506 (1990) 593.
754. C. Petterson and G. Schill, J. Chromatogr., 9 (1986) 269.
755. C. Petterson and M. Josefsson, Chromatographia, 21 (1986) 321.
756. F. Bruner (Ed.) "The Science of Chromatography", Elsevier, Amsterdam, 1985.
757. R. H. Liu and W. W. Ku, J. Chromatogr., 271 (1983) 309.
758. H. Frank, J. High Resolut. Chromatogr., 11 (1988) 787.
759. T. Saeed, P. Sandra, and M. Verzele, J. Chromatogr., 186 (1980) 611.
760. W. Roder, F.-J. Ruffing, G. Schomburg, and W. H. Pikle, J. High Resolut. Chromatogr., 10 (1987) 665.
761. G. Schomburg, I. Benecke, and G. Severin, J. High Resolut. Chromatogr., 8 (1985) 391.
762. W. A. Konig and I. Benecke, J. Chromatogr., 269 (1983) 19.
763. B. Koppenhoefer and E. Bayer, Chromatographia, 19 (1985) 123.
764. H. Frank, G. Nicholson, and E. Bayer, J. Chromatogr., 167 (1978) 187.
765. W. A. Konig and U. Sturm, J. Chromatogr., 328 (1985) 357.
766. W. A. Konig, K. Ernst, and J. Vessman, J. Chromatogr., 294 (1984) 423.

944

767. W. A. Konig and K. Ernst, J. Chromatogr., 280 (1983) 135.
768. K. Nakamura, S. Hara, and Y. Dobashi, Anal. Chem., 61 (1989) 2121.
769. W. A. Konig, "Enantioselective Gas Chromatography With Modified Cyclodextrins", Huethig, Heidelberg, 1991.
770. D. W. Armstrong, W. Li, A. M. Stalcup, H. V. Secor, R. R. Izac, and J. I. Seeman, Anal. Chim. Acta, 234 (1990) 365.
771. D. W. Armstrong, W. Li, C.-D. Chang, and J. Pitha, Anal. Chem., 62 (1990) 994.
772. W. A. Konig, R. Krebber, and P. Mischnick, J. High Resolut. Chromatogr., 12 (1989) 732.
773. W. A. Konig, R. Krebber, and G. Wenz, J. High Resolut. Chromatogr., 12 (1989) 790.
774. W. A. Konig, R. Krebber, P. Evers, and G. Bruhn, J. High Resolut. Chromatogr., 13 (1990) 328.
775. V. Schuring, M. Jung, D. Schmalzing, M. Scleimer, J. Duvelzot, J. C. Buyten, J. A. Peene, and P. Mussche, J. High Resolut. Chromatogr., 13 (1990) 470.
776. W. Blum and R. Aichholz, J. High Resolut. Chromatogr., 13 (1990) 515.
777. H.-P. Nowotny, D. Schmalzing, D. Wistuba, and V. Schuring, J. High Resolut. Chromatogr., 12 (1989) 383.
778. A. Mosandl, K. Rettinger, K. Fischer, V. Schubert, H.-G. Schmarr, and B. Maas, J. High Resolut. Chromatogr., 13 (1990) 382.
779. V. Schuring, D. Schmalzing, U. Muhleck, M. Jung, M. Schleimer, P. Mussche, C. Duvekot, and J. C. Buyten, J. High Resolut. Chromatogr., 13 (1990) 713.
780. A. C. Mehta, J. Chromatogr., 426 (1988) 1.
781. D. W. Armstrong, Anal. Chem., 59 (1987) 84A.
782. W. Lindner, Chromatographia, 24 (1987) 97.
783. W. H. Pirkle and T. C. Pochapsky, Adv. Chromatogr., 27 (1987) 73.
784. I. W. Wainer, Trends. Anal. Chem., 6 (1987) 125.
785. R. Dappen, H. Arm, and V. R. Meyer, J. Chromatogr., 373 (1986) 1.
786. A. M. Krstulovic (Ed.), "Chiral Separations in HPLC", Ellis Horwood, Chichester, UK, 1989.
787. W. J. Lough, "Chiral Liquid Chromatography", Blackie, Glasgow, UK, 1989.
788. S. G. Allenmark, "Chromatographic Enantioseparations: Methods and Applications", Ellis Horwood, Chichester, UK, 1988.
789. D. W. Armstrong and S. M. Han, CRC Crit. Revs. Anal. Chem., 19 (1988) 175.
790. V. A. Davankov, Chromatogrphia, 27 (1989) 475.
791. G. Gubitz, Chromatographia, 30 (1990) 555.
792. W. H. Pirkle and T. C. Pochapsky, Chem. Revs., 89 (1989) 347.
793. M. H. Hyun and W. H. Pirkle, J. Chromatogr., 393 (1987) 357.
794. W. H. Pirkle and T. J. Sowin, J. Chromatogr., 396 (1987) 83.
795. W. H. Pirkle and T. J. Sowin, J. Chromatogr., 387 (1987) 313.
796. I. W. Wainer and M. C. Alembik, J. Chromatogr., 367 (1986) 59.
797. S. C. Dhanesar and D. J. Gisch, J. Chromatogr., 461 (1988) 407.
798. T. Shibata, I. Okamoto, and K. Ishii, J. Liq. Chromatogr., 9 (1986) 313.
799. G. Blaschke, J. Liq. Chromatogr., 9 (1986) 341.
800. A. M. Rizzi, J. Chromatogr., 478 (1989) 71, 87, and 101.
801. R. Isaksson, P. Erlandsson, L. Hanson, A. Holmberg, and S. Berner, J. Chromatogr., 498 (1990) 257.
802. Y. Okamoto and Y. Kaida, J. High Resolut. Chromatogr., 13 (1990) 708.

803. Y. Fukui, A. Ichida, T. Shibata, and K. Mori, J. Chromatogr., 515 (1990) 85.
804. Y. Okamoto and K. Hatada, J. Liq Chromatogr., 9 (1986) 369.
805. T. J. Ward and D. W. Armstrong, J. Liq. Chromatogr., 9 (1986) 407.
806. D. W. Armstrong, A. M. Stalcup, M. L. Hilton, J. D. Duncan, J. R. Faulkner, and S.-C. Chang, Anal. Chem., 62 (1990) 1610.
807. S. Allenmark, J. Liq. Chromatogr., 9 (1986) 425.
808. J. Hermansson and M. Eriksson, J. Liq Chromatogr., 9 (1986) 621.
809. J. Hermasson, K. Strom, and R. Sundberg, Chromatographia, 24 (1987) 520.
810. J. Hermansson, Trends Anal. Chem., 8 (1989) 251.
811. V. Schurig, R. C. Chang, A. Zlatkis, E. Gil-Av, and F. Mikes, Chromatographia, 6 (1973) 223.
812. W. Szczepaniak, J. Nawrocki, and W. Wasiak, Chromatographia, 12 (1979) 484.
813. B. Vonach and G. Schomburg, J. Chromatogr., 149 (1978) 417.
814. M. G. M. de Ruyter and A. P. de Leenheer, Anal. Chem., 51 (1979) 43.
815. P. L. Phelan and J. R. Miller, J. Chromatogr. Sci., 19 (1981) 13.
816. L. D. Kissinger and R. H. Robins, J. Chromatogr., 321 (1985) 353.
817. V. Schurig and R. Weber, J. Chromatogr., 289 (1984) 321.
818. V. Schurig, U. Leyrer, and R. Weber, J. High Resolut. Chromatogr., 8 (1985) 459.
819. V. Schuring, W. Burkle, K. Hintzer, and R. Weber, J. Chromatogr., 475 (1989) 23.
820. V. A. Davankov, Adv. Chromatogr., 18 (1980) 139.
821. S. Lam and A. Karmen, J. Liq Chromatogr., 9 (1986) 291.
822. V. A. Davankov, J. D. Navratil, and H. F. Walton, "Ligand Exchange Chromatography", CRC Press, Boca Raton, FL, 1988.
823. A. Duchateau, M. Crombach, M. Ausserns, and J. Bongers, J. Chromatogr., 461 (1989) 419.
824. S. Lam and A. Karmen, J. Chromatogr., 289 (1984) 339.
825. E. Grushka and F. K. Chow, J. Chromatogr., 199 (1980) 283.
826. H. Kelker, Adv. Liq. Cryst., 3 (1978) 237.
827. G. M. Janini, Adv. Chromatogr., 17 (1979) 231.
828. Z. Witkiewicz, J. Chromatogr., 251 (1982) 311.
829. Z. Witkiewicz, J. Chromatogr., 466 (1989) 37.
830. J. Szulc and Z. Witkiewicz, J. Chromatogr., 262 (1983) 141.
831. W. Marciniak and Z. Witkiewicz, J. Chromatogr., 324 (1985) 299 and 309.
832. Z. Suprynowicz, W. M. Buda, M. Mardarowicz, and Z. Witkiewicz, Chromatographia, 19 (1985) 418.
833. J. Szulc, Z. Witkiewicz, and A. Ziolek, J. Chromatogr., 262 (1983) 161.
834. M. A. Apfel, H. Fikelmann, G. M. Janini, R. J. Laub, B.-H. Luhmann, A. Price, W. L. Robert, T. J. Shaw, and C. A. Smith, Anal. Chem., 57 (1985) 651.
835. S. Rokushika, K. P. Naikwadi, A. L. Jadhav, and H. Hatano, J. High Resolut. Chromatogr., 8 (1985) 480.
836. K. E. Markides, H.-C. Chang, C. M. Schregenberger, B. J. Tarbet, J. S. Bradshaw, and M. L. Lee, J. High Resolut. Chromatogr., 8 (1985) 516.
837. J. S. Bradshaw, C. M. Schregenberger, H.-C. Chang, K. E. Markides, and M. L. Lee, J. Chromatogr., 358 (1986) 95.
838. N. Nishioka, B. A. Jones, B. J. Tarbet, J. S. Bradshaw, and M. L. Lee, J. Chromatogr., 357 (1986) 79.

839. K. E. Markides, M. Nishioka, B. J. Tarbet, J. S. Bradshaw, and M. L. Lee, Anal. Chem., 57 (1985) 1296.
840. V. G. Berezkin, CRC Crit. Revs. Anal. Chem., 11 (1981) 1.
841. W. J. Irwin, "Analytical Pyrolysis: A Comprehensive Guide", Dekker, New York, NY, 1982.
842. S. A. Liebman and E. J. Levy, "Pyrolsis and GC in Polymer Analysis", Dekker, New York, NY, 1985.
843. K. J. Vooehees, "Analytical Pyrolysis: Techniques and Application", Butterworth, London, UK, 1984.
844. R. S. Lehrle, J. Anal. Appl. Pyrol., 11 (1987) 55.
845. S. T. Jones, Analyst, 109 (1984) 823.
846. S. A. Liebman, T. P. Wampler, and E. J. Levy, in P. Sandra (Ed.), Sample Introduction in Capillary Gas Chromatography, Huthig, Heidelberg, Vol. 1, 1985, p. 165.
847. E. L. Colling, B. H. Burda, and P. A. Kelley, J. Chromatogr. Sci., 24 (1986) 7.
848. C. J. Wolf, M. A. Grayson, and D. L. Fanter, Anal. Chem., 52 (1980) 348A.
849. W. G. Fischer and P. Kuseh, J. Chromatogr., 518 (1990) 9.
850. J. W. Washall and T. P. Wampler, J. Chromatogr. Sci., 27 (1989) 144.
851. A. Venema and J. Veurink, J. Anal. Appl. Pyrol., 7 (1985) 207.
852. T. P. Wampler and E. J. Levy, J. Anal. Appl. Pyrol., 12 (1987) 75.
853. E. J. Levy and J. Q. Walker, J. Chromatogr. Sci., 22 (1984) 49.
854. T. P. Wampler and E. J. Levy, J. Anal. Appl. Pyrol., 8 (1985) 65.
855. R. S. Whiton and S. L. Morgan, Anal. Chem., 59 (1985) 778.
856. S. A. Liebman. T. P. Wamper, and E. J. Levy, J. High Resolut. Chromatogr., 7 (1984) 172.
857. R. Milina, N. Dimov, and M. D. Dimitrova, Chromatographia, 17 (1983) 29.
858. J. J. R. Mertens, E. Jacobs, A. J. A. Callaerts, and A. Buekens, Anal. Chem., 54 (1982) 2620.
859. H. Engman, H. T. Mayfield, T. Mar, and W. Bertsch, J. Anal. Appl. Pyrol., 6 (1984) 137.
860. C. S. Smith, S. L. Morgan, C. D. Parks, A. Fox, and D. G. Pritchard, Anal. Chem., 59 (1987) 1410.

CHAPTER 9

HYPHENATED METHODS FOR IDENTIFICATION AFTER CHROMATOGRAPHIC
SEPARATION

9.1 INTRODUCTION

A chromatogram provides information regarding the complexity (number of components), quantity (peak height or area) and identity (retention parameter) of the components in a mixture. Of these parameters the certainty of identification based solely on retention is considered very suspect, even for simple mixtures. When the identity can be firmly established the quantitative information from the chromatogram is very good. The reverse situation applies to spectroscopic techniques which provide a rich source of qualitative information from which substance identity may be inferred with a reasonable degree of certainty. Spectroscopic instruments have, however, two practical limitations: it is often difficult to extract quantitative information from their signals and, for reliable identification, they require pure samples. Chromatographic and spectroscopic techniques thus provide complementary information about the identity of the components and their concentration in a sample. This provides the driving force for the inception of combined instruments, often referred to as "hyphenated" systems. The principal hyphenated techniques are gas chromatography interfaced to mass spectrometry (GC/MS) and Fourier transform infrared spectrometry (GC/FTIR); liquid chromatography interfaced to mass spectrometry (LC/MS), Fourier transform infrared spectrometry (LC/FTIR) and nuclear magnetic resonance spectroscopy (LC/NMR); supercritical fluid chromatography interfaced to mass spectrometry (SFC/MS) and Fourier transform infrared spectrometry (SFC/FTIR); and thin-layer chromatography interfaced with mass spectrometry (TLC/MS), Fourier transform infrared spectrometry (TLC/FTIR) and surface enhanced Raman spectroscopy (TLC/SERS). In recent times serial flow GC/FTIR/MS instruments have reached commercial maturity. All the above techniques are data intensive and require the use of computers to control the acquisition, storage, display and manipulation of data. Computers have also been important in reducing the difficulty of data interpretation bringing acceptance of these instruments to non-spectroscopists.

9.2 INSTRUMENTAL REQUIREMENTS FOR MASS SPECTROMETRY

A mass spectrum is a histogram of the relative abundance of individual ions having different mass-to-charge ratios (m/z) generated from a sample of, in most cases, neutral molecules. The mass spectrum is a molecular fingerprint of the molecule conveying information about the molecular weight of the sample and, if fragmentation occurs during ionization, structurally useful products yielding information characteristic of the position and bonding order of the molecular substructures. Through interpretation, reassembly of the molecular substructures using sets of rules and intuition allows a molecule to be identified even if its mass spectrum was previously unknown, at least in the most favorable case.

A mass spectrometer is a sophisticated measurement instrument and the processes of ionization, separation of the ions in a vacuum according to their m/z ratio, and ion detection are complex processes [1-8]. Only the briefest of reviews will be given here to enable the reader to appreciate the problems of interfacing a mass spectrometer to chromatographic instruments.

The ionization of organic molecules may be achieved in many ways; those of principal importance for use with chromatographic inlets include electron impact, chemical ionization, atmospheric pressure ionization, fast atom bombardment, thermospray and electrospray. Sources for electron impact ionization consist of a heated, evacuated chamber in which a beam of electrons with a narrow energy distribution is generated from a heated metal filament. The energy of the ionizing electrons is controlled normally by the accelerating voltage established between the cathode and the source housing. Electrons with energies in the range of 5 to 100 eV may be used but, unless otherwise stated, a value of 70 eV is standard practice. Since most organic compounds have ionization potentials of 7 to 20 eV, the energy transferred on collision between the electron and a neutral molecule is sufficient to cause both ionization and extensive fragmentation. The majority of ions formed by this process are singly-charged parent ions or molecular fragments, as well as a few multiply-charged ions and some negatively-charged ions. In most cases the number of negatively-charged ions is only a small fraction of the total number of ions formed. Ionization takes place at a temperature sufficient to maintain the sample in the vapor phase at a pressure below 10^{-5} Torr, which is sufficient to ensure that the average mean free path of the ion is large enough for it to escape the source without undergoing a significant number of ion-molecule collisions. The positive ions are extracted from the ion source by a repeller electrode having a small positive potential which directs them towards the analyzer slits. Alternatively, they may be extracted by the field penetration of the high voltage on the focusing electrodes. In both instances the ion beam is usually focused, collimated and accelerated to provide a beam of narrow energy dispersion that is capable of traversing the analyzer section of the mass spectrometer. In modern mass spectrometers the ionization source and analyzer sections are usually differentially pumped, allowing the source to operate at a distinctly higher pressure than the analyzer unit, for which pressures below 10^{-7} Torr are common.

The bombardment of a neutral molecule with a beam of high energy electrons results in the formation of a molecular (or parent) ion radical with excess energy. The molecular ion provides useful information concerning the identity of the molecule as it defines the molecular weight. Unfortunately, under electron impact conditions this ion may be too unstable to be present in the spectrum even when beams of comparatively low energy (e.g., 15-20 eV) are used. In these circumstances a softer ionization method, such as chemical ionization, may be used to obtain molecular weight information. As well as providing information complementary to electron impact generated mass spectra, chemical ionization methods are blessed with other characteristics which makes them suitable for the trace analysis of organic compounds.

Chemical ionization mass spectra are produced by ion-molecule reactions between neutral sample molecules and a high pressure (0.2-2 Torr) reagent gas ion plasma [10-12]. Fundamentally and practically several differences exist between chemical and electron impact ionization. In the chemical ionization source the concentration of reagent gas molecules exceeds that of the sample by several orders of magnitude. Thus, the electron beam ionizes the reagent gas with little direct ionization of the sample molecules. As the source is operated at high pressures compared to electron impact sources, ion-molecule reactions are now favored and promote sample ionization. These collisional processes are gentler than electron impact ionization and produce stable molecular ions or molecular ion adducts with little additional fragmentation. Chemical ionization sources are operated at much higher electron energies (200-500 eV) to ensure penetration of the ionizing electrons into the active source volume and require a "tight source", which is differentially pumped, because of the large pressure difference between the source and analyzer sections of the mass spectrometer. The high pressure conditions in the chemical ionization source also favor the production of thermal electrons which may be captured by molecules with a high electron affinity to form negatively charged molecular ions or molecular ion fragments after dissociation [11,13]. The negative ion population may be two or three orders of magnitude greater than that generated under electron impact conditions, making negative chemical ionization mass spectrometry a viable analytical technique. Since many compounds which readily form negative ions are biologically active or toxic substances, negative chemical ionization mass spectrometry is an important technique in biomedical and environmental studies. Its areas of general application parallel those of the electron-capture detector used in gas chromatography. Whereas all mass spectrometers are designed for the separation of positive ions, not all instruments support the separation of negative ions or are easily converted to do so, particularly among the low-cost instruments.

Several reagent gases are used in chemical ionization mass spectrometry. A combination of electron impact and ion-molecule reactions results in the formation of a plasma containing a steady-state concentration of ions. These ions can ionize neutral molecules by collision induced processes involving proton transfer, hydride abstraction, charge exchange and adduct formation. For methane, one of the most widely used reagent gases, at a source pressure of approximately one Torr the dominant plasma ions consist mainly of $[CH_5]^+$ (48%), $[C_2H_5]^+$ (40%) and $[C_3H_5]^+$ (6%). In the gas phase, $[CH_5]^+$ and $[C_2H_5]^+$ ions function as strong acids and can transfer a proton to the sample as shown below.

$$[CH_5]^+ + M \longrightarrow [MH]^+ + CH_4$$

$$[C_2H_5]^+ + M \longrightarrow [MH]^+ + C_2H_4$$

The $[C_2H_5]^+$ ion can also function as a Lewis acid, forming collision-stabilized complexes or molecular ion adducts:

$$[C_2H_5]^+ + M \longrightarrow [M(C_2H_5)]^+$$

The product ions generated by chemical ionization are stable even-electron species with relatively little excess energy compared to those generated by electron impact. The chemical ionization mass

TABLE 9.1

PRINCIPAL REAGENT IONS FORMED UNDER CHEMICAL IONIZATION CONDITIONS

Reagent Gas	Predominant Reagent Ions at ca. One Torr
Methane	CH_5^+, $C_2H_5^+$, $C_3H_5^+$
Propane	$C_3H_7^+$, $C_3H_8^+$
Isobutane	$C_4H_9^+$
Hydrogen	H_3^+
Ammonia	NH_4^+, $(NH_3)_2H^+$, $(NH_3)_3H^+$
Water	H_3O^+
Tetramethylsilane	$(CH_3)_3Si^+$
Dimethylamine	$(CH_3)_2NH_2^+$, $[(CH_3)_2NH]_2H^+$, $C_3H_8N^+$
Helium	He^+

spectrum is characterized, therefore, by the presence of a few intense molecular ion adducts with very little further fragmentation. With some experience or knowledge of the sample type, the sample molecular weight is readily identified from the m/z value of the molecular ion adducts. The nature of the molecular ion adduct formed depends primarily on the sample composition and the identity of the reagent gas. Some typical reagent gases and the predominant reactive ions formed in the chemical ionization plasma are summarized in Table 9.1 [4]. Water and ammonia are relatively mild protonating reagents. With tetramethylsilane, trimethylsilylation rather than protonation occurs while dimethylamine appears to selectively ionize carbonyl compounds, with protonated dimethylamine apparently attached to the carbonyl group. Nitric oxide reacts with ketones, esters and carboxylic acids to give primarily $[M + NO]^+$ ions. With aldehydes and esters hydride abstraction occurs. The rare gases, such as He, Ar and Xe cause ionization by a charge transfer mechanism. The reagent ion abstracts an electron from the sample which results in the formation of an odd electron molecular ion, the same product as would be expected from electron impact, except that it is formed under less energetic conditions.

$$[He]^+ + M \longrightarrow [M]^+ + He$$

The appearance and reproducibility of chemical ionization mass spectra depends on the ionizing conditions, principally the source temperature and pressure and the purity of the reagent gas. Chemical ionization mass spectra are generally not as reproducible as electron impact spectra.

Atmospheric pressure ionization sources differ from the other ionization methods discussed so far in that ionization occurs at atmospheric pressure under conditions where equilibrium between ions and neutral molecules is likely to occur [15]. Reagent ions are generated by bombarding an inert gas such as nitrogen with electrons generated from a nickel-63 foil or from a corona discharge device. This results in the formation of a plasma consisting of electrons and ions formed by ionization of the carrier gas and by ion-molecule reactions involving trace impurities in the carrier gas, such as water and oxygen, or solvent molecules admitted to the source. Ionization of sample molecules results from ion-molecule reactions of these impurity ions and cluster ions as well as from direct electron attachment reactions. Both positive and negative ions are generated by hydride abstraction or addition, charge exchange, electron capture and adduct formation with cluster ions. Sample and plasma ions are sampled through a pinhole separating the source from the analyzer section of the mass spectrometer. Atmospheric pressure ionization sources are not available for most commercial mass spectrometers and, although capable of use for both gas and liquid chromatography, most recent research has focused on their use as an interface for LC/MS.

The above ionization techniques require that the sample can be vaporized prior to ionization precluding the analysis of many labile, high molecular weight or ionic samples. Alternative ionization techniques using thermospray, electrospray or fast atom bombardment are needed for these samples. In the case of fast atom bombardment (FAB), the sample is dissolved in a viscous solvent (e.g., glycerol) and then bombarded by a beam of high velocity particles, usually rare gas atoms with an energy between 3-8 KeV [16]. This results in the expulsion of material into the gas phase by momentum transfer (sputtering), in which some of the sputtered material will be in the form of both positively and negatively charged ions. These ions are then extracted by a slit lens system and directed to the mass analyzer. The mass spectrum is dominated by the formation of even electron product molecular ions of the type $[M + H]^+$ and $[M - H]^-$ sometimes accompanied by cluster ions involving the psuedomolecular ion combined with various numbers of unionized parent molecules or solvent matrix molecules. In most cases sputtering and ionization occurs with the formation of a small number of fragment ions sufficient to yield structural as well as molecular weight information about the sample. Thermospray (TSP) is a combined liquid introduction and sample ionization method for LC/MS [17-19]. Thermospray results in the formation of a supersonic jet of small droplets, at least some of which are electrically charged, when the mobile phase is partly aqueous and a volatile buffer, such as ammonium acetate, is also present. An equal number of positively and negatively charged droplets are formed leading to the production of both positive and negative sample ions. As the droplets evaporate the electric field at the liquid surface increases until ions present in the liquid phase are ejected from the droplet (ion-evaporation). Alternatively, ions can be generated in a two-step process similar to conventional chemical ionization, whereby an ion of the electrolyte, e.g. NH_4^+, is ejected from the droplet, reacts with a sample molecule in the gas phase and generates a sample ion. The types of ions observed by either process are usually molecular adduct ions such as $[M + H]^+$, $[M + NH_4]^+$, etc., accompanied by a few low intensity, fragment ions. This ionization process is by no means universal and in unfavorable cases the number of ions generated can be increased by combining thermospray with a filament or discharge ionization source. In this case, ionization occurs under conditions similar to conventional chemical ionization in which the solvent vapor functions as the chemical ionization reagent gas.

Ions leaving the source of the mass spectrometer must be separated according to their mass-to-charge ratio prior to detection. This is achieved by imposing an external electric or magnetic field on the ion beam to affect dispersion (resolution). The resolving power of a mass spectrometer is a measure of its ability to distinguish between two neighboring masses. In general, mass spectrometers operate in two resolution ranges, low to medium resolution with values of R = 500-2000 and high resolution with values of R = 10,000-75,000. Low resolution mass spectrometry is routinely used in GC/MS and LC/MS applications. High resolution measurements provide accurate elemental composition data for all ions in the spectra or for a single ion such as the molecular ion. High resolution mass spectrometers are often referred to as double focusing, since the ion beam from the source is first focused by an electrostatic analyzer which diminishes the energy dispersion of the ion beam, and then by a magnetic analyzer which provides dispersion of the ion beam according to the mass-to-charge ratio of the ions. The combined process permits very accurate mass measurement (to within 10 ppm). Double-focusing mass spectrometers are more expensive and, when operated at high resolution, less sensitive and slower scanning than single-focusing instruments.

Rapid scanning mass spectrometers providing unit resolution are routinely used as chromatographic detectors. Ion separation is accomplished using either a magnetic sector, quadrupole filter or ion trap device. Ions can also be separated by time-of-flight or ion cyclotron resonance mass analyzers but these devices are not widely used with chromatographic inlets and will not be discussed here [20].

In a magnetic field, ions of different mass-to-charge ratio will be focused at the detector according to the conditions represented by equation (9.1)

$$m/z = (H^2r^2)/2V \qquad\qquad (9.1)$$

Where m is the mass of the ion, z the charge on the ion, H the magnetic field strength, r the radius of the circular path along which the ions are deflected, and V the ion accelerating voltage. The position of the detector and the radius of curvature of the flight tube are usually fixed so that the complete mass spectrum is recorded by scanning (i.e., sequentially bringing all ions of different m/z into focus at the detector). Scanning is accomplished by changing the magnetic field strength (H) at a constant accelerating voltage (V) or by holding the magnetic field constant and varying the accelerating voltage. For most applications it is preferable to vary the magnetic field at a constant accelerating potential to record the mass spectrum. This is partly because the efficiency with which ions are transmitted through the analyzer depends on their momentum; at lower accelerating potentials fewer ions are transmitted than at higher accelerating potentials, and thus scanning by varying the accelerating potential results in mass discrimination. Practically, this results in fewer ions of high m/z being recorded compared to those of low m/z. Scan times of 1-2 s per decade of mass (e.g., 50-500 amu) are required to accurately record the mass spectrum of early eluting peaks from an open tubular column in GC/MS. This is possible with modern laminated magnets, although it should be kept in mind that because of magnetic hysteresis, a certain amount of time is required prior to each scan for the magnetic field to reset and stabilize. Typical conditions might involve a scan time of 1.5 s and a reset time of 1.5 s. Effectively, therefore, scans can only be initiated every 3 s and chromatographic peaks with peak widths less than about 15 s may be erroneously recorded under these conditions due to the rapidly changing sample concentration over the scan period.

The quadrupole mass filter consists of four parallel hyperbolic rods in a square array such that the inside radius of the array (field radius) is equal to the smallest radius of curvature of the hyperbola. Diagonally opposite rods are electrically connected to radio frequency and direct current voltages [7,21]. For a given radio frequency/direct current voltage ratio only ions of a specific m/z value are transmitted by the filter and reach the detector. Ions with a m/z different from the transmitted ion are deflected away from the principal axis of the system and strike the rods; these ions are not transmitted by the quadrupole filter. To scan the mass spectrum the frequency of the radio frequency voltage and the ratio of the alternating current/direct current voltages are held constant while the magnitude of the alternating current and direct current voltages are varied. The transmitted ions of mass m/z are then linearly dependent on the voltage applied to the quadrupoles, producing a m/z scale which is linear with time. The voltages applied to the rods are usually chosen to give equal peak widths over the entire mass range.

Quadrupole mass spectrometers have several attractive features when compared to magnetic sector instruments. They are more compact, less expensive, and easier to operate. Electric fields can be precisely controlled more easily than magnetic fields so that tuning desired m/z values for selected ion monitoring is easily accomplished. Quadrupoles can scan at faster rates, 0.1 s/decade, with a delay time of 3 ms between scans, and are suitable for recording fast eluting peaks from open tubular columns in GC/MS (to obtain reasonable ion statistics scan times of about 1 s/decade are more common in practice). Also, quadrupole mass analyzer can be rapidly pulsed to facilitate simultaneous recording of positive and negative ion mass spectra, which is not possible with magnetic sector instruments. On the other hand, magnetic sector instruments are capable of higher resolution and operation in higher mass ranges without loss in sensitivity, which can be of overriding importance for some applications.

The introduction of low cost, bench top mass spectrometers and mass selective detectors taking advantage of the favorable operating characteristics of quadrupole mass analyzers, in large measure contributed to the wider acceptance of the mass spectrometer as a routine chromatographic detector [6,8,22]. Mounting the ion source, quadrupole analyzer, and ion detector within the diffusion pump housing of the mass spectrometer vacuum system enabled compact instruments to be designed, Figure 9.1 [2,7]. The whole instrument is operated by a desk-top computer which performs control functions as well as a full array of data processing tasks without delays. The ion trap detector, also shown in Figure 9.1, is the most recent low cost mass analyzer introduced specifically for GC/MS and SFC/MS applications [7,23,24]. Unlike conventional quadrupole analyzers ion formation, storage and scanning are performed in the same chamber. The ion trap also uses pulsed ionization (ionization is continuous for other mass spectrometers) and in its current version is capable of electron impact or chemical ionization ion formation, scan rates from 0.125-2s per mass decade and has a mass range from 10 to 650 amu. Although the ion trap detector is based on the quadrupole mass analyzer its orientation has been completely changed and the radio frequency signal controlling scanning, instead of being applied to the conventional rods, is now applied to a central, circular ring electrode situated between two end-caps held at ground potential. Sample entering the detector is ionized by thermal electrons generated from a metal filament and accelerated through an electron gate which can be opened and closed by application of an appropriate voltage. The ions formed are stored in the ion trap by the application of a radio frequency voltage to the ring electrode. The ion trap detector is operated at the comparatively high pressure of 10^{-2} to 10^{-3} Torr to dampen the motion of the stored ions and prevent their immediate loss to the end caps. Analysis of the ions is performed by increasing the radio frequency voltage in steps. This makes the ions successively unstable, such that the ions exit the trap and are detected by an electron multiplier. A unique feature of the ion trap detector is the use of an automatic gain control to simultaneously minimize the influence of space charge effects at high sample concentrations and to maximize sensitivity at low sample levels. Too many ions in the trap can result in distortion of the mass spectrum through collisions. The automatic gain control adjusts

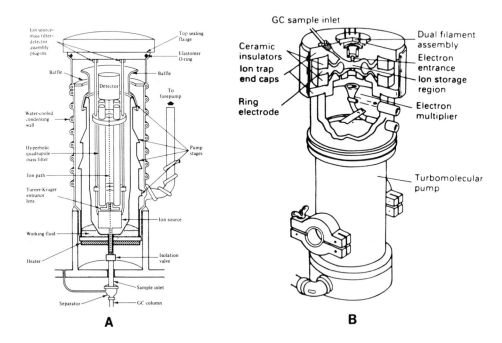

Figure 9.1 Schematic view of (A) the bench-top quadrupole mass spectrometer (Hewlett-Packard) and (B) the ion trap detector (Finnigan MAT).

the ionization time for each scan according to the product of the sample size and ionization efficiency to avoid overloading the trap at high sample concentration while maximizing the number of ions in the trap at low sample concentrations. This is accomplished by a preliminary short scan of 0.2 ms in which all ions above a certain threshold value are collected. From the preliminary scan the full scan acquisition time is estimated and the ionization time set. The ion trap detector is very sensitive and capable of providing full scan mass spectra from 2-5 pg. of sample with a dynamic range of about 10^6.

For ion detection an electron multiplier is commonly used or less frequently a Faraday cage collector. The Faraday cage collector consists of a collector plate housed inside a cup at an angle such that impinging ions cannot escape. The total ion current measured is proportional to the total number of ions collected and their charge. The response is independent of the mass and energy of the ions and mass discrimination is low. Obtaining measurable currents requires large amplification circuits which are slow and prevent fast scanning. The electron multiplier works on the principle of secondary electron emission, in which an ion impinging on a conversion dynode ejects several electrons which are then accelerated to another electrode (dynode) releasing additional electrons. The sequence is repeated through a number of stages until the final current is detected. The electron multiplier is capable of current amplification of 10^3 to 10^8 with negligible background noise. Its response is very fast so signal distortion is generally eliminated. However, electron multiplier detectors show poor gain stability, a gain that is dependent upon the charge, mass, angle of incidence, and nature of the impacting ion, as well as the work function and surface condition of the impacted surface. Saturation effects limit the dynamic range of the detector and are usually significant when the output current exceeds 10^{-8} A. These characteristics are obviously not desirable in an ion detector but the high gain of the electron multiplier makes it indispensable and its short-comings must be tolerated.

9.3 INTERFACE REQUIREMENTS FOR CHROMATOGRAPHIC SAMPLE INTRODUCTION SYSTEMS IN MASS SPECTROMETRY

Problems arise in interfacing column chromatographic techniques to a mass spectrometer from the difference in material flow requirements between the two instruments and the desire to generate information about the sample without interference from the mobile phase in which it is diluted. The most favorable case occurs for gas chromatography where the mobile phases commonly used do not generally influence the spectra observed and the sample, being in the vapor phase, is compatible with the widest range of mass spectral ionization techniques. The primary incompatibility in this case is the difference in operating pressure for the two instruments. The column outlet in gas chromatography is typically at atmospheric pressure while source pressures in the mass spectrometer are in the range of 2 to 10^{-5} Torr for chemical and electron impact ionization, respectively. The interface must be capable of providing an adequate pressure drop between the two instruments and should also maximize the throughput of sample while maintaining a gas flow rate compatible with the source operating pressure. Further, the interface should not introduce excessive dead volume at the column exit and should not degrade or modify the chemical constitution of the sample.

Modern differentially pumped mass spectrometers can accept column flow rates of 1-2 ml/min (adjusted to STP) into the ion source. This is not incompatible with the optimum flow rates of some types of open tubular columns in gas chromatography but is much less than typical flow rates for packed columns. The gas burden of the source becomes an even bigger problem for liquid and supercritical fluid chromatography. Even with additional pumping capacity and the use of chemical ionization the gas burden is reached by the evaporation of about 5-50 microliter/min of liquid into the ion source. Flow rates in this range are compatible with the requirements of open tubular columns and some packed capillary columns but not those used with conventional packed columns. A liquid chromatograph operating at a flow rate of 1.0 ml/min, for example, would produce nearly 200-1000 ml/min of vapor at STP, of which the mass spectrometer could accept 1-2 ml/min. Also, liquid chromatography is frequently selected for the separation of samples which are thermally sensitive and cannot be volatilized without pyrolysis. Those ionization techniques which require volatilization in a heated source prior to ionization are not suitable for these compounds. Some mobile phases, such as those containing inorganic salts or buffers, may present additional problems such as clogging of the interface, as they are generally not volatile or volatilizable substances.

9.3.1 Interfaces for Gas Chromatography/Mass Spectrometery

Several interface designs are available for GC/MS and selection depends on the circumstances of the experiment [3-6,8,25,26]. Column flow rates of 1-2 ml/min (adjusted to STP) into the ion source are compatible with modern mass spectrometer vacuum systems. This is also the optimum flow rate range for open tubular capillary columns of conventional dimensions. Coupling such columns to a modern mass spectrometer, therefore, presents the fewest problems. Because of their flexibility, inertness and low bleed characteristics fused silica columns with immobilized phases are often connected directly to the ion source with a simple vacuum-tight flange coupling and column support. This allows the position of the column end within the ion source to be adjusted for maximum ion yield [26-30]. The sample is quantitatively transferred to the ion source but the performance of the capillary column may be compromised to some extent by the large pressure drop placed across the column and the retention time of components will vary with changes in the source pressure [31,32]. Another direct coupling device employs a fixed inlet restrictor to the mass spectrometer and a low-dead volume needle valve that acts as a flow diverter. This interface is more versatile as it allows a wider selection of column flow rates, permits columns to be changed without instrument shutdown, and provides a by-pass by which large solvent volumes or corrosive derivatizing reagents can be vented away from the mass spectrometer. Disadvantages are the introduction of dead volume and the possibility of transforming labile compounds on the hot metal surfaces of the valve.

In many respects the use of an open split coupling is preferred over the valve interface [33-37]. The open split interface essentially consists of a low-dead volume tube swept by a stream of gas into which the column end or transfer capillary and a capillary restrictor tube from the mass spectrometer are inserted from either end leaving a small gap between them. In some cases the restrictor capillary for the mass spectrometer is made narrower than the internal diameter of the column so that it can be inserted directly into the column end without completely sealing the column. This minimizes possible adsorption problems in the interface and avoids diluting the column effluent since the mass spectrometer restriction capillary pulls a constant flow from the column independent of the column flow rate. Returning to the conventional gap design, the amount of sample entering the mass spectrometer depends on the ratio of the inlet flow rate of the ion source (which is fixed for a given inlet capillary, source conductance, temperature and carrier gas) to the column flow rate. The total sample is transferred to the mass spectrometer only if the outlet flow rate of the column equals the inlet flow rate of the ion source; hence, no split occurs. If the flow rate of the column is larger than that of the source inlet, part of the sample is split off at the interface and does not reach the mass spectrometer. If the column flow rate is less than the inlet flow rate of the source, the sample is diluted by makeup gas and the mass flow of sample to the ion source is reduced. Activating a valve in the makeup gas line provides a means of diverting solvent or reagent away from the mass spectrometer and then rapidly restoring the analytical conditions after a preset time. The interface can be operated as a split with connection to a second detector. Other advantages of the open split interface are that the column end is always at atmospheric pressure so that retention times are reproducible, the interface is versatile with respect to the use of different column types and flow rates and the interface permits easy change of the analytical column without affecting the operation of the mass spectrometer. It should be noted that the yield for the interface will remain constant throughout a temperature programmed separation only if the carrier gas is controlled by a mass flow controller or if the interface is heated at the same rate as the column (i.e., the interface is installed in the GC oven) [37].

The problems associated with coupling packed columns to a mass spectrometer are more severe than those encountered with capillary columns. Conventional packed columns are operated at much higher flow rates, 20 to 60 ml/min, and although this diminishes the influence of dead volumes in the interface on sample resolution, it poses a problem due to the pressure and volume flow rate restrictions of the mass spectrometer. The interface must provide a pressure drop between column and mass spectrometer source on the order of 10^4 to 10^6, it must reduce the volumetric flow of gas into the mass spectrometer without diminishing the mass flow of sample by the same amount, and it must retain the integrity of the sample eluting from the column in terms of the separation obtained and its chemical constitution [3,25,26]. To meet the above requirements the interface must function as a molecular separator.

The performance of any type of molecular separator is characterized in terms of its separation factor (enrichment) N and separator yield (efficiency) Y [8]. The separator yield is defined as the ratio of the amount of sample entering the mass spectrometer to that entering the separator, usually expressed as a percentage. It represents the ability of the device to allow organic material to pass into the ion source of the mass spectrometer. The separation factor is defined as the ratio of sample concentration in the carrier gas entering the mass spectrometer to the sample concentration in the carrier gas entering the separator. The separation factor varies greatly depending on the type of separator, the gas chromatographic flow rate, the vacuum system efficiency of the source, and the molecular weight of the sample. The separation factor and the separator yield are algebraically related by

$$N = (Y/100)(V_{GC}/V_{MS}) \qquad\qquad (9.2)$$

where V_{GC} is the volume of carrier gas entering the separator and V_{MS} the volume of carrier gas entering the mass spectrometer.

Many types of molecular separators have been described; the most common being the effusion separator, the jet separator and the membrane separator, Figure 9.2. The Watson-Biemann effusion separator consists of an ultrafine-porosity, sintered glass tube enclosed in a thermostated vacuum envelope with glass capillaries at the entrance and exit to provide flow restriction. The surrounding vacuum pressure is on the order of a few Torr and the rate at which sample and carrier gas effuse through the pores of the frit is inversely proportional to the square root of the molecular weight and directly proportional to the partial pressure of each component. This differential flow of carrier gas and sample through the porous fritted tube is responsible for the enrichment of the sample in the carrier gas reaching the mass spectrometer. The enrichment obtained is both flow and temperature dependent and these parameters must be optimized empirically.

Probably the most popular separator for use with packed columns is the jet separator. The effluent from the gas chromatograph is throttled through a fine orifice where it rapidly expands into a heated vacuum chamber. During this expansion the lighter helium gas molecules rapidly diffuse away from the core of the supersonic jet which becomes enriched in the heavier sample molecules. The core is received by a second jet and enters the mass spectrometer. The alignment, orifice diameter and relative spacing of the expansion and collection jets are crucial. Normally these are fixed by the manufacturer so that maximum enrichment will only be obtained over a narrow flow rate range.

The silicone membrane separator works on the principle of differential permeability for the transmission of organic solutes compared to carrier gas molecules [38]. The amount of sample

964

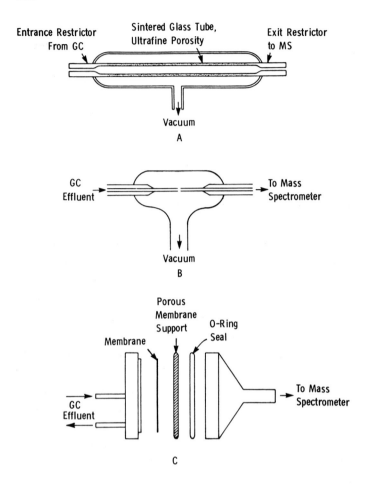

Figure 9.2 Common interfaces for GC/MS. Effusion separator (A), jet separator (B) and a membrane separator (C).

transmitted by the membrane is directly proportional to the solubility and diffusivity of the solute in the membrane, the exposed area of the membrane, and the pressure differential across the membrane. It is inversely proportional to the membrane thickness. The transmission of organic molecules may be up to two orders of magnitude greater than that of an inert gas. The effluent from the gas chromatograph is exposed to the high pressure side of the membrane. It is either channeled along a short (several cm) spiral path over the membrane or simply allowed to flow into a small, dead-volume chamber in contact with the membrane. To maintain column resolution the dead volume of the cavity on the high pressure side of the membrane must be

TABLE 9.2

COMPARISON OF MOLECULAR SEPARATORS

Typical Properties	Effusion (Watson-Bieman)	Jet	Membrane
Yield (%)	20-30	30-70	30-80
Enrichment	4-7	6-14	10-30
Carrier gas flow range (ml/min)	10-30	15-25	1-60
Temperature (approximate upper limit, °C)	350	300	255

minimized. The low pressure side of the membrane leads directly to the mass spectrometer source. The surface area, thickness, and contact time are critical design parameters and temperature is an important operating parameter. The highest enrichment is generally obtained at the lowest separator temperature commensurate with maintaining the integrity of the column separation.

No single separator is ideal for all situations. As can be seen from Table 9.2, at first sight the membrane separator would seem to be the most useful. It provides the highest sample enrichment and essentially allows the column to be operated separately from the mass spectrometer so that both systems can be optimized individually. Unfortunately, the performance of the membrane separator is critically influenced by temperature. There is an optimum temperature for the transmission of each organic compound across the membrane; at less than optimum temperature conditions peak distortion may occur and the relatively low upper operating temperature precludes its use with organic compounds of low volatility. The jet and effusion separators require auxiliary vacuum equipment and only operate efficiently over a narrow range of carrier gas flow rates. The jet separator is fairly rugged and reliable and is the standard packed column interface offered by most instrument manufacturers. Careful deactivation of the effusion separator may be required to avoid sample decomposition at high temperature.

9.3.2 Interfaces for Liquid Chromatography/Mass Spectrometry

Interfacing a gas chromatograph to a mass spectrometer is relatively simple compared to the problems encountered in LC/MS. Here the mass flow created by vaporizing the eluent from the liquid chromatograph is several orders of magnitude greater than the volume conductance of the mass spectrometer. This situation is compounded by the fact that wide bore columns operating at comparatively high liquid flow rates are the norm for liquid chromatography while capillary columns are more popular for gas chromatography. Also, liquid chromatography is often selected for the separation of involatile and thermally labile samples requiring alternative ionization methods to vapor phase techniques that are predominantly used in GC/MS. Consequently, the mixture of interface designs and ionization sources is more varied for LC/MS than for GC/MS, with many interface designs still somewhat short of maturity [22,39-47]. While one might consider the current practice of GC/MS as being reliable and quite straightforward this is not the case for LC/MS. So far, no single interface has proven reliable for all LC/MS applications and different interfaces are required to tackle the varied problems that might be experienced in a research laboratory.

The principal methods of interfacing a liquid chromatograph to a mass spectrometer are direct liquid introduction, continuous flow FAB, moving belt, thermospray, electrospray and various particle beam approaches. Open tubular columns with internal diameters less than 20 micrometers can be operated at flow rates below 100 nl/min which are compatible with the pumping capacity of both electron impact and chemical ionization sources [48,49]. Tapering and heating the end of the column in the ion source was found to be important for the detection of compounds of low volatility and for minimizing band broadening. The limited sample capacity of these columns results in concentration detection limits that are inadequate for most practical problems. With a cryopumped chemical ionization source the maximum liquid flow that can be tolerated is increased to about 50-100 microliters/min. This is compatible with the normal flow rates used with small bore columns and for conventional packed columns if a flow splitter is used [43,44]. In most direct liquid introduction interfaces the column is connected to a narrow bore capillary terminated at a metal diaphragm with a pinhole (2-25 micrometers in diameter), all housed in a probe that makes a vacuum seal with the source housing. The probe may be cooled by circulating water to minimize vaporization of the column eluent up to the diaphragm while inside the heated source block. Liquid flowing through the orifice in the diaphragm forms a stable liquid jet of fine droplets which are caused to traverse a desolvation chamber to intensify the evaporation of solvent from the droplets. The desolvation chamber is usually a length of tubing independently heated or heated by heat transfer from the ion source. After the desolvation process the mixture of solvent vapors, desolvated solutes and tiny droplets (solvent-solute clusters) enters the ion source, in which the vapors from the mobile phase act as reagent gases in the chemical ionization process. Diminished column resolution due to external peak broadening in the interface, clogging of the diaphragm orifice by particles, etc., and the need to "deice" the cold trap at regular intervals make the above approach far from ideal.

An alternative type of direct liquid introduction interface is represented by the continuous-flow FAB source. In this case column flow rates are restricted to about 5-10 microliters/min by the rate of solvent removal from the target. These flow rates are compatible with the normal operation of open tubular columns [50] and packed capillary columns or small bore columns with a flow splitter [51-57]. The interfaces are comparatively simple consisting, for example, of a hollow shaft with an angled probe tip through which a fused silica capillary is passed with its end terminated just above the probe tip. As the column eluent emerges from the column tip it is bombarded by high energy inert gas atoms producing ions which are subsequently mass analyzed. In another design the interface consisted of a length of fused silica tubing terminated in a stainless steel frit which was bombarded by the atom beam. In both cases glycerol (ca. 15 %v/v) or some similar matrix solvent, was added to the mobile phase to create the correct conditions for the ionization process. For a particular application the observed sensitivity was found to depend on the wettability of the probe tip, the temperature of the probe, and the composition and flow rate of the mobile phase. Major advantages of the continuous-flow FAB interface for LC/MS are high sensitivity, high mass capability (> 6,000 amu) and applicability to thermally unstable and/or involatile polar compounds, especially biopolymers.

As there often exists a substantial difference in volatility between the solutes typically analyzed by liquid chromatography and the solvents used for their separation, it should be possible to use selective solvent vaporization as an enrichment technique in LC/MS. This principle is employed on-line in the moving belt

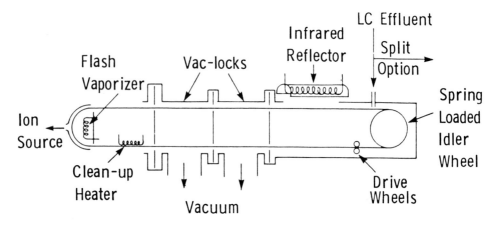

Figure 9.3   Schematic diagram of a moving belt interface for LC/MS.

interface, Figure 9.3. [25,39,42,55,58]. Three distinct steps are involved; deposition of the column eluent, removal of solvent in vacuum, and the volatilization or sputtering of the sample into the ion source. The column eluent flow that can be accommodated by the belt depends on the dimensions of the belt and its material of construction, the composition of the mobile phase and the method of eluent deposition. The belts are typically made of a metal mesh or polyimide material about 0.3 cm wide and have a capacity of about 1.5 ml/min for volatile nonpolar solvents and about 0.1-0.2 ml/min for water-containing mobile phases. Spray deposition allows higher flow rates and application to eluents of higher water content [58]. Spray deposition also reduces mixing of the analytes within the film deposited on the belt and yields a more even deposition of the eluent as small droplets on the belt surface aiding solvent removal. The thin film of eluent on the belt is carried along under a focused infrared lamp which causes a rapid preliminary vaporization of the solvent. The solute and residual solvent then pass through two serial vacuum locks and into the mass spectrometer vacuum envelope. Placing the terminus of the belt inside the ion source increases the sensitivity of the method by causing the sample to be volatilized directly into the electron beam. On the return journey a cleanup heater (and a solvent wash bath in some cases) is provided to remove solute residues that might cause ghost peaks on recycle. The main advantage of the moving belt interface is that it is compatible with normal column flow rates and solvents and permits the free choice of electron impact, chemical ionization and FAB ion sources with a free choice of reactant gases in chemical ionization. On the debit side it is a fairly complex and technically demanding device that may alter sample composition by surface adsorption/decomposition and, if the cleaning process is inefficient, a high chemical background can develop.

The thermospray interface, Figure 9.4, is one of the most popular interfaces in use today because it is rugged, fully compatible with the flow rates commonly used with conventional liquid chromatographic columns (1-2 ml/min) and because interfaces are commercially available for most of the common quadrupole and magnetic sector mass spectrometers [17-19,59]. The thermospray interface consists of a probe, source and vacuum system combined in a single unit that can be attached to most mass spectrometers with little further modification. The vaporizer probe, which connects the liquid chromatograph to the mass spectrometer, contains a resistively heated capillary tube that generates a supersonic jet of vapor containing a mist of fine droplets and particles. In addition the source block is heated to continue the solvent vaporization process as the liquid jet traverses the source region. A thermocouple downstream senses the vapor temperature and controls the heat input to the source to maintain a constant vapor temperature. Ions generated in the source are transmitted to the mass analyzer through a sampling cone placed in the path of the vapor jet. In addition to thermospray ionization an additional filament or discharge device is provided to assist in the ionization of samples introduced in nonaqueous or aqueous mobile phases containing a high percentage of organic modifier. Under these conditions many compounds show insufficient ion yields by normal thermospray. A mechanical pump with a cold trap is coupled directly to the ion source to provide the additional pumping capacity needed to remove excess vapor delivered to the source by the probe. The degree of vaporization of the eluent is a critical experimental parameter which depends largely on the eluent composition and flow rate, the source pressure and temperature of the vaporizer and source. For this reason, to obtain maximum ion current throughout a gradient elution separation it is usually necessary to simultaneously program changes in the vaporizer temperature.

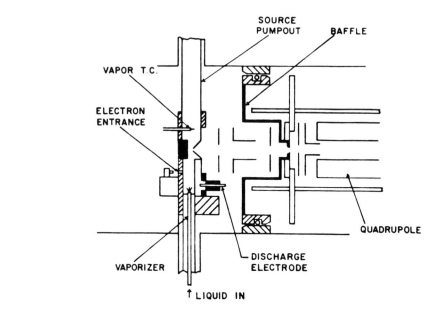

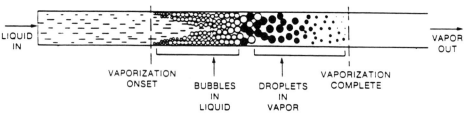

Figure 9.4 Schematic diagram of a thermospray interface ionization source connected to a quadrupole mass analyzer and insert indicating the mechanism of vapor production by the heated thermospray.

The monodisperse aerosol generation interface for LC/MS, MAGIC, enables both electron impact and chemical ionization mass spectra to be obtained from any compound sufficiently volatile to generate an adequate vapor pressure in the ion source [60-62]. In this interface the solvent is pumped away on the inlet side of the source allowing virtual independent operation of the liquid chromatograph and mass spectrometer. In practice, maximum liquid flow rates are limited to about 0.1 to 0.6 ml/min, compatible with the use of small bore columns and conventional packed columns with a moderate split flow. The interface is shown in Figure 9.5. The

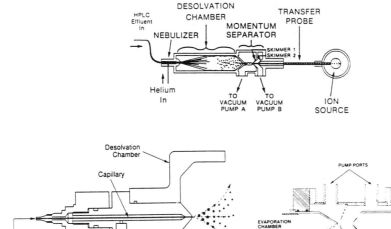

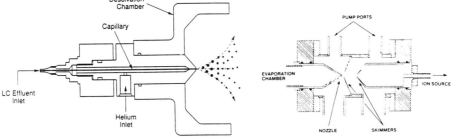

Figure 9.5 Schematic diagram of the monodisperse aerosol generation interface (MAGIC) for LC/MS with insert providing a more detailed view of the eluent nebulizer and particle beam momentum separator.

eluent from the column is pumped through a small (typically 25 micrometer) orifice resulting in the formation of a liquid jet of closely spaced droplets that is then subjected to a perpendicular flow of helium gas to aid dispersion and to prevent agglomeration of the droplets. The aerosol formed is then directed through a glass evaporation chamber, maintained at slightly below atmospheric pressure, which continues the solvent evaporation process. As solvent evaporates, any analyte present in the drops forms a solid residue (unless the analyte is a liquid at room temperature when the final state would be that of a droplet), thus becoming a high velocity particle beam directed at the momentum separator portion of the interface. On leaving the evaporation chamber the mixture of dispersal gas, solvent vapor and analyte particles expands through a nozzle, undergoing supersonic expansion into the first low-pressure region (ca. 10 Torr). The high mass of the analyte particles relative to the gas stream provides them with much greater momentum than the solvent vapor molecules and the helium atoms. Consequently, the particles do not expand from the core of the expansion jet as rapidly as do the gases. Skimming the core of the expansion jet, therefore, results in a substantial enrichment of the analyte relative to helium and the solvent vapors which are removed by the vacuum line. This process is repeated in a second chamber of the momentum separator, this time at a chamber pressure of about 1 Torr. After leaving the momentum separator the analyte particles continue in a straight line entering the ion source where they are flash vaporized and ionized by electron impact or chemical ionization processes.

A number of interfaces based on heated gas nebulizers and electrospray probes for eluent introduction into atmospheric pressure chemical ionization sources have been described [42,63-69]. Since the ion source is at atmospheric pressure operation of the liquid chromatograph is not compromised by the mass spectrometer and up to about 2 ml/min of eluent can be introduced into the source, although the optimum flow rate for maximizing the ion current may be a good deal lower depending on the source design. The nebulizer consists of three concentric tubes (in the case of the nebulizer-assisted electrospray interface the inner tube is held at a high voltage with respect to the outer tube), the eluent being pumped through the inner most tube and nebulizer gas and makeup gas through the outer tubes, Figure 9.6. The combination of heat and gas flow desolvates the nebulized droplets producing a dry vapor of solvent and analyte molecules. The solvent molecules are then ionized by a corona discharge and the solvent ions formed ionize the analyte molecules by atmospheric pressure chemical ionization. The ions formed are extracted through a small (100 micrometers) orifice at the tip of a cone pointing towards the source region. The area in front of the orifice is flushed with dry nitrogen gas which acts as a

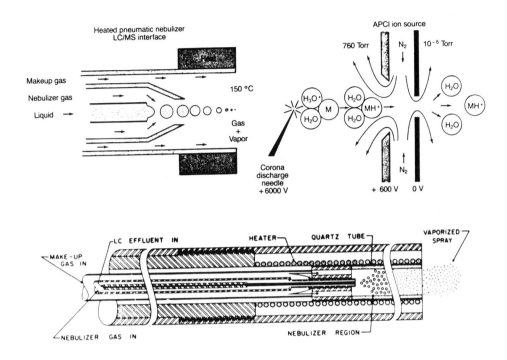

Figure 9.6 Schematic diagram of a heated pneumatic nebulizer LC/MS interface combined with an APCI ion source and cross-sectional view of the nebulizer probe.

curtain to keep solvent vapors away from the orifice and minimizes clustering of the analyte ions with water and other polar molecules. The mass analyzer section of the mass spectrometer is maintained at high vacuum using cylindrical cryosurfaces located around the quadrupoles. One limitation of this approach is that few instrument manufacturers offer atmospheric pressure ionization sources as standard or optional equipment.

Electrospray ionization processes are also performed at atmospheric pressure with sampling of the ions through a small orifice by the mass spectrometer [66-69]. An electrospray is generally produced by application of a high electric filed to a slow flow of a conducting liquid (generally 1-10 microliters/min) from a capillary tube. The high electric field results in disruption of the liquid surface and the formation of a stream of highly charged liquid droplets. Desolvation by collision and

TABLE 9.3

CHARACTERISTICS OF LC/MS INTERFACES

Property	Direct Liquid Introduction	Moving Belt	Thermo -spray	MAGIC	APCI	Continuous -Flow FAB
Accepted flow of solvent (ml/min)	0.01-0.05 0.2-2.0 (with splitter)	1-2	1-2	0.1-0.6	1-2	0.005-0.01
Mode of ioniz- ation	CI	EI/CI/ FAB	TSP CI	EI/CI	APCI	FAB
Mass range (minimum)	130	50	120	40	150	
Identification limit (full scan) ng	1-10	10-20	5-10	10	1-10	10-50

thermal means results in the formation of analyte ions largely by ion evaporation. A characteristic of electrospray ionization is the formation of multiply charged ions which have been exploited for the identification of high molecular weight biopolymers on conventional quadrupole mass spectrometers with a limited mass scan range. For LC/MS the main limitation is the low flow rates that have to be used, which are only appropriate for capillary and microcolumn operation without splitting the eluent flow. Electrospray is also the most recent of the ionization/interface methods developed for LC/MS and further improvements may provide a solution to the limitations existing at present..

The operating characteristics of the LC/MS interfaces discussed above are compared in Table 9.3. No single interface is ideal for all sample types and separation conditions, or is necessarily compatible with all existing mass spectrometers. The moving belt interface is perhaps the most universal and provides reasonable enrichment for samples with boiling points greater than about 170°C but is technically complex and demanding to operate. Thermospray is probably the most popular interface as it can be retrofitted to most mass spectrometers and the interface is rugged and reliable in routine use. Not all analytes are suitable for ionization by this method and only molecular weight information with very little fragmentation is generally obtained. This same deficiency is found for direct liquid introduction and atmospheric pressure chemical ionization methods as well. Fragmentation information is usually required in combination with molecular weight information to identify unknowns. One solution, available with tandem mass spectrometers, is collisionally activated dissociation (CAD), in which a selected ion is dissociated into daughter ions in a collision cell and then mass analyzed to produce a full scan mass spectrum of the isolated ion. The cost of such equipment, however, is too high for most laboratories at present. The MAGIC interface solves this problem in a different manner and provides clean library-searchable electron impact and chemical ionization mass spectra but is only applicable to thermally stable molecules with a reasonable vapor pressure at ion source temperatures. The recent introduction of continuous-flow FAB techniques offers a potential analytical method for those analytes that do not yield electron impact, chemical ionization, or thermospray mass spectra and were often difficult to analyze previously. None of the interfaces discussed in this section are as reliable, convenient or simple as those employed for GC/MS

9.3.3 Interfaces for Supercritical Fluid Chromatography/Mass Spectrometry

The basic problems of interfacing supercritical fluid chromatography to mass spectrometry are intermediate between those of gas and liquid chromatography [70-74]. Nearly all current research in SFC utilizes either open tubular columns of very small internal diameter, 25-100 micrometers, or slurry packed capillary or small bore columns prepared for use in liquid chromatography. The low flow rates of open tubular columns, 1-5 microliters/min (as liquid) are compatible with the pumping capacity of most mass spectrometers with chemical ionization sources and, therefore, the most common type of interface for these columns is direct fluid introduction [71,73-79]. The most critical component of the interface is the restrictor (a short length of fused silica tubing of 5-20 micrometers internal diameter, tapered tube, capillary with a fritted end, etc.) which is required to maintain supercritical fluid conditions inside the column while allowing rapid decompression of the fluid in the low-pressure region of the ionization source. The tip of the restrictor should be heated to compensate for cooling during the fluid expansion facilitating the transfer of high molecular weight solutes while diminishing solvent cluster formation. In the chemical ionization mode the higher source pressure also serves to minimize cluster formation which is a much more serious problem for low-pressure electron impact sources [76,80]. Full scan mass spectra in the chemical ionization mode are possible from a few nanograms of sample but in the electron impact mode the sensitivity can be poorer by up to two orders of magnitude. With a conventional electron impact source, ionization occurs by both charge exchange and electron impact; both processes producing similar mass spectra. Under typical operating conditions the pressure in the ion source is higher than under normal electron impact operating conditions. As a result, electrons generated by the filament have a much shorter range lowering significantly the ionization efficiency in both mechanisms. To obtain true electron impact mass spectra a heated desolvation chamber at a pressure higher than the source pressure and situated before the source ionization region is required for efficient cluster breakup prior to ionization.

For small bore packed column SFC/MS direct fluid introduction into a modified chemical ionization source, a particle beam interface, a modified filament on thermospray interface, and the moving belt interface have been used successfully [71-73,81-83]. The moving belt interface used in liquid chromatography requires only minor modification to the spray deposition device to avoid freezing of the nozzle for SFC/MS and can be used with mixed mobile phases [82]. It enables either electron impact or chemical ionization mass spectra with free choice of reagent gas to be obtained for those analytes that can be vaporized from the belt. The particle beam separator is similar in construction to the MAGIC interface used for LC/MS but optimum operating conditions are different to the liquid chromatography version [81]. Flow rates for small bore columns with an internal diameter of 1 mm are typically about 60-100 microliters/min (liquid) requiring a modified source for direct fluid introduction, Figure 9.7 [83]. The small bore column is connected to the restrictor via a short length of fused silica capillary tubing housed in a cylindrical probe maintained at the same temperature as the column oven using forced air circulation. The restrictor is housed in a separate stainless steel capillary which is independently heated to the optimum temperature for sample transmission. Expansion of the supercritical fluid occurs into a heated chamber behind the ion source which is separately pumped to maintain a relatively high pressure of 5-50 Torr to complete the

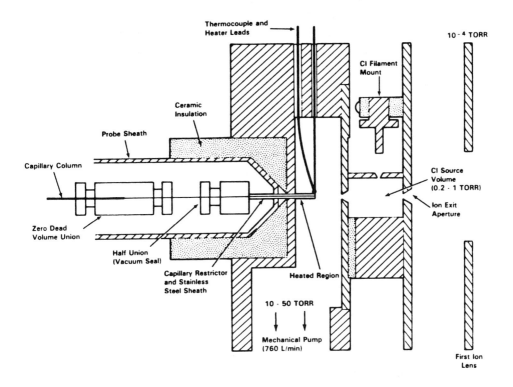

Figure 9.7 Schematic diagram of a high flow rate interface for direct fluid introduction into a modified chemical ionization source for SFC/MS. (Reproduced with permission from ref. 83. Copyright American Chemical Society).

desolvation process. The exit orifice is aligned with the capillary restrictor and provides flow into the chemical ionization region where ionization takes place at a pressure of 0.2-1 Torr in the usual manner.

9.3.4 Interfaces for Thin-Layer Chromatography/Mass Spectrometry

The problems encountered in interfacing thin-layer chromatography to mass spectrometry are entirely different to those of liquid chromatography and other column based separation methods. At the completion of the separation the chromatogram is fixed in time and space with the major portion of the TLC solvent eliminated by evaporation. The sample is thus compatible with the vacuum requirements of the mass spectrometer and can be analyzed in an off-line manner so that time constraints are not a problem. However, no commercial interface for TLC/MS is presently available and the most common method of analysis involves excising chromatographic zones from the plate followed by thermal desorption of the sample from its matrix in the ion source or similar treatment of the sample residue obtained by solvent elution from the sorbent matrix [84,85]. A direct interface for obtaining chemical ionization mass spectra from in situ spots employed a scanning stage and either a laser or incandescent lamp to desorb samples into a gas stream which then swept the desorbed materials into the ion source of the mass spectrometer. This method was suitable for relatively volatile and nonpolar compounds but poor sensitivity was observed in other cases [85,86]. It has been shown that the laser microprobe mass analyzer can be used to obtain mass spectra by laser ablation and ionization of chromatographic zones on TLC plates but this approach is limited both by the availability of the rather specialized instrumentation and the requirement of very low sample volatility [85,87]. However, sputtering techniques, which could be said to include laser ablation, represent the most sensitive and versatile approach to an integrated system incorporating scanning and continuous spectral analysis. FAB mass spectra have been routinely obtained by plate scanning or simply using a strip of doubled-sided masking tape to attach excised areas of the chromatographic zones to a standard FAB probe [84,85]. A few microliters of FAB matrix solution is then added to the sorbent and the probe introduced into the mass spectrometer to acquire spectra in the usual way from samples in the 10s of nanograms to microgram range. Busch and co-workers have developed several scanning TLC/MS systems based on a liquid surface ionization mass spectrometry (liquid SIMS) approach [84,88-90]. The mass spectra obtained by

SIMS are typically even-electron ions of the same type observed in chemical ionization mass spectra with similar modes of fragmentation in many cases. In the SIMS experiment the dried plate in a vacuum chamber is scanned through an ion beam using a translational device to position the plate in the x and y direction. The primary ion beam, usually cesium or gallium, is accelerated to high energy and collides with the surface transferring its energy in a collision cascade to molecules in the surface region. Neutral molecules, ions and electrons are ejected from the surface. The ions of interest are drawn into the mass spectrometer by an extraction potential where they are mass analyzed. The sensitivity of the technique can be improved by using a phase transition matrix and time averaging. Since ions are removed only from the surface, the bulk of the samples is not ionized by the primary ion beam unless a mechanism can be found to extract and cycle the major portion of the sample to the surface in a continuous manner. Impregnating the chromatogram with a viscous liquid or low melting point solid fulfills this role without destroying the integrity of the chromatogram through zone broadening. By using a phase transition matrix a steady signal can be obtained for several hours permitting extended data collection. For a full scan mass spectra sample sizes in the tens of nanograms to microgram range are required. A CCD camera can be used in combination with the mass spectrometer to locate the coordinates of sample zones in the chromatogram to eliminate time lost while scanning areas of the plate that do not contain sample.

9.4 DATA MANAGEMENT AND QUANTITATIVE ION MONITORING IN MASS SPECTROMETRY

Virtually all modern mass spectrometers with chromatographic sample inlets employ computers and/or microprocessors for several purposes. The most important of these are data acquisition and instrument control; data processing, including background subtraction, mass calibration, normalization, formatting, printing and display; and identification of mass spectra by library search methods [4-6,91-93]. The computer acquires sample-dependent information from the detector and the ion source. The total ion current monitor is positioned between the ion source and mass analyzer in such a way that it intercepts a few percent of the total ion beam. It thus provides a continuous monitor of the rate of ion production in the source which can be correlated with the rate of sample mass entering the source. The total ion current monitor, then, provides a representation of the chromatogram that can be used to identify chromatographic peaks of interest. Other methods for reconstructing the chromatogram under continuous scan control are available and these will be discussed later.

From the detector the computer receives information about the detected ion current as a function of time. The conversion of the time axis into mass increments is usually achieved through a calibration of the instrument based on the position of peaks of known mass from a standard compound such as perfluorokerosene or heptacosafluorotributylamine. For low resolution mass spectrometry external calibration is used to establish the calibration at the start of the day and then repeated as required to check for drift. In addition to time, the Hall voltage in magnetic scanning instruments or the applied voltage in quadrupole instruments can be used to define the mass scale. A Hall effect probe is used to sense the magnetic field during scanning and provides a time base for mass marking the spectrum. Whichever method is used, reproducible scan behavior is required, as succeeding spectra are mass converted by interpolation of the calibration data. The mass spectra are usually normalized so that either the most intense peak or the sum of all mass peaks is assigned a value of 100%. If the total ion current signal is recorded simultaneously with mass data it can be used to correct for fluctuations in ion current during the course of a scan. Other algorithms are available to subtract out background contributions, for example, from column bleed, or to eliminate interferences from overlapping chromatographic peaks.

A further function of the computer is to continuously monitor and modify the operating status of the instrument by actuating electronic switches and adjusting voltages in accordance with a menu of preset operating instructions, selected at the beginning of the analysis. Control of the scanning function of the mass spectrometer is one variable that has a direct bearing on the methods used for data processing. Two scanning techniques, continuous scanning and selected ion monitoring, are used routinely. In the continuous scan mode the computer initiates a scan of the mass spectrum at fixed time intervals and repeats this process throughout the analysis. An advantage of this technique is that a complete record of the sample analysis is available that may be searched retrospectively if the object of the analysis should change at a future date. The immediate problem is to decide which of the many recorded spectra are most relevant to the problem at hand. In most cases a summary of the total analysis in the form of a total ion current profile (TICP), also called a reconstructed chromatogram, is generated. The TICP is defined as a normalized plot of the sum of the ion abundance measurements in each member of a series of mass spectra as a function of the serially indexed spectrum number. A scan containing significant information about the sample is then easily located from the TICP and a copy of the mass spectrum of interest requested by identifying the appropriate scan number. When specific information is required about the analysis of a single component or a few components whose mass spectral properties are known or can be estimated, then mass chromatography may be used. In this case a mass chromatogram, a display against time of the ion intensity recorded at a specific mass value, is used to identify the substances of interest. The specificity is increased by simultaneously displaying the mass chromatograms of several characteristic ions of the substance of interest. The mass chromatograms should maximize in the same scan sequence if the desired substance is present in the sample. The scan number at which this occurs can be identified and the full mass spectrum obtained to ensure the substance is correctly identified.

When the object of the analysis can be clearly defined, it is not necessary to record the full mass spectra of all substances present in the sample. As discussed above, a few ions characteristic of the substance or substances of interest may be monitored throughout the experiment. This approach is known as selected ion monitoring (SIM), and is defined as the dedication of the mass spectrometer to the acquisition of ion abundance data at only selected masses in real time as components emerge from the chromatographic system. It differs from the continuous scan mode in that only preselected ions are recorded during the analysis by rapid switching of the accelerating voltage for magnetic sector mass spectrometers or the quadrupole voltage for quadrupole mass spectrometers. Its principal advantage is high sensitivity, as only the ions of interest are measured at the detector and the dwell time over which the ion current is recorded for each ion is increased compared to the continuous scan mode. On the other hand, since the full range of ions are not recorded, if at a later date information different from that originally sought is required, it will not be available and the analysis must be repeated.

Selected ion monitoring lends itself to problems in quantitative trace organic analysis of complex mixtures and is now a routine tool in many environmental and biomedical laboratories [8,94-98]. Selectivity is obtained because mass is a substance-characteristic parameter. When several masses are recorded sequentially the specificity is increased as the probability of two different substances having in common the same retention time and the same characteristic mass ion ratios is very small. The sensitivity of selected ion monitoring is from 100 to 1000 times greater than is obtained in the full scan mode, providing detection limits in the range 10^{-9} to 10^{-15}g. This limit will depend on the ionization efficiency of the compound, the fraction of the total ion current carried by the ion monitored, the contribution of background ions to the signal, mass spectrometer resolution, and the sensitivity of the detector. To improve the accuracy of analytical methods stable isotope analogs of the substances of interest are used as internal standards and can be added to the sample at an early stage to correct for losses at all stages of the sample workup and analysis. Since they differ from the substance to be determined only by mass, they are ideally suited to the multiple ion monitoring techniques. When a stable isotope analog is unavailable then a homologous compound containing the same m/z ions as the substance to be determined is the next best choice and may be more readily available or easier to synthesize.

As well as processing data into a form convenient for interpretation the computer can also help to identify the mass spectrum of an unknown compound by retrieval or interpretive procedures [6,93,99]. Retrieval processes using library search procedures are available on most mass spectrometer data systems. These libraries contain from 25,000-80,000 reference mass spectra of a general nature or for specific compound types such as EPA priority pollutants, common drugs of abuse, etc. Most data systems also have software for construction and searching of specialized libraries collected individually. Searching involves comparing the library spectra in turn with the unknown spectrum and reporting a degree of fit. The computer should correctly identify the authentic spectrum if present, and otherwise should pick similar spectra, guiding the analyst towards the correct identification. Selection and encoding of certain mass peaks or features from a

TABLE 9.4

COMPARISON OF COMPUTER SERACH METHODS

Method of Search	Basis of Comparison	Reliability of Identification	Other Features
N most intense peaks	Match largest peaks in unknown with corresponding peaks in each reference spectrum.	Fair if unknown has limited fragmentation and is in the reference file.	Fast, easy to program, good for systems with limited computing facilities.
Biemann search	Divide spectrum into 14 amu divisions, take 2 largest peaks each division beginning at amu = 6.	Excellent if unknown is in reference file. Compounds of closest match may have similar structure to unknown.	About 25% of peaks used for search. Molecular ion must be chosen if present.
Whole spectrum search	Uses every peak in unknown for search.	Excellent if unknown is in reference file. Poor if unknown spectrum contains contaminant peaks.	Slow, efficient programing required and large data storage for reference files.
Reverse search	Whole spectrum search but treat reference spectra as unknown. Each comparison based on peaks in the refernce spectrum only.	Better than whole spectrum search for impure spectra.	Identification of multiple components of mixtures can be achieved by using residual spectrum.
Combined match factor-retention time/ index	Retention time or retention index of unknown used in combination with a search procedure.	Good, can be used to differentiate between compounds having similar mass spectra.	System must be setup for specific chromatographic condition.

spectrum reduces both the storage and time requirements of a search system. Some common approaches are summarized in Table 9.4. These approaches seek to abbreviate the mass spectrum while retaining the maximum amount of structurally useful information. The degree of correspondence between the library and unknown spectra are then calculated and ranked according to the degree of fit using one of several pattern recognition methods. Those methods based on the selection of the N most intense peaks (N = 3-10; commonly 8) are the easiest to implement but since intense peaks generally represent the most stable and common fragment ions, only a low degree of discrimination is obtained. Giving extra weight to high mass ions or choosing the two most abundant peaks present every fourteen mass units (Biemann search) improves the discrimination. For reliable high precision matching and ranking of matches, it is probably best to view the library search process as a two-stage process. In the first stage, globally significant peaks are used to select a small set of spectra from the data base most like the unknown. Then in a second stage precision matching and ranking are carried out based on a model that uses locally significant peaks.

Since there are many more organic compounds than there are reference mass spectra, an unknown compound may remain unknown after a library search. Interpretive procedures using artificial intelligence methods may be useful. In this case a range of empirical rules for characterizing mass spectral features are used, and probability estimates generated that certain substructures may occur in a given unknown. The programs for this process are rather large and elaborate requiring a main frame computer for their efficient execution. They are usually accessed from one of the regional mass spectrometry centers by a telephone hookup.

9.5   INSTRUMENTAL REQUIREMENTS FOR FOURIER TRANSFORM INFRARED SPECTROSCOPY

Infrared spectra are particularly valuable for the identification of functional groups. Also, they are often very sensitive to small changes in the molecular architecture and can be used to distinguish different isomers. Except by comparison with an authentic spectrum, however, it is rarely possible to identify an unknown compound on the basis of its infrared spectrum alone [100]. Infrared spectrometry is particularly poor for distinguishing between homologs without additional molecular weight information. This information may be obtained by mass spectrometry. The two techniques are complementary since a knowledge of the infrared spectra of a substance helps in the interpretation of the mass fragmentation pattern and fully integrated GC/FTIR/MS instruments are commercially available.

Until a few years ago dispersive (grating or prism) instruments were used to acquire infrared spectra. These instruments were relatively slow and insensitive, which made recording spectral information on-the-fly from chromatographic instruments all but impossible except under conditions which severely compromised the performance of the chromatograph. This situation improved dramatically with the advent of interferometric recording instruments and algorithms for the fast Fourier transform of the interferogram into an IR absorption or transmission spectrum [101-104]. The infrared spectrum is recorded by a Michelson interferometer which consists of two mirrors at right angles to one another and a beam splitter which bisects the angle between the two mirrors. One mirror is stationary and the other may be moved at a constant velocity. After reflection the two beams recombine at the beam splitter and, for any particular wavelength, constructively or destructively interfere depending on the difference in optical paths between the two branches of the interferometer. For a broad band emission source each frequency comes in and out of phase at a characteristic optical path difference, and the superposition of all frequencies produces the observed interferogram. The combined beam then traverses the sample compartment and is focused on the detector, usually a small area liquid-nitrogen cooled mercury-cadmium-telluride (MCT) photoconductivity detector. The detector signal is sampled at very precise intervals during the mirror scan. Both the sampling rate and mirror velocity are precisely controlled by modulation of an auxiliary reference beam from a helium-neon laser. After data collection the dedicated computer executes a Fourier transform of the data to provide a single beam spectrum which may then be ratioed against a background spectrum to provide the customary transmittance (or absorbance) versus wavenumber infrared spectrum. The advantages of Fourier transform infrared spectrometry (FTIR) vis. a vis. dispersive measurements are that all frequencies are recorded at all times throughout the experiment (Fellgett advantage), as programmable slits are not required the energy throughput of the instrument is higher (Jacquinot advantage), higher frequency accuracy (0.01 cm^{-1}) is possible because of the use of the frequency lock mechanism provided by the reference source, and there is no stray light contribution to background noise as in dispersive instruments. For on-line coupling to chromatographic instruments the principal advantages are those of greater sensitivity and rapid scanning capability. The infrared spectrum may be scanned in less than a second and individual scans summed during the elution of a chromatographic peak to provide sensitivity two orders of magnitude greater than those obtained with dispersive instruments.

Computer acquisition provides attendant advantages, such as real time chromatogram plots, background subtraction, peak deconvolution and library search facilities as well as calculation and storage of the infrared spectra. Three techniques are commonly used for reconstructing the chromatographic information from the absorbance data [105-108]. The Gram-Schmidt orthogonalization method is a normalization routine employing interferometric data without prior Fourier transformation. It is the preferred method for chromatogram reconstruction using a wide frequency window because it is computationally faster and provides greater sensitivity than integrated absorbance methods. The Gram-Schmidt orthogonalization process is a method used by mathematicians to generate coordinate systems in multidimensional space. To use this method a series of base line interferograms are collected that contain no sample information. These interferograms are then used to construct a set of orthogonal basis vectors, using a number of consecutive data points from each of the original interferograms. For chromatogram reconstruction, each interferogram collected following the start of the separation is used to compute a resultant vector orthogonal to the basis vectors. The magnitude of the difference between the computed vectors and the basis vectors is proportional to the infrared absorbance of the sample. Alternative approaches use either the integrated absorbance or maximum absorbance obtained from the Fourier transform of the interferograms for chromatogram reconstruction. Both methods are generally used to create selective chromatograms by integrating over a narrow frequency range, characteristic of a particular functional group. The integrated absorbance method calculates the total absorbance in a selected frequency range of each transformed spectra. In the maximum absorbance method the maximum absorbance value within the selected frequency range of the transformed interferogram is used as the Y-coordinate value for corresponding time (X-coordinate) value in the reconstruction. For base line regions, this procedure results in low values and, as absorbing compounds elute, the values rise and fall to reflect their absorbance behavior, producing the desired reconstructed chromatogram. The maximum absorbance method is more sensitive than the integrated absorbance method and does not decline if the frequency window is much larger than the bandwidth of the peak within the window. All three of the above methods provide reconstructed chromatograms in real time and can also be used for quantitative analysis after calibration [101,105]. The slope of the calibration curves from the Gram-Schmidt plots is influenced by changes in background contributions from contaminants, particularly water vapor, requiring that the carrier gas and spectrometer purge gas are rigorously dried. The integrated or maximum absorbance values are usually more reliable but are complicated by the uncertainties associated with estimating the percentage of injected material in the detector cell during the data acquisition period.

9.6 INTERFACE REQUIREMENTS FOR CHROMATOGRAPHIC SAMPLE INTRODUCTION SYSTEMS IN FOURIER TRANSFORM INFRARED SPECTROSCOPY

Three techniques have been predominantly used as interfaces for column chromatographic techniques; the flow-through cell (lightpipe), matrix isolation and cold trapping [101,109,110]. In GC/FTIR the mobile phases commonly used are transparent in the mid-infrared region favoring direct coupling via a flow-through cell. Mobile phase elimination techniques are used primarily to increase sensitivity. Supercritical fluids, such as carbon dioxide absorb weakly and selectively in the mid-infrared region making flow cells acceptable for many applications but when a complete spectrum and/or solvent-modified mobile phases are used, mobile phase elimination interfaces are preferred. Most common solvents used in liquid chromatography absorb infrared radiation at certain wavelengths in the mid-infrared region and some, such as water, have intense absorption across most of the spectrum. This effect completely obscures sample bands in several spectral regions and limits the useful path length of the cell and, therefore, sensitivity. Spectral subtraction of solvent bands is almost impossible when the composition of the mobile phase is changing, for example, during gradient elution. The flow cell approach is thus less attractive for liquid chromatography and, in general, mobile phase elimination interfaces are preferred.

9.6.1 Interfaces for Gas Chromatography/FTIR Spectroscopy

The simplest interface for GC/FTIR is a flow-through cell, lightpipe, connected to the gas chromatograph by a length of heated capillary tubing, Figure 9.8 [101-103,109-114]. The lightpipe usually consists of an internally, gold-coated, heated glass tube with potassium bromide or zinc selenide windows affixed

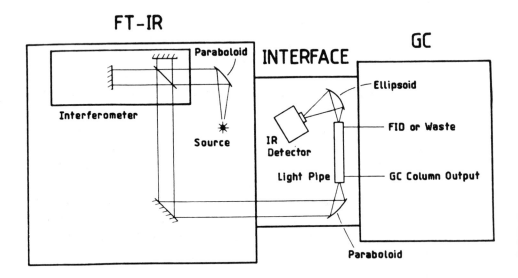

Figure 9.8   Schematic diagram of an integrated GC/FTIR system with a lightpipe interface.

to either end. The lightpipe is located in the optical path of the spectrometer. For optimum performance the volume of the lightpipe is determined by the width of the peaks in the chromatogram [111]. To avoid severe degradation of the chromatographic resolution the lightpipe volume must be no larger than the volume of carrier gas between the half-height points of the main peaks in the chromatogram. Makeup gas added to the effluent from the column can be used to preserve resolution at the expense of sensitivity. On the other hand, if the lightpipe volume is much less than the half-height peak volume sensitivity is again reduced since the fraction of the sample peak present in the lightpipe at any instant during detection is small. Chromatographic peak widths are determined by the column dimensions, the temperature program rate, and the carrier gas flow rate. Since it is not convenient to change lightpipes to match each chromatogram, it is standard practice to select certain lightpipe volumes and then set the chromatographic conditions so that the half-peak height volumes of each (or as many as possible) peaks are equal to, or slightly greater than, the lightpipe volume. For capillary columns lightpipes of 1 mm internal diameter and 10-20 cm long are used giving volumes between 50-200 microliters. For packed columns typical dimensions are 1 to 3 mm internal diameter and 20 to 100 cm in length, giving volumes of approximately 0.8 to 5.0 ml. For full spectra about 5-25 ng per component for strong absorbers and about 100 ng for weak absorbers is required. The popularity of the lightpipe interface is due to its simple design, provision for real-time measurements, and the availability of a fairly large library of vapor phase spectra for search purposes. The principal problems are a lack of sensitivity, particularly for use with thin-film open tubular columns, and dead volume limitations for fast separations with narrow bore open tubular columns.

For matrix isolation the effluent from the gas chromatograph is mixed with a small quantity of argon (ca. 1%) or helium containing a small quantity of argon is used as the carrier gas, and is directed under vacuum by a short length of heated capillary tubing onto the surface of a gold-plated collection disk maintained at about 12 K [109,110,115,116]. The collection disk is continuously rotated and can store up to about 5 h of chromatography frozen at one time. Under the prevailing conditions the helium remains gaseous and is pumped away, while the sample components remain trapped within a matrix of solid argon. The strip of argon is subsequently rotated through the infrared beam and a series of absorption spectra are measured by reflectance. In the matrix of solid argon the sample molecules do not interact with each other and interact only slightly with the inert argon atoms surrounding them. This lack of intermolecular interactions and absence of molecular rotation leads to infrared spectra having narrow lines and detailed fine structure. Since the separation and measurement process are separated in time, signal averaging techniques can be applied to enhance sensitivity compared to lightpipe measurements. An improvement in sensitivity of at least one order of magnitude is not unusual with the capability of recording full spectra from about 0.01-0.10 ng of strong absorbers and 0.20-0.50 ng of average absorbers. The high cost and complex apparatus combined with the fact that data is not provided in real time and the need for special matrix isolation libraries for computer searching have limited the general acceptance of this approach [117]. A similar approach that uses a cryocollector at liquid nitrogen temperatures for cold trapping but does not utilize a condensable matrix was described recently [118]. In this case the transfer line was terminated in a 5 cm piece of 50 micrometer internal diameter tubing located about 75-100 micrometers from the collector surface. The narrow bore restrictor prevents evacuation of the column and reduces the size of the deposited track increasing sensitivity. The sensitivity was not quite as good as the Cryolect matrix isolation interface but, presumably, would be less costly to construct. A similar, and somewhat simpler device, uses a zinc selenide window maintained at 77 K [119-121]. Gas chromatographic eluent was passed through a heated 50 micrometer internal diameter fused silica restrictor and deposited onto the moving window which transported the condensed sample into the focus of an FTIR microscope, used to match the FTIR beam diameter to the size of the sample spot. This approach has the advantage of providing near real-time chromatogram reconstruction and spectra, provides spectra which can be identified using available condensed phase infrared libraries, and offers sensitivity approaching that of the matrix isolation method with a similar experimental design. However, volatile compounds will not be efficiently trapped unless much lower temperatures are used, in which case a vacuum housing is also needed to avoid co-condensation of water [110,113].

9.6.2 Interfaces for Gas Chromatography/FTIR/MS

Interfacing of Fourier transform infrared and mass spectrometers to a gas chromatograph has been achieved by either splitting of the column effluent between the infrared spectrometer and mass spectrometer in a parallel configuration or by using a serial configuration where the column effluent passes through the lightpipe and then directly to the mass spectrometer [105,112,122-124]. In the parallel configuration the carrier gas flow (and makeup gas flow if used) is split between the two instruments such that about 90-99% of the flow goes to the infrared spectrometer. In this way the different sensitivity of the two instruments can be matched and optimum flow provided to the mass spectrometer and lightpipe used for infrared spectroscopy. For serial coupling makeup gas is added to the column effluent to match the requirements of the lightpipe for optimum sensitivity and the exit flow is diverted to the mass spectrometer using a splitter and a jet separator or an open split coupling to reduce the gas flow into the mass spectrometer to its optimum value. The open split interface is the most versatile as it is easy to optimize for different flow rates and can function as a splitter when high flow rates are used in the interface to minimize losses in resolution and to maximize sensitivity in the lightpipe. Some problems remain in harmonizing the use of the mass spectral and infrared data to reduce the ambiguity of spectral interpretation using library search routines [125]. A considerable reduction in the cost of spectroscopic equipment and computers used in GC/FTIR/MS combination has improved the likelihood that such hyphenated systems will be more widely employed in the future. The low sensitivity of the dynamic lightpipe infrared interface compared to the sensitivity of mass spectrometry remains the primary limitation in problem selection, since the compromises made in the chromatography are done to accommodate the circumstances of infrared detection and are unrelated to the capabilities of the mass spectrometer.

9.6.3 Interfaces for Supercritical Fluid Chromatography/FTIR Spectroscopy

Small bore packed columns and narrow bore open tubular columns have been used for SFC/FTIR analysis using a pressure-stable, thermostated, flow-through cell or solvent elimination interface in which the sample is deposited on an infrared transparent substrate after evaporation of the mobile phase, Figure 9.9 [110,126-129]. Carbon dioxide is the most widely used supercritical fluid mobile phase in current practice and has many favorable infrared spectral properties when compared with the solvents commonly employed in liquid chromatography. Only the absorption region between 3475 to 3850 cm^{-1} and 2040 to 2575 cm^{-1} are completely inaccessible due to strong absorption by carbon dioxide. Information may be lost or reduced in the region between 1200 and 1475 cm^{-1} due to absorption resulting from Fermi resonance whose magnitude is a function of carbon dioxide density. For short path length flow-through cells, the background may be effectively subtracted from the spectrum. FTIR spectra recorded in supercritical fluid carbon dioxide differ from those recorded in the vapor or condensed phase due to shifts of the maximum absorbances, variation of band widths, and modification of the

992

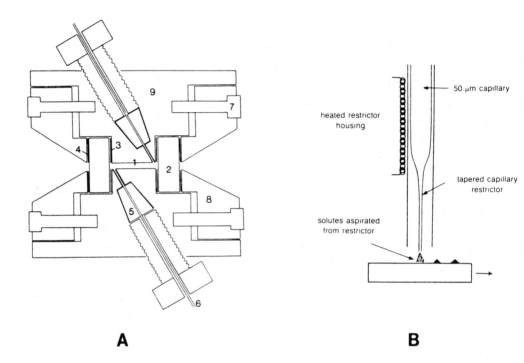

heated restrictor housing

50 µm capillary

tapered capillary restrictor

solutes aspirated from restrictor

A   **B**

Figure 9.9 Schematic diagrams of flow-through cell, A, and solvent elimination interface, B, for SFC/FTIR. For A: (1) polished stainless steel lightpipe; (2) zinc selenide window; (3) PTFE spacer; (4) viton rubber o-ring; (5) graphitized Vespel microferrule; (6) deactivated fused-silica capillary tubing; (7) bolt with Allen nut; (8) stainless steel end-fitting; and (9) stainless steel body of flow cell.

intensity distribution [130]. The influence of mobile phase density on infrared spectra has been attributed to changes in dipole-induced dipole interactions between the solute and fluid because of changes in the polarizability of the supercritical fluid. Stretching vibrations are more sensitive to carbon dioxide density variation than bending vibrations; furthermore wavenumber shifts are relatively moderate for apolar functional groups (< 2 cm^{-1}), but more important for polar functional groups (up to 10 cm^{-1}). When organic solvent modifiers are used with carbon dioxide to improve sample separation direct infrared recording may no longer be possible due to the difficulty of subtracting the background contribution from the modifier. Besides carbon dioxide, supercritical fluid xenon has been proposed as a solvent for infrared spectroscopy due to its favorable critical constants and transparency throughout the mid-infrared region [131]. High cost and uncertainties regarding its solubilizing properties for solutes other than hydrocarbons have limited its use so far.

Typical flow cells are cylindrical in shape and polished internally to maximize reflection with infrared transparent end-windows made from calcium fluoride, zinc selenide or some similar material [128,132,133]. The cells are positioned between the column and restrictor or backpressure regulator and must be capable of withstanding the operating pressure and temperature required to maintain the separation conditions (or different conditions if these are selected to improve sample detectability). Typical cell volumes vary from 0.8 to 8.0 microliters representing a compromise between maintaining chromatographic resolution and providing reasonable sample mass detectability. The cell diameter is dictated by the optics of the spectrometer and cannot be reduced easily below about 0.5 mm without a significant increase in noise. Similarly, the cell path length, which should be long to increase sample absorption, cannot be increased easily beyond about 5 mm due to excessive absorption of the infrared radiation by the mobile phase (carbon dioxide). These cell volumes are compatible with typical peak volumes of retained solutes on packed small bore columns operated at normal flow rates but are too large to maintain chromatographic integrity for typical separation conditions using 50 and 100 micrometer internal diameter open tubular columns. In the latter case a cell volume less than about 0.1 microliters is required, and even if makeup flow is added prior to the cell to reduce dead volumes (and also sensitivity), some loss in chromatographic resolution must be accepted. Sample sizes at least in the range of 10-100 ng are required for spectral identification.

The solvent elimination approach is quite straightforward for supercritical fluids which are often gases at atmospheric pressure. Each chromatographic peak is deposited from the end of a restrictor, connected to the end of the column by a heated transfer line, onto a small area of infrared-transparent support [110,128,129,134]. The support can be moved manually to collect each peak at a new position or stepped continuously to record the full chromatogram. The support is then positioned in the infrared beam and spectra collected from as many scans as are necessary to obtain adequate signal-to-noise ratios. There is a potential for sub-nanogram identification limits using the solvent elimination interface, it is comparatively easy to construct, and the spectra recorded can be referenced to conventional condensed-phase libraries for identification. Also, with some modification to the experimental conditions it may be possible to obtain spectra from mobile phases containing volatile organic solvents.

9.6.4 Interfaces for Liquid Chromatography/ FTIR Spectroscopy

The solvents commonly used in liquid chromatography almost always show significant absorption in the mid-infrared region and, consequently, solvent elimination interfaces are required for many applications. The flow cell interface has been used to monitor certain selected frequency windows to provide functional group detection more often than to record major portions of the complete spectrum (full spectral recording is rarely possible due to solvent bands dominating certain regions of the spectrum).

The flow cell approach is the most straightforward, the eluent from the column is passed through the flow cell and interferograms are continuously measured and stored during the entire chromatographic run [135-138]. At the end of the run, the solution spectra are computed and the absorption bands due to the mobile phase are subtracted out. In order to prevent large portions of the spectrum from being blocked out by solvent bands the path length of the cell has to be kept short, generally less than 0.2 mm for organic solvents and less than 0.03 mm for water-organic solvent mixtures. Sensitivity is thus quite low compared to gas and supercritical fluid flow cell interfaces and sample sizes in the 0.1-1.0 microgram range may be required to give useful information. Even with the above constraints there are several further disadvantages. No information about the solute can be obtained in spectral regions where the mobile phase shows appreciable absorption. Subtraction of the mobile phase contribution to the signal is very difficult for some solvents, especially water (reversed-phase chromatography is barely compatible with the flow cell approach). Solvent subtraction cannot be satisfactorily performed under gradient elution conditions, limiting the method to those samples that can be conveniently separated isocratically. Whereas simplicity and cost considerations maintain an interest in the flow cell interface, in practice, the information that can be obtained from real samples is often very limited.

The solvent elimination interface offers three primary advantages over the flow cell approach. Firstly, the full amount of spectral information from the sample can be obtained; secondly, the mobile phase composition and elution method can be freely selected and optimized virtually independently of spectrometer requirements; and finally, since each chromatographically separated component is stored in the interface, recording can be performed off-line and free of time constraints, allowing sensitivity to be enhanced by spectrum accumulation methods. However, the common solvents used in liquid chromatography are not as easily removed as are gases and, consequently, solvent elimination interfaces are more complex and difficult to use than those discussed earlier [136-140]. Since enrichment is achieved by selective evaporation, the solute must be significantly less volatile than the mobile phase. Early interfaces employed a carousel arrangement of collection cups holding a few milligrams of potassium chloride supported on a metal screen. The eluent from a conventional column was passed through a heated concentrator tube and mixed with a flow of nitrogen to evaporate the bulk of the solvent. When an upstream UV detector signals the elution of a solute peak the effluent from the concentrator tube is allowed to drip into one of the cups on the carousel. After solute collection the carousel cup is automatically advanced to the second position where a slow stream of air is drawn through the cup to eliminate the remaining solvent and a fresh cup is simultaneously located beneath the concentrator tube. A further movement positions the sample in the spectrometer and an infrared spectrum is measured by diffuse reflectance. Identifiable spectra can be obtained from submicrogram samples of low volatility and chromatographic resolution was maintained even for closely separated peaks. With small bore columns operating at low eluent flow rates a drop monitor suffices for the interface. However, there are certain disadvantages to this interface besides its complexity. It is not easy to obtain a continuous chromatogram of the separation, late eluting peaks tend to become collected over several cups, and the method is not suitable for use with aqueous mobile phases due to the high surface tension and latent heat of evaporation for water and because of the solubility of the potassium chloride substrate in water. The eluent from a microcolumn can be deposited directly on a moving crystal plate or cylinder as a continuous narrow band [137]. Aqueous eluents can be used by replacing the alkali salt plate/cylinder with a stainless steel wire net as the transfer medium. Flow rates were typically less than 10 microliters/min which can be easily evaporated from the transfer medium by a stream of warm nitrogen gas applied at the point of eluent contact. The transfer medium is then automatically brought into the infrared beam of the spectrometer giving a near real time chromatographic record of the separation. To obtain identifiable spectra, 50-250 ng of sample is required, which is close to the sample capacity of the microcolumn used for the separation. The method is restricted to major sample components and the choice of transfer medium precludes the use of aqueous eluents. An interface employing a rotating reflective aluminum mirror has been described for use with small bore columns with both organic and aqueous solvents [140]. The column eluent is applied to the mirror as a damp aerosol from a heated, nitrogen gas nebulizer as a continuous track about 0.5-1.5 mm wide which rotates continuously into the beam of the infrared spectrometer for measurement in the reflectance mode. As well as providing a complete chromatographic record of the separation in near real time, compatibility with most mobile phases and gradient separation is maintained with improved sensitivity over other interfaces. Full spectra can be recorded from samples in the 30-150 ng range in most cases. Other methods of interfacing for aqueous solvents employ postcolumn treatment of the column eluent to remove most of the water, for example, by continuous extraction of the separated solute into an organic solvent or by removal of water by reaction with 2,2-dimethyoxypropane [139]. Mobile phase flow rates up to about 0.5 ml/min (higher with an effluent splitter) can be employed in this case. A heated tube operated at reduced pressure is used as a concentrator to reduce the volume of solvent which is applied to a sampling train, such as a series of cups containing potassium chloride, for measurement by diffuse reflectance infrared spectroscopy. Sample sizes in the low microgram range are usually required to obtain full spectra by this approach. At present most of the studies employing solvent elimination interfaces are at the research stage and will undoubtedly undergo further evolution before they become widely used as routine devices.

9.6.5 Interfaces for Thin-Layer Chromatography/FTIR Spectroscopy

Infrared spectra of sample zones on a thin-layer chromatographic plate can be made in situ using diffuse reflectance measurements or after transfer to an infrared-transparent substrate [68,69,104,141-143]. In situ measurements are limited by poor sensitivity and by the strong infrared absorption by the chromatographic support. Spectral information is usually unavailable between 3700-3100 and 1650-800 cm^{-1} for silica (and bonded phases) and 4000-3500 and 2700-1500 cm^{-1} for cellulose [142]. Although silica has been the most widely used substrate in the past alumina shows weaker absorption than silica over the region 3200-1300 cm^{-1} and would be preferred from a spectroscopic point of view [144]. Corrected spectra are obtained by ratio-recording or background subtraction using an identical plate without sample, developed and treated in the same way as that used to separate the sample, from which a background spectra is recorded at the same position on the reference plate as the sample zone on the analytical plate. Normally at least 1 microgram of material is required to detect individual functional groups and at least ten times that amount for partial spectral recording with about 30 min accumulation of spectral data. These sample sizes would be compatible with conventional TLC layers but would cause overloading and loss of resolution for high performance layers. Spectral identification is made more difficult in the absence of reference spectra of standards measured under the same conditions due to changes in band position that result from sorbent-analyte interactions.

Many of the above problems can be eliminated by transferring the TLC chromatogram to an infrared transparent substrate without loss of the chromatographic resolution [143,145]. In this way the amount of sample required for identification can be lowered by an order of magnitude and the information lost due to absorption by the sorbent eliminated. The transfer is achieved by eluting the sample in the orthogonal direction to the direction of development with a stronger solvent than that used for development, and towards a metal strip placed along one edge of the TLC plate. The strip contains a series of sample cups filled with a suitable infrared substrate and a wick to aid transfer of the sample to the infrared substrate. A controlled flow of air over the surface of the powder results in evaporation of the solvent, leaving the analyte deposited on the surface. Spectra can then be recorded by placing the metal strip in an automated diffuse reflectance sampler for measurement.

In situ Raman spectra are limited in sensitivity and spectral range by the same factors that affect in situ infrared spectra. Unlike infrared, however, Raman spectra can be intensified by surface interactions. A recent study demonstrated that spectral information could be obtained from sub-picogram amounts of sample adsorbed onto silica gel TLC plates using sliver colloid activated surface enhanced Raman microspectroscopy [146,147]. The silver colloid was applied to the plate by a light spraying after development and does not interfere in the chromatography. Surface enhanced Raman spectroscopy (SERS) has the potential to become an important structural identification tool in modern TLC.

9.7 LIQUID CHROMATOGRAPHY/NUCLEAR MAGNETIC RESONANCE SPECTRO-SCOPY

Nuclear magnetic resonance (NMR) spectroscopy is one of the most widely used methods for structural elucidation and stereochemical assignments in organic chemistry. This has provided the impetus for exploring the direct coupling of liquid chromatography to nuclear magnetic resonance spectrometric detection (LC/NMR). Although several continuous or stopped flow interfaces have been devised, they are not, however, without significant problems that have delayed the routine implementation of this technique so far [148-151]. Primary problems are poor sensitivity and limited dynamic range when protonated solvents are used to record proton spectra. The best limits of detection that have been demonstrated in a continuous flow experiment are 5-10 micrograms while 50-250 micrograms is a more general requirement to obtain useful spectra. These are obtained with flow cells, straight bore glass tubes of 1-3 mm internal diameter, with volumes of about 25 to 125 microliters. When connecting tubing volumes are also considered, these extracolumn volumes are substantial from a chromatographic point of view. Compatibility is limited to conventional packed columns with some degradation of resolution and semipreparative columns. The use of deuterated solvents with these high-flow rate columns is not economically feasible on a routine basis so that protonated solvents must be frequently used with selective pulse excitation techniques to achieve solvent suppression. The dynamic range achievable, however, is usually less than two orders of magnitude. As a result of the limited residence time of a nucleus in the flow cell both the spin-lattice and spin-spin relaxation times are reduced compared to static measurements. This leads to a favorable increase in signal intensity which increases with flow rate but is accompanied by increasing spectral line broadening requiring the selection of compromise conditions for the spectral acquisition parameters and the ratio of the column eluent flow rate to detector cell volume. At the present state of maturity the advantages of on-line coupling are not so well founded in practice and off-line techniques present fewer compromises in the operation of both the chromatograph and the spectrometer. The current practice of LC/NMR is restricted to ^{1}H and ^{19}F nuclei, and might possibly be extended to ^{31}P nuclei, but for other less sensitive nuclei a substantial increase in spectrometer sensitivity would be needed.

9.8 REFERENCES

1. R. D. Craig, R. H. Bateman, B. N. Green, and D. S. Millington, Phil. Trans. R. Soc. Lond. A, 293 (1979) 135.
2. N. Gochman, L. J. Bowie, and D. N. Bailey, Anal. Chem., 51 (1979) 525A.
3. W. H. McFadden, "Techniques of Combined Gas Chromatography-Mass Spectrometry: Applications in Organic Analysis", Wiley, New York, NY, 1973.
4. B. J. Gudzinowicz, M. J. Gudzinowicz and H. F. Martin, "Fundamentals of Integrated Gas Chromatography-Mass Spectrometry", Dekker, New York, NY, Part II (1976) and Part III (1977).
5. G. M. Message. "Practical Aspects of Gas Chromatography-Mass Spectrometry", Wiley, New York, NY, 1984.
6. F. W. Karasek and R. E. Clement,: Basic Gas Chromatography-Mass Spectrometry. Principles and Techniques", Elsevier, Amsterdam, 1988.
7. R. E. March and R. J. Hughes, "Quadrupole Storage Mass Spectrometry", Wiley, New York, NY, 1989.
8. M. A. Grayson, J. Chromatogr. Sci., 24 (1986) 529.
9. R. M. Milberg and J. C. Cook, J. Chromatogr. Sci., 17 (1979) 43.
10. W. J. Richter and H. Schwarz, Angew. Chem. Int. Ed. Engl., 17 (1978) 424.
11. R. E. Mather and J. F. J. Todd, Int. J. Mass Spectrom. Ion Phys., 30 (1979) 1.
12. A. G. Harrison, "Chemical Ionization Mass Spectrometry", CRC Press, Boca Raton, FL, 1983.
13. R. C. Dougherty, Anal. Chem., 53 (1981) 625A.

14. E. A. Stemmler, R. A. Hites, B. Arbogast, W. L. Budde, M. L. Deinzer, R. C. Dougherty, J. W. Eichelberger, R. L. Foltz, C. Grimm, E. P. Grimsrud, C. Sakashita, and L. J. Sears, Anal. Chem., 60 (1988) 731.

15. R. K. Mitchum and W. A. Korfmacher, Anal. Chem., 55 (1983) 1485A.

16. M. Barber, R. S. Bordoli, G. J. Elliot, R. D. Sedgwick, and A. N. Tyler, Anal. Chem., 54 (1982) 645A.

17. M. L. Vestel and G. J. Fergusson, Anal. Chem., 57 (1985) 2373.

18. W. H. McFadden and S. A. Lammert, J. Chromatogr., 385 (1987) 201.

19. C. E. M. Heeremans, R. A. Van Der Hoeven, W. M. A. Niessen, U. R. Tjaden, and J. Van Der Greef, J. Chromatogr., 474 (1989) 149.

20. C. L. Wilkins and M. L. Gross, Anal. Chem., 53 (1981) 1661A.

21. K. Feser and W. Kogler, J. Chromatogr. Sci., 17 (1979) 57.

22. E. R. Schmid, Chromatographia, 30 (1990) 573.

23. R. P. Adams, "Identification of Essential Oils by Ion Trap Mass Spectroscopy", Academic Press, San Diego, CA, 1989.

24. B. D. Nourse and R. G. Cooks, Anal. Chim. Acta, 228 (1990) 1.

25. M. C. Ten Noever De Brauw, J. Chromatogr., 165 (1979) 207.

26. W. H. McFadden, J. Chromatogr. Sci., 17 (1979) 2.

27. T. E. Jensen, R. Kaminsky, B. D. Veety, T. J. Wozniak and R. A. Hites, Anal. Chem., 54 (1982) 2388.

28. K. Rose, J. Chromatogr., 259 (1983) 445.

29. R. Tiebach and W. Blaas, J. Chromatogr., 454 (1988) 372.

30. H. J. Stan, J. Chromatogr., 467 (1989) 85.

31. N. W. Davies, J. Chromatogr., 325 (1988) 23.

32. C. A. Cramers and P. A. Leclercq, CRC Crit. Revs. Anal. Chem., 20 (1988) 117.

33. D. Henneberg, U. Henrichs, H. Husmann, and G. Schomburg, J. Chromatogr., 167 (1978) 139.

34. J. F. K. Huber, E. Matisova, and E. Kenndler, Anal. Chem., 54 (1982) 1297.

35. E. Wetzel, Th. Kuster, and H.-Ch. Curtius, J. Chromatogr., 239 (1982) 107.

36. J. F. Pankow and L. M. Isabelle, J. High Resolut. Chromatogr., 10 (1987) 617.

37. N. W. Davies, J. Chromatogr., 450 (1988) 388.

38. C. C. Greenwalt, K. J. Voorhees, and J. H. Futrell, Anal. Chem., 55 (1983) 468.

39. D. E. Games, Adv. Chromatogr., 21 (1983) 1.

40. P. J. Arpino, J. Chromatogr., 323 (1985) 3.

41. A. P. Bruins, J. Chromatogr., 323 (1985) 99.

42. T. R. Covet, E. D. Lee, A. P. Bruins, and J. D. Henion, Anal. Chem., 58 (1986) 1451A.

43. W. M. A. Niessen, Chromatographia, 21 (1986) 277 and 342.

44. E. D. Lee and J. D. Henion, J. Chromatogr. Sci., 23 (1985) 253.

45. A. L. Yergey, C. G. Edmonds, I. A. S. Lewis, and M. L. Vestel, "Liquid Chromatography/Mass Spectrometry. Techniques and Applications", Plenum Press, New York, NY, 1989.

46. K. Vekey, D. Edwards, and L. F. Zerilli, J. Chromatogr., 488 (1989) 73.

47. K. B. Tomer and C. E. Parker, J. Chromatogr., 492 (1989) 189.

48. J. S. M. de Wit, C. E. Parker, K. B. Tomer, and J. W. Jorgenson, Anal. Chem., 59 (1987) 2400.

49. W. M. A. Niessen and H. Poppe, J. Chromatogr., 394 (1987) 21.

50. M. A. Moseley, L. J. Deterding, J. S. M. de Wit, K. B. Tomer, R. T. Kennedy, N. Bragg, and J. W. Jorgenson, Anal. Chem., 61 (1989) 1577.

51. R. M. Caprioli, Trends Anal. Chem., 7 (1988) 328.
52. T. Takeuchi, S. Watanabe, N. Kondo, D. Ishii, and M. Goto, J. Chromatogr., 435 (1988) 482.
53. D. Ishii and T. Takeuchi, Trends Anal. Chem., 8 (1989) 25.
54. J. A. Page, M. T. Beer, and R. Lauber, J. Chromatogr., 474 (1989) 51.
55. A. C. Barefoot, R. W. Reiser, and S. A. Cousins, J. Chromatogr., 474 (1989) 39.
56. P. Kokkonen, E. Schroder, W. M. A. Niessen, U. R. Tjaden, and J. Van Der Greef, J. Chromatogr., 511 (1990) 35.
57. R. M. Caprioli, Anal. Chem., 62 (1990) 477A.
58. G. M. Kresbach, T. R. Baker, R. J. Nelson, J. Wronka, B. L. Karger, and P. Vouros, J. Chromatogr., 394 (1987) 89.
59. J. R. Chapman and J. A. E. Pratt, J. Chromatogr., 394 (1987) 231.
60. P. C. Winkler, D. D. Perkins, W. K. Williams, and R. F. Browner, Anal. Chem., 60 (1988) 489.
61. T. D. Behymer, T. A. Bellar, and W. L. Budde, Anal. Chem., 62 (1990) 1686.
62. W. V. Ligon and S. B. Dorn, Anal. Chem., 62 (1990) 2573.
63. A. P. Bruins, T. R. Covey, and J. D. Henion, Anal. Chem., 59 (1987) 2642.
64. M. Sakairi and H. Kambara, Anal. Chem., 60 (1988) 774.
65. E. C. Huang, J. Wachs, J. J. Conboy, and J. D. Henion, Anal. Chem., 62 (1990) 713A.
66. G. J. Van Berkel, G. L. Glish, and S. A. Mcluckey, Anal. Chem., 62 (1990) 1284.
67. R. D. Smith, Z. A. Loo, C. G. Edmonds, C. J. Barinaga, and H. R. Udseth, Anal. Chem., 62 (1990) 882.
68. M. G. Ikonomou, A. T. Blades, and P. Kebarle, Anal. Chem., 62 (1990) 957.
69. J. B. Fenn, M. Mann, C. K. Meng, S. F. Wong, and C. M. Whitehouse, Mass Spectrom. Rev., 9 (1990) 37.
70. P. J. Arpino, J. Cousin, and J. Higgins, Trends Anal. Chem., 6 (1987) 69.
71. R. D. Smith, H. T. Kalinoski, and H. R. Udseth, Mass Spectrom. Rev., 6 (1987) 445.
72. A. J. Berry, D. E. Games, I. C. Mylchreest, J. R. Perkins, and S. Pleasance, J. High Resolut. Chromatogr., 11 (1988) 61.
73. D. E. Games, A. J. Beryy, I. C. Mylchreest, J. R. Perkins, and S. Pleasance, in R. M. Smith (Ed.), "Supercritical Fluid Chromatography", Royal Society of Chemistry, Letchworth, UK, 1988, p. 159.
74. D. M. Sheeley and V. N. Reinhold, J. Chromatogr., 474 (1989) 83.
75. J. Cousin and P. J. Arpino, J. Chromatogr., 398 (1987) 125.
76. B. W. Wright, H. T. Kalinoski, H. R. Udseth, and R. D. Smith, J. High Resolut. Chromatogr., 9 (1986) 145.
77. H. T. Kalinoski, H. R. Udseth, E. K. Chess, and R. D. Smith, J. Chromatogr., 394 (1987) 3.
78. S. B. Hawthorne and D. J. Miller, J. Chromatogr., 468 (1989) 115.
79. E. C. Huang, B. J. Jackson, K. E. Markides, and M. L. Lee, Anal. Chem., 60 (1988) 2715.
80. G. Holzer, S. Deluca, and K. J. Voorhees, J. High Resolut. Chromatogr., 8 (1985) 528.
81. P. O. Edlund and J. D. Henion, J. Chromatogr. Sci., 27 (1989) 274.
82. A. J. Berry, D. E. Games, and J. R. Perknis, J. Chromatogr., 363 (1986) 147.
83. R. D. Smith and H. R. Udseth, Anal. Chem., 59 (1987) 13.

84. J. Sherma and B. Fried (Eds.), "Handbook of Thin Layer Chromatography", Dekker, New York, NY, 1990.
85. C. F. Poole and S. K. Poole, J. Chromatogr., 492 (1989) 539.
86. L. Ramley, M. A. Vaughn, and W. D. Jamieson, Anal. Chem., 57 (1985) 353.
87. A. J. Kubis, K. V. Somayajula, A. G. Sharkey, and D. M. Hercules, Anal. Chem., 61 (1989) 2516.
88. K. L. Busch, Trends Anal. Chem., 6 (1987) 95.
89. K. L. Bush, J. Planar Chromatogr., 2 (1989) 355.
90. S. M. Brown, H. Schurz, and K. L. Busch, J. Planar Chromatogr., 3 (1990) 222.
91. J. R. Chapman, "Computers in Mass Spectrometry", Academic Press, New York, NY, 1978.
92. J. R. Chapman, J. Phys. E, 13 (1980) 365.
93. R. G. Dromey, Spectra, 10 (1984) 3.
94. B. J. Millard, "Quantitative Mass Spectrometry", Heyden, London, UK, 1978.
95. W. A. Garland and M. L. Powell, J. Chromatogr. Sci., 19 (1981) 392.
96. T. A. Baillie, Pharmacol. Rev., 33 (1981) 81.
97. S. J. Gaskell, Trends Anal. Chem., 1 (1982) 110.
98. W. A. Garland and M. P. Barbalas, J. Clin. Pharmacol., 26 (1986) 412.
99. F. W. McLafferty and R. Venkataraghavan, J. Chromatogr. Sci., 17 (1979) 24.
100. H. J. Luinge, Vibrational Spectroscopy, 1 (1990) 3.
101. P. R. Griffiths and J. A. de Haseth, "Fourier Transform Infrared Spectrometry", Wiley, New York, NY, 1986.
102. W. Herres, "HRGC-FTIR: Capillary Gas Chromatography-Fourier Transform Infrared Spectrometry. Theory and Applications", Huethig, Heidelberg, 1987.
103. M. W. Mackenzie (Ed.), "Advances in Applied Fourier Transform Infrared Spectroscopy", Wiley, NY, 1988.
104. R. White, "Chromatography/Fourier Transform Infrared Spectroscopy and Its Applications", Dekker, New York, NY, 1990.
105. D. F. Gurka, I. Farnham, B. B. Potter, S. Pyle, R. Titus, and W. Duncan, Anal. Chem., 61 (1989) 1584.
106. I. C. Bowater, R. S. Brown, J. R. Cooper, and C. L. Wilkins, Anal. Chem., 58 (1986) 2195.
107. R. L. White, G. N. Giss, G. M. Brissey, and C. L. Wilkins, Anal. Chem., 53 (1981) 1778.
108. J. M. Bjerga and G. W. Small, Anal. Chem., 61 (1989) 1073.
109. P. R. Griffiths, S. L. Pentoney, A. Giorgetti, and K. H. Shafer, Anal. Chem., 58 (1986) 1349A.
110. P. R. Griffiths, A. M. Haefner, K. L. Norton, D. J. J. Fraser, D. Pyo, and H. Markishima, J. High Resolut. Chromatogr., 12 (1989) 119.
111. P. R. Griffiths, J. A. de Haseth, and L. V. Azarraga, Anal. Chem., 55 (1983) 1361A.
112. R. L. White, Appl. Spectrosc. Revs., 23 (1987) 165.
113. C. J. Wurrey, Trends Anal. Chem., 8 (1989) 52.
114. B. Lacroix, J. P. Huvenne, and M. Deveaux, J. Chromatogr., 492 (1989) 109.
115. G. T. Reedy, D. G. Ettinger, J. F. Schneider, and S. Bourne, Anal. Chem., 57 (1985) 1602.
116. T. T. Holloway, B. F. Fairless, C. E. Freidline, H. E. Kimball, R. D. Kloepfer, C. J. Wurrey, L. A. Jonooby, and H. G. Palmer, Appl. Spectrosc., 42 (1988) 359.
117. S. Jagannathan, J. R. Cooper, and C. L. Wilkins, Appl. Spectrosc., 43 (1989) 781.
118. R. S. Brown and C. L. Wilkins, Anal. Chem., 60 (1988) 1483.

119. R. Fuoco, K. H. Shafer, and P. R. Griffiths, Anal. Chem., 58 (1986) 3249.
120. S. Bourne, A. M. Haefner, K. L. Norton, and P. R. Griffiths, Anal. Chem., 62 (1990) 2448.
121. T. Visser and M. J. Vredenbregt, Vibrational Spectroscopy, 1 (1990) 205.
122. C. L. Wilkins, Anal. Chem., 59 (1987) 571A.
123. E. S. Olson and J. W. Diehl, Anal. Chem., 59 (1987) 443.
124. J. C. Demirgian, Trends Anal. Chem., 6 (1987) 58.
125. J. R. Cooper and C. L. Wilkins, Anal. Chem., 61 (1989) 1571.
126. K. Jinno, Chromatographia, 23 (1987) 55.
127. L. T. Taylor and E. M. Calvey, Chem. Revs., 89 (1989) 321.
128. K. D. Bartle, M. W. Raynor, A. A. Clifford, I. L. Davies, J. P. Kithinji, G. F. Shilstone, J. M. Chalmers, and B. W. Cook, J. Chromatogr. Sci., 27 (1989) 283.
129. R. Fuoco, S. L. Pentoney, and P. R. Griffiths, Anal. Chem., 61 (1989) 2212.
130. Ph. Morin, B. Beccard, M. Caude, and R. Rosset, J. High Resolut. Chromatogr., 11 (1988) 697.
131. M. W. Raynor, G. F. Shilstone, K. D. Bartle, A. A. Clifford, M. Cleary, and B. W. Cook, J. High Resolut. Chromatogr., 12 (1989) 300.
132. R. C. Wieboldt, G. E. Adams, and D. W. Later, Anal. Chem., 60 (1988) 2422.
133. M. W. Raynor, A. A. Clifford, K. D. Bartle, C. Reyner, A. Williams, and B. W. Cook, J. Microcol. Sepns, 1 (1989) 101.
134. M. W. Raynor, K. D. Bartle, I. L. Davies, A. Williams, A. A. Clifford, J. M. Chalmers, and B. W. Cook, Anal. Chem., 60 (1988) 427.
135. L. T. Taylor, J. Chromatogr. Sci., 23 (1985) 265.
136. J. W. Hellgeth and L. T. Taylor, J. Chromatogr. Sci., 24 (1986) 519.
137. C. Fujimoto and K. Jinno, Trends Anal. Chem., 8 (1989) 90.
138. K. Jinno and C. Fujimoto, J. Chromatogr., 506 (1990) 443.
139. V. F. Kalasinsky, K. G. Whitehead, R. C. Kenton, J. A. S. Smith, and K. S. Kalasinsky, J. Chromatogr. Sci., 25 (1987) 273.
140. J. J. Gagel and K. Biemann, Anal. Chem., 59 (1987) 1266.
141. P. R. Brown and B. T. Beauchemin, J. Liquid Chromatogr., 11 (1988) 1001.
142. G. E. Zuber, R. J. Warner, P. P. Begosh, and E. L. O' Donnell, Anal. Chem., 56 (1984) 2935.
143. K. H. Shafer, P. R. Griffiths, and W. Shu-Qin, Anal. Chem., 58 (1986) 2708.
144. C. Fujimoto, T. Morita, K. Jinno, and K. H. Shafer, J. High Resolut. Chromatogr., 11 (1988) 810.
145. J. M. Chalmers, M. W. Mackenzie, and J. L. Sharp, Anal. Chem., 59 (1987) 415.
146. E. Koglin, J. Plannar Chromatogr., 2 (1989) 194.
147. E. Koglin, J. Planar Chromatogr., 3 (1990) 117.
148. H. C. Dorn, Anal. Chem., 56 (1984) 747A.
149. D. A. Laude and C. L. Wilkins, Trends Anal. Chem., 5 (1986) 230.
150. K. Albert and E. Bayer, Trends Anal. Chem., 7 (1988) 288.
151. K. Albert, M. Kunst, E. Bayer, M. Spraul, and W. Bermel, J. Chromatogr., 463 (1989) 355.

SUBJECT INDEX

CGE = Capillary gel
electrophoresis
CZE = Capillary zone
electrophoresis
ECD = Electron-capture
detector
FAB = Fast atom bombardment
FID = Flame ionization
detector
FTIR = Fourier transform
infrared spectroscopy
GC = Gas chromatography
LC = Liquid chromatography
LSC = Liquid-solid
chromatography
MS = Mass spectrometry
NMR = Nuclear magnetic
resonance spectroscopy
OTC = Open tubular columns
SEC = Size-exclusion
chromatography
SERS = Surface enhanced Raman
spectroscopy
SFC = Supercritical fluid
chromatography
SFE = Supercritical fluid
extraction
TLC = Thin-layer
chromatography

A

Absorption detectors
derivatizing reagents 873
LC 573, 589
SFC 642
TLC 720
Activated carbon 783
applications 784, 828, 833
preparation 784
properties 784
Activity (sorbents) 384, 774
Activity coefficient (GC) 7
Activity tests OTC (GC) 159
Acylating reagents (GC) 858
Adjusted retention time/
(volume) 4
Adsorption chromatography
[see gas-solid (GC)]
[see liquid-solid (LC)]
[see sample cleanup by (LC)]
Air sampling 831
cold trapping 833
filtration 840
grab 832
impactors 840

preparation of standards 844
sorbent trapping 833
Alkali flame ionization
detector
[see thermionic ionization
detector]
Alkydimethylsilyl reagents 855
Alkylating reagents (GC) 861
alkyl halide/catalyst 861
dialkyl acetals 863
diazoalkanes 862
extractive alkylation 863
pyrolytic 865
Alumina
GC 200
ion exchanger 388
LC 323
preparation 323
sample cleanup 771
TLC 673
Amperometric detectors
CZE 524
LC 589
Analysis time
[see separation time]
Anticircular development
(TLC) 684,716
Apiezon phases (GC) 109
Apolane-87 phase (GC) 100
Applicators
[see sample application]
Assisted distillation 748
Asymmetric peak models 26
Asymmetry function 28, 359
Atmospheric pressure ionization
(MS) 953
interface (LC) 972
Atomic emission detector (GC)
293
elemental composition 293
operation 294
plasma sources 293
Automated multiple development
(TLC) 687, 717
Axial compression columns (LC)
502

B

Backflushing 791, 814
Balanced density slurry, column
packing (LC) 348
Band applicators (TLC) 711
Band broadening
coupled plate height equation
19
eddy diffusion 15
effective theoretical plate
12
extracolumn effects (LC) 74,
548
frictional heat (LC) 69

1018

BC